ERROR CONTROL CODING

Fundamentals and Applications

PRENTICE-HALL COMPUTER APPLICATIONS IN ELECTRICAL ENGINEERING SERIES

FRANKLIN F. KUO, editor

ABRAMSON and KUO, Computer-Communication Networks
BOWERS and SEDORE, Sceptre: A Computer Program for Circuit and Systems Analysis
CADZOW, Discrete Time Systems: An Introduction with Interdisciplinary Applications
CADZOW and MARTENS, Discrete-Time and Computer Control Systems
DAVIS, Computer Data Displays
FRIEDMAN and MENON, Fault Detection in Digital Circuits
HUELSMAN, Basic Circuit Theory
JENSEN and LIEBERMAN, IBM Circuit Analysis Program: Techniques and Applications
JENSEN and WATKINS, Network Analysis: Theory and Computer Methods
KLINE, Digital Computer Design
KOCHENBURGER, Computer Simulation of Dynamic Systems
KUO, (ed.) Protocols and Techniques for Data Communication Networks
KUO and MAGNUSON, Computer Oriented Circuit Design
LIN, An Introduction to Error-Correcting Codes
LIN and COSTELLO, Error Control Coding: Fundamentals and Applications
NAGLE, CARROLL, and IRWIN, An Introduction to Computer Logic
RHYNE, Fundamentals of Digitals Systems Design
SIFFERLEN and VARTANIAN, Digital Electronics with Engineering Applications
STAUDHAMMER, Circuit Analysis by Digital Computer
STOUTEMYER, PL/1 Programming for Engineering and Science

ERROR CONTROL CODING

Fundamentals and Applications

SHU LIN

University of Hawaii
Texas A&M University

DANIEL J. COSTELLO, JR.

Illinois Institute of Technology

Prentice-Hall, Inc. Englewood Cliffs, New Jersey 07632

Library of Congress Cataloging in Publication Data

Lin, Shu.
Error control coding.

(Prentice-Hall computer applications in
electrical engineering series)
Includes bibliographical references and index.
1. Error-correcting codes (Information theory)
1. Costello, Daniel J. . II. Title.
III. Series.
QA268.L55 001.53'9 82-5255
ISBN 0-13-283796-X AACR2

Editorial/production supervision and interior design by Anne Simpson

Cover design by Marvin Warshaw

Manufacturing buyer: Joyce Levatino

Printed in the United States of America

10 9 8 7

ISBN 0-13-283796-X

PRENTICE-HALL INTERNATIONAL, INC., *London*
PRENTICE-HALL OF AUSTRALIA PTY. LIMITED, *Sydney*
EDITORA PRENTICE-HALL DO BRAZIL, LTDA, *Rio de Janeiro*
PRENTICE-HALL CANADA INC., *Toronto*
PRENTICE-HALL OF INDIA PRIVATE LIMITED, *New Delhi*
PRENTICE-HALL OF JAPAN, INC., *Tokyo*
PRENTICE-HALL OF SOUTHEAST ASIA PTE. LTD., *Singapore*
WHITEHALL BOOKS LIMITED, *Wellington, New Zealand*

With Love and Affection for
Ivy,
Julian, Patrick, and Michelle Lin
and
Lucretia,
Kevin, Nick, Daniel, and Anthony Costello

Contents

PREFACE xiii

CHAPTER 1 CODING FOR RELIABLE DIGITAL TRANSMISSION AND STORAGE 1

1.1 Introduction 1
1.2 Types of Codes 3
1.3 Modulation and Demodulation 5
1.4 Maximum Likelihood Decoding 8
1.5 Types of Errors 11
1.6 Error Control Strategies 12
References 14

CHAPTER 2 INTRODUCTION TO ALGEBRA 15

2.1 Groups 15
2.2 Fields 19
2.3 Binary Field Arithmetic 24
2.4 Construction of Galois Field GF(2^m) 29
2.5 Basic Properties of Galois Field GF(2^m) 34
2.6 Computations Using Galois Field GF(2^m) Arithmetic 39
2.7 Vector Spaces 40
2.8 Matrices 46
Problems 48
References 50

CHAPTER 3 LINEAR BLOCK CODES 51

3.1 Introduction to Linear Block Codes 51
3.2 Syndrome and Error Detection 58
3.3 Minimum Distance of a Block Code 63
3.4 Error-Detecting and Error-Correcting Capabilities of a Block Code 65
3.5 Standard Array and Syndrome Decoding 68
3.6 Probability of an Undetected Error for Linear Codes over a BSC 76
3.7 Hamming Codes 79
Problems 82
References 84

CHAPTER 4 CYCLIC CODES 85

4.1 Description of Cyclic Codes 85
4.2 Generator and Parity-Check Matrices of Cyclic Codes 92
4.3 Encoding of Cyclic Codes 95
4.4 Syndrome Computation and Error Detection 98
4.5 Decoding of Cyclic Codes 103
4.6 Cyclic Hamming Codes 111
4.7 Shortened Cyclic Codes 116
Problems 121
References 123

CHAPTER 5 ERROR-TRAPPING DECODING FOR CYCLIC CODES 125

5.1 Error-Trapping Decoding 125
5.2 Improved Error-Trapping Decoding 131
5.3 The Golay Code 134
Problems 139
References 139

CHAPTER 6 BCH CODES 141

6.1 Description of the Codes 142
6.2 Decoding of the BCH Codes 151
6.3 Implementation of Galois Field Arithmetic 161
6.4 Implementation of Error Correction 167
6.5 Nonbinary BCH Codes and Reed–Solomon Codes 170
6.6 Weight Distribution and Error Detection of Binary BCH Codes 177
Problems 180
References 182

CHAPTER 7 MAJORITY-LOGIC DECODING FOR CYCLIC CODES 184

7.1 One-Step Majority-Logic Decoding 184
7.2 Class of One-Step Majority-Logic Decodable Codes 194
7.3 Other One-Step Majority-Logic Decodable Codes 201
7.4 Multiple-Step Majority-Logic Decoding 209
Problems 219
References 221

CHAPTER 8 FINITE GEOMETRY CODES 223

8.1 Euclidean Geometry 223
8.2 Majority-Logic Decodable Cyclic Codes Based on Euclidean Geometry 227
8.3 Projective Geometry and Projective Geometry Codes 240
8.4 Modifications of Majority-Logic Decoding 245
Problems 253
References 255

CHAPTER 9 BURST-ERROR-CORRECTING CODES 257

9.1 Introduction 257
9.2 Decoding of Single-Burst-Error-Correcting Cyclic Codes 259
9.3 Single-Burst-Error-Correcting Codes 261
9.4 Interleaved Codes 271
9.5 Phased-Burst-Error-Correcting Codes 272
9.6 Burst-and-Random-Error-Correcting Codes 274
9.7 Modified Fire Codes for Simultaneous Correction of Burst and Random Errors 280
Problems 282
References 284

CHAPTER 10 CONVOLUTIONAL CODES 287

10.1 Encoding of Convolutional Codes 288
10.2 Structural Properties of Convolutional Codes 295
10.3 Distance Properties of Convolutional Codes 308
Problems 312
References 313

CHAPTER 11 MAXIMUM LIKELIHOOD DECODING OF CONVOLUTIONAL CODES 315

11.1 The Viterbi Algorithm 315
11.2 Performance Bounds for Convolutional Codes 322

11.3 Construction of Good Convolutional Codes 329
11.4 Implementation of the Viterbi Algorithm 337
11.5 Modifications of the Viterbi Algorithm 345
Problems 346
References 348

CHAPTER 12 SEQUENTIAL DECODING OF CONVOLUTIONAL CODES 350

12.1 The Stack Algorithm 351
12.2 The Fano Algorithm 360
12.3 Performance Characteristics of Sequential Decoding 364
12.4 Code Construction for Sequential Decoding 374
12.5 Other Approaches to Sequential Decoding 380
Problems 384
References 386

CHAPTER 13 MAJORITY-LOGIC DECODING OF CONVOLUTIONAL CODES 388

13.1 Feedback Decoding 389
13.2 Error Propagation and Definite Decoding 406
13.3 Distance Properties and Code Performance 408
13.4 Code Construction for Majority-Logic Decoding 414
13.5 Comparison with Probabilistic Decoding 424
Problems 426
References 428

CHAPTER 14 BURST-ERROR-CORRECTING CONVOLUTIONAL CODES 429

14.1 Bounds on Burst-Error-Correcting Capability 430
14.2 Burst-Error-Correcting Convolutional Codes 430
14.3 Interleaved Convolutional Codes 441
14.4 Burst-and-Random-Error-Correcting Convolutional Codes 442
Problems 455
References 456

CHAPTER 15 AUTOMATIC-REPEAT-REQUEST STRATEGIES 458

15.1 Basic ARQ Schemes 459
15.2 Selective-Repeat ARQ System with Finite Receiver Buffer 465
15.3 ARQ Schemes with Mixed Modes of Retransmission 474
15.4 Hybrid ARQ Schemes 477
15.5 Class of Half-Rate Invertible Codes 481

15.6 Type II Hybrid Selective-Repeat ARQ with Finite Receiver Buffer 483
Problems 494
References 495

CHAPTER 16 APPLICATIONS OF BLOCK CODES FOR ERROR CONTROL IN DATA STORAGE SYSTEMS 498

16.1 Error Control for Computer Main Processor and Control Storages 498
16.2 Error Control for Magnetic Tapes 503
16.3 Error Control in IBM 3850 Mass Storage System 516
16.4 Error Control for Magnetic Disks 525
16.5 Error Control in Other Data Storage Systems 531
Problems 532
References 532

CHAPTER 17 PRACTICAL APPLICATIONS OF CONVOLUTIONAL CODES 533

17.1 Applications of Viterbi Decoding 533
17.2 Applications of Sequential Decoding 539
17.3 Applications of Majority-Logic Decoding 543
17.4 Applications to Burst-Error Correction 547
17.5 Applications of Convolutional Codes in ARQ Systems 551
Problems 556
References 557

Appendix A Tables of Galois Fields 561

Appendix B Minimal Polynomials of Elements in GF(2^m) 579

Appendix C Generator Polynomials of Binary Primitive BCH Codes of Length up to $2^{10} - 1$ 583

INDEX 599

Preface

This book owes its beginnings to the pioneering work of Claude Shannon in 1948 on achieving reliable communication over a noisy transmission channel. Shannon's central theme was that if the signaling rate of the system is less than the channel capacity, reliable communication can be achieved if one chooses proper encoding and decoding techniques. The design of good codes and of efficient decoding methods, initiated by Hamming, Slepian, and others in the early 1950s, has occupied the energies of many researchers since then. Much of this work is highly mathematical in nature, and requires an extensive background in modern algebra and probability theory to understand. This has acted as an impediment to many practicing engineers and computer scientists, who are interested in applying these techniques to real systems. One of the purposes of this book is to present the essentials of this highly complex material in such a manner that it can be understood and applied with only a minimum of mathematical background.

Work on coding in the 1950s and 1960s was devoted primarily to developing the theory of efficient encoders and decoders. In 1970, the first author published a book entitled *An Introduction to Error-Correcting Codes*, which presented the fundamentals of the previous two decades of work covering both block and convolutional codes. The approach was to explain the material in an easily understood manner, with a minimum of mathematical rigor. The present book takes the same approach to covering the fundamentals of coding. However, the entire manuscript has been rewritten and much new material has been added. In particular, during the 1970s the emphasis in coding research shifted from theory to practical applications. Consequently, three completely new chapters on the applications of coding to digital transmission and storage systems have been added. Other major additions include a comprehensive treatment of the error-detecting capabilities of block codes, and an emphasis on probabilistic decoding methods for convolutional codes. A brief description of each chapter follows.

Chapter 1 presents an overview of coding for error control in data transmission and storage systems. A brief discussion of modulation and demodulation serves to place coding in the context of a complete system. Chapter 2 develops those concepts from modern algebra that are necessary to an understanding of the material in later chapters. The presentation is at a level that can be understood by students in the senior year as well as by practicing engineers and computer scientists.

Chapters 3 through 8 cover in detail block codes for random-error correction. The fundamentals of linear codes are presented in Chapter 3. Also included is an extensive section on error detection with linear codes, an important topic which is discussed only briefly in most other books on coding. Most linear codes used in practice are cyclic codes. The basic structure and properties of cyclic codes are presented in Chapter 4. A simple way of decoding some cyclic codes, known as error-trapping decoding, is covered in Chapter 5. The important class of BCH codes for multiple-error correction is presented in detail in Chapter 6. A discussion of hardware and software implementation of BCH decoders is included, as well as the use of BCH codes for error detection. Chapters 7 and 8 provide detailed coverage of majority-logic decoding and majority-logic decodable codes. The material on fundamentals of block codes concludes with Chapter 9 on burst-error correction. This discussion includes codes for correcting a combination of burst and random errors.

Chapters 10 through 14 are devoted to the presentation of the fundamentals of convolutional codes. Convolutional codes are introduced in Chapter 10, with the encoder state diagram serving as the basis for studying code structure and distance properties. The Viterbi decoding algorithm for both hard and soft demodulator decisions is covered in Chapter 11. A detailed performance analysis based on code distance properties is also included. Chapter 12 presents the basics of sequential decoding using both the stack and Fano algorithms. The difficult problem of the computational performance of sequential decoding is discussed without including detailed proofs. Chapter 13 covers majority-logic decoding of convolutional codes. The chapter concludes with a comparison of the three primary decoding methods for convolutional codes. Burst-error-correcting convolutional codes are presented in Chapter 14. A section is included on convolutional codes that correct a combination of burst and random errors. Burst-trapping codes, which embed a block code in a convolutional code, are also covered here.

Chapters 15 through 17 cover a variety of applications of coding to modern day data communication and storage systems. Although they are not intended to be comprehensive, they are representative of the many different ways in which coding is used as a method of error control. This emphasis on practical applications makes the book unique in the coding literature. Chapter 15 is devoted to automatic-repeat-request (ARQ) error control schemes used for data communications. Both pure ARQ (error detection with retransmission) and hybrid ARQ (a combination of error correction and error detection with retransmission) are discussed. Chapter 16 covers the application of block codes for error control in data storage systems. Coding techniques for computer memories, magnetic tape, magnetic disk, and optical storage systems are included. Finally, Chapter 17 presents a wide range of applications of convolutional codes to digital communication systems. Codes actually used on many space and satellite systems are included, as well as a section on using convolutional codes in a hybrid ARQ system.

Several additional features are included to make the book useful both as a classroom text and as a comprehensive reference for engineers and computer scientists involved in the design of error control systems. Three appendices are given which include details of algebraic structure used in the construction of block codes. Many tables of the best known codes for a given decoding structure are presented throughout the book. These should prove valuable to designers looking for the best code for a particular application. A set of problems is given at the end of each chapter. Most of the problems are relatively straightforward applications of material covered in the text, although some more advanced problems are also included. There are a total of over 250 problems. A solutions manual will be made available to instructors using the text. Over 300 references are also included. Although no attempt was made to compile a complete bibliography on coding, the references listed serve to provide additional detail on topics covered in the book.

The book can be used as a text for an introductory course on error-correcting codes and their applications at the senior or beginning graduate level. It can also be used as a self-study guide for engineers and computer scientists in industry who want to learn the fundamentals of coding and how they can be applied to the design of error control systems.

As a text, the book can be used as the basis for a two-semester sequence in coding theory and applications, with Chapters 1 through 9 on block codes covered in one semester and Chapters 10 through 17 on convolutional codes and applications in a second semester. Alternatively, portions of the book can be covered in a one-semester course. One possibility is to cover Chapters 1 through 6 and 10 through 12, which include the basic fundamentals of both block and convolutional codes. A course on block codes and applications can be comprised of Chapters 1 through 6, 9, 15, and 16, whereas Chapters 1 through 3, 10 through 14, and 17 include convolutional codes and applications as well as the rudiments of block codes. Preliminary versions of the notes on which the book is based have been classroom tested by both authors for university courses and for short courses in industry, with very gratifying results.

It is difficult to identify the many individuals who have influenced this work over the years. Naturally, we both owe a great deal of thanks to our thesis advisors, Professors Paul E. Pfeiffer and James L. Massey. Without their stimulating our interest in this exciting field and their constant encouragement and guidance through the early years of our research, this book would not have been possible.

Much of the material in the first half of the book on block codes owes a great deal to Professors W. Wesley Peterson and Tadao Kasami. Their pioneering work in algebraic coding and their valuable discussions and suggestions had a major impact on the writing of this material. The second half of the book on convolutional codes was greatly influenced by Professor James L. Massey. His style of clarifying the basic elements in highly complex subject matter was instrumental throughout the preparation of this material. In particular, most of Chapter 14 was based on a set of notes that he prepared.

We are grateful to the National Science Foundation, and to Mr. Elias Schutzman, for their continuing support of our research in the coding field. Without this assistance, our interest in coding could never have developed to the point of writing this book. We thank the University of Hawaii and Illinois Institute of Technology for their support of our efforts in writing this book and for providing facilities. We also owe thanks to Professor Franklin F. Kuo for suggesting that we write this book, and for his constant encouragement and guidance during the preparation of the manuscript. Another major source of stimulation for this effort came from our graduate students, who have provided a continuing stream of new ideas and insights. Those who have made contributions directly reflected in this book include Drs. Pierre Chevillat, Farhad Hemmati, Alexander Drukarev, and Michael J. Miller.

We would like to express our special appreciation to Professors Tadao Kasami, Michael J. Miller, and Yu-ming Wang, who read the first draft very carefully and made numerous corrections and suggestions for improvements. We also wish to thank our secretaries for their dedication and patience in typing this manuscript. Deborah Waddy and Michelle Masumoto deserve much credit for their perseverence in preparing drafts and redrafts of this work.

Finally, we would like to give special thanks to our parents, wives, and children for their continuing love and affection throughout this project.

Shu Lin
Daniel J. Costello, Jr.

ERROR CONTROL CODING

Fundamentals and Applications

1

Coding for Reliable Digital Transmission and Storage

1.1 INTRODUCTION

In recent years, there has been an increasing demand for efficient and reliable digital data transmission and storage systems. This demand has been accelerated by the emergence of large-scale, high-speed data networks for the exchange, processing, and storage of digital information in the military, governmental, and private spheres. A merging of communications and computer technology is required in the design of these systems. A major concern of the designer is the control of errors so that reliable reproduction of data can be obtained.

In 1948, Shannon [1] demonstrated in a landmark paper that, by proper encoding of the information, errors induced by a noisy channel or storage medium can be reduced to any desired level without sacrificing the rate of information transmission or storage. Since Shannon's work, a great deal of effort has been expended on the problem of devising efficient encoding and decoding methods for error control in a noisy environment. Recent developments have contributed toward achieving the reliability required by today's high-speed digital systems, and the use of coding for error control has become an integral part in the design of modern communication systems and digital computers.

The transmission and storage of digital information have much in common. They both transfer data from an information source to a destination (or user). A typical transmission (or storage) system may be represented by the block diagram shown in Figure 1.1. The *information source* can be either a person or a machine (e.g., a digital computer). The source output, which is to be communicated to the destination, can be either a continuous waveform or a sequence of discrete symbols.

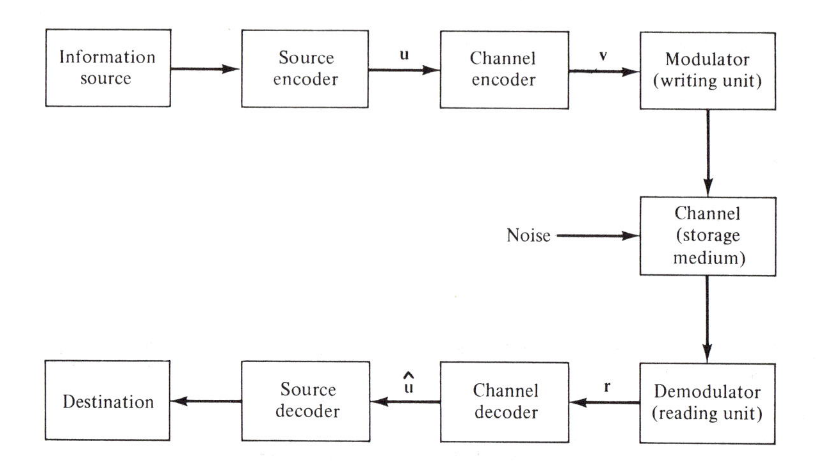

Figure 1.1 Block diagram of a typical data transmission or storage system.

The *source encoder* transforms the source output into a sequence of binary digits (bits) called the *information sequence* **u**. In the case of a continuous source, this involves analog-to-digital (A/D) conversion. The source encoder is ideally designed so that (1) the number of bits per unit time required to represent the source output is minimized, and (2) the source output can be reconstructed from the information sequence **u** without ambiguity. The subject of source coding is not discussed in this book. For a thorough treatment of this important topic, see References 2 and 3.

The *channel encoder* transforms the information sequence **u** into a discrete *encoded sequence* **v** called a *code word.* In most instances **v** is also a binary sequence, although in some applications nonbinary codes have been used. The design and implementation of channel encoders to combat the noisy environment in which code words must be transmitted or stored is one of the major topics of this book.

Discrete symbols are not suitable for transmission over a physical channel or recording on a digital storage medium. The *modulator* (or *writing unit*) transforms each output symbol of the channel encoder into a waveform of duration T seconds which is suitable for transmission (or recording). This waveform enters the *channel* (or *storage medium*) and is corrupted by noise. Typical transmission channels include telephone lines, high-frequency radio links, telemetry links, microwave links, satellite links, and so on. Typical storage media include core and semiconductor memories, magnetic tapes, drums, disk files, optical memory units, and so on. Each of these examples is subject to various types of noise disturbances. On a telephone line, the disturbance may come from switching impulse noise, thermal noise, crosstalk from other lines, or lightning. On magnetic tape, surface defects are regarded as a noise disturbance. The *demodulator* (or *reading unit*) processes each received waveform of duration T and produces an output that may be discrete (quantized) or continuous (unquantized). The sequence of demodulator outputs corresponding to the encoded sequence **v** is called the *received sequence* **r**.

The *channel decoder* transforms the received sequence **r** into a binary sequence $\hat{\mathbf{u}}$ called the *estimated sequence*. The decoding strategy is based on the rules of channel encoding and the noise characteristics of the channel (or storage medium). Ideally, $\hat{\mathbf{u}}$

will be a replica of the information sequence **u**, although the noise may cause some *decoding errors*. Another major topic of this book is the design and implementation of channel decoders that minimize the probability of decoding error.

The *source decoder* transforms the *estimated sequence* $\hat{\mathbf{u}}$ into an *estimate* of the source output and delivers this estimate to the *destination*. When the source is continuous, this involves digital-to-analog (D/A) conversion. In a well-designed system, the estimate will be a faithful reproduction of the source output except when the channel (or storage medium) is very noisy.

To focus attention on the channel encoder and channel decoder, (1) the information source and source encoder are combined into a *digital source* with output **u**; (2) the modulator (or writing unit), the channel (or storage medium), and the demodulator (or reading unit) are combined into a *coding channel* with input **v** and output **r**; and (3) the source decoder and destination are combined into a *digital sink* with input $\hat{\mathbf{u}}$. A simplified block diagram is shown in Figure 1.2.

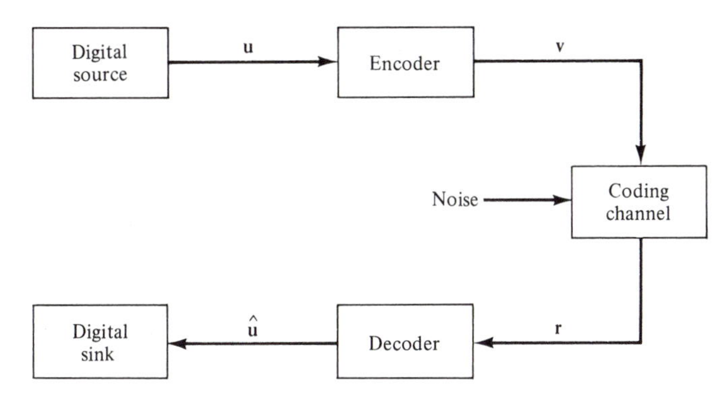

Figure 1.2 Simplified model of a coded system.

The major engineering problem that is addressed in this book is to design and implement the channel encoder/decoder pair such that (1) information can be transmitted (or recorded) in a noisy environment as fast as possible, (2) reliable reproduction of the information can be obtained at the output of the channel decoder, and (3) the cost of implementing the encoder and decoder falls within acceptable limits.

1.2 TYPES OF CODES

There are two different types of codes in common use today, block codes and convolutional codes. The encoder for a block code divides the information sequence into message blocks of k information bits each. A message block is represented by the binary k-tuple $\mathbf{u} = (u_1, u_2, \ldots, u_k)$ called a *message*. (In block coding, the symbol **u** is used to denote a k-bit message rather than the entire information sequence.) There are a total of 2^k different possible messages. The encoder transforms each message **u** independently into an n-tuple $\mathbf{v} = (v_1, v_2, \ldots, v_n)$ of discrete symbols called a *code word*. (In block coding, the symbol **v** is used to denote an n-symbol block rather than the entire encoded sequence.) Therefore, corresponding to the 2^k different possible messages, there are 2^k different possible code words at the encoder output. This set of 2^k code words of length n is called an (n, k) *block code*. The ratio $R = k/n$ is called the *code rate*, and can be interpreted as the number of information bits entering the encoder per transmitted symbol. Since the n-symbol output code word depends only on the corresponding k-bit input message, the encoder is memoryless, and can be implemented with a combinational logic circuit.

In a binary code, each code word $\mathbf{v}$ is also binary. Hence, for a binary code to be useful (i.e., to have a different code word assigned to each message), $k \leq n$ or $R \leq 1$. When $k < n$, $n - k$ redundant bits can be added to each message to form a code word. These redundant bits provide the code with the capability of combating the channel noise. For a fixed code rate R, more redundant bits can be added by increasing the block length n of the code while holding the ratio k/n constant. How to choose these redundant bits to achieve reliable transmission over a noisy channel is the major problem in designing the encoder. An example of a binary block code with $k = 4$ and $n = 7$ is shown in Table 1.1. Chapters 3 through 9 are devoted to the analysis and design of block codes for controlling errors in a noisy environment.

TABLE 1.1 BINARY BLOCK CODE WITH $k = 4$ AND $n = 7$

Messages	Code words
(0 0 0 0)	(0 0 0 0 0 0 0)
(1 0 0 0)	(1 1 0 1 0 0 0)
(0 1 0 0)	(0 1 1 0 1 0 0)
(1 1 0 0)	(1 0 1 1 1 0 0)
(0 0 1 0)	(1 1 1 0 0 1 0)
(1 0 1 0)	(0 0 1 1 0 1 0)
(0 1 1 0)	(1 0 0 0 1 1 0)
(1 1 1 0)	(0 1 0 1 1 1 0)
(0 0 0 1)	(1 0 1 0 0 0 1)
(1 0 0 1)	(0 1 1 1 0 0 1)
(0 1 0 1)	(1 1 0 0 1 0 1)
(1 1 0 1)	(0 0 0 1 1 0 1)
(0 0 1 1)	(0 1 0 0 0 1 1)
(1 0 1 1)	(1 0 0 1 0 1 1)
(0 1 1 1)	(0 0 1 0 1 1 1)
(1 1 1 1)	(1 1 1 1 1 1 1)

The encoder for a convolutional code also accepts k-bit blocks of the information sequence $\mathbf{u}$ and produces an encoded sequence (code word) $\mathbf{v}$ of n-symbol blocks. (In convolutional coding, the symbols $\mathbf{u}$ and $\mathbf{v}$ are used to denote sequences of blocks rather than a single block.) However, each encoded block depends not only on the corresponding k-bit message block at the same time unit, but also on m previous message blocks. Hence, the encoder has a *memory order* of m. The set of encoded sequences produced by a k-input, n-output encoder of memory order m is called an (n, k, m) *convolutional code*. The ratio $R = k/n$ is called the *code rate*. Since the encoder contains memory, it must be implemented with a sequential logic circuit.

In a binary convolutional code, redundant bits for combating the channel noise can be added to the information sequence when $k < n$ or $R < 1$. Typically, k and n are small integers and more redundancy is added by increasing the memory order m of the code while holding k and n, and hence the code rate R, fixed. How to use the memory to achieve reliable transmission over a noisy channel is the major problem in designing the encoder. An example of a binary convolutional encoder with $k = 1$, $n = 2$, and $m = 2$ is shown in Figure 1.3. As an illustration of how code words are generated, consider the information sequence $\mathbf{u} = (1\ 1\ 0\ 1\ 0\ 0\ 0\ \ldots)$, where the leftmost bit is assumed to enter the encoder first. Using the rules of exclusive-or addition, and assuming that the multiplexer takes the first encoded bit from the top output, it is easy to see that the encoded sequence is $\mathbf{v} = (1\ 1,\ 1\ 0,\ 1\ 0,\ 0\ 0,\ 0\ 1,\ 1\ 1,\ 0\ 0,\ 0\ 0,\ 0\ 0,\ \ldots)$. Chapters 10 through 14 are devoted to the analysis and design of convolutional codes for controlling errors in a noisy environment.

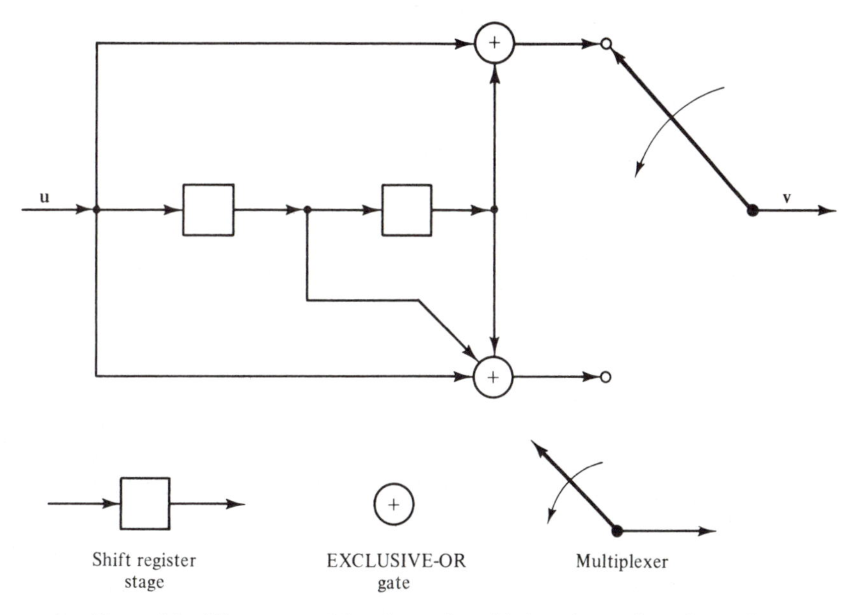

Figure 1.3 Binary convultional encoder with $k = 1$, $n = 2$, and $m = 2$.

1.3 MODULATION AND DEMODULATION

The modulator in a communication system must select a waveform of duration T seconds, which is suitable for transmission, for each encoder output symbol. In the case of a binary code, the modulator must generate one of two signals, $s_0(t)$ for an encoded "0" or $s_1(t)$ for an encoded "1." For a wideband channel, the optimum choice of signals is

$$\begin{aligned} s_0(t) &= \sqrt{\frac{2E}{T}} \sin\left(2\pi f_0 t + \frac{\pi}{2}\right), \qquad 0 \le t \le T \\ s_1(t) &= \sqrt{\frac{2E}{T}} \sin\left(2\pi f_0 t - \frac{\pi}{2}\right), \qquad 0 \le t \le T, \end{aligned} \tag{1.1}$$

where f_0 is a multiple of $1/T$ and E is the energy of each signal. This is called *binary-phase-shift-keyed* (BPSK) modulation, since the transmitted signal is a sine-wave pulse whose phase is either $+\pi/2$ or $-\pi/2$, depending on the encoder output. The BPSK modulated waveform corresponding to the code word $\mathbf{v} = (1\ 1\ 0\ 1\ 0\ 0\ 0)$ in the code of Table 1.1 is shown in Figure 1.4.

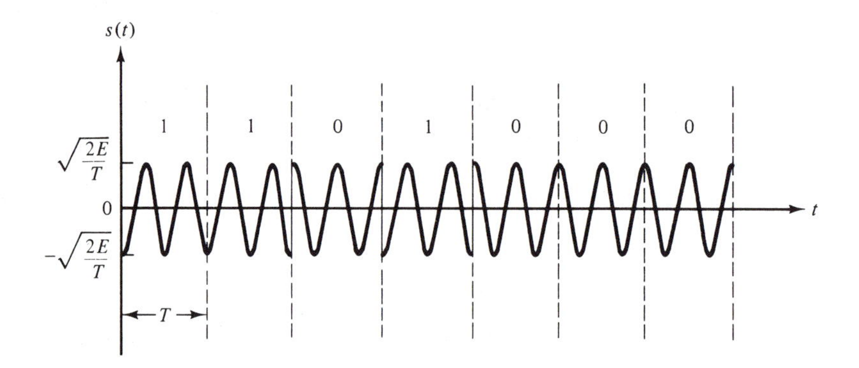

Figure 1.4 BPSK modulated waveform corresponding to the code word $\mathbf{v} = (1\ 1\ 0\ 1\ 0\ 0\ 0)$.

A common form of noise disturbance present in any communication system is *additive white Gaussian noise* (AWGN). If the transmitted signal is $s(t)$ [$= s_0(t)$ or $s_1(t)$], the received signal is

$$r(t) = s(t) + n(t), \tag{1.2}$$

where $n(t)$ is a Gaussian random process with one-sided *power spectral density* (PSD) N_0. Other forms of noise are also present in many systems. For example, in a communication system subject to multipath transmission, the received signal is observed to fade (lose strength) during certain time intervals. This fading can be modeled as a multiplicative noise component on the signal $s(t)$.

The demodulator must produce an output corresponding to the received signal in each T-second interval. This output may be a real number or one of a discrete set of preselected symbols, depending on the demodulator design. An optimum demodulator always includes a matched filter or correlation detector followed by a sampling switch. For BPSK modulation with coherent detection the sampled output is a real number,

$$\rho = \int_0^T r(t)\sqrt{\frac{2E}{T}} \sin\left(2\pi f_0 t + \frac{\pi}{2}\right) dt. \tag{1.3}$$

The sequence of unquantized demodulator outputs can be passed on directly to the channel decoder for processing. In this case, the channel decoder must be capable of handling analog inputs; that is, it must be an *analog decoder*. A much more common approach to decoding is to quantize the continuous detector output ρ into one of a finite number Q of discrete output symbols. In this case, the channel decoder has discrete inputs; that is, it must be a *digital decoder*. Almost all coded communication systems use some form of digital decoding.

If the detector output in a given interval depends only on the transmitted signal in that interval, and not on any previous transmission, the channel is said to be *memoryless*. In this case, the combination of an M-ary input modulator, the physical channel, and a Q-ary output demodulator can be modeled as a *discrete memoryless channel* (DMC). A DMC is completely described by a set of *transition probabilities* $P(j|i)$, $0 \leq i \leq M-1$, $0 \leq j \leq Q-1$, where i represents a modulator input symbol, j represents a demodulator output symbol, and $P(j|i)$ is the probability of receiving j given that i was transmitted. As an example, consider a communication system in which (1) binary modulation is used ($M = 2$), (2) the amplitude distribution of the noise is symmetric, and (3) the demodulator output is quantized to $Q = 2$ levels. In this case a particularly simple and practically important channel model, called the *binary symmetric channel* (BSC), results. The transition probability diagram for a BSC is shown in Figure 1.5(a). Note that the transition probability p completely describes the channel.

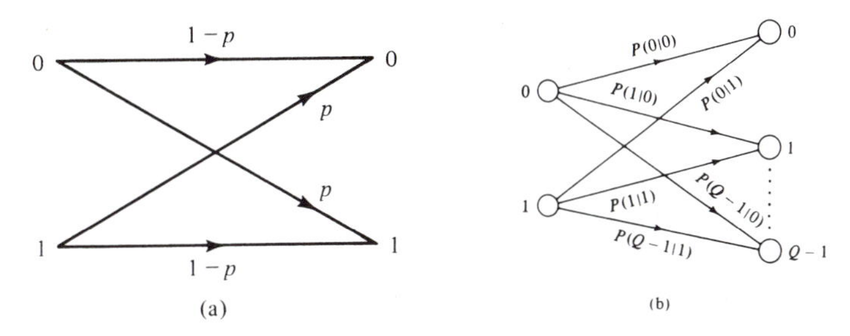

Figure 1.5 Transition probability diagrams: (a) binary symmetric channel; (b) binary-input, Q-ary-output discrete memoryless channel.

The transition probability p can be calculated from a knowledge of the signals used, the probability distribution of the noise, and the output quantization threshold of the demodulator. When BPSK modulation is used on an AWGN channel with optimum coherent detection and binary output quantization, the BSC transition probability is just the BPSK bit error probability for equally likely signals given by

$$p = Q\left(\sqrt{\frac{2E}{N_0}}\right), \tag{1.4}$$

where $Q(x) \triangleq (1/\sqrt{2\pi}) \int_x^{\infty} e^{-y^2/2}\, dy$ is the *complementary error function* of Gaussian statistics. An upper bound on $Q(x)$ which will be used later in evaluating the error performance of codes on a BSC is

$$Q(x) \leq \tfrac{1}{2} e^{-x^2/2}, \qquad x \geq 0. \tag{1.5}$$

When binary coding is used, the modulator has only binary inputs ($M = 2$). Similarly, when binary demodulator output quantization is used ($Q = 2$), the decoder has only binary inputs. In this case, the demodulator is said to make *hard decisions*. Most coded digital communication systems, whether block or convolutional, use binary coding with hard-decision decoding, owing to the resulting simplicity of implementation compared to nonbinary systems. However, some binary coded systems do not use hard decisions at the demodulator output. When $Q > 2$ (or the output is left unquantized) the demodulator is said to make *soft decisions.* In this case the decoder must accept multilevel (or analog) inputs. Although this makes the decoder more difficult to implement, soft-decision decoding offers significant performance improvement over hard-decision decoding, as discussed in Chapter 11. A transition probability diagram for a soft-decision DMC with $M = 2$ and $Q > 2$ is shown in Figure 1.5(b). This is the appropriate model for a binary-input AWGN channel with finite output quantization. The transition probabilities can be calculated from a knowledge of the signals used, the probability distribution of the noise, and the output quantization thresholds of the demodulator in a manner similar to the calculation of the BSC transition probability p. For a more thorough treatment of the calculation of DMC transition probabilities, see References 4 and 5.

If the detector output in a given interval depends on the transmitted signal in previous intervals as well as the transmitted signal in the present interval, the channel is said to have *memory.* A fading channel is a good example of a channel with memory, since the multipath transmission destroys the independence from interval to interval. Appropriate models for channels with memory are difficult to construct, and coding for these channels is normally done on an ad hoc basis.

Two important and related parameters in any digital communication system are the speed of information transmission and the bandwidth of the channel. Since one encoded symbol is transmitted every T seconds, the *symbol transmission rate* (baud rate) is $1/T$. In a coded system, if the code rate is $R = k/n$, k information bits correspond to the transmission of n symbols, and the *information transmission rate* (data rate) is R/T bits per second (bps). In addition to signal modification due to the effects of noise, all communication channels are subject to signal distortion due to bandwidth limitations. To minimize the effect of this distortion on the detection process, the channel should have a *bandwidth* W of roughly $1/2T$ hertz (Hz).[1] In an uncoded system ($R = 1$), the data rate is $1/T = 2W$, and is limited by the channel bandwidth. In a binary-coded system, with a code rate $R < 1$, the data rate is $R/T = 2RW$, and is reduced by the factor R compared to an uncoded system. Hence, to maintain the same data rate as the uncoded system, the coded system requires a *bandwidth expansion* by a factor of $1/R$. This is characteristic of binary-coded systems: they require some bandwidth expansion to maintain a constant data rate. If no additional bandwidth is available without undergoing severe signal distortion, binary coding is not feasible, and other means of reliable communication must be sought.[2]

1.4 MAXIMUM LIKELIHOOD DECODING

A block diagram of a coded system on an AWGN channel with finite output quantization is shown in Figure 1.6. In a block-coded system, the source output $\mathbf{u}$ represents a k-bit message, the encoder output $\mathbf{v}$ represents an n-symbol code word, the demodu-

[1]The exact bandwidth required depends on the shape of the signal waveform, the acceptable limits of distortion, and the definition of bandwidth.

[2]This does not preclude the use of coding, but requires only that a larger set of signals be found. See References 4 to 6.

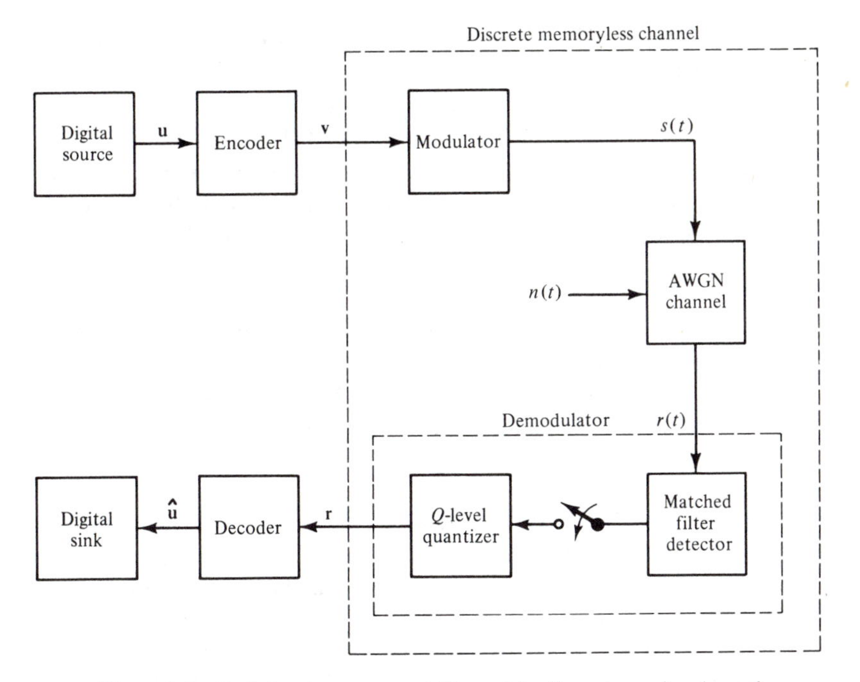

Figure 1.6 Coded system on an additive white Gaussian noise channel.

lator output $\mathbf{r}$ represents the corresponding Q-ary received n-tuple, and the decoder output $\hat{\mathbf{u}}$ represents the k-bit estimate of the encoded message. In a convolutional coded system, $\mathbf{u}$ represents a sequence of kL information bits and $\mathbf{v}$ represents a code word containing $N \triangleq nL + nm = n(L + m)$ symbols, where kL is the length of the information sequence and N is the length of the code word. The additional nm encoded symbols are produced after the last block of information bits has entered the encoder. This is due to the m time unit memory of the encoder, and is discussed more fully in Chapter 10. The demodulator output $\mathbf{r}$ is a Q-ary received N-tuple, and the decoder output $\hat{\mathbf{u}}$ is a kL-bit estimate of the information sequence.

The decoder must produce an estimate $\hat{\mathbf{u}}$ of the information sequence $\mathbf{u}$ based on the received sequence $\mathbf{r}$. Equivalently, since there is a one-to-one correspondence between the information sequence $\mathbf{u}$ and the code word $\mathbf{v}$, the decoder can produce an estimate $\hat{\mathbf{v}}$ of the code word $\mathbf{v}$. Clearly, $\hat{\mathbf{u}} = \mathbf{u}$ if and only if $\hat{\mathbf{v}} = \mathbf{v}$. A *decoding rule* is a strategy for choosing an estimated code word $\hat{\mathbf{v}}$ for each possible received sequence $\mathbf{r}$. If the code word $\mathbf{v}$ was transmitted, a *decoding error* occurs if and only if $\hat{\mathbf{v}} \neq \mathbf{v}$. Given that $\mathbf{r}$ is received, the *conditional error probability of the decoder* is defined as

$$P(E|\mathbf{r}) \triangleq P(\hat{\mathbf{v}} \neq \mathbf{v}|\mathbf{r}). \tag{1.6}$$

The *error probability of the decoder* is then given by

$$P(E) = \sum_{\mathbf{r}} P(E|\mathbf{r})P(\mathbf{r}). \tag{1.7}$$

$P(\mathbf{r})$ is independent of the decoding rule used since $\mathbf{r}$ is produced prior to decoding. Hence, an optimum decoding rule [i.e., one that minimizes $P(E)$] must minimize $P(E|\mathbf{r}) = P(\hat{\mathbf{v}} \neq \mathbf{v}|\mathbf{r})$ for all $\mathbf{r}$. Since minimizing $P(\hat{\mathbf{v}} \neq \mathbf{v}|\mathbf{r})$ is equivalent to maximizing $P(\hat{\mathbf{v}} = \mathbf{v} | \mathbf{r})$, $P(E|\mathbf{r})$ is minimized for a given $\mathbf{r}$ by choosing $\hat{\mathbf{v}}$ as the code word $\mathbf{v}$ which maximizes

$$P(\mathbf{v}|\mathbf{r}) = \frac{P(\mathbf{r}|\mathbf{v})P(\mathbf{v})}{P(\mathbf{r})}, \tag{1.8}$$

that is, $\hat{\mathbf{v}}$ is chosen as the most likely code word given that $\mathbf{r}$ is received. If all information sequences, and hence all code words, are equally likely [i.e., $P(\mathbf{v})$ is the same for all $\mathbf{v}$], maximizing (1.8) is equivalent to maximizing $P(\mathbf{r}|\mathbf{v})$. For a DMC

$$P(\mathbf{r}|\mathbf{v}) = \prod_i P(r_i | v_i), \tag{1.9}$$

since for a memoryless channel each received symbol depends only on the corresponding transmitted symbol. A decoder that chooses its estimate to maximize (1.9) is called a *maximum likelihood decoder* (MLD). Since $\log x$ is a monotone increasing function of x, maximizing (1.9) is equivalent to maximizing the *log-likelihood function*

$$\log P(\mathbf{r}|\mathbf{v}) = \sum_i \log P(r_i | v_i). \tag{1.10}$$

An MLD for a DMC then chooses $\hat{\mathbf{v}}$ as the code word $\mathbf{v}$ that maximizes the sum in (1.10). If the code words are not equally likely, an MLD is not necessarily optimum, since the conditional probabilities $P(\mathbf{r}|\mathbf{v})$ must be weighted by the code word probabilities $P(\mathbf{v})$ to determine which code word maximizes $P(\mathbf{v}|\mathbf{r})$. However, in many systems, the code word probabilities are not known exactly at the receiver, making optimum decoding impossible, and an MLD then becomes the best feasible decoding rule.

Now consider specializing the MLD decoding rule to the BSC. In this case $\mathbf{r}$ is a binary sequence which may differ from the transmitted code word $\mathbf{v}$ in some positions because of the channel noise. When $r_i \neq v_i$, $P(r_i | v_i) = p$, and when $r_i = v_i$, $P(r_i | v_i) = 1 - p$. Let $d(\mathbf{r}, \mathbf{v})$, be the distance between $\mathbf{r}$ and $\mathbf{v}$ (i.e., the number of positions in which $\mathbf{r}$ and $\mathbf{v}$ differ). For a block code of length n, (1.10) becomes

$$\begin{aligned} \log P(\mathbf{r}|\mathbf{v}) &= d(\mathbf{r}, \mathbf{v}) \log p + [n - d(\mathbf{r}, \mathbf{v})] \log (1 - p) \\ &= d(\mathbf{r}, \mathbf{v}) \log \frac{p}{1-p} + n \log (1 - p). \end{aligned} \tag{1.11}$$

[For a convolutional code, n in (1.11) is replaced by N.] Since $\log [p/(1 - p)] < 0$ for $p < \frac{1}{2}$ and $n \log (1 - p)$ is a constant for all $\mathbf{v}$, the MLD decoding rule for the BSC *chooses* $\hat{\mathbf{v}}$ *as the code word* $\mathbf{v}$ *which minimizes the distance* $d(\mathbf{r}, \mathbf{v})$ *between* $\mathbf{r}$ *and* $\mathbf{v}$; *that is, it chooses the code word that differs from the received sequence in the fewest number of positions.* Hence, an MLD for the BSC is sometimes called a *minimum distance decoder*.

The capability of a noisy channel to transmit information reliably was determined by Shannon [1] in his original work. This result, called the *noisy channel coding theorem*, states that every channel has a *channel capacity* C, and that for any rate $R < C$, there exists codes of rate R which, with maximum likelihood decoding, have an arbitrarily small decoding error probability $P(E)$. In particular, for any $R < C$, there exists block codes of length n such that

$$P(E) \leq 2^{-nE_b(R)}, \tag{1.12}$$

and there exists convolutional codes of *memory order* m such that

$$P(E) \leq 2^{-(m+1)nE_c(R)} = 2^{-n_A E_c(R)}, \tag{1.13}$$

where $n_A \triangleq (m + 1)n$ is called the code *constraint length*. $E_b(R)$ and $E_c(R)$ are positive functions of R for $R < C$ and are completely determined by the channel characteristics. The bound of (1.12) implies that arbitrarily small error probabilities are achievable with block coding for any fixed $R < C$ by increasing the block length n while holding the ratio k/n constant. The bound of (1.13) implies that arbitrarily small error probabilities are achievable with convolutional coding for any fixed $R < C$ by increasing the constraint length n_A (i.e., by increasing the memory order m while holding k and n constant).

The noisy channel coding theorem is based on an argument called *random coding*. The bound obtained is actually on the average error probability of the ensemble of all codes. Since some codes must perform better than the average, the noisy channel coding theorem guarantees the existence of codes satisfying (1.12) and (1.13), but does not indicate how to construct these codes. Furthermore, to achieve very low error probabilities for block codes of fixed rate $R < C$, long block lengths are needed. This requires that the number of code words $2^k = 2^{nR}$ must be very large. Since a MLD must compute $\log P(\mathbf{r} \mid \mathbf{v})$ for each code word, and then choose the code word that gives the maximum, the number of computations that must be performed by a MLD becomes excessively large. For convolutional codes, low error probabilities require a large memory order m. As will be seen in Chapter 11, a MLD for convolutional codes requires approximately 2^{km} computations to decode each block of k information bits. This, too, becomes excessively large as m increases. Hence, it is impractical to achieve very low error probabilities with maximum likelihood decoding. Therefore, two major problems are encountered when designing a coded system to achieve low error probabilities: (1) to construct good long codes whose performance with maximum likelihood decoding would satisfy (1.12) and (1.13), and (2) to find easily implementable methods of encoding and decoding these codes such that their actual performance is close to what could be achieved with maximum likelihood decoding. The remainder of this book is devoted to finding solutions to these two problems.

1.5 TYPES OF ERRORS

On memoryless channels, the noise affects each transmitted symbol independently. As an example, consider the BSC whose transition diagram is shown in Figure 1.5(a). Each transmitted bit has a probability p of being received incorrectly and a probability $1 - p$ of being received correctly, independently of other transmitted bits. Hence transmission errors occur randomly in the received sequence, and memoryless channels are called *random-error channels*. Good examples of random-error channels are the deep-space channel and many satellite channels. Most line-of-sight transmission facilities, as well, are affected primarily by random errors. The codes devised for correcting random errors are called *random-error-correcting codes*. Most of the codes presented in this book are random-error-correcting codes. In particular, Chapters 3 through 8 and 10 through 13 are devoted to codes of this type.

On channels with memory, the noise is not independent from transmission to transmission. A simplified model of a channel with memory is shown in Figure 1.7. This model contains two states, a "good state," in which transmission errors occur infrequently, $p_1 \approx 0$, and a "bad state," in which transmission errors are highly

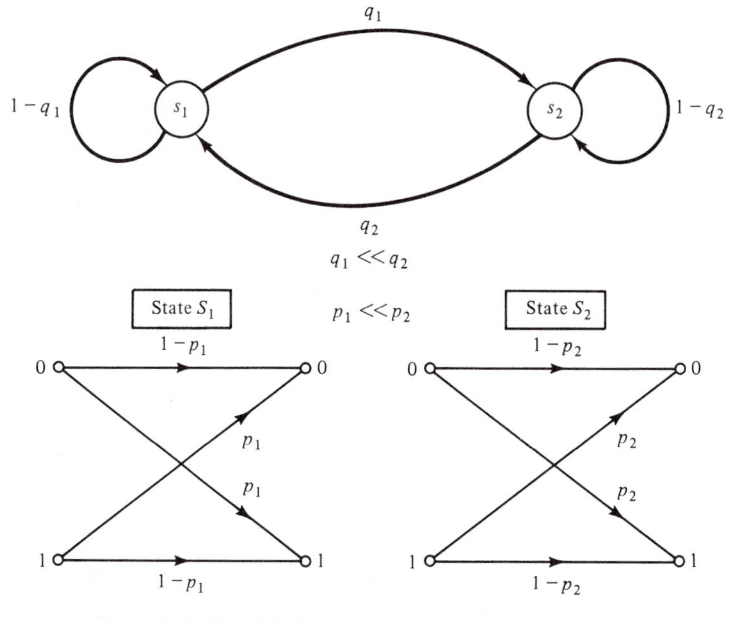

Figure 1.7 Simplified model of a channel with memory.

probable, $p_2 \approx 0.5$. The channel is in the good state most of the time, but on occasion shifts to the bad state due to a change in the transmission characteristic of the channel (e.g., a "deep fade" caused by multipath transmission). As a consequence, transmission errors occur in clusters or bursts because of the high transition probability in the bad state, and channels with memory are called *burst-error channels.* Examples of burst-error channels are radio channels, where the error bursts are caused by signal fading due to multipath transmission, wire and cable transmission, which is affected by impulsive switching noise and crosstalk, and magnetic recording, which is subject to tape dropouts due to surface defects and dust particles. The codes devised for correcting burst errors are called *burst-error-correcting codes.* Sections 9.1 to 9.5 and 14.1 to 14.3 are devoted to codes of this type.

Finally, some channels contain a combination of both random and burst errors. These are called *compound channels*, and codes devised for correcting errors on these channels are called *burst-and-random-error-correcting codes.* Sections 9.6, 9.7, and 14.4 are devoted to codes of this type.

1.6 ERROR CONTROL STRATEGIES

The block diagram shown in Figure 1.1 represents a one-way system. The transmission (or recording) is strictly in one direction, from transmitter to receiver. Error control for a one-way system must be accomplished using *forward error correction* (FEC), that is, by employing error-correcting codes that automatically correct errors detected at the receiver. Examples are magnetic tape storage systems, in which the information recorded on tape may be replayed weeks or even months after it is recorded, and deep-space communication systems, where the relatively simple encoding equipment can be placed aboard the spacecraft, but the much more complex decoding procedure must be performed on earth. Most of the coded systems in use today employ some form of FEC, even if the channel is not strictly one-way. This book is devoted mostly to the analysis and design of FEC systems. Applications of FEC to storage and communication systems are presented in Chapter 16 and Sections 17.1 to 17.4.

In some cases, a transmission system can be two-way; that is, information can be sent in both directions and the transmitter also acts as a receiver (a transceiver), and vice versa. Examples of two-way systems are telephone channels and some satellite communication systems. Error control for a two-way system can be accomplished using error detection and retransmission, called *automatic repeat request* (ARQ). In an ARQ system, when errors are detected at the receiver, a request is sent for the transmitter to repeat the message, and this continues until the message is received correctly.

There are two types of ARQ systems: stop-and-wait ARQ and continuous ARQ. With *stop-and-wait ARQ*, the transmitter sends a code word to the receiver and waits for a positive (ACK) or negative (NAK) acknowledgment from the receiver. If ACK is received (no errors detected), the transmitter sends the next code word. If NAK is received (errors detected), it resends the preceding code word. When the noise is persistent, the same code word may be retransmitted several times before it is correctly received and acknowledged.

With *continuous ARQ*, the transmitter sends code words to the receiver continuously and receives acknowledgments continuously. When a NAK is received, the transmitter begins a retransmission. It may back up to the code word in error and resend that word plus the words that follow it. This is called *go-back-N ARQ*. Alternatively, the transmitter may simply resend only those code words that are acknowledged negatively. This is known as *selective-repeat ARQ*. Selective-repeat ARQ is more efficient than go-back-N ARQ, but requires more logic and buffering.

Continuous ARQ is more efficient than stop-and-wait ARQ, but it is also more expensive. In a satellite communication system where the transmission rate is high and the round-trip delay is long, continuous ARQ is normally used. Stop-and-wait ARQ is used in systems where the time taken to transmit a code word is long compared to the time taken to receive an acknowledgment. Stop-and-wait ARQ is designed for use on half-duplex channels, whereas continuous ARQ is designed for use on full-duplex channels.

The major advantage of ARQ over FEC is that error detection requires much simpler decoding equipment than does error correction. Also, ARQ is adaptive in the sense that information is retransmitted only when errors occur. On the other hand, when the channel error rate is high, retransmissions must be sent too frequently, and the system throughput, the rate at which newly generated messages are correctly received, is lowered by ARQ. In this situation, a combination of FEC for the most frequent error patterns, together with error detection and retransmission for the less likely error patterns, is more efficient than ARQ alone. Although this hybrid ARQ error control strategy has not been implemented in many systems, it clearly carries the potential for improving throughput in two-way systems subject to a high channel error rate. Various types of ARQ and hybrid ARQ schemes are discussed in Chapter 15 and Section 17.5.

REFERENCES

1. C. E. Shannon, "A Mathematical Theory of Communication," *Bell Syst. Tech. J.*, 27, pp. 379–423 (Part I), 623–656 (Part II), July 1948.
2. T. Berger, *Rate Distortion Theory*, Prentice-Hall, Englewood Cliffs, N.J., 1971.
3. L. Davisson and R. Gray, eds., *Data Compression*, Dowden, Hutchinson, & Ross, Stroudsburg, Pa., 1976.
4. J. M. Wozencraft and I. M. Jacobs, *Principles of Communication Engineering*, Wiley, New York, 1965.
5. A. J. Viterbi and J. K. Omura, *Principles of Digital Communication and Coding*, McGraw-Hill, New York, 1979.
6. R. G. Gallager, *Information Theory and Reliable Communication*, Wiley, New York, 1968.

2

Introduction to Algebra

The purpose of this chapter is to provide the reader with an elementary knowledge of algebra that will aid in the understanding of the material in the following chapters. The treatment is basically descriptive and no attempt is made to be mathematically rigorous. There are many good textbooks on algebra. The reader who is interested in more advance algebraic coding theory is referred to the textbooks listed at the end of the chapter. Birkhoff and MacLane [2] is probably the most easily understood text on modern algebra. Fraleigh [4] is also a good and fairly simple text.

2.1 GROUPS

Let G be a set of elements. A *binary operation* $*$ on G is a *rule* that assigns to each pair of elements a and b a uniquely defined third element $c = a * b$ in G. When such a binary operation $*$ is defined on G, we say that G is *closed* under $*$. For example, let G be the set of all integers and let the binary operation on G be real addition $+$. We all know that, for any two integers i and j in G, $i + j$ is a uniquely defined integer in G. Hence, the set of integers is closed under real addition. A binary operation $*$ on G is said to be *associative* if, for any a, b, and c in G,

$$a * (b * c) = (a * b) * c.$$

Now, we introduce a useful algebraic system called a *group*.

Definition 2.1. A set G on which a binary operation $*$ is defined is called a *group* if the following conditions are satisfied:

(i) The binary operation $*$ is associative.

(ii) G contains an element e such that, for any a in G,

$$a * e = e * a = a.$$

This element e is called an *identity* element of G.

(iii) For any element a in G, there exists another element a' in G such that

$$a * a' = a' * a = e.$$

The element a' is called an *inverse* of a (a is also an inverse of a').

A group G is said to be *commutative* if its binary operation $*$ also satisfies the following condition: For any a and b in G,

$$a * b = b * a.$$

Theorem 2.1. The identity element in a group G is unique.

Proof. Suppose that there exist two identity elements e and e' in G. Then $e' = e' * e = e$. This implies that e and e' are identical. Therefore, there is one and only one identity element. Q.E.D.

Theorem 2.2. The inverse of a group element is unique.

Proof. Suppose that there exist two inverses a' and a'' for a group element a. Then

$$a' = a' * e = a' * (a * a'') = (a' * a) * a'' = e * a'' = a''.$$

This implies that a' and a'' are identical and there is only one inverse for a.

Q.E.D.

The set of all integers is a commutative group under real addition. In this case, the integer 0 is the identity element and the integer $-i$ is the inverse of integer i. The set of all rational numbers excluding zero is a commutative group under real multiplication. The integer 1 is the identity element with respect to real multiplication, and the rational number b/a is the multiplicative inverse of a/b. The groups noted above contain infinite numbers of elements. Groups with finite numbers of elements do exist, as we shall see in the next example.

Example 2.1

Consider the set of two integers, $G = \{0, 1\}$. Let us define a binary operation, denoted by $\oplus$, on G as follows:

$$0 \oplus 0 = 0, \qquad 0 \oplus 1 = 1, \qquad 1 \oplus 0 = 1, \qquad 1 \oplus 1 = 0.$$

This binary operation is called *modulo-2* addition. The set $G = \{0, 1\}$ is a group under modulo-2 addition. It follows from the definition of modulo-2 addition $\oplus$ that G is closed under $\oplus$ and $\oplus$ is commutative. We can easily check that $\oplus$ is associative. The element 0 is the identity element. The inverse of 0 is itself and the inverse of 1 is also itself. Thus, G together with $\oplus$ is a commutative group.

The number of elements in a group is called the *order* of the group. A group of finite order is called a *finite* group. For any positive integer m, it is possible to construct a group of order m under a binary operation which is very similar to real addition. This is shown in the next example.

Example 2.2

Let m be a positive integer. Consider the set of integers $G = \{0, 1, 2, \ldots, m-1\}$. Let $+$ denote real addition. Define a binary operation $\boxplus$ on G as follows: For any integers i and j in G,

$$i \boxplus j = r,$$

where r is the *remainder* resulting from dividing $i + j$ by m. The remainder r is an integer between 0 and $m - 1$ (Euclid's division algorithm) and is therefore in G. Hence, G is closed under the binary operation $\boxplus$, which is called *modulo-m addition.* The set $G = \{0, 1, \ldots, m-1\}$ is a group under modulo-m addition. First we see that 0 is the identity element. For $0 < i < m$, i and $m - i$ are both in G. Since

$$i + (m - i) = (m - i) + i = m,$$

it follows from the definition of modulo-m addition that

$$i \boxplus (m - i) = (m - i) \boxplus i = 0.$$

Therefore, i and $m - i$ are inverses to each other with respect to $\boxplus$. It is also clear that the inverse of 0 is itself. Since real addition is commutative, it follows from the definition of modulo-m addition that, for any i and j in G, $i \boxplus j = j \boxplus i$. Therefore, modulo-$m$ addition is commutative. Next, we show that modulo-m addition is also associative. Let i, j, and k be three integers in G. Since real addition is associative, we have

$$i + j + k = (i + j) + k = i + (j + k).$$

Dividing $i + j + k$ by m, we obtain

$$i + j + k = qm + r,$$

where q and r are the quotient and the remainder, respectively, and $0 \leq r < m$. Now, dividing $i + j$ by m, we have

$$i + j = q_1 m + r_1 \tag{2.1}$$

with $0 \leq r_1 < m$. Therefore, $i \boxplus j = r_1$. Dividing $r_1 + k$ by m, we obtain

$$r_1 + k = q_2 m + r_2 \tag{2.2}$$

with $0 \leq r_2 < m$. Hence, $r_1 \boxplus k = r_2$ and

$$(i \boxplus j) \boxplus k = r_2.$$

Combining (2.1) and (2.2), we have

$$i + j + k = (q_1 + q_2)m + r_2.$$

This implies that r_2 is also the remainder when $i + j + k$ is divided by m. Since the remainder resulting from dividing an integer by another integer is unique, we must have $r_2 = r$. As a result, we have

$$(i \boxplus j) \boxplus k = r.$$

Similarly, we can show that

$$i \boxplus (j \boxplus k) = r.$$

Therefore, $(i \boxplus j) \boxplus k = i \boxplus (j \boxplus k)$ and modulo-m addition is associative. This concludes our proof that the set $G = \{0, 1, 2, \ldots, m-1\}$ is a group under modulo-

m addition. We shall call this group an *additive* group. For $m = 2$, we obtain the binary group given in Example 2.1.

The additive group under modulo-5 addition is given by Table 2.1.

TABLE 2.1 MODULO-5 ADDITION

$\boxplus$	0	1	2	3	4
0	0	1	2	3	4
1	1	2	3	4	0
2	2	3	4	0	1
3	3	4	0	1	2
4	4	0	1	2	3

Finite groups with a binary operation similar to real multiplication can also be constructed.

Example 2.3

Let p be a prime (e.g., $p = 2, 3, 5, 7, 11, \ldots$). Consider the set of integers, $G = \{1, 2, 3, \ldots, p - 1\}$. Let $\cdot$ denote real multiplication. Define a binary operation $\boxdot$ on G as follows: For i and j in G,

$$i \boxdot j = r,$$

where r is the remainder resulting from dividing $i \cdot j$ by p. First we note that $i \cdot j$ is not divisible by p. Hence, $0 < r < p$ and r is an element in G. Therefore, the set G is closed under the binary operation $\boxdot$, which is referred to as *modulo-p multiplication*. The set $G = \{1, 2, \ldots, p - 1\}$ is a group under modulo-p multiplication. We can easily check that modulo-p multiplication is commutative and associative. The identity element is 1. The only thing left to be proved is that every element in G has an inverse. Let i be an element in G. Since p is a prime and $i < p$, i and p must be relatively prime (i.e., i and p do not have any common factor greater than 1). It is well known that there exist two integers a and b such that

$$a \cdot i + b \cdot p = 1 \tag{2.3}$$

and a and p are relatively prime (Euclid's theorem). Rearranging (2.3), we have

$$a \cdot i = -b \cdot p + 1. \tag{2.4}$$

This says that when $a \cdot i$ is divided by p, the remainder is 1. If $0 < a < p$, a is in G and it follows from (2.4) and the definition of modulo-p multiplication that

$$a \boxdot i = i \boxdot a = 1.$$

Therefore, a is the inverse of i. However, if a is not in G, we divide a by p,

$$a = q \cdot p + r. \tag{2.5}$$

Since a and p are relatively prime, the remainder r cannot be 0 and it must be between 1 and $p - 1$. Therefore, r is in G. Now, combining (2.4) and (2.5), we obtain

$$r \cdot i = -(b + qi)p + 1.$$

Therefore, $r \boxdot i = i \boxdot r = 1$ and r is the inverse of i. Hence, any element i in G has an inverse with respect to modulo-p multiplication. The group $G = \{1, 2, \ldots, p - 1\}$

under modulo-p multiplication is called a *multiplicative* group. For $p = 2$, we obtain a group $G = \{1\}$ with only one element under modulo-2 multiplication.

If p is *not* a prime, the set $G = \{1, 2, \ldots, p - 1\}$ is not a group under modulo-p multiplication (see Problem 2.3). Table 2.2 illustrates the group $G = \{1, 2, 3, 4\}$ under modulo-5 multiplication.

TABLE 2.2 MODULO-5 MULTIPLICATION

$\boxdot$	1	2	3	4
1	1	2	3	4
2	2	4	1	3
3	3	1	4	2
4	4	3	2	1

Let H be a nonempty subset of G. The subset H is said to be a *subgroup* of G if H is closed under the group operation of G and satisfies all the conditions of a group. For example, the set of all rational numbers is a group under real addition. The set of all integers is a subgroup of the group of rational numbers under real addition.

2.2 FIELDS

Now, we use the group concepts to introduce another algebraic system, called a *field.* Roughly speaking, a field is a set of elements in which we can do addition, subtraction, multiplication, and division without leaving the set. Addition and multiplication must satisfy the commutative, associative, and distributive laws. A formal definition of a field is given below.

Definition 2.2. Let F be a set of elements on which two binary operations, called addition "$+$" and multiplication "$\cdot$," are defined. The set F together with the two binary operations $+$ and $\cdot$ is a field if the following conditions are satisfied:

(i) F is a commutative group under addition $+$. The identity element with respect to addition is called the *zero* element or the additive identity of F and is denoted by 0.

(ii) The set of nonzero elements in F is a commutative group under multiplication $\cdot$. The identity element with respect to multiplication is called the *unit* element or the multiplicative identity of F and is denoted by 1.

(iii) Multiplication is *distributive* over addition; that is, for any three elements a, b, and c in F,

$$a \cdot (b + c) = a \cdot b + a \cdot c.$$

It follows from the definition that a field consists of at least two elements, the additive identity and the multiplicative identity. Later, we will show that a field of two elements does exist. The number of elements in a field is called the *order* of the field. A field with finite number of elements is called a *finite field.* In a field, the additive inverse of an element a is denoted by $-a$, and the multiplicative inverse of a is denoted by a^{-1}, provided that $a \neq 0$. Subtracting a field element b from another field element a is defined as adding the additive inverse $-b$ of b to a [i.e., $a - b \triangleq a + (-b)$]. If b is a nonzero element, dividing a by b is defined as multiplying a by the multiplicative inverse b^{-1} of b [i.e., $a \div b \triangleq a \cdot b^{-1}$].

A number of basic properties of fields can be derived from the definition of a field.

Property I. For every element a in a field, $a \cdot 0 = 0 \cdot a = 0$.

Proof. First we note that

$$a = a \cdot 1 = a \cdot (1 + 0) = a + a \cdot 0.$$

Adding $-a$ to both sides of the equality above, we have

$$-a + a = -a + a + a \cdot 0$$
$$0 = 0 + a \cdot 0$$
$$0 = a \cdot 0.$$

Similarly, we can show that $0 \cdot a = 0$. Therefore, we obtain $a \cdot 0 = 0 \cdot a = 0$.

Q.E.D.

Property II. For any two nonzero elements a and b in a field, $a \cdot b \neq 0$.

Proof. This is a direct consequence of the fact that the nonzero elements of a field are closed under multiplication. Q.E.D.

Property III. $a \cdot b = 0$ and $a \neq 0$ imply that $b = 0$.

Proof. This is a direct consequence of Property II. Q.E.D.

Property IV. For any two elements a and b in a field,

$$-(a \cdot b) = (-a) \cdot b = a \cdot (-b).$$

Proof. $0 = 0 \cdot b = (a + (-a)) \cdot b = a \cdot b + (-a) \cdot b$. Therefore, $(-a) \cdot b$ must be the additive inverse of $a \cdot b$ and $-(a \cdot b) = (-a) \cdot b$. Similarly, we can prove that $-(a \cdot b) = a \cdot (-b)$. Q.E.D.

Property V. For $a \neq 0$, $a \cdot b = a \cdot c$ implies that $b = c$.

Proof. Since a is a nonzero element in the field, it has a multiplicative inverse a^{-1}. Multiplying both side of $a \cdot b = a \cdot c$ by a^{-1}, we obtain

$$a^{-1} \cdot (a \cdot b) = a^{-1} \cdot (a \cdot c)$$
$$(a^{-1} \cdot a) \cdot b = (a^{-1} \cdot a) \cdot c$$
$$1 \cdot b = 1 \cdot c.$$

Thus, $b = c$. Q.E.D.

We can verify readily that the set of real numbers is a field under real number addition and multiplication. This field has an infinite number of elements. Fields with finite number of elements can be constructed and are illustrated in the next two examples and in Section 2.4.

Example 2.4

Consider the set {0, 1} together with modulo-2 addition and multiplication shown in Tables 2.3 and 2.4. In Example 2.1 we have shown that {0, 1} is a commutative group under modulo-2 addition; and in Example 2.3, we have shown that {1} is a group under modulo-2 multiplication. We can easily check that modulo-2 multiplication is distributive over modulo-2 addition by simply computing $a \cdot (b + c)$ and $a \cdot b + a \cdot c$ for eight possible combinations of a, b and c ($a = 0$ or 1, $b = 0$ or 1 and $c = 0$ or 1). Therefore, the set {0, 1} is a field of two elements under modulo-2 addition and modulo-2 multiplication.

TABLE 2.3
MODULO-2 ADDITION

+	0	1
0	0	1
1	1	0

TABLE 2.4
MODULO-2 MULTIPLICATION

·	0	1
0	0	0
1	0	1

The field given in Example 2.4 is usually called a *binary* field and is denoted by GF(2). The binary field GF(2) plays an important role in coding theory and is widely used in digital computers and digital data transmission (or storage) systems.

Example 2.5

Let p be a prime. We have shown in Example 2.2 that the set of integers $\{0, 1, 2, \ldots, p-1\}$ is a commutative group under modulo-p addition. We have also shown in Example 2.3 that the nonzero elements, $\{1, 2, \ldots, p-1\}$ form a commutative group under modulo-p multiplication. Following the definitions of modulo-p addition and multiplication and the fact that real number multiplication is distributive over real number addition, we can show that modulo-p multiplication is distributive over modulo-p addition. Therefore, the set $\{0, 1, 2, \ldots, p-1\}$ is a field of order p under modulo-p addition and multiplication. Since this field is constructed from a prime p, it is called a *prime field* and is denoted by GF(p). For $p = 2$, we obtain the binary field GF(2).

Let $p = 7$. Modulo-7 addition and multiplication are given by Tables 2.5 and 2.6, respectively. The set of integers {0, 1, 2, 3, 4, 5, 6} is a field of seven elements, denoted by GF(7), under modulo-7 addition and multiplication. The addition table is also used for subtraction. For example, if we want to subtract 6 from 3, we first use the addition table to find the additive inverse of 6, which is 1. Then we add 1 to 3 to obtain the result [i.e., $3 - 6 = 3 + (-6) = 3 + 1 = 4$]. For division, we use the multiplication table. Suppose that we divide 3 by 2. We first find the multiplicative inverse of 2, which is 4, and then we multiply 3 by 4 to obtain the result [i.e., $3 \div 2 = 3 \cdot (2^{-1}) = 3 \cdot 4 = 5$]. Here we have demonstrated that, in a finite field, addition, subtraction, multiplication, and division can be carried out in a manner similar to ordinary arithmetic, with which we are quite familiar.

In Example 2.5 we have shown that, for any prime p, there exists a finite field

TABLE 2.5
MODULO-7 ADDITION

+	0	1	2	3	4	5	6
0	0	1	2	3	4	5	6
1	1	2	3	4	5	6	0
2	2	3	4	5	6	0	1
3	3	4	5	6	0	1	2
4	4	5	6	0	1	2	3
5	5	6	0	1	2	3	4
6	6	0	1	2	3	4	5

TABLE 2.6
MODULO-7 MULTIPLICATION

·	0	1	2	3	4	5	6
0	0	0	0	0	0	0	0
1	0	1	2	3	4	5	6
2	0	2	4	6	1	3	5
3	0	3	6	2	5	1	4
4	0	4	1	5	2	6	3
5	0	5	3	1	6	4	2
6	0	6	5	4	3	2	1

of p elements. In fact, for any positive integer m, it is possible to extend the prime field GF(p) to a field of p^m elements which is called an *extension field* of GF(p) and is denoted by GF(p^m). Furthermore, it has been proved that the order of any finite field is a power of a prime. Finite fields are also called *Galois* fields, in honor of their discoverer. A large portion of algebraic coding theory, code construction, and decoding is built around finite fields. In the rest of this section and in the next two sections we examine some basic structures of finite fields, their arithmetic, and the construction of extension fields from prime fields. Our presentation will be mainly descriptive and no attempt is made to be mathematically rigorous. Since finite-field arithmetic is very similar to ordinary arithmetic, most of the rules of ordinary arithmetic apply to finite-field arithmetic. Therefore, it is possible to utilize most of the techniques of algebra in the computations over finite fields.

Consider a finite field of q elements, GF(q). Let us form the following sequence of sums of the unit element 1 in GF(q):

$$\sum_{i=1}^{1} 1 = 1, \quad \sum_{i=1}^{2} 1 = 1 + 1, \quad \sum_{i=1}^{3} 1 = 1 + 1 + 1, \quad \ldots,$$

$$\sum_{i=1}^{k} 1 = 1 + 1 + \cdots + 1 \ (k \text{ times}), \quad \ldots$$

Since the field is closed under addition, these sums must be elements in the field. Since the field has finite number of elements, these sums cannot be all distinct. Therefore, at some point of the sequence of sums, there must be a repetition; that is, there must exist two positive integers m and n such that $m < n$ and

$$\sum_{i=1}^{m} 1 = \sum_{i=1}^{n} 1.$$

This implies that $\sum_{i=1}^{n-m} 1 = 0$. Therefore, there must exist a *smallest positive integer* λ such that $\sum_{i=1}^{\lambda} 1 = 0$. This integer λ is called the *characteristic* of the field GF(q). The characteristic of the binary field GF(2) is 2, since $1 + 1 = 0$. The characteristic of the prime field GF(p) is p, since $\sum_{i=1}^{k} 1 = k \neq 0$ for $1 \leq k < p$ and $\sum_{i=1}^{p} 1 = 0$.

Theorem 2.3. The characteristic λ of a finite field is prime.

Proof. Suppose that λ is not a prime and is equal to the product of two smaller integers k and m (i.e., $\lambda = km$). Since the field is closed under multiplication,

$$\left(\sum_{i=1}^{k} 1\right) \cdot \left(\sum_{i=1}^{m} 1\right)$$

is also a field element. It follows from the distributive law that

$$\left(\sum_{i=1}^{k} 1\right) \cdot \left(\sum_{i=1}^{m} 1\right) = \sum_{i=1}^{km} 1.$$

Since $\sum_{i=1}^{km} 1 = 0$, then either $\sum_{i=1}^{k} 1 = 0$ or $\sum_{i=1}^{m} 1 = 0$. However, this contradicts the definition that λ is the smallest positive integer such that $\sum_{i=1}^{\lambda} 1 = 0$. Therefore, we conclude that λ is prime. Q.E.D.

It follows from the definition of the characteristic of a finite field that for any two distinct positive integers k and m less than λ,

$$\sum_{i=1}^{k} 1 \neq \sum_{i=1}^{m} 1.$$

Suppose that $\sum_{i=1}^{k} 1 = \sum_{i=1}^{m} 1$. Then we have

$$\sum_{i=1}^{m-k} 1 = 0$$

(assuming that $m > k$). However, this is impossible since $m - k < \lambda$. Therefore, the sums

$$1 = \sum_{i=1}^{1} 1, \quad \sum_{i=1}^{2} 1, \quad \sum_{i=1}^{3} 1, \quad \ldots, \quad \sum_{i=1}^{\lambda-1} 1, \quad \sum_{i=1}^{\lambda} 1 = 0$$

are λ distinct elements in GF(q). In fact, this set of sums itself is a field of λ elements, GF(λ), under the addition and multiplication of GF(q) (see Problem 2.6). Since GF(λ) is a subset of GF(q), GF(λ) is called a *subfield* of GF(q). Therefore, any finite field GF(q) of characteristic λ contains a subfield of λ elements. It can be proved that if $q \neq \lambda$, then q is a power of λ.

Now let a be a nonzero element in GF(q). Since the set of nonzero elements of GF(q) is closed under multiplication, the following powers of a,

$$a^1 = a, \quad a^2 = a \cdot a, \quad a^3 = a \cdot a \cdot a, \quad \ldots$$

must also be nonzero elements in GF(q). Since GF(q) has only a finite number of elements, the powers of a given above cannot all be distinct. Therefore, at some point of the sequence of powers of a, there must be a repetition; that is, there must exist two positive integers k and m such that $m > k$ and $a^k = a^m$. Let a^{-1} be the multiplicative inverse of a. Then $(a^{-1})^k = a^{-k}$ is the multiplicative inverse of a^k. Multiplying both sides of $a^k = a^m$ by a^{-k}, we obtain

$$1 = a^{m-k}.$$

This implies that there must exist a *smallest positive integer* n such that $a^n = 1$. This integer n is called the *order* of the field element a. Therefore, the sequence $a^1, a^2, a^3, \ldots$ repeats itself after $a^n = 1$. Also, the powers $a^1, a^2, \ldots, a^{n-1}, a^n = 1$ are all distinct. In fact, they form a group under the multiplication of GF(q). First we see that they contain the unit element 1. Consider $a^i \cdot a^j$. If $i + j \leq n$,

$$a^i \cdot a^j = a^{i+j}.$$

If $i + j > n$, we have $i + j = n + r$, where $0 < r \leq n$. Hence,

$$a^i \cdot a^j = a^{i+j} = a^n \cdot a^r = a^r.$$

Therefore, the powers $a^1, a^2, \ldots, a^{n-1}, a^n = 1$ are closed under the multiplication of GF(q). For $1 \leq i < n$, a^{n-i} is the multiplicative inverse of a^i. Since the powers of a are nonzero elements in GF(q), they satisfy the associative and commutative laws. Therefore, we conclude that $a^n = 1, a^1, a^2, \ldots, a^{n-1}$ form a group under the multiplication of GF(q). A group is said to be *cyclic* if there exists an element in the group whose powers constitute the whole group.

Theorem 2.4. Let a be a nonzero element of a finite field GF(q). Then $a^{q-1} = 1$.

Proof. Let $b_1, b_2, \ldots, b_{q-1}$ be the $q - 1$ nonzero elements of GF(q). Clearly, the $q - 1$ elements, $a \cdot b_1, a \cdot b_2, \ldots, a \cdot b_{q-1}$, are nonzero and distinct. Thus,

$$(a \cdot b_1) \cdot (a \cdot b_2) \cdots (a \cdot b_{q-1}) = b_1 \cdot b_2 \cdots b_{q-1}$$

$$a^{q-1} \cdot (b_1 \cdot b_2 \cdots b_{q-1}) = b_1 \cdot b_2 \cdots b_{q-1}.$$

Since $a \neq 0$ and $(b_1 \cdot b_2 \cdots b_{q-1}) \neq 0$, we must have $a^{q-1} = 1$. Q.E.D.

Theorem 2.5. Let a be a nonzero element in a finite field GF(q). Let n be the order of a. Then n divides $q - 1$.

Proof. Suppose that n does not divide $q - 1$. Dividing $q - 1$ by n, we obtain

$$q - 1 = kn + r,$$

where $0 < r < n$. Then

$$a^{q-1} = a^{kn+r} = a^{kn} \cdot a^r = (a^n)^k \cdot a^r.$$

Since $a^{q-1} = 1$ and $a^n = 1$, we must have $a^r = 1$. This is impossible since $0 < r < n$ and n is the smallest integer such that $a^n = 1$. Therefore, n must divide $q - 1$. Q.E.D.

In a finite field GF(q), a nonzero element a is said to be *primitive* if the order of a is $q - 1$. Therefore, the powers of a primitive element generate all the nonzero elements of GF(q). Every finite field has a primitive element (see Problem 2.7).

Consider the prime field GF(7) illustrated by Tables 2.5 and 2.6. The characteristic of this field is 7. If we take the powers of the integer 3 in GF(7) using the multiplication table, we obtain

$$3^1 = 3, \quad 3^2 = 3 \cdot 3 = 2, \quad 3^3 = 3 \cdot 3^2 = 6,$$

$$3^4 = 3 \cdot 3^3 = 4, \quad 3^5 = 3 \cdot 3^4 = 5, \quad 3^6 = 3 \cdot 3^5 = 1.$$

Therefore, the order of the integer 3 is 6 and the integer 3 is a primitive element of GF(7). The powers of the integer 4 in GF(7) are

$$4^1 = 4, \quad 4^2 = 4 \cdot 4 = 2, \quad 4^3 = 4 \cdot 4^2 = 1.$$

Clearly, the order of the integer 4 is 3, which is a factor of 6.

2.3 BINARY FIELD ARITHMETIC

In general, we can construct code with symbols from any Galois field GF(q), where q is either a prime p or a power of p. However, codes with symbols from the binary field GF(2) or its extension GF(2^m) are most widely used in digital data transmission and storage systems because information in these systems is universally coded in binary form for practical reasons. In this book we are concerned only with binary codes and codes with symbols from the field GF(2^m). Most of the results presented in this book can be generalized to codes with symbols from any finite field GF(q) with $q \neq 2$ or 2^m. In this section we discuss arithmetic over the binary field GF(2), which will be used in the rest of this book.

In binary arithmetic we use modulo-2 addition and multiplication, which are defined by Tables 2.3 and 2.4, respectively. This arithmetic is actually equivalent to ordinary arithmetic, except that we consider 2 to be equal to 0 (i.e., $1 + 1 = 2 = 0$). Note that since $1 + 1 = 0$, $1 = -1$. Hence, in binary arithmetic, subtraction is the same as addition. To illustrate how the ideas of ordinary algebra can be used with the binary arithmetic, we consider the following sets of equations:

$$X + Y = 1$$
$$X + Z = 0$$
$$X + Y + Z = 1.$$

These can be solved by adding the first equation to the third, giving $Z = 0$. Then from the second equation, since $Z = 0$ and $X + Z = 0$, we obtain $X = 0$. From the first equation, since $X = 0$ and $X + Y = 1$, we have $Y = 1$. We can substitute these solutions back into the original set of equations and verify that they are correct.

Since we were able to solve the equations shown above, they must be linearly independent, and the determinant of the coefficients on the left side must be nonzero. If the determinant is nonzero, it must be 1. This can be verified as follows:

$$\begin{vmatrix} 1 & 1 & 0 \\ 1 & 0 & 1 \\ 1 & 1 & 1 \end{vmatrix} = 1 \cdot \begin{vmatrix} 0 & 1 \\ 1 & 1 \end{vmatrix} - 1 \cdot \begin{vmatrix} 1 & 1 \\ 1 & 1 \end{vmatrix} + 0 \cdot \begin{vmatrix} 1 & 0 \\ 1 & 1 \end{vmatrix}$$
$$= 1 \cdot 1 - 1 \cdot 0 + 0 \cdot 1 = 1.$$

We could have solved the equations by Cramer's rule:

$$X = \frac{\begin{vmatrix} 1 & 1 & 0 \\ 0 & 0 & 1 \\ 1 & 1 & 1 \end{vmatrix}}{\begin{vmatrix} 1 & 1 & 0 \\ 1 & 0 & 1 \\ 1 & 1 & 1 \end{vmatrix}} = \frac{0}{1} = 0, \qquad Y = \frac{\begin{vmatrix} 1 & 1 & 0 \\ 1 & 0 & 1 \\ 1 & 1 & 1 \end{vmatrix}}{\begin{vmatrix} 1 & 1 & 0 \\ 1 & 0 & 1 \\ 1 & 1 & 1 \end{vmatrix}} = \frac{1}{1} = 1,$$

$$Z = \frac{\begin{vmatrix} 1 & 1 & 1 \\ 1 & 0 & 0 \\ 1 & 1 & 1 \end{vmatrix}}{\begin{vmatrix} 1 & 1 & 0 \\ 1 & 0 & 1 \\ 1 & 1 & 1 \end{vmatrix}} = \frac{0}{1} = 0.$$

Next we consider computations with polynomials whose coefficients are from the binary field GF(2). A polynomial $f(X)$ with one *variable* X and with coefficients from GF(2) is of the following form:

$$f(X) = f_0 + f_1 X + f_2 X^2 + \cdots + f_n X^n,$$

where $f_i = 0$ or 1 for $0 \leq i \leq n$. The *degree* of a polynomial is the largest power of X with a nonzero coefficient. For the polynomial above, if $f_n = 1$, $f(X)$ is a polynomial of degree n; if $f_n = 0$, $f(X)$ is a polynomial of degree less than n. The degree of $f(X) = f_0$ is zero. In the following we use the phrase "a polynomial over GF(2)" to mean "a polynomial with coefficients from GF(2)." There are two polynomials over GF(2) with degree 1: X and $1 + X$. There are four polynomials over GF(2) with degree 2: X^2, $1 + X^2$, $X + X^2$, and $1 + X + X^2$. In general, there are 2^n polynomials over GF(2) with degree n.

Polynomials over GF(2) can be added (or subtracted), multiplied, and divided in the usual way. Let

$$g(X) = g_0 + g_1 X + g_2 X^2 + \cdots + g_m X^m$$

be another polynomial over GF(2). To add $f(X)$ and $g(X)$, we simply add the coefficients of the same power of X in $f(X)$ and $g(X)$ as follows (assuming that $m \leq n$):

$$f(X) + g(X) = (f_0 + g_0) + (f_1 + g_1)X + \cdots + (f_m + g_m)X^m + f_{m+1}X^{m+1} + \cdots + f_n X^n,$$

where $f_i + g_i$ is carried out in modulo-2 addition. For example, adding $a(X) = 1 + X + X^3 + X^5$ and $b(X) = 1 + X^2 + X^3 + X^4 + X^7$, we obtain the following sum:

$$\begin{aligned} a(X) + b(X) &= (1 + 1) + X + X^2 + (1 + 1)X^3 + X^4 + X^5 + X^7 \\ &= X + X^2 + X^4 + X^5 + X^7. \end{aligned}$$

When we multiply $f(X)$ and $g(X)$, we obtain the following product:

$$f(X) \cdot g(X) = c_0 + c_1 X + c_2 X^2 + \cdots + c_{n+m} X^{n+m},$$

where

$$\begin{aligned} c_0 &= f_0 g_0 \\ c_1 &= f_0 g_1 + f_1 g_0 \\ c_2 &= f_0 g_2 + f_1 g_1 + f_2 g_0 \\ &\vdots \\ c_i &= f_0 g_i + f_1 g_{i-1} + f_2 g_{i-2} + \cdots + f_i g_0 \\ &\vdots \\ c_{n+m} &= f_n g_m. \end{aligned} \tag{2.6}$$

(Multiplication and addition of coefficients are modulo-2.) It is clear from (2.6) that if $g(X) = 0$, then

$$f(X) \cdot 0 = 0. \tag{2.7}$$

We can readily verify that the polynomials over GF(2) satisfy the following conditions:

(i) Commutative:

$$a(X) + b(X) = b(X) + a(X)$$
$$a(X) \cdot b(X) = b(X) \cdot a(X).$$

(ii) Associative:

$$a(X) + [b(X) + c(X)] = [a(X) + b(X)] + c(X)$$
$$a(X) \cdot [b(X) \cdot c(X)] = [a(X) \cdot b(X)] \cdot c(X).$$

(iii) Distributive:

$$a(X) \cdot [b(X) + c(X)] = [a(X) \cdot b(X)] + [a(X) \cdot c(X)]. \tag{2.8}$$

Suppose that the degree of $g(X)$ is *not* zero. When $f(X)$ is divided by $g(X)$, we obtain a unique pair of polynomials over GF(2)—$q(X)$, called the quotient, and $r(X)$, called the remainder—such that

$$f(X) = q(X)g(X) + r(X)$$

and the degree of $r(X)$ is less than that of $g(X)$. This is known as Euclid's division algorithm. As an example, we divide $f(X) = 1 + X + X^4 + X^5 + X^6$ by $g(X) = 1 + X + X^3$. Using the long-division technique, we have

$$\begin{array}{r l}
 & X^3 + X^2 \quad \text{(quotient)} \\
X^3 + X + 1 \,\big) & X^6 + X^5 + X^4 \qquad\qquad\qquad + X + 1 \\
 & X^6 \qquad\quad + X^4 + X^3 \\
 & \qquad X^5 \qquad\quad + X^3 \qquad\quad + X + 1 \\
 & \qquad X^5 \qquad\quad + X^3 + X^2 \\
 & \qquad\qquad\qquad\qquad X^2 + X + 1 \quad \text{(remainder)}.
\end{array}$$

We can easily verify that

$$X^6 + X^5 + X^4 + X + 1 = (X^3 + X^2)(X^3 + X + 1) + X^2 + X + 1.$$

When $f(X)$ is divided by $g(X)$, if the remainder $r(X)$ is identical to zero $[r(X) = 0]$, we say that $f(X)$ is divisible by $g(X)$ and $g(X)$ is a factor of $f(X)$.

For real numbers, if a is a *root* of a polynomial $f(X)$ [i.e., $f(a) = 0$], $f(X)$ is divisible by $x - a$. (This fact follows from Euclid's division algorithm.) This is still true for $f(X)$ over GF(2). For example, let $f(X) = 1 + X^2 + X^3 + X^4$. Substituting $X = 1$, we obtain

$$f(1) = 1 + 1^2 + 1^3 + 1^4 = 1 + 1 + 1 + 1 = 0.$$

Thus, $f(X)$ has 1 as a root and it should be divisible by $X + 1$.

$$\begin{array}{r l}
 & X^3 + X + 1 \\
X + 1 \,\big) & X^4 + X^3 + X^2 \qquad + 1 \\
 & X^4 + X^3 \\
 & \qquad\qquad X^2 \qquad + 1 \\
 & \qquad\qquad X^2 + X \\
 & \qquad\qquad\qquad X + 1 \\
 & \qquad\qquad\qquad X + 1 \\
 & \qquad\qquad\qquad\quad 0
\end{array}$$

For a polynomial $f(X)$ over GF(2), if it has an even number of terms, it is divisible by $X + 1$. A polynomial $p(X)$ over GF(2) of degree m is said to be *irreducible* over GF(2) if $p(X)$ is not divisible by any polynomial over GF(2) of degree less than m but greater than zero. Among the four polynomials of degree 2, X^2, $X^2 + 1$ and $X^2 + X$ are not irreducible since they are either divisible by X or $X + 1$. However, $X^2 + X + 1$ does not have either "0" or "1" as a root and so is not divisible by any polynomial of degree 1. Therefore, $X^2 + X + 1$ is an irreducible polynomial of degree 2. The polynomial $X^3 + X + 1$ is an irreducible polynomial of degree 3. First we note that $X^3 + X + 1$ does not have either 0 or 1 as a root. Therefore, $X^3 + X + 1$ is not divisible by X or $X + 1$. Since it is not divisible by any polynomial of degree 1, it cannot be divisible by a polynomial of degree 2. Consequently, $X^3 + X + 1$ is irreducible over GF(2). We may verify that $X^4 + X + 1$ is an irreducible polynomial of degree 4. It has been proved that, for any $m \geq 1$, there exists an irreducible polynomial of degree m. An important theorem regarding irreducible polynomials over GF(2) is given below without a proof.

Theorem 2.6. Any irreducible polynomial over GF(2) of degree m divides $X^{2^m - 1} + 1$.

As an example of Theorem 2.6, we can check that $X^3 + X + 1$ divides $X^{2^3 - 1} + 1 = X^7 + 1$:

$$
\begin{array}{r}
X^4 + X^2 + X + 1 \qquad\qquad\qquad\qquad \\
X^3 + X + 1 \overline{)X^7 \qquad\qquad\qquad\qquad\qquad\qquad + 1} \\
\underline{X^7 \qquad + X^5 + X^4 \qquad\qquad\qquad\quad} \\
X^5 + X^4 \qquad\qquad\quad + 1 \\
\underline{X^5 \qquad + X^3 + X^2 \qquad} \\
X^4 + X^3 + X^2 \quad + 1 \\
\underline{X^4 \qquad + X^2 + X \quad} \\
X^3 \qquad + X + 1 \\
\underline{X^3 \qquad + X + 1} \\
0.
\end{array}
$$

An irreducible polynomial $p(X)$ of degree m is said to be *primitive* if the smallest positive integer n for which $p(X)$ divides $X^n + 1$ is $n = 2^m - 1$. We may check that $p(X) = X^4 + X + 1$ divides $X^{15} + 1$ but does not divide any $X^n + 1$ for $1 \leq n < 15$. Hence, $X^4 + X + 1$ is a primitive polynomial. The polynomial $X^4 + X^3 + X^2 + X + 1$ is irreducible but it is not primitive, since it divides $X^5 + 1$. It is not easy to recognize a primitive polynomial. However, there are tables of irreducible polynomials in which primitive polynomials are indicated [6,7]. For a given m, there may be more than one primitive polynomial of degree m. A list of primitive polynomials is given in Table 2.7. For each degree m, we list only a primitive polynomial with the smallest number of terms.

Before leaving this section, we derive another useful property of polynomials over GF(2). Consider

TABLE 2.7 LIST OF PRIMITIVE POLYNOMIALS

m		m	
3	$1 + X + X^3$	14	$1 + X + X^6 + X^{10} + X^{14}$
4	$1 + X + X^4$	15	$1 + X + X^{15}$
5	$1 + X^2 + X^5$	16	$1 + X + X^3 + X^{12} + X^{16}$
6	$1 + X + X^6$	17	$1 + X^3 + X^{17}$
7	$1 + X^3 + X^7$	18	$1 + X^7 + X^{18}$
8	$1 + X^2 + X^3 + X^4 + X^8$	19	$1 + X + X^2 + X^5 + X^{19}$
9	$1 + X^4 + X^9$	20	$1 + X^3 + X^{20}$
10	$1 + X^3 + X^{10}$	21	$1 + X^2 + X^{21}$
11	$1 + X^2 + X^{11}$	22	$1 + X + X^{22}$
12	$1 + X + X^4 + X^6 + X^{12}$	23	$1 + X^5 + X^{23}$
13	$1 + X + X^3 + X^4 + X^{13}$	24	$1 + X + X^2 + X^7 + X^{24}$

$$\begin{aligned}
f^2(X) &= (f_0 + f_1X + \cdots + f_nX^n)^2 \\
&= [f_0 + (f_1X + f_2X^2 + \cdots + f_nX^n)]^2 \\
&= f_0^2 + f_0\cdot(f_1X + f_2X^2 + \cdots + f_nX^n) \\
&\qquad + f_0\cdot(f_1X + f_2X^2 + \cdots + f_nX^n) + (f_1X + f_2X^2 + \cdots + f_nX^n)^2 \\
&= f_0^2 + (f_1X + f_2X^2 + \cdots + f_nX^n)^2.
\end{aligned}$$

Expanding the equation above repeatedly, we eventually obtain

$$f^2(X) = f_0^2 + (f_1X)^2 + (f_2X^2)^2 + \cdots + (f_nX^n)^2.$$

Since $f_i = 0$ or 1, $f_i^2 = f_i$. Hence, we have

$$\begin{aligned}
f^2(X) &= f_0 + f_1X^2 + f_2(X^2)^2 + \cdots + f_n(X^2)^n \\
&= f(X^2).
\end{aligned} \tag{2.9}$$

It follows from (2.9) that, for any $l \geq 0$,

$$[f(X)]^{2^l} = f(X^{2^l}). \tag{2.10}$$

2.4 CONSTRUCTION OF GALOIS FIELD GF(2^m)

In this section we present a method for constructing the Galois field of 2^m elements ($m > 1$) from the binary field GF(2). We begin with the two elements 0 and 1, from GF(2) and a new symbol α. Then we define a multiplication "$\cdot$" to introduce a sequence of powers of α as follows:

$$\begin{aligned}
0\cdot 0 &= 0, \\
0\cdot 1 &= 1\cdot 0 = 0, \\
1\cdot 1 &= 1, \\
0\cdot\alpha &= \alpha\cdot 0 = 0, \\
1\cdot\alpha &= \alpha\cdot 1 = \alpha,
\end{aligned} \tag{2.11}$$

$$
\begin{aligned}
\alpha^2 &= \alpha \cdot \alpha, \\
\alpha^3 &= \alpha \cdot \alpha \cdot \alpha, \\
&\ \ \vdots \\
\alpha^j &= \alpha \cdot \alpha \cdot \cdots \cdot \alpha \quad (j \text{ times}), \\
&\ \ \vdots
\end{aligned} \tag{2.11}
$$

It follows from the definition of multiplication above that

$$
\begin{aligned}
0 \cdot \alpha^j &= \alpha^j \cdot 0 = 0, \\
1 \cdot \alpha^j &= \alpha^j \cdot 1 = \alpha^j, \\
\alpha^i \cdot \alpha^j &= \alpha^j \cdot \alpha^i = \alpha^{i+j}.
\end{aligned} \tag{2.12}
$$

Now, we have the following set of elements on which a multiplication operation "$\cdot$" is defined:

$$F = \{0, 1, \alpha, \alpha^2, \ldots, \alpha^j, \ldots\}.$$

The element 1 is sometimes denoted α^0.

Next we put a condition on the element α so that the set F contains only 2^m elements and is closed under the multiplication "$\cdot$" defined by (2.11). Let $p(X)$ be a primitive polynomial of degree m over GF(2). We assume that $p(\alpha) = 0$. Since $p(X)$ divides $X^{2^m-1} + 1$ (Theorem 2.6), we have

$$X^{2^m-1} + 1 = q(X)p(X). \tag{2.13}$$

If we replace X by α in (2.13), we obtain

$$\alpha^{2^m-1} + 1 = q(\alpha)p(\alpha).$$

Since $p(\alpha) = 0$, we have

$$\alpha^{2^m-1} + 1 = q(\alpha) \cdot 0.$$

If we regard $q(\alpha)$ as a polynomial of α over GF(2), it follows from (2.7) that $q(\alpha) \cdot 0 = 0$. As a result, we obtain the following equality:

$$\alpha^{2^m-1} + 1 = 0.$$

Adding 1 to both sides of $\alpha^{2^m-1} + 1 = 0$ (use modulo-2 addition) results in the following equality:

$$\alpha^{2^m-1} = 1. \tag{2.14}$$

Therefore, under the condition that $p(\alpha) = 0$, the set F becomes finite and contains the following elements:

$$F^* = \{0, 1, \alpha, \alpha^2, \ldots, \alpha^{2^m-2}\}.$$

The nonzero elements of F^* are closed under the multiplication operation "$\cdot$" defined by (2.11). To see this, let i and j be two integers such that $0 \leq i, j < 2^m - 1$. If $i + j < 2^m - 1$, then $\alpha^i \cdot \alpha^j = \alpha^{i+j}$, which is obviously a nonzero element in F^*. If $i + j \geq 2^m - 1$, we can express $i + j$ as follows: $i + j = (2^m - 1) + r$, where $0 \leq r < 2^m - 1$.

Then

$$\alpha^i \cdot \alpha^j = \alpha^{i+j} = \alpha^{(2^m-1)+r} = \alpha^{2^m-1} \cdot \alpha^r = 1 \cdot \alpha^r = \alpha^r,$$

which is also a nonzero element in F^*. Hence, we conclude that the nonzero elements of F^* are closed under the multiplication "$\cdot$" defined by (2.11). In fact, these nonzero elements form a commutative group under "$\cdot$". First, we see that the element 1 is the unit element. From (2.11) and (2.12) we see readily that the multiplication operation "$\cdot$" is commutative and associative. For $0 < i < 2^m - 1$, α^{2^m-i-1} is the multiplicative inverse of α^i since

$$\alpha^{2^m-i-1} \cdot \alpha^i = \alpha^{2^m-1} = 1.$$

(Note that $\alpha^0 = \alpha^{2^m-1} = 1$.) It will be clear in what follows that $1, \alpha, \alpha^2, \ldots, \alpha^{2^m-2}$ represent $2^m - 1$ distinct elements. Therefore, the nonzero elements of F^* form a group of order $2^m - 1$ under the multiplication operation "$\cdot$" defined by (2.11).

Our next step is to define an addition operation "$+$" on F^* so that F^* forms a commutative group under "$+$." For $0 \leq i < 2^m - 1$, we divide the polynomial X^i by $p(X)$ and obtain the following:

$$X^i = q_i(X)p(X) + a_i(X), \tag{2.15}$$

where $q_i(X)$ and $a_i(X)$ are the quotient and the remainder, respectively. The remainder $a_i(X)$ is a polynomial of degree $m - 1$ or less over GF(2) and is of the following form:

$$a_i(X) = a_{i0} + a_{i1}X + a_{i2}X^2 + \cdots + a_{i,m-1}X^{m-1}.$$

Since X and $p(X)$ are relatively prime (i.e., they do not have any common factor except 1), X^i is not divisible by $p(X)$. Therefore, for any $i \geq 0$,

$$a_i(X) \neq 0. \tag{2.16}$$

For $0 \leq i, j < 2^m - 1$, and $i \neq j$, we can also show that

$$a_i(X) \neq a_j(X). \tag{2.17}$$

Suppose that $a_i(X) = a_j(X)$. Then it follows from (2.15) that

$$\begin{aligned} X^i + X^j &= [q_i(X) + q_j(X)]p(X) + a_i(X) + a_j(X) \\ &= [q_i(X) + q_j(X)]p(X). \end{aligned}$$

This implies that $p(X)$ divides $X^i + X^j = X^i(1 + X^{j-i})$ (assuming that $j > i$). Since X^i and $p(X)$ are relatively prime, $p(X)$ must divide $X^{j-i} + 1$. However, this is impossible since $j - i < 2^m - 1$ and $p(X)$ is a primitive polynomial of degree m which does not divide $X^n + 1$ for $n < 2^m - 1$. Therefore, our hypothesis that $a_i(X) = a_j(X)$ is invalid. As a result, for $0 \leq i, j < 2^m - 1$ and $i \neq j$, we must have $a_i(X) \neq a_j(X)$. Hence, for $i = 0, 1, 2, \ldots, 2^m - 2$, we obtain $2^m - 1$ distinct nonzero polynomials $a_i(X)$ of degree $m - 1$ or less. Now, replacing X by α in (2.15) and using the equality that $q_i(\alpha) \cdot 0 = 0$ [see (2.7)], we obtain the following polynomial expression for α^i:

$$\alpha^i = a_i(\alpha) = a_{i0} + a_{i1}\alpha + a_{i2}\alpha^2 + \cdots + a_{i,m-1}\alpha^{m-1}. \tag{2.18}$$

From (2.16), (2.17), and (2.18), we see that the $2^m - 1$ nonzero elements, $\alpha^0, \alpha^1, \ldots, \alpha^{2^m-2}$ in F^*, are represented by $2^m - 1$ *distinct nonzero polynomials* of α over GF(2) with degree $m - 1$ or less. The zero element 0 in F^* may be represented by the *zero* *polynomial.* As a result, the 2^m elements in F^* are represented by 2^m *distinct polynomials* of α over GF(2) with degree $m-1$ or less and are regarded as 2^m distinct elements.

Now, we define an addition "+" on F^* as follows:

$$0 + 0 = 0 \tag{2.19a}$$

and, for $0 \leq i, j < 2^m - 1$,

$$0 + \alpha^i = \alpha^i + 0 = \alpha^i, \tag{2.19b}$$

$$\begin{aligned} \alpha^i + \alpha^j &= (a_{i0} + a_{i1}\alpha + \cdots + a_{i,m-1}\alpha^{m-1}) + (a_{j0} + a_{j1}\alpha + \cdots + a_{j,m-1}\alpha^{m-1}) \\ &= (a_{i0} + a_{j0}) + (a_{i1} + a_{j1})\alpha + \cdots + (a_{i,m-1} + a_{j,m-1})\alpha^{m-1}, \end{aligned} \tag{2.19c}$$

where $a_{i,l} + a_{j,l}$ is carried out in modulo-2 addition. From (2.19c) we see that, for $i = j$,

$$\alpha^i + \alpha^i = 0 \tag{2.20}$$

and for $i \neq j$,

$$(a_{i0} + a_{j0}) + (a_{i1} + a_{j1})\alpha + \cdots + (a_{i,m-1} + a_{j,m-1})\alpha^{m-1}$$

is nonzero and must be the polynomial expression for some α^k in F^*. Hence, the set F^* is closed under the addition "+" defined by (2.19). We can immediately verify that F^* is a commutative group under "+." First, we see that 0 is the additive identity. Using the fact that modulo-2 addition is commutative and associative, the addition defined on F^* is also commutative and associative. From (2.19a) and (2.20) we see that the additive inverse of any element in F^* is itself.

Up to this point, we have shown that the set $F^* = \{0, 1, \alpha, \alpha^2, \ldots, \alpha^{2^m-2}\}$ is a commutative group under an addition operation "+" and the nonzero elements of F^* form a commutative group under a multiplication operation "$\cdot$." Using the polynomial representation for the elements in F^* and (2.8) (polynomial multiplication satisfies distributive law), we readily see that the multiplication on F^* is distributive over the addition on F^*. Therefore, the set $F^* = \{0, 1, \alpha, \alpha^2, \ldots, \alpha^{2^m-2}\}$ is a Galois field of 2^m elements, GF(2^m). We notice that the addition and multiplication defined on $F^* =$ GF(2^m) imply modulo-2 addition and multiplication. Hence, the subset $\{0, 1\}$ forms a subfield of GF(2^m) [i.e., GF(2) is a subfield of GF(2^m)]. The binary field GF(2) is usually called the *ground* field of GF(2^m). The characteristic of GF(2^m) is 2.

In our process of constructing GF(2^m) from GF(2), we have developed two representations for the nonzero elements of GF(2^m): the power representation and the polynomial representation. The power representation is convenient for multiplication and the polynomial representation is convenient for addition.

Example 2.6

Let $m = 4$. The polynomial $p(X) = 1 + X + X^4$ is a primitive polynomial over GF(2). Set $p(\alpha) = 1 + \alpha + \alpha^4 = 0$. Then $\alpha^4 = 1 + \alpha$. Using this, we can construct GF(2^4). The elements of GF(2^4) are given in Table 2.8. The identity $\alpha^4 = 1 + \alpha$ is used repeatedly to form the polynomial representations for the elements of GF(2^4). For example,

$$\alpha^5 = \alpha \cdot \alpha^4 = \alpha(1 + \alpha) = \alpha + \alpha^2,$$

$$\alpha^6 = \alpha \cdot \alpha^5 = \alpha(\alpha + \alpha^2) = \alpha^2 + \alpha^3,$$

$$\alpha^7 = \alpha \cdot \alpha^6 = \alpha(\alpha^2 + \alpha^3) = \alpha^3 + \alpha^4 = \alpha^3 + 1 + \alpha = 1 + \alpha + \alpha^3.$$

TABLE 2.8 THREE REPRESENTATIONS FOR THE ELEMENTS OF GF(2^4) GENERATED BY $p(X) = 1 + X + X^4$

Power representation	Polynomial representation	4-Tuple representation
0	0	(0 0 0 0)
1	1	(1 0 0 0)
α	α	(0 1 0 0)
α^2	α^2	(0 0 1 0)
α^3	α^3	(0 0 0 1)
α^4	$1 + \alpha$	(1 1 0 0)
α^5	$\alpha + \alpha^2$	(0 1 1 0)
α^6	$\alpha^2 + \alpha^3$	(0 0 1 1)
α^7	$1 + \alpha + \alpha^3$	(1 1 0 1)
α^8	$1 + \alpha^2$	(1 0 1 0)
α^9	$\alpha + \alpha^3$	(0 1 0 1)
α^{10}	$1 + \alpha + \alpha^2$	(1 1 1 0)
α^{11}	$\alpha + \alpha^2 + \alpha^3$	(0 1 1 1)
α^{12}	$1 + \alpha + \alpha^2 + \alpha^3$	(1 1 1 1)
α^{13}	$1 + \alpha^2 + \alpha^3$	(1 0 1 1)
α^{14}	$1 + \alpha^3$	(1 0 0 1)

To multiply two elements α^i and α^j, we simply add their exponents and use the fact that $\alpha^{15} = 1$. For example, $\alpha^5 \cdot \alpha^7 = \alpha^{12}$ and $\alpha^{12} \cdot \alpha^7 = \alpha^{19} = \alpha^4$. Dividing α^j by α^i, we simply multiply α^j by the multiplicative inverse α^{15-i} of α^i. For example, $\alpha^4/\alpha^{12} = \alpha^4 \cdot \alpha^3 = \alpha^7$ and $\alpha^{12}/\alpha^5 = \alpha^{12} \cdot \alpha^{10} = \alpha^{22} = \alpha^7$. To add α^i and α^j, we use their polynomial representations in Table 2.8. Thus,

$$\alpha^5 + \alpha^7 = (\alpha + \alpha^2) + (1 + \alpha + \alpha^3) = 1 + \alpha^2 + \alpha^3 = \alpha^{13}$$

$$1 + \alpha^5 + \alpha^{10} = 1 + (\alpha + \alpha^2) + (1 + \alpha + \alpha^2) = 0.$$

There is another useful representation for the field elements in GF(2^m). Let $a_0 + a_1\alpha + a_2\alpha^2 + \cdots + a_{m-1}\alpha^{m-1}$ be the polynomial representation of a field element β. Then we can represent β by an ordered sequence of m components, called an *m-tuple*, as follows:

$$(a_0, a_1, a_2, \ldots, a_{m-1}),$$

where the m components are simply the m coefficients of the polynomial representation of β. Clearly, we see that there is one-to-one correspondence between this m-tuple and the polynomial representation of β. The zero element 0 of GF(2^m) is represented by the zero m-tuple $(0, 0, \ldots, 0)$. Let $(b_0, b_1, \ldots, b_{m-1})$ be the m-tuple representation of γ in GF(2^m). Adding β and γ, we simply add the corresponding components of their m-tuple representations as follows:

$$(a_0 + b_0, a_1 + b_1, \ldots, a_{m-1} + b_{m-1}),$$

where $a_i + b_i$ is carried out in modulo-2 addition. Obviously, the components of the resultant m-tuple are the coefficients of the polynomial representation for $\beta + \gamma$. All three representations for the elements of GF(2^4) are given in Table 2.8.

Galois fields of 2^m elements with $m = 3$ to 10 are given in Appendix A.

2.5 BASIC PROPERTIES OF GALOIS FIELD GF(2^m)

In ordinary algebra we often see that a polynomial with real coefficients has roots not from the field of real numbers but from the field of complex numbers that contains the field of real numbers as a subfield. For example, the polynomial $X^2 + 6X + 25$ does not have roots from the field of real numbers but has two complex conjugate roots, $-3 + 4i$ and $-3 - 4i$, where $i = \sqrt{-1}$. This is also true for polynomials with coefficients from GF(2). In this case, a polynomial with coefficients from GF(2) may not have roots from GF(2) but has roots from an extension field of GF(2). For example, $X^4 + X^3 + 1$ is irreducible over GF(2) and therefore it does not have roots from GF(2). However, it has four roots from the field GF(2^4). If we substitute the elements of GF(2^4) given by Table 2.8 into $X^4 + X^3 + 1$, we find that α^7, α^{11}, α^{13}, and α^{14} are the roots of $X^4 + X^3 + 1$. We may verify this as follows:

$$(\alpha^7)^4 + (\alpha^7)^3 + 1 = \alpha^{28} + \alpha^{21} + 1 = (1 + \alpha^2 + \alpha^3) + (\alpha^2 + \alpha^3) + 1 = 0.$$

Indeed, α^7 is a root for $X^4 + X^3 + 1$. Similarly, we may verify that α^{11}, α^{13}, and α^{14} are the other three roots. Since α^7, α^{11}, α^{13}, and α^{14} are all roots of $X^4 + X^3 + 1$, then $(X + \alpha^7)(X + \alpha^{11})(X + \alpha^{13})(X + \alpha^{14})$ must be equal to $X^4 + X^3 + 1$. To see this, we multiply out the product above using Table 2.8:

$$\begin{aligned}
(X + \alpha^7)&(X + \alpha^{11})(X + \alpha^{13})(X + \alpha^{14}) \\
&= [X^2 + (\alpha^7 + \alpha^{11})X + \alpha^{18}][X^2 + (\alpha^{13} + \alpha^{14})X + \alpha^{27}] \\
&= (X^2 + \alpha^8 X + \alpha^3)(X^2 + \alpha^2 X + \alpha^{12}) \\
&= X^4 + (\alpha^8 + \alpha^2)X^3 + (\alpha^{12} + \alpha^{10} + \alpha^3)X^2 + (\alpha^{20} + \alpha^5)X + \alpha^{15} \\
&= X^4 + X^3 + 1.
\end{aligned}$$

Let $f(X)$ be a polynomial with coefficients from GF(2). If β, an element in GF(2^m), is a root of $f(X)$, the polynomial $f(X)$ may have other roots from GF(2^m). Then, what are these roots? This is answered by the following theorem.

Theorem 2.7. Let $f(X)$ be a polynomial with coefficients from GF(2). Let β be an element in an extension field of GF(2). If β is a root of $f(X)$, then for any $l \geq 0$, β^{2^l} is also a root of $f(X)$.

Proof. From (2.10), we have

$$[f(X)]^{2^l} = f(X^{2^l}).$$

Substituting β into the equation above, we obtain

$$[f(\beta)]^{2^l} = f(\beta^{2^l}).$$

Since $f(\beta) = 0$, $f(\beta^{2^l}) = 0$. Therefore, β^{2^l} is also a root of $f(X)$. Q.E.D.

The element β^{2^l} is called a *conjugate* of β. Theorem 2.7 says that if β, an element in GF(2^m), is a root of a polynomial $f(X)$ over GF(2), then all the distinct conjugates of β, also elements in GF(2^m), are roots of $f(X)$. For example, the polynomial $f(X) = 1 + X^3 + X^4 + X^5 + X^6$ has α^4, an element in GF(2^4) given by Table 2.8, as a root.

To verify this, we use Table 2.8 and the fact that $\alpha^{15} = 1$,

$$f(\alpha^4) = 1 + \alpha^{12} + \alpha^{16} + \alpha^{20} + \alpha^{24} = 1 + \alpha^{12} + \alpha + \alpha^5 + \alpha^9$$
$$= 1 + (1 + \alpha + \alpha^2 + \alpha^3) + \alpha + (\alpha + \alpha^2) + (\alpha + \alpha^3) = 0.$$

The conjugates of α^4 are

$$(\alpha^4)^2 = \alpha^8, \qquad (\alpha^4)^{2^2} = \alpha^{16} = \alpha, \qquad (\alpha^4)^{2^3} = \alpha^{32} = \alpha^2.$$

[Note that $(\alpha^4)^{2^4} = \alpha^{64} = \alpha^4$.] It follows from Theorem 2.7 that α^8, α, and α^2 must be also roots of $f(X) = 1 + X^3 + X^4 + X^5 + X^6$. We can check that α^5 and its conjugate α^{10} are roots of $f(X) = 1 + X^3 + X^4 + X^5 + X^6$.

Let β be a nonzero element in the field GF(2^m). It follows from Theorem 2.4 that

$$\beta^{2^m-1} = 1.$$

Adding 1 to both sides of $\beta^{2^m-1} = 1$, we obtain

$$\beta^{2^m-1} + 1 = 0.$$

This says that β is a root of the polynomial $X^{2^m-1} + 1$. Hence, every nonzero element of GF(2^m) is a root of $X^{2^m-1} + 1$. Since the degree of $X^{2^m-1} + 1$ is $2^m - 1$, the $2^m - 1$ nonzero elements of GF(2^m) form all the roots of $X^{2^m-1} + 1$. Summarizing the result above, we obtain Theorem 2.8.

Theorem 2.8. The $2^m - 1$ nonzero elements of GF(2^m) form all the roots of $X^{2^m-1} + 1$.

Since the zero element 0 of GF(2^m) is the root of X, Theorem 2.8 has the following corollary:

Corollary 2.8.1. The elements of GF(2^m) form all the roots of $X^{2^m} + X$.

Since any element β in GF(2^m) is a root of the polynomial $X^{2^m} + X$, β may be a root of a polynomial over GF(2) with a degree less than 2^m. Let $\phi(X)$ be the polynomial of *smallest degree* over GF(2) such that $\phi(\beta) = 0$. [We can easily prove that $\phi(X)$ is unique.] This polynomial $\phi(X)$ is called the *minimal polynomial* of β. For example, the minimal polynomial of the zero element 0 of GF(2^m) is X and the minimal polynomial of the unit element 1 is $X + 1$. Next, a number of properties of minimal polynomials are derived.

Theorem 2.9. The minimal polynomial $\phi(X)$ of a field element β is irreducible.

Proof. Suppose that $\phi(X)$ is not irreducible and that $\phi(X) = \phi_1(X)\phi_2(X)$, where both $\phi_1(X)$ and $\phi_2(X)$ have degrees greater than 0 and less than the degree of $\phi(X)$. Since $\phi(\beta) = \phi_1(\beta)\phi_2(\beta) = 0$, either $\phi_1(\beta) = 0$ or $\phi_2(\beta) = 0$. This contradicts the hypothesis that $\phi(X)$ is a polynomial of smallest degree such that $\phi(\beta) = 0$. Therefore, $\phi(X)$ must be irreducible. Q.E.D.

Theorem 2.10. Let $f(X)$ be a polynomial over GF(2). Let $\phi(X)$ be the minimal polynomial of a field element β. If β is a root of $f(X)$, then $f(X)$ is divisible by $\phi(X)$.

Proof. Dividing $f(X)$ by $\phi(X)$, we obtain

$$f(X) = a(X)\phi(X) + r(X),$$

where the degree of the remainder $r(X)$ is less than the degree of $\phi(X)$. Substituting β into the equation above and using the fact that $f(\beta) = \phi(\beta) = 0$, we have $r(\beta) = 0$. If $r(X) \neq 0$, $r(X)$ would be a polynomial of lower degree than $\phi(X)$, which has β as a root. This is a contradiction to the fact that $\phi(X)$ is the minimal polynomial of β. Hence, $r(X)$ must be identical to 0 and $\phi(X)$ divides $f(X)$. Q.E.D.

It follows from Corollary 2.8.1 and Theorem 2.10 that we have the following result:

Theorem 2.11. The minimal polynomial $\phi(X)$ of an element β in GF(2^m) divides $X^{2^m} + X$.

Theorem 2.11 says that all the roots of $\phi(X)$ are from GF(2^m). Then, what are the roots of $\phi(X)$? This will be answered by the next two theorems.

Theorem 2.12. Let $f(X)$ be an irreducible polynomial over GF(2). Let β be an element in GF(2^m). Let $\phi(X)$ be the minimal polynomial of β. If $f(\beta) = 0$, then $\phi(X) = f(X)$.

Proof. It follows from Theorem 2.10 that $\phi(X)$ divides $f(X)$. Since $\phi(X) \neq 1$ and $f(X)$ is irreducible, we must have $\phi(X) = f(X)$. Q.E.D.

Theorem 2.12 says that if an irreducible polynomial has β as a root, it is the minimal polynomial $\phi(X)$ of β. It follows from Theorem 2.7 that β and its conjugates $\beta^2, \beta^{2^2}, \ldots, \beta^{2^l}, \ldots$ are roots of $\phi(X)$. Let e be the smallest integer such that $\beta^{2^e} = \beta$. Then $\beta^2, \beta^{2^2}, \ldots, \beta^{2^{e-1}}$ are all the distinct conjugates of β (see Problem 2.14). Since $\beta^{2^m} = \beta$, $e \leq m$.

Theorem 2.13. Let β be an element in GF(2^m) and let e be the smallest nonnegative integer such that $\beta^{2^e} = \beta$. Then

$$f(X) = \prod_{i=0}^{e-1} (X + \beta^{2^i})$$

is an irreducible polynomial over GF(2).

Proof. Consider

$$[f(X)]^2 = \left[\prod_{i=0}^{e-1} (X + \beta^{2^i})\right]^2 = \prod_{i=0}^{e-1} (X + \beta^{2^i})^2.$$

Since $(X + \beta^{2^i})^2 = X^2 + (\beta^{2^i} + \beta^{2^i})X + \beta^{2^{i+1}} = X^2 + \beta^{2^{i+1}}$,

$$[f(X)]^2 = \prod_{i=0}^{e-1} (X^2 + \beta^{2^{i+1}}) = \prod_{i=1}^{e} (X^2 + \beta^{2^i})$$

$$= \left[\prod_{i=1}^{e-1} (X^2 + \beta^{2^i})\right](X^2 + \beta^{2^e}).$$

Since $\beta^{2^e} = \beta$, then

$$[f(X)]^2 = \prod_{i=0}^{e-1} (X^2 + \beta^{2^i}) = f(X^2). \tag{2.21}$$

Let $f(X) = f_0 + f_1X + \cdots + f_eX^e$, where $f_e = 1$. Expand

$$[f(X)]^2 = (f_0 + f_1X + \cdots + f_eX^e)^2$$
$$= \sum_{i=0}^{e} f_i^2 X^{2i} + (1+1)\sum_{i=0}^{e}\sum_{\substack{j=0 \\ i\neq j}}^{e} f_i f_j X^{i+j} = \sum_{i=0}^{e} f_i^2 X^{2i}. \tag{2.22}$$

From (2.21) and (2.22), we obtain

$$\sum_{i=0}^{e} f_i X^{2i} = \sum_{i=0}^{e} f_i^2 X^{2i}.$$

Then, for $0 \leq i \leq e$, we must have

$$f_i = f_i^2.$$

This holds only when $f_i = 0$ or 1. Therefore, $f(X)$ has coefficients from GF(2).

Now suppose that $f(X)$ is not irreducible over GF(2) and $f(X) = a(X)b(X)$. Since $f(\beta) = 0$, either $a(\beta) = 0$ or $b(\beta) = 0$. If $a(\beta) = 0$, $a(X)$ has $\beta, \beta^2, \ldots, \beta^{2^{e-1}}$ as roots, so $a(X)$ has degree e and $a(X) = f(X)$. Similarly, if $b(\beta) = 0$, $b(X) = f(X)$. Therefore, $f(X)$ must be irreducible. Q.E.D.

A direct consequence of Theorems 2.12 and 13 is Theorem 2.14.

Theorem 2.14. Let $\phi(X)$ be the minimal polynomial of an element β in GF(2^m). Let e be the smallest integer such that $\beta^{2^e} = \beta$. Then

$$\phi(X) = \prod_{i=0}^{e-1} (X + \beta^{2^i}). \tag{2.23}$$

Example 2.7

Consider the Galois field GF(2^4) given by Table 2.8. Let $\beta = \alpha^3$. The conjugates of β are

$$\beta^2 = \alpha^6, \qquad \beta^{2^2} = \alpha^{12}, \qquad \beta^{2^3} = \alpha^{24} = \alpha^9.$$

The minimal polynomial of $\beta = \alpha^3$ is then

$$\phi(X) = (X + \alpha^3)(X + \alpha^6)(X + \alpha^{12})(X + \alpha^9).$$

Multiplying out the right-hand side of the equation above with the aid of Table 2.8, we obtain

$$\phi(X) = [X^2 + (\alpha^3 + \alpha^6)X + \alpha^9][X^2 + (\alpha^{12} + \alpha^9)X + \alpha^{21}]$$
$$= (X^2 + \alpha^2X + \alpha^9)(X^2 + \alpha^8X + \alpha^6)$$
$$= X^4 + (\alpha^2 + \alpha^8)X^3 + (\alpha^6 + \alpha^{10} + \alpha^9)X^2 + (\alpha^{17} + \alpha^8)X + \alpha^{15}$$
$$= X^4 + X^3 + X^2 + X + 1.$$

There is another way of finding the minimal polynomial of a field element, which is illustrated by the following example.

Example 2.8

Suppose that we want to determine the minimal polynomial $\phi(X)$ of $\gamma = \alpha^7$ in GF(2^4). The distinct conjugates of γ are

$$\gamma^2 = \alpha^{14}, \qquad \gamma^{2^2} = \alpha^{28} = \alpha^{13}, \qquad \gamma^{2^3} = \alpha^{56} = \alpha^{11}.$$

Hence, $\phi(X)$ has degree 4 and must be of the following form:

$$\phi(X) = a_0 + a_1X + a_2X^2 + a_3X^3 + X^4.$$

Substituting γ into $\phi(X)$, we have

$$\phi(\gamma) = a_0 + a_1\gamma + a_2\gamma^2 + a_3\gamma^3 + \gamma^4 = 0.$$

Using the polynomial representations for $\gamma, \gamma^2, \gamma^3$, and γ^4 in the equation above, we obtain the following:

$$a_0 + a_1(1 + \alpha + \alpha^3) + a_2(1 + \alpha^3) + a_3(\alpha^2 + \alpha^3) + (1 + \alpha^2 + \alpha^3) = 0$$

$$(a_0 + a_1 + a_2 + 1) + a_1\alpha + (a_3 + 1)\alpha^2 + (a_1 + a_2 + a_3 + 1)\alpha^3 = 0.$$

For the equality above to be true, we must have the coefficients equal to zero,

$$\begin{aligned} a_0 + a_1 + a_2 \qquad + 1 &= 0, \\ a_1 \qquad\qquad &= 0, \\ a_3 + 1 &= 0, \\ a_1 + a_2 + a_3 + 1 &= 0. \end{aligned}$$

Solving the linear equations above, we obtain $a_0 = 1$, $a_1 = a_2 = 0$, and $a_3 = 1$. Therefore, the minimal polynomial of $\gamma = \alpha^7$ is $\phi(X) = 1 + X^3 + X^4$. All the minimal polynomials of the elements in GF(2^4) are given by Table 2.9.

TABLE 2.9 MINIMAL POLYNOMIALS OF THE ELEMENTS IN GF(2^4) GENERATED BY $p(X) = X^4 + X + 1$

Conjugate roots	Minimal polynomials
0	X
1	$X + 1$
$\alpha, \alpha^2, \alpha^4, \alpha^8$	$X^4 + X + 1$
$\alpha^3, \alpha^6, \alpha^9, \alpha^{12}$	$X^4 + X^3 + X^2 + X + 1$
α^5, α^{10}	$X^2 + X + 1$
$\alpha^7, \alpha^{11}, \alpha^{13}, \alpha^{14}$	$X^4 + X^3 + 1$

A direct consequence of Theorem 2.14 is Theorem 2.15.

Theorem 2.15. Let $\phi(X)$ be the minimal polynomial of an element β in GF(2^m). Let e be the degree of $\phi(X)$. Then e is the smallest integer such that $\beta^{2^e} = \beta$. Moreover, $e \leq m$.

In particular, the degree of the minimal polynomial of any element in GF(2^m) divides m. The proof of this property is omitted here. Table 2.9 shows that the degree of the minimal polynomial of each element in GF(2^4) divides 4. Minimal polynomials of the elements in GF(2^m) for $m = 2$ to 10 are given in Appendix B.

In the construction of the Galois field GF(2^m), we use a primitive polynomial $p(X)$ of degree m and require that the element α be a root of $p(X)$. Since the powers of α generate all the nonzero elements of GF(2^m), α is a primitive element. In fact, all the conjugates of α are primitive elements of GF(2^m). To see this, let n be the order of α^{2^l} for $l > 0$. Then

$$(\alpha^{2^l})^n = \alpha^{n2^l} = 1.$$

Also, it follows from Theorem 2.5 that n divides $2^m - 1$,

$$2^m - 1 = k \cdot n. \tag{2.24}$$

Since α is a primitive element of GF(2^m), its order is $2^m - 1$. For $\alpha^{n2^l} = 1$, $n2^l$ must be a multiple of $2^m - 1$. Since 2^l and $2^m - 1$ are relatively prime, n must be divisible by $2^m - 1$, say

$$n = q \cdot (2^m - 1). \tag{2.25}$$

From (2.24) and (2.25) we conclude that $n = 2^m - 1$. Consequently, α^{2^l} is also a primitive element of GF(2^m). In general, we have the following theorem:

Theorem 2.16. If β is a primitive element of GF(2^m), all its conjugates β^2, $\beta^{2^2}, \ldots$ are also primitive elements of GF(2^m).

Example 2.9

Consider the field GF(2^4) given by Table 2.8. The powers of $\beta = \alpha^7$ are

$$\beta^0 = 1, \quad \beta^1 = \alpha^7, \quad \beta^2 = \alpha^{14}, \quad \beta^3 = \alpha^{21} = \alpha^6, \quad \beta^4 = \alpha^{28} = \alpha^{13},$$
$$\beta^5 = \alpha^{35} = \alpha^5, \quad \beta^6 = \alpha^{42} = \alpha^{12}, \quad \beta^7 = \alpha^{49} = \alpha^4, \quad \beta^8 = \alpha^{56} = \alpha^{11},$$
$$\beta^9 = \alpha^{63} = \alpha^3, \quad \beta^{10} = \alpha^{70} = \alpha^{10}, \quad \beta^{11} = \alpha^{77} = \alpha^2, \quad \beta^{12} = \alpha^{84} = \alpha^9,$$
$$\beta^{13} = \alpha^{91} = \alpha, \quad \beta^{14} = \alpha^{98} = \alpha^8, \quad \beta^{15} = \alpha^{105} = 1.$$

Clearly, the powers of $\beta = \alpha^7$ generate all the nonzero elements of GF(2^4), so $\beta = \alpha^7$ is a primitive element of GF(2^7). The conjugates of $\beta = \alpha^7$ are

$$\beta^2 = \alpha^{14}, \qquad \beta^{2^2} = \alpha^{13}, \qquad \beta^{2^3} = \alpha^{11}.$$

We may readily check that they are all primitive elements of GF(2^m).

A more general form of Theorem 2.16 is Theorem 2.17.

Theorem 2.17. If β is an element of order n in GF(2^m), all its conjugates have the same order n. (The proof is left as an exercise.)

Example 2.10

Consider the element α^5 in GF(2^4) given by Table 2.8. Since $(\alpha^5)^{2^2} = \alpha^{20} = \alpha^5$, the only conjugate of α^5 is α^{10}. Both α^5 and α^{10} have order $n = 3$. The minimal polynomial of α^5 and α^{10} is $X^2 + X + 1$, whose degree is a factor of $m = 4$. The conjugates of α^3 are α^6, α^9, and α^{12}. They all have order $n = 5$.

2.6 COMPUTATIONS USING GALOIS FIELD GF(2^m) ARITHMETIC

Here we perform some example computations using arithmetic over GF(2^m). Consider the following linear equations over GF(2^4) (see Table 2.8):

$$\begin{aligned} X + \alpha^7 Y &= \alpha^2, \\ \alpha^{12} X + \alpha^8 Y &= \alpha^4. \end{aligned} \tag{2.26}$$

Multiplying the second equation by α^3 gives

$$X + \alpha^7 Y = \alpha^2,$$
$$X + \alpha^{11} Y = \alpha^7.$$

By adding the two equations above, we get

$$(\alpha^7 + \alpha^{11})Y = \alpha^2 + \alpha^7,$$
$$\alpha^8 Y = \alpha^{12},$$
$$Y = \alpha^4.$$

Substituting $Y = \alpha^4$ into the first equation of (2.26), we obtain $X = \alpha^9$. Thus, the solution for the equations of (2.26) is $X = \alpha^9$ and $Y = \alpha^4$.

Alternatively, the equations of (2.26) could be solved by using Cramer's rule:

$$X = \frac{\begin{vmatrix} \alpha^2 & \alpha^7 \\ \alpha^4 & \alpha^8 \end{vmatrix}}{\begin{vmatrix} 1 & \alpha^7 \\ \alpha^{12} & \alpha^8 \end{vmatrix}} = \frac{\alpha^{10} + \alpha^{11}}{\alpha^8 + \alpha^{19}} = \frac{1 + \alpha^3}{\alpha + \alpha^2} = \frac{\alpha^{14}}{\alpha^5} = \alpha^9,$$

$$Y = \frac{\begin{vmatrix} 1 & \alpha^2 \\ \alpha^{12} & \alpha^4 \end{vmatrix}}{\begin{vmatrix} 1 & \alpha^7 \\ \alpha^{12} & \alpha^8 \end{vmatrix}} = \frac{\alpha^4 + \alpha^{14}}{\alpha^8 + \alpha^{19}} = \frac{\alpha + \alpha^3}{\alpha + \alpha^2} = \frac{\alpha^9}{\alpha^5} = \alpha^4.$$

As one more example, suppose that we want to solve the equation

$$f(X) = X^2 + \alpha^7 X + \alpha = 0$$

over GF(2^4). The quadratic formula will not work because it requires dividing by 2, and in this field, $2 = 0$. If $f(X) = 0$ has any solutions in GF(2^4), the solutions can be found simply by substituting all the elements of Table 2.8 for X. By doing so, we would find that $f(\alpha^6) = 0$ and $f(\alpha^{10}) = 0$, since

$$f(\alpha^6) = (\alpha^6)^2 + \alpha^7 \cdot \alpha^6 + \alpha = \alpha^{12} + \alpha^{13} + \alpha = 0,$$
$$f(\alpha^{10}) = (\alpha^{10})^2 + \alpha^7 \cdot \alpha^{10} + \alpha = \alpha^5 + \alpha^2 + \alpha = 0.$$

Thus, α^6 and α^{10} are the roots of $f(X)$ and $f(X) = (X + \alpha^6)(X + \alpha^{10})$.

The computations above are typical of those required for decoding a class of block codes, known as BCH codes, and they can be programmed quite easily on a general-purpose computer. It is also a simple matter to build a computer that can do this kind of arithmetic.

2.7 VECTOR SPACES

Let V be a set of elements on which a binary operation called addition $+$ is defined. Let F be a field. A multiplication operation, denoted by $\cdot$, between the elements in F and elements in V is also defined. The set V is called a *vector space* over the field F if it satisfies the following conditions:

(i) V is a commutative group under addition.

(ii) For any element a in F and any element $\mathbf{v}$ in V, $a \cdot \mathbf{v}$ is an element in V.

(iii) (Distributive Laws) For any elements $\mathbf{u}$ and $\mathbf{v}$ in V and any elements a and b in F,

$$a\cdot(\mathbf{u}+\mathbf{v}) = a\cdot\mathbf{u} + a\cdot\mathbf{v},$$
$$(a+b)\cdot\mathbf{v} = a\cdot\mathbf{v} + b\cdot\mathbf{v}.$$

(iv) (Associative Law) For any $\mathbf{v}$ in V and any a and b in F,

$$(a\cdot b)\cdot\mathbf{v} = a\cdot(b\cdot\mathbf{v}).$$

(v) Let 1 be the unit element of F. Then, for any $\mathbf{v}$ in V, $1\cdot\mathbf{v} = \mathbf{v}$.

The elements of V are called *vectors* and the elements of the field F are called *scalars*. The addition on V is called a *vector addition* and the multiplication that combines a scalar in F and a vector in V into a vector in V is referred to as *scalar multiplication* (or *product*). The additive identity of V is denoted by $\mathbf{0}$.

Some basic properties of a vector space V over a field F can be derived from the definition above.

Property I. Let 0 be the zero element of the field F. For any vector $\mathbf{v}$ in V, $0\cdot\mathbf{v} = \mathbf{0}$.

Proof. Since $1 + 0 = 1$ in F, we have $1\cdot\mathbf{v} = (1+0)\cdot\mathbf{v} = 1\cdot\mathbf{v} + 0\cdot\mathbf{v}$. Using condition (v) of the definition of a vector space given above, we obtain $\mathbf{v} = \mathbf{v} + 0\cdot\mathbf{v}$. Let $-\mathbf{v}$ be the additive inverse of $\mathbf{v}$. Adding $-\mathbf{v}$ to both sides of $\mathbf{v} = \mathbf{v} + 0\cdot\mathbf{v}$, we have

$$\mathbf{0} = \mathbf{0} + 0\cdot\mathbf{v}$$
$$\mathbf{0} = 0\cdot\mathbf{v}.$$

Q.E.D.

Property II. For any scalar c in F, $c\cdot\mathbf{0} = \mathbf{0}$. (The proof is left as an exercise.)

Property III. For any scalar c in F and any vector $\mathbf{v}$ in V,

$$(-c)\cdot\mathbf{v} = c\cdot(-\mathbf{v}) = -(c\cdot\mathbf{v})$$

That is, $(-c)\cdot\mathbf{v}$ or $c\cdot(-\mathbf{v})$ is the additive inverse of the vector $c\cdot\mathbf{v}$. (The proof is left as an exercise.)

Next, we present a very useful vector space over GF(2) which plays a central role in coding theory. Consider an ordered sequence of n components,

$$(a_0, a_1, \ldots, a_{n-1}),$$

where each component a_i is an element from the binary field GF(2) (i.e., $a_i = 0$ or 1). This sequence is generally called an *n-tuple* over GF(2). Since there are two choices for each a_i, we can construct 2^n distinct n-tuples. Let V_n denote this set of 2^n distinct n-tuples. Now, we define an addition $+$ on V_n as the following: For any $\mathbf{u} = (u_0, u_1, \ldots, u_{n-1})$ and $\mathbf{v} = (v_0, v_1, \ldots, v_{n-1})$ in V_n,

$$\mathbf{u} + \mathbf{v} = (u_0 + v_0, u_1 + v_1, \ldots, u_{n-1} + v_{n-1}), \tag{2.27}$$

where $u_i + v_i$ is carried out in modulo-2 addition. Clearly, $\mathbf{u} + \mathbf{v}$ is also an n-tuple over GF(2). Hence, V_n is closed under the addition defined by (2.27). We can readily verify that V_n is a commutative group under the addition defined by (2.27). First we see that the all-zero n-tuple $\mathbf{0} = (0, 0, \ldots, 0)$ is the additive identity. For any $\mathbf{v}$ in V_n,

$$\mathbf{v} + \mathbf{v} = (v_0 + v_0, v_1 + v_1, \ldots, v_{n-1} + v_{n-1})$$
$$= (0, 0, \ldots, 0) = \mathbf{0}.$$

Hence, the additive inverse of each n-tuple in V_n is itself. Since modulo-2 addition is commutative and associative, we can easily check that the addition defined by (2.27) is also commutative and associative. Therefore, V_n is a commutative group under the addition defined by (2.27).

Next we define scalar multiplication of an n-tuple $\mathbf{v}$ in V_n by an element a from GF(2) as follows:

$$a \cdot (v_0, v_1, \ldots, v_{n-1}) = (a \cdot v_0, a \cdot v_1, \ldots, a \cdot v_{n-1}), \tag{2.28}$$

where $a \cdot v_i$ is carried out in modulo-2 multiplication. Clearly, $a \cdot (v_0, v_1, \ldots, v_{n-1})$ is also an n-tuple in V_n. If $a = 1$,

$$1 \cdot (v_0, v_1, \ldots, v_{n-1}) = (1 \cdot v_0, 1 \cdot v_1, \ldots, 1 \cdot v_{n-1})$$
$$= (v_0, v_1, \ldots, v_{n-1}).$$

We can easily show that the vector addition and scalar multiplication defined by (2.27) and (2.28), respectively, satisfy the distributive and associative laws. Therefore, the set V_n of all n-tuples over GF(2) forms a vector space over GF(2).

Example 2.11

Let $n = 5$. The vector space V_5 of all 5-tuples over GF(2) consists of the following 32 vectors:

(0 0 0 0 0), (0 0 0 0 1), (0 0 0 1 0), (0 0 0 1 1),
(0 0 1 0 0), (0 0 1 0 1), (0 0 1 1 0), (0 0 1 1 1),
(0 1 0 0 0), (0 1 0 0 1), (0 1 0 1 0), (0 1 0 1 1),
(0 1 1 0 0), (0 1 1 0 1), (0 1 1 1 0), (0 1 1 1 1),
(1 0 0 0 0), (1 0 0 0 1), (1 0 0 1 0), (1 0 0 1 1),
(1 0 1 0 0), (1 0 1 0 1), (1 0 1 1 0), (1 0 1 1 1),
(1 1 0 0 0), (1 1 0 0 1), (1 1 0 1 0), (1 1 0 1 1),
(1 1 1 0 0), (1 1 1 0 1), (1 1 1 1 0), (1 1 1 1 1).

The vector sum of (1 0 1 1 1) and (1 1 0 0 1) is

$$(1\ 0\ 1\ 1\ 1) + (1\ 1\ 0\ 0\ 1) = (1 + 1, 0 + 1, 1 + 0, 1 + 0, 1 + 1) = (0\ 1\ 1\ 1\ 0).$$

Using the rule of scalar multiplication defined by (2.28), we obtain

$$0 \cdot (1\ 1\ 0\ 1\ 0) = (0 \cdot 1, 0 \cdot 1, 0 \cdot 0, 0 \cdot 1, 0 \cdot 0) = (0\ 0\ 0\ 0\ 0),$$
$$1 \cdot (1\ 1\ 0\ 1\ 0) = (1 \cdot 1, 1 \cdot 1, 1 \cdot 0, 1 \cdot 1, 1 \cdot 0) = (1\ 1\ 0\ 1\ 0).$$

The vector space of all n-tuples over any field F can be constructed in a similar manner. However, in this book, we are concerned only with the vector space of all n-tuples over GF(2) or over an extension field of GF(2) [e.g., GF(2^m)].

V being a vector space over a field F, it may happen that a subset S of V is also a vector space over F. Such a subset is called a *subspace* of V.

Theorem 2.18. Let S be a nonempty subset of a vector space V over a field F. Then S is a subspace of V if the following conditions are satisfied:

(i) For any two vectors $\mathbf{u}$ and $\mathbf{v}$ in S, $\mathbf{u} + \mathbf{v}$ is also a vector in S.
(ii) For any element a in F and any vector $\mathbf{u}$ in S, $a \cdot \mathbf{u}$ is also in S.

Proof. Conditions (i) and (ii) say simply that S is closed under vector addition and scalar multiplication of V. Condition (ii) ensures that, for any vector $\mathbf{v}$ in S, its additive inverse $(-1) \cdot \mathbf{v}$ is also in S. Then, $\mathbf{v} + (-1) \cdot \mathbf{v} = \mathbf{0}$ is also in S. Therefore, S is a subgroup of V. Since the vectors of S are also vectors of V, the associative and distributive laws must hold for S. Hence, S is a vector space over F and is a subspace of V.
Q.E.D.

Example 2.12

Consider the vector space V_5 of all 5-tuples over GF(2) given in Example 2.11. The set

$$\{(0\ 0\ 0\ 0\ 0),\ (0\ 0\ 1\ 1\ 1),\ (1\ 1\ 0\ 1\ 0),\ (1\ 1\ 1\ 0\ 1)\}$$

satisfies both conditions of Theorem 2.18, so it is a subspace of V_5.

Let $\mathbf{v}_1, \mathbf{v}_2, \ldots, \mathbf{v}_k$ be k vectors in a vector space V over a field F. Let $a_1, a_2, \ldots, a_k$ be k scalars from F. The sum

$$a_1\mathbf{v}_1 + a_2\mathbf{v}_2 + \cdots + a_k\mathbf{v}_k$$

is called a *linear combination* of $\mathbf{v}_1, \mathbf{v}_2, \ldots, \mathbf{v}_k$. Clearly, the sum of two linear combinations of $\mathbf{v}_1, \mathbf{v}_2, \ldots, \mathbf{v}_k$,

$$(a_1\mathbf{v}_1 + a_2\mathbf{v}_2 + \cdots + a_k\mathbf{v}_k) + (b_1\mathbf{v}_1 + b_2\mathbf{v}_2 + \cdots + b_k\mathbf{v}_k)$$
$$= (a_1 + b_1)\mathbf{v}_1 + (a_2 + b_2)\mathbf{v}_2 + \cdots + (a_k + b_k)\mathbf{v}_k,$$

is also a linear combination of $\mathbf{v}_1, \mathbf{v}_2, \ldots, \mathbf{v}_k$, and the product of a scalar c in F and a linear combination of $\mathbf{v}_1, \mathbf{v}_2, \ldots, \mathbf{v}_k$,

$$c \cdot (a_1\mathbf{v}_1 + a_2\mathbf{v}_2 + \cdots + a_k\mathbf{v}_k) = (c \cdot a_1)\mathbf{v}_1 + (c \cdot a_2)\mathbf{v}_2 + \cdots + (c \cdot a_k)\mathbf{v}_k,$$

is also a linear combination of $\mathbf{v}_1, \mathbf{v}_2, \ldots, \mathbf{v}_k$. It follows from Theorem 2.18 that we have the following result.

Theorem 2.19. Let $\mathbf{v}_1, \mathbf{v}_2, \ldots, \mathbf{v}_k$ be k vectors in a vector space V over a field F. The set of all linear combinations of $\mathbf{v}_1, \mathbf{v}_2, \ldots, \mathbf{v}_k$ forms a subspace of V.

Example 2.13

Consider the vector space V_5 of all 5-tuples over GF(2) given by Example 2.11. The linear combinations of (0 0 1 1 1) and (1 1 1 0 1) are

$$0 \cdot (0\ 0\ 1\ 1\ 1) + 0 \cdot (1\ 1\ 1\ 0\ 1) = (0\ 0\ 0\ 0\ 0),$$
$$0 \cdot (0\ 0\ 1\ 1\ 1) + 1 \cdot (1\ 1\ 1\ 0\ 1) = (1\ 1\ 1\ 0\ 1),$$
$$1 \cdot (0\ 0\ 1\ 1\ 1) + 0 \cdot (1\ 1\ 1\ 0\ 1) = (0\ 0\ 1\ 1\ 1),$$
$$1 \cdot (0\ 0\ 1\ 1\ 1) + 1 \cdot (1\ 1\ 1\ 0\ 1) = (1\ 1\ 0\ 1\ 0).$$

These four vectors form the same subspace given by Example 2.12.

A set of vectors $\mathbf{v}_1, \mathbf{v}_2, \ldots, \mathbf{v}_k$ in a vector space V over a field F is said to be *linearly dependent* if and only if there exist k scalars $a_1, a_2, \ldots, a_k$ from F, *not all zero*, such that

$$a_1\mathbf{v}_1 + a_2\mathbf{v}_2 + \cdots + a_k\mathbf{v}_k = \mathbf{0}.$$

A set of vectors, $\mathbf{v}_1, \mathbf{v}_2, \ldots, \mathbf{v}_k$, is said to be *linearly independent* if it is not linearly dependent. That is, if $\mathbf{v}_1, \mathbf{v}_2, \ldots, \mathbf{v}_k$ are linearly independent, then

$$a_1\mathbf{v}_1 + a_2\mathbf{v}_2 + \cdots + a_k\mathbf{v}_k \neq \mathbf{0}$$

unless $a_1 = a_2 = \cdots = a_k = 0$.

Example 2.14

The vectors (1 0 1 1 0), (0 1 0 0 1), and (1 1 1 1 1) are linearly dependent since

$$1\cdot(1\ 0\ 1\ 1\ 0) + 1\cdot(0\ 1\ 0\ 0\ 1) + 1\cdot(1\ 1\ 1\ 1\ 1) = (0\ 0\ 0\ 0\ 0).$$

However, (1 0 1 1 0), (0 1 0 0 1), and (1 1 0 1 1) are linearly independent. All eight linear combinations of these three vectors are given below:

$$0\cdot(1\ 0\ 1\ 1\ 0) + 0\cdot(0\ 1\ 0\ 0\ 1) + 0\cdot(1\ 1\ 0\ 1\ 1) = (0\ 0\ 0\ 0\ 0),$$
$$0\cdot(1\ 0\ 1\ 1\ 0) + 0\cdot(0\ 1\ 0\ 0\ 1) + 1\cdot(1\ 1\ 0\ 1\ 1) = (1\ 1\ 0\ 1\ 1),$$
$$0\cdot(1\ 0\ 1\ 1\ 0) + 1\cdot(0\ 1\ 0\ 0\ 1) + 0\cdot(1\ 1\ 0\ 1\ 1) = (0\ 1\ 0\ 0\ 1),$$
$$0\cdot(1\ 0\ 1\ 1\ 0) + 1\cdot(0\ 1\ 0\ 0\ 1) + 1\cdot(1\ 1\ 0\ 1\ 1) = (1\ 0\ 0\ 1\ 0),$$
$$1\cdot(1\ 0\ 1\ 1\ 0) + 0\cdot(0\ 1\ 0\ 0\ 1) + 0\cdot(1\ 1\ 0\ 1\ 1) = (1\ 0\ 1\ 1\ 0),$$
$$1\cdot(1\ 0\ 1\ 1\ 0) + 0\cdot(0\ 1\ 0\ 0\ 1) + 1\cdot(1\ 1\ 0\ 1\ 1) = (0\ 1\ 1\ 0\ 1),$$
$$1\cdot(1\ 0\ 1\ 1\ 0) + 1\cdot(0\ 1\ 0\ 0\ 1) + 0\cdot(1\ 1\ 0\ 1\ 1) = (1\ 1\ 1\ 1\ 1),$$
$$1\cdot(1\ 0\ 1\ 1\ 0) + 1\cdot(0\ 1\ 0\ 0\ 1) + 1\cdot(1\ 1\ 0\ 1\ 1) = (0\ 0\ 1\ 0\ 0).$$

A set of vectors is said to *span* a vector space V if every vector in V is a linear combination of the vectors in the set. In any vector space or subspace there exists at least one set B of linearly independent vectors which span the space. This set is called a *basis* (or *base*) of the vector space. The number of vectors in a basis of a vector space is called the *dimension* of the vector space. (Note that the number of vectors in any two bases are the same.)

Consider the vector space V_n of all n-tuples over GF(2). Let us form the following n n-tuples:

$$\mathbf{e}_0 = (1, 0, 0, 0, \ldots, 0, 0)$$
$$\mathbf{e}_1 = (0, 1, 0, 0, \ldots, 0, 0)$$
$$\vdots$$
$$\mathbf{e}_{n-1} = (0, 0, 0, 0, \ldots, 0, 1),$$

where the n-tuple $\mathbf{e}_i$ has only one nonzero component at ith position. Then every n-tuple $(a_0, a_1, a_2, \ldots, a_{n-1})$ in V_n can be expressed as a linear combination of $\mathbf{e}_0, \mathbf{e}_1, \ldots, \mathbf{e}_{n-1}$ as follows:

$$(a_0, a_1, a_2, \ldots, a_{n-1}) = a_0\mathbf{e}_0 + a_1\mathbf{e}_1 + a_2\mathbf{e}_2 + \cdots + a_{n-1}\mathbf{e}_{n-1}.$$

Therefore, $\mathbf{e}_0, \mathbf{e}_1, \ldots, \mathbf{e}_{n-1}$ span the vector space V_n of all n-tuples over GF(2). From the equation above, we also see that $\mathbf{e}_0, \mathbf{e}_1, \ldots, \mathbf{e}_{n-1}$ are linearly independent. Hence, they form a basis for V_n and the dimension of V_n is n. If $k < n$ and $\mathbf{v}_1, \mathbf{v}_2, \ldots, \mathbf{v}_k$ are k linearly independent vectors in V_n, then all the linear combinations of $\mathbf{v}_1, \mathbf{v}_2, \ldots, \mathbf{v}_k$ of the form

$$\mathbf{u} = c_1\mathbf{v}_1 + c_2\mathbf{v}_2 + \cdots + c_k\mathbf{v}_k$$

form a k-dimensional subspace S of V_n. Since each c_i has two possible values, 0 or 1, there are 2^k possible distinct linear combinations of $\mathbf{v}_1, \mathbf{v}_2, \ldots, \mathbf{v}_k$. Thus, S consists of 2^k vectors and is a k-dimensional subspace of V_n.

Let $\mathbf{u} = (u_0, u_1, \ldots, u_{n-1})$ and $\mathbf{v} = (v_0, v_1, \ldots, v_{n-1})$ be two n-tuples in V_n. We define the *inner product* (or *dot product*) of $\mathbf{u}$ and $\mathbf{v}$ as

$$\mathbf{u}\cdot\mathbf{v} = u_0\cdot v_0 + u_1\cdot v_1 + \cdots + u_{n-1}\cdot v_{n-1}, \tag{2.29}$$

where $u_i\cdot v_i$ and $u_i\cdot v_i + u_{i+1}\cdot v_{i+1}$ are carried out in modulo-2 multiplication and addition. Hence, the inner product $\mathbf{u}\cdot\mathbf{v}$ is a scalar in GF(2). If $\mathbf{u}\cdot\mathbf{v} = 0$, $\mathbf{u}$ and $\mathbf{v}$ are said to be *orthogonal* to each other. The inner product has the following properties:

(i) $\mathbf{u}\cdot\mathbf{v} = \mathbf{v}\cdot\mathbf{u}$.

(ii) $\mathbf{u}\cdot(\mathbf{v} + \mathbf{w}) = \mathbf{u}\cdot\mathbf{v} + \mathbf{u}\cdot\mathbf{w}$.

(iii) $(a\mathbf{u})\cdot\mathbf{v} = a(\mathbf{u}\cdot\mathbf{v})$.

(The concept of inner product can be generalized to any Galois field.)

Let S be a k-dimensional subspace of V_n and let S_d be the set of vectors in V_n such that, for any $\mathbf{u}$ in S and $\mathbf{v}$ in S_d, $\mathbf{u}\cdot\mathbf{v} = 0$. The set S_d contains at least the all-zero n-tuple $\mathbf{0} = (0, 0, \ldots, 0)$, since for any $\mathbf{u}$ in S, $\mathbf{0}\cdot\mathbf{u} = 0$. Thus, S_d is nonempty. For any element a in GF(2) and any $\mathbf{v}$ in S_d,

$$a\cdot\mathbf{v} = \begin{cases} \mathbf{0} & \text{if } a = 0 \\ \mathbf{v} & \text{if } a = 1. \end{cases}$$

Therefore, $a\cdot\mathbf{v}$ is also in S_d. Let $\mathbf{v}$ and $\mathbf{w}$ be any two vectors in S_d. For any vector $\mathbf{u}$ in S, $\mathbf{u}\cdot(\mathbf{v} + \mathbf{w}) = \mathbf{u}\cdot\mathbf{v} + \mathbf{u}\cdot\mathbf{w} = 0 + 0 = 0$. This says that if $\mathbf{v}$ and $\mathbf{w}$ are orthogonal to $\mathbf{u}$, the vector sum $\mathbf{v} + \mathbf{w}$ is also orthogonal to $\mathbf{u}$. Consequently, $\mathbf{v} + \mathbf{w}$ is a vector in S_d. It follows from Theorem 2.18 that S_d is also a subspace of V_n. This subspace S_d is called the *null* (or *dual*) *space* of S. Conversely, S is also the null space of S_d. The dimension of S_d is given by Theorem 2.20, whose proof is omitted here [2].

Theorem 2.20. Let S be a k-dimensional space of the vector space V_n of all n-tuples over GF(2). The dimension of its null space S_d is $n - k$. In other words, $\dim(S) + \dim(S_d) = n$.

Example 2.15

Consider the vector space V_5 of all 5-tuples over GF(2) given by Example 2.11. The following eight vectors form a three-dimensional subspace S of V_5:

$$(0\ 0\ 0\ 0\ 0),\quad (1\ 1\ 1\ 0\ 0),\quad (0\ 1\ 0\ 1\ 0),\quad (1\ 0\ 0\ 0\ 1).$$
$$(1\ 0\ 1\ 1\ 0),\quad (0\ 1\ 1\ 0\ 1),\quad (1\ 1\ 0\ 1\ 1),\quad (0\ 0\ 1\ 1\ 1).$$

The null space S_d of S consists of the following 4-vectors:

$$(0\ 0\ 0\ 0\ 0),\quad (1\ 0\ 1\ 0\ 1),\quad (0\ 1\ 1\ 1\ 0),\quad (1\ 1\ 0\ 1\ 1).$$

S_d is spanned by (1 0 1 0 1) and (0 1 1 1 0), which are linearly independent. Thus, the dimension of S_d is 2.

All the results presented in this section can be generalized in a straightforward manner to the vector space of all n-tuples over GF(q), where q is a power of prime.

2.8 MATRICES

A $k \times n$ matrix over GF(2) (or over any other field) is a rectangular array with k rows and n columns,

$$\mathbf{G} = \begin{bmatrix} g_{00} & g_{01} & g_{02} & \cdots & g_{0,n-1} \\ g_{10} & g_{11} & g_{12} & \cdots & g_{1,n-1} \\ \cdot & & & & \\ \cdot & & & & \\ \cdot & & & & \\ g_{k-1,0} & g_{k-1,1} & g_{k-1,2} & \cdots & g_{k-1,n-1} \end{bmatrix}, \tag{2.30}$$

where each entry g_{ij} with $0 \leq i < k$ and $0 \leq j < n$ is an element from the binary field GF(2). Observe that the first index i indicates the row containing g_{ij} and the second index j tells which column g_{ij} is in. We shall sometimes abbreviate the matrix of (2.30) by the notation $[g_{ij}]$. We also observe that each row of $\mathbf{G}$ is an n-tuple over GF(2) and each column is a k-tuple over GF(2). The matrix $\mathbf{G}$ can also be represented by its k rows $\mathbf{g}_0, \mathbf{g}_1, \ldots, \mathbf{g}_{k-1}$ as follows:

$$\mathbf{G} = \begin{bmatrix} \mathbf{g}_0 \\ \mathbf{g}_1 \\ \cdot \\ \cdot \\ \cdot \\ \mathbf{g}_{k-1} \end{bmatrix}.$$

If the k ($k \leq n$) rows of $\mathbf{G}$ are linearly independent, then the 2^k linear combinations of these rows form a k-dimensional subspace of the vector space V_n of all the n-tuples over GF(2). This subspace is called the *row space* of $\mathbf{G}$. We may interchange any two rows of $\mathbf{G}$ or add one row to another. These are called *elementary row operations.* Performing elementary row operations on $\mathbf{G}$, we obtain another matrix $\mathbf{G}'$ over GF(2). However, both $\mathbf{G}$ and $\mathbf{G}'$ gave the same row space.

Example 2.16

Consider a 3×6 matrix $\mathbf{G}$ over GF(2),

$$\mathbf{G} = \begin{bmatrix} 1 & 1 & 0 & 1 & 1 & 0 \\ 0 & 0 & 1 & 1 & 1 & 0 \\ 0 & 1 & 0 & 0 & 1 & 1 \end{bmatrix}.$$

Adding the third row to the first row and interchanging the second and third rows, we obtain the following matrix:

$$\mathbf{G}' = \begin{bmatrix} 1 & 0 & 0 & 1 & 0 & 1 \\ 0 & 1 & 0 & 0 & 1 & 1 \\ 0 & 0 & 1 & 1 & 1 & 0 \end{bmatrix}.$$

Both $\mathbf{G}$ and $\mathbf{G}'$ give the following row space:

$$(0\ 0\ 0\ 0\ 0\ 0),\quad (1\ 0\ 0\ 1\ 0\ 1),\quad (0\ 1\ 0\ 0\ 1\ 1),\quad (0\ 0\ 1\ 1\ 1\ 0),$$
$$(1\ 1\ 0\ 1\ 1\ 0),\quad (1\ 0\ 1\ 0\ 1\ 1),\quad (0\ 1\ 1\ 1\ 0\ 1),\quad (1\ 1\ 1\ 0\ 0\ 0).$$

This is a three-dimensional subspace of the vector space V_6 of all the 6-tuples over GF(2).

Let S be the row space of a $k \times n$ matrix $\mathbf{G}$ over GF(2) whose k rows $\mathbf{g}_0, \mathbf{g}_1, \ldots, \mathbf{g}_{k-1}$ are linearly independent. Let S_d be the null space of S. Then the dimension of S_d is $n - k$. Let $\mathbf{h}_0, \mathbf{h}_1, \ldots, \mathbf{h}_{n-k-1}$ be $n - k$ linearly independent vectors in S_d. Clearly, these vectors span S_d. We may form an $(n - k) \times n$ matrix $\mathbf{H}$ using $\mathbf{h}_0, \mathbf{h}_1, \ldots, \mathbf{h}_{n-k-1}$ as rows:

$$\mathbf{H} = \begin{bmatrix} \mathbf{h}_0 \\ \mathbf{h}_1 \\ \cdot \\ \cdot \\ \cdot \\ \mathbf{h}_{n-k-1} \end{bmatrix} = \begin{bmatrix} h_{00} & h_{01} & \cdots & h_{0,n-1} \\ h_{10} & h_{11} & \cdots & h_{1,n-1} \\ \cdot & \cdot & & \cdot \\ \cdot & \cdot & & \cdot \\ \cdot & \cdot & & \cdot \\ h_{n-k-1,0} & h_{n-k-1,1} & \cdots & h_{n-k-1,n-1} \end{bmatrix}.$$

The row space of $\mathbf{H}$ is S_d. Since each row $\mathbf{g}_i$ of $\mathbf{G}$ is a vector in S and each row $\mathbf{h}_j$ of $\mathbf{H}$ is a vector in S_d, the inner product of $\mathbf{g}_i$ and $\mathbf{h}_j$ must be zero (i.e., $\mathbf{g}_i \cdot \mathbf{h}_j = 0$). Since the row space S of $\mathbf{G}$ is the null space of the row space S_d of $\mathbf{H}$, we call S the null (or dual) space of $\mathbf{H}$. Summarizing the results above, we have:

Theorem 2.21. For any $k \times n$ matrix $\mathbf{G}$ over GF(2) with k linearly independent rows, there exists an $(n - k) \times n$ matrix $\mathbf{H}$ over GF(2) with $n - k$ linearly independent rows such that for any row $\mathbf{g}_i$ in $\mathbf{G}$ and any $\mathbf{h}_j$ in $\mathbf{H}$, $\mathbf{g}_i \cdot \mathbf{h}_j = 0$. The row space of $\mathbf{G}$ is the null space of $\mathbf{H}$, and vice versa.

Example 2.17

Consider the following 3×6 matrix over GF(2):

$$\mathbf{G} = \begin{bmatrix} 1 & 1 & 0 & 1 & 1 & 0 \\ 0 & 0 & 1 & 1 & 1 & 0 \\ 0 & 1 & 0 & 0 & 1 & 1 \end{bmatrix}.$$

The row space of this matrix is the null space of

$$\mathbf{H} = \begin{bmatrix} 1 & 0 & 1 & 1 & 0 & 0 \\ 0 & 1 & 1 & 0 & 1 & 0 \\ 1 & 1 & 0 & 0 & 0 & 1 \end{bmatrix}.$$

We can easily check that each row of $\mathbf{G}$ is orthogonal to each row of $\mathbf{H}$.

Two matrices can be added if they have the same number of rows and the same number of columns. Adding two $k \times n$ matrices $\mathbf{A} = [a_{ij}]$ and $\mathbf{B} = [b_{ij}]$, we simply add their corresponding entries a_{ij} and b_{ij} as follows:

$$[a_{ij}] + [b_{ij}] = [a_{ij} + b_{ij}].$$

Hence, the resultant matrix is also a $k \times n$ matrix. Two matrices can be multiplied provided that the number of columns in the first matrix is equal to the number of rows in the second matrix. Multiplying a $k \times n$ matrix $\mathbf{A} = [a_{ij}]$ by an $n \times l$ matrix $\mathbf{B} = [b_{ij}]$, the product

$$\mathbf{C} = \mathbf{A} \times \mathbf{B} = [c_{ij}]$$

is a $k \times l$ matrix where the entry c_{ij} is equal to the inner product of the ith row $\mathbf{a}_i$ in $\mathbf{A}$ and the jth column $\mathbf{b}_j$ in $\mathbf{B}$, that is,

$$c_{ij} = \mathbf{a}_i \cdot \mathbf{b}_j = \sum_{t=0}^{n-1} a_{it} b_{tj}.$$

Let $\mathbf{G}$ be a $k \times n$ matrix over GF(2). The *transpose* of $\mathbf{G}$, denoted by $\mathbf{G}^T$, is an $n \times k$ matrix whose rows are columns of $\mathbf{G}$ and whose columns are rows of $\mathbf{G}$. A $k \times k$ matrix is called an *identity* matrix if it has 1's on the main diagonal and 0's elsewhere. This matrix is usually denoted by $\mathbf{I}_k$. A *submatrix* of a matrix $\mathbf{G}$ is a matrix that is obtained by striking out given rows or columns of $\mathbf{G}$.

It is straightforward to generalize the concepts and results presented in this section to matrices with entries from GF(q) with q as a power of a prime.

PROBLEMS

2.1. Construct the group under modulo-6 addition.

2.2. Construct the group under modulo-3 multiplication.

2.3. Let m be a positive integer. If m is not a prime, prove that the set $\{1, 2, \ldots, m-1\}$ is not a group under modulo-m multiplication.

2.4. Construct the prime field GF(11) with modulo-11 addition and multiplication. Find all the primitive elements and determine the orders of other elements.

2.5. Let m be a positive integer. If m is not prime, prove that the set $\{0, 1, 2, \ldots, m-1\}$ is not a field under modulo-m addition and multiplication.

2.6. Let λ be the characteristic of a Galois field GF(q). Let 1 be the unit element of GF(q). Show that the sums

$$1, \ \sum_{i=1}^{2} 1, \ \sum_{i=1}^{3} 1, \ \ldots, \ \sum_{i=1}^{\lambda-1} 1, \ \sum_{i=1}^{\lambda} 1 = 0$$

form a subfield of GF(q).

2.7. Prove that every finite field has a primitive element.

2.8. Solve the following simultaneous equations of X, Y, Z, and W with modulo-2 arithmetic:

$$\begin{aligned} X + Y + W &= 1, \\ X + Z + W &= 0, \\ X + Y + Z + W &= 1, \\ Y + Z + W &= 0. \end{aligned}$$

2.9. Show that $X^5 + X^3 + 1$ is irreducible over GF(2).

2.10. Let $f(X)$ be a polynomial of degree n over GF(2). The reciprocal of $f(X)$ is defined as

$$f^*(X) = X^n f\left(\frac{1}{X}\right).$$

(a) Prove that $f^*(X)$ is irreducible over GF(2) if and only if $f(X)$ is irreducible over GF(2).

(b) Prove that $f^*(X)$ is primitive if and only if $f(X)$ is primitive.

2.11. Find all the irreducible polynomials of degree 5 over GF(2).

2.12. Construct a table for GF(2^3) based on the primitive polynomial $p(X) = 1 + X + X^3$. Display the power, polynomial, and vector representations of each element. Determine the order of each element.

2.13. Construct a table for GF(2^5) based on the primitive polynomial $p(X) = 1 + X^2 + X^5$. Let α be a primitive element of GF(2^5). Find the minimal polynomials of α^3 and α^7.

2.14. Let β be an element in GF(2^m). Let e be the smallest nonnegative integer such that $\beta^{2^e} = \beta$. Prove that $\beta^2, \beta^{2^2}, \ldots, \beta^{2^{e-1}}$ are all the distinct conjugates of β.

2.15. Prove Theorem 2.17.

2.16. Let α be a primitive element in GF(2^4). Use Table 2.8 to find the roots of $f(X) = X^3 + \alpha^6 X^2 + \alpha^9 X + \alpha^9$.

2.17. Let α be a primitive element in GF(2^4). Use Table 2.8 to solve the following simultaneous equations for X, Y, and Z:

$$\begin{aligned} X + \alpha^5 Y + \quad Z &= \alpha^7, \\ X + \alpha Y + \alpha^7 Z &= \alpha^9, \\ \alpha^2 X + \quad Y + \alpha^6 Z &= \alpha. \end{aligned}$$

2.18. Let V be a vector space over a field F. For any element c in F, prove that $c \cdot \mathbf{0} = \mathbf{0}$.

2.19. Let V be a vector space over a field F. Prove that, for any c in F and any $\mathbf{v}$ in V, $(-c) \cdot \mathbf{v} = c \cdot (-\mathbf{v}) = -(c \cdot \mathbf{v})$.

2.20. Let S be a subset of the vector space V_n of all n-tuples over GF(2). Prove that S is a subspace if for any $\mathbf{u}$ and $\mathbf{v}$ in S, $\mathbf{u} + \mathbf{v}$ is in S.

2.21. Prove that GF(2^m) is a vector space over GF(2).

2.22. Construct the vector space V_5 of all 5-tuples over GF(2). Find a three-dimensional subspace and determine its null space.

2.23. Given the matrices

$$\mathbf{G} = \begin{bmatrix} 1 & 1 & 0 & 1 & 1 & 0 & 0 \\ 1 & 1 & 1 & 0 & 0 & 1 & 0 \\ 0 & 1 & 1 & 1 & 0 & 0 & 1 \end{bmatrix}, \qquad \mathbf{H} = \begin{bmatrix} 1 & 0 & 0 & 0 & 1 & 1 & 0 \\ 0 & 1 & 0 & 0 & 1 & 1 & 1 \\ 0 & 0 & 1 & 0 & 0 & 1 & 1 \\ 0 & 0 & 0 & 1 & 1 & 0 & 1 \end{bmatrix},$$

show that the row space of $\mathbf{G}$ is the null space of $\mathbf{H}$, and vice versa.

2.24. Let S_1 and S_2 be two subspaces of a vector V. Show that the intersection of S_1 and S_2 is also a subspace in V.

2.25. Construct the vector space of all 3-tuples over GF(3). Form a two-dimensional subspace and its dual space.

REFERENCES

1. A. A. Albert, *Modern Higher Algebra*, The University of Chicago Press, Chicago, 1937.
2. G. Birkhoff and S. MacLane, *A Survey of Modern Algebra*, Macmillan, New York, 1953.
3. R. D. Carmichael, *Introduction to the Theory of Group of Finite Order*, Ginn & Company, Boston, 1937.
4. J. B. Fraleigh, *A First Course in Abstract Algebra*, 2nd ed., Addison-Wesley, Reading, Mass., 1976.
5. N. Jacobson, *Lectures in Abstract Algebra*, Van Nostrand, Princeton, N.J., Vol. 1, 1951; Vol. 2, 1953; Vol. 3, 1964.
6. R. W. Marsh, *Table of Irreducible Polynomials over GF(2) through Degree 19*, NSA, Washington, D.C., 1957.
7. W. W. Peterson, *Error-Correcting Codes*, MIT Press, Cambridge, Mass., 1961.
8. B. L. Van der Waerden, *Modern Algebra*, Vols. 1 and 2, Ungar, New York, 1949.

3

Linear Block Codes

In this chapter basic concepts of block codes are introduced. For ease of code synthesis and implementation, we restrict our attention to a subclass of the class of all block codes, the *linear block codes*. Since in most present digital computers and digital data communication systems, information is coded in binary digits "0" or "1," we discuss only the linear block codes with symbols from the binary field GF(2). The theory developed for the binary codes can be generalized to codes with symbols from a nonbinary field in a straightforward manner.

First, linear block codes are defined and described in terms of *generator* and *parity-check* matrices. The parity-check equations for a *systematic* code are derived. Encoding of linear block codes is discussed. In Section 3.2 the concept of *syndrome* is introduced. The use of syndrome for error detection and correction is discussed. In Sections 3.3 and 3.4 we define the *minimum distance* of a block code and show that the random-error-detecting and random-error-correcting capabilities of a code are determined by its minimum distance. Probabilities of a decoding error are discussed. In Section 3.5 the *standard array* and its application to the decoding of linear block codes are presented. A general decoder based on the *syndrome decoding* scheme is given. Finally, we conclude the chapter by presenting a class of single-error-correcting linear codes.

References 1 to 4 contain excellent treatments of linear block codes.

3.1 INTRODUCTION TO LINEAR BLOCK CODES

We assume that the output of an information source is a sequence of binary digits "0" or "1." In block coding, this binary information sequence is segmented into *message* blocks of fixed length; each message block, denoted by $\mathbf{u}$, consists of k

information digits. There are a total of 2^k distinct messages. The encoder, *according to certain rules*, transforms each input message $\mathbf{u}$ into a binary n-tuple $\mathbf{v}$ with $n > k$. This binary n-tuple $\mathbf{v}$ is referred to as the *code word* (or *code vector*) of the message $\mathbf{u}$. Therefore, corresponding to the 2^k possible messages, there are 2^k code words. This set of 2^k code words is called a *block* code. For a block code to be useful, the 2^k code words must be distinct. Therefore, there should be a one-to-one correspondence between a message $\mathbf{u}$ and its code word $\mathbf{v}$.

For a block code with 2^k code words and length n, unless it has a certain special structure, the encoding apparatus would be prohibitively complex for large k and n since it has to store the 2^k code words of length n in a dictionary. Therefore, we must restrict our attention to block codes that can be mechanized in a practical manner. A desirable structure for a block code to possess is the *linearity*. With this structure in a block code, the encoding complexity will be greatly reduced, as we will see.

Definition 3.1. A block code of length n and 2^k code words is called a *linear* (n, k) code if and only if its 2^k code words form a k-dimensional subspace of the vector space of all the n-tuples over the field GF(2).

In fact, a binary block code is linear if and only if the modulo-2 sum of two code words is also a code word. The block code given in Table 3.1 is a (7, 4) linear code. One can easily check that the sum of any two code words in this code is also a code word.

Since an (n, k) linear code C is a k-dimensional subspace of the vector space V_n of all the binary n-tuples, it is possible to find k linearly independent code words,

TABLE 3.1 LINEAR BLOCK CODE WITH $k = 4$ AND $n = 7$

Messages	Code words
(0 0 0 0)	(0 0 0 0 0 0 0)
(1 0 0 0)	(1 1 0 1 0 0 0)
(0 1 0 0)	(0 1 1 0 1 0 0)
(1 1 0 0)	(1 0 1 1 1 0 0)
(0 0 1 0)	(1 1 1 0 0 1 0)
(1 0 1 0)	(0 0 1 1 0 1 0)
(0 1 1 0)	(1 0 0 0 1 1 0)
(1 1 1 0)	(0 1 0 1 1 1 0)
(0 0 0 1)	(1 0 1 0 0 0 1)
(1 0 0 1)	(0 1 1 1 0 0 1)
(0 1 0 1)	(1 1 0 0 1 0 1)
(1 1 0 1)	(0 0 0 1 1 0 1)
(0 0 1 1)	(0 1 0 0 0 1 1)
(1 0 1 1)	(1 0 0 1 0 1 1)
(0 1 1 1)	(0 0 1 0 1 1 1)
(1 1 1 1)	(1 1 1 1 1 1 1)

$\mathbf{g}_0, \mathbf{g}_1, \ldots, \mathbf{g}_{k-1}$ in C such that every code word $\mathbf{v}$ in C is a linear combination of these k code words, that is,

$$\mathbf{v} = u_0\mathbf{g}_0 + u_1\mathbf{g}_1 + \cdots + u_{k-1}\mathbf{g}_{k-1}, \tag{3.1}$$

where $u_i = 0$ or 1 for $0 \leq i < k$. Let us arrange these k linearly independent code words as the rows of a $k \times n$ matrix as follows:

$$\mathbf{G} = \begin{bmatrix} \mathbf{g}_0 \\ \mathbf{g}_1 \\ \cdot \\ \cdot \\ \cdot \\ \mathbf{g}_{k-1} \end{bmatrix} = \begin{bmatrix} g_{00} & g_{01} & g_{02} & \cdots & g_{0,n-1} \\ g_{10} & g_{11} & g_{12} & \cdots & g_{1,n-1} \\ \cdot & \cdot & \cdot & & \cdot \\ \cdot & \cdot & \cdot & & \cdot \\ \cdot & \cdot & \cdot & & \cdot \\ g_{k-1,0} & g_{k-1,1} & g_{k-1,2} & \cdots & g_{k-1,n-1} \end{bmatrix}, \tag{3.2}$$

where $\mathbf{g}_i = (g_{i0}, g_{i1}, \ldots, g_{i,n-1})$ for $0 \leq i < k$. If $\mathbf{u} = (u_0, u_1, \ldots, u_{k-1})$ is the message to be encoded, the corresponding code word can be given as follows:

$$\begin{aligned} \mathbf{v} &= \mathbf{u} \cdot \mathbf{G} \\ &= (u_0, u_1, \ldots, u_{k-1}) \cdot \begin{bmatrix} \mathbf{g}_0 \\ \mathbf{g}_1 \\ \cdot \\ \cdot \\ \cdot \\ \mathbf{g}_{k-1} \end{bmatrix} \\ &= u_0\mathbf{g}_0 + u_1\mathbf{g}_1 + \cdots + u_{k-1}\mathbf{g}_{k-1}. \end{aligned} \tag{3.3}$$

Clearly, the rows of $\mathbf{G}$ *generate* (or *span*) the (n, k) linear code C. For this reason, the matrix $\mathbf{G}$ is called a *generator* matrix for C. Note that any k linearly independent code words of an (n, k) linear code can be used to form a generator matrix for the code. It follows from (3.3) that an (n, k) linear code is completely specified by the k rows of a generator matrix $\mathbf{G}$. Therefore, the encoder has only to store the k rows of $\mathbf{G}$ and to form a linear combination of these k rows based on the input message $\mathbf{u} = (u_0, u_1, \ldots, u_{k-1})$.

Example 3.1

The (7, 4) linear code given in Table 3.1 has the following matrix as a generator matrix:

$$\mathbf{G} = \begin{bmatrix} \mathbf{g}_0 \\ \mathbf{g}_1 \\ \mathbf{g}_2 \\ \mathbf{g}_3 \end{bmatrix} = \begin{bmatrix} 1 & 1 & 0 & 1 & 0 & 0 & 0 \\ 0 & 1 & 1 & 0 & 1 & 0 & 0 \\ 1 & 1 & 1 & 0 & 0 & 1 & 0 \\ 1 & 0 & 1 & 0 & 0 & 0 & 1 \end{bmatrix}.$$

If $\mathbf{u} = (1 \ 1 \ 0 \ 1)$ is the message to be encoded, its corresponding code word, according to (3.3), would be

$$\begin{aligned} \mathbf{v} &= 1 \cdot \mathbf{g}_0 + 1 \cdot \mathbf{g}_1 + 0 \cdot \mathbf{g}_2 + 1 \cdot \mathbf{g}_3 \\ &= (1 \ 1 \ 0 \ 1 \ 0 \ 0 \ 0) + (0 \ 1 \ 1 \ 0 \ 1 \ 0 \ 0) + (1 \ 0 \ 1 \ 0 \ 0 \ 0 \ 1) \\ &= (0 \ 0 \ 0 \ 1 \ 1 \ 0 \ 1). \end{aligned}$$

A desirable property for a linear block code to possess is the *systematic structure* of the code words as shown in Figure 3.1, where a code word is divided into two parts, the message part and the redundant checking part. The message part consists of k unaltered information (or message) digits and the redundant checking part consists of $n - k$ *parity-check* digits, which are *linear sums* of the information digits. A linear block code with this structure is referred to as a *linear systematic block code*. The (7, 4) code given in Table 3.1 is a linear systematic block code, the rightmost four digits of each code word are identical to the corresponding information digits.

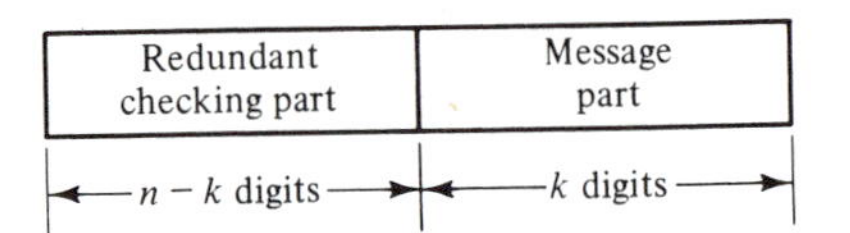

Figure 3.1 Systematic format of a code word.

A linear systematic (n, k) code is completely specified by a $k \times n$ matrix $\mathbf{G}$ of the following form:

$$\mathbf{G} = \begin{bmatrix} \mathbf{g}_0 \\ \mathbf{g}_1 \\ \mathbf{g}_2 \\ \cdot \\ \cdot \\ \cdot \\ \mathbf{g}_{k-1} \end{bmatrix} = \left[\underbrace{\begin{matrix} p_{00} & p_{01} & \cdots & p_{0,n-k-1} \\ p_{10} & p_{11} & \cdots & p_{1,n-k-1} \\ p_{20} & p_{21} & \cdots & p_{2,n-k-1} \\ & & & \\ & & & \\ p_{k-1,0} & p_{k-1,1} & \cdots & p_{k-1,n-k-1} \end{matrix}}_{\mathbf{P} \text{ matrix}} \left| \underbrace{\begin{matrix} 1 & 0 & 0 & \cdots & 0 \\ 0 & 1 & 0 & \cdots & 0 \\ 0 & 0 & 1 & \cdots & 0 \\ & & & & \\ & & & & \\ 0 & 0 & 0 & \cdots & 1 \end{matrix}}_{k \times k \text{ identity matrix}} \right. \right], \tag{3.4}$$

where $p_{ij} = 0$ or 1. Let $\mathbf{I}_k$ denote the $k \times k$ identity matrix. Then $\mathbf{G} = [\mathbf{P} \; \mathbf{I}_k]$. Let $\mathbf{u} = (u_0, u_1, \ldots, u_{k-1})$ be the message to be encoded. The corresponding code word is

$$\begin{aligned} \mathbf{v} &= (v_0, v_1, v_2, \ldots, v_{n-1}) \\ &= (u_0, u_1, \ldots, u_{k-1}) \cdot \mathbf{G}. \end{aligned} \tag{3.5}$$

It follows from (3.4) and (3.5) that the components of $\mathbf{v}$ are

$$v_{n-k+i} = u_i \qquad \text{for } 0 \leq i < k \tag{3.6a}$$

and

$$v_j = u_0 p_{0j} + u_1 p_{1j} + \cdots + u_{k-1} p_{k-1,j} \tag{3.6b}$$

for $0 \leq j < n - k$. Equation (3.6a) shows that the rightmost k digits of a code word $\mathbf{v}$ are identical to the information digits $u_0, u_1, \ldots, u_{k-1}$ to be encoded, and (3.6b) shows that the leftmost $n - k$ redundant digits are linear sums of the information digits. The $n - k$ equations given by (3.6b) are called *parity-check equations* of the code.

Example 3.2

The matrix $\mathbf{G}$ given in Example 3.1 is in systematic form. Let $\mathbf{u} = (u_0, u_1, u_2, u_3)$ be the message to be encoded and let $\mathbf{v} = (v_0, v_1, v_2, v_3, v_4, v_5, v_6)$ be the corresponding code word. Then

$$\mathbf{v} = (u_0, u_1, u_2, u_3) \cdot \begin{bmatrix} 1 & 1 & 0 & 1 & 0 & 0 & 0 \\ 0 & 1 & 1 & 0 & 1 & 0 & 0 \\ 1 & 1 & 1 & 0 & 0 & 1 & 0 \\ 1 & 0 & 1 & 0 & 0 & 0 & 1 \end{bmatrix}.$$

By matrix multiplication, we obtain the following digits of the code word $\mathbf{v}$:

$$\begin{aligned} v_6 &= u_3 \\ v_5 &= u_2 \\ v_4 &= u_1 \\ v_3 &= u_0 \\ v_2 &= u_1 + u_2 + u_3 \\ v_1 &= u_0 + u_1 + u_2 \\ v_0 &= u_0 + u_2 + u_3. \end{aligned}$$

The code word corresponding to the message (1 0 1 1) is (1 0 0 1 0 1 1).

There is another useful matrix associated with every linear block code. As stated in Chapter 2, for any $k \times n$ matrix $\mathbf{G}$ with k linearly independent rows, there exists an $(n - k) \times n$ matrix $\mathbf{H}$ with $n - k$ linearly independent rows such that any vector in the row space of $\mathbf{G}$ is orthogonal to the rows of $\mathbf{H}$ and any vector that is orthogonal to the rows of $\mathbf{H}$ is in the row space of $\mathbf{G}$. Hence, we can describe the (n, k) linear code generated by $\mathbf{G}$ in an alternate way as follows: *An n-tuple* $\mathbf{v}$ *is a code word in the code generated by* $\mathbf{G}$ *if and only if* $\mathbf{v} \cdot \mathbf{H}^T = \mathbf{0}$. This matrix $\mathbf{H}$ is called a *parity-check matrix* of the code. The 2^{n-k} linear combinations of the rows of matrix $\mathbf{H}$ form an $(n, n - k)$ linear code C_d. This code is the null space of the (n, k) linear code C generated by matrix $\mathbf{G}$ (i.e., for any $\mathbf{v} \in C$ and any $\mathbf{w} \in C_d$, $\mathbf{v} \cdot \mathbf{w} = 0$). C_d is called the *dual* code of C. Therefore, a parity-check matrix for a linear code C is a generator matrix for its dual code C_d.

If the generator matrix of an (n, k) linear code is in the systematic form of (3.4), the parity-check matrix may take the following form:

$$\mathbf{H} = [\mathbf{I}_{n-k} \quad \mathbf{P}^T]$$

$$= \begin{bmatrix} 1 & 0 & 0 & \cdots & 0 & p_{00} & p_{10} & \cdots & p_{k-1,0} \\ 0 & 1 & 0 & \cdots & 0 & p_{01} & p_{11} & \cdots & p_{k-1,1} \\ 0 & 0 & 1 & \cdots & 0 & p_{02} & p_{12} & \cdots & p_{k-1,2} \\ \cdot & & & & & & & & \\ \cdot & & & & & & & & \\ \cdot & & & & & & & & \\ 0 & 0 & 0 & \cdots & 1 & p_{0,n-k-1} & p_{1,n-k-1} & \cdots & p_{k-1,n-k-1} \end{bmatrix}, \tag{3.7}$$

where $\mathbf{P}^T$ is the transpose of the matrix $\mathbf{P}$. Let $\mathbf{h}_j$ be the jth row of $\mathbf{H}$. We can check readily that the inner product of the ith row of $\mathbf{G}$ given by (3.4) and the jth row of $\mathbf{H}$ given by (3.7) is

$$\mathbf{g}_i \cdot \mathbf{h}_j = p_{ij} + p_{ij} = 0$$

for $0 \le i < k$ and $0 \le j < n - k$. This implies that $\mathbf{G} \cdot \mathbf{H}^T = \mathbf{0}$. Also, the $n - k$ rows of $\mathbf{H}$ are linearly independent. Therefore, the $\mathbf{H}$ matrix of (3.7) is a parity-check matrix of the (n, k) linear code generated by the matrix $\mathbf{G}$ of (3.4).

The parity-check equations given by (3.6b) can also be obtained from the parity-check matrix $\mathbf{H}$ of (3.7). Let $\mathbf{u} = (u_0, u_1, \ldots, u_{k-1})$ be the message to be encoded. In systematic form the corresponding code word would be

$$\mathbf{v} = (v_0, v_1, \ldots, v_{n-k-1}, u_0, u_1, \ldots, u_{k-1}).$$

Using the fact that $\mathbf{v} \cdot \mathbf{H}^T = \mathbf{0}$, we obtain

$$v_j + u_0 p_{0j} + u_1 p_{1j} + \cdots + u_{k-1} p_{k-1,j} = 0 \tag{3.8}$$

for $0 \le j < n - k$. Rearranging the equations of (3.8), we obtain the same parity-check equations of (3.6b). Therefore, an (n, k) linear code is completely specified by its parity-check matrix.

Example 3.3

Consider the generator matrix of a (7, 4) linear code given in Example 3.1. The corresponding parity-check matrix is

$$\mathbf{H} = \begin{bmatrix} 1 & 0 & 0 & 1 & 0 & 1 & 1 \\ 0 & 1 & 0 & 1 & 1 & 1 & 0 \\ 0 & 0 & 1 & 0 & 1 & 1 & 1 \end{bmatrix}.$$

At this point, let us summarize the foregoing results: For any (n, k) linear block code C, there exists a $k \times n$ matrix $\mathbf{G}$ whose row space gives C. Furthermore, there exists an $(n - k) \times n$ matrix $\mathbf{H}$ such that an n-tuple $\mathbf{v}$ is a code word in C if and only if $\mathbf{v} \cdot \mathbf{H}^T = \mathbf{0}$. If $\mathbf{G}$ is of the form given by (3.4), then $\mathbf{H}$ may take the form given by (3.7), and vice versa.

Based on the equations of (3.6a) and (3.6b), the encoding circuit for an (n, k) linear systematic code can be implemented easily. The encoding circuit is shown in Figure 3.2, where →□→ denotes a shift-register stage (e.g., a flip-flop), $\oplus$ denotes a modulo-2 adder, and →(p_{ij})→ denotes a connection if $p_{ij} = 1$ and no connection if $p_{ij} = 0$. The encoding operation is very simple. The message $\mathbf{u} = (u_0, u_1, \ldots, u_{k-1})$ to be encoded is shifted into the message register and simultaneously into the channel. As soon as the entire message has entered the message register, the $n - k$ parity-check digits are formed at the outputs of the $n - k$ modulo-2 adders. These parity-check digits are then serialized and shifted into the channel. We see that the complexity of the encoding circuit is linearly proportional to the block length of the code. The encoding circuit for the (7, 4) code given in Table 3.1 is shown in Figure 3.3, where the connection is based on the parity-check equations given in Example 3.2.

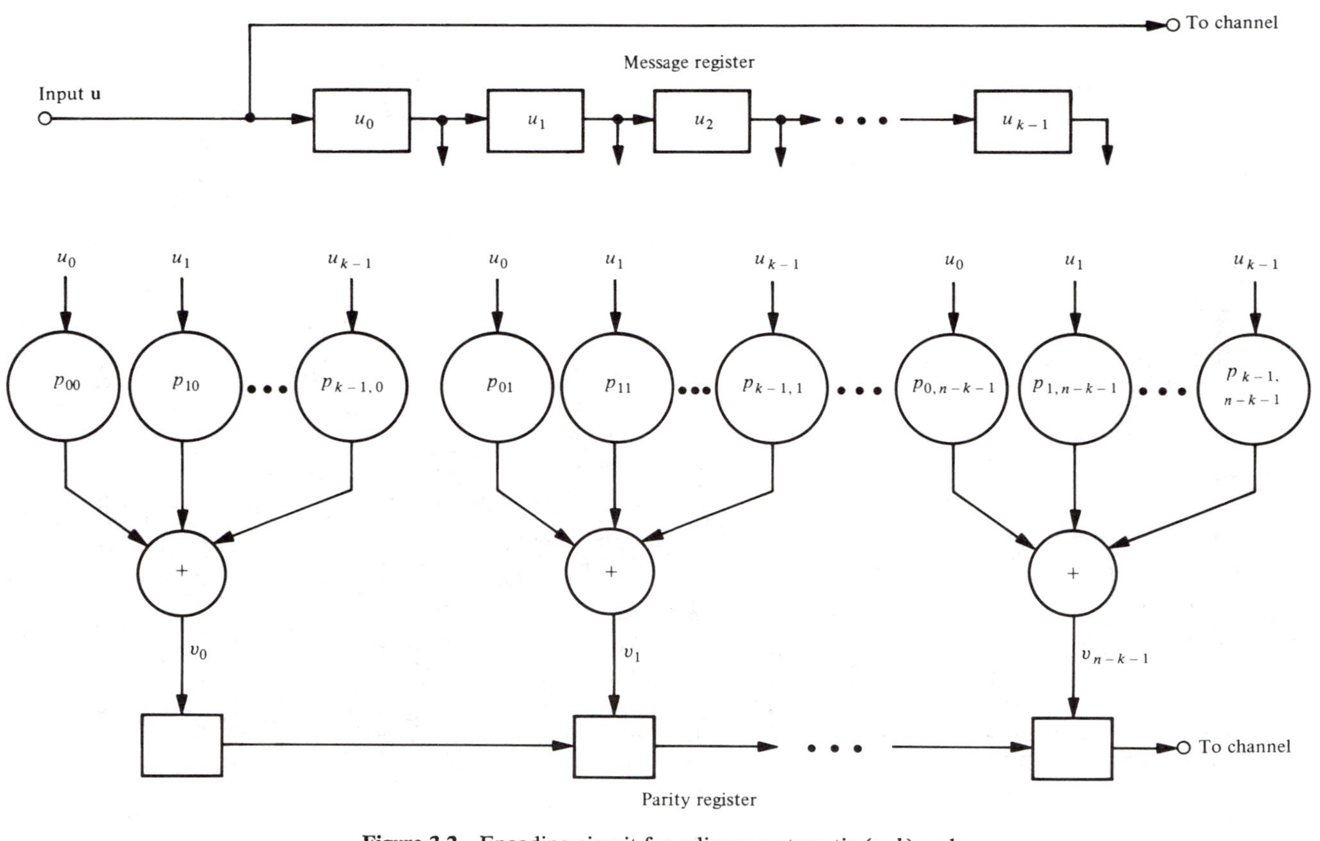

Figure 3.2 Encoding circuit for a linear systematic (n, k) code.

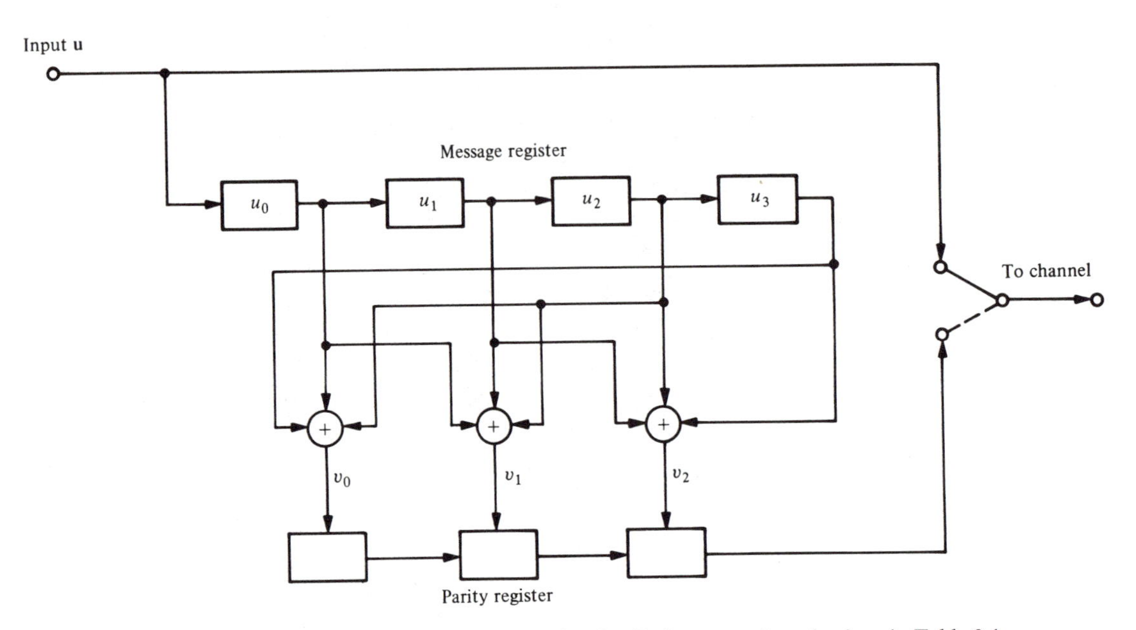

Figure 3.3 Encoding circuit for the (7, 4) systematic code given in Table 3.1.

3.2 SYNDROME AND ERROR DETECTION

Consider an (n, k) linear code with generator matrix $\mathbf{G}$ and parity-check matrix $\mathbf{H}$. Let $\mathbf{v} = (v_0, v_1, \ldots, v_{n-1})$ be a code word that was transmitted over a noisy channel. Let $\mathbf{r} = (r_0, r_1, \ldots, r_{n-1})$ be the received vector at the output of the channel. Because of the channel noise, $\mathbf{r}$ may be different from $\mathbf{v}$. The vector sum

$$\begin{aligned}\mathbf{e} &= \mathbf{r} + \mathbf{v} \\ &= (e_0, e_1, \ldots, e_{n-1})\end{aligned} \tag{3.9}$$

is an n-tuple where $e_i = 1$ for $r_i \neq v_i$ and $e_i = 0$ for $r_i = v_i$. This n-tuple is called the *error vector* (or *error pattern*). The 1's in $\mathbf{e}$ are the *transmission errors* caused by the channel noise. It follows from (3.9) that the received vector $\mathbf{r}$ is the vector sum of the transmitted code word and the error vector, that is,

$$\mathbf{r} = \mathbf{v} + \mathbf{e}.$$

Of course, the receiver does not know either $\mathbf{v}$ or $\mathbf{e}$. Upon receiving $\mathbf{r}$, the decoder must first determine whether $\mathbf{r}$ contains transmission errors. If the presence of errors is detected, the decoder will either take actions to locate the errors and correct them (FEC) or request for a retransmission of $\mathbf{v}$(ARQ).

When $\mathbf{r}$ is received, the decoder computes the following $(n - k)$-tuple:

$$\begin{aligned}\mathbf{s} &= \mathbf{r} \cdot \mathbf{H}^T \\ &= (s_0, s_1, \ldots, s_{n-k-1}).\end{aligned} \tag{3.10}$$

which is called the *syndrome* of $\mathbf{r}$. Then $\mathbf{s} = \mathbf{0}$ if and only if $\mathbf{r}$ is a code word, and $\mathbf{s} \neq \mathbf{0}$

if and only if $\mathbf{r}$ is not a code word. Therefore, when $\mathbf{s} \neq \mathbf{0}$, we know that $\mathbf{r}$ is not a code word and the presence of errors has been detected. When $\mathbf{s} = \mathbf{0}$, $\mathbf{r}$ is a code word and the receiver accepts $\mathbf{r}$ as the transmitted code word. It is possible that the errors in certain error vectors are not detectable (i.e., $\mathbf{r}$ contains errors but $\mathbf{s} = \mathbf{r} \cdot \mathbf{H}^T = \mathbf{0}$). This happens when the error pattern $\mathbf{e}$ is identical to a nonzero code word. In this event, $\mathbf{r}$ is the sum of two code words which is a code word, and consequently $\mathbf{r} \cdot \mathbf{H}^T = \mathbf{0}$. Error patterns of this kind are called *undetectable* error patterns. Since there are $2^k - 1$ nonzero code words, there are $2^k - 1$ undetectable error patterns. When an undetectable error pattern occurs, the decoder makes a *decoding error*. In a later section of the chapter we derive the probability of an undetected error for a BSC and show that this error probability can be made very small.

Based on (3.7) and (3.10), the syndrome digits are as follows:

$$\begin{aligned}
s_0 &= r_0 + r_{n-k}p_{00} + r_{n-k+1}p_{10} + \cdots + r_{n-1}p_{k-1,0} \\
s_1 &= r_1 + r_{n-k}p_{01} + r_{n-k+1}p_{11} + \cdots + r_{n-1}p_{k-1,1} \\
&\;\;\vdots \\
s_{n-k-1} &= r_{n-k-1} + r_{n-k}p_{0,n-k-1} + r_{n-k+1}p_{1,n-k-1} + \cdots + r_{n-1}p_{k-1,n-k-1}.
\end{aligned} \tag{3.11}$$

If we examine the equations above carefully, we find that the syndrome $\mathbf{s}$ is simply the vector sum of the received parity digits $(r_0, r_1, \ldots, r_{n-k-1})$ and the parity-check digits recomputed from the received information digits $r_{n-k}, r_{n-k+1}, \ldots, r_{n-1}$. Therefore, the syndrome can be formed by a circuit similar to the encoding circuit. A general syndrome circuit is shown in Figure 3.4.

Example 3.4

Consider the (7, 4) linear code whose parity-check matrix is given in Example 3.3. Let $\mathbf{r} = (r_0, r_1, r_2, r_3, r_4, r_5, r_6)$ be the received vector. Then the syndrome is given by

$$\mathbf{s} = (s_0, s_1, s_2)$$

$$= (r_0, r_1, r_2, r_3, r_4, r_5, r_6)\begin{bmatrix} 1 & 0 & 0 \\ 0 & 1 & 0 \\ 0 & 0 & 1 \\ 1 & 1 & 0 \\ 0 & 1 & 1 \\ 1 & 1 & 1 \\ 1 & 0 & 1 \end{bmatrix}.$$

The syndrome digits are

$$\begin{aligned}
s_0 &= r_0 + r_3 + r_5 + r_6 \\
s_1 &= r_1 + r_3 + r_4 + r_5 \\
s_2 &= r_2 + r_4 + r_5 + r_6.
\end{aligned}$$

The syndrome circuit for this code is shown in Figure 3.5.

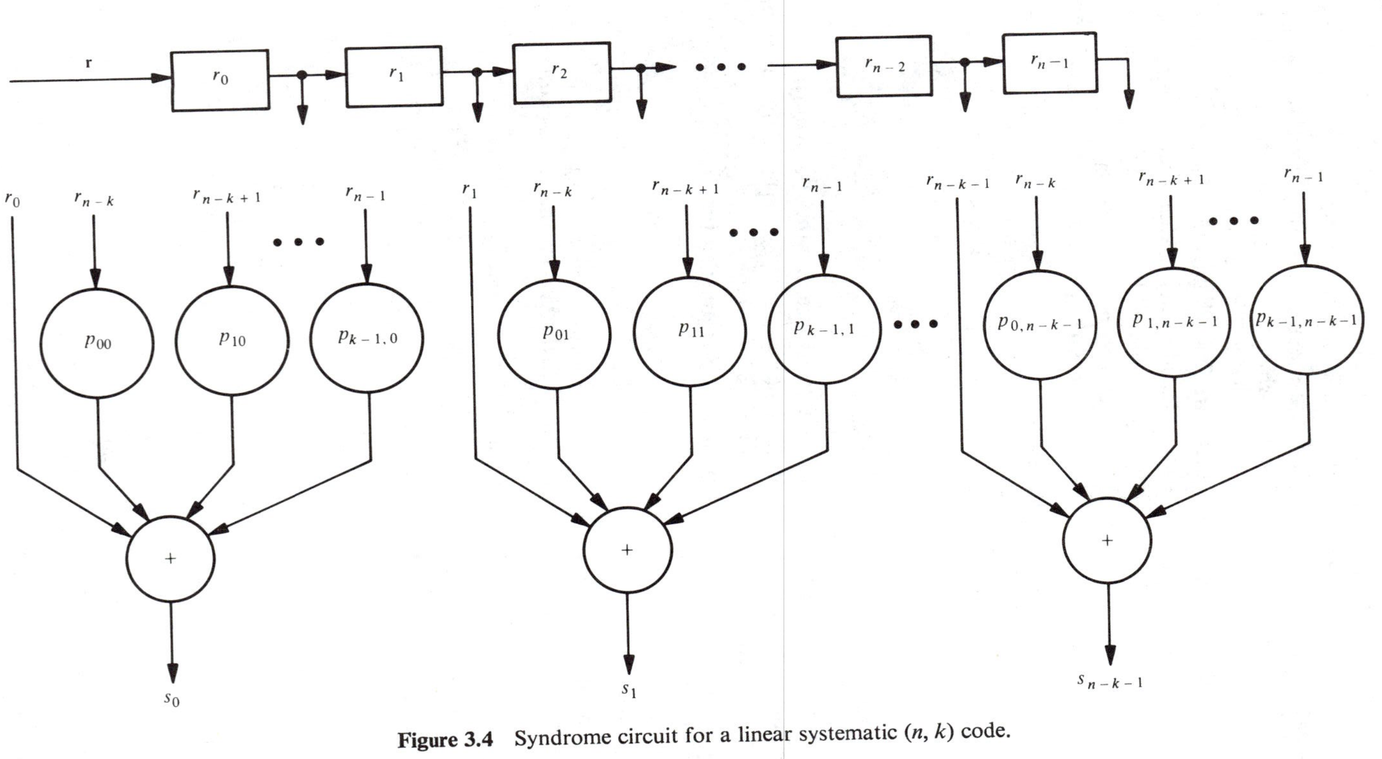

Figure 3.4 Syndrome circuit for a linear systematic (n, k) code.

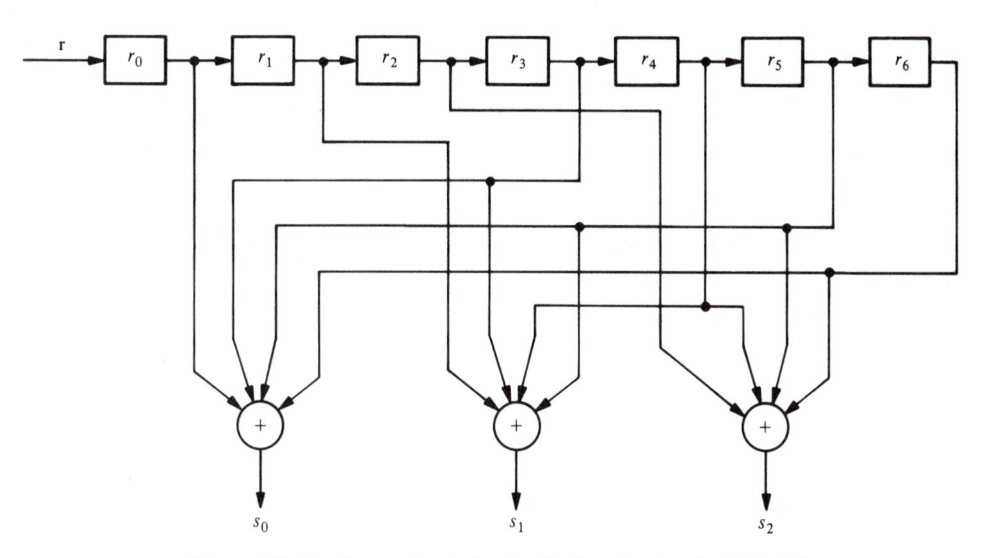

Figure 3.5 Syndrome circuit for the (7, 4) code given in Table 3.1.

The syndrome **s** computed from the received vector **r** actually depends only on the error pattern **e**, and not on the transmitted code word **v**. Since **r** is the vector sum of **v** and **e**, it follows from (3.10) that

$$\mathbf{s} = \mathbf{r} \cdot \mathbf{H}^T = (\mathbf{v} + \mathbf{e})\mathbf{H}^T = \mathbf{v} \cdot \mathbf{H}^T + \mathbf{e} \cdot \mathbf{H}^T.$$

However, $\mathbf{v} \cdot \mathbf{H}^T = \mathbf{0}$. Consequently, we obtain the following relation between the syndrome and the error pattern:

$$\mathbf{s} = \mathbf{e} \cdot \mathbf{H}^T. \tag{3.12}$$

If the parity-check matrix **H** is expressed in the systematic form as given by (3.7), multiplying out $\mathbf{e} \cdot \mathbf{H}^T$ yields the following linear relationship between the syndrome digits and the error digits:

$$\begin{aligned}
s_0 &= e_0 + e_{n-k}p_{00} + e_{n-k+1}p_{10} + \cdots + e_{n-1}p_{k-1,0} \\
s_1 &= e_1 + e_{n-k}p_{01} + e_{n-k+1}p_{11} + \cdots + e_{n-1}p_{k-1,1} \\
&\vdots \\
s_{n-k-1} &= e_{n-k-1} + e_{n-k}p_{0,n-k-1} + e_{n-k+1}p_{1,n-k-1} + \cdots + e_{n-1}p_{k-1,n-k-1}.
\end{aligned} \tag{3.13}$$

The syndrome digits are simply linear combinations of the error digits. Clearly, they provide information about the error digits and therefore can be used for error correction.

At this point, one would feel that any error correction scheme is a method of solving the $n - k$ linear equations of (3.13) for the error digits. Once the error pattern **e** has been found, the vector $\mathbf{r} + \mathbf{e}$ is taken as the actual transmitted code word. Unfortunately, determining the true error vector **e** is not a simple matter. This is because the $n - k$ linear equations of (3.13) do not have a unique solution but have 2^k solutions (this will be proved in Theorem 3.6). In other words, there are 2^k error patterns that result in the same syndrome, and the true error pattern $\mathbf{e}$ is just one of them. Therefore, the decoder has to determine the true error vector from a set of 2^k candidates. To minimize the probability of a decoding error, the most *probable* error pattern that satisfies the equations of (3.13) is chosen as the true error vector. If the channel is a BSC, the most probable error pattern is the one that has the smallest number of nonzero digits.

The notion of using syndrome for error correction may be clarified by an example.

Example 3.5

Again, we consider the (7, 4) code whose parity-check matrix is given in Example 3.3. Let $\mathbf{v} = (1\ 0\ 0\ 1\ 0\ 1\ 1)$ be the transmitted code word and $\mathbf{r} = (1\ 0\ 0\ 1\ 0\ 0\ 1)$ be the received vector. Upon receiving $\mathbf{r}$, the receiver computes the syndrome:

$$\mathbf{s} = \mathbf{r} \cdot \mathbf{H}^T = (1 \quad 1 \quad 1).$$

Next, the receiver attempts to determine the true error vector $\mathbf{e} = (e_0, e_1, e_2, e_3, e_4, e_5$ $e_6)$, which yields the syndrome above. It follows from (3.12) or (3.13) that the error digits are related to the syndrome digits by the following linear equations:

$$1 = e_0 + e_3 + e_5 + e_6$$
$$1 = e_1 + e_3 + e_4 + e_5$$
$$1 = e_2 + e_4 + e_5 + e_6.$$

There are $2^4 = 16$ error patterns that satisfy the equations above. They are

$$\begin{array}{ll}
(0\ 0\ 0\ 0\ 0\ 1\ 0), & (1\ 0\ 1\ 0\ 0\ 1\ 1),\\
(1\ 1\ 0\ 1\ 0\ 1\ 0), & (0\ 1\ 1\ 1\ 0\ 1\ 1),\\
(0\ 1\ 1\ 0\ 1\ 1\ 0), & (1\ 1\ 0\ 0\ 1\ 1\ 1),\\
(1\ 0\ 1\ 1\ 1\ 1\ 0), & (0\ 0\ 0\ 1\ 1\ 1\ 1),\\
(1\ 1\ 1\ 0\ 0\ 0\ 0), & (0\ 1\ 0\ 0\ 0\ 0\ 1),\\
(0\ 0\ 1\ 1\ 0\ 0\ 0), & (1\ 0\ 0\ 1\ 0\ 0\ 1),\\
(1\ 0\ 0\ 0\ 1\ 0\ 0), & (0\ 0\ 1\ 0\ 1\ 0\ 1),\\
(0\ 1\ 0\ 1\ 1\ 0\ 0), & (1\ 1\ 1\ 1\ 1\ 0\ 1).
\end{array}$$

The error vector $\mathbf{e} = (0\ 0\ 0\ 0\ 0\ 1\ 0)$ has the smallest number of nonzero components. If the channel is a BSC, $\mathbf{e} = (0\ 0\ 0\ 0\ 0\ 1\ 0)$ is the most probable error vector that satisfies the equations above. Taking $\mathbf{e} = (0\ 0\ 0\ 0\ 0\ 1\ 0)$ as the true error vector, the receiver decodes the received vector $\mathbf{r} = (1\ 0\ 0\ 1\ 0\ 0\ 1)$ into the following code word:

$$\begin{aligned}
\mathbf{v}^* &= \mathbf{r} + \mathbf{e}\\
&= (1\ \ 0\ \ 0\ \ 1\ \ 0\ \ 0\ \ 1) + (0\ \ 0\ \ 0\ \ 0\ \ 0\ \ 1\ \ 0)\\
&= (1\ \ 0\ \ 0\ \ 1\ \ 0\ \ 1\ \ 1).
\end{aligned}$$

We see that $\mathbf{v}^*$ is the actual transmitted code word. Hence, the receiver has made a correct decoding. Later we show that the (7, 4) linear code considered in this example is capable of correcting any single error over a span of seven digits; that is, if a code word is transmitted and if only one digit is changed by the channel noise, the receiver will be able to determine the true error vector and to perform a correct decoding.

More discussion on error correction based on syndrome is given in Section 3.5. Various methods of determining the true error pattern from the $n - k$ linear equations of (3.13) are presented in later chapters.

3.3 THE MINIMUM DISTANCE OF A BLOCK CODE

In this section an important parameter of a block code called the *minimum distance* is introduced. This parameter determines the random-error-detecting and random-error-correcting capabilities of a code. Let $\mathbf{v} = (v_0, v_1, \ldots, v_{n-1})$ be a binary n-tuple. The *Hamming weight* (or simply weight) of $\mathbf{v}$, denoted by $w(\mathbf{v})$, is defined as the number of nonzero components of $\mathbf{v}$. For example, the Hamming weight of $\mathbf{v} = (1\ 0\ 0\ 1\ 0\ 1\ 1)$ is 4. Let $\mathbf{v}$ and $\mathbf{w}$ be two n-tuples. The *Hamming distance* (or simply distance) between $\mathbf{v}$ and $\mathbf{w}$, denoted $d(\mathbf{v}, \mathbf{w})$, is defined as the number of places where they differ. For example, the Hamming distance between $\mathbf{v} = (1\ 0\ 0\ 1\ 0\ 1\ 1)$ and $\mathbf{w} = (0\ 1\ 0\ 0\ 0\ 1\ 1)$ is 3; they differ in the zeroth, first, and third places. The Hamming distance is a metric function that satisfies the *triangle inequality*. Let $\mathbf{v}$, $\mathbf{w}$, and $\mathbf{x}$ be three n-tuples. Then

$$d(\mathbf{v}, \mathbf{w}) + d(\mathbf{w}, \mathbf{x}) \geq d(\mathbf{v}, \mathbf{x}). \tag{3.14}$$

(The proof of this inequality is left as a problem.) It follows from the definition of Hamming distance and the definition of modulo-2 addition that the Hamming distance between two n-tuples, $\mathbf{v}$ and $\mathbf{w}$, is equal to the Hamming weight of the sum of $\mathbf{v}$ and $\mathbf{w}$, that is,

$$d(\mathbf{v}, \mathbf{w}) = w(\mathbf{v} + \mathbf{w}). \tag{3.15}$$

For example, the Hamming distance between $\mathbf{v} = (1\ 0\ 0\ 1\ 0\ 1\ 1)$ and $\mathbf{w} = (1\ 1\ 1\ 0\ 0\ 1\ 0)$ is 4 and the weight of $\mathbf{v} + \mathbf{w} = (0\ 1\ 1\ 1\ 0\ 0\ 1)$ is also 4.

Given a block code C, one can compute the Hamming distance between any two distinct code words. The *minimum distance* of C, denoted $d_{\min}$, is defined as

$$d_{\min} = \min\{d(\mathbf{v}, \mathbf{w}) : \mathbf{v}, \mathbf{w} \in C, \mathbf{v} \neq \mathbf{w}\}. \tag{3.16}$$

If C is a linear block code, the sum of two vectors is also a code vector. It follows from (3.15) that the Hamming distance between two code vectors in C is equal to the Hamming weight of a third code vector in C. Then it follows from (3.16) that

$$\begin{aligned} d_{\min} &= \min\{w(\mathbf{v} + \mathbf{w}) : \mathbf{v}, \mathbf{w} \in C, \mathbf{v} \neq \mathbf{w}\} \\ &= \min\{w(\mathbf{x}) : \mathbf{x} \in C, \mathbf{x} \neq \mathbf{0}\} \\ &\triangleq w_{\min}. \end{aligned} \tag{3.17}$$

The parameter $w_{\min} \triangleq \{w(\mathbf{x}) : \mathbf{x} \in C, \mathbf{x} \neq \mathbf{0}\}$ is called the *minimum weight* of the linear code C. Summarizing the result above, we have the following theorem.

Theorem 3.1. The minimum distance of a linear block code is equal to the minimum weight of its nonzero code words.

Therefore, for a linear block code, to determine the minimum distance of the code is equivalent to determining its minimum weight. The (7, 4) code given in Table 3.1 has minimum weight 3; thus, its minimum distance is 3. Next, we prove a number of theorems that relate the weight structure of a linear block code to its parity-check matrix.

Theorem 3.2. Let C be an (n, k) linear code with parity-check matrix $\mathbf{H}$. For each code vector of Hamming weight l, there exist l columns of $\mathbf{H}$ such that the vector sum of these l columns is equal to the zero vector. Conversely, if there exist l columns of $\mathbf{H}$ whose vector sum is the zero vector, there exists a code vector of Hamming weight l in C.

Proof. Let us express the parity-check matrix in the following form:

$$\mathbf{H} = [\mathbf{h}_0, \mathbf{h}_1, \ldots, \mathbf{h}_{n-1}],$$

where $\mathbf{h}_i$ represents the ith column of $\mathbf{H}$. Let $\mathbf{v} = (v_0, v_1, \ldots, v_{n-1})$ be a code vector of weight l. Then $\mathbf{v}$ has l nonzero components. Let $v_{i_1}, v_{i_2}, \ldots, v_{i_l}$ be the l nonzero components of $\mathbf{v}$, where $0 \leq i_1 < i_2 < \cdots < i_l \leq n - 1$. Then $v_{i_1} = v_{i_2} = \cdots = v_{i_l} = 1$. Since $\mathbf{v}$ is a code vector, we must have

$$\begin{aligned}\mathbf{0} &= \mathbf{v} \cdot \mathbf{H}^T \\ &= v_0\mathbf{h}_0 + v_1\mathbf{h}_1 + \cdots + v_{n-1}\mathbf{h}_{n-1} \\ &= v_{i_1}\mathbf{h}_{i_1} + v_{i_2}\mathbf{h}_{i_2} + \cdots + v_{i_l}\mathbf{h}_{i_l} \\ &= \mathbf{h}_{i_1} + \mathbf{h}_{i_2} + \cdots + \mathbf{h}_{i_l}.\end{aligned}$$

This proves the first part of the theorem.

Now suppose that $\mathbf{h}_{i_1}, \mathbf{h}_{i_2}, \ldots, \mathbf{h}_{i_l}$ are l columns of $\mathbf{H}$ such that

$$\mathbf{h}_{i_1} + \mathbf{h}_{i_2} + \cdots + \mathbf{h}_{i_l} = \mathbf{0}. \tag{3.18}$$

Let us form a binary n-tuple $\mathbf{x} = (x_1, x_2, \ldots, x_{n-1})$ whose nonzero components are $x_{i_1}, x_{i_2}, \ldots, x_{i_l}$. The Hamming weight of $\mathbf{x}$ is l. Consider the product

$$\begin{aligned}\mathbf{x} \cdot \mathbf{H}^T &= x_0\mathbf{h}_0 + x_1\mathbf{h}_1 + \cdots + x_{n-1}\mathbf{h}_{n-1} \\ &= x_{i_1}\mathbf{h}_{i_1} + x_{i_2}\mathbf{h}_{i_2} + \cdots + x_{i_l}\mathbf{h}_{i_l} \\ &= \mathbf{h}_{i_1} + \mathbf{h}_{i_2} + \cdots + \mathbf{h}_{i_l}.\end{aligned}$$

It follows from (3.18) that $\mathbf{x} \cdot \mathbf{H}^T = \mathbf{0}$. Thus, $\mathbf{x}$ is a code vector of weight l in C. This proves the second part of the theorem. Q.E.D.

It follows from Theorem 3.2 that we have the following two corollaries.

Corollary 3.2.1. Let C be a linear block code with parity-check matrix $\mathbf{H}$. If no $d - 1$ or fewer columns of $\mathbf{H}$ add to $\mathbf{0}$, the code has minimum weight at least d.

Corollary 3.2.2. Let C be a linear code with parity-check matrix $\mathbf{H}$. The minimum weight (or the minimum distance) of C is equal to the smallest number of columns of $\mathbf{H}$ that sum to $\mathbf{0}$.

Consider the (7, 4) linear code given in Table 3.1. The parity-check matrix of this code is

$$\mathbf{H} = \begin{bmatrix} 1 & 0 & 0 & 1 & 0 & 1 & 1 \\ 0 & 1 & 0 & 1 & 1 & 1 & 0 \\ 0 & 0 & 1 & 0 & 1 & 1 & 1 \end{bmatrix}.$$

We see that all columns of **H** are nonzero and that no two of them are alike. Therefore, no two or fewer columns sum to **0**. Hence, the minimum weight of this code is at least 3. However, the zeroth, second and sixth columns sum to **0**. Thus, the minimum weight of the code is 3. From Table 3.1 we see that the minimum weight of the code is indeed 3. It follows from Theorem 3.1 that the minimum distance is 3.

Corollaries 3.2.1 and 3.2.2 are generally used to determine the minimum distance or to establish a lower bound on the minimum distance of a linear block code.

3.4 ERROR-DETECTING AND ERROR-CORRECTING CAPABILITIES OF A BLOCK CODE

When a code vector $\mathbf{v}$ is transmitted over a noisy channel, an error pattern of l errors will result in a received vector $\mathbf{r}$ which differs from the transmitted vector $\mathbf{v}$ in l places [i.e., $d(\mathbf{v}, \mathbf{r}) = l$]. If the minimum distance of a block code C is $d_{\min}$, any two distinct code vectors of C differ in at least $d_{\min}$ places. For this code C, no error pattern of $d_{\min} - 1$ or fewer errors can change one code vector into another. Therefore, any error pattern of $d_{\min} - 1$ or fewer errors will result in a received vector $\mathbf{r}$ that is not a code word in C. When the receiver detects that the received vector is not a code word of C, we say that errors are detected. Hence, a block code with minimum distance $d_{\min}$ is capable of detecting all the error patterns of $d_{\min} - 1$ or fewer errors. However, it cannot detect all the error patterns of $d_{\min}$ errors because there exists at least one pair of code vectors that differ in $d_{\min}$ places and there is an error pattern of $d_{\min}$ errors that will carry one into the other. The same argument applies to error patterns of more than $d_{\min}$ errors. For this reason, we say that the random-error-detecting capability of a block code with minimum distance $d_{\min}$ is $d_{\min} - 1$.

Even though a block code with minimum distance $d_{\min}$ guarantees detecting all the error patterns of $d_{\min} - 1$ or fewer errors, it is also capable of detecting a large fraction of error patterns with $d_{\min}$ or more errors. In fact, an (n, k) linear code is capable of detecting $2^n - 2^k$ error patterns of length n. This can be shown as follows. Among the $2^n - 1$ possible nonzero error patterns, there are $2^k - 1$ error patterns that are identical to the $2^k - 1$ nonzero code words. If any of these $2^k - 1$ error patterns occurs, it alters the transmitted code word $\mathbf{v}$ into another code word $\mathbf{w}$. Thus, $\mathbf{w}$ will be received and its syndrome is zero. In this case, the decoder accepts $\mathbf{w}$ as the transmitted code word and thus commits an incorrect decoding. Therefore, there are $2^k - 1$ *undetectable* error patterns. If an error pattern is not identical to a nonzero code word, the received vector $\mathbf{r}$ will not be a code word and the syndrome will not be zero. In this case, error will be detected. There are exactly $2^n - 2^k$ error patterns that are not identical to the code words of an (n, k) linear code. These $2^n - 2^k$ error patterns are *detectable* error patterns. For large n, $2^k - 1$ is in general much smaller than 2^n. Therefore, only a small fraction of error patterns pass through the decoder without being detected.

Let C be an (n, k) linear code. Let A_i be the number of code vectors of weight i in C. The numbers $A_0, A_1, \ldots, A_n$ are called the *weight distribution* of C. If C is used only for error detection on a BSC, the probability that the decoder fails to detect the presence of errors can be computed from the weight distribution of C. Let $P_u(E)$ denote the probability of an undetected error. Since an undetected error occurs only when the error pattern is identical to a nonzero code vector of C,

$$P_u(E) = \sum_{i=1}^{n} A_i p^i (1-p)^{n-i}, \tag{3.19}$$

where p is the transition probability of the BSC. If the minimum distance of C is $d_{\min}$, then A_1 to $A_{d_{\min}-1}$ are zero.

Consider the (7, 4) code given in Table 3.1. The weight distribution of this code is $A_0 = 1$, $A_1 = A_2 = 0$, $A_3 = 7$, $A_4 = 7$, $A_5 = A_6 = 0$, and $A_7 = 1$. The probability of an undetected error is

$$P_u(E) = 7p^3(1-p)^4 + 7p^4(1-p)^3 + p^7.$$

If $p = 10^{-2}$, this probability is approximately 7×10^{-6}. In other words, if 1 million code words are transmitted over a BSC with $p = 10^{-2}$, there are on the average seven erroneous code words passing through the decoder without being detected.

If a block code C with minimum distance $d_{\min}$ is used for random-error correction, one would like to know how many errors that the code is able to correct. The minimum distance $d_{\min}$ is either odd or even. Let t be a positive integer such that

$$2t + 1 \leq d_{\min} \leq 2t + 2. \tag{3.20}$$

Next, we show that the code C is capable of correcting all the error patterns of t or fewer errors. Let $\mathbf{v}$ and $\mathbf{r}$ be the transmitted code vector and the received vector, respectively. Let $\mathbf{w}$ be any other code vector in C. The Hamming distances among $\mathbf{v}$, $\mathbf{r}$, and $\mathbf{w}$ satisfy the triangle inequality:

$$d(\mathbf{v}, \mathbf{r}) + d(\mathbf{w}, \mathbf{r}) \geq d(\mathbf{v}, \mathbf{w}). \tag{3.21}$$

Suppose that an error pattern of t' errors occurs during the transmission of $\mathbf{v}$. Then the received vector $\mathbf{r}$ differs from $\mathbf{v}$ in t' places and therefore $d(\mathbf{v}, \mathbf{r}) = t'$. Since $\mathbf{v}$ and $\mathbf{w}$ are code vectors in C, we have

$$d(\mathbf{v}, \mathbf{w}) \geq d_{\min} \geq 2t + 1. \tag{3.22}$$

Combining (3.21) and (3.22) and using the fact that $d(\mathbf{v}, \mathbf{r}) = t'$, we obtain the following inequality:

$$d(\mathbf{w}, \mathbf{r}) \geq 2t + 1 - t'.$$

If $t' \leq t$, then

$$d(\mathbf{w}, \mathbf{r}) > t.$$

The inequality above says that if an error pattern of t or fewer errors occurs, the received vector $\mathbf{r}$ is closer (in Hamming distance) to the transmitted code vector $\mathbf{v}$ than to any other code vector $\mathbf{w}$ in C. For a BSC, this means that the conditional probability $P(\mathbf{r} \mid \mathbf{v})$ is greater than the conditional probability $P(\mathbf{r} \mid \mathbf{w})$ for $\mathbf{w} \neq \mathbf{v}$. Based on the maximum likelihood decoding scheme, $\mathbf{r}$ is decoded into $\mathbf{v}$, which is the actual transmitted code vector. This results in a correct decoding and thus errors are corrected.

On the other hand, the code is not capable of correcting all the error patterns of l errors with $l > t$, for there is at least one case where an error pattern of l errors results in a received vector which is closer to an incorrect code vector than to the actual transmitted code vector. To show this, let $\mathbf{v}$ and $\mathbf{w}$ be two code vectors in C such that

$$d(\mathbf{v}, \mathbf{w}) = d_{\min}.$$

Let $\mathbf{e}_1$ and $\mathbf{e}_2$ be two error patterns that satisfy the following conditions:

(i) $\mathbf{e}_1 + \mathbf{e}_2 = \mathbf{v} + \mathbf{w}$.

(ii) $\mathbf{e}_1$ and $\mathbf{e}_2$ do not have nonzero components in common places.

Obviously, we have

$$w(\mathbf{e}_1) + w(\mathbf{e}_2) = w(\mathbf{v} + \mathbf{w}) = d(\mathbf{v}, \mathbf{w}) = d_{\text{min}}. \tag{3.23}$$

Now suppose that $\mathbf{v}$ is transmitted and is corrupted by the error pattern $\mathbf{e}_1$. Then the received vector is

$$\mathbf{r} = \mathbf{v} + \mathbf{e}_1.$$

The Hamming distance between $\mathbf{v}$ and $\mathbf{r}$ is

$$d(\mathbf{v}, \mathbf{r}) = w(\mathbf{v} + \mathbf{r}) = w(\mathbf{e}_1). \tag{3.24}$$

The Hamming distance between $\mathbf{w}$ and $\mathbf{r}$ is

$$d(\mathbf{w}, \mathbf{r}) = w(\mathbf{w} + \mathbf{r}) = w(\mathbf{w} + \mathbf{v} + \mathbf{e}_1) = w(\mathbf{e}_2). \tag{3.25}$$

Now suppose that the error pattern $\mathbf{e}_1$ contains more than t errors [i.e., $w(\mathbf{e}_1) > t$]. Since $2t + 1 \leq d_{\text{min}} \leq 2t + 2$, it follows from (3.23) that

$$w(\mathbf{e}_2) \leq t + 1.$$

Combining (3.24) and (3.25) and using the fact that $w(\mathbf{e}_1) > t$ and $w(\mathbf{e}_2) \leq t + 1$, we obtain the following inequality:

$$d(\mathbf{v}, \mathbf{r}) \geq d(\mathbf{w}, \mathbf{r}).$$

This inequality says that there exists an error pattern of l ($l > t$) errors which results in a received vector that is closer to an incorrect code vector than to the transmitted code vector. Based on the maximum likelihood decoding scheme, an incorrect decoding would be committed.

Summarizing the results above, a block code with minimum distance d_{min} *guarantees* correcting all the error patterns of $t = \lfloor(d_{\text{min}} - 1)/2\rfloor$ or fewer errors, where $\lfloor(d_{\text{min}} - 1)/2\rfloor$ denotes the largest integer no greater than $(d_{\text{min}} - 1)/2$. The parameter $t = \lfloor(d_{\text{min}} - 1)/2\rfloor$ is called the *random-error-correcting capability* of the code. The code is referred to as a t-error-correcting code. The (7, 4) code given in Table 3.1 has minimum distance 3 and thus $t = 1$. It is capable of correcting any error pattern of single error over a block of seven digits.

A block code with random-error-correcting capability t is usually capable of correcting many error patterns of $t + 1$ or more errors. For a t-error-correcting (n, k) linear code, it is capable of correcting a total 2^{n-k} error patterns, including those with t or fewer errors (this will be seen in the next section). If a t-error-correcting block code is used strictly for error correction on a BSC with transition probability p, the probability that the decoder commits an erroneous decoding is upper bounded by

$$P(E) \leq \sum_{i=t+1}^{n} \binom{n}{i} p^i (1 - p)^{n-i}. \tag{3.26}$$

In practice, a code is often used for correcting λ or fewer errors and simultaneously detecting l ($l > \lambda$) or fewer errors. That is, when λ or fewer errors occur, the code is capable of correcting them; when more than λ but fewer than $l + 1$ errors occur, the code is capable of detecting their presence without making a decoding error. For this purpose, the minimum distance $d_{\min}$ of the code is at least $\lambda + l + 1$ (left as a problem). Thus, a block code with $d_{\min} = 10$ is capable of correcting three or fewer errors and simultaneously detecting six or fewer errors.

From the discussion above, we see that random-error-detecting and random-error-correcting capabilities of a block code are determined by the code's minimum distance. Clearly, for given n and k, one would like to construct a block code with minimum distance as large as possible, in addition to the implementation considerations.

3.5 STANDARD ARRAY AND SYNDROME DECODING

In this section a scheme for decoding linear block codes is presented. Let C be an (n, k) linear code. Let $\mathbf{v}_1, \mathbf{v}_2, \ldots, \mathbf{v}_{2^k}$ be the code vectors of C. No matter which code vector is transmitted over a noisy channel, the received vector $\mathbf{r}$ may be any of the 2^n n-tuples over GF(2). Any decoding scheme used at the receiver is a rule to partition the 2^n possible received vectors into 2^k disjoint subsets $D_1, D_2, \ldots, D_{2^k}$ such that the code vector $\mathbf{v}_i$ is contained in the subset D_i for $1 \leq i \leq 2^k$. Thus, each subset D_i is one-to-one correspondence to a code vector $\mathbf{v}_i$. If the received vector $\mathbf{r}$ is found in the subset D_i, $\mathbf{r}$ is decoded into $\mathbf{v}_i$. Correct decoding is made if and only if the received vector $\mathbf{r}$ is in the subset D_i that corresponds to the actual code vector transmitted.

A method to partition the 2^n possible received vectors into 2^k disjoint subsets such that each subset contains one and only one code vector is described here. The partition is based on the linear structure of the code. First, the 2^k code vectors of C are placed in a row with the all-zero code vector $\mathbf{v}_1 = (0, 0, \ldots, 0)$ as the first (leftmost) element. From the remaining $2^n - 2^k$ n-tuples, an n-tuple $\mathbf{e}_2$ is chosen and is placed under the zero vector $\mathbf{v}_1$. Now, we form a second row by adding $\mathbf{e}_2$ to each code vector $\mathbf{v}_i$ in the first row and placing the sum $\mathbf{e}_2 + \mathbf{v}_i$ under $\mathbf{v}_i$. Having completed the second row, an unused n-tuple $\mathbf{e}_3$ is chosen from the remaining n-tuples and is placed under $\mathbf{v}_1$. Then a third row is formed by adding $\mathbf{e}_3$ to each code vector $\mathbf{v}_i$ in the first row and placing $\mathbf{e}_3 + \mathbf{v}_i$ under $\mathbf{v}_i$. We continue this process until all the n-tuples are used. Then we have an array of rows and columns as shown in Figure 3.6. This array is called a *standard array* of the given linear code C.

It follows from the construction rule of a standard array that the sum of any two vectors in the same row is a code vector in C. Next, we prove some important properties of a standard array.

Theorem 3.3. No two n-tuples in the same row of a standard array are identical. Every n-tuple appears in one and only one row.

Proof. The first part of the theorem follows from the fact that all the code vectors of C are distinct. Suppose that two n-tuples in the lth rows are identical, say $\mathbf{e}_l + \mathbf{v}_i = \mathbf{e}_l + \mathbf{v}_j$ with $i \neq j$. This means that $\mathbf{v}_i = \mathbf{v}_j$, which is impossible. Therefore, no two n-tuples in the same row are identical.

It follows from the construction rule of the standard array that every n-tuple appears at least once. Now suppose that an n-tuple appears in both lth row and the

$$
\begin{array}{cccccccc}
\mathbf{v}_1 = 0 & \mathbf{v}_2 & \cdots & \mathbf{v}_i & \cdots & \mathbf{v}_{2^k} \\
\mathbf{e}_2 & \mathbf{e}_2 + \mathbf{v}_2 & \cdots & \mathbf{e}_2 + \mathbf{v}_i & \cdots & \mathbf{e}_2 + \mathbf{v}_{2^k} \\
\mathbf{e}_3 & \mathbf{e}_3 + \mathbf{v}_2 & \cdots & \mathbf{e}_3 + \mathbf{v}_i & \cdots & \mathbf{e}_3 + \mathbf{v}_{2^k} \\
\vdots & & & & & \vdots \\
\mathbf{e}_l & \mathbf{e}_l + \mathbf{v}_2 & \cdots & \mathbf{e}_l + \mathbf{v}_i & \cdots & \mathbf{e}_l + \mathbf{v}_{2^k} \\
\vdots & & & & & \vdots \\
\mathbf{e}_{2^{n-k}} & \mathbf{e}_{2^{n-k}} + \mathbf{v}_2 & \cdots & \mathbf{e}_{2^{n-k}} + \mathbf{v}_i & \cdots & \mathbf{e}_{2^{n-k}} + \mathbf{v}_{2^k}
\end{array}
$$

Figure 3.6 Standard array for an (n, k) linear code.

mth row with $l < m$. Then this n-tuple must be equal to $\mathbf{e}_l + \mathbf{v}_i$ for some i and equal to $\mathbf{e}_m + \mathbf{v}_j$ for some j. As a result, $\mathbf{e}_l + \mathbf{v}_i = \mathbf{e}_m + \mathbf{v}_j$. From this equality we obtain $\mathbf{e}_m = \mathbf{e}_l + (\mathbf{v}_i + \mathbf{v}_j)$. Since $\mathbf{v}_i$ and $\mathbf{v}_j$ are code vectors in C, $\mathbf{v}_i + \mathbf{v}_j$ is also a code vector in C, say $\mathbf{v}_s$. Then $\mathbf{e}_m = \mathbf{e}_l + \mathbf{v}_s$. This implies that the n-tuple $\mathbf{e}_m$ is in the lth row of the array, which contradicts the construction rule of the array that $\mathbf{e}_m$, the first element of the mth row, should be unused in any previous row. Therefore, no n-tuple can appear in more than one row of the array. This concludes the proof of the second part of the theorem. Q.E.D.

From Theorem 3.3 we see that there are $2^n/2^k = 2^{n-k}$ disjoint rows in the standard array, and that each row consists of 2^k distinct elements. The 2^{n-k} rows are called the *cosets* of the code C and the first n-tuple $\mathbf{e}_j$ of each coset is called a *coset leader*. Any element in a coset can be used as its coset leader. This does not change the elements of the coset; it simply permutes them.

Example 3.6

Consider the (6, 3) linear code generated by the following matrix:

$$
\mathbf{G} = \begin{bmatrix} 0 & 1 & 1 & 1 & 0 & 0 \\ 1 & 0 & 1 & 0 & 1 & 0 \\ 1 & 1 & 0 & 0 & 0 & 1 \end{bmatrix}.
$$

The standard array of this code is shown in Figure 3.7.

A standard array of an (n, k) linear code C consists of 2^k disjoint columns. Each column consists of 2^{n-k} n-tuples with the topmost one as a code vector in C. Let D_j denote the jth column of the standard array. Then

$$
D_j = \{\mathbf{v}_j, \mathbf{e}_2 + \mathbf{v}_j, \mathbf{e}_3 + \mathbf{v}_j, \ldots, \mathbf{e}_{2^{n-k}} + \mathbf{v}_j\}, \tag{3.27}
$$

where $\mathbf{v}_j$ is a code vector of C and $\mathbf{e}_2, \mathbf{e}_3, \ldots, \mathbf{e}_{2^{n-k}}$ are the coset leaders. The 2^k disjoint columns $D_1, D_2, \ldots, D_{2^k}$ can be used for decoding the code C as described earlier in this section. Suppose that the code vector $\mathbf{v}_j$ is transmitted over a noisy channel. From (3.27) we see that the received vector $\mathbf{r}$ is in D_j if the error pattern caused by the channel is a coset leader. In this event, the received vector $\mathbf{r}$ will be decoded

Coset leader 000000	011100	101010	110001	110110	101101	011011	000111
100000	111100	001010	010001	010110	001101	111011	100111
010000	001100	111010	100001	100110	111101	001011	010111
001000	010100	100010	111001	111110	100101	010011	001111
000100	011000	101110	110101	110010	101001	011111	000011
000010	011110	101000	110011	110100	101111	011001	000101
000001	011101	101011	110000	110111	101100	011010	000110
100100	111000	001110	010101	010010	001001	111111	100011

Figure 3.7 Standard array for the (6, 3) code.

correctly into the transmitted code vector $\mathbf{v}_j$. On the other hand, if the error pattern caused by the channel is not a coset leader, an erroneous decoding will result. This can be seen as follows. The error pattern $\mathbf{x}$ caused by the channel must be in some coset and under some nonzero code vector, say in the lth coset and under the code vector $\mathbf{v}_i \neq \mathbf{0}$. Then $\mathbf{x} = \mathbf{e}_l + \mathbf{v}_i$ and the received vector is

$$\mathbf{r} = \mathbf{v}_j + \mathbf{x} = \mathbf{e}_l + (\mathbf{v}_i + \mathbf{v}_j) = \mathbf{e}_l + \mathbf{v}_s.$$

The received vector $\mathbf{r}$ is thus in D_s and is decoded into $\mathbf{v}_s$, which is not the transmitted code vector. This results in an erroneous decoding. Therefore, the decoding is correct if and only if the error pattern caused by the channel is a coset leader. For this reason, the 2^{n-k} coset leaders (including the zero vector $\mathbf{0}$) are called the *correctable error patterns*. Summarizing the results above, we have the following theorem:

Theorem 3.4. Every (n, k) linear block code is capable of correcting 2^{n-k} error patterns.

To minimize the probability of a decoding error, the error patterns that are most likely to occur for a given channel should be chosen as the coset leaders. For a BSC, an error pattern of smaller weight is more probable than an error pattern of larger weight. Therefore, when a standard array is formed, each coset leader should be chosen to be a vector of *least weight* from the remaining available vectors. Choosing coset leaders in this manner, each coset leader has minimum weight in its coset. As a result, the decoding based on the standard array is the minimum distance decoding (i.e., the maximum likelihood decoding). To see this, let $\mathbf{r}$ be the received vector. Suppose that $\mathbf{r}$ is found in the ith column D_i and lth coset of the standard array. Then $\mathbf{r}$ is decoded into the code vector $\mathbf{v}_i$. Since $\mathbf{r} = \mathbf{e}_l + \mathbf{v}_i$, the distance between $\mathbf{r}$ and $\mathbf{v}_i$ is

$$d(\mathbf{r}, \mathbf{v}_i) = w(\mathbf{r} + \mathbf{v}_i) = w(\mathbf{e}_l + \mathbf{v}_i + \mathbf{v}_i) = w(\mathbf{e}_l). \tag{3.28}$$

Now, consider the distance between $\mathbf{r}$ and any other code vector, say $\mathbf{v}_j$,

$$d(\mathbf{r}, \mathbf{v}_j) = w(\mathbf{r} + \mathbf{v}_j) = w(\mathbf{e}_l + \mathbf{v}_i + \mathbf{v}_j).$$

Since $\mathbf{v}_i$ and $\mathbf{v}_j$ are two different code vectors, their vector sum, $\mathbf{v}_i + \mathbf{v}_j$, is a nonzero code vector, say $\mathbf{v}_s$. Thus,

$$d(\mathbf{r}, \mathbf{v}_j) = w(\mathbf{e}_l + \mathbf{v}_s). \tag{3.29}$$

Since, $\mathbf{e}_l$ and $\mathbf{e}_l + \mathbf{v}_s$ are in the same coset and since $w(\mathbf{e}_l) \leq w(\mathbf{e}_l + \mathbf{v}_s)$, it follows from (3.28) and (3.29) that

$$d(\mathbf{r}, \mathbf{v}_i) \leq d(\mathbf{r}, \mathbf{v}_j).$$

This says that the received vector is decoded into a closest code vector. Hence, if each coset leader is chosen to have minimum weight in its coset, the decoding based on the standard array is the minimum distance decoding or MLD.

Let α_i denote the number of coset leaders of weight i. The numbers $\alpha_0, \alpha_1, \ldots, \alpha_n$ are called the *weight distribution* of the coset leaders. Knowing these numbers, we can compute the probability of a decoding error. Since a decoding error occurs if and only if the error pattern is not a coset leader, the error probability for a BSC with transition probability p is

$$P(E) = 1 - \sum_{i=0}^{n} \alpha_i p^i (1-p)^{n-i}. \quad (3.30)$$

Example 3.7

Consider the (6, 3) code given in Example 3.6. The standard array for this code is shown in Figure 3.7. The weight distribution of the coset leaders is $\alpha_0 = 1, \alpha_1 = 6, \alpha_2 = 1$, and $\alpha_3 = \alpha_4 = \alpha_5 = \alpha_6 = 0$. Thus,

$$P(E) = 1 - (1-p)^6 - 6p(1-p)^5 - p^2(1-p)^4.$$

For $p = 10^{-2}$, we have $P(E) \approx 1.37 \times 10^{-3}$.

An (n, k) linear code is capable of detecting $2^n - 2^k$ error patterns; however, it is capable of correcting only 2^{n-k} error patterns. For large n, 2^{n-k} is a small fraction of $2^n - 2^k$. Therefore, the probability of a decoding error is much higher than the probability of an undetected error.

Theorem 3.5. For an (n, k) linear code C with minimum distance $d_{\min}$, all the n-tuples of weight of $t = \lfloor (d_{\min} - 1)/2 \rfloor$ or less can be used as coset leaders of a standard array of C. If all the n-tuples of weight t or less are used as coset leaders, there is at least one n-tuple of weight $t + 1$ that cannot be used as a coset leader.

Proof. Since the minimum distance of C is $d_{\min}$, the minimum weight of C is also $d_{\min}$. Let $\mathbf{x}$ and $\mathbf{y}$ be two n-tuples of weight t or less. Clearly, the weight of $\mathbf{x} + \mathbf{y}$ is

$$w(\mathbf{x} + \mathbf{y}) \leq w(\mathbf{x}) + w(\mathbf{y}) \leq 2t < d_{\min}.$$

Suppose that $\mathbf{x}$ and $\mathbf{y}$ are in the same coset; then $\mathbf{x} + \mathbf{y}$ must be a nonzero code vector in C. This is impossible because the weight of $\mathbf{x} + \mathbf{y}$ is less than the minimum weight of C. Therefore, no two n-tuples of weight t or less can be in the same coset of C, and all the n-tuples of weight t or less can be used as coset leaders.

Let $\mathbf{v}$ be a minimum weight code vector of C [i.e., $w(\mathbf{v}) = d_{\min}$]. Let $\mathbf{x}$ and $\mathbf{y}$ be two n-tuples which satisfy the following two conditions:

(i) $\mathbf{x} + \mathbf{y} = \mathbf{v}$.

(ii) $\mathbf{x}$ and $\mathbf{y}$ do not have nonzero components in common places.

It follows from the definition that $\mathbf{x}$ and $\mathbf{y}$ must be in the same coset and

$$w(\mathbf{x}) + w(\mathbf{y}) = w(\mathbf{v}) = d_{\min}.$$

Suppose we choose $\mathbf{y}$ such that $w(\mathbf{y}) = t + 1$. Since $2t + 1 \leq d_{\min} \leq 2t + 2$, we have $w(\mathbf{x}) = t$ or $t + 1$. If $\mathbf{x}$ is used as a coset leader, then $\mathbf{y}$ cannot be a coset leader.
Q.E.D.

Theorem 3.5 reconfirms the fact that an (n, k) linear code with minimum distance $d_{\min}$ is capable of correcting all the error patterns of $\lfloor (d_{\min} - 1)/2 \rfloor$ or fewer errors, but it is not capable of correcting all the error patterns of weight $t + 1$.

A standard array has an important property that can be used to simplify the decoding process. Let $\mathbf{H}$ be the parity-check matrix of the given (n, k) linear code C.

Theorem 3.6. All the 2^k n-tuples of a coset have the same syndrome. The syndromes for different cosets are different.

Proof. Consider the coset whose coset leader is $\mathbf{e}_l$. A vector in this coset is the sum of $\mathbf{e}_l$ and some code vector $\mathbf{v}_i$ in C. The syndrome of this vector is

$$(\mathbf{e}_l + \mathbf{v}_i)\mathbf{H}^T = \mathbf{e}_l\mathbf{H}^T + \mathbf{v}_i\mathbf{H}^T = \mathbf{e}_l\mathbf{H}^T$$

(since $\mathbf{v}_i\mathbf{H}^T = \mathbf{0}$). The equality above says that the syndrome of any vector in a coset is equal to the syndrome of the coset leader. Therefore, all the vectors of a coset have the same syndrome.

Let $\mathbf{e}_j$ and $\mathbf{e}_l$ be the coset leaders of the jth and lth cosets, respectively, where $j < l$. Suppose that the syndromes of these two cosets are equal. Then

$$\mathbf{e}_j\mathbf{H}^T = \mathbf{e}_l\mathbf{H}^T,$$

$$(\mathbf{e}_j + \mathbf{e}_l)\mathbf{H}^T = \mathbf{0}.$$

This implies that $\mathbf{e}_j + \mathbf{e}_l$ is a code vector in C, say $\mathbf{v}_i$. Thus, $\mathbf{e}_j + \mathbf{e}_l = \mathbf{v}_i$ and $\mathbf{e}_l = \mathbf{e}_j + \mathbf{v}_i$. This implies that $\mathbf{e}_l$ is in the jth coset, which contradicts the construction rule of a standard array that a coset leader should be previously unused. Therefore, no two cosets have the same syndrome.
Q.E.D.

We recall that the syndrome of an n-tuple is an $(n - k)$-tuple and there are 2^{n-k} distinct $(n - k)$-tuples. It follows from Theorem 3.6 that there is a one-to-one correspondence between a coset and an $(n - k)$-tuple syndrome. Or, there is a one-to-one correspondence between a coset leader (a correctable error pattern) and a syndrome. Using this one-to-one correspondence relationship, we can form a decoding table, which is much simpler to use than a standard array. The table consists of 2^{n-k} coset leaders (the correctable error patterns) and their corresponding syndromes. This table is either stored or wired in the receiver. The decoding of a received vector consists of three steps:

Step 1. Compute the syndrome of $\mathbf{r}$, $\mathbf{r} \cdot \mathbf{H}^T$.

Step 2. Locate the coset leader $\mathbf{e}_l$ whose syndrome is equal to $\mathbf{r} \cdot \mathbf{H}^T$. Then $\mathbf{e}_l$ is assumed to be the error pattern caused by the channel.

Step 3. Decode the received vector $\mathbf{r}$ into the code vector $\mathbf{v} = \mathbf{r} + \mathbf{e}_l$.

The decoding scheme described above is called the *syndrome decoding* or *table-lookup decoding*. In principle, table-lookup decoding can be applied to any (n, k) linear code. It results in minimum decoding delay and minimum error probability. However, for large $n - k$, the implementation of this decoding scheme becomes impractical, and either a large storage or a complicated logic circuitry is needed. Several practical decoding schemes which are variations of table-lookup decoding are discussed in subsequent chapters. Each of these decoding schemes requires additional properties in a code other than the linear structure.

Example 3.8

Consider the (7, 4) linear code given in Table 3.1. The parity-check matrix, as given in Example 3.3, is

$$\mathbf{H} = \begin{bmatrix} 1 & 0 & 0 & 1 & 0 & 1 & 1 \\ 0 & 1 & 0 & 1 & 1 & 1 & 0 \\ 0 & 0 & 1 & 0 & 1 & 1 & 1 \end{bmatrix}.$$

The code has $2^3 = 8$ cosets and, therefore, there are eight correctable error patterns (including the all-zero vector). Since the minimum distance of the code is 3, it is capable of correcting all the error patterns of weight 1 or 0. Hence, all the 7-tuples of weight 1 or 0 can be used as coset leaders. There are $\binom{7}{0} + \binom{7}{1} = 8$ such vectors. We see that, for the (7, 4) linear code considered in this example, the number of correctable error patterns guaranteed by the minimum distance is equal to the total number of correctable error patterns. The correctable error patterns and their corresponding syndromes are given in Table 3.2.

TABLE 3.2 DECODING TABLE FOR THE (7, 4) LINEAR CODE GIVEN IN TABLE 3.1

Syndrome	Coset leaders
(1 0 0)	(1 0 0 0 0 0 0)
(0 1 0)	(0 1 0 0 0 0 0)
(0 0 1)	(0 0 1 0 0 0 0)
(1 1 0)	(0 0 0 1 0 0 0)
(0 1 1)	(0 0 0 0 1 0 0)
(1 1 1)	(0 0 0 0 0 1 0)
(1 0 1)	(0 0 0 0 0 0 1)

Suppose that the code vector $\mathbf{v} = (1\ 0\ 0\ 1\ 0\ 1\ 1)$ is transmitted and $\mathbf{r} = (1\ 0\ 0\ 1\ 1\ 1\ 1)$ is received. For decoding $\mathbf{r}$, we compute the syndrome of $\mathbf{r}$,

$$\mathbf{s} = (1 \quad 0 \quad 0 \quad 1 \quad 1 \quad 1 \quad 1) \begin{bmatrix} 1 & 0 & 0 \\ 0 & 1 & 0 \\ 0 & 0 & 1 \\ 1 & 1 & 0 \\ 0 & 1 & 1 \\ 1 & 1 & 1 \\ 1 & 0 & 1 \end{bmatrix} = (0 \quad 1 \quad 1).$$

From Table 3.2 we find that $(0\ 1\ 1)$ is the syndrome of the coset leader $\mathbf{e} = (0\ 0\ 0\ 0\ 1\ 0\ 0)$. Thus, $(0\ 0\ 0\ 0\ 1\ 0\ 0)$ is assumed to be the error pattern caused by the channel, and $\mathbf{r}$ is decoded into

$$\begin{aligned} \mathbf{v}^* &= \mathbf{r} + \mathbf{e} \\ &= (1\ 0\ 0\ 1\ 1\ 1\ 1) + (0\ 0\ 0\ 0\ 1\ 0\ 0) \\ &= (1\ 0\ 0\ 1\ 0\ 1\ 1), \end{aligned}$$

which is the actual code vector transmitted. The decoding is correct since the error pattern caused by the channel is a coset leader.

Now suppose that $\mathbf{v} = (0\ 0\ 0\ 0\ 0\ 0\ 0)$ is transmitted and $\mathbf{r} = (1\ 0\ 0\ 0\ 1\ 0\ 0)$ is received. We see that two errors have occurred during the transmission of $\mathbf{v}$. The error pattern is not correctable and will cause a decoding error. When $\mathbf{r}$ is received, the receiver computes the syndrome

$$\mathbf{s} = \mathbf{r} \cdot \mathbf{H}^T = (1\ 1\ 1).$$

From the decoding table we find that the coset leader $\mathbf{e} = (0\ 0\ 0\ 0\ 0\ 1\ 0)$ corresponds to the syndrome $\mathbf{s} = (1\ 1\ 1)$. As a result, $\mathbf{r}$ is decoded into the code vector

$$\begin{aligned} \mathbf{v}^* &= \mathbf{r} + \mathbf{e} \\ &= (1\ 0\ 0\ 0\ 1\ 0\ 0) + (0\ 0\ 0\ 0\ 0\ 1\ 0) \\ &= (1\ 0\ 0\ 0\ 1\ 1\ 0). \end{aligned}$$

Since $\mathbf{v}^*$ is not the actual code vector transmitted, a decoding error is committed.

Using Table 3.2, the code is capable of correcting any single error over a block of seven digits. When two or more errors occur, a decoding error will be committed.

The table-lookup decoding of an (n, k) linear code may be implemented as follows. The decoding table is regarded as the truth table of n switching functions:

$$\begin{aligned} e_0 &= f_0(s_0, s_1, \ldots, s_{n-k-1}), \\ e_1 &= f_1(s_0, s_1, \ldots, s_{n-k-1}), \\ &\vdots \\ e_{n-1} &= f_{n-1}(s_0, s_1, \ldots, s_{n-k-1}), \end{aligned}$$

where $s_0, s_1, \ldots, s_{n-k-1}$ are the syndrome digits, which are regarded as switching variables, and $e_0, e_1, \ldots, e_{n-1}$ are the estimated error digits. When these n switching functions are derived and simplified, a combinational logic circuit with the $n - k$ syndrome digits as inputs and the estimated error digits as outputs can be realized. The implementation of the syndrome circuit has been discussed in Section 3.2. The general decoder for an (n, k) linear code based on the table-lookup scheme is shown in Figure 3.8. The cost of this decoder depends primarily on the complexity of the combinational logic circuit.

Example 3.9

Again, we consider the (7, 4) code given in Table 3.1. The syndrome circuit for this code is shown in Figure 3.5. The decoding table is given by Table 3.2. From this table we form the truth table (Table 3.3). The switching expressions for the seven error digits are

$$\begin{aligned} e_0 &= s_0 \wedge s'_1 \wedge s'_2, & e_1 &= s'_0 \wedge s_1 \wedge s'_2, \\ e_2 &= s'_0 \wedge s'_1 \wedge s_2, & e_3 &= s_0 \wedge s_1 \wedge s'_2, \\ e_4 &= s'_0 \wedge s_1 \wedge s_2, & e_5 &= s_0 \wedge s_1 \wedge s_2, \\ e_6 &= s_0 \wedge s'_1 \wedge s_2, \end{aligned}$$

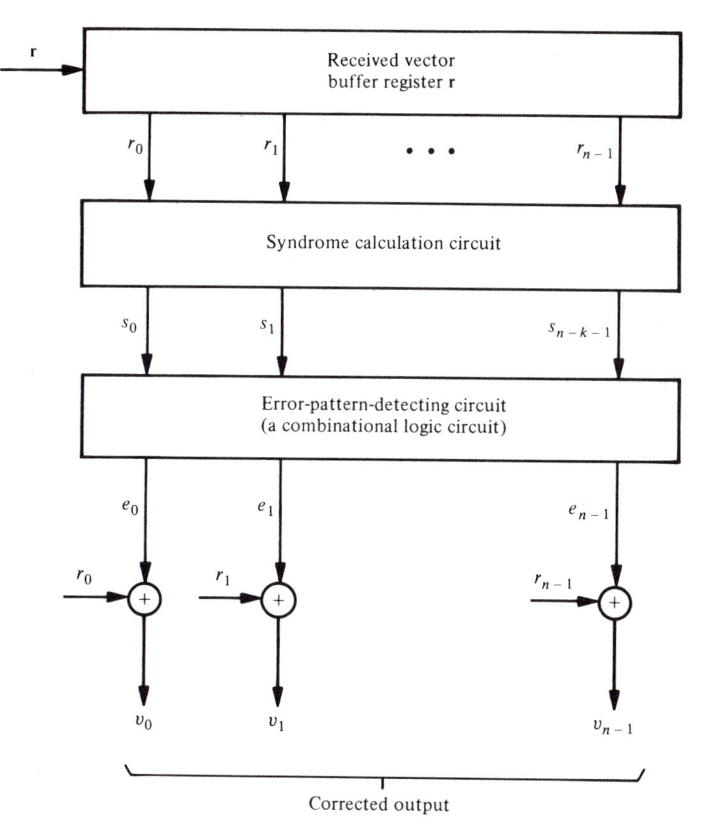

Figure 3.8 General decoder for a linear block code.

TABLE 3.3 TRUTH TABLE FOR THE ERROR DIGITS OF THE CORRECTABLE ERROR PATTERNS OF THE (7, 4) LINEAR CODE GIVEN IN TABLE 3.1

Syndromes			Correctable error patterns (coset leaders)						
s_0	s_1	s_2	e_0	e_1	e_2	e_3	e_4	e_5	e_6
0	0	0	0	0	0	0	0	0	0
1	0	0	1	0	0	0	0	0	0
0	1	0	0	1	0	0	0	0	0
0	0	1	0	0	1	0	0	0	0
1	1	0	0	0	0	1	0	0	0
0	1	1	0	0	0	0	1	0	0
1	1	1	0	0	0	0	0	1	0
1	0	1	0	0	0	0	0	0	1

where Λ denotes the logic-AND operation and s' denotes the logic-COMPLEMENT of s. These seven switching expressions can be realized by seven 3-input AND gates. The complete circuit of the decoder is shown in Figure 3.9.

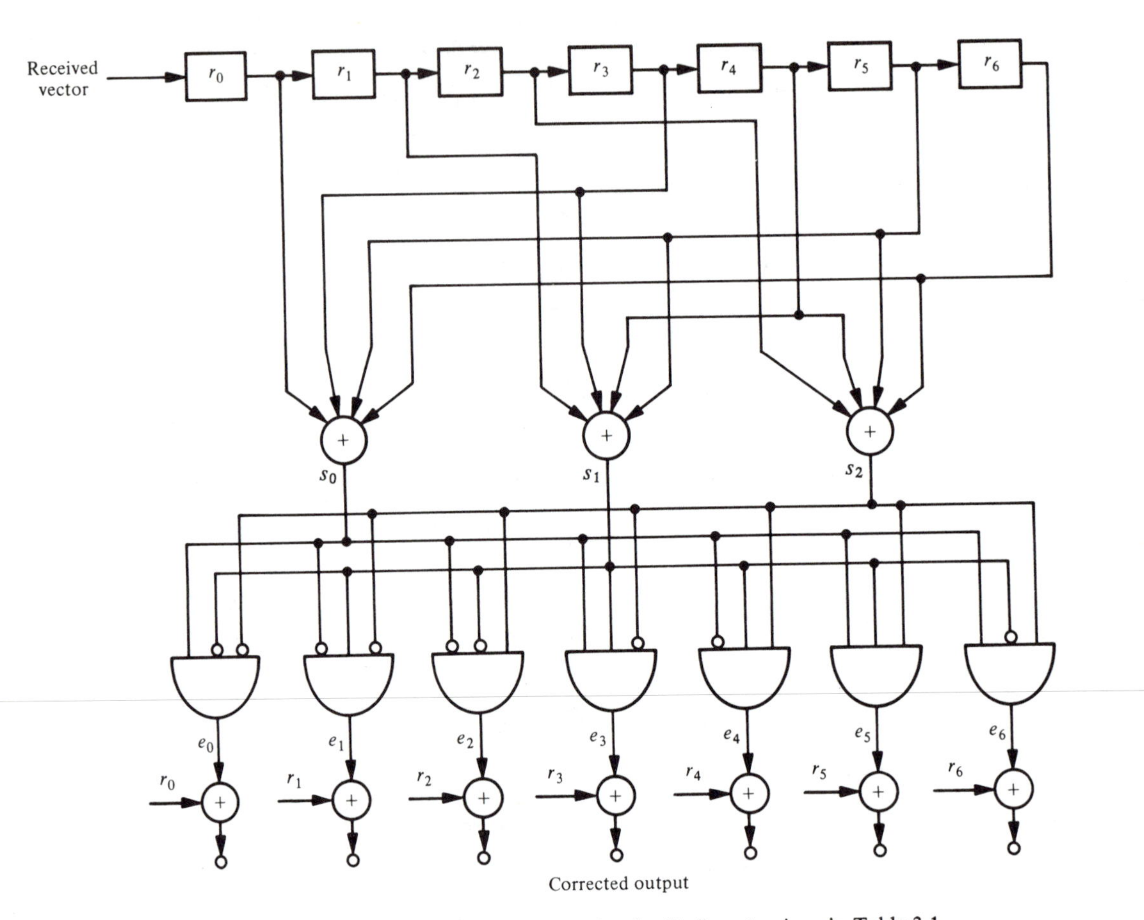

Figure 3.9 Decoding circuit for the (7, 4) code given in Table 3.1.

3.6 PROBABILITY OF AN UNDETECTED ERROR FOR LINEAR CODES OVER A BSC

If an (n, k) linear code is used only for error detection over a BSC, the probability of an undetected error, $P_u(E)$, can be computed from (3.19) if the weight distribution of the code is known. There exists an interesting relationship between the weight distribution of a linear code and the weight distribution of its dual code. This relationship often makes the computation of $P_u(E)$ much easier. Let $\{A_0, A_1, \ldots, A_n\}$ be the weight distribution of an (n, k) linear code C and let $\{B_0, B_1, \ldots, B_n\}$ be the weight distribution of its dual code C_d. Now we represent these two weight distributions in polynomial form as follows:

$$\begin{aligned} A(z) &= A_0 + A_1 z + \cdots + A_n z^n, \\ B(z) &= B_0 + B_1 z + \cdots + B_n z^n. \end{aligned} \tag{3.31}$$

Then $A(z)$ and $B(z)$ are related by the following identity:

$$A(z) = 2^{-(n-k)}(1 + z)^n B\left(\frac{1 - z}{1 + z}\right). \tag{3.32}$$

This identity is known as the *MacWilliams identity* [5]. The polynomials $A(z)$ and $B(z)$ are called the *weight enumerators* for the (n, k) linear code C and its dual C_d. From the MacWilliams identity, we see that if the weight distribution of the dual of a linear code is known, the weight distribution of the code itself can be determined. As a result, this gives us more flexibility of computing the weight distribution of a linear code.

Using the MacWilliams identity, we can compute the probability of an undetected error for an (n, k) linear code from the weight distribution of its dual. First, we put the expression of (3.19) into the following form:

$$\begin{aligned} P_u(E) &= \sum_{i=1}^{n} A_i p^i (1-p)^{n-i} \\ &= (1-p)^n \sum_{i=1}^{n} A_i \left(\frac{p}{1-p}\right)^i. \end{aligned} \tag{3.33}$$

Substituting $z = p/(1-p)$ in $A(z)$ of (3.31) and using the fact that $A_0 = 1$, we obtain the following identity:

$$A\left(\frac{p}{1-p}\right) - 1 = \sum_{i=1}^{n} A_i \left(\frac{p}{1-p}\right)^i. \tag{3.34}$$

Combining (3.33) and (3.34), we have the following expression for the probability of an undetected error:

$$P_u(E) = (1-p)^n \left[A\left(\frac{p}{1-p}\right) - 1 \right]. \tag{3.35}$$

From (3.35) and the MacWilliams identity of (3.32), we finally obtain the following expression for $P_u(E)$:

$$P_u(E) = 2^{-(n-k)} B(1-2p) - (1-p)^n, \tag{3.36}$$

where

$$B(1-2p) = \sum_{i=0}^{n} B_i (1-2p)^i.$$

Hence, there are two ways for computing the probability of an undetected error for a linear code; often one is easier than the other. If $n - k$ is smaller than k, it is much easier to compute $P_u(E)$ from (3.36); otherwise, it is easier to use (3.35).

Example 3.10

Consider the (7, 4) linear code given in Table 3.1. The dual of this code is generated by its parity-check matrix,

$$\mathbf{H} = \begin{bmatrix} 1 & 0 & 0 & 1 & 0 & 1 & 1 \\ 0 & 1 & 0 & 1 & 1 & 1 & 0 \\ 0 & 0 & 1 & 0 & 1 & 1 & 1 \end{bmatrix}$$

(see Example 3.3). Taking the linear combinations of the rows of **H**, we obtain the following eight vectors in the dual code:

$$\begin{array}{ll} (0\ 0\ 0\ 0\ 0\ 0\ 0), & (1\ 1\ 0\ 0\ 1\ 0\ 1), \\ (1\ 0\ 0\ 1\ 0\ 1\ 1), & (1\ 0\ 1\ 1\ 1\ 0\ 0), \\ (0\ 1\ 0\ 1\ 1\ 1\ 0), & (0\ 1\ 1\ 1\ 0\ 0\ 1), \\ (0\ 0\ 1\ 0\ 1\ 1\ 1), & (1\ 1\ 1\ 0\ 0\ 1\ 0). \end{array}$$

Thus, the weight enumerator for the dual code is $B(z) = 1 + 7z^4$. Using (3.36), we obtain the probability of an undetected error for the (7, 4) linear code given in Table 3.1,

$$P_u(E) = 2^{-3}[1 + 7(1 - 2p)^4] - (1 - p)^7.$$

This probability was also computed in Section 3.4 using the weight distribution of the code itself.

Theoretically, we can compute the weight distribution of an (n, k) linear code by examining its 2^k code words or by examining the 2^{n-k} code words of its dual and then applying the MacWilliams identity. However, for large n, k, and $n - k$, the computation becomes practically impossible. Except for some short linear codes and a few small classes of linear codes, the weight distributions for many known linear codes are still unknown. Consequently, it is very difficult, if not impossible, to compute their probability of an undetected error.

Although it is difficult to compute the probability of an undetected error for a specific (n, k) linear code for large n and k, it is quite easy to derive an upper bound on the average probability of an undetected error for the ensemble of all (n, k) linear systematic codes. As we have shown earlier, an (n, k) linear systematic code is completely specified by a matrix **G** of the form given by (3.4). The submatrix **P** consists of $k(n - k)$ entries. Since each entry p_{ij} can be either a 0 or a 1, there are $2^{k(n-k)}$ distinct matrices **G**'s of the form given by (3.4). Let Γ denote the ensemble of codes generated by these $2^{k(n-k)}$ matrices. Suppose that we choose a code randomly from Γ and use it for error detection. Let C_j be the chosen code. Then the probability of C_j being chosen is

$$P(C_j) = 2^{-k(n-k)}. \tag{3.37}$$

Let A_{ji} denote the number of code words in C_j with weight i. It follows from (3.19) that probability of an undetected error for C_j is given by

$$P_u(E \mid C_j) = \sum_{i=1}^{n} A_{ji} p^i (1 - p)^{n-i}. \tag{3.38}$$

The average probability of an undetected error for a linear code in Γ is defined as

$$\mathbf{P_u(E)} = \sum_{j=1}^{|\Gamma|} P(C_j) P_u(E \mid C_j), \tag{3.39}$$

where $|\Gamma|$ denotes the number of codes in Γ. Substituting (3.37) and (3.38) into (3.39), we obtain

$$\mathbf{P_u(E)} = 2^{-k(n-k)} \sum_{i=1}^{n} p^i (1 - p)^{n-i} \sum_{j=1}^{|\Gamma|} A_{ji}. \tag{3.40}$$

A nonzero n-tuple is either contained in exactly $2^{(k-1)(n-k)}$ codes in Γ or contained in none of the codes (left as a problem). Since there are $\binom{n}{i}$ n-tuples of weight i, we have

$$\sum_{j=1}^{|\Gamma|} A_{ji} \leq \binom{n}{i} 2^{(k-1)(n-k)}. \tag{3.41}$$

Substituting (3.41) into (3.40), we obtain the following upper bound on the average probability of an undetected error for an (n, k) linear systematic code:

$$\begin{aligned}\mathbf{P_u(E)} &\leq 2^{-(n-k)} \sum_{i=1}^{n} \binom{n}{i} p^i (1-p)^{n-i} \\ &= 2^{-(n-k)}[1-(1-p)^n].\end{aligned} \tag{3.42}$$

Since $[1-(1-p)^n] \leq 1$, it is clear that $\mathbf{P_u(E)} \leq 2^{-(n-k)}$.

The result above says that there exist (n, k) linear codes with probability of an undetected error, $P_u(E)$, upper bounded by $2^{-(n-k)}$. In other words, there exist (n, k) linear codes with $P_u(E)$ *decreasing exponentially with the number of parity-check digits*, $n-k$. Even for moderate $n-k$, these codes have a very small probability of an undetected error. For example, let $n-k=30$. There exist (n, k) linear codes for which $P_u(E)$ is upper bounded by $2^{-30} \approx 10^{-9}$. Many classes of linear codes have been constructed for the past three decades. However, only a few small classes of linear codes have been proved to have $P_u(E)$ satisfying the upper bound $2^{-(n-k)}$. It is still not known whether the other known linear codes satisfy this upper bound. A class of linear codes that satisfies this upper bound is presented in the next section. Other codes with probability of an undetected error decreasing exponentially with $n-k$ are presented in subsequent chapters.

3.7 HAMMING CODES

Hamming codes are the first class of linear codes devised for error correction [6]. These codes and their variations have been widely used for error control in digital communication and data storage systems.

For any positive integer $m \geq 3$, there exists a Hamming code with the following parameters:

Code length:	$n = 2^m - 1$
Number of information symbols:	$k = 2^m - m - 1$
Number of parity-check symbols:	$n - k = m$
Error-correcting capability:	$t = 1 (d_{\min} = 3)$.

The parity-check matrix $\mathbf{H}$ of this code consists of all the nonzero m-tuples as its columns. In systematic form, the columns of $\mathbf{H}$ are arranged in the following form:

$$\mathbf{H} = [\mathbf{I}_m \quad \mathbf{Q}],$$

where $\mathbf{I}_m$ is an $m \times m$ identity matrix and the submatrix $\mathbf{Q}$ consists of $2^m - m - 1$ columns which are the m-tuples of weight 2 or more. For example, let $m = 3$. The parity-check matrix of a Hamming code of length 7 can be put in the form

$$\mathbf{H} = \begin{bmatrix} 1 & 0 & 0 & 1 & 0 & 1 & 1 \\ 0 & 1 & 0 & 1 & 1 & 1 & 0 \\ 0 & 0 & 1 & 0 & 1 & 1 & 1 \end{bmatrix},$$

which is the parity-check matrix of the (7, 4) linear code given in Table 3.1 (see Example 3.3). Hence, the code given in Table 3.1 is a Hamming code. The columns of $\mathbf{Q}$ may be arranged in any order without affecting the distance property and weight distribution of the code. In systematic form, the generator matrix of the code is

$$\mathbf{G} = [\mathbf{Q}^T \quad \mathbf{I}_{2^m-m-1}],$$

where $\mathbf{Q}^T$ is the transpose of $\mathbf{Q}$ and $\mathbf{I}_{2^m-m-1}$ is an $(2^m - m - 1) \times (2^m - m - 1)$ identity matrix.

Since the columns of $\mathbf{H}$ are nonzero and distinct, no two columns add to zero. It follows from Corollary 3.2.1 that the minimum distance of a Hamming code is at least 3. Since $\mathbf{H}$ consists of all the nonzero m-tuples as its columns, the vector sum of any two columns, say $\mathbf{h}_i$ and $\mathbf{h}_j$, must also be a column in $\mathbf{H}$, say $\mathbf{h}_l$. Thus,

$$\mathbf{h}_i + \mathbf{h}_j + \mathbf{h}_l = \mathbf{0}.$$

It follows from Corollary 3.2.2 that the minimum distance of a Hamming code is exactly 3. Hence, the code is capable of correcting all the error patterns with a single error or of detecting all the error patterns of two or fewer errors.

If we form the standard array for the Hamming code of length $2^m - 1$, all the $(2^m - 1)$-tuples of weight 1 can be used as coset leaders (Theorem 3.5). The number of $(2^m - 1)$-tuples of weight 1 is $2^m - 1$. Since $n - k = m$, the code has 2^m cosets. Thus, the zero vector $\mathbf{0}$ and the $(2^m - 1)$-tuples of weight 1 form all the coset leaders of the standard array. This says that a Hamming code corrects only the error patterns of single error and no others. This is a very interesting structure. A t-error-correcting code is called a *perfect code* if its standard array has all the error patterns of t or fewer errors and no others as coset leaders. Thus, Hamming codes form a class of single-error-correcting perfect codes. Perfect codes are rare [3]. Besides the Hamming codes, the only other nontrivial binary perfect code is the (23, 12) Golay code (see Section 5.3).

Decoding of Hamming codes can be accomplished easily with the table-lookup scheme described in Section 3.5. The decoder for a Hamming code of length $2^m - 1$ can be implemented in the same manner as that for the (7,4) Hamming code given in Example 3.9.

We may delete any l columns from the parity-check matrix $\mathbf{H}$ of a Hamming code. This deletion results in an $m \times (2^m - l - 1)$ matrix $\mathbf{H}'$. Using $\mathbf{H}'$ as a parity-check matrix, we obtain a shortened Hamming code with the following parameters:

Code length:	$n = 2^m - l - 1$
Number of information symbols:	$k = 2^m - m - l - 1$
Number of parity-check symbols:	$n - k = m$
Minimum distance:	$d_{\min} \geq 3.$

If we delete columns from $\mathbf{H}$ properly, we may obtain a shortened Hamming code with minimum distance 4. For example, if we delete from the submatrix $\mathbf{Q}$ all the columns of even weight, we obtain an $m \times 2^{m-1}$ matrix

$$\mathbf{H}' = [\mathbf{I}_m \quad \mathbf{Q}'],$$

where $\mathbf{Q}'$ consists of $2^{m-1} - m$ columns of odd weight. Since all the columns of $\mathbf{H}'$ have odd weight, no three columns add to zero. However, for a column $\mathbf{h}_i$ of weight 3 in $\mathbf{Q}'$, there exists three columns $\mathbf{h}_j$, $\mathbf{h}_l$, and $\mathbf{h}_s$ in $\mathbf{I}_m$ such that $\mathbf{h}_i + \mathbf{h}_j + \mathbf{h}_l + \mathbf{h}_s = \mathbf{0}.$

Thus, the shortened Hamming code with $\mathbf{H}'$ as a parity-check matrix has minimum distance exactly 4.

The distance 4 shortened Hamming code can be used for correcting all error patterns of single error and simultaneously detecting all error patterns of double errors. When a single error occurs during the transmission of a code vector, the resultant syndrome is nonzero and it contains an odd number of 1's. However, when double errors occur, the syndrome is also nonzero, but it contains even number of 1's. Based on these facts, decoding can be accomplished in the following manner:

1. If the syndrome $\mathbf{s}$ is zero, we assume that no error occurred.
2. If $\mathbf{s}$ is nonzero and it contains odd number of 1's, we assume that a single error occurred. The error pattern of a single error that corresponds to $\mathbf{s}$ is added to the received vector for error correction.
3. If $\mathbf{s}$ is nonzero and it contains even number of 1's, an uncorrectable error pattern has been detected.

A class of single-error-correcting and double-error-detecting shortened Hamming codes which is widely used for error control in computer main/or control storages is presented in Chapter 16.

The weight distribution of a Hamming code of length $n = 2^m - 1$ is known [1–4]. The number of code vectors of weight i, A_i, is simply the coefficient of z^i in the expansion of the following polynomial:

$$A(z) = \frac{1}{n+1}\{(1+z)^n + n(1-z)(1-z^2)^{(n-1)/2}\}. \tag{3.43}$$

This polynomial is the weight enumerator for the Hamming codes.

Example 3.11

Let $m = 3$. Then $n = 2^3 - 1 = 7$ and the weight enumerator for the (7, 4) Hamming code is

$$A(z) = \tfrac{1}{8}\{(1+z)^7 + 7(1-z)(1-z^2)^3\} = 1 + 7z^3 + 7z^4 + z^7.$$

Hence, the weight distribution for the (7, 4) Hamming code is $A_0 = 1$, $A_3 = A_4 = 7$, and $A_7 = 1$.

The dual code of a $(2^m - 1, 2^m - m - 1)$ Hamming code is a $(2^m - 1, m)$ linear code. This code has a very simple weight distribution; it consists of the all-zero code word and $2^m - 1$ code words of weight 2^{m-1}. Thus, its weight enumerator is

$$B(z) = 1 + (2^m - 1)z^{2^{m-1}}. \tag{3.44}$$

The duals of Hamming codes are discussed further in Chapter 7.

If a Hamming code is used for error detection over a BSC, its probability of an undetected error, $P_u(E)$, can be computed either from (3.35) and (3.43) or from (3.36) and (3.44). Computing $P_u(E)$ from (3.36) and (3.44) is easier. Combining (3.36) and (3.44), we obtain

$$P_u(E) = 2^{-m}\{1 + (2^m - 1)(1-2p)^{2^{m-1}}\} - (1-p)^{2^m - 1}. \tag{3.45}$$

The probability $P_u(E)$ for Hamming codes does satisfy the upper bound $2^{-(n-k)} = 2^{-m}$ for $p \leq \frac{1}{2}$ [i.e., $P_u(E) \leq 2^{-m}$] [7]. This can be shown by using the expression of (3.45) (see Problem 3.21).

PROBLEMS

3.1. Consider a systematic (8, 4) code whose parity-check equations are

$$\begin{aligned} v_0 &= u_1 + u_2 + u_3, \\ v_1 &= u_0 + u_1 + u_2, \\ v_2 &= u_0 + u_1 + u_3, \\ v_3 &= u_0 + u_2 + u_3. \end{aligned}$$

where u_0, u_1, u_2, and u_3 are message digits and v_0, v_1, v_2, and v_3 are parity-check digits. Find the generator and parity-check matrices for this code. Show analytically that the minimum distance of this code is 4.

3.2. Construct an encoder for the code given in Problem 3.1.

3.3. Construct a syndrome circuit for the code given in Problem 3.1.

3.4. Let $\mathbf{H}$ be the parity-check matrix of an (n, k) linear code C that has both odd- and even-weight code vectors. Construct a new linear code C_1 with the following parity-check matrix:

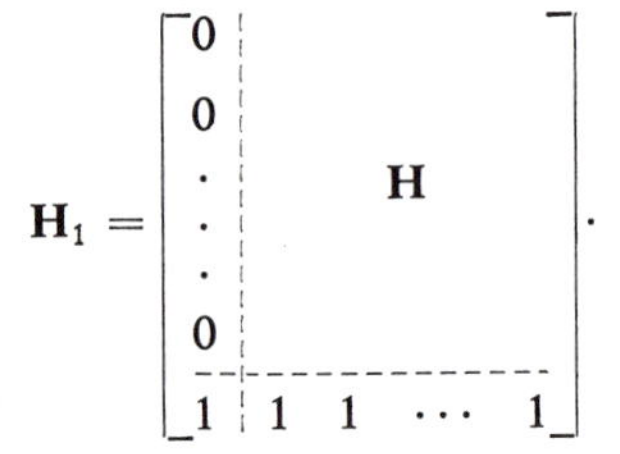

$$\mathbf{H}_1 = \left[\begin{array}{c|cccc} 0 & & & & \\ 0 & & & & \\ \cdot & & \mathbf{H} & & \\ \cdot & & & & \\ \cdot & & & & \\ 0 & & & & \\ \hline 1 & 1 & 1 & \cdots & 1 \end{array}\right].$$

(Note that the last row of $\mathbf{H}_1$ consists of all 1's)

(a) Show that C_1 is an $(n + 1, k)$ linear code. C_1 is called an *extension* of C.

(b) Show that every code vector of C_1 has even weight.

(c) Show that C_1 can be obtained from C by adding an extra parity-check digit, denoted v_∞, to the left of each code vector $\mathbf{v}$ as follows: (1) if $\mathbf{v}$ has odd weight, then $v_\infty = 1$, and (2) if $\mathbf{v}$ has even weight, then $v_\infty = 0$. The parity-check digit v_∞ is called an *overall parity-check* digit.

3.5. Let C be a linear code with both even-weight and odd-weight code vectors. Show that the number of even-weight code vectors is equal to the number of odd-weight code vectors.

3.6. Consider an (n, k) linear code C whose generator matrix $\mathbf{G}$ contains no zero column. Arrange all the code vectors of C as rows of a 2^k-by-n array.

(a) Show that no column of the array contains only zeros.

(b) Show that each column of the array consists of 2^{k-1} zeros and 2^{k-1} ones.

(c) Show that the set of all code vectors with zeros in a particular component forms a subspace of C. What is the dimension of this subspace?

3.7. Prove that the Hamming distance satisfies the triangle inequality; that is, let $\mathbf{x}$, $\mathbf{y}$, and $\mathbf{z}$ be three n-tuples over GF(2), and show that

$$d(\mathbf{x}, \mathbf{y}) + d(\mathbf{y}, \mathbf{z}) \geq d(\mathbf{x}, \mathbf{z}).$$

3.8. Prove that a linear code is capable of correcting λ or fewer errors and simultaneously detecting $l(l > \lambda)$ or fewer errors if its minimum distance $d_{\min} \geq \lambda + l + 1$.

3.9. Determine the weight distribution of the (8, 4) linear code given in Problem 3.1. Let the transition probability of a BSC be $p = 10^{-2}$. Compute the probability of an undetected error of this code.

3.10. Since the (8, 4) linear code given in Problem 3.1 has minimum distance 4, it is capable of correcting all the single-error patterns and simultaneously detecting any combination of double errors. Construct a decoder for this code. The decoder must be capable of correcting any single error and detecting any double errors.

3.11. Let Γ be the ensemble of all the binary systematic (n, k) linear codes. Prove that a nonzero binary n-tuple $\mathbf{v}$ is either contained in exactly $2^{(k-1)(n-k)}$ codes in Γ or contained in none of the codes in Γ.

3.12. The (8, 4) linear code given in Problem 3.1 is capable of correcting 16 error patterns (the coset leaders of a standard array). Suppose that this code is used for a BSC. Devise a decoder for this code based on the table-lookup decoding scheme. The decoder is designed to correct the 16 most probable error patterns.

3.13. Let C_1 be an (n_1, k) linear systematic code with minimum distance d_1 and generator matrix $\mathbf{G}_1 = [\mathbf{P}_1 \ \mathbf{I}_k]$. Let C_2 be an (n_2, k) linear systematic code with minimum distance d_2 and generator matrix $\mathbf{G}_2 = [\mathbf{P}_2 \ \mathbf{I}_k]$. Consider an $(n_1 + n_2, k)$ linear code with the following parity-check matrix:

$$\mathbf{H} = \left[\begin{array}{c|c} & \mathbf{P}_1^T \\ \mathbf{I}_{n_1+n_2-k} & \mathbf{I}_k \\ & \mathbf{P}_2^T \end{array} \right].$$

Show that this code has minimum distance at least $d_1 + d_2$.

3.14. Show that the dual code of the (8, 4) linear code C given in Problem 3.1 is identical to C. C is said to be *self-dual.*

3.15. Form a parity-check matrix for a (15, 11) Hamming code. Devise a decoder for this code.

3.16. For any binary (n, k) linear code with minimum distance (or minimum weight) $2t + 1$ or greater, show that the number of parity-check digits satisfies the following inequality:

$$n - k \geq \log_2 \left[1 + \binom{n}{1} + \binom{n}{2} + \cdots + \binom{n}{t} \right].$$

The inequality above gives an upper bound on the random error-correcting capability t of an (n, k) linear code. This bound is known as the *Hamming bound* [5]. [*Hint:* For an (n, k) linear code with minimum distance $2t + 1$ or greater, all the n-tuples of weight t or less can be used as coset leaders in a standard array.]

3.17. Show that the Hamming codes achieve the Hamming bound.

3.18. Show that the minimum distance $d_{\min}$ of an (n, k) linear code satisfies the following inequality:

$$d_{\min} \leq \frac{n \cdot 2^{k-1}}{2^k - 1}.$$

(*Hint:* Use the result of Problem 3.6(b). The bound above is known as the *Plotkin bound* [1–3].)

3.19. Show that there exists an (n, k) linear code with minimum distance at least d if

$$\sum_{i=1}^{d-1} \binom{n}{i} < 2^{n-k}.$$

[*Hint:* Use the result of Problem 3.11 and the fact that the nonzero n-tuples of weight $d-1$ or less can be at most in

$$\left\{\sum_{i=1}^{d-1}\binom{n}{i}\right\}\cdot 2^{(k-1)(n-k)}$$

(n, k) systematic linear codes.]

3.20. Show that there exists an (n, k) linear code with minimum distance at least $d_{\min}$ which satisfies the following inequality:

$$\sum_{i=1}^{d_{\min}-1}\binom{n}{i} < 2^{n-k} \leq \sum_{i=1}^{d_{\min}}\binom{n}{i}.$$

(*Hint:* See Problem 3.19. The second inequality provides a lower bound on the minimum distance attainable with an (n, k) linear code. This bound is known as *Varsharmov–Gilbert* bound [1–3].)

3.21. Show that the probability of an undetected error for Hamming codes on a BSC with transition probability p satisfies the upper bound 2^{-m} for $p \leq \frac{1}{2}$. [*Hint:* Use the inequality $(1-2p) \leq (1-p)^2$.]

3.22. Compute the probability of an undetected error for a (15, 11) Hamming code on a BSC with transition probability $p = 10^{-2}$.

REFERENCES

1. E. R. Berlekamp, *Algebraic Coding Theory*, McGraw-Hill, New York, 1968.
2. W. W. Peterson and E. J. Weldon, Jr., *Error-Correcting Codes*, 2nd ed., MIT Press, Cambridge, Mass., 1972.
3. F. J. MacWilliams and J. J. A. Sloane, *The Theory of Error-Correcting Codes*, North-Holland, Amsterdam, 1977.
4. R. J. McEliece, *The Theory of Information and Coding*, Addison-Wesley, Reading, Mass., 1977.
5. F. J. MacWilliams, "A Theorem on the Distribution of Weights in a Systematic Code," *Bell Syst. Tech. J.*, 42, pp. 79–94, 1963.
6. R. W. Hamming, "Error Detecting and Error Correcting Codes," *Bell Syst. Tech. J.*, 29, pp. 147–160, April 1950.
7. S. K. Leung-Yan-Cheong and M. E. Hellman, "Concerning a Bound on Undetected Error Probability," *IEEE Trans. Inf. Theory*, IT-22, pp. 235–237, March 1976.

4

Cyclic Codes

Cyclic codes form an important subclass of linear codes. These codes are attractive for two reasons: first, encoding and syndrome computation can be implemented easily by employing shift registers with feedback connections (or linear sequential circuits); and second, because they have considerable inherent algebraic structure, it is possible to find various practical methods for decoding them.

Cyclic codes were first studied by Prange in 1957 [1]. Since then, progress in the study of cyclic codes for both random-error correction and burst-error correction has been spurred by many algebraic coding theorists. References 2 to 7 contain excellent expositions of cyclic codes.

4.1 DESCRIPTION OF CYCLIC CODES

If the components of an n-tuple $\mathbf{v} = (v_0, v_1, \ldots, v_{n-1})$ are cyclically shifted one place to the right, we obtain another n-tuple,

$$\mathbf{v}^{(1)} = (v_{n-1}, v_0 \ldots, v_{n-2}),$$

which is called a cyclic shift of $\mathbf{v}$. If the components of $\mathbf{v}$ are cyclically shifted i places to the right, the resultant n-tuple would be

$$\mathbf{v}^{(i)} = (v_{n-i}, v_{n-i+1}, \ldots, v_{n-1}, v_0, v_1, \ldots, v_{n-i-1}).$$

Clearly, cyclically shifting $\mathbf{v}$ i places to the right is equivalent to cyclically shifting $\mathbf{v}$ $n - i$ places to the left.

Definition 4.1. An (n, k) linear code C is called a *cyclic code* if every cyclic shift of a code vector in C is also a code vector in C.

The (7, 4) linear code given in Table 4.1 is a cyclic code. Cyclic codes form an important subclass of the linear codes and they possess many algebraic properties that simplify the encoding and the decoding implementations.

TABLE 4.1 A (7, 4) CYCLIC CODE GENERATED BY $\mathbf{g}(X) = 1 + X + X^3$

Messages	Code Vectors	Code polynomials
(0 0 0 0)	0 0 0 0 0 0 0	$\mathbf{0} = 0 \cdot \mathbf{g}(X)$
(1 0 0 0)	1 1 0 1 0 0 0	$1 + X + X^3 = 1 \cdot \mathbf{g}(X)$
(0 1 0 0)	0 1 1 0 1 0 0	$X + X^2 + X^4 = X \cdot \mathbf{g}(X)$
(1 1 0 0)	1 0 1 1 1 0 0	$1 + X^2 + X^3 + X^4 = (1 + X) \cdot \mathbf{g}(X)$
(0 0 1 0)	0 0 1 1 0 1 0	$X^2 + X^3 + X^5 = X^2 \cdot \mathbf{g}(X)$
(1 0 1 0)	1 1 1 0 0 1 0	$1 + X + X^2 + X^5 = (1 + X^2) \cdot \mathbf{g}(X)$
(0 1 1 0)	0 1 0 1 1 1 0	$X + X^3 + X^4 + X^5 = (X + X^2) \cdot \mathbf{g}(X)$
(1 1 1 0)	1 0 0 0 1 1 0	$1 + X^4 + X^5 = (1 + X + X^2) \cdot \mathbf{g}(X)$
(0 0 0 1)	0 0 0 1 1 0 1	$X^3 + X^4 + X^6 = X^3 \cdot \mathbf{g}(X)$
(1 0 0 1)	1 1 0 0 1 0 1	$1 + X + X^4 + X^6 = (1 + X^3) \cdot \mathbf{g}(X)$
(0 1 0 1)	0 1 1 1 0 0 1	$X + X^2 + X^3 + X^6 = (X + X^3) \cdot \mathbf{g}(X)$
(1 1 0 1)	1 0 1 0 0 0 1	$1 + X^2 + X^6 = (1 + X + X^3) \cdot \mathbf{g}(X)$
(0 0 1 1)	0 0 1 0 1 1 1	$X^2 + X^4 + X^5 + X^6 = (X^2 + X^3) \cdot \mathbf{g}(X)$
(1 0 1 1)	1 1 1 1 1 1 1	$1 + X + X^2 + X^3 + X^4 + X^5 + X^6 = (1 + X^2 + X^3) \cdot \mathbf{g}(X)$
(0 1 1 1)	0 1 0 0 0 1 1	$X + X^5 + X^6 = (X + X^2 + X^3) \cdot \mathbf{g}(X)$
(1 1 1 1)	1 0 0 1 0 1 1	$1 + X^3 + X^5 + X^6 = (1 + X + X^2 + X^3) \cdot \mathbf{g}(X)$

To develop the algebraic properties of a cyclic code, we treat the components of a code vector $\mathbf{v} = (v_0, v_1, \ldots, v_{n-1})$ as the coefficients of a polynomial as follows:

$$\mathbf{v}(X) = v_0 + v_1 X + v_2 X^2 + \cdots + v_{n-1} X^{n-1}.$$

Thus, each code vector corresponds to a polynomial of degree $n - 1$ or less. If $v_{n-1} \neq 0$, the degree of $\mathbf{v}(X)$ is $n - 1$; if $v_{n-1} = 0$, the degree of $\mathbf{v}(X)$ is less than $n - 1$. The correspondence between the vector $\mathbf{v}$ and the polynomial $\mathbf{v}(X)$ is one-to-one. We shall call $\mathbf{v}(X)$ the code polynomial of $\mathbf{v}$. Hereafter, we use the terms "code vector" and "code polynomial" interchangeably. The code polynomial that corresponds to the code vector $\mathbf{v}^{(i)}$ is

$$\mathbf{v}^{(i)}(X) = v_{n-i} + v_{n-i+1} X + \cdots + v_{n-1} X^{i-1} + v_0 X^i + v_1 X^{i+1} + \cdots + v_{n-i-1} X^{n-1}.$$

There exists an interesting algebraic relationship between $\mathbf{v}(X)$ and $\mathbf{v}^{(i)}(X)$. Multiplying $\mathbf{v}(X)$ by X^i, we obtain

$$X^i \mathbf{v}(X) = v_0 X^i + v_1 X^{i+1} + \cdots + v_{n-i-1} X^{n-1} + \cdots + v_{n-1} X^{n+i-1}.$$

The equation above can be manipulated into the following form:

$$\begin{aligned} X^i \mathbf{v}(X) &= v_{n-i} + v_{n-i+1} X + \cdots + v_{n-1} X^{i-1} + v_0 X^i + \cdots + v_{n-i-1} X^{n-1} \\ &\quad + v_{n-i}(X^n + 1) + v_{n-i+1} X(X^n + 1) + \cdots + v_{n-1} X^{i-1}(X^n + 1) \\ &= \mathbf{q}(X)(X^n + 1) + \mathbf{v}^{(i)}(X), \end{aligned} \tag{4.1}$$

where $\mathbf{q}(X) = v_{n-i} + v_{n-i+1}X + \cdots + v_{n-1}X^{i-1}$. From (4.1) we see that the code polynomial $\mathbf{v}^{(i)}(X)$ is simply the remainder resulting from dividing the polynomial $X^i\mathbf{v}(X)$ by $X^n + 1$.

Next, we prove a number of important algebraic properties of a cyclic code which make possible the simple implementation of encoding and syndrome computation.

Theorem 4.1. The nonzero code polynomial of minimum degree in a cyclic code C is unique.

Proof. Let $\mathbf{g}(X) = g_0 + g_1X + \cdots + g_{r-1}X^{r-1} + X^r$ be a nonzero code polynomial of minimum degree in C. Suppose that $\mathbf{g}(X)$ is not unique. Then there exists another code polynomial of degree r, say $\mathbf{g}'(X) = g'_0 + g'_1X + \cdots + g'_{r-1}X^{r-1} + X^r$. Since C is linear, $\mathbf{g}(X) + \mathbf{g}'(X) = (g_0 + g'_0) + (g_1 + g'_1)X + \cdots + (g_{r-1} + g'_{r-1})X^{r-1}$ is also a code polynomial which has degree less than r. If $\mathbf{g}(X) + \mathbf{g}'(X) \neq 0$, then $\mathbf{g}(X) + \mathbf{g}'(X)$ is a nonzero code polynomial with degree less than the minimum degree r. This is impossible. Therefore, $\mathbf{g}(X) + \mathbf{g}'(X) = 0$. This implies that $\mathbf{g}'(X) = \mathbf{g}(X)$. Hence, $\mathbf{g}(X)$ is unique.

Q.E.D.

Theorem 4.2. Let $\mathbf{g}(X) = g_0 + g_1X + \cdots + g_{r-1}X^{r-1} + X^r$ be the nonzero code polynomial of minimum degree in an (n, k) cyclic code C. Then the constant term g_0 must be equal to 1.

Proof: Suppose that $g_0 = 0$. Then

$$\begin{aligned}\mathbf{g}(X) &= g_1X + g_2X^2 + \cdots + g_{r-1}X^{r-1} + X^r \\ &= X(g_1 + g_2X + \cdots + g_{r-1}X^{r-2} + X^{r-1}).\end{aligned}$$

If we shift $\mathbf{g}(X)$ cyclically $n - 1$ places to the right (or one place to the left), we obtain a nonzero code polynomial, $g_1 + g_2X + \cdots + g_{r-1}X^{r-2} + X^{r-1}$, which has a degree less than r. This is a contradiction to the assumption that $\mathbf{g}(X)$ is the nonzero code polynomial with minimum degree. Thus, $g_0 \neq 0$.

Q.E.D.

It follows from Theorem 4.2 that the nonzero code polynomial of minimum degree in an (n, k) cyclic code C is of the following form:

$$\mathbf{g}(X) = 1 + g_1X + g_2X^2 + \cdots + g_{r-1}X^{r-1} + X^r. \tag{4.2}$$

Consider the (7, 4) cyclic code given in Table 4.1. The nonzero code polynomial with minimum degree is $\mathbf{g}(X) = 1 + X + X^3$.

Consider the polynomials $X\mathbf{g}(X)$, $X^2\mathbf{g}(X)$, . . ., $X^{n-r-1}\mathbf{g}(X)$, which have degrees $r + 1, r + 2, \ldots, n - 1$, respectively. It follows from (4.1) that $X\mathbf{g}(X) = \mathbf{g}^{(1)}(X)$, $X^2\mathbf{g}(X) = \mathbf{g}^{(2)}(X), \ldots, X^{n-r-1}\mathbf{g}(X) = \mathbf{g}^{(n-r-1)}(X)$; that is, they are cyclic shifts of the code polynomial $\mathbf{g}(X)$. Therefore, they are code polynomials in C. Since C is linear, a linear combination of $\mathbf{g}(X)$, $X\mathbf{g}(X)$, . . ., $X^{n-r-1}\mathbf{g}(X)$,

$$\begin{aligned}\mathbf{v}(X) &= u_0\mathbf{g}(X) + u_1X\mathbf{g}(X) + \cdots + u_{n-r-1}X^{n-r-1}\mathbf{g}(X) \\ &= (u_0 + u_1X + \cdots + u_{n-r-1}X^{n-r-1})\mathbf{g}(X),\end{aligned} \tag{4.3}$$

is also a code polynomial where $u_i = 0$ or 1. The following theorem characterizes an important property of a cyclic code.

Theorem 4.3. Let $\mathbf{g}(X) = 1 + g_1 X + \cdots + g_{r-1}X^{r-1} + X^r$ be the nonzero code polynomial of minimum degree in an (n, k) cyclic code C. A binary polynomial of degree $n - 1$ or less is a code polynomial if and only if it is a multiple of $\mathbf{g}(X)$.

Proof: Let $\mathbf{v}(X)$ be a binary polynomial of degree $n - 1$ or less. Suppose that $\mathbf{v}(X)$ is a multiple of $\mathbf{g}(X)$. Then

$$\begin{aligned}\mathbf{v}(X) &= (a_0 + a_1 X + \cdots + a_{n-r-1}X^{n-r-1})\mathbf{g}(X)\\ &= a_0\mathbf{g}(X) + a_1 X\mathbf{g}(X) + \cdots + a_{n-r-1}X^{n-r-1}\mathbf{g}(X).\end{aligned}$$

Since $\mathbf{v}(X)$ is a linear combination of the code polynomials, $\mathbf{g}(X)$, $X\mathbf{g}(X)$, . . ., $X^{n-r-1}\mathbf{g}(X)$, it is a code polynomial in C. This proves the first part of the theorem—that if a polynomial of degree $n - 1$ or less is a multiple of $\mathbf{g}(X)$, it is a code polynomial.

Now let $\mathbf{v}(X)$ be a code polynomial in C. Dividing $\mathbf{v}(X)$ by $\mathbf{g}(X)$, we obtain

$$\mathbf{v}(X) = \mathbf{a}(X)\mathbf{g}(X) + \mathbf{b}(X),$$

where either $\mathbf{b}(X)$ is identical to zero or the degree of $\mathbf{b}(X)$ is less than the degree of $\mathbf{g}(X)$. Rearranging the equation above, we have

$$\mathbf{b}(X) = \mathbf{v}(X) + \mathbf{a}(X)\mathbf{g}(X).$$

It follows from the first part of the theorem that $\mathbf{a}(X)\mathbf{g}(X)$ is a code polynomial. Since both $\mathbf{v}(X)$ and $\mathbf{a}(X)\mathbf{g}(X)$ are code polynomials, $\mathbf{b}(X)$ must also be a code polynomial. If $\mathbf{b}(X) \neq 0$, then $\mathbf{b}(X)$ is a nonzero code polynomial whose degree is less than the degree of $\mathbf{g}(X)$. This contradicts the assumption that $\mathbf{g}(X)$ is the nonzero code polynomial of minimum degree. Thus, $\mathbf{b}(X)$ must be identical to zero. This proves the second part of the theorem—that a code polynomial is a multiple of $\mathbf{g}(X)$.

Q.E.D.

The number of binary polynomials of degree $n - 1$ or less that are multiples of $\mathbf{g}(X)$ is 2^{n-r}. It follows from Theorem 4.3 that these polynomials form all the code polynomials of the (n, k) cyclic code C. Since there are 2^k code polynomials in C, then 2^{n-r} must be equal to 2^k. As a result, we have $r = n - k$ [i.e., the degree of $\mathbf{g}(X)$ is $n - k$]. Hence, the nonzero code polynomial of minimum degree in an (n, k) cyclic code is of the following form:

$$\mathbf{g}(X) = 1 + g_1 X + g_2 X^2 + \cdots + g_{n-k-1}X^{n-k-1} + X^{n-k}.$$

Summarizing the results above, we have the following theorem:

Theorem 4.4. In an (n, k) cyclic code, there exists one and only one code polynomial of degree $n - k$,

$$\mathbf{g}(X) = 1 + g_1 X + g_2 X^2 + \cdots + g_{n-k-1}X^{n-k-1} + X^{n-k}. \tag{4.4}$$

Every code polynomial is a multiple of $\mathbf{g}(X)$ and every binary polynomial of degree $n - 1$ or less that is a multiple of $\mathbf{g}(X)$ is a code polynomial.

It follows from Theorem 4.4 that every code polynomial $\mathbf{v}(X)$ in an (n, k) cyclic code can be expressed in the following form:

$$\begin{aligned}\mathbf{v}(X) &= \mathbf{u}(X)\mathbf{g}(X) \\ &= (u_0 + u_1 X + \cdots + u_{k-1}X^{k-1})\mathbf{g}(X).\end{aligned}$$

If the coefficients of $\mathbf{u}(X)$, $u_0, u_1, \ldots, u_{k-1}$, are the k information digits to be encoded, $\mathbf{v}(X)$ is the corresponding code polynomial. Hence, the encoding can be achieved by multiplying the message $\mathbf{u}(X)$ by $\mathbf{g}(X)$. Therefore, an (n, k) cyclic code is completely specified by its nonzero code polynomial of minimum degree, $\mathbf{g}(X)$, given by (4.4). The polynomial $\mathbf{g}(X)$ is called the *generator polynomial* of the code. The degree of $\mathbf{g}(X)$ is equal to the number of parity-check digits of the code. The generator polynomial of the (7, 4) cyclic code given in Table 4.1 is $\mathbf{g}(X) = 1 + X + X^3$. We see that each code polynomial is a multiple of $\mathbf{g}(X)$.

The next important property of a cyclic code is given in the following theorem.

Theorem 4.5. The generator polynomial $\mathbf{g}(X)$ of an (n, k) cyclic code is a factor of $X^n + 1$.

Proof: Multiplying $\mathbf{g}(X)$ by X^k results in a polynomial $X^k\mathbf{g}(X)$ of degree n. Dividing $X^k\mathbf{g}(X)$ by $X^n + 1$, we obtain

$$X^k\mathbf{g}(X) = (X^n + 1) + \mathbf{g}^{(k)}(X), \tag{4.5}$$

where $\mathbf{g}^{(k)}(X)$ is the remainder. It follows from (4.1) that $\mathbf{g}^{(k)}(X)$ is the code polynomial obtained by shifting $\mathbf{g}(X)$ to the right cyclically k times. Hence, $\mathbf{g}^{(k)}(X)$ is a multiple of $\mathbf{g}(X)$, say $\mathbf{g}^{(k)}(X) = \mathbf{a}(X)\mathbf{g}(X)$. From (4.5) we obtain

$$X^n + 1 = \{X^k + \mathbf{a}(X)\}\mathbf{g}(X).$$

Thus, $\mathbf{g}(X)$ is a factor of $X^n + 1$. Q.E.D.

At this point, a natural question is whether, for any n and k, there exists an (n, k) cyclic code. This is answered by the following theorem.

Theorem 4.6. If $\mathbf{g}(X)$ is a polynomial of degree $n - k$ and is a factor of $X^n + 1$, then $\mathbf{g}(X)$ generates an (n, k) cyclic code.

Proof. Consider the k polynomials $\mathbf{g}(X), X\mathbf{g}(X), \ldots, X^{k-1}\mathbf{g}(X)$, which all have degree $n - 1$ or less. A linear combination of these k polynomials,

$$\begin{aligned}\mathbf{v}(X) &= a_0\mathbf{g}(X) + a_1 X\mathbf{g}(X) + \cdots + a_{k-1}X^{k-1}\mathbf{g}(X) \\ &= (a_0 + a_1 X + \cdots + a_{k-1}X^{k-1})\mathbf{g}(X),\end{aligned}$$

is also a polynomial of degree $n - 1$ or less and is a multiple of $\mathbf{g}(X)$. There are a total of 2^k such polynomials and they form an (n, k) linear code.

Let $\mathbf{v}(X) = v_0 + v_1 X + \cdots + v_{n-1}X^{n-1}$ be a code polynomial in this code. Multiplying $\mathbf{v}(X)$ by X, we obtain

$$
\begin{aligned}
X\mathbf{v}(X) &= v_0 X + v_1 X^2 + \cdots + v_{n-2}X^{n-1} + v_{n-1}X^n \\
&= v_{n-1}(X^n + 1) + (v_{n-1} + v_0 X + \cdots + v_{n-2}X^{n-1}) \\
&= v_{n-1}(X^n + 1) + \mathbf{v}^{(1)}(X),
\end{aligned}
$$

where $\mathbf{v}^{(1)}(X)$ is a cyclic shift of $\mathbf{v}(X)$. Since both $X\mathbf{v}(X)$ and $X^n + 1$ are divisible by $\mathbf{g}(X)$, $\mathbf{v}^{(1)}(X)$ must be divisible by $\mathbf{g}(X)$. Thus, $\mathbf{v}^{(1)}(X)$ is a multiple of $\mathbf{g}(X)$ and is a linear combination of $\mathbf{g}(X)$, $X\mathbf{g}(X)$, . . ., $X^{k-1}\mathbf{g}(X)$. Hence, $\mathbf{v}^{(1)}(X)$ is also a code polynomial. It follows from Definition 4.1 that the linear code generated by $\mathbf{g}(X)$, $X\mathbf{g}(X)$, . . ., $X^{k-1}\mathbf{g}(X)$ is an (n, k) cyclic code. Q.E.D.

Theorem 4.6 actually says that any factor of $X^n + 1$ with degree $n - k$ generates an (n, k) cyclic code. For large n, $X^n + 1$ may have many factors of degree $n - k$. Some of these polynomials generate good codes and some generate bad codes. How to select generator polynomials to produce good cyclic codes is a very difficult problem. For the past two decades, coding theorists have expended much effort in searching for good cyclic codes. Several classes of good cyclic codes have been discovered and they can be practically implemented.

Example 4.1

The polynomial $X^7 + 1$ can be factored as follows:

$$X^7 + 1 = (1 + X)(1 + X + X^3)(1 + X^2 + X^3).$$

There are two factors of degree 3; each generates a (7, 4) cyclic code. The (7, 4) cyclic code given by Table 4.1 is generated by $\mathbf{g}(X) = 1 + X + X^3$. This code has minimum distance 3 and it is a single-error-correcting code. Notice that the code is not in systematic form. Each code polynomial is the product of a message polynomial of degree 3 or less and the generator polynomial $\mathbf{g}(X) = 1 + X + X^3$. For example, let $\mathbf{u} = (1\ 0\ 1\ 0)$ be the message to be encoded. The corresponding message polynomial is $\mathbf{u}(X) = 1 + X^2$. Multiplying $\mathbf{u}(X)$ by $\mathbf{g}(X)$ results in the following code polynomial:

$$
\begin{aligned}
\mathbf{v}(X) &= (1 + X^2)(1 + X + X^3) \\
&= 1 + X + X^2 + X^5,
\end{aligned}
$$

or the code vector $(1\ 1\ 1\ 0\ 0\ 1\ 0)$.

Given the generator polynomials $\mathbf{g}(X)$ of an (n, k) cyclic code, the code can be put into systematic form (i.e., the rightmost k digits of each code vector are the unaltered information digits and the leftmost $n - k$ digits are parity-check digits). Suppose that the message to be encoded is $\mathbf{u} = (u_0, u_1, \ldots, u_{k-1})$. The corresponding message polynomial is

$$\mathbf{u}(X) = u_0 + u_1 X + \cdots + u_{k-1}X^{k-1}.$$

Multiplying $\mathbf{u}(X)$ by X^{n-k}, we obtain a polynomial of degree $n - 1$ or less,

$$X^{n-k}\mathbf{u}(X) = u_0 X^{n-k} + u_1 X^{n-k+1} + \cdots + u_{k-1}X^{n-1}.$$

Dividing $X^{n-k}\mathbf{u}(X)$ by the generator polynomial $\mathbf{g}(X)$, we have

$$X^{n-k}\mathbf{u}(X) = \mathbf{a}(X)\mathbf{g}(X) + \mathbf{b}(X) \tag{4.6}$$

where $\mathbf{a}(X)$ and $\mathbf{b}(X)$ are the quotient and the remainder, respectively. Since the degree of $\mathbf{g}(X)$ is $n - k$, the degree of $\mathbf{b}(X)$ must be $n - k - 1$ or less, that is,

$$\mathbf{b}(X) = b_0 + b_1 X + \cdots + b_{n-k-1} X^{n-k-1}.$$

Rearranging (4.6), we obtain the following polynomial of degree $n - 1$ or less:

$$\mathbf{b}(X) + X^{n-k}\mathbf{u}(X) = \mathbf{a}(X)\mathbf{g}(X). \tag{4.7}$$

This polynomial is a multiple of the generator polynomial $\mathbf{g}(X)$ and therefore it is a code polynomial of the cyclic code generated by $\mathbf{g}(X)$. Writing out $\mathbf{b}(X) + X^{n-k}\mathbf{u}(X)$, we have

$$\begin{aligned} \mathbf{b}(X) + X^{n-k}\mathbf{u}(X) = b_0 + b_1 X + \cdots + b_{n-k-1} X^{n-k-1} \\ + u_0 X^{n-k} + u_1 X^{n-k+1} + \cdots + u_{k-1} X^{n-1}, \end{aligned} \tag{4.8}$$

which corresponds to the code vector

$$(b_0, b_1, \ldots, b_{n-k-1}, u_0, u_1, \ldots, u_{k-1}).$$

We see that the code vector consists of k unaltered information digits $(u_0, u_1, \ldots, u_{k-1})$ followed by $n - k$ parity-check digits. The $n - k$ parity-check digits are simply the coefficients of the remainder resulting from dividing the message polynomial $X^{n-k}\mathbf{u}(X)$ by the generator polynomial $\mathbf{g}(X)$. The process above yields an (n, k) cyclic code in systematic form. In connection with cyclic codes in systematic form, the following convention is used: The first $n - k$ symbols, the coefficients of $1, X, \ldots, X^{n-k-1}$, are taken as parity-check digits and the last k symbols, the coefficients of $X^{n-k}, X^{n-k+1}, \ldots, X^{n-1}$, are taken as the information digits. In summary, encoding in systematic form consists of three steps:

Step 1. Premultiply the message $\mathbf{u}(X)$ by X^{n-k}.

Step 2. Obtain the remainder $\mathbf{b}(X)$ (the parity-check digits) from dividing $X^{n-k}\mathbf{u}(X)$ by the generator polynomial $\mathbf{g}(X)$.

Step 3. Combine $\mathbf{b}(X)$ and $X^{n-k}\mathbf{u}(X)$ to obtain the code polynomial $\mathbf{b}(X) + X^{n-k}\mathbf{u}(X)$.

Example 4.2

Consider the (7, 4) cyclic code generated by $\mathbf{g}(X) = 1 + X + X^3$. Let $\mathbf{u}(X) = 1 + X^3$ be the message to be encoded. Dividing $X^3\mathbf{u}(X) = X^3 + X^6$ by $\mathbf{g}(X)$,

$$\begin{array}{r|l} & X^3 + X \quad \text{(quotient)} \\ X^3 + X + 1 & X^6 \qquad\qquad + X^3 \\ & X^6 \quad + X^4 + X^3 \\ \hline & \qquad X^4 \\ & \qquad X^4 \quad + X^2 + X \\ \hline & \qquad\qquad X^2 + X \quad \text{(remainder)}, \end{array}$$

we obtain the remainder $\mathbf{b}(X) = X + X^2$. Thus, the code polynomial is $\mathbf{v}(X) = \mathbf{b}(X) + X^3\mathbf{u}(X) = X + X^2 + X^3 + X^6$ and the corresponding code vector is $\mathbf{v} = (0\ 1\ 1\ 1\ 0\ 0\ 1)$, where the four rightmost digits are the information digits. The 16 code vectors in systematic form are listed in Table 4.2.

TABLE 4.2 A (7, 4) CYCLIC CODE GENERATED BY $\mathbf{g}(X) = 1 + X + X^3$

Message	Code word	
(0 0 0 0)	(0 0 0 0 0 0 0)	$0 = 0 \cdot \mathbf{g}(X)$
(1 0 0 0)	(1 1 0 1 0 0 0)	$1 + X + X^3 = \mathbf{g}(X)$
(0 1 0 0)	(0 1 1 0 1 0 0)	$X + X^2 + X^4 = X\mathbf{g}(X)$
(1 1 0 0)	(1 0 1 1 1 0 0)	$1 + X^2 + X^3 + X^4 = (1 + X)\mathbf{g}(X)$
(0 0 1 0)	(1 1 1 0 0 1 0)	$1 + X + X^2 + X^5 = (1 + X^2)\mathbf{g}(X)$
(1 0 1 0)	(0 0 1 1 0 1 0)	$X^2 + X^3 + X^5 = X^2\mathbf{g}(X)$
(0 1 1 0)	(1 0 0 0 1 1 0)	$1 + X^4 + X^5 = (1 + X + X^2)\mathbf{g}(X)$
(1 1 1 0)	(0 1 0 1 1 1 0)	$X + X^3 + X^4 + X^5 = (X + X^2)\mathbf{g}(X)$
(0 0 0 1)	(1 0 1 0 0 0 1)	$1 + X^2 + X^6 = (1 + X + X^3)\mathbf{g}(X)$
(1 0 0 1)	(0 1 1 1 0 0 1)	$X + X^2 + X^3 + X^6 = (X + X^3)\mathbf{g}(X)$
(0 1 0 1)	(1 1 0 0 1 0 1)	$1 + X + X^4 + X^6 = (1 + X^3)\mathbf{g}(X)$
(1 1 0 1)	(0 0 0 1 1 0 1)	$X^3 + X^4 + X^6 = X^3\mathbf{g}(X)$
(0 0 1 1)	(0 1 0 0 0 1 1)	$X + X^5 + X^6 = (X + X^2 + X^3)\mathbf{g}(X)$
(1 0 1 1)	(1 0 0 1 0 1 1)	$1 + X^3 + X^5 + X^6 = (1 + X + X^2 + X^3)\mathbf{g}(X)$
(0 1 1 1)	(0 0 1 0 1 1 1)	$X^2 + X^4 + X^5 + X^6 = (X^2 + X^3)\mathbf{g}(X)$
(1 1 1 1)	(1 1 1 1 1 1 1)	$1 + X + X^2 + X^3 + X^4 + X^5 + X^6 = (1 + X^2 + X^5)\mathbf{g}(X)$

4.2 GENERATOR AND PARITY-CHECK MATRICES OF CYCLIC CODES

Consider an (n, k) cyclic code C with generator polynomial $\mathbf{g}(X) = g_0 + g_1X + \cdots + g_{n-k}X^{n-k}$. In Section 4.1 we have shown that the k code polynomials $\mathbf{g}(X)$, $X\mathbf{g}(X)$, $\ldots$, $X^{k-1}\mathbf{g}(X)$ span C. If the k n-tuples corresponding to these k code polynomials are used as the rows of an $k \times n$ matrix, we obtain the following generator matrix for C:

$$\mathbf{G} = \begin{bmatrix} g_0 & g_1 & g_2 & \cdot & \cdot & \cdot & \cdot & \cdot & g_{n-k} & 0 & 0 & 0 & \cdot & \cdot & 0 \\ 0 & g_0 & g_1 & g_2 & \cdot & \cdot & \cdot & \cdot & \cdot & g_{n-k} & 0 & 0 & \cdot & \cdot & 0 \\ 0 & 0 & g_0 & g_1 & g_2 & \cdot & \cdot & \cdot & \cdot & \cdot & g_{n-k} & 0 & \cdot & \cdot & 0 \\ \cdot & & & & & & & & & & & & & & \cdot \\ \cdot & & & & & & & & & & & & & & \cdot \\ \cdot & & & & & & & & & & & & & & \cdot \\ 0 & 0 & \cdot & \cdot & \cdot & 0 & g_0 & g_1 & g_2 & \cdot & \cdot & \cdot & \cdot & \cdot & g_{n-k} \end{bmatrix} \quad (4.9)$$

(Note that $g_0 = g_{n-k} = 1$.) In general, $\mathbf{G}$ is not in systematic form. However, it can be put into systematic form with row operations. For example, the (7, 4) cyclic code given in Table 4.1 with generator polynomial $\mathbf{g}(X) = 1 + X + X^3$ has the following matrix as a generator matrix:

$$\mathbf{G} = \begin{bmatrix} 1 & 1 & 0 & 1 & 0 & 0 & 0 \\ 0 & 1 & 1 & 0 & 1 & 0 & 0 \\ 0 & 0 & 1 & 1 & 0 & 1 & 0 \\ 0 & 0 & 0 & 1 & 1 & 0 & 1 \end{bmatrix}.$$

Clearly, $\mathbf{G}$ is not in systematic form. If the first row is added to the third row and the sum of the first two rows is added to the fourth row, we obtain the following matrix:

$$\mathbf{G}' = \begin{bmatrix} 1 & 1 & 0 & 1 & 0 & 0 & 0 \\ 0 & 1 & 1 & 0 & 1 & 0 & 0 \\ 1 & 1 & 1 & 0 & 0 & 1 & 0 \\ 1 & 0 & 1 & 0 & 0 & 0 & 1 \end{bmatrix},$$

which is in systematic form. This matrix generates the same code as **G**.

Recall that the generator polynomial $\mathbf{g}(X)$ is a factor of $X^n + 1$, say

$$X^n + 1 = \mathbf{g}(X)\mathbf{h}(X), \tag{4.10}$$

where the polynomial $\mathbf{h}(X)$ has the degree k and is of the following form:

$$\mathbf{h}(X) = h_0 + h_1 X + \cdots + h_k X^k$$

with $h_0 = h_k = 1$. Next we want to show that a parity-check matrix of C may be obtained from $\mathbf{h}(X)$. Let $\mathbf{v} = (v_0, v_1, \ldots, v_{n-1})$ be a code vector in C. Then $\mathbf{v}(X) = \mathbf{a}(X)\mathbf{g}(X)$. Multiplying $\mathbf{v}(X)$ by $\mathbf{h}(X)$, we obtain

$$\begin{aligned} \mathbf{v}(X)\mathbf{h}(X) &= \mathbf{a}(X)\mathbf{g}(X)\mathbf{h}(X) \\ &= \mathbf{a}(X)(X^n + 1) \\ &= \mathbf{a}(X) + X^n\mathbf{a}(X). \end{aligned} \tag{4.11}$$

Since the degree of $\mathbf{a}(X)$ is $k - 1$ or less, the powers $X^k, X^{k+1}, \ldots, X^{n-1}$ do not appear in $\mathbf{a}(X) + X^n\mathbf{a}(X)$. If we expand the product $\mathbf{v}(X)\mathbf{h}(X)$ on the left-hand side of (4.11), the coefficients of $X^k, X^{k+1}, \ldots, X^{n-1}$ must be equal to zero. Therefore, we obtain the following $n - k$ equalities:

$$\sum_{i=0}^{k} h_i v_{n-i-j} = 0 \qquad \text{for } 1 \leq j \leq n - k. \tag{4.12}$$

Now, we take the *reciprocal* of $\mathbf{h}(X)$, which is defined as follows:

$$X^k\mathbf{h}(X^{-1}) = h_k + h_{k-1}X + h_{k-2}X^2 + \cdots + h_0 X^k. \tag{4.13}$$

We can see easily that $X^k\mathbf{h}(X^{-1})$ is also a factor of $X^n + 1$. The polynomial $X^k\mathbf{h}(X^{-1})$ generates an $(n, n - k)$ cyclic code with the following $(n - k) \times n$ matrix as a generator matrix:

$$\mathbf{H} = \begin{bmatrix} h_k & h_{k-1} & h_{k-2} & \cdot & \cdot & \cdot & \cdot & \cdot & h_0 & 0 & \cdot & \cdot & \cdot & \cdot & 0 \\ 0 & h_k & h_{k-1} & h_{k-2} & \cdot & \cdot & \cdot & \cdot & \cdot & h_0 & 0 & \cdot & \cdot & \cdot & 0 \\ 0 & 0 & h_k & h_{k-1} & h_{k-2} & \cdot & \cdot & \cdot & \cdot & \cdot & h_0 & \cdot & \cdot & \cdot & 0 \\ \cdot & & & & & & & & & & & & & & \cdot \\ \cdot & & & & & & & & & & & & & & \cdot \\ \cdot & & & & & & & & & & & & & & \cdot \\ 0 & 0 & \cdot & \cdot & \cdot & 0 & h_k & h_{k-1} & h_{k-2} & \cdot & \cdot & \cdot & \cdot & \cdot & h_0 \end{bmatrix}. \tag{4.14}$$

It follows from the $n - k$ equalities of (4.12) that any code vector $\mathbf{v}$ in C is orthogonal to every row of **H**. Therefore, **H** is a parity-check matrix of the cyclic code C, and the row space of **H** is the dual code of C. Since the parity-check matrix **H** is obtained from the polynomial $\mathbf{h}(X)$, we call $\mathbf{h}(X)$ the *parity polynomial* of C. Hence, a cyclic code is also uniquely specified by its parity polynomial.

Besides deriving a parity-check matrix for a cyclic code, we have also proved another important property, which is stated in the following theorem.

Theorem 4.7. Let C be an (n, k) cyclic code with generator polynomial $\mathbf{g}(X)$. The dual code of C is also cyclic and is generated by the polynomial $X^k\mathbf{h}(X^{-1})$, where $\mathbf{h}(X) = (X^n + 1)/\mathbf{g}(X)$.

Example 4.3

Consider the (7, 4) cyclic code given in Table 4.1 with generator polynomial $\mathbf{g}(X) = 1 + X + X^3$. The parity polynomial is

$$\mathbf{h}(X) = \frac{X^7 + 1}{\mathbf{g}(X)}$$
$$= 1 + X + X^2 + X^4.$$

The reciprocal of $\mathbf{h}(X)$ is

$$X^4\mathbf{h}(X^{-1}) = X^4(1 + X^{-1} + X^{-2} + X^{-4}).$$
$$= 1 + X^2 + X^3 + X^4.$$

The polynomial $X^4\mathbf{h}(X^{-1})$ divides $X^7 + 1$, $(X^7 + 1)/X^4\mathbf{h}(X^{-1}) = 1 + X^2 + X^3$. If we construct all the vectors of the (7, 3) code generated by $X^4\mathbf{h}(X^{-1}) = 1 + X^2 + X^3 + X^4$, we will find that it has minimum distance 4. Hence, it is capable of correcting any single error and simultaneously detecting any combination of double errors.

The generator matrix in systematic form can also be formed easily. Dividing X^{n-k+i} by the generator polynomial $\mathbf{g}(X)$ for $i = 0, 1, \ldots, k - 1$, we obtain

$$X^{n-k+i} = \mathbf{a}_i(X)\mathbf{g}(X) + \mathbf{b}_i(X), \tag{4.15}$$

where $\mathbf{b}_i(X)$ is the remainder with the following form:

$$\mathbf{b}_i(X) = b_{i0} + b_{i1}X + \cdots + b_{i,n-k-1}X^{n-k-1}.$$

Since $\mathbf{b}_i(X) + X^{n-k+i}$ for $i = 0, 1, \ldots, k - 1$ are multiples of $\mathbf{g}(X)$, they are code polynomials. Arranging these k code polynomials as rows of a $k \times n$ matrix, we obtain

$$\mathbf{G} = \begin{bmatrix} b_{00} & b_{01} & b_{02} & \cdots & b_{0,n-k-1} & 1 & 0 & 0 & \cdots & 0 \\ b_{10} & b_{11} & b_{12} & \cdots & b_{1,n-k-1} & 0 & 1 & 0 & \cdots & 0 \\ b_{20} & b_{21} & b_{22} & \cdots & b_{2,n-k-1} & 0 & 0 & 1 & \cdots & 0 \\ & \vdots & & & \vdots & & & & & \vdots \\ b_{k-1,0} & b_{k-1,1} & b_{k-1,2} & \cdots & b_{k-1,n-k-1} & 0 & 0 & 0 & \cdots & 1 \end{bmatrix}, \tag{4.16}$$

which is the generator matrix of C in systematic form. The corresponding parity-check matrix for C is

$$\mathbf{H} = \begin{bmatrix} 1 & 0 & 0 & \cdots & 0 & b_{00} & b_{10} & b_{20} & \cdots & b_{k-1,0} \\ 0 & 1 & 0 & \cdots & 0 & b_{01} & b_{11} & b_{21} & \cdots & b_{k-1,1} \\ 0 & 0 & 1 & \cdots & 0 & b_{02} & b_{12} & b_{22} & \cdots & b_{k-1,2} \\ \vdots & & & & & & \vdots & & & \vdots \\ 0 & 0 & 0 & \cdots & 1 & b_{0,n-k-1} & b_{1,n-k-1} & b_{2,n-k-1} & \cdots & b_{k-1,n-k-1} \end{bmatrix}. \tag{4.17}$$

Example 4.4

Again, let us consider the (7, 4) cyclic code generated by $\mathbf{g}(X) = 1 + X + X^3$. Dividing X^3, X^4, X^5, and X^6 by $\mathbf{g}(X)$, we have

$$\begin{aligned} X^3 &= \mathbf{g}(X) + (1 + X), \\ X^4 &= X\mathbf{g}(X) + (X + X^2), \\ X^5 &= (X^2 + 1)\mathbf{g}(X) + (1 + X + X^2), \\ X^6 &= (X^3 + X + 1)\mathbf{g}(X) + (1 + X^2). \end{aligned}$$

Rearranging the equations above, we obtain the following four code polynomials:

$$\begin{aligned} \mathbf{v}_0(X) &= 1 + X + X^3, \\ \mathbf{v}_1(X) &= X + X^2 + X^4, \\ \mathbf{v}_2(X) &= 1 + X + X^2 + X^5, \\ \mathbf{v}_3(X) &= 1 + X^2 + X^6. \end{aligned}$$

Taking these four code polynomials as rows of a 4×7 matrix, we obtain the following generator matrix in systematic form for the (7, 4) cyclic code:

$$\mathbf{G} = \begin{bmatrix} 1 & 1 & 0 & 1 & 0 & 0 & 0 \\ 0 & 1 & 1 & 0 & 1 & 0 & 0 \\ 1 & 1 & 1 & 0 & 0 & 1 & 0 \\ 1 & 0 & 1 & 0 & 0 & 0 & 1 \end{bmatrix},$$

which is identical to the matrix $\mathbf{G}'$ obtain earlier in this section.

4.3 ENCODING OF CYCLIC CODES

We have shown in Section 4.1 that encoding of an (n, k) cyclic code in systematic form consists of three steps; (1) multiply the message polynomial $\mathbf{u}(X)$ by X^{n-k}; (2) divide $X^{n-k}\mathbf{u}(X)$ by $\mathbf{g}(X)$ to obtain the remainder $\mathbf{b}(X)$; and (3) form the code word $\mathbf{b}(X) + X^{n-k}\mathbf{u}(X)$. All these three steps can be accomplished with a division circuit which is a linear $(n - k)$-stage shift register with feedback connections based on the generator polynomial $\mathbf{g}(X) = 1 + g_1X + g_2X^2 + \cdots + g_{n-k-1}X^{n-k-1} + X^{n-k}$. Such a circuit is shown in Fig. 4.1. The encoding operation is carried out as follows:

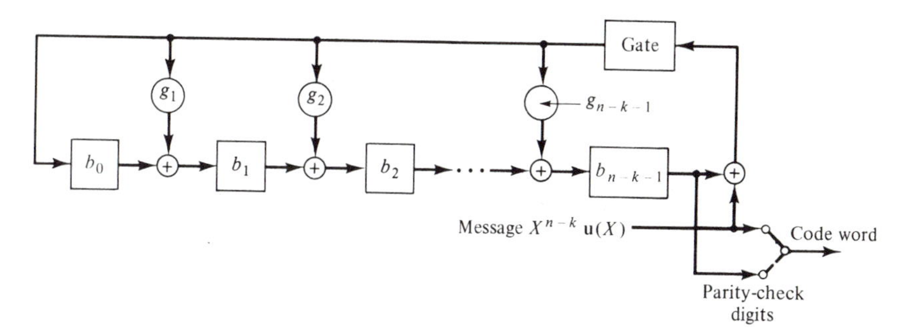

Figure 4.1 Encoding circuit for an (n, k) cyclic code with generator polynomial $\mathbf{g}(X) = 1 + g_1X + g_2X^2 + \cdots + g_{n-k-1}X^{n-k-1} + X^{n-k}$.

Step 1. With the gate turned on, the k information digits $u_0, u_1, \ldots, u_{k-1}$ [or $\mathbf{u}(X) = u_0 + u_1X + \cdots + u_{k-1}X^{k-1}$ in polynomial form] are shifted into the circuit and simultaneously into the communication channel. Shifting the message $\mathbf{u}(X)$ into the circuit from the front end is equivalent to premultiplying $\mathbf{u}(X)$ by X^{n-k}. As soon as the complete message has entered the circuit, the $n - k$ digits in the register form the remainder and thus they are the parity-check digits.

Step 2. Break the feedback connection by turning off the gate.

Step 3. Shift the parity-check digits out and send them into the channel. These $n - k$ parity-check digits $b_0, b_1, \ldots, b_{n-k-1}$, together with the k information digits, form a complete code vector.

Example 4.5

Consider the (7, 4) cyclic code generated by $\mathbf{g}(X) = 1 + X + X^3$. The encoding circuit based on $\mathbf{g}(X)$ is shown in Figure 4.2. Suppose that the message $\mathbf{u} = (1\ 0\ 1\ 1)$ is to be encoded. As the message digits are shifted into the register, the contents in the register are as follows:

Input	Register contents	
	0 0 0	(initial state)
1	1 1 0	(first shift)
1	1 0 1	(second shift)
0	1 0 0	(third shift)
1	1 0 0	(fourth shift)

After four shifts, the contents of the register are (1 0 0). Thus, the complete code vector is (1 0 0 1 0 1 1) and the code polynomial is $1 + X^3 + X^5 + X^6$.

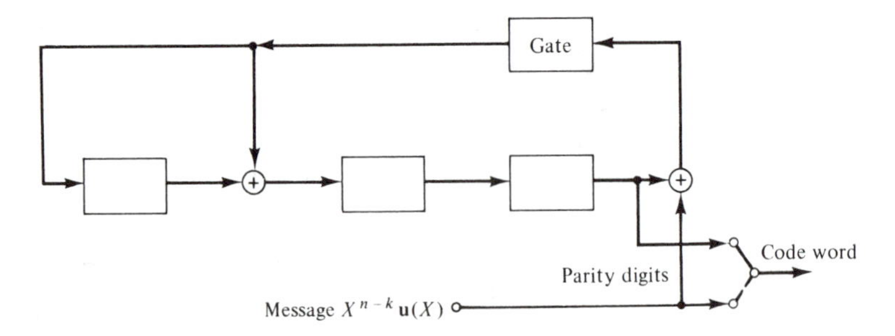

Figure 4.2 Encoder for the (7, 4) cyclic code generated by $\mathbf{g}(X) = 1 + X + X^3$.

Encoding of a cyclic code can also be accomplished by using its parity polynomial $\mathbf{h}(X) = h_0 + h_1X + \cdots + h_kX^k$. Let $\mathbf{v} = (v_0, v_1, \ldots, v_{n-1})$ be a code vector. We have shown in Section 4.2 that the components of $\mathbf{v}$ satisfy the $n - k$ equalities of (4.12). Since $h_k = 1$, the equalities of (4.12) can be put into the following form:

$$v_{n-k-j} = \sum_{i=0}^{k-1} h_i v_{n-i-j} \qquad \text{for } 1 \leq j \leq n-k \tag{4.18}$$

which is known as a *difference equation*. For a cyclic code in systematic form, the components $v_{n-k}, v_{n-k+1}, \ldots, v_{n-1}$ of each code vector are the information digits. Given these k information digits, (4.18) is a rule to determine the $n-k$ parity-check digits, $v_0, v_1, \ldots, v_{n-k-1}$. An encoding circuit based on (4.18) is shown in Figure 4.3.

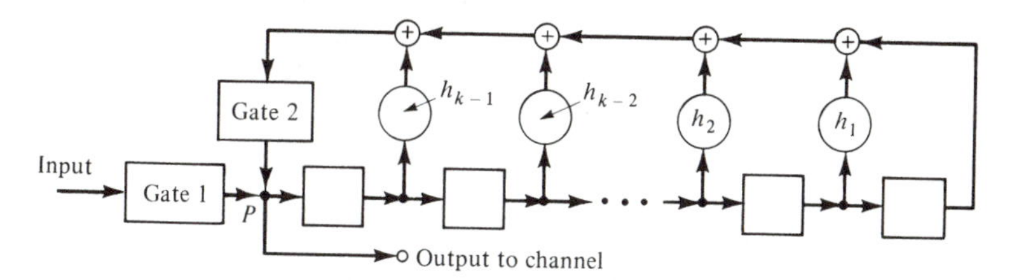

Figure 4.3 Encoding circuit for an (n, k) cyclic code based on the parity polynomial $\mathbf{h}(X) = 1 + h_1X + \cdots + X^k$.

The feedback connections are based on the coefficients of the parity polynomial $\mathbf{h}(X)$. (Note that $h_0 = h_k = 1$.) The encoding operation can be described in the following steps:

Step 1. Initially, gate 1 is turned on and gate 2 is turned off. The k information digits $\mathbf{u}(X) = u_0 + u_1X + \cdots + u_{k-1}X^{k-1}$ are shifted into the register and the communication channel simultaneously.

Step 2. As soon as the k information digits have entered the shift register, gate 1 is turned off and gate 2 is turned on. The first parity-check digit,

$$\begin{aligned} v_{n-k-1} &= h_0v_{n-1} + h_1v_{n-2} + \cdots + h_{k-1}v_{n-k} \\ &= u_{k-1} + h_1u_{k-2} + \cdots + h_{k-1}u_0, \end{aligned}$$

is formed and appears at point P.

Step 3. The register is shifted once. The first parity-check digit is shifted into the channel and is also shifted into the register. Now, the second parity-check digit,

$$\begin{aligned} v_{n-k-2} &= h_0v_{n-2} + h_1v_{n-3} + \cdots + h_{k-1}v_{n-k-1} \\ &= u_{k-2} + h_1u_{k-3} + \cdots + h_{k-2}u_0 + h_{k-1}v_{n-k-1}, \end{aligned}$$

is formed at P.

Step 4. Step 3 is repeated until $n-k$ parity-check digits have been formed and shifted into the channel. Then gate 1 is turned on and gate 2 is turned off. The next message is now ready to be shifted into the register.

The encoding circuit above employs a k-stage shift register. Comparing the two encoding circuits presented in this section, we can make the following remark: For codes with more parity-check digits than the message digits, the k-stage encoding circuit is more economical; otherwise, the $(n-k)$-stage encoding circuit is preferable.

Example 4.6

The parity polynomial of the (7, 4) cyclic code generated by $\mathbf{g}(X) = 1 + X + X^3$ is

$$\mathbf{h}(X) = \frac{X^7 + 1}{1 + X + X^3} = 1 + X + X^2 + X^4.$$

The encoding circuit based on $\mathbf{h}(X)$ is shown in Figure 4.4. Each code vector is of the form $\mathbf{v} = (v_0, v_1, v_2, v_3, v_4, v_5, v_6)$, where v_3, v_4, v_5, and v_6 are message digits and v_0, v_1, and v_2 are parity-check digits. The difference equation that determines the parity-check digits is

$$\begin{aligned} v_{3-j} &= 1 \cdot v_{7-j} + 1 \cdot v_{6-j} + 1 \cdot v_{5-j} + 0 \cdot v_{4-j} \\ &= v_{7-j} + v_{6-j} + v_{5-j} \qquad \text{for } 1 \leq j \leq 3. \end{aligned}$$

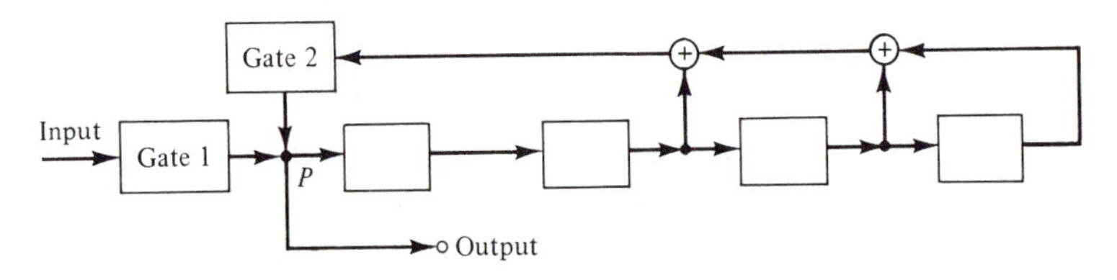

Figure 4.4 Encoding circuit for the (7, 4) cyclic code based on its parity polynomial $\mathbf{h}(X) = 1 + X + X^2 + X^4$.

Suppose that the message to be encoded is (1 0 1 1). Then $v_3 = 1, v_4 = 0, v_5 = 1, v_6 = 1$. The first parity-check digit is

$$v_2 = v_6 + v_5 + v_4 = 1 + 1 + 0 = 0.$$

The second parity-check digit is

$$v_1 = v_5 + v_4 + v_3 = 1 + 0 + 1 = 0.$$

The third parity-check digit is

$$v_0 = v_4 + v_3 + v_2 = 0 + 1 + 0 = 1.$$

Thus, the code vector that corresponds to the message (1 0 1 1) is (1 0 0 1 0 1 1).

4.4 SYNDROME COMPUTATION AND ERROR DETECTION

Suppose that a code vector is transmitted. Let $\mathbf{r} = (r_0, r_1, \ldots, r_{n-1})$ be the received vector. Because of the channel noise, the received vector may not be the same as the transmitted code vector. In the decoding of a linear code, the first step is to compute the syndrome $\mathbf{s} = \mathbf{r} \cdot \mathbf{H}^T$, where $\mathbf{H}$ is the parity-check matrix. If the syndrome is zero, $\mathbf{r}$ is a code vector and the decoder accepts $\mathbf{r}$ as the transmitted code vector. If the syndrome is not identical to zero, $\mathbf{r}$ is not a code vector and the presence of errors has been detected.

We have shown that for a linear systematic code, the syndrome is simply the vector sum of the received parity digits and the parity-check digits recomputed from the received information digits. For a cyclic code in systematic form, the syndrome can be computed easily. The received vector $\mathbf{r}$ is treated as a polynomial of degree $n - 1$ or less,

$$\mathbf{r}(X) = r_0 + r_1 X + r_2 X^2 + \cdots + r_{n-1} X^{n-1}.$$

Dividing $\mathbf{r}(X)$ by the generator polynomial $\mathbf{g}(X)$, we obtain

$$\mathbf{r}(X) = \mathbf{a}(X)\mathbf{g}(X) + \mathbf{s}(X). \tag{4.19}$$

The remainder $\mathbf{s}(X)$ is a polynomial of degree $n - k - 1$ or less. The $n - k$ coefficients of $\mathbf{s}(X)$ form the syndrome $\mathbf{s}$. It is clear from Theorem 4.4 that $\mathbf{s}(X)$ is identical to zero if and only if the received polynomial $\mathbf{r}(X)$ is a code polynomial. Hereafter, we will simply call $\mathbf{s}(X)$ the syndrome. The syndrome computation can be accomplished with a division circuit as shown in Fig. 4.5, which is identical to the $(n - k)$-stage encoding circuit except that the received polynomial $\mathbf{r}(X)$ is shifted into the register from the left end. The received polynomial $\mathbf{r}(X)$ is shifted into the register with all stages initially set to 0. As soon as the entire $\mathbf{r}(X)$ has been shifted into the register, the contents in the register form the syndrome $\mathbf{s}(X)$.

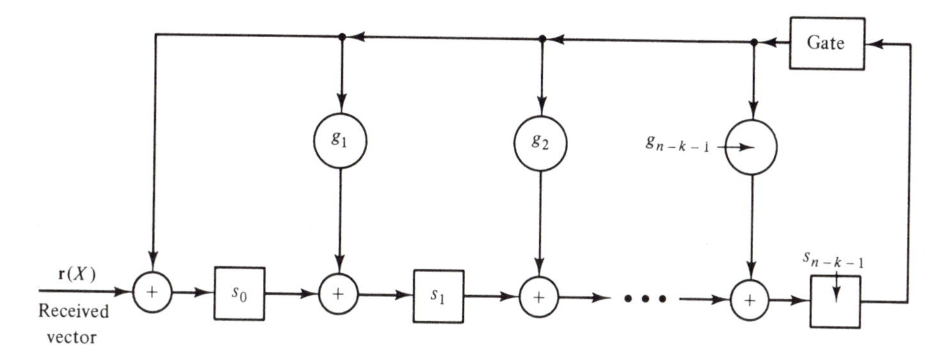

Figure 4.5 An $(n - k)$-stage syndrome circuit with input from the left end.

Because of the cyclic structure of the code, the syndrome $\mathbf{s}(X)$ has the following property.

Theorem 4.8. Let $\mathbf{s}(X)$ be the syndrome of a received polynomial $\mathbf{r}(X) = r_0 + r_1X + \cdots + r_{n-1}X^{n-1}$. Then the remainder $\mathbf{s}^{(1)}(X)$ resulting from dividing $X\mathbf{s}(X)$ by the generator polynomial $\mathbf{g}(X)$ is the syndrome of $\mathbf{r}^{(1)}(X)$, which is a cyclic shift of $\mathbf{r}(X)$.

Proof. It follows from (4.1) that $\mathbf{r}(X)$ and $\mathbf{r}^{(1)}(X)$ satisfy the following relationship:

$$X\mathbf{r}(X) = r_{n-1}(X^n + 1) + \mathbf{r}^{(1)}(X). \tag{4.20}$$

Rearranging (4.20), we have

$$\mathbf{r}^{(1)}(X) = r_{n-1}(X^n + 1) + X\mathbf{r}(X). \tag{4.21}$$

Dividing both sides of (4.21) by $\mathbf{g}(X)$ and using the fact that $X^n + 1 = \mathbf{g}(X)\mathbf{h}(X)$, we obtain

$$\mathbf{c}(X)\mathbf{g}(X) + \boldsymbol{\rho}(X) = r_{n-1}\mathbf{g}(X)\mathbf{h}(X) + X[\mathbf{a}(X)\mathbf{g}(X) + \mathbf{s}(X)], \tag{4.22}$$

where $\boldsymbol{\rho}(X)$ is the remainder resulting from dividing $\mathbf{r}^{(1)}(X)$ by $\mathbf{g}(X)$. Then $\boldsymbol{\rho}(X)$ is the syndrome of $\mathbf{r}^{(1)}(X)$.

Rearranging (4.22), we obtain the following relationship between $\mathbf{p}(X)$ and $X\mathbf{s}(X)$:

$$X\mathbf{s}(X) = [\mathbf{c}(X) + r_{n-1}\mathbf{h}(X) + X\mathbf{a}(X)]\mathbf{g}(X) + \mathbf{p}(X). \tag{4.23}$$

From (4.23) we see that $\mathbf{p}(X)$ is also the remainder resulting from dividing $X\mathbf{s}(X)$ by $\mathbf{g}(X)$. Therefore, $\mathbf{p}(X) = \mathbf{s}^{(1)}(X)$. This completes the proof. Q.E.D.

It follows from Theorem 4.8 that the remainder $\mathbf{s}^{(i)}(X)$ resulting from dividing $X^i\mathbf{s}(X)$ by the generator polynomial $\mathbf{g}(X)$ is the syndrome of $\mathbf{r}^{(i)}(X)$, which is the ith cyclic shift of $\mathbf{r}(X)$. This property is useful in decoding of cyclic codes. The syndrome $\mathbf{s}^{(1)}(X)$ of $\mathbf{r}^{(1)}(X)$ can be obtained by shifting (or clocking) the syndrome register once with $\mathbf{s}(X)$ as the initial contents and with the input gate disabled. This is due to the fact that shifting the syndrome register once with $\mathbf{s}(X)$ as the initial contents is equivalent to dividing $X\mathbf{s}(X)$ by $\mathbf{g}(X)$. Thus, after the shift, the register contains $\mathbf{s}^{(1)}(X)$. To obtain the syndrome $\mathbf{s}^{(i)}(X)$ of $\mathbf{r}^{(i)}(X)$, we simply shift the syndrome register i times with $\mathbf{s}(X)$ as the initial contents.

Example 4.7

A syndrome circuit for the (7, 4) cyclic code generated by $\mathbf{g}(X) = 1 + X + X^3$ is shown in Figure 4.6. Suppose that the received vector is $\mathbf{r} = (0\ 0\ 1\ 0\ 1\ 1\ 0)$. The syndrome of $\mathbf{r}$ is $\mathbf{s} = (1\ 0\ 1)$. As the received vector is shifted into the circuit, the contents in the register are given in Table 4.3. At the end of the seventh shift, the reg-

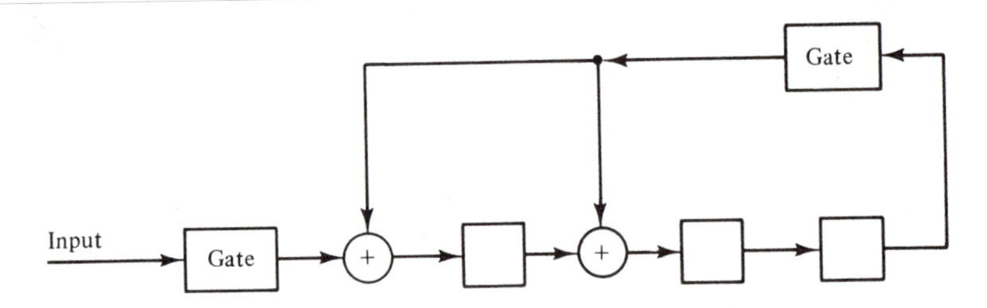

Figure 4.6 Syndrome circuit for the (7, 4) cyclic code generated by $\mathbf{g}(X) = 1 + X + X^3$.

TABLE 4.3 CONTENTS OF THE SYNDROME REGISTER SHOWN IN FIGURE 4.6 WITH $\mathbf{r} = (0\ 0\ 1\ 0\ 1\ 1\ 0)$ AS INPUT

Shift	Input	Register contents
		0 0 0 (initial state)
1	0	0 0 0
2	1	1 0 0
3	1	1 1 0
4	0	0 1 1
5	1	0 1 1
6	0	1 1 1
7	0	1 0 1 (syndrome $\mathbf{s}$)
8	—	1 0 0 (syndrome $\mathbf{s}^{(1)}$)
9	—	0 1 0 (syndrome $\mathbf{s}^{(2)}$)

ister contains the syndrome $\mathbf{s} = (1\ 0\ 1)$. If the register is shifted once more with the input gate disabled, the new contents will be $\mathbf{s}^{(1)} = (1\ 0\ 0)$, which is the syndrome of $\mathbf{r}^{(1)} = (0\ 0\ 0\ 1\ 0\ 1\ 1)$, a cyclic shift of $\mathbf{r}$.

We may shift the received vector $\mathbf{r}(X)$ into the syndrome register from the right end, as shown in Figure 4.7. However, after the entire $\mathbf{r}(X)$ has been shifted into the register, the contents in the register do not form the syndrome of $\mathbf{r}(X)$; rather, they form the syndrome $\mathbf{s}^{(n-k)}(X)$ of $\mathbf{r}^{(n-k)}(X)$, which is the $(n-k)$th cyclic shift of $\mathbf{r}(X)$.

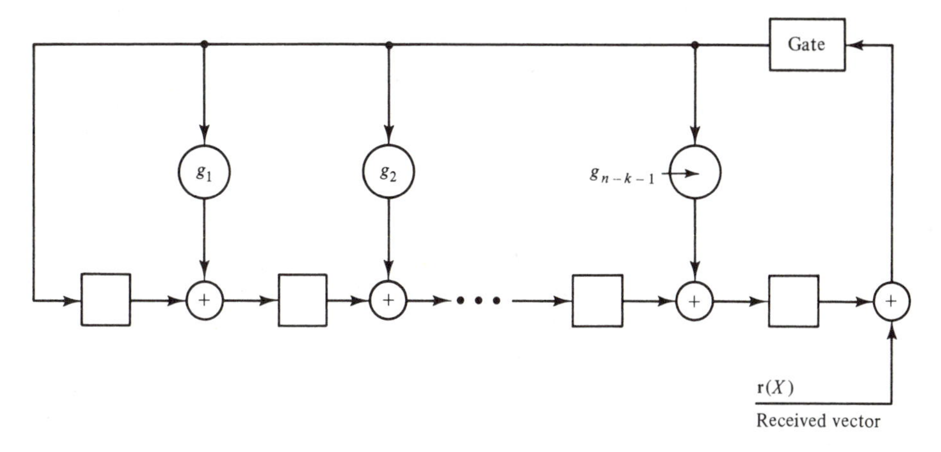

Figure 4.7 An $(n-k)$-stage syndrome circuit with input from the right end.

To show this, we notice that shifting $\mathbf{r}(X)$ from the right end is equivalent to premultiplying $\mathbf{r}(X)$ by X^{n-k}. When the entire $\mathbf{r}(X)$ has entered the register, the register contains the remainder $\boldsymbol{\rho}(X)$ resulting from dividing $X^{n-k}\mathbf{r}(X)$ by the generator polynomial $\mathbf{g}(X)$. Thus, we have

$$X^{n-k}\mathbf{r}(X) = \mathbf{a}(X)\mathbf{g}(X) + \boldsymbol{\rho}(X). \tag{4.24}$$

It follows from (4.1) that $\mathbf{r}(X)$ and $\mathbf{r}^{(n-k)}(X)$ satisfy the following relation:

$$X^{n-k}\mathbf{r}(X) = \mathbf{b}(X)(X^n + 1) + \mathbf{r}^{(n-k)}(X). \tag{4.25}$$

Combining (4.24) and (4.25) and using the fact that $X^n + 1 = \mathbf{g}(X)\mathbf{h}(X)$, we have

$$\mathbf{r}^{(n-k)}(X) = [\mathbf{b}(X)\mathbf{h}(X) + \mathbf{a}(X)]\mathbf{g}(X) + \boldsymbol{\rho}(X).$$

This says that, when $\mathbf{r}^{(n-k)}(X)$ is divided by $\mathbf{g}(X)$, $\boldsymbol{\rho}(X)$ is also the remainder. Therefore, $\boldsymbol{\rho}(X)$ is indeed the syndrome of $\mathbf{r}^{(n-k)}(X)$.

Let $\mathbf{v}(X)$ be the transmitted code word and let $\mathbf{e}(X) = e_0 + e_1X + \cdots + e_{n-1}X^{n-1}$ be the error pattern. Then the received polynomial is

$$\mathbf{r}(X) = \mathbf{v}(X) + \mathbf{e}(X). \tag{4.26}$$

Since $\mathbf{v}(X)$ is a multiple of the generator polynomial $\mathbf{g}(X)$, combining (4.19) and (4.26), we have the following relationship between the error pattern and the syndrome:

$$\mathbf{e}(X) = [\mathbf{a}(X) + \mathbf{b}(X)]\mathbf{g}(X) + \mathbf{s}(X), \tag{4.27}$$

where $\mathbf{b}(X)\mathbf{g}(X) = \mathbf{v}(X)$. This shows that the syndrome is actually equal to the remainder resulting from dividing the error pattern by the generator polynomial. The syndrome can be computed from the received vector; however, the error pattern $\mathbf{e}(X)$ is unknown to the decoder. Therefore, the decoder has to estimate $\mathbf{e}(X)$ based on the syndrome $\mathbf{s}(X)$. If $\mathbf{e}(X)$ is a coset leader in the standard array and if table-lookup decoding is used, $\mathbf{e}(X)$ can be correctly determined from the syndrome.

From (4.27), we see that $\mathbf{s}(X)$ is identical to zero if and only if either the error pattern $\mathbf{e}(X) = \mathbf{0}$ or it is identical to a code vector. If $\mathbf{e}(X)$ is identical to a code vector, $\mathbf{e}(X)$ is an undetectable error pattern. Cyclic codes are very effective for detecting errors, random or burst. The error-detection circuit is simply a syndrome circuit with an OR gate with the syndrome digits as inputs. If the syndrome is not zero, the output of the OR gate is "1" and the presence of errors has been detected.

Now, we investigate the error-detecting capability of an (n, k) cyclic code. Suppose that the error pattern $\mathbf{e}(X)$ is a burst of length $n - k$ or less (i.e., errors are confined to $n - k$ or fewer consecutive positions). Then $\mathbf{e}(X)$ can be expressed in the following form:

$$\mathbf{e}(X) = X^j\mathbf{B}(X),$$

where $0 \leq j \leq n - 1$ and $\mathbf{B}(X)$ is a polynomial of degree $n - k - 1$ or less. Since the degree of $\mathbf{B}(X)$ is less than the degree of the generator polynomial $\mathbf{g}(X)$, $\mathbf{B}(X)$ is not divisible by $\mathbf{g}(X)$. Since $\mathbf{g}(X)$ is a factor of $X^n + 1$ and X is not a factor of $\mathbf{g}(X)$, $\mathbf{g}(X)$ and X^j must be relatively prime. Therefore, $\mathbf{e}(X) = X^j\mathbf{B}(X)$ is not divisible by $\mathbf{g}(X)$. As a result, the syndrome caused by $\mathbf{e}(X)$ is not equal to zero. This implies that an (n, k) cyclic code is capable of detecting any error burst of length $n - k$ or less. For a cyclic code, an error pattern with errors confined to i high-order positions and $l - i$ low-order positions is also regarded as a burst of length l or less. Such a burst is called *end-around* burst. For example,

$$e = (\overleftarrow{1\ 0\ 1}\ 0\ 0\ 0\ 0\ 0\ 0\ 0\ 0\ \overrightarrow{1\ 1\ 0\ 1})$$

is an end-around burst of length 7. An (n, k) cyclic code is also capable of detecting all the end-around error bursts of length $n - k$ or less (the proof of this is left as a problem). Summarizing the results above, we have the following property:

Theorem 4.9. An (n, k) cyclic code is capable of detecting any error burst of length $n - k$ or less, including the end-around bursts.

In fact, a large percentage of error bursts of length $n - k + 1$ or longer can be detected. Consider the bursts of length $n - k + 1$ starting from the ith digit position and ending at the $(i + n - k)$th digit position (i.e., errors are confined to digits $e_i, e_{i+1}, \ldots, e_{i+n-k}$ with $e_i = e_{i+n-k} = 1$). There are 2^{n-k-1} such bursts. Among these bursts, the only one that cannot be detected is

$$\mathbf{e}(X) = X^i\mathbf{g}(X).$$

Therefore, the fraction of undetectable bursts of length $n - k + 1$ starting from the ith digit position is $2^{-(n-k-1)}$. This fraction applies to bursts of length $n - k + 1$ starting from any digit position (including the end-around case). Therefore, we have the following result:

Theorem 4.10. The fraction of undetectable bursts of length $n - k + 1$ is $2^{-(n-k-1)}$.

For $l > n - k + 1$, there are 2^{l-2} bursts of length l starting from the ith digit position and ending at the $(i + l - 1)$th digit position. Among these bursts, the undetectable ones must be of the following form:

$$\mathbf{e}(X) = X^i\mathbf{a}(X)\mathbf{g}(X),$$

where $\mathbf{a}(X) = a_0 + a_1X + \cdots + a_{l-(n-k)-1}X^{l-(n-k)-1}$, with $a_0 = a_{l-(n-k)-1} = 1$. The number of such bursts is $2^{l-(n-k)-2}$. Therefore, the fraction of undetectable bursts of length l starting from the ith digit position is $2^{-(n-k)}$. Again this fraction applies to bursts of length l starting from any digit position (including the end-around case). This leads to the following conclusion:

Theorem 4.11. For $l > n - k + 1$, the fraction of undetectable error bursts of length l is $2^{-(n-k)}$.

The analysis above shows that cyclic codes are very effective for burst-error detection.

Example 4.8

The (7, 4) cyclic code generated by $\mathbf{g}(X) = 1 + X + X^3$ has minimum distance 3. It is capable of detecting any combination of two or fewer random errors or any burst of length 3 or less. It also detects many bursts of length greater than 3.

4.5 DECODING OF CYCLIC CODES

Decoding of cyclic codes consists of the same three steps as for decoding linear codes: syndrome computation, association of the syndrome to an error pattern, and error correction. We have shown in Section 4.4 that syndrome computation for cyclic codes can be accomplished with a division circuit whose complexity is linearly proportional to the number of parity-check digits (i.e., $n - k$). The error correction step is simply adding (modulo-2) the error pattern to the received vector. This can be achieved with a single EXCLUSIVE-OR gate if correction is carried out in serial manner (i.e., one digit at a time); n EXCLUSIVE-OR gates are required if correction is carried out in parallel manner, as shown in Figure 3.8. The association of the syndrome to an error pattern can be completely specified by a decoding table. A straightforward approach to the design of a decoding circuit is via a combinational logic circuit that implements the table-lookup procedure. However, the limit to this approach is that the complexity of the decoding circuit tends to grow exponentially with the code length and the number of errors that we intend to correct. Cyclic codes have considerable algebraic and geometric properties. If these properties are properly used, simplification in the decoding circuit is possible.

The cyclic structure of a cyclic code allows us to decode a received vector $\mathbf{r}(X) = r_0 + r_1X + r_2X^2 + \cdots + r_{n-1}X^{n-1}$ in serial manner. The received digits are decoded one at a time and each digit is decoded with the same circuitry. As soon as the syndrome has been computed, the decoding circuit checks whether the syndrome

$\mathbf{s}(X)$ corresponds to a correctable error pattern $\mathbf{e}(X) = e_0 + e_1X + \cdots + e_{n-1}X^{n-1}$ with an error at the highest-order position X^{n-1} (i.e., $e_{n-1} = 1$). If $\mathbf{s}(X)$ does not correspond to an error pattern with $e_{n-1} = 1$, the received polynomial (stored in a buffer register) and the syndrome register are cyclically shifted once simultaneously. By doing this, we obtain $\mathbf{r}^{(1)}(X) = r_{n-1} + r_0X + \cdots + r_{n-2}X^{n-1}$ and the new contents in the syndrome register form the syndrome $\mathbf{s}^{(1)}(X)$ of $\mathbf{r}^{(1)}(X)$. Now, the second digit r_{n-2} of $\mathbf{r}(X)$ becomes the first digit of $\mathbf{r}^{(1)}(X)$. The same decoding circuit will check whether $\mathbf{s}^{(1)}(X)$ corresponds to an error pattern with an error at location X^{n-1}.

If the syndrome $\mathbf{s}(X)$ of $\mathbf{r}(X)$ does correspond to an error pattern with an error at the location X^{n-1}(i.e., $e_{n-1} = 1$), the first received digit r_{n-1} is an erroneous digit and it must be corrected. The correction is carried out by taking the sum $r_{n-1} \oplus e_{n-1}$. This correction results in a modified received polynomial, denoted by $\mathbf{r}_1(X) = r_0 + r_1X + \cdots + r_{n-2}X^{n-2} + (r_{n-1} \oplus e_{n-1})X^{n-1}$. The effect of the error digit e_{n-1} on the syndrome is then removed from the syndrome $\mathbf{s}(X)$. This can be achieved by adding the syndrome of $\mathbf{e}'(X) = X^{n-1}$ to $\mathbf{s}(X)$. This sum is the syndrome of the modified received polynomial $\mathbf{r}_1(X)$. Now cyclically shift $\mathbf{r}_1(X)$ and the syndrome register once simultaneously. This shift results in a received polynomial $\mathbf{r}_1^{(1)}(X) = (r_{n-1} \oplus e_{n-1}) + r_0X + \cdots + r_{n-2}X^{n-1}$. The syndrome $\mathbf{s}_1^{(1)}(X)$ of $\mathbf{r}_1^{(1)}(X)$ is the remainder resulting from dividing $X[\mathbf{s}(X) + X^{n-1}]$ by the generator polynomial $\mathbf{g}(X)$. Since the remainders resulting from dividing $X\mathbf{s}(X)$ and X^n by $\mathbf{g}(X)$ are $\mathbf{s}^{(1)}(X)$ and 1, respectively, we have

$$\mathbf{s}_1^{(1)}(X) = \mathbf{s}^{(1)}(X) + 1.$$

Therefore, if 1 is added to the left end of the syndrome register while it is shifted, we obtain $\mathbf{s}_1^{(1)}(X)$. The decoding circuitry proceeds to decode the received digit r_{n-2}. The decoding of r_{n-2} and the other received digits is identical to the decoding of r_{n-1}. Whenever an error is detected and corrected, its effect on the syndrome is removed. The decoding stops after a total of n shifts. If $\mathbf{e}(X)$ is a correctable error pattern, the contents of the syndrome register should be zero at the end of the decoding operation, and the received vector $\mathbf{r}(X)$ has been correctly decoded. If the syndrome register does not contain all 0's at the end of the decoding process, an uncorrectable error pattern has been detected.

A general decoder for an (n, k) cyclic code is shown in Figure 4.8. It consists of three major parts: (1) a syndrome register, (2) an error-pattern detector, and (3) a buffer register to hold the received vector. The received polynomial is shifted into the syndrome register from the left end. To remove the effect of an error digit on the syndrome, we simply feed the error digit into the shift register from the left end through an EXCLUSIVE-OR gate. The decoding operation is described as follows:

Step 1. The syndrome is formed by shifting the entire received vector into the syndrome register. At the same time the received vector is stored into the buffer register.

Step 2. The syndrome is read into the detector and is tested for the corresponding error pattern. The detector is a combinational logic circuit which is designed in such a way that its output is 1 if and only if the syndrome in the syndrome register corresponds to a correctable error pattern with an error at the highest-order position X^{n-1}. That is, if a "1" appears at the output of the

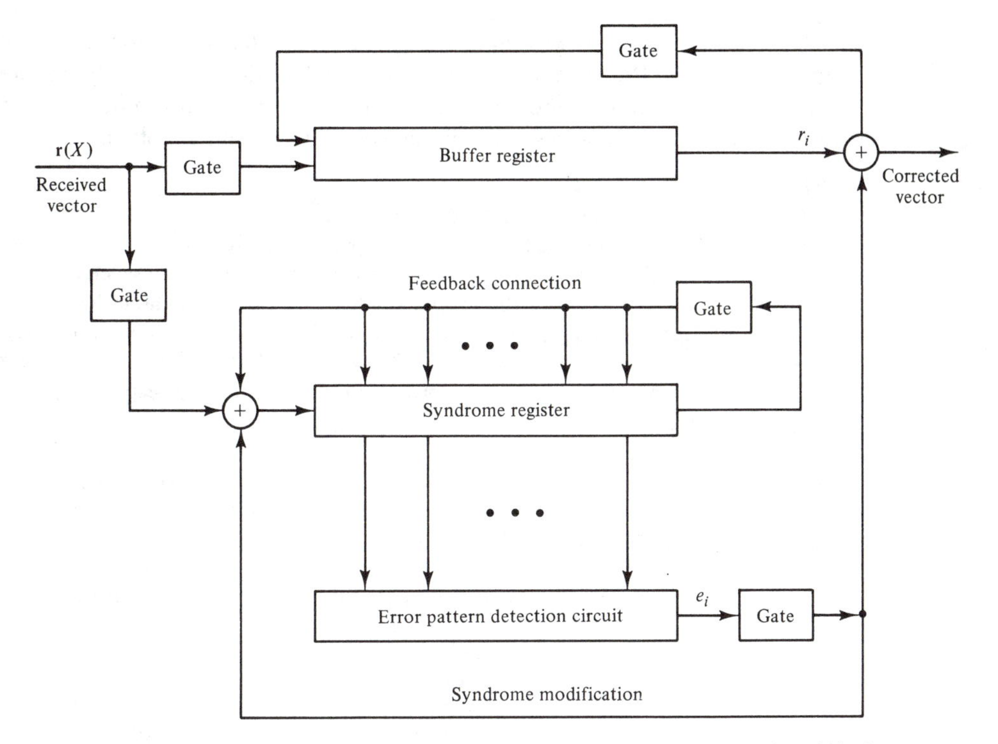

Figure 4.8 General cyclic code decoder with received polynomial $\mathbf{r}(X)$ shifted into the syndrome register from the left end.

detector, the received symbol in the rightmost stage of the buffer register is assumed to be erroneous and must be corrected; if a "0" appears at the output of the detector, the received symbol at the rightmost stage of the buffer register is assumed to be correct and no correction is necessary. Thus, the output of the detector is the estimated error value for the symbol to come out of the buffer.

Step 3. The first received symbol is read out of the buffer. At the same time, the syndrome register is shifted once. If the first received symbol is detected to be an erroneous symbol, it is then corrected by the output of the detector. The output of the detector is also fed back to the syndrome register to modify the syndrome (i.e., to remove the error effect from the syndrome). This results in a new syndrome, which corresponds to the altered received vector shifted one place to the right.

Step 4. The new syndrome formed in step 3 is used to detect whether or not the second received symbol (now at the rightmost stage of the buffer register) is an erroneous symbol. The decoder repeats steps 2 and 3. The second received symbol is corrected in exactly the same manner as the first received symbol was corrected.

Step 5. The decoder decodes the received vector symbol by symbol in the manner outlined above until the entire received vector is read out of the buffer register.

The decoder above is known as Meggitt decoder [8], which applies in principle to any cyclic code. But whether or not it is practical depends entirely on its error-pattern detection circuit. There are cases in which the error-pattern detection circuits are simple. Several of these cases are discussed in subsequent chapters.

Example 4.9

Consider the decoding of the (7, 4) cyclic code generated by $\mathbf{g}(X) = 1 + X + X^3$. This code has minimum distance 3 and is capable of correcting any single error over a block of seven digits. There are seven single-error patterns. These seven error patterns and the all-zero vector form all the coset leaders of the decoding table. Thus, they form all the correctable error patterns. Suppose that the received polynomial $\mathbf{r}(X) = r_0 + r_1X + r_2X^2 + r_3X^3 + r_4X^4 + r_5X^5 + r_6X^6$ is shifted into the syndrome register from the left end. The seven single-error patterns and their corresponding syndromes are listed in Table 4.4.

TABLE 4.4 ERROR PATTERNS AND THEIR SYNDROMES WITH THE RECEIVED POLYNOMIAL $\mathbf{r}(X)$ SHIFTED INTO THE SYNDROME REGISTER FROM THE LEFT END

Error pattern $\mathbf{e}(X)$	Syndrome $\mathbf{s}(X)$	Syndrome vector (s_0, s_1, s_2)
$\mathbf{e}_6(X) = X^6$	$\mathbf{s}(X) = 1 + X^2$	(1 0 1)
$\mathbf{e}_5(X) = X^5$	$\mathbf{s}(X) = 1 + X + X^2$	(1 1 1)
$\mathbf{e}_4(X) = X^4$	$\mathbf{s}(X) = X + X^2$	(0 1 1)
$\mathbf{e}_3(X) = X^3$	$\mathbf{s}(X) = 1 + X$	(1 1 0)
$\mathbf{e}_2(X) = X^2$	$\mathbf{s}(X) = X^2$	(0 0 1)
$\mathbf{e}_1(X) = X^1$	$\mathbf{s}(X) = X$	(0 1 0)
$\mathbf{e}_0(X) = X^0$	$\mathbf{s}(X) = 1$	(1 0 0)

We see that $\mathbf{e}_6(X) = X^6$ is the only error pattern with an error at location X^6. When this error pattern occurs, the syndrome in the syndrome register will be (1 0 1) after the entire received polynomial $\mathbf{r}(X)$ has entered the syndrome register. The detection of this syndrome indicates that r_6 is an erroneous digit and must be corrected. Suppose that the single error occurs at location X^i [i.e., $\mathbf{e}_i(X) = X^i$]. After the entire received polynomial has been shifted into the syndrome register, the syndrome in the register is not (1 0 1). However, after another $6 - i$ shifts, the contents in the syndrome register will be (1 0 1) and the next received digit to come out of the buffer register is the erroneous digit. Therefore, only the syndrome (1 0 1) needs to be detected. This can be accomplished with a single three-input AND gate. The complete decoding circuit is shown in Figure 4.9. Figure 4.10 illustrates the decoding process. Suppose that the code vector $\mathbf{v} = (1\ 0\ 0\ 1\ 0\ 1\ 1)$ [or $\mathbf{v}(X) = 1 + X^3 + X^5 + X^6$] is transmitted and $\mathbf{r} = (1\ 0\ 1\ 1\ 0\ 1\ 1)$ [or $\mathbf{r}(X) = 1 + X^2 + X^3 + X^5 + X^6$] is received. A single error occurs at location X^2. When the entire received polynomial has been shifted into the syndrome and buffer registers, the syndrome register contains (0 0 1). In Figure 4.10, the contents in the syndrome register and the contents in the buffer register are recorded after each shift. Also, there is a pointer to indicate the error location after each shift. We see that, after four more shifts, the contents in the syndrome register are (1 0 1) and the erroneous digit r_2 is the next digit to come out from the buffer register.

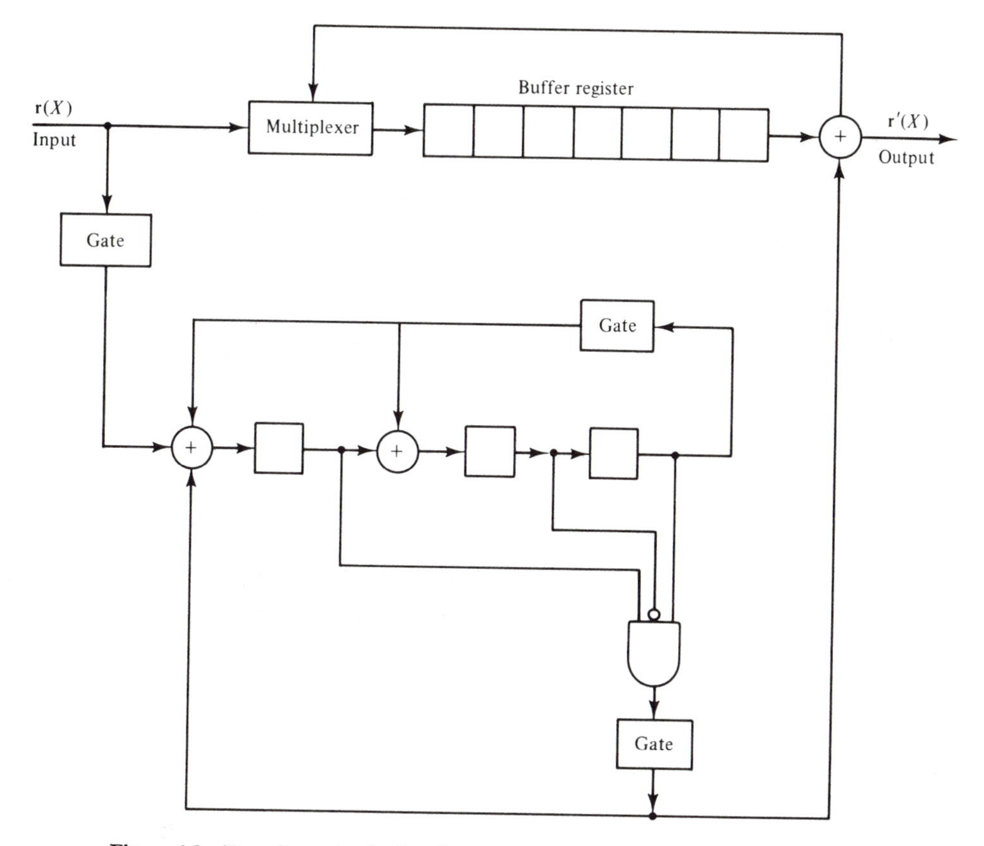

Figure 4.9 Decoding circuit for the (7, 4) cyclic code generated by $\mathbf{g}(X) = 1 + X + X^3$.

The (7, 4) cyclic code considered in Example 4.9 is actually the same code considered in Example 3.9. Comparing the decoding circuit shown in Figure 3. 9 and the decoding circuit shown in Figure 4.9, we see that the circuit shown in Figure 4.9 is simpler than the circuit shown in Figure 3.9. Thus, the cyclic structure does simplify the decoding circuit. However, the circuit shown in Figure 4.9 takes a longer time to decode a received vector because the decoding is carried out in serial manner. In general, speed and simplicity cannot be achieved at the same time, and a trade-off between them must be made.

The Meggitt decoder described above decodes a received polynomial $\mathbf{r}(X) = r_0 + r_1X + \cdots + r_{n-1}X^{n-1}$ from the highest-order received digit r_{n-1} to the lowest-order received digit r_0. After decoding the received digit r_i, both the buffer and syndrome registers are shifted once to the right. The next received digit to be decoded is r_{i-1}. It is possible to implement a Meggitt decoder to decode a received polynomial in the reverse order (i.e., to decode a received polynomial from the lowest-order received digit r_0 to the highest-order received digit r_{n-1}). After decoding the received digit r_i, both the buffer and syndrome registers are shifted once to the left. The next received digit to be decoded is r_{i+1}. The details of this decoding of a received polynomial in the reverse order are left as an exercise.

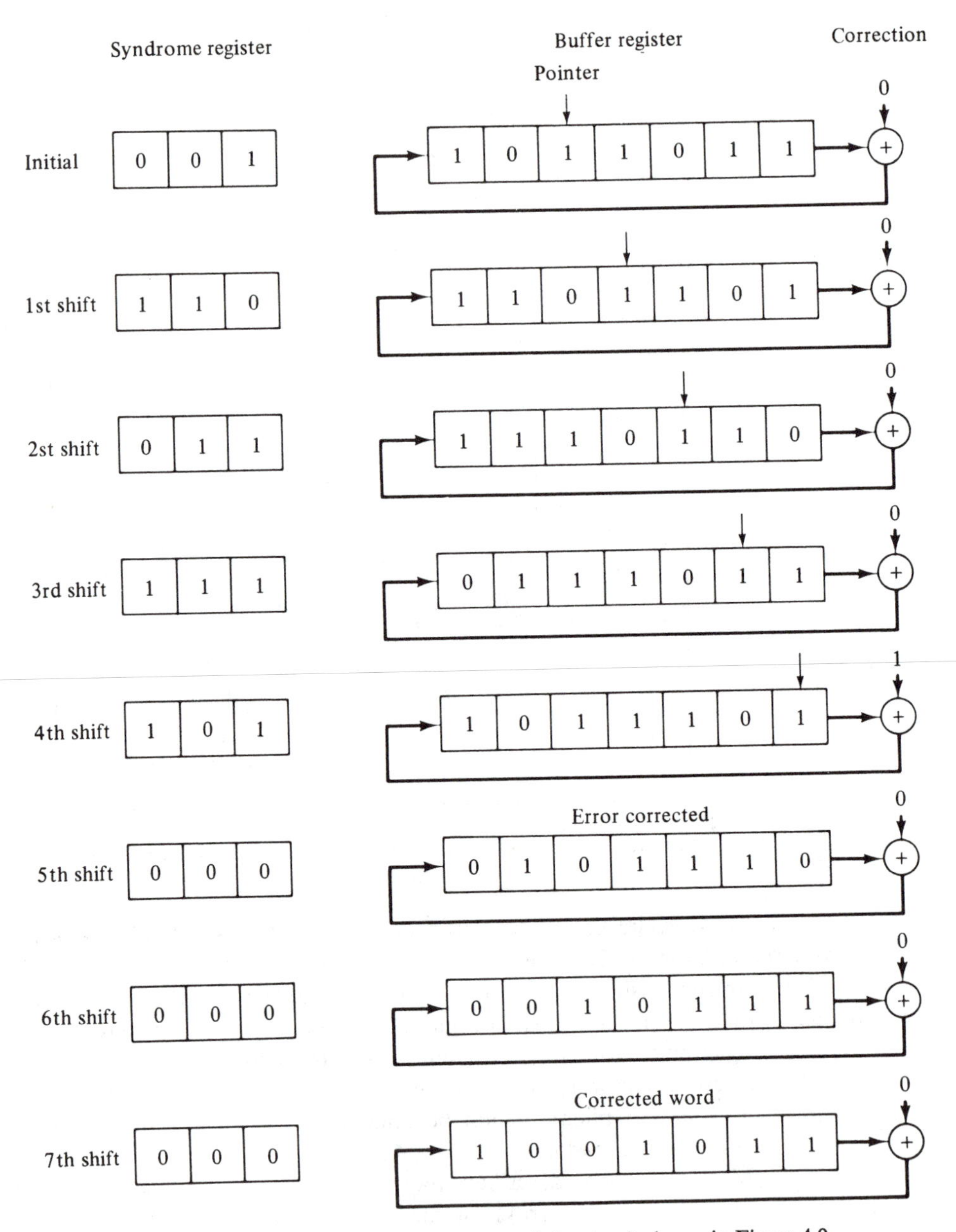

Figure 4.10 Error-correction process of the circuit shown in Figure 4.9.

To decode a cyclic code, the received polynomial $\mathbf{r}(X)$ may be shifted into the syndrome register from the right end for computing the syndrome. When $\mathbf{r}(X)$ has been shifted into the syndrome register, the register contains $\mathbf{s}^{(n-k)}(X)$, which is the syndrome of $\mathbf{r}^{(n-k)}(X)$, the $(n-k)$th cyclic shift of $\mathbf{r}(X)$. If $\mathbf{s}^{(n-k)}(X)$ corresponds to an error pattern $\mathbf{e}(X)$ with $e_{n-1} = 1$, the highest-order digit r_{n-1} of $\mathbf{r}(X)$ is erroneous and must be corrected. In $\mathbf{r}^{(n-k)}(X)$, the digit r_{n-1} is at the location X^{n-k-1}. When

r_{n-1} is corrected, the error effect must be removed from $\mathbf{s}^{(n-k)}(X)$. The new syndrome, denoted $\mathbf{s}_1^{(n-k)}(X)$, is the sum of $\mathbf{s}^{(n-k)}(X)$ and the remainder $\boldsymbol{\rho}(X)$ resulting from dividing X^{n-k-1} by the generator polynomial $\mathbf{g}(X)$. Since the degree of X^{n-k-1} is less than the degree of $\mathbf{g}(X)$,

$$\boldsymbol{\rho}(X) = X^{n-k-1}.$$

Therefore

$$\mathbf{s}_1^{(n-k)}(X) = \mathbf{s}^{(n-k)}(X) + X^{n-k-1}.$$

This indicates that the effect of an error at the location X^{n-1} on the syndrome can be removed by feeding the error digit into the syndrome register from the right end through an EXCLUSIVE-OR gate as shown in Figure 4.11. The decoding process of the decoder shown in Figure 4.11 is identical to the decoding process of the decoder shown in Figure 4.8.

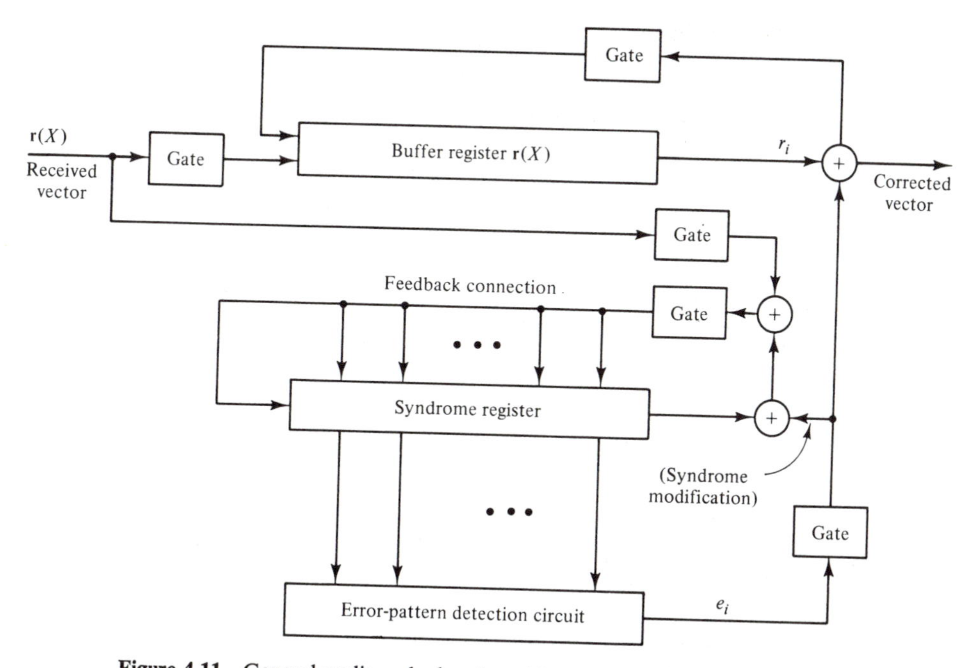

Figure 4.11 General cyclic code decoder with received polynomial $\mathbf{r}(X)$ shifted into the syndrome register from the right end.

Example 4.10

Again, we consider the decoding of the (7, 4) cyclic code generated by $\mathbf{g}(X) = 1 + X + X^3$. Suppose that the received polynomial $\mathbf{r}(X)$ is shifted into the syndrome register from the right end. The seven single-error patterns and their corresponding syndromes are listed in Table 4.5.

We see that only when $\mathbf{e}(X) = X^6$ occurs, the syndrome is (0 0 1) after the entire received polynomial $\mathbf{r}(X)$ has been shifted into the syndrome register. If the single error occurs at the location X^i with $i \neq 6$, the syndrome in the register will not be (0 0 1) after the entire received polynomial $\mathbf{r}(X)$ has been shifted into the

TABLE 4.5 ERROR PATTERNS AND THEIR SYNDROMES WITH THE RECEIVED POLYNOMIAL $\mathbf{r}(X)$ SHIFTED INTO THE SYNDROME REGISTER FROM THE RIGHT END

Error pattern $\mathbf{e}(X)$	Syndrome $\mathbf{s}^{(3)}(X)$	Syndrome vector (s_0, s_1, s_2)
$\mathbf{e}(X) = X^6$	$\mathbf{s}^{(3)}(X) = X^2$	(0 0 1)
$\mathbf{e}(X) = X^5$	$\mathbf{s}^{(3)}(X) = X$	(0 1 0)
$\mathbf{e}(X) = X^4$	$\mathbf{s}^{(3)}(X) = 1$	(1 0 0)
$\mathbf{e}(X) = X^3$	$\mathbf{s}^{(3)}(X) = 1 + X^2$	(1 0 1)
$\mathbf{e}(X) = X^2$	$\mathbf{s}^{(3)}(X) = 1 + X + X^2$	(1 1 1)
$\mathbf{e}(X) = X$	$\mathbf{s}^{(3)}(X) = X + X^2$	(0 1 1)
$\mathbf{e}(X) = X^0$	$\mathbf{s}^{(3)}(X) = 1 + X$	(1 1 0)

syndrome register. However, after another $6 - i$ shifts, the syndrome register will contain (0 0 1). Based on this fact, we obtain another decoding circuit for the (7, 4) cyclic code generated by $\mathbf{g}(X) = 1 + X + X^3$, as shown in Figure 4.12. We see that the circuit shown in Figure 4.9 and the circuit shown in Figure 4.12 have the same complexity.

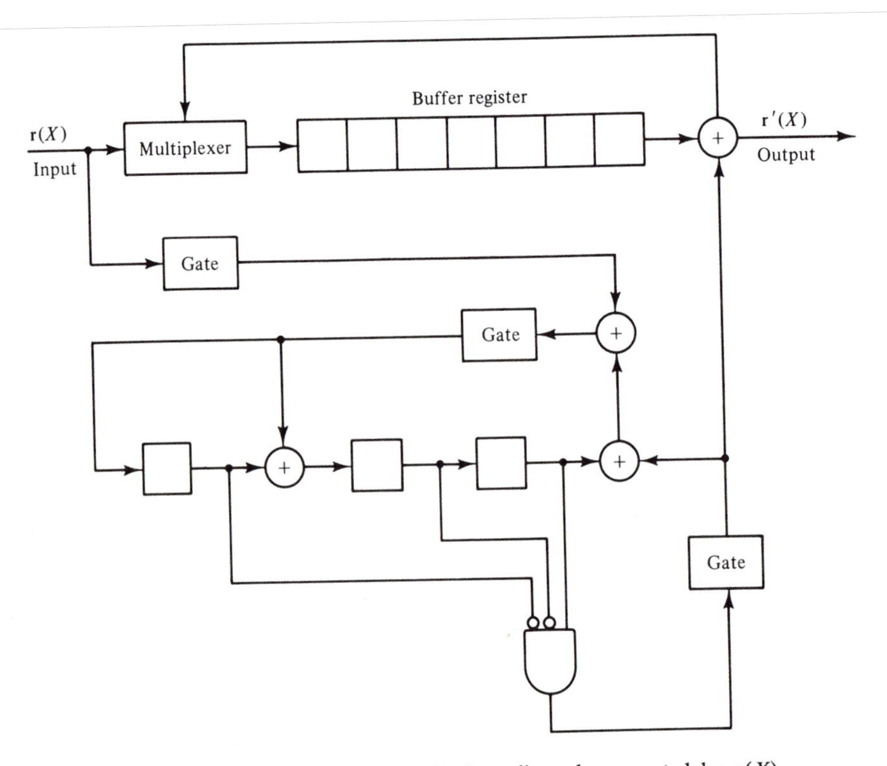

Figure 4.12 Decoding circuit for the (7, 4) cyclic code generated by $\mathbf{g}(X) = 1 + X + X^3$.

4.6 CYCLIC HAMMING CODES

The Hamming codes presented in Section 3.7 can be put in cyclic form. A cyclic Hamming code of length $2^m - 1$ with $m \geq 3$ is generated by a primitive polynomial $\mathbf{p}(X)$ of degree m.

In the following we show that the cyclic code defined above is indeed a Hamming code. For this purpose, we examine its parity-check matrix in systematic form. The method presented in Section 4.2 is used to form the parity-check matrix. Dividing X^{m+i} by the generator polynomial $\mathbf{p}(X)$ for $0 \leq i < 2^m - m - 1$, we obtain

$$X^{m+i} = \mathbf{a}_i(X)\mathbf{p}(X) + \mathbf{b}_i(X), \tag{4.28}$$

where the remainder $\mathbf{b}_i(X)$ is of the form

$$\mathbf{b}_i(X) = b_{i0} + b_{i1}X + \cdots + b_{i,m-1}X^{m-1}.$$

Since X is not a factor of the primitive polynomial $\mathbf{p}(X)$, X^{m+i} and $\mathbf{p}(X)$ must be relatively prime. As a result, $\mathbf{b}_i(X) \neq 0$. Moreover, $\mathbf{b}_i(X)$ consists of at least two terms. Suppose that $\mathbf{b}_i(X)$ has only one term, say X^j with $0 \leq j < m$. It follows from (4.28) that

$$X^{m+i} = \mathbf{a}_i(X)\mathbf{p}(X) + X^j.$$

Rearranging the equation above, we have

$$X^j(X^{m+i-j} + 1) = \mathbf{a}_i(X)\mathbf{p}(X).$$

Since X^j and $\mathbf{p}(X)$ are relatively prime, the equation above implies that $\mathbf{p}(X)$ divides $X^{m+i-j} + 1$. However, this is impossible since $m + i - j < 2^m - 1$ and $\mathbf{p}(X)$ is a primitive polynomial of degree m. [Recall that the smallest positive integer n such that $\mathbf{p}(X)$ divides $X^n + 1$ is $2^m - 1$.] Therefore, for $0 \leq i < 2^m - m - 1$, the remainder $\mathbf{b}_i(X)$ contains at least two terms. Next we show that, for $i \neq j$, $\mathbf{b}_i(X) \neq \mathbf{b}_j(X)$. From (4.28) we obtain

$$\mathbf{b}_i(X) + X^{m+i} = \mathbf{a}_i(X)\mathbf{p}(X),$$

$$\mathbf{b}_j(X) + X^{m+j} = \mathbf{a}_j(X)\mathbf{p}(X).$$

Suppose that $\mathbf{b}_i(X) = \mathbf{b}_j(X)$. Assuming that $i < j$ and combining the two equations above, the following relation is obtained:

$$X^{m+i}(X^{j-i} + 1) = [\mathbf{a}_i(X) + \mathbf{a}_j(X)]\mathbf{p}(X).$$

This equation implies that $\mathbf{p}(X)$ divides $X^{j-i} + 1$. This is impossible since $i \neq j$ and $j - i < 2^m - 1$. Therefore, $\mathbf{b}_i(X) \neq \mathbf{b}_j(X)$.

Let $\mathbf{H} = [\mathbf{I}_m \quad \mathbf{Q}]$ be the parity-check matrix of the cyclic code generated by $\mathbf{p}(X)$ where $\mathbf{I}_m$ is an $m \times m$ identity matrix and $\mathbf{Q}$ is an $m \times (2^m - m - 1)$ matrix. Let $\mathbf{b}_i = (b_{i0}, b_{i1}, \ldots, b_{i,m-1})$ be the m-tuple corresponding to $\mathbf{b}_i(X)$. It follows from (4.17) that the matrix $\mathbf{Q}$ has the $2^m - m - 1$ $\mathbf{b}_i$'s with $0 \leq i < 2^m - m - 1$ as all its columns. It follows from the analysis above that no two columns of $\mathbf{Q}$ are alike and each column has at least two 1's. Therefore, the matrix $\mathbf{H}$ is indeed a parity-check matrix of a Hamming code, and $\mathbf{p}(X)$ generates this code.

The polynomial $\mathbf{p}(X) = 1 + X + X^3$ is a primitive polynomial. Therefore, the (7, 4) cyclic code generated by $\mathbf{p}(X) = 1 + X + X^3$ is a Hamming code. A list of primitive polynomials with degree ≥ 3 is given in Table 2.7.

Decoding of cyclic Hamming codes can be easily implemented. To devise the decoding circuit, all we need to know is how to decode the first received digit. All the other received digits will be decoded in the same manner and with the same circuitry. Suppose that a single error has occurred at the highest-order position, X^{2^m-2}, of the received vector $\mathbf{r}(X)$ [i.e., the error polynomial is $\mathbf{e}(X) = X^{2^m-2}$]. Suppose that $\mathbf{r}(X)$ is shifted into the syndrome register from the right end. After the entire $\mathbf{r}(X)$ has entered the register, the syndrome in the register is equal to the remainder resulting from dividing $X^m \cdot X^{2^m-2}$ (the error polynomial preshifted m times) by the generator polynomial $\mathbf{p}(X)$. Since $\mathbf{p}(X)$ divides $X^{2^m-1} + 1$, the syndrome is of the following form:

$$\mathbf{s}(X) = X^{m-1}.$$

Therefore, if a single error occurs at the highest-order location of $\mathbf{r}(X)$, the resultant syndrome is (0, 0, . . ., 0, 1). If a single error occurs at any other location of $\mathbf{r}(X)$, the resultant syndrome will be different from (0, 0, . . ., 0, 1). Based on this fact, only a single m-input AND gate is needed to detect the syndrome pattern (0, 0, . . ., 0, 1). The inputs to this AND gate are $s'_0, s'_1, \ldots, s'_{m-2}$ and s_{m-1}, where s_i is a syndrome digit and s'_i denotes its complement. A complete decoding circuit for a cyclic Hamming code is shown in Figure 4.13. The decoding operation is described in the following steps:

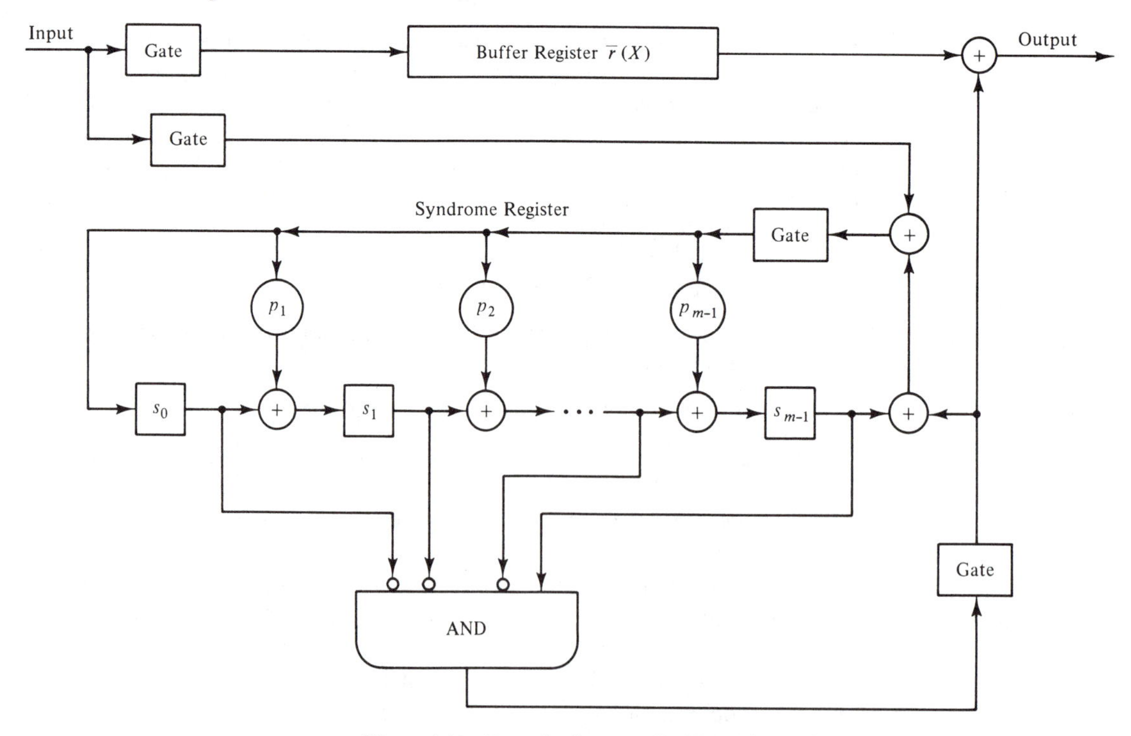

Figure 4.13 Decoder for a cyclic Hamming code.

Step 1. The syndrome is obtained by shifting the entire received vector into the syndrome register. At the same time, the received vector is stored into the buffer register. If the syndrome is zero, the decoder assumes that no error has occurred, and no correction is necessary. If the syndrome is not zero, the decoder assumes that a single error has occurred.

Step 2. The received word is read out of the buffer register digit by digit. As each digit is read out of the buffer register, the syndrome register is shifted cyclically once. As soon as the syndrome in the register is $(0, 0, 0, \ldots, 0, 1)$, the next digit to come out of the buffer is the erroneous digit, and the output of the m-input AND gate is 1.

Step 3. The erroneous digit is read out of the buffer register and is corrected by the output of the m-input AND gate. The correction is accomplished by an EXCLUSIVE-OR gate.

Step 4. The syndrome register is reset to zero after the entire received vector is read out of the buffer.

The cyclic Hamming code presented above can be modified to correct any single error and simultaneously to detect any combination of double errors. Let $\mathbf{g}(X) = (X + 1)\mathbf{p}(X)$, where $\mathbf{p}(X)$ is a primitive polynomial of degree m. Since both $X + 1$ and $\mathbf{p}(X)$ divide $X^{2^m-1} + 1$ and since they are relatively prime, $\mathbf{g}(X)$ must also divide $X^{2^m-1} + 1$. A single-error-correcting and double-error-detecting cyclic Hamming code of length $2^m - 1$ is generated by $\mathbf{g}(X) = (X + 1)\mathbf{p}(X)$. The code has $m + 1$ parity-check digits. We show next that the minimum distance of this code is 4.

For convenience, we denote the single-error-correcting cyclic Hamming code by C_1 and denote the cyclic code generated by $\mathbf{g}(X) = (X + 1)\mathbf{p}(X)$ by C_2. Since $\mathbf{p}(X)$ is a proper factor of $\mathbf{g}(X)$, C_1 contains C_2 as a proper subcode. In fact, C_2 consists of the even-weight code vectors of C_1 as all its vectors. This is due to the fact that any odd-weight code polynomial in C_1 does not have $X + 1$ as a factor. Therefore, an odd-weight code polynomial of C_1 is not divisible by $\mathbf{g}(X) = (X + 1)\mathbf{p}(X)$, and it is not a code polynomial of C_2. However, an even-weight code polynomial of C_1 has $X + 1$ as a factor. Therefore, it is divisible by $\mathbf{g}(X) = (X + 1)\mathbf{p}(X)$ and it is also a code polynomial in C_2. As a result, the minimum weight of C_2 is at least 4.

Next, we show that the minimum weight of C_2 is exactly 4. Let i, j, and k be three distinct nonnegative integers less than $2^m - 1$ such that $X^i + X^j + X^k$ is not divisible by $\mathbf{p}(X)$. Such integers do exist. For example, we first choose i and j. Dividing $X^i + X^j$ by $\mathbf{p}(X)$, we obtain

$$X^i + X^j = \mathbf{a}(X)\mathbf{p}(X) + \mathbf{b}(X),$$

where $\mathbf{b}(X)$ is the remainder with degree $m - 1$ or less. Since $X^i + X^j$ is not divisible by $\mathbf{p}(X)$, $\mathbf{b}(X) \neq 0$. Now, we choose an integer k such that, when X^k is divided by $\mathbf{p}(X)$, the remainder is not equal to $\mathbf{b}(X)$. Therefore, $X^i + X^j + X^k$ is not divisible by $\mathbf{p}(X)$. Dividing this polynomial by $\mathbf{p}(X)$, we have

$$X^i + X^j + X^k = \mathbf{c}(X)\mathbf{p}(X) + \mathbf{d}(X). \tag{4.29}$$

Next, we choose a nonnegative integer l less than $2^m - 1$ such that, when X^l is divided by $\mathbf{p}(X)$, the remainder is $\mathbf{d}(X)$, that is,

$$X^l = \mathbf{f}(X)\mathbf{p}(X) + \mathbf{d}(X). \tag{4.30}$$

The integer l cannot be equal to any of the three integers $i, j,$ and k. Suppose that $l = i$. From (4.29) and (4.30) we would obtain

$$X^j + X^k = [\mathbf{c}(X) + \mathbf{f}(X)]\mathbf{p}(X).$$

This implies that $\mathbf{p}(X)$ divides $X^{k-j} + 1$ (assuming that $j < k$), which is impossible since $k - j < 2^m - 1$ and $\mathbf{p}(X)$ is a primitive polynomial. Therefore, $l \neq i$. Similarly, we can show that $l \neq j$ and $l \neq k$. Using this fact and combining (4.29) and (4.30), we obtain

$$X^i + X^j + X^k + X^l = [\mathbf{c}(X) + \mathbf{f}(X)]\mathbf{p}(X).$$

Since $X + 1$ is a factor of $X^i + X^j + X^k + X^l$ and it is not a factor of $\mathbf{p}(X)$, $\mathbf{c}(X) + \mathbf{f}(X)$ must be divisible by $X + 1$. As a result, $X^i + X^j + X^k + X^l$ is divisible by $\mathbf{g}(X) = (X + 1)\mathbf{p}(X)$. Therefore, it is a code vector in the code generated by $\mathbf{g}(X)$. It has weight 4. This proves that the cyclic code C_2 generated by $\mathbf{g}(X) = (X + 1)\mathbf{p}(X)$ has minimum weight (or distance) 4. Hence it is capable of correcting any single error and simultaneously detecting any combination of double errors.

The decoding circuit for the single-error-correcting Hamming code shown in Figure 4.13 can be modified to decode the single-error-correcting and double-error-detecting Hamming code. Let $\mathbf{r}(X)$ be the received polynomial. Dividing $X^m\mathbf{r}(X)$ by $\mathbf{p}(X)$ and $\mathbf{r}(X)$ by $(X + 1)$, respectively, we have

$$X^m\mathbf{r}(X) = \mathbf{a}_1(X)\mathbf{p}(X) + \mathbf{s}_p(X)$$

and

$$\mathbf{r}(X) = \mathbf{a}_2(X)(X + 1) + \sigma,$$

where $\mathbf{s}_p(X)$ is of degree $m - 1$ or less and σ is either 0 or 1. If $\mathbf{s}_p(X) = 0$ and $\sigma = 0$, $\mathbf{r}(X)$ is divisible by $(1 + X)\mathbf{p}(X)$ and is a code polynomial; otherwise, $\mathbf{r}(X)$ is not a code polynomial. We define the syndrome of $\mathbf{r}(X)$ as

$$\mathbf{s}(X) = X\mathbf{s}_p(X) + \sigma.$$

If a single error occurs, $\mathbf{s}_p(X) \neq 0$ and $\sigma = 1$. However, when an error pattern with double errors occurs, we would have $\mathbf{s}_p(X) \neq 0$ and $\sigma = 0$. Based on these facts, we may implement a decoder for a single-error-correcting and double-error-detecting cyclic Hamming code as shown in Figure 4.14. The error-correction and error-detection operations are described as follows:

1. For $\sigma = 0$ and $\mathbf{s}_p(X) = 0$, the decoder assumes that there is no error in the received polynomial. No corrective action takes place.
2. For $\sigma = 1$ and $\mathbf{s}_p(X) \neq 0$, the decoder assumes that a single error has occurred and proceeds with the error-correction process as described in the decoding of a single-error-correcting cyclic Hamming code.
3. For $\sigma = 0$ and $\mathbf{s}_p(X) \neq 0$, the decoder assumes that double errors have occurred. The error alarm is turned on.

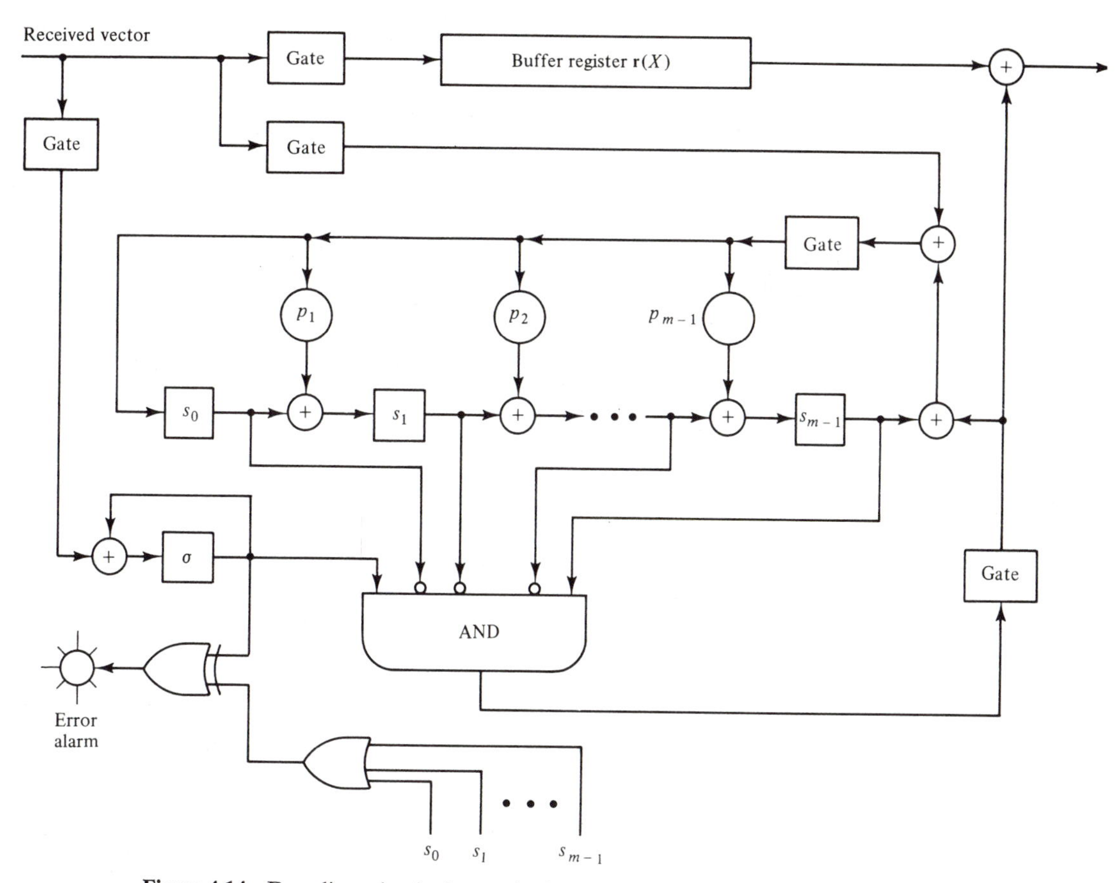

Figure 4.14 Decoding circuit for a single-error-correcting and double-error-detecting cyclic Hamming code.

4. For $\sigma = 1$ and $\mathbf{s}_p(X) = 0$, the error alarm is also turned on. This happens when an error pattern with odd number (greater than 1) of errors has occurred and the error pattern is divisible by $\mathbf{p}(X)$.

Since the distance 4 Hamming code C_2 of length $2^m - 1$ consists of the even-weight code vectors of the distance 3 Hamming code C_1 of length $2^m - 1$ as its code vectors, the weight enumerator $A_2(z)$ for C_2 can be determined from the weight enumerator $A_1(z)$ for C_1. $A_2(z)$ consists of only the even power terms of $A_1(z)$. Therefore,

$$A_2(z) = \tfrac{1}{2}[A_1(z) + A_1(-z)] \tag{4.31}$$

(see Problem 4.8). Since $A_1(z)$ is known and is given by (3.43), $A_2(z)$ can be determined from (3.43) and (4.31):

$$A_2(z) = \frac{1}{2(n+1)}\{(1+z)^n + (1-z)^n + 2n(1-z^2)^{(n-1)/2}\}, \tag{4.32}$$

where $n = 2^m - 1$. The dual of a distance 4 cyclic Hamming code is a $(2^m - 1, m + 1)$ cyclic code which has the following weight distribution:

$$B_0 = 1, \quad B_{2^{m-1}-1} = 2^m - 1, \quad B_{2^{m-1}} = 2^m - 1, \quad B_{2^m-1} = 1.$$

Therefore, the weight enumerator for the dual of a distance 4 cyclic Hamming code is

$$B_2(z) = 1 + (2^m - 1)z^{2^{m-1}-1} + (2^m - 1)z^{2^{m-1}} + z^{2^m-1}. \tag{4.33}$$

If a distance 4 cyclic Hamming code is used for error detection on a BSC, its probability of an undetected error, $P_u(E)$, can be computed either from (3.33) and (4.32) or from (3.36) and (4.33). Computing $P_u(E)$ from (3.36) and (4.33), we obtain the following expression:

$$P_u(E) = 2^{-(m+1)}\{1 + 2(2^m - 1)(1 - p)(1 - 2p)^{2^{m-1}-1} + (1 - 2p)^{2^m-1}\} - (1 - p)^{2^m-1}. \tag{4.34}$$

Again, from (4.34), we can show that the probability of an undetected error for the distance 4 cyclic Hamming codes satisfies the upper bound $2^{-(n-k)} = 2^{-(m+1)}$ (see Problem 4.21).

Distance 3 and distance 4 cyclic Hamming codes are often used in communication systems for error detection.

4.7 SHORTENED CYCLIC CODES

In system design, if a code of suitable natural length or suitable number of information digits cannot be found, it may be desirable to shorten a code to meet the requirements. A technique for shortening a cyclic code is presented in this section. This technique leads to simple implementation of the encoding and decoding for the shortened code.

Given an (n, k) cyclic code C consider the set of code vectors for which the l leading high-order information digits are identical to zero. There are 2^{k-l} such code vectors and they form a linear subcode of C. If the l zero information digits are deleted from each of these code vectors, we obtain a set of 2^{k-l} vectors of length $n - l$. These 2^{k-l} shortened vectors form an $(n - l, k - l)$ linear code. This code is called a *shortened* cyclic code (or *polynomial* code) and it is not cyclic. A shortened cyclic code has at least the same error-correcting capability as the code from which it is derived.

The encoding and decoding for a shortened cyclic code can be accomplished by the same circuits as those employed by the original cyclic code. This is so because the deleted l leading-zero information digits do not affect the parity-check and syndrome computations. However, in decoding the shortened cyclic code after the entire received vector has been shifted into the syndrome register, the syndrome register must be cyclically shifted l times to generate the proper syndrome for decoding the first received digit r_{n-l-1}. For large l, these extra l shifts of the syndrome register cause undesirable decoding delay; they can be eliminated by modifying either the connections of the syndrome register or the error-pattern detection circuit.

Let $\mathbf{r}(X) = r_0 + r_1X + \cdots + r_{n-l-1}X^{n-l-1}$ be the received polynomial. Suppose that $\mathbf{r}(X)$ is shifted into the syndrome register from the right end. If the decoding circuit for the original cyclic code is used for decoding the shortened code, the proper syndrome for decoding the received digit r_{n-l-1} is equal to the remainder resulting from dividing $X^{n-k+l}\mathbf{r}(X)$ by the generator polynomial $\mathbf{g}(X)$. Since shifting $\mathbf{r}(X)$ into the syndrome register from the right end is equivalent to premultiplying $\mathbf{r}(X)$ by X^{n-k}, the syndrome register must be cyclically shifted for another l times after the entire $\mathbf{r}(X)$ has been shifted into the register. Now, we want to show how these extra l shifts can be eliminated by modifying the connections of the syndrome register. Dividing $X^{n-k+l}\mathbf{r}(X)$ by $\mathbf{g}(X)$, we obtain

$$X^{n-k+l}\mathbf{r}(X) = \mathbf{a}_1(X)\mathbf{g}(X) + \mathbf{s}^{(n-k+l)}(X), \tag{4.35}$$

where $\mathbf{s}^{(n-k+l)}(X)$ is the remainder and is the syndrome for decoding the received digit r_{n-l-1}. Next, we divide X^{n-k+l} by $\mathbf{g}(X)$. Let $\boldsymbol{\rho}(X) = \rho_0 + \rho_1 X + \cdots + \rho_{n-k-1} X^{n-k-1}$ be the remainder resulting from this division. Then we have the following relation:

$$\boldsymbol{\rho}(X) = X^{n-k+l} + \mathbf{a}_2(X)\mathbf{g}(X). \tag{4.36}$$

Multiplying both sides of (4.36) by $\mathbf{r}(X)$ and using the equality of (4.35), we obtain the following relation between $\boldsymbol{\rho}(X)\mathbf{r}(X)$ and $\mathbf{s}^{(n-k+l)}(X)$:

$$\boldsymbol{\rho}(X)\mathbf{r}(X) = [\mathbf{a}_1(X) + \mathbf{a}_2(X)\mathbf{r}(X)]\mathbf{g}(X) + \mathbf{s}^{(n-k+l)}(X). \tag{4.37}$$

The equality above suggests that we can obtain the syndrome $\mathbf{s}^{(n-k+l)}(X)$ by multiplying $\mathbf{r}(X)$ by $\boldsymbol{\rho}(X)$ and dividing the product $\boldsymbol{\rho}(X)\mathbf{r}(X)$ by $\mathbf{g}(X)$. Computing $\mathbf{s}^{(n-k+l)}(X)$ this way, the extra l shifts of the syndrome register can be avoided. Simultaneously multiplying $\mathbf{r}(X)$ by $\boldsymbol{\rho}(X)$ and dividing $\boldsymbol{\rho}(X)\mathbf{r}(X)$ by $\mathbf{g}(X)$ can be accomplished by a circuit as shown in Figure 4.15. As soon as the received polynomial $\mathbf{r}(X)$ has been shifted into the register, the contents in the register form the syndrome $\mathbf{s}^{(n-k+l)}(X)$ and the first received digit is ready to be decoded.

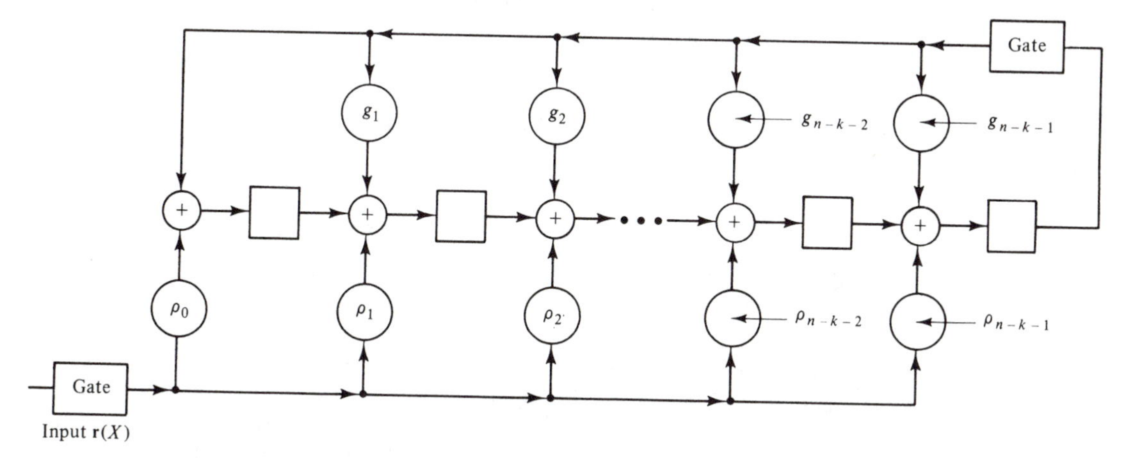

Figure 4.15 Circuit for multiplying $\mathbf{r}(X)$ by $\boldsymbol{\rho}(X) = \rho_0 + \rho_1 X + \cdots + \rho_{n-k-1}X^{n-k-1}$ and dividing $\boldsymbol{\rho}(X)\mathbf{r}(X)$ by $\mathbf{g}(X) = 1 + g_1 X + \cdots + X^{n-k}$.

Example 4.11

For $m = 5$, there exists a (31, 26) cyclic Hamming code generated by $\mathbf{g}(X) = 1 + X^2 + X^5$. Suppose that it is shortened by three digits. The resultant shortened code is a (28, 23) linear code. The decoding circuit for the (31, 26) cyclic code is shown in Figure

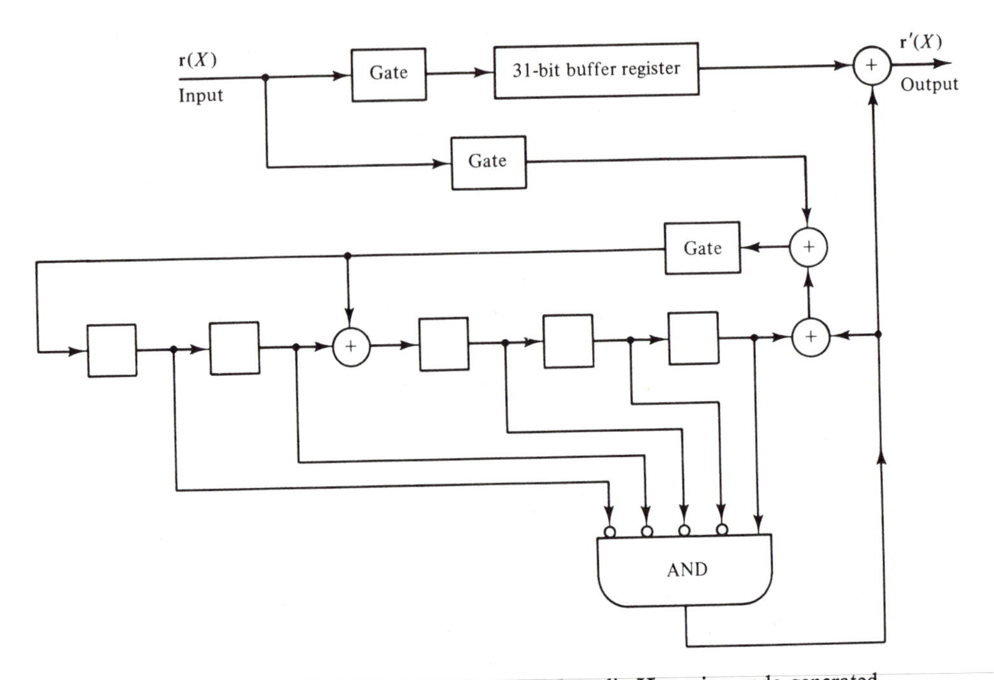

Figure 4.16 Decoding circuit for the (31, 26) cyclic Hamming code generated by $\mathbf{g}(X) = 1 + X^2 + X^5$.

4.16. This circuit can be used to decode the (28, 23) shortened code. To eliminate the extra shifts of the syndrome register, we need to modify the connections of the syndrome register. First, we need to determine the polynomial $\boldsymbol{\rho}(X)$. Dividing X^{n-k+3} by $\mathbf{g}(X) = 1 + X^2 + X^5$, we have

$$\begin{array}{r|l} & X^3 + 1 \\ \hline X^5 + X^2 + 1 & X^8 \\ & X^8 + X^5 + X^3 \\ \hline & X^5 + X^3 \\ & X^5 \qquad + X^2 + 1 \\ \hline & X^3 + X^2 + 1 \end{array}$$

and $\boldsymbol{\rho}(X) = 1 + X^2 + X^3$. The modified decoding circuit for the (28, 23) shortened code is shown in Figure 4.17.

The extra l shifts of the syndrome register for decoding the shortened cyclic code can also be avoided by modifying the error-pattern detection circuit of the decoder for the original cyclic code. The error-pattern detection circuit is redesigned to check whether the syndrome in the syndrome register corresponds to a correctable error pattern $\mathbf{e}(X)$ with an error at position X^{n-l-1} (i.e., $e_{n-l-1} = 1$). When the received digit r_{n-l-1} is corrected the effect of the error digit e_{n-l-1} on the syndrome should be removed. Suppose that the received vector is shifted into the syndrome register from the right end. Let $\boldsymbol{\rho}(X) = \rho_0 + \rho_1 X + \cdots + \rho_{n-k-1}X^{n-k-1}$ be the remainder resulting from dividing $X^{n-l-1} \cdot X^{n-k} = X^{2n-k-l-1}$ by the generator poly-

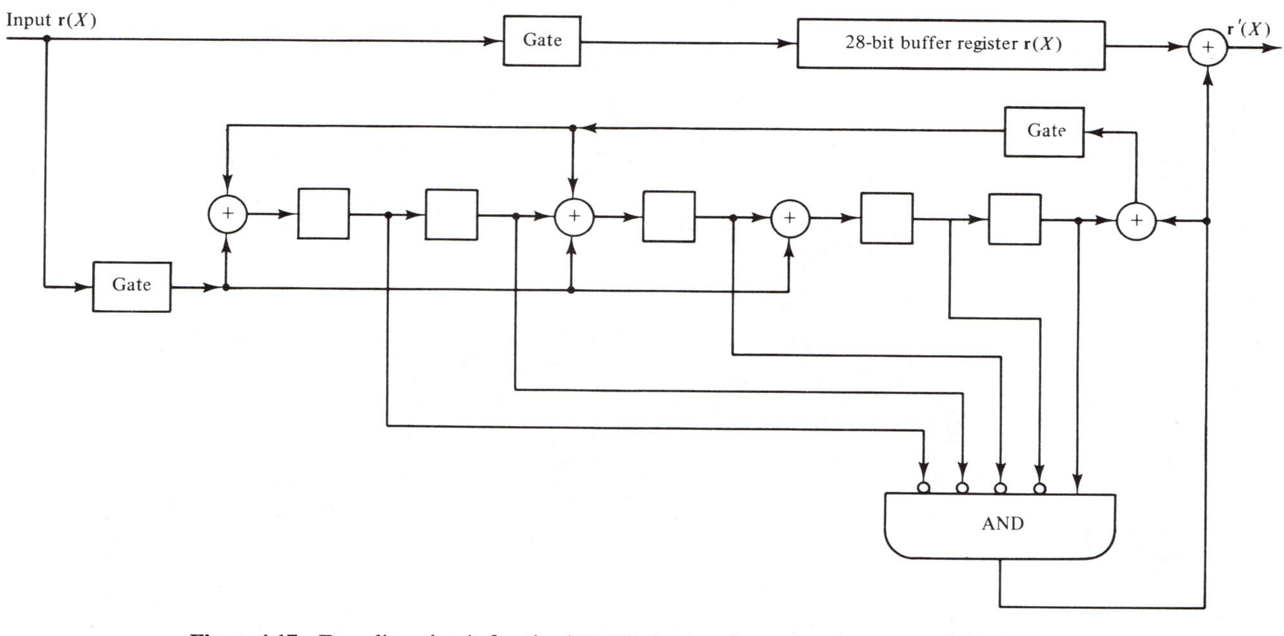

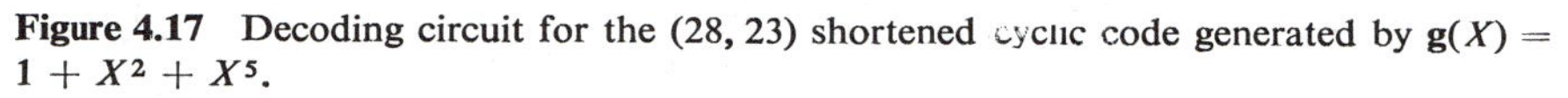
Figure 4.17 Decoding circuit for the (28, 23) shortened cyclic code generated by $\mathbf{g}(X) = 1 + X^2 + X^5$.

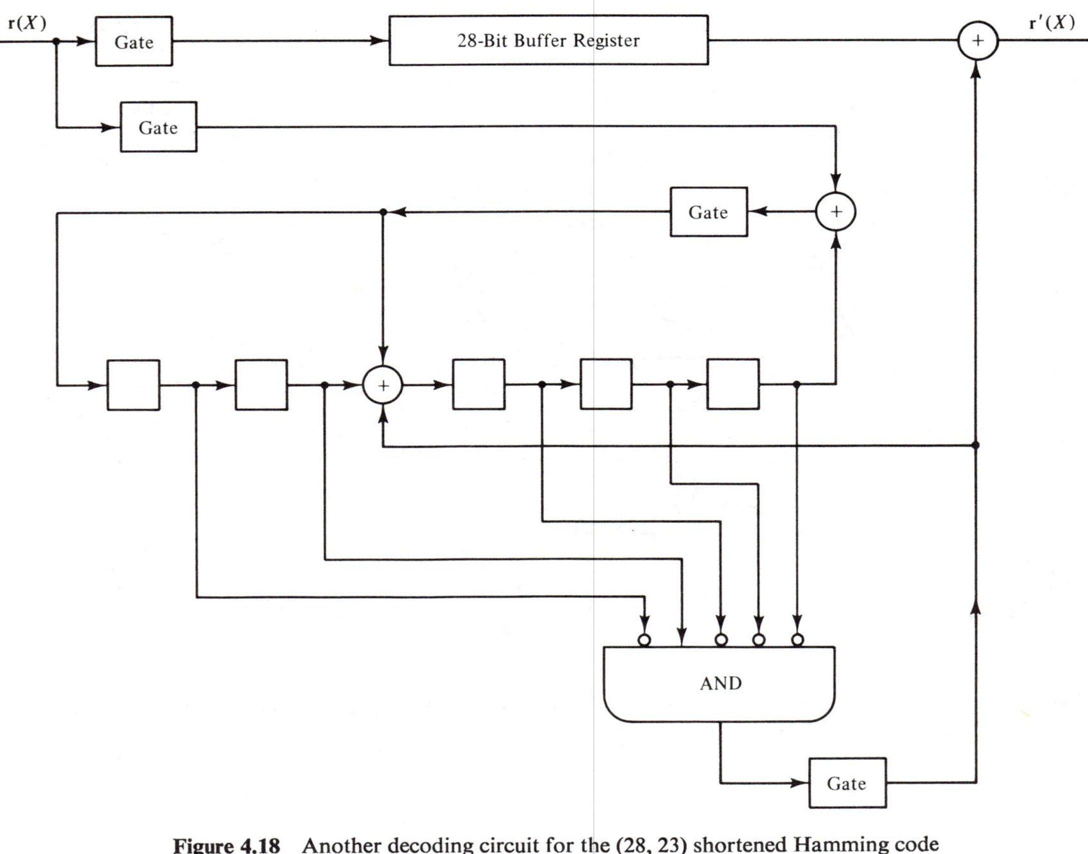

Figure 4.18 Another decoding circuit for the (28, 23) shortened Hamming code generated by $\mathbf{g}(X) = 1 + X^2 + X^5$.

nomial $\mathbf{g}(X)$. Then the effect of the error digit e_{n-l-1} on the syndrome is removed by adding $\boldsymbol{\rho}(X)$ to the syndrome in the syndrome register.

Example 4.12

Consider the (28, 23) shortened cyclic code obtained by deleting three digits from the (31, 26) cyclic Hamming code generated by $\mathbf{g}(X) = 1 + X^2 + X^5$. Suppose that, in decoding this code, the received vector is shifted into the syndrome register from the right end. If a single error occurs at the position X^{27} [or $\mathbf{e}(X) = X^{27}$], the syndrome corresponding to this error pattern is the remainder resulting from dividing $X^5\mathbf{e}(X) = X^{32}$ by $\mathbf{g}(X) = 1 + X^2 + X^5$. This resultant syndrome is (0 1 0 0 0). Thus, in decoding the (28, 23) shortened Hamming code, the error-pattern detection circuit may be designed to check whether the syndrome in the syndrome register is (0 1 0 0 0). By doing this, the extra three shifts of the syndrome register can be avoided. The resultant decoding circuit with syndrome resetting is shown in Figure 4.18.

PROBLEMS

4.1. Consider the (15, 11) cyclic Hamming code generated by $\mathbf{g}(X) = 1 + X + X^4$.
  (a) Determine the parity polynomial $\mathbf{h}(X)$ of this code.
  (b) Determine the generator polynomial of its dual code.
  (c) Find the generator and parity matrices in systematic form for this code.

4.2. Devise an encoder and a decoder for the (15, 11) cyclic Hamming code generated by $\mathbf{g}(X) = 1 + X + X^4$.

4.3. Show that $\mathbf{g}(X) = 1 + X^2 + X^4 + X^6 + X^7 + X^{10}$ generates a (21, 11) cyclic code. Devise a syndrome computation circuit for this code. Let $\mathbf{r}(X) = 1 + X^5 + X^{17}$ be a received polynomial. Compute the syndrome of $\mathbf{r}(X)$. Display the contents of the syndrome register after each digit of $\mathbf{r}$ has been shifted into the syndrome computation circuit.

4.4. Shorten the (15, 11) cyclic Hamming by deleting the seven leading high-order message digits. The resultant code is a (8, 4) shortened cyclic code. Design a decoder for this code which eliminates the extra shifts of the syndrome register.

4.5. Shorten the (31, 26) cyclic Hamming code by deleting the 11 leading high-order message digits. The resultant code is a (20, 15) shortened cyclic code. Devise a decoding circuit for this code which requires no extra shifts of the syndrome register.

4.6. Let $\mathbf{g}(X)$ be the generator polynomial of a binary cyclic code of length n.
  (a) Show that, if $\mathbf{g}(X)$ has $X + 1$ as a factor, the code contains no code vectors of odd weight.
  (b) If n is odd and $X + 1$ is not a factor of $\mathbf{g}(X)$, show that the code contains a code vector consisting of all 1's.
  (c) Show that the code has minimum weight at least 3 if n is the smallest integer such that $\mathbf{g}(X)$ divides $X^n + 1$.

4.7. Consider a binary (n, k) cyclic code C generated by $\mathbf{g}(X)$. Let

$$\mathbf{g}^*(X) = X^{n-k}\mathbf{g}(X^{-1})$$

be the reciprocal polynomial of $\mathbf{g}(X)$.
  (a) Show that $\mathbf{g}^*(X)$ also generates an (n, k) cyclic code.

(b) Let C^* denote the cyclic code generated by $\mathbf{g}^*(X)$. Show that C and C^* have the same weight distribution. [*Hint:* Show that

$$\mathbf{v}(X) = v_0 + v_1 X + \cdots + v_{n-2}X^{n-2} + v_{n-1}X^{n-1}$$

is a code polynomial in C if and only if

$$X^{n-1}\mathbf{v}(X^{-1}) = v_{n-1} + v_{n-2}X + \cdots + v_1 X^{n-2} + v_0 X^{n-1}$$

is a code polynomial in C^*.]

4.8. Consider a cyclic code C of length n which consists of both odd-weight and even-weight code vectors. Let $\mathbf{g}(X)$ and $A(z)$ be the generator polynomial and weight enumerator for this code. Show that the cyclic code generated by $(X + 1)\mathbf{g}(X)$ has weight enumerator

$$A_1(z) = \tfrac{1}{2}[A(z) + A(-z)].$$

4.9. Suppose that the (15, 10) cyclic Hamming code of minimum distance 4 is used for error detection over a BSC with transition probability $p = 10^{-2}$. Compute the probability of an undetected error $P_u(E)$ for this code.

4.10. Consider the $(2^m - 1, 2^m - m - 2)$ cyclic Hamming code C generated by $\mathbf{g}(X) = (X + 1)\mathbf{p}(X)$, where $\mathbf{p}(X)$ is a primitive polynomial of degree m. An error pattern of the form

$$\mathbf{e}(X) = X^i + X^{i+1}$$

is called a *double-adjacent-error pattern.* Show that no two double-adjacent-error patterns can be in the same coset of a standard array for C. Therefore, the code is capable of correcting all the single-error patterns and all the double-adjacent-error patterns.

4.11. Devise a decoding circuit for the (7, 3) Hamming code generated by $\mathbf{g}(X) = (X + 1)(X^3 + X + 1)$. The decoding circuit corrects all the single-error patterns and all the double-adjacent-error patterns. (See Problem 4.10.)

4.12. For a cyclic code, if an error pattern $\mathbf{e}(X)$ is detectable, show that its ith cyclic shift $\mathbf{e}^{(i)}(X)$ is also detectable.

4.13. In decoding an (n, k) cyclic code, suppose that the received polynomial $\mathbf{r}(X)$ is shifted into the syndrome register from the right end as shown in Figure 4.11. Show that when a received digit r_i is detected in error and is corrected, the effect of error digit e_i on the syndrome can be removed by feeding e_i into the syndrome register from the right end as shown in Figure 4.8.

4.14. Let $\mathbf{v}(X)$ be a code polynomial in a cyclic code of length n. Let l be the smallest integer such that

$$\mathbf{v}^{(l)}(X) = \mathbf{v}(X).$$

Show that if $l \neq 0$, l is a factor of n.

4.15. Let $\mathbf{g}(X)$ be the generator polynomial of an (n, k) cyclic code C.

(a) Show that $\mathbf{g}(X^\lambda)$ generates a $(\lambda n, \lambda k)$ cyclic code.

(b) Show that the code generated by $\mathbf{g}(X^\lambda)$ has the same minimum weight as that of the code generated by $\mathbf{g}(X)$.

4.16. Construct all the binary cyclic codes of length 15. [*Hint:* Using the fact that $X^{15} + 1$ has all the nonzero elements of GF(2^4) as roots and using Table 2.9, factor $X^{15} + 1$ as a product of irreducible polynomials.]

4.17. Let β be a nonzero element in the Galois field GF(2^m) and $\beta \neq 1$. Let $\phi(X)$ be the minimum polynomial of β. Is there a cyclic code with $\phi(X)$ as the generator polynomial?

If your answer is "yes," find the shortest cyclic code with $\phi(X)$ as the generator polynomial.

4.18. Let β_1 and β_2 be two distinct nonzero elements in GF(2^m). Let $\phi_1(X)$ and $\phi_2(X)$ be the minimal polynomials of β_1 and β_2, respectively. Is there a cyclic code with $\mathbf{g}(X) = \phi_1(X)\cdot\phi_2(X)$ as the generator polynomial? If your answer is yes, find the shortest cyclic code with $\mathbf{g}(X) = \phi_1(X)\cdot\phi_2(X)$ as the generator polynomial.

4.19. Consider the Galois field GF(2^m), which is constructed based on the the primitive polynomial $\mathbf{p}(X)$ of degree m. Let α be a primitive element of GF(2^m) whose minimal polynomial is $\mathbf{p}(X)$. Show that every code polynomial in the Hamming code generated by $\mathbf{p}(X)$ has α and its conjugates as roots. Show that any binary polynomial of degree $2^m - 2$ or less which has α as a root is a code polynomial in the Hamming code generated by $\mathbf{p}(X)$.

4.20. Let C_1 and C_2 be two cyclic codes of length n which are generated by $\mathbf{g}_1(X)$ and $\mathbf{g}_2(X)$, respectively. Show that the code polynomials common to both C_1 and C_2 also form a cyclic code C_3. Determine the generator polynomial of C_3. If d_1 and d_2 are the minimum distances of C_1 and C_2, respectively, what can you say about the minimum distance of C_3?

4.21. Show that the probability of an undetected error for the distance 4 cyclic Hamming codes is upper bounded by $2^{-(m+1)}$.

4.22. Let C be a $(2^m - 1, 2^m - m - 1)$ Hamming code generated by a primitive polynomial $\mathbf{p}(X)$ of degree m. Let C_d be the dual code of C. Then C_d is a $(2^m - 1, m)$ cyclic code generated by

$$\mathbf{h}^*(X) = X^{2^m - m - 1}\mathbf{h}(X^{-1}),$$

where

$$\mathbf{h}(X) = \frac{X^{2^m - 1} + 1}{\mathbf{p}(X)}.$$

(a) Let $\mathbf{v}(X)$ be a code vector in C_d and let $\mathbf{v}^{(i)}(X)$ be the ith cyclic shift of $\mathbf{v}(X)$. Show that, for $1 \leq i \leq 2^m - 2$, $\mathbf{v}^{(i)}(X) \neq \mathbf{v}(X)$.

(b) Show that C_d contains the all-zero code vector and $2^m - 1$ code vectors of weight 2^{m-1}.

[*Hint:* For part (a), use (4.1) and the fact that the smallest integer n such that $X^n + 1$ is divisible by $\mathbf{p}(X)$ is $2^m - 1$. For part (b), use the result of Problem 3.6(b).]

4.23. For an (n, k) cyclic code, show that the syndrome of an end-around burst of length $n - k$ cannot be zero.

4.24. Design a Meggitt decoder that decodes a received polynomial $\mathbf{r}(X) = r_0 + r_1X + \cdots + r_{n-1}X^{n-1}$ from the lowest-order received digit r_0 to the highest-order received digit r_{n-1}. Describe the decoding operation and the syndrome modification after each correction.

REFERENCES

1. E. Prange, "Cyclic Error-Correcting Codes in Two Symbols," AFCRC-TN-57, 103, Air Force Cambridge Research Center, Cambridge, Mass., September 1957.
2. E. R. Berlekamp, *Algebraic Coding Theory*, McGraw-Hill, New York, 1968.
3. W. W. Peterson and E. J. Weldon, Jr., *Error-Correcting Codes*, 2nd ed., MIT Press, Cambridge, Mass., 1972.

4. T. Kasami, N. Tokura, Y. Iwadare, and Y. Inagaki, *Coding Theory*, Corona, Tokyo, 1974 (in Japanese).
5. I. F. Blake and R. C. Mullin, *The Mathematical Theory of Coding*, Academic Press, New York, 1975.
6. F. J. MacWilliams and N. J. A. Sloane, *The Theory of Error-Correcting Codes*, North-Holland, Amsterdam, 1977.
7. R. J. McEliece, *The Theory of Information and Coding*, Addison-Wesley, Reading, Mass., 1977.
8. J. E. Meggitt, "Error Correcting Codes and Their Implementation," *IRE Trans. Inf. Theory*, IT-7, pp. 232–244, October 1961.

5

Error-Trapping Decoding for Cyclic Codes

In principle, the general decoding method of Meggitt's applies to any cyclic code, but refinements are necessary for practical implementation. In this chapter a practical variation of Meggitt decoding, called *error-trapping decoding*, is presented. A decoder based on this decoding technique employs a very simple combinational logic circuit for error detection and correction. Error-trapping decoding was devised independently by Kasami [1], Mitchell [2, 3], and Rudolph [4]. It is most effective for decoding single-error-correcting codes, some short double-error-correcting codes, and burst-error-correcting codes. When it is applied to long and high rate codes with large error-correcting capability, it becomes very ineffective and much error-correcting capability will be sacrificed. Several improved error-trapping methods for decoding multiple-error-correcting codes have been devised. One such method due to Kasami [5] is presented in this chapter.

Error-trapping decoding is particularly effective for decoding burst-error-correcting codes, discussed in Chapter 9.

5.1 ERROR-TRAPPING DECODING

If we put some restrictions on the error patterns that we intend to correct, the Meggitt decoder can be practically implemented. Consider an (n, k) cyclic code with generator polynomial $\mathbf{g}(X)$. Suppose that a code vector $\mathbf{v}(X)$ is transmitted and is corrupted by an error pattern $\mathbf{e}(X)$. Then the received polynomial is $\mathbf{r}(X) = \mathbf{v}(X) + \mathbf{e}(X)$. We have shown in Section 4.4 that the syndrome $\mathbf{s}(X)$ computed from $\mathbf{r}(X)$ is equal to the remainder resulting from dividing the error pattern $\mathbf{e}(X)$ by the generator $\mathbf{g}(X)$

[i.e. $\mathbf{e}(X) = \mathbf{a}(X)\mathbf{g}(X) + \mathbf{s}(X)$]. Suppose that errors are confined to the $n - k$ high-order positions, $X^k, X^{k+1}, \ldots, X^{n-1}$ of $\mathbf{r}(X)$ [i.e., $\mathbf{e}(X) = e_k X^k + e_{k+1}X^{k+1} + \cdots + e_{n-1}X^{n-1}$]. If $\mathbf{r}(X)$ is cyclically shifted $n - k$ times, the errors will be confined to $n - k$ low-order parity positions, $X^0, X^1, \ldots, X^{n-k-1}$ of $\mathbf{r}^{(n-k)}(X)$. The corresponding error pattern is then

$$\mathbf{e}^{(n-k)}(X) = e_k + e_{k+1}X + \cdots + e_{n-1}X^{n-k-1}.$$

Since the syndrome $\mathbf{s}^{(n-k)}(X)$ of $\mathbf{r}^{(n-k)}(X)$ is equal to the remainder resulting from dividing $\mathbf{e}^{(n-k)}(X)$ by $\mathbf{g}(X)$ and since the degree of $\mathbf{e}^{(n-k)}(X)$ is less than $n - k$, we obtain the following equality:

$$\mathbf{s}^{(n-k)}(X) = \mathbf{e}^{(n-k)}(X) = e_k + e_{k+1}X + \cdots + e_{n-1}X^{n-k-1}.$$

Multiplying $\mathbf{s}^{(n-k)}(X)$ by X^k, we have

$$\begin{aligned} X^k\mathbf{s}^{(n-k)}(X) &= \mathbf{e}(X) \\ &= e_kX^k + r_{k+1}X^{k+1} + \cdots + e_{n-1}X^{n-1}. \end{aligned}$$

This says that, if errors are confined to the $n - k$ high-order positions of the received polynomial $\mathbf{r}(X)$, *the error pattern* $\mathbf{e}(X)$ *is identical to* $X^k\mathbf{s}^{(n-k)}(X)$, *where* $\mathbf{s}^{(n-k)}(X)$ *is the syndrome of* $\mathbf{r}^{(n-k)}(X)$, *the* $(n - k)$th *cyclic shift of* $\mathbf{r}(X)$. When this event occurs, we simply compute $\mathbf{s}^{(n-k)}(X)$ and add $X^k\mathbf{s}^{(n-k)}(X)$ to $\mathbf{r}(X)$. The resultant vector is the transmitted code vector.

Suppose that errors are not confined to the $n - k$ high-order positions but are confined to $n - k$ consecutive positions, say $X^i, X^{i+1}, \ldots, X^{(n-k)+i-1}$, of $\mathbf{r}(X)$ (including the end-around case). If $\mathbf{r}(X)$ is cyclically shifted $n - i$ times to the right, errors will be confined to the $n - k$ low-order position of $\mathbf{r}^{(n-i)}(X)$ and the error pattern will be identical to $X^i\mathbf{s}^{(n-i)}(X)$, where $\mathbf{s}^{(n-i)}(X)$ is the syndrome of $\mathbf{r}^{(n-i)}(X)$.

Now suppose that we shift the received polynomial $\mathbf{r}(X)$ into the syndrome register from the right end. Shifting $\mathbf{r}(X)$ into the syndrome register from the right end is equivalent to premultiplying $\mathbf{r}(X)$ by X^{n-k}. After the entire $\mathbf{r}(X)$ has been shifted into the syndrome register, the contents of the syndrome register form the syndrome $\mathbf{s}^{(n-k)}(X)$ of $\mathbf{r}^{(n-k)}(X)$. If the errors are confined to $n - k$ high-order positions, $X^k, X^{k+1}, \ldots, X^{n-1}$ of $\mathbf{r}(X)$, they are identical to $\mathbf{s}^{(n-k)}(X)$. However, if the errors are confined to $n - k$ consecutive positions (including end-around) other than the $n - k$ high-order positions of $\mathbf{r}(X)$, after the entire $\mathbf{r}(X)$ has been shifted into the syndrome register, the syndrome register must be shifted a certain number of times before its contents are identical to the error digits. This shifting of the syndrome register until its contents are identical to the error digits is called *error trapping*. If errors are confined to $n - k$ consecutive positions of $\mathbf{r}(X)$ and if we can detect when the errors are trapped in the syndrome register, error correction can be accomplished by simply adding the contents of the syndrome register to the received digits at the $n - k$ proper positions.

Suppose that a t-error-correcting cyclic code is used. To detect the event that the errors are trapped in the syndrome register, we may simply test the *weight* of the syndrome after each shift of the syndrome register. As soon as the weight of the syndrome becomes t or less, we assume that errors are trapped in the syndrome register. *If the number of errors in* $\mathbf{r}(X)$ *is t or less and if they are confined to $n - k$*

consecutive positions, the errors are trapped in the syndrome register only when the weight of the syndrome in the register becomes t or less. This can be shown as follows. An error pattern $\mathbf{e}(X)$ with t or fewer errors which are confined to $n - k$ consecutive positions must be of the form $\mathbf{e}(X) = X^j B(X)$, where $B(X)$ has t or fewer terms and has degree $n - k - 1$ or less. [For the end-around case, the same form would be obtained after certain number of cyclic shifts of $\mathbf{e}(X)$.] Dividing $\mathbf{e}(X)$ by the generator polynomial $\mathbf{g}(X)$, we have

$$X^j B(X) = \mathbf{a}(X)\mathbf{g}(X) + \mathbf{s}(X),$$

where $\mathbf{s}(X)$ is the syndrome of $X^j B(X)$. Since $\mathbf{s}(X) + X^j B(X)$ is a multiple of $\mathbf{g}(X)$, it is a code polynomial. The syndrome $\mathbf{s}(X)$ cannot have weight t or less unless $\mathbf{s}(X) = X^j B(X)$. Suppose that the weight of $\mathbf{s}(X)$ is t or less and $\mathbf{s}(X) \neq X^j B(X)$. Then $\mathbf{s}(X) + X^j B(X)$ is a nonzero code vector with weight less than $2t + 1$. This is impossible since a t-error-correcting code must have a minimum weight of at least $2t + 1$. Therefore we conclude that the errors are trapped in the syndrome register only when the weight of the syndrome becomes t or less.

Based on the error-trapping concept and the test described above, an error-trapping decoder can be implemented as shown in Figure 5.1. The decoding operation can be described in the following steps:

Step 1. The received polynomial $\mathbf{r}(X)$ is shifted into the buffer and syndrome registers simultaneously with gates 1 and 3 turned on and all the other gates turned off. Since we are only interested in the recovery of the k information digits, the buffer register has only to store the k received information digits.

Step 2. As soon as the entire $\mathbf{r}(X)$ has been shifted into the syndrome register, the weight of the syndrome in the register is tested by an $(n - k)$-input *threshold* gate whose output is 1 when t or fewer of its inputs are 1; otherwise, it is zero. (a) If the weight of the syndrome is t or less, the syndrome digits in the syndrome register are identical to the error digits at the $n - k$ high-order positions X^k,

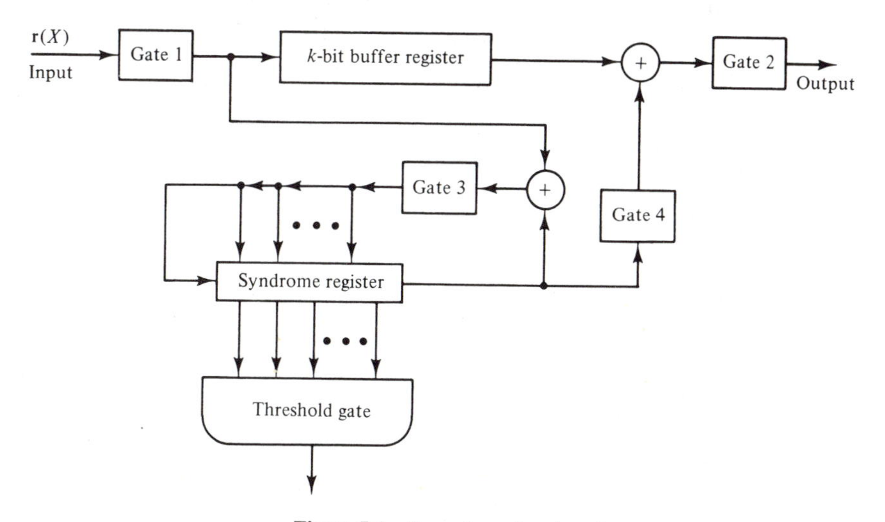

Figure 5.1 Error-trapping decoder.

$X^{k+1}, \ldots, X^{n-1}$ of $\mathbf{r}(X)$. Now, gates 2 and 4 are turned on and the other gates turned off. The received vector is read out of the buffer register one digit at a time and is corrected by the error digits shifted out from the syndrome register. (b) If the weight of the syndrome is greater than t, the errors are *not* confined to the $n-k$ high-order positions of $\mathbf{r}(X)$ and they have not been trapped in the syndrome register. Go to step 3.

Step 3. Cyclically shift the syndrome register once with gate 3 turned on and other gates turned off. The weight of the new syndrome is tested. (a) If it is t or less, the errors are confined to the locations $X^{k-1}, X^k, \ldots, X^{n-2}$ of $\mathbf{r}(X)$ and the contents in the syndrome register are identical to the errors at these locations. Since the first received digit r_{n-1} is *error-free*, it is read out of the buffer register with gate 2 turned on. As soon as r_{n-1} has been read out, gate 4 is turned on and gate 3 is turned off. The contents in the syndrome register are shifted out and are used to correct the next $n-k$ received digits to come out from the buffer register. (b) If the weight of the syndrome is greater than t, the syndrome register is shifted once more with gate 3 turned on.

Step 4. The syndrome register is continuously shifted until the weight of its contents goes down to t or less. If the weight goes down to t or less at the end of the ith shift, for $1 \leq i \leq k$, the first i received digits, $r_{n-i}, r_{n-i+1}, \ldots, r_{n-1}$, in the buffer register are *error-free* and the contents in the syndrome register are identical to the errors at the locations of $X^{k-i}, X^{k-i+1}, \ldots, X^{n-i-1}$. As soon as the i error-free received digits have been read out of the buffer register, the contents in the syndrome register are shifted out and are used to correct the next $n-k$ received digit to come out from the buffer register. When the k received information digits have been read out of the buffer register and have been corrected, gate 2 will be turned off. Any nonzero digits left in the syndrome register are errors in the parity part of $\mathbf{r}(X)$ and they will be ignored.

Step 5. If the weight of the syndrome never goes down to t or less by the time that the syndrome register has been shifted k times, either an error pattern with errors confined to $n-k$ consecutive end-around locations has occurred or an uncorrectable error pattern has occurred. We keep on shifting the syndrome register. Suppose that the weight of its contents becomes t or less at the end of $k+l$ shifts with $1 \leq l \leq n-k$. Then errors are confined to the $n-k$ consecutive end-around locations, $X^{n-l}, X^{n-l+1}, \ldots, X^{n-1}, X^0, X^1, \ldots, X^{n-k-l-1}$ of $\mathbf{r}(X)$. The l digits in the l leftmost stages of the syndrome register match the errors at the l high-order locations $X^{n-l}, X^{n-l+1}, \ldots, X^{n-1}$ of $\mathbf{r}(X)$. Since we do not need the errors at the $n-k-l$ parity locations, we shift the syndrome register $n-k-l$ times with all the gates turned off. Now, the l errors at the locations $X^{n-l}, X^{n-l+1}, \ldots, X^{n-1}$ of $\mathbf{r}(X)$ are contained in the l rightmost stages of the syndrome register. With gates 2 and 4 turned on and other gates turned off, the received digits in the buffer register are read out and are corrected by the corresponding error digits shifted out from the syndrome register. This completes the decoding operation.

If the weight of the syndrome never goes down to t or less by the time the syndrome register has been shifted a total of n times, either an uncorrectable error pattern has occurred or errors are not confined to $n-k$ consecutive positions. In either case, errors are detected. Except when errors are confined to the $n-k$ consecutive end-around positions of $\mathbf{r}(X)$, the received information digits can be read out of the buffer register, corrected, and delivered to the data sink after at most k cyclic shifts of the syndrome register. When an error pattern with errors confined to $n-k$ consecutive end-around locations of $\mathbf{r}(X)$, a total of n cyclic shifts of the syndrome register is required before the received message can be read out of the buffer register for corrections. For large n and $n-k$, the number of correctable end-around error patterns becomes big and it causes undesirable long decoding delay.

It is possible to implement the error-trapping decoding in a different manner so that the error patterns with errors confined to $n-k$ consecutive end-around locations can be corrected as fast as possible. This can be achieved by shifting the received vector $\mathbf{r}(X)$ into the syndrome register from the *left* end as shown in Figure 5.2. This variation is based on the following facts. If the errors are confined to $n-k$

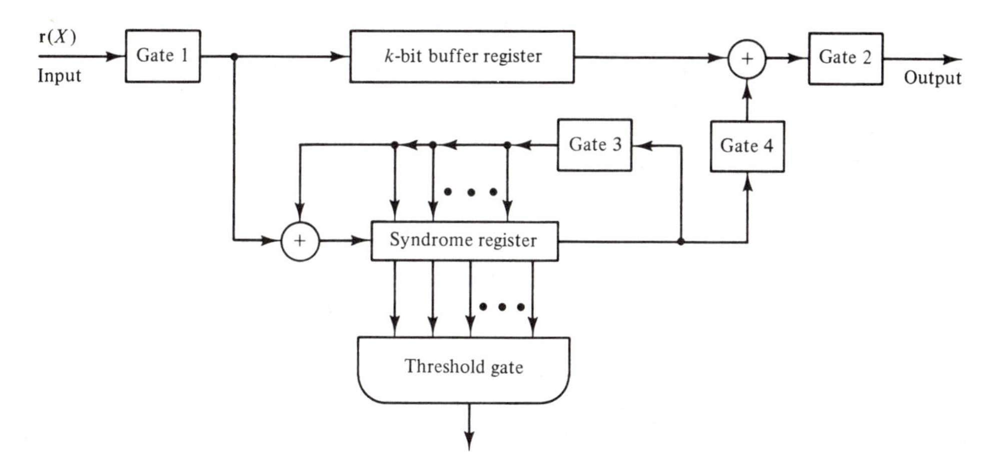

Figure 5.2 Another error-trapping decoder.

low-order parity positions $X^0, X^1, \ldots, X^{n-k-1}$ of $\mathbf{r}(X)$, then after the entire $\mathbf{r}(X)$ has entered the syndrome register, the contents in the register are identical to the error digits at the locations $X^0, X^1, \ldots, X^{n-k-1}$ of $\mathbf{r}(X)$. Suppose that the errors are not confined to the $n-k$ low-order positions of $\mathbf{r}(X)$ but are confined to $n-k$ consecutive locations (including the end-around case), say $X^i, X^{i+1}, \ldots, X^{(n-k)+i-1}$. After $n-i$ cyclic shifts of $\mathbf{r}(X)$, the errors will be shifted to the $n-k$ low-order positions of $\mathbf{r}^{(n-i)}(X)$, and the syndrome of $\mathbf{r}^{(n-i)}(X)$ will be identical to the errors confined to positions, $X^i, X^{i+1}, \ldots, X^{(n-k)+i-1}$ of $\mathbf{r}(X)$. The operation of the decoder shown in Figure 5.2 is described as follows:

Step 1. Gates 1 and 3 turned on and the other gates are turned off. The received vector $\mathbf{r}(X)$ is shifted into the syndrome register and simultaneously into the buffer register (only the k received information digits are stored in the buffer register). As soon as the entire $\mathbf{r}(X)$ has been shifted into the syndrome register, the contents of the register form the syndrome $\mathbf{s}(X)$ of $\mathbf{r}(X)$.

Step 2. The weight of the syndrome is tested. (a) If the weight is t or less, the errors are confined to the $(n-k)$ low-order parity positions $X^0, X^1, \ldots, X^{n-k-1}$ of $\mathbf{r}(X)$. Thus, the k received information digits in the buffer register are *error-free*. Gate 2 is then turned on and the error-free information digits are read out of the buffer with gate 4 turned off. (b) If the weight of the syndrome is greater than t, the syndrome register is then shifted once with gate 3 turned on and the other gates turned off. Go to step 3.

Step 3. The weight of the new contents in the syndrome register is tested. (a) If the weight is t or less, the errors are confined to the positions $X^{n-1}, X^0, X^1, \ldots, X^{n-k-2}$ of $\mathbf{r}(X)$ (end-around case). The leftmost digit in the syndrome register is identical to the error at the position X^{n-1} of $\mathbf{r}(X)$; the other $n-k-1$ digits in the syndrome register match the errors at parity positions $X^0, X^1, \ldots, X^{n-k-2}$ of $\mathbf{r}(X)$. The output of the threshold gate turns gate 3 off and sets a clock to count from 2. The syndrome register is then shifted (in step with the clock) with gate 3 turned off. As soon as the clock has counted to $n-k$, the contents of the syndrome register will be $(0 \ 0 \ \cdots \ 0 \ 1)$. The rightmost digit matches the error at position X^{n-1} of $\mathbf{r}(X)$. The k received information digits are then read out of the buffer and the first received information digit is corrected by the 1 coming out from the syndrome register. The decoding is thus completed. (b) If the weight of the contents in the syndrome register is greater than t, the syndrome register is shifted once again with gate 3 turned on and other gates turned off. Go to step 4.

Step 4. Step 3(b) repeats until the weight of the contents of the syndrome register goes down to t or less. If the weight goes down to t or less after the ith shift, for $1 \leq i \leq n-k$, the clock starts to count from $i+1$. At the same time, the syndrome register is shifted with gate 3 turned off. As soon as the clock has counted to $n-k$, the rightmost i digits in the syndrome register match the errors in the first i received information digits in the buffer register. The other information digits are error-free. Gates 2 and 4 are then turned on. The received information digits are read out of the buffer for correction.

Step 5. If the weight of the contents of the syndrome register never goes down to t or less by the time that the syndrome register has been shifted $n-k$ times (with gate 3 turned on), gate 2 is then turned on and the received information digits are read out of the buffer one at a time. At the same time the syndrome register is shifted with gate 3 turned on. As soon as the weight of the contents of the syndrome register goes down to t or less, the contents match the errors in the next $n-k$ digits to come out of the buffer. Gate 4 is then turned on and the erroneous information digits are corrected by the digits coming out from the syndrome register with gate 3 turned off. Gate 2 is turned off as soon as k information digits have been read out of the buffer.

With the implementation of error-trapping decoding described above, the received information digits can be read out of the buffer register after at most $n-k$ shifts of the syndrome register. For large $n-k$, this implementation provides faster decoding than the decoder shown in Figure 5.1. However, when $n-k$ is much smaller than k, the first implementation of error trapping is more advantageous in decoding speed than the one shown in Figure 5.2.

The decoding of cyclic Hamming codes presented in Chapter 4 is actually an error-trapping decoding. The syndrome register is cyclically shifted until the single error is trapped in the rightmost stage of the register. Error-trapping decoding is most effective for decoding single-error-correcting codes and burst-error-correcting codes (decoding of burst-error-correcting codes is discussed in Chapter 9). It is also effective for decoding some short double-error-correcting codes. When it is applied to long and high rate codes (small $n - k$) with large error-correcting capability, it becomes very ineffective and much of the error-correcting capability will be sacrificed. Several refinements of this simple decoding technique [1–9] have been devised to extend its application to multiple-error-correcting codes. One of the refinements is presented in the next section.

Example 5.1

Consider the (15, 7) cyclic code generated by $\mathbf{g}(X) = 1 + X^4 + X^6 + X^7 + X^8$. This code has minimum distance $d_{\min} = 5$, which will be proved in Chapter 6. Hence, the code is capable of correcting any combination of two or fewer errors over a block of 15 digits. Suppose that we decode this code with an error-trapping decoder. Clearly, any single error is confined to $n - k = 8$ consecutive positions. Therefore, any single error can be trapped and corrected. Now consider any double errors over a span of 15 digits. If we arrange the 15 digit positions X^0 to X^{14} as a ring, as shown in Figure 5.3, we see that any double errors are confined to eight consecutive positions. Hence, any double errors can be trapped and corrected. An error-trapping decoder for the (15, 7) generated by $\mathbf{g}(X) = 1 + X^4 + X^6 + X^7 + X^8$ is shown in Figure 5.4.

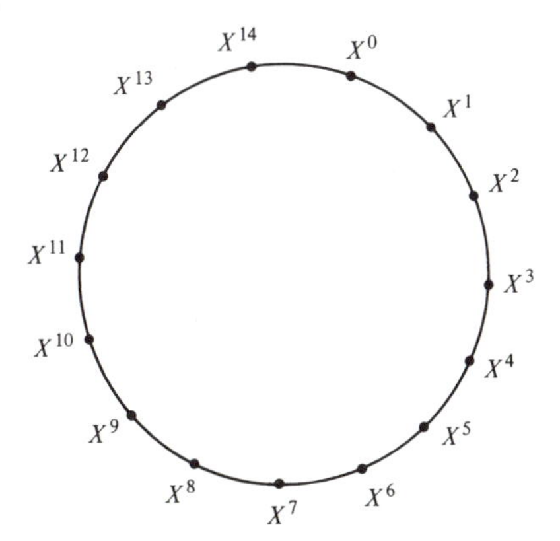

Figure 5.3 Ring arrangement of code digit positions.

5.2 IMPROVED ERROR-TRAPPING DECODING

The error-trapping decoding discussed in Section 5.1 can be improved to correct error patterns such that, for each error pattern, most errors are confined to $n - k$ consecutive positions and fewer errors are outside the $(n - k)$-digit span. This improvement needs additional equipment. The complexity of the additional equipment depends on how many errors outside an $(n - k)$-digit span are to be corrected. An improvement proposed by Kasami [5] is discussed here.

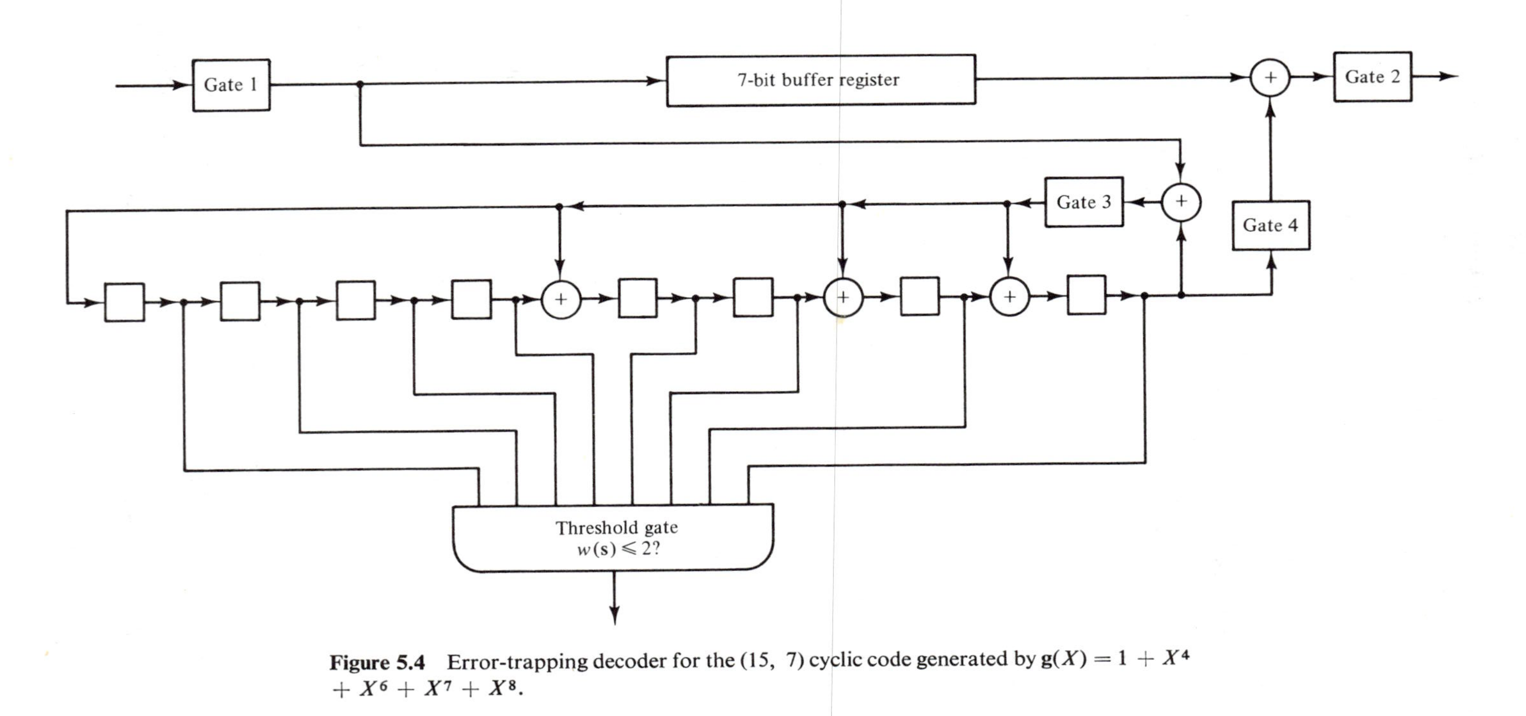

Figure 5.4 Error-trapping decoder for the (15, 7) cyclic code generated by $\mathbf{g}(X) = 1 + X^4 + X^6 + X^7 + X^8$.

The error pattern $\mathbf{e}(X) = e_0 + e_1 X + e_2 X^2 + \cdots + e_{n-1} X^{n-1}$, which corrupted the transmitted code vector, can be divided into two parts:

$$\mathbf{e}_p(X) = e_0 + e_1 X + \cdots + e_{n-k-1} X^{n-k-1}$$
$$\mathbf{e}_I(X) = e_{n-k} X^{n-k} + \cdots + e_{n-1} X^{n-1},$$

where $\mathbf{e}_I(X)$ contains the errors in the message section of the received vector and $\mathbf{e}_p(X)$ contains the errors in the parity section of the received vector. Dividing $\mathbf{e}_I(X)$ by the code-generator polynomial $\mathbf{g}(X)$, we obtain

$$\mathbf{e}_I(X) = \mathbf{q}(X)\mathbf{g}(X) + \boldsymbol{\rho}(X), \tag{5.1}$$

where $\boldsymbol{\rho}(X)$ is the remainder with degree $n - k - 1$ or less. Adding $\mathbf{e}_p(X)$ to both sides of (5.1), we obtain

$$\mathbf{e}(X) = \mathbf{e}_p(X) + \mathbf{e}_I(X) = \mathbf{q}(X)\mathbf{g}(X) + \boldsymbol{\rho}(X) + \mathbf{e}_p(X). \tag{5.2}$$

Since $\mathbf{e}_p(X)$ has degree $n - k - 1$ or less, $\boldsymbol{\rho}(X) + \mathbf{e}_p(X)$ must be the remainder resulting from dividing the error pattern $\mathbf{e}(X)$ by the generator polynomial. Thus, $\boldsymbol{\rho}(X) + \mathbf{e}_p(X)$ is equal to the syndrome of the received vector $\mathbf{r}(X)$,

$$\mathbf{s}(X) = \boldsymbol{\rho}(X) + \mathbf{e}_p(X). \tag{5.3}$$

Rearranging (5.3), we have

$$\mathbf{e}_p(X) = \mathbf{s}(X) + \boldsymbol{\rho}(X). \tag{5.4}$$

That is, if the error pattern $\mathbf{e}_I(X)$ in the message positions is known, the error pattern $\mathbf{e}_p(X)$ in the parity positions can be found.

Kasami's error-trapping decoding requires finding a set of polynomials $[\boldsymbol{\phi}_j(X)]_{j=1}^{N}$ of degree $k - 1$ or less, such that, for any correctable error pattern $\mathbf{e}(X)$, there is one polynomial $\boldsymbol{\phi}_j(X)$ such that $X^{n-k}\boldsymbol{\phi}_j(X)$ matches the message section of $\mathbf{e}(X)$ or the message section of a cyclic shift of $\mathbf{e}(X)$. The polynomials $\boldsymbol{\phi}_j(X)$'s are called the *covering polynomials*. Let $\boldsymbol{\rho}_j(X)$ be the remainder resulting from dividing $X^{n-k}\boldsymbol{\phi}_j(X)$ by the generator polynomial $\mathbf{g}(X)$ of the code.

The decoding procedure can be described in the following steps:

Step 1. Calculate the syndrome $\mathbf{s}(X)$ by entering the entire received vector into the syndrome register.

Step 2. Calculate the weight of the sum $\mathbf{s}(X) + \boldsymbol{\rho}_j(X)$ for each $j = 1, 2, \ldots, N$ (i.e., $w[\mathbf{s}(X) + \boldsymbol{\rho}_j(X)]$ for $j = 1, 2, \ldots, N$).

Step 3. If, for some l,

$$w[\mathbf{s}(X) + \boldsymbol{\rho}_l(X)] \leq t - w[\boldsymbol{\phi}_l(X)],$$

then $X^{n-k}\boldsymbol{\phi}_l(X)$ matches the error pattern in the message section of $\mathbf{e}(X)$ and $\mathbf{s}(X) + \boldsymbol{\rho}_l(X)$ matches the error pattern in the parity section of $\mathbf{e}(X)$. Thus,

$$\mathbf{e}(X) = \mathbf{s}(X) + \boldsymbol{\rho}_l(X) + X^{n-k}\boldsymbol{\phi}_l(X).$$

Correction is then accomplished by taking the modulo-2 sum $\mathbf{r}(X) + \mathbf{e}(X)$. This step requires N $(n - k)$-input threshold gates to test the weights of $\mathbf{s}(X) + \boldsymbol{\rho}_j(X)$ for $j = 1, 2, \ldots, N$.

Step 4. If $w[\mathbf{s}(X) + \boldsymbol{\rho}_j(X)] > t - w[\boldsymbol{\phi}_j(X)]$ for all $j = 1, 2, \ldots, N$, both syndrome and buffer registers are shifted cyclically once. Then the new contents $\mathbf{s}^{(1)}(X)$ of the syndrome register is the syndrome corresponding to $\mathbf{e}^{(1)}(X)$ which is obtained by shifting the error pattern $\mathbf{e}(X)$ cyclically one place to the right.

Step 5. The weight of $\mathbf{s}^{(1)}(X) + \boldsymbol{\rho}_j(X)$ is computed for $j = 1, 2, \ldots, N$. If, for some l,

$$w[\mathbf{s}^{(1)}(X) + \boldsymbol{\rho}_l(X)] \leq t - w[\boldsymbol{\phi}_l(X)],$$

then $X^{n-k}\boldsymbol{\phi}_l(X)$ matches the errors in the message section of $\mathbf{e}^{(1)}(X)$ and $\mathbf{s}^{(1)}(X) + \boldsymbol{\rho}_l(X)$ matches the errors in the parity section of $\mathbf{e}^{(1)}(X)$. Thus,

$$\mathbf{e}^{(1)}(X) = \mathbf{s}^{(1)}(X) + \boldsymbol{\rho}_l(X) + X^{n-k}\boldsymbol{\phi}_l(X).$$

Correction is then accomplished by taking the modulo-2 sum $\mathbf{r}^{(1)}(X) + \mathbf{e}^{(1)}(X)$. If

$$w[\mathbf{s}^{(1)}(X) + \boldsymbol{\rho}_j(X)] > t - w[\boldsymbol{\phi}_j(X)]$$

for all $j = 1, 2, \ldots, N$, both syndrome and buffer registers are shifted cyclically once again.

Step 6. The syndrome and buffer registers are continuously shifted until $\mathbf{s}^{(i)}(X)$ (the syndrome after the ith shift) is found such that, for some l,

$$w[\mathbf{s}^{(i)}(X) + \boldsymbol{\rho}_l(X)] \leq t - w[\boldsymbol{\phi}_l(X)].$$

Then

$$\mathbf{e}^{(i)}(X) = \mathbf{s}^{(i)}(X) + \boldsymbol{\rho}_l(X) + X^{n-k}\boldsymbol{\phi}_l(X),$$

where $\mathbf{e}^{(i)}(X)$ is the ith cyclic shift of $\mathbf{e}(X)$. If the weight $w[\mathbf{s}^{(i)}(X) + \boldsymbol{\rho}_j(X)]$ never goes down to $t - w[\boldsymbol{\phi}_j(X)]$ or less for all j by the time that the syndrome and buffer registers have been cyclically shifted $n - 1$ times, an uncorrectable error pattern is detected.

The complexity of a decoder that employs the decoding method described above depends on N, the number of covering polynomials in $\{\boldsymbol{\phi}_j(X)\}_{j=1}^{N}$. The combinational logical circuitry consists of N $(n - k)$-input threshold gates. To find the set of covering polynomials $\{\boldsymbol{\phi}_j(X)\}_{j=1}^{N}$ for a specific code is not an easy problem. Several methods for finding this set can be found in References 5, 10, and 11.

This improved error-trapping method is applicable to many double- and triple-error-correcting codes. However, it is still only applicable to relatively short and low rate codes. When the code length n and error-correcting capability t become large, the number of threshold gates required in the error-detecting logical circuitry becomes very large and impractical.

Other variations of error trapping decoding can be found in References 4, 7, and 8.

5.3 THE GOLAY CODE

The (23, 12) Golay code is the only known multiple-error-correcting binary perfect code which is capable of correcting any combination of three or fewer random errors in a block of 23 digits. This code has abundant and beautiful algebraic structure. Since its discovery by Golay in 1949 [12], it has become a subject of study by many coding theorists and mathematicians. Many research papers have been published on its structure and decoding. The recent book by MacWilliams and Sloane [13] presents a thorough coverage of this code and its cousins [the extended (23, 12) Golay code and two ternary Golay codes], where a whole chapter is devoted to their algebraic structure. Besides its beautiful structure, the (23, 12) Golay code has been used in several real communication systems.

The (23, 12) Golay code is either generated by

$$\mathbf{g}_1(X) = 1 + X^2 + X^4 + X^5 + X^6 + X^{10} + X^{11}$$

or by

$$\mathbf{g}_2(X) = 1 + X + X^5 + X^6 + X^7 + X^9 + X^{11}.$$

Both $\mathbf{g}_1(X)$ and $\mathbf{g}_2(X)$ are factors of $X^{23} + 1$ and $X^{23} + 1 = (1 + X)\mathbf{g}_1(X)\mathbf{g}_2(X)$. The encoding can be accomplished by an 11-stage shift register with feedback connections according to either $\mathbf{g}_1(X)$ or $\mathbf{g}_2(X)$. If the simple error-trapping scheme described in Section 5.1 is used for decoding this code, some of the double-error patterns and many of the triple-error patterns cannot be trapped. For example, consider the double-error pattern $\mathbf{e}(X) = X^{11} + X^{22}$. The two errors are never confined to $n - k = 11$ consecutive positions, no matter how many times we cyclically shift $\mathbf{e}(X)$. Therefore, they can never be trapped in the syndrome register and cannot be corrected. We can also readily see that the triple-error pattern $\mathbf{e}(X) = X^5 + X^{11} + X^{22}$ cannot be trapped. Therefore, using the simple error-trapping scheme for decoding the Golay code, some of its error-correcting capability will be lost. However, the decoding circuitry is simple.

There are several practical ways to decode the (23, 12) Golay code up to its error-correcting capability $t = 3$. Two of the best are discussed in this section. Both are refined error-trapping schemes.

Kasami Decoder [5]

The Golay code can be easily decoded by Kasami's error-trapping technique. The set of polynomials $\{\boldsymbol{\phi}_j(X)\}_{j=1}^{N}$ is chosen as follows:

$$\boldsymbol{\phi}_1(X) = 0, \qquad \boldsymbol{\phi}_2(X) = X^5, \qquad \boldsymbol{\phi}_3(X) = X^6.$$

Let $\mathbf{g}_1(X) = 1 + X^2 + X^4 + X^5 + X^6 + X^{10} + X^{11}$ be the generator polynomial. Dividing $X^{11}\boldsymbol{\phi}_j(X)$ by $\mathbf{g}_1(X)$ for $j = 1, 2, 3,$ we obtain the following remainders:

$$\begin{aligned}\boldsymbol{\rho}_1(X) &= 0,\\ \boldsymbol{\rho}_2(X) &= X + X^2 + X^5 + X^6 + X^8 + X^9,\\ \boldsymbol{\rho}_3(X) &= X\boldsymbol{\rho}_2(X) = X^2 + X^3 + X^6 + X^7 + X^9 + X^{10}.\end{aligned}$$

A decoder based on the Kasami's error-trapping scheme is shown in Figure 5.5. The received vector $\mathbf{r}(X) = r_0 + r_1X + r_2X^2 + \cdots + r_{22}X^{22}$ is shifted into the syndrome register from the rightmost stage; this is equivalent to preshifting the received vector 11 times cyclically. After the entire received vector has entered the syndrome register, the syndrome in the register corresponds to $\mathbf{r}^{(11)}(X)$ which is the eleventh cyclic shift of $\mathbf{r}(X)$. In this case, if the errors are confined to the first 11

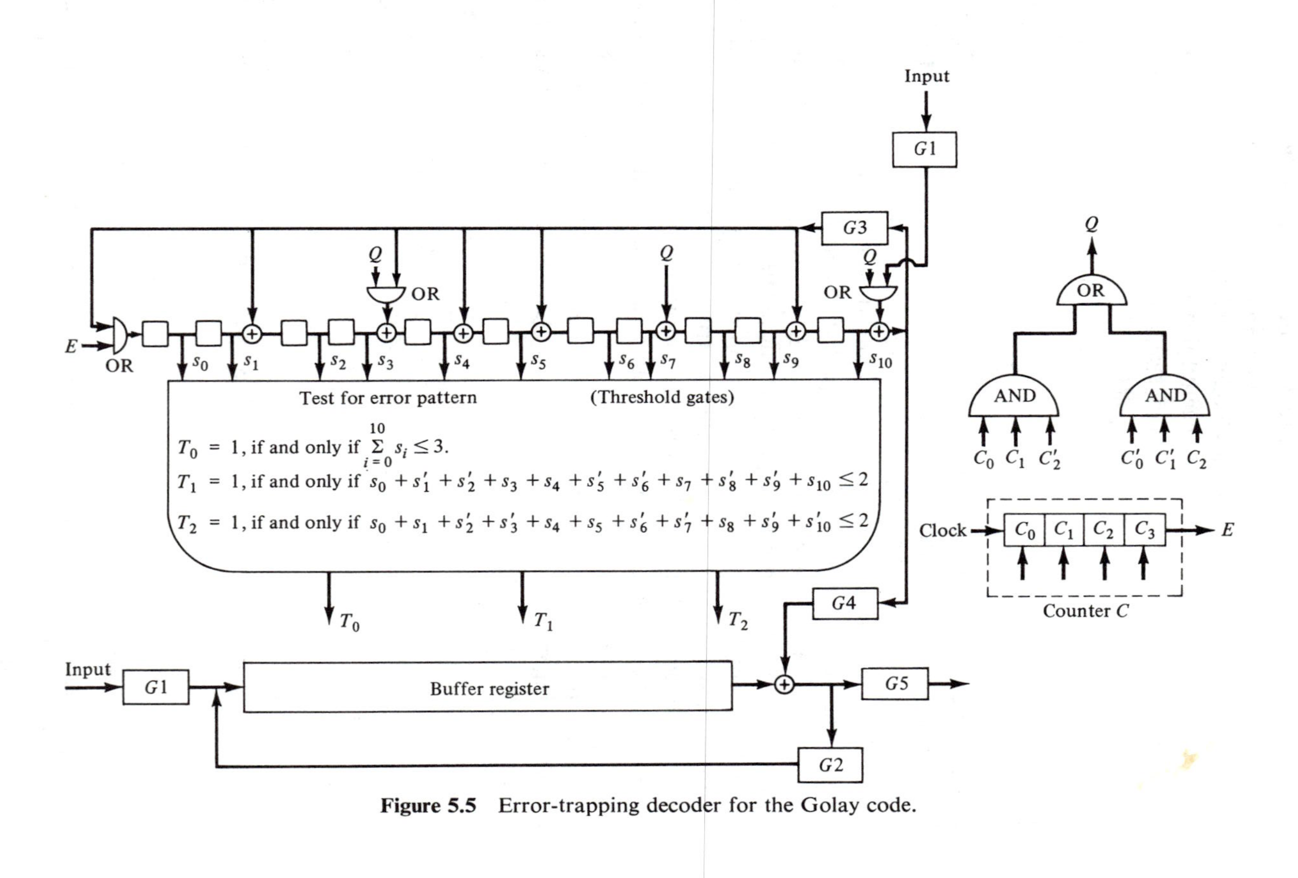

Figure 5.5 Error-trapping decoder for the Golay code.

high-order positions $X^{12}, X^{13}, \ldots, X^{22}$ of $\mathbf{r}(X)$, the syndrome matches the errors in those positions. The error-correction procedure of this decoder is described in the following steps:

Step 1. Gates 1, 3, and 5 are turned on; gates 2 and 4 are turned off. The received vector $\mathbf{r}(X)$ is read into the syndrome register and simultaneously into the buffer register. The syndrome $\mathbf{s}(X) = s_0 + s_1X + \cdots + s_{10}X^{10}$ is formed and is read into three threshold gates.

Step 2. Gates 1, 4, and 5 are turned off; gates 2 and 3 are turned on. The syndrome is tested for correctable error patterns as follows:

(a) If the weight $w[\mathbf{s}(X)] \leq 3$, all the errors are confined to the 11 high-order positions of $\mathbf{r}(X)$ and $\mathbf{s}(X)$ matches the errors. Thus, the erroneous symbols are the next 11 digits to come out of the buffer register. The output of the threshold gate T_0 turns gate 4 on and gate 3 off. Digits are read out one at a time from the buffer register. The digit coming out of the syndrome register is added (modulo-2) to the digit coming out of the buffer. This corrects the errors.

(b) If $w[\mathbf{s}(X)] > 3$, the weight of $\mathbf{s}(X) + \boldsymbol{\rho}_2(X)$ is tested. If $w[\mathbf{s}(X) + \boldsymbol{\rho}_2(X)] \leq 2$, then $\mathbf{s}(X) + \boldsymbol{\rho}_2(X) = s_0 + s_1'X + s_2'X^2 + s_3X^3 + s_4X^4 + s_5'X^5 + s_6'X^6 + s_7X^7 + s_8'X^8 + s_9'X^9 + s_{10}X^{10}$ is identical to the error pattern in the 11 high-order positions of the received word and a single error occurs at location X^5, where s_i' is the complement of s_i. Gate 4 is turned on, and gate 3 is turned off. The counter C starts to count from 2. At the same time, the syndrome register is shifted without feedback. The output Q, which is 1 when and only when C counts 3 and 4, is fed into the syndrome register to form the error pattern $\mathbf{s}(X) + \boldsymbol{\rho}_2(X)$. When the counter C counts 8, its output E is 1 and the leftmost stage of the syndrome register is set to 1. This 1 is used for correcting the error at location X^5 in the received vector $\mathbf{r}(X)$. The digits coming out of the buffer are then corrected by the digits coming out of the syndrome register.

(c) If $w[\mathbf{s}(X)] > 3$ and $w[\mathbf{s}(X) + \boldsymbol{\rho}_2(X)] > 2$, the weight of $\mathbf{s}(X) + \boldsymbol{\rho}_3(X)$ is tested. If $w[\mathbf{s}(X) + \boldsymbol{\rho}_3(X)] \leq 2$, then $\mathbf{s}(X) + \boldsymbol{\rho}_3(X) = s_0 + s_1X + s_2'X^2 + s_3'X^3 + s_4X^4 + s_5X^5 + s_6'X^6 + s_7'X^7 + s_8X^8 + s_9'X^9 + s_{10}'X^{10}$ is identical to the error pattern in the 11 high-order positions of the received word and a single error occurs at positions X^6. The correction is the same as step (b), except that counter C starts to count from 3. If $w[\mathbf{s}(X)] > 3$, $w[\mathbf{s}(X) + \boldsymbol{\rho}_2(X)] > 2$, and $w[\mathbf{s}(X) + \boldsymbol{\rho}_3(X)] > 2$, then the decoder moves to step 3.

Step 3. Both the syndrome and buffer registers are cyclically shifted once with gates 1, 4, and 5 turned off and gates 2 and 3 turned on. The new contents of the syndrome register are $\mathbf{s}^{(1)}(X)$. Step 2 is then repeated.

Step 4. The decoding operation is completed as soon as the buffer register has been cyclically shifted 46 times. Gate 5 is then turned on and the vector in the buffer is shifted out to the data sink.

If there are three or fewer errors in the received vector, the vector in the buffer at the end of decoding will be the transmitted code vector. If there are more than three errors in the received vector, the vector in the buffer at the end of decoding will not be the transmitted code vector.

Systematic Search Decoder [14]

This decoding method is based on the fact that every pattern of three or fewer errors in a block of 23 digits can be cyclically shifted so that at most one of the errors lies outside a specified 11-digit section of the word. The decoding procedure can be described as follows:

Step 1. Compute the syndrome from the received vector.

Step 2. Shift the syndrome and the received vector 23 times, checking whether the weight of the syndrome ever falls to 3 or less. If it does, the syndrome with weight 3 or less matches the error pattern and correction can be made.

Step 3. If it does not, the first received information digit is inverted and step 2 is repeated, checking for a syndrome of weight of 2 or less. If one is found, the first received information digit was incorrect and the other two errors are specified by the syndrome. This completes the decoding.

Step 4. If no syndrome of weight 2 or less is found in step 3, the first information digit was originally correct. In this case, this bit must be reinverted.

Step 5. Repeat step 3 by inverting the second, third, . . . , and twelfth information digits. Since not all the errors are in the parity-check section, an error must be corrected in this manner.

In every pattern of 3 or fewer errors, there is at least one error which, if corrected, will leave the remaining error or errors within 11 successive positions. When the digit corresponding to this error is inverted, the remaining errors are corrected as in ordinary error trapping.

Compared to the Kasami decoder, the systematic search decoder has the advantage that only one weight sensing (threshold) gate is required. However, it has the disadvantage that the clock and timing circuitry is more complex than the Kasami decoder since 12 different digits must be inverted sequentially. Also, the Kasami decoder operates faster than the systematic search decoder.

This systematic search technique can be generalized for decoding other multiple-error-correcting cyclic codes.

The weight enumerator for the (23, 12) Golay code is [15]

$$A(z) = 1 + 253z^7 + 506z^8 + 1288z^{11} + 1288z^{12} + 506z^{15} + 253z^{16} + z^{23}.$$

If this code is used for error detection on a BSC, its probability of an undetected error $P_u(E)$ can be computed from (3.19). Moreover, $P_u(E)$ satisfies the upper bound 2^{-11} [16] [i.e., $P_u(E) \leq 2^{-11}$].

PROBLEMS

5.1. Consider the (15, 5) cyclic code generated by the following polynomial:

$$\mathbf{g}(X) = 1 + X + X^2 + X^4 + X^5 + X^8 + X^{10}.$$

This code has been proved to be capable of correcting any combination of three or fewer errors. Suppose that this code is to be decoded by the simple error-trapping decoding scheme.

(a) Show that all the double errors can be trapped.

(b) Can all the error patterns of three errors be trapped? If not, how many error patterns of three errors cannot be trapped?

(c) Devise a simple error-trapping decoder for this code.

5.2. **(a)** Devise a simple error-trapping decoder for the (23, 12) Golay code.

(b) How many error patterns of double errors cannot be trapped?

(c) How many error patterns of three errors cannot be trapped?

5.3. Suppose that the (23, 12) Golay code is used only for error correction on a BSC with transition probability p. If Kasami's decoder of Figure 5.5 is used for decoding this code, what is the probability of a decoding error? [*Hint:* Use the fact that the (23, 12) Golay code is a perfect code.]

5.4. Use the decoder of Figure 5.5 to decode the following received polynomials:

(a) $\mathbf{r}(X) = X^5 + X^{19}$

(b) $\mathbf{r}(X) = X^4 + X^{11} + X^{21}$

At each step in the decoding process, write down the contents in the syndrome register.

5.5. Consider the following binary polynomial:

$$\mathbf{g}(X) = (X^3 + 1)\mathbf{p}(X),$$

where $(X^3 + 1)$ and $\mathbf{p}(X)$ are relatively prime and $\mathbf{p}(X)$ is an irreducible polynomial of degree m with $m \geq 3$. Let n be the smallest integer such that $\mathbf{g}(X)$ divides $X^n + 1$. Thus, $\mathbf{g}(X)$ generates a cyclic code of length n.

(a) Show that this code is capable of correcting all the single-error, double-adjacent-error, and triple-adjacent-error patterns. (*Hint:* Show that these error patterns can be used as coset leaders of a standard array for the code.)

(b) Devise an error-trapping decoder for this code. The decoder must be capable of correcting all the single-error, double-adjacent-error, and triple-adjacent-error patterns. Design a combinational logic circuit whose output is 1 when the errors are trapped in the appropriate stages of the syndrome register.

(c) Suppose that $\mathbf{p}(X) = 1 + X + X^4$, which is a primitive polynomial of degree 4. Determine the smallest integer n such that $\mathbf{g}(X) = (X^3 + 1)\mathbf{p}(X)$ divides $X^n + 1$.

REFERENCES

1. T. Kasami, "A Decoding Method for Multiple-Error-Correcting Cyclic Codes by Using Threshold Logics," Conv. Rec. Inf. Process. Soc. Jap. (in Japanese), Tokyo, November 1961.
2. M. E. Mitchell et al., "Coding and Decoding Operation Research," G. E. Advanced

Electronics Final Report on Contract AF 19 (604)-6183, Air Force Cambridge Research Labs., Cambridge, Mass., 1961.

3. M. E. Mitchell, "Error-Trap Decoding of Cyclic Codes," G. E. Report No. 62MCD3, General Electric Military Communications Dep., Oklahoma City, Okla., December 1962.
4. L. Rudolph, "Easily Implemented Error-Correction Encoding–Decoding," G. E. Report No. 62MCD2, General Electric Corporation, Oklahoma City, Okla., December 1962.
5. T. Kasami, "A Decoding Procedure For Multiple-Error-Correction Cyclic Codes," *IEEE Trans. Inf. Theory*, IT-10, pp. 134–139, April 1964.
6. E. Prange, "The Use of Coset Equivalence in the Analysis and Decoding of Group Codes," AFCRC-TR-59-164, Air Force Cambridge Research Labs., Cambridge, Mass., June 1959.
7. E. Prange, "The Use of Information Sets in Decoding Cyclic Codes," *IEEE Trans. Inf. Theory*, IT-8, pp. 85–89, September 1962.
8. F. J. MacWilliams, "Permutation Decoding of Systematic Codes," *Bell Syst. Tech. J.*, 43, Part I, pp. 485–505, January 1964.
9. L. Rudolph and M. E. Mitchell, "Implementation of Decoders for Cyclic Codes," *IEEE Trans. Inf. Theory*, IT-10, pp. 259–260, July 1964.
10. D. C. Foata, "On a Program for Ray-Chaudhuri's Algorithm for a Minimum Cover of an Abstract Complex," *Commun. ACM*, 4, pp. 504–506, November 1961.
11. I. B. Pyne and E. J. McCluskey, "The Reduction of Redundancy in Solving Prime Implicant Tables," *IRE Trans. Electron. Comput.*, EC-11, pp. 473–482, August 1962.
12. M. J. E. Golay, "Notes on Digital Coding," *Proc. IRE*, 37, p. 657, June 1949.
13. F.J. MacWilliams and N. J. A. Sloane, *The Theory of Error-Correcting Codes*, North-Holland, Amsterdam, 1977.
14. E. J. Weldon, Jr., "A Comparison of an Interleaved Golay Code and a Three-Dimensional Product Code," Final Report, USNELC Contract N0095368M5345, August 1968.
15. W. W. Peterson and E. J. Weldon, Jr., *Error-Correcting Codes*, 2nd ed., MIT Press, Cambridge, Mass., 1972.
16. S. K. Leung-Yan-Cheong, E. R. Barnes, and D. U. Friedman, "On Some Properties of the Undetected Error Probability of Linear Codes," *IEEE Trans. Inf. Theory*, IT-25 (1), pp. 110–112, January 1979.

6

BCH Codes

The Bose, Chaudhuri, and Hocquenghem (BCH) codes form a large class of powerful random error-correcting cyclic codes. This class of codes is a remarkable generalization of the Hamming codes for multiple-error correction. Binary BCH codes were discovered by Hocquenghem in 1959 [1] and independently by Bose and Chaudhuri in 1960 [2]. The cyclic structure of these codes was proved by Peterson in 1960 [3]. Generalization of the binary BCH codes to codes in p^m symbols (where p is a prime) was obtained by Gorenstein and Zierler in 1961 [4]. Among the nonbinary BCH codes, the most important subclass is the class of Reed–Solomon (RS) codes. The RS codes were introduced by Reed and Solomon in 1960 [5] independently of the works by Hocquenghem, Bose, and Chaudhuri.

The first decoding algorithm for binary BCH codes was devised by Peterson in 1960 [3]. Since then, Peterson's algorithm has been generalized and refined by Gorenstein and Zierler [4], Chien [6], Forney [7], Berlekamp [8, 9], Massey [10, 11], Burton [12], and others. Among all the decoding algorithms for BCH codes, Berlekamp's iterative algorithm, and Chien's search algorithm are the most efficient ones.

In this chapter we consider primarily a subclass of the binary BCH codes which is the most important subclass from the standpoint of both theory and implementation. For nonbinary BCH codes, we discuss only the Reed–Solomon codes. For a detailed description of the BCH codes, their algebraic properties and decoding algorithms, the reader is referred to References 9 and 13 to 15.

6.1 DESCRIPTION OF THE CODES

For any positive integers $m(m \geq 3)$ and $t(t < 2^{m-1})$, there exists a binary BCH code with the following parameters:

Block length:	$n = 2^m - 1$
Number of parity-check digits:	$n - k \leq mt$
Minimum distance:	$d_{\min} \geq 2t + 1.$

Clearly, this code is capable of correcting any combination of t or fewer errors in a block of $n = 2^m - 1$ digits. We call this code a t-error-correcting BCH code. The generator polynomial of this code is specified in terms of its roots from the Galois field GF(2^m). Let α be a primitive element in GF(2^m). The generator polynomial $\mathbf{g}(X)$ of the t-error-correcting BCH code of length $2^m - 1$ is the *lowest-degree polynomial* over GF(2) which has

$$\alpha, \alpha^2, \alpha^3, \ldots, \alpha^{2t} \tag{6.1}$$

as its roots [i.e., $\mathbf{g}(\alpha^i) = 0$ for $1 \leq i \leq 2t$]. It follows from Theorem 2.7 that $\mathbf{g}(X)$ has $\alpha, \alpha^2, \ldots, \alpha^{2t}$ and their conjugates as all its roots. Let $\phi_i(X)$ be the minimal polynomial of α^i. Then $\mathbf{g}(X)$ must be the *least common multiple* of $\phi_1(X), \phi_2(X), \ldots, \phi_{2t}(X)$, that is,

$$\mathbf{g}(X) = \text{LCM}\,\{\phi_1(X), \phi_2(X), \ldots, \phi_{2t}(X)\}. \tag{6.2}$$

If i is an even integer, it can be expressed as a product of the following form:

$$i = i'2^l,$$

where i' is an odd number and $l \geq 1$. Then $\alpha^i = (\alpha^{i'})^{2^l}$ is a conjugate of $\alpha^{i'}$ and therefore α^i and $\alpha^{i'}$ have the same minimal polynomial, that is,

$$\phi_i(X) = \phi_{i'}(X).$$

Hence, every even power of α in the sequence of (6.1) has the same minimal polynomial as some preceding odd power of α in the sequence. As a result, the generator polynomial $\mathbf{g}(X)$ of the binary t-error-correcting BCH code of length $2^m - 1$ given by (6.2) can be reduced to

$$\mathbf{g}(X) = \text{LCM}\,\{\phi_1(X), \phi_3(X), \ldots, \phi_{2t-1}(X)\}. \tag{6.3}$$

Since the degree of each minimal polynomial is m or less, the degree of $\mathbf{g}(X)$ is at most mt. That is, the number of parity-check digits, $n - k$, of the code is at most equal to mt. There is no simple formula for enumerating $n - k$, but if t is small, $n - k$ is exactly equal to mt [9, 16]. The parameters for all binary BCH codes of length $2^m - 1$ with $m \leq 10$ are given in Table 6.1. The BCH codes defined above are usually called *primitive* (or *narrow-sense*) BCH codes.

From (6.3), we see that the single-error-correcting BCH code of length $2^m - 1$ is generated by

$$\mathbf{g}(X) = \phi_1(X).$$

Since α is a primitive element of GF(2^m), $\phi_1(X)$ is a primitive polynomial of degree m. Therefore, the single-error-correcting BCH code of length $2^m - 1$ is a Hamming code.

TABLE 6.1 BCH CODES GENERATED BY PRIMITIVE ELEMENTS OF ORDER LESS THAN 2^{10}

n	k	t	n	k	t	n	k	t
7	4	1	255	163	12	511	268	29
15	11	1		155	13		259	30
	7	2		147	14		250	31
	5	3		139	15		241	36
31	26	1		131	18		238	37
	21	2		123	19		229	38
	16	3		115	21		220	39
	11	5		107	22		211	41
	6	7		99	23		202	42
63	57	1		91	25		193	43
	51	2		87	26		184	45
	45	3		79	27		175	46
	39	4		71	29		166	47
	36	5		63	30		157	51
	30	6		55	31		148	53
	24	7		47	42		139	54
	18	10		45	43		130	55
	16	11		37	45		121	58
	10	13		29	47		112	59
	7	15		21	55		103	61
127	120	1		13	59		94	62
	113	2		9	63		85	63
	106	3	511	502	1		76	85
	99	4		493	2		67	87
	92	5		484	3		58	91
	85	6		475	4		49	93
	78	7		466	5		40	95
	71	9		457	6		31	109
	64	10		448	7		28	111
	57	11		439	8		19	119
	50	13		430	9		10	121
	43	14		421	10	1023	1013	1
	36	15		412	11		1003	2
	29	21		403	12		993	3
	22	23		394	13		983	4
	15	27		385	14		973	5
	8	31		376	15		963	6
255	247	1		367	16		953	7
	239	2		358	18		943	8
	231	3		349	19		933	9
	223	4		340	20		923	10
	215	5		331	21		913	11
	207	6		322	22		903	12
	199	7		313	23		893	13
	191	8		304	25		883	14
	187	9		295	26		873	15
	179	10		286	27		863	16
	171	11		277	28		858	17

TABLE 6.1 Continued.

n	k	t	n	k	t	n	k	t
1023	848	18	1023	553	52	1023	268	103
	838	19		543	53		258	106
	828	20		533	54		248	107
	818	21		523	55		238	109
	808	22		513	57		228	110
	798	23		503	58		218	111
	788	24		493	59		208	115
	778	25		483	60		203	117
	768	26		473	61		193	118
	758	27		463	62		183	119
	748	28		453	63		173	122
	738	29		443	73		163	123
	728	30		433	74		153	125
	718	31		423	75		143	126
	708	34		413	77		133	127
	698	35		403	78		123	170
	688	36		393	79		121	171
	678	37		383	82		111	173
	668	38		378	83		101	175
	658	39		368	85		91	181
	648	41		358	86		86	183
	638	42		348	87		76	187
	628	43		338	89		66	189
	618	44		328	90		56	191
	608	45		318	91		46	219
	598	46		308	93		36	223
	588	47		298	94		26	239
	578	49		288	95		16	147
	573	50		278	102		11	255
	563	51						

Example 6.1

Let α be a primitive element of the Galois field GF(2^4) given by Table 2.8 such that $1 + \alpha + \alpha^4 = 0$. From Table 2.9 we find that the minimal polynomials of α, α^3, and α^5 are

$$\phi_1(X) = 1 + X + X^4,$$
$$\phi_3(X) = 1 + X + X^2 + X^3 + X^4,$$

and

$$\phi_5(X) = 1 + X + X^2,$$

respectively. It follows from (6.3) that the double-error-correcting BCH code of length $n = 2^4 - 1 = 15$ is generated by

$$\mathbf{g}(X) = \text{LCM}\,\{\phi_1(X), \phi_3(X)\}.$$

Since $\phi_1(X)$ and $\phi_3(X)$ are two distinct irreducible polynomials,

$$\begin{aligned}\mathbf{g}(X) &= \phi_1(X)\phi_3(X)\\ &= (1 + X + X^4)(1 + X + X^2 + X^3 + X^4)\\ &= 1 + X^4 + X^6 + X^7 + X^8.\end{aligned}$$

Thus, the code is a (15, 7) cyclic code with $d_{\min} \geq 5$. Since the generator polynomial is code polynomial of weight 5, the minimum distance of this code is exactly 5.

The triple-error-correcting BCH code of length 15 is generated by

$$\begin{aligned}\mathbf{g}(X) &= \text{LCM}\,\{\phi_1(X), \phi_3(X), \phi_5(X)\} \\ &= (1 + X + X^4)(1 + X + X^2 + X^3 + X^4)(1 + X + X^2) \\ &= 1 + X + X^2 + X^4 + X^5 + X^8 + X^{10}.\end{aligned}$$

This triple-error-correcting BCH code is a (15, 5) cyclic code with $d_{\min} \geq 7$. Since the weight of the generator polynomial is 7, the minimum distance of this code is exactly 7.

Using the primitive polynomial $\mathbf{p}(X) = 1 + X + X^6$, we may construct the Galois field GF(2^6) as shown in Table 6.2. The minimal polynomials of the elements in GF(2^6) are listed in Table 6.3. Using (6.3), we find the generator polynomials of all the BCH codes of length 63 as shown in Table 6.4. The generator polynomials of all binary primitive BCH codes of length $2^m - 1$ with $m \leq 10$ are given in Appendix C.

TABLE 6.2 GALOIS FIELD GF(2^6) WITH $p(\alpha) = 1 + \alpha + \alpha^6 = 0$

0	0	(0 0 0 0 0 0)
1	1	(1 0 0 0 0 0)
α	α	(0 1 0 0 0 0)
α^2	α^2	(0 0 1 0 0 0)
α^3	α^3	(0 0 0 1 0 0)
α^4	α^4	(0 0 0 0 1 0)
α^5	α^5	(0 0 0 0 0 1)
α^6	$1 + \alpha$	(1 1 0 0 0 0)
α^7	$\alpha + \alpha^2$	(0 1 1 0 0 0)
α^8	$\alpha^2 + \alpha^3$	(0 0 1 1 0 0)
α^9	$\alpha^3 + \alpha^4$	(0 0 0 1 1 0)
α^{10}	$\alpha^4 + \alpha^5$	(0 0 0 0 1 1)
α^{11}	$1 + \alpha + \alpha^5$	(1 1 0 0 0 1)
α^{12}	$1 + \alpha^2$	(1 0 1 0 0 0)
α^{13}	$\alpha + \alpha^3$	(0 1 0 1 0 0)
α^{14}	$\alpha^2 + \alpha^4$	(0 0 1 0 1 0)
α^{15}	$\alpha^3 + \alpha^5$	(0 0 0 1 0 1)
α^{16}	$1 + \alpha + \alpha^4$	(1 1 0 0 1 0)
α^{17}	$\alpha + \alpha^2 + \alpha^5$	(0 1 1 0 0 1)
α^{18}	$1 + \alpha + \alpha^2 + \alpha^3$	(1 1 1 1 0 0)
α^{19}	$\alpha + \alpha^2 + \alpha^3 + \alpha^4$	(0 1 1 1 1 0)
α^{20}	$\alpha^2 + \alpha^3 + \alpha^4 + \alpha^5$	(0 0 1 1 1 1)
α^{21}	$1 + \alpha + \alpha^3 + \alpha^4 + \alpha^5$	(1 1 0 1 1 1)
α^{22}	$1 + \alpha^2 + \alpha^4 + \alpha^5$	(1 0 1 0 1 1)
α^{23}	$1 + \alpha^3 + \alpha^5$	(1 0 0 1 0 1)
α^{24}	$1 + \alpha^4$	(1 0 0 0 1 0)
α^{25}	$\alpha + \alpha^5$	(0 1 0 0 0 1)
α^{26}	$1 + \alpha + \alpha^2$	(1 1 1 0 0 0)
α^{27}	$\alpha + \alpha^2 + \alpha^3$	(0 1 1 1 0 0)
α^{28}	$\alpha^2 + \alpha^3 + \alpha^4$	(0 0 1 1 1 0)
α^{29}	$\alpha^3 + \alpha^4 + \alpha^5$	(0 0 0 1 1 1)
α^{30}	$1 + \alpha + \alpha^4 + \alpha^5$	(1 1 0 0 1 1)

TABLE 6.2 Continued.

α^{31}	$1 + \alpha^2 + \alpha^5$	(1 0 1 0 0 1)
α^{32}	$1 + \alpha^3$	(1 0 0 1 0 0)
α^{33}	$\alpha + \alpha^4$	(0 1 0 0 1 0)
α^{34}	$\alpha^2 + \alpha^5$	(0 0 1 0 0 1)
α^{35}	$1 + \alpha + \alpha^3$	(1 1 0 1 0 0)
α^{36}	$\alpha + \alpha^2 + \alpha^4$	(0 1 1 0 1 0)
α^{37}	$\alpha^2 + \alpha^3 + \alpha^5$	(0 0 1 1 0 1)
α^{38}	$1 + \alpha + \alpha^3 + \alpha^4$	(1 1 0 1 1 0)
α^{39}	$\alpha + \alpha^2 + \alpha^4 + \alpha^5$	(0 1 1 0 1 1)
α^{40}	$1 + \alpha + \alpha^2 + \alpha^3 + \alpha^5$	(1 1 1 1 0 1)
α^{41}	$1 + \alpha^2 + \alpha^3 + \alpha^4$	(1 0 1 1 1 0)
α^{42}	$\alpha + \alpha^3 + \alpha^4 + \alpha^5$	(0 1 0 1 1 1)
α^{43}	$1 + \alpha + \alpha^2 + \alpha^4 + \alpha^5$	(1 1 1 0 1 1)
α^{44}	$1 + \alpha^2 + \alpha^3 + \alpha^5$	(1 0 1 1 0 1)
α^{45}	$1 + \alpha^3 + \alpha^4$	(1 0 0 1 1 0)
α^{46}	$\alpha + \alpha^4 + \alpha^5$	(0 1 0 0 1 1)
α^{47}	$1 + \alpha + \alpha^2 + \alpha^5$	(1 1 1 0 0 1)
α^{48}	$1 + \alpha^2 + \alpha^3$	(1 0 1 1 0 0)
α^{49}	$\alpha + \alpha^3 + \alpha^4$	(0 1 0 1 1 0)
α^{50}	$\alpha^2 + \alpha^4 + \alpha^5$	(0 0 1 0 1 1)
α^{51}	$1 + \alpha + \alpha^3 + \alpha^5$	(1 1 0 1 0 1)
α^{52}	$1 + \alpha^2 + \alpha^4$	(1 0 1 0 1 0)
α^{53}	$\alpha + \alpha^3 + \alpha^5$	(0 1 0 1 0 1)
α^{54}	$1 + \alpha + \alpha^2 + \alpha^4$	(1 1 1 0 1 0)
α^{55}	$\alpha + \alpha^2 + \alpha^3 + \alpha^5$	(0 1 1 1 0 1)
α^{56}	$1 + \alpha + \alpha^2 + \alpha^3 + \alpha^4$	(1 1 1 1 1 0)
α^{57}	$\alpha + \alpha^2 + \alpha^3 + \alpha^4 + \alpha^5$	(0 1 1 1 1 1)
α^{58}	$1 + \alpha + \alpha^2 + \alpha^3 + \alpha^4 + \alpha^5$	(1 1 1 1 1 1)
α^{59}	$1 + \alpha^2 + \alpha^3 + \alpha^4 + \alpha^5$	(1 0 1 1 1 1)
α^{60}	$1 + \alpha^3 + \alpha^4 + \alpha^5$	(1 0 0 1 1 1)
α^{61}	$1 + \alpha^4 + \alpha^5$	(1 0 0 0 1 1)
α^{62}	$1 + \alpha^5$	(1 0 0 0 0 1)

$\alpha^{63} = 1$

TABLE 6.3 MINIMAL POLYNOMIALS OF THE ELEMENTS IN GF(2^6)

Elements	Minimal polynomials
$\alpha, \alpha^2, \alpha^4, \alpha^8, \alpha^{16}, \alpha^{32}$	$1 + X + X^6$
$\alpha^3, \alpha^6, \alpha^{12}, \alpha^{24}, \alpha^{48}, \alpha^{33}$	$1 + X + X^2 + X^4 + X^6$
$\alpha^5, \alpha^{10}, \alpha^{20}, \alpha^{40}, \alpha^{17}, \alpha^{34}$	$1 + X + X^2 + X^5 + X^6$
$\alpha^7, \alpha^{14}, \alpha^{28}, \alpha^{56}, \alpha^{49}, \alpha^{35}$	$1 + X^3 + X^6$
$\alpha^9, \alpha^{18}, \alpha^{36}$	$1 + X^2 + X^3$
$\alpha^{11}, \alpha^{22}, \alpha^{44}, \alpha^{25}, \alpha^{50}, \alpha^{37}$	$1 + X^2 + X^3 + X^5 + X^6$
$\alpha^{13}, \alpha^{26}, \alpha^{52}, \alpha^{41}, \alpha^{19}, \alpha^{38}$	$1 + X + X^3 + X^4 + X^6$
$\alpha^{15}, \alpha^{30}, \alpha^{60}, \alpha^{57}, \alpha^{51}, \alpha^{39}$	$1 + X^2 + X^4 + X^5 + X^6$
α^{21}, α^{42}	$1 + X + X^2$
$\alpha^{23}, \alpha^{46}, \alpha^{29}, \alpha^{58}, \alpha^{53}, \alpha^{43}$	$1 + X + X^4 + X^5 + X^6$
$\alpha^{27}, \alpha^{54}, \alpha^{45}$	$1 + X + X^3$
$\alpha^{31}, \alpha^{62}, \alpha^{61}, \alpha^{59}, \alpha^{55}, \alpha^{47}$	$1 + X^5 + X^6$

TABLE 6.4 GENERATOR POLYNOMIALS OF ALL THE BCH CODES OF LENGTH 63

n	k	t	$\mathbf{g}(X)$
63	57	1	$\mathbf{g}_1(X) = 1 + X + X^6$
	51	2	$\mathbf{g}_2(X) = (1 + X + X^6)(1 + X + X^2 + X^4 + X^6)$
	45	3	$\mathbf{g}_3(X) = (1 + X + X^2 + X^5 + X^6)\mathbf{g}_2(X)$
	39	4	$\mathbf{g}_4(X) = (1 + X^3 + X^6)\mathbf{g}_3(X)$
	36	5	$\mathbf{g}_5(X) = (1 + X^2 + X^3)\mathbf{g}_4(X)$
	30	6	$\mathbf{g}_6(X) = (1 + X^2 + X^3 + X^5 + X^6)\mathbf{g}_5(X)$
	24	7	$\mathbf{g}_7(X) = (1 + X + X^3 + X^4 + X^6)\mathbf{g}_6(X)$
	18	10	$\mathbf{g}_{10}(X) = (1 + X^2 + X^4 + X^5 + X^6)\mathbf{g}_7(X)$
	16	11	$\mathbf{g}_{11}(X) = (1 + X + X^2)\mathbf{g}_{10}(X)$
	10	13	$\mathbf{g}_{13}(X) = (1 + X + X^4 + X^5 + X^6)\mathbf{g}_{11}(X)$
	7	15	$\mathbf{g}_{15}(X) = (1 + X + X^3)\mathbf{g}_{13}(X)$

It follows from the definition of a t-error-correcting BCH code of length $n = 2^m - 1$ that each code polynomial has $\alpha, \alpha^2, \ldots, \alpha^{2t}$ and their conjugates as roots. Now, let $\mathbf{v}(X) = v_0 + v_1X + \cdots + v_{n-1}X^{n-1}$ be a polynomial with coefficients from GF(2). If $\mathbf{v}(X)$ has $\alpha, \alpha^2, \ldots, \alpha^{2t}$ as roots, it follows from Theorem 2.10 that $\mathbf{v}(X)$ is divisible by the minimal polynomials $\phi_1(X), \phi_2(X), \ldots, \phi_{2t}(X)$ of $\alpha, \alpha^2, \ldots, \alpha^{2t}$. Obviously, $\mathbf{v}(X)$ is divisible by their least common multiple (the generator polynomial),

$$\mathbf{g}(X) = \text{LCM}\,\{\phi_1(X), \phi_2(X), \ldots, \phi_{2t}(X)\}.$$

Hence, $\mathbf{v}(X)$ is a code polynomial. Consequently, we may define a t-error-correcting BCH code of length $n = 2^m - 1$ in the following manner: A binary n-tuple $\mathbf{v} = (v_0, v_1, v_2, \ldots, v_{n-1})$ is a code word if and only if the polynomial $\mathbf{v}(X) = v_0 + v_1X + \cdots + v_{n-1}X^{n-1}$ has $\alpha, \alpha^2, \ldots, \alpha^{2t}$ as roots. This definition is useful in proving the minimum distance of the code.

Let $\mathbf{v}(X) = v_0 + v_1X + \cdots + v_{n-1}X^{n-1}$ be a code polynomial in a t-error-correcting BCH code of length $n = 2^m - 1$. Since α^i is a root of $\mathbf{v}(X)$ for $1 \le i \le 2t$, then

$$\mathbf{v}(\alpha^i) = v_0 + v_1\alpha^i + v_2\alpha^{2i} + \cdots + v_{n-1}\alpha^{(n-1)i} = 0 \tag{6.4}$$

This equality can be written as a matrix product as follows:

$$(v_0, v_1, \ldots, v_{n-1}) \cdot \begin{bmatrix} 1 \\ \alpha^i \\ \alpha^{2i} \\ \cdot \\ \cdot \\ \cdot \\ \alpha^{(n-1)i} \end{bmatrix} = 0 \tag{6.5}$$

for $1 \le i \le 2t$. The condition given by (6.5) simply says that the inner product of $(v_0, v_1, \ldots, v_{n-1})$ and $(1, \alpha^i, \alpha^{2i}, \ldots, \alpha^{(n-1)i})$ is equal to zero. Now we form the following matrix:

$$\mathbf{H} = \begin{bmatrix} 1 & \alpha & \alpha^2 & \alpha^3 & \cdots & \alpha^{n-1} \\ 1 & (\alpha^2) & (\alpha^2)^2 & (\alpha^2)^3 & \cdots & (\alpha^2)^{n-1} \\ 1 & (\alpha^3) & (\alpha^3)^2 & (\alpha^3)^3 & \cdots & (\alpha^3)^{n-1} \\ \cdot & & & & & \cdot \\ \cdot & & & & & \cdot \\ \cdot & & & & & \cdot \\ 1 & (\alpha^{2t}) & (\alpha^{2t})^2 & (\alpha^{2t})^3 & \cdots & (\alpha^{2t})^{n-1} \end{bmatrix}. \tag{6.6}$$

It follows from (6.5) that if $\mathbf{v} = (v_0, v_1, \ldots, v_{n-1})$ is a code word in the t-error-correcting BCH code, then

$$\mathbf{v} \cdot \mathbf{H}^T = \mathbf{0}. \tag{6.7}$$

On the other hand, if an n-tuple $\mathbf{v} = (v_0, v_1, \ldots, v_{n-1})$ satisfies the condition of (6.7), it follows from (6.5) and (6.4) that, for $1 \leq i \leq 2t$, α^i is a root of the polynomial $\mathbf{v}(X)$. Therefore, $\mathbf{v}$ must be a code word in the t-error-correcting BCH code. Hence, the code is the null space of the matrix $\mathbf{H}$, and $\mathbf{H}$ is the parity-check matrix of the code. If for some i and j, α^j is a conjugate of α^i, then $\mathbf{v}(\alpha^j) = 0$ if and only if $\mathbf{v}(\alpha^i) = 0$ (see Theorem 2.7). This says that if the inner product of $\mathbf{v} = (v_0, v_1, \ldots, v_{n-1})$ and the ith row of $\mathbf{H}$ is zero, the inner product of $\mathbf{v}$ and the jth row of $\mathbf{H}$ is also zero. For this reason, the jth row of $\mathbf{H}$ can be omitted. As a result, the $\mathbf{H}$ matrix given by (6.6) can be reduced to the following form:

$$\mathbf{H} = \begin{bmatrix} 1 & \alpha & \alpha^2 & \alpha^3 & \cdots & \alpha^{n-1} \\ 1 & (\alpha^3) & (\alpha^3)^2 & (\alpha^3)^3 & \cdots & (\alpha^3)^{n-1} \\ 1 & (\alpha^5) & (\alpha^5)^2 & (\alpha^5)^3 & \cdots & (\alpha^5)^{n-1} \\ \cdot & & & & & \cdot \\ \cdot & & & & & \cdot \\ \cdot & & & & & \cdot \\ 1 & (\alpha^{2t-1}) & (\alpha^{2t-1})^2 & (\alpha^{2t-1})^3 & \cdots & (\alpha^{2t-1})^{n-1} \end{bmatrix}. \tag{6.8}$$

Note that the entries of $\mathbf{H}$ are elements in GF(2^m). Each element in GF(2^m) can be represented by a m-tuple over GF(2). If each entry of $\mathbf{H}$ is replaced by its corresponding m-tuple over GF(2) arranged in column form, we obtain a binary parity-check matrix for the code.

Example 6.2

Consider the double-error-correcting BCH code of length $n = 2^4 - 1 = 15$. From Example 6.1 we know that this is a (15, 7) code. Let α be a primitive element in GF(2^4). Then the parity-check matrix of this code is

$$\mathbf{H} = \begin{bmatrix} 1 & \alpha & \alpha^2 & \alpha^3 & \alpha^4 & \alpha^5 & \alpha^6 & \alpha^7 & \alpha^8 & \alpha^9 & \alpha^{10} & \alpha^{11} & \alpha^{12} & \alpha^{13} & \alpha^{14} \\ 1 & \alpha^3 & \alpha^6 & \alpha^9 & \alpha^{12} & \alpha^{15} & \alpha^{18} & \alpha^{21} & \alpha^{24} & \alpha^{27} & \alpha^{30} & \alpha^{33} & \alpha^{36} & \alpha^{39} & \alpha^{42} \end{bmatrix}$$

[use (6.8)]. Using Table 2.8, the fact that $\alpha^{15} = 1$, and representing each entry of $\mathbf{H}$ by its corresponding 4-tuple, we obtain the following binary parity-check matrix for the code:

$$
\mathbf{H} = \left[\begin{array}{ccccccccccccccc}
1 & 0 & 0 & 0 & 1 & 0 & 0 & 1 & 1 & 0 & 1 & 0 & 1 & 1 & 1 \\
0 & 1 & 0 & 0 & 1 & 1 & 0 & 1 & 0 & 1 & 1 & 1 & 1 & 0 & 0 \\
0 & 0 & 1 & 0 & 0 & 1 & 1 & 0 & 1 & 0 & 1 & 1 & 1 & 1 & 0 \\
0 & 0 & 0 & 1 & 0 & 0 & 1 & 1 & 0 & 1 & 0 & 1 & 1 & 1 & 1 \\
\hdashline
1 & 0 & 0 & 0 & 1 & 1 & 0 & 0 & 0 & 1 & 1 & 0 & 0 & 0 & 1 \\
0 & 0 & 0 & 1 & 1 & 0 & 0 & 0 & 1 & 1 & 0 & 0 & 0 & 1 & 1 \\
0 & 0 & 1 & 0 & 1 & 0 & 0 & 1 & 0 & 1 & 0 & 0 & 1 & 0 & 1 \\
0 & 1 & 1 & 1 & 1 & 0 & 1 & 1 & 1 & 1 & 0 & 1 & 1 & 1 & 1
\end{array}\right].
$$

Now we are ready to prove that the t-error-correcting BCH code defined above indeed has minimum distance at least $2t + 1$. To prove this, we need to show that no $2t$ or fewer columns of $\mathbf{H}$ given by (6.6) sum to zero. Suppose that there exists a nonzero code vector $\mathbf{v} = (v_0, v_1, \ldots, v_{n-1})$ with weight $\delta \leq 2t$. Let $v_{j_1}, v_{j_2}, \ldots, v_{j_\delta}$ be the nonzero components of $\mathbf{v}$ (i.e., $v_{j_1} = v_{j_2} = \ldots = v_{j_\delta} = 1$). Using (6.6) and (6.7), we have

$$
\mathbf{0} = \mathbf{v} \cdot \mathbf{H}^T
$$

$$
= (v_{j_1}, v_{j_2}, \ldots, v_{j_\delta}) \cdot \begin{bmatrix}
\alpha^{j_1} & (\alpha^2)^{j_1} & \cdots & (\alpha^{2t})^{j_1} \\
\alpha^{j_2} & (\alpha^2)^{j_2} & \cdots & (\alpha^{2t})^{j_2} \\
\alpha^{j_3} & (\alpha^2)^{j_3} & \cdots & (\alpha^{2t})^{j_3} \\
\vdots & \vdots & & \vdots \\
\alpha^{j_\delta} & (\alpha^2)^{j_\delta} & \cdots & (\alpha^{2t})^{j_\delta}
\end{bmatrix}
$$

$$
= (1, 1, \ldots, 1) \cdot \begin{bmatrix}
\alpha^{j_1} & (\alpha^{j_1})^2 & \cdots & (\alpha^{j_1})^{2t} \\
\alpha^{j_2} & (\alpha^{j_2})^2 & \cdots & (\alpha^{j_2})^{2t} \\
\alpha^{j_3} & (\alpha^{j_3})^2 & \cdots & (\alpha^{j_3})^{2t} \\
\vdots & \vdots & & \vdots \\
\alpha^{j_\delta} & (\alpha^{j_\delta})^2 & \cdots & (\alpha^{j_\delta})^{2t}
\end{bmatrix}.
$$

The equality above implies the following equality:

$$
(1, 1, \ldots, 1) \cdot \begin{bmatrix}
\alpha^{j_1} & (\alpha^{j_1})^2 & \cdots & (\alpha^{j_1})^{\delta} \\
\alpha^{j_2} & (\alpha^{j_2})^2 & \cdots & (\alpha^{j_2})^{\delta} \\
\alpha^{j_3} & (\alpha^{j_3})^2 & \cdots & (\alpha^{j_3})^{\delta} \\
\vdots & \vdots & & \vdots \\
\alpha^{j_\delta} & (\alpha^{j_\delta})^2 & \cdots & (\alpha^{j_\delta})^{\delta}
\end{bmatrix} = \mathbf{0}, \tag{6.9}
$$

where the second matrix on the left is a $\delta \times \delta$ square matrix. To satisfy the equality of (6.9), the *determinant* of the $\delta \times \delta$ matrix must be zero, that is,

$$\begin{vmatrix} \alpha^{j_1} & (\alpha^{j_1})^2 & \cdots & (\alpha^{j_1})^\delta \\ \alpha^{j_2} & (\alpha^{j_2})^2 & \cdots & (\alpha^{j_2})^\delta \\ \alpha^{j_3} & (\alpha^{j_3})^2 & \cdots & (\alpha^{j_3})^\delta \\ \vdots & \vdots & & \vdots \\ \alpha^{j_\delta} & (\alpha^{j_\delta})^2 & \cdots & (\alpha^{j_\delta})^\delta \end{vmatrix} = 0.$$

Taking out the common factor from each row of the determinant above, we obtain

$$\alpha^{(j_1+j_2+\cdots+j_\delta)} \cdot \begin{vmatrix} 1 & \alpha^{j_1} & \cdots & \alpha^{(\delta-1)j_1} \\ 1 & \alpha^{j_2} & \cdots & \alpha^{(\delta-1)j_2} \\ 1 & \alpha^{j_3} & \cdots & \alpha^{(\delta-1)j_3} \\ \vdots & \vdots & & \vdots \\ 1 & \alpha^{j_\delta} & \cdots & \alpha^{(\delta-1)j_\delta} \end{vmatrix} = 0. \tag{6.10}$$

The determinant in the equality above is a *Vandermonde determinant* which is *nonzero*. Therefore, the product on the left-hand side of (6.10) cannot be zero. This is a contradiction and hence our assumption that there exists a nonzero code vector $\mathbf{v}$ of weight $\delta \leq 2t$ is *invalid*. This implies that the minimum weight of the t-error-correcting BCH code defined above is at least $2t + 1$. Consequently, the minimum distance of the code is at least $2t + 1$.

The parameter $2t + 1$ is usually called the *designed distance* of the t-error-correcting BCH code. The true minimum distance of a BCH code may or may not be equal to its designed distance. There are many cases where the true minimum distance of a BCH code is equal to its designed distance. However, there are also cases where the true minimum distance is greater than the designed distance.

Binary BCH codes with length $n \neq 2^m - 1$ can be constructed in the same manner as for the case $n = 2^m - 1$. Let β be an element of order n in the field GF(2^m). We know that n is a factor of $2^m - 1$. Let $\mathbf{g}(X)$ be the binary polynomial of minimum degree that has

$$\beta, \beta^2, \ldots, \beta^{2t}$$

as roots. Let $\psi_1(X), \psi_2(X), \ldots, \psi_{2t}(X)$ be the minimal polynomials of $\beta, \beta^2, \ldots,$ β^{2t}, respectively. Then

$$\mathbf{g}(X) = \text{LCM}\,\{\psi_1(X), \psi_2(X), \ldots, \psi_{2t}(X)\}.$$

Since $\beta^n = 1$, $\beta, \beta^2, \ldots, \beta^{2t}$ are roots of $X^n + 1$. We see that $\mathbf{g}(X)$ is a factor of $X^n + 1$. The cyclic code generated by $\mathbf{g}(X)$ is a t-error-correcting BCH code of length n. In a manner similar to the case $n = 2^m - 1$, we can prove that the number of parity-check digits of this code is at most mt and the minimum distance of the code is at least $2t + 1$. If β is not a primitive element of GF(2^m), $n \neq 2^m - 1$ and the code is called a *nonprimitive* BCH code.

Example 6.3

Consider the Galois field GF(2^6) given in Table 6.2. The element $\beta = \alpha^3$ has order $n = 21$. Let $t = 2$. Let $\mathbf{g}(X)$ be the binary polynomial of minimum degree that has

$$\beta, \beta^2, \beta^3, \beta^4$$

as roots. The elements β, β^2, and β^4 have the same minimal polynomial, which is

$$\psi_1(X) = 1 + X + X^2 + X^4 + X^6.$$

The minimal polynomial of β^3 is

$$\psi_3(X) = 1 + X^2 + X^3.$$

Therefore,

$$\begin{aligned}\mathbf{g}(X) &= \psi_1(X)\psi_3(X) \\ &= 1 + X + X^4 + X^5 + X^7 + X^8 + X^9.\end{aligned}$$

We can check easily that $\mathbf{g}(X)$ divides $X^{21} + 1$. The (21, 12) code generated by $\mathbf{g}(X)$ is a double-error-correcting nonprimitive BCH code.

Now, we give a general definition of binary BCH codes. Let β be an element of GF(2^m). Let l_0 be any nonnegative integer. Then a binary BCH code with designed distance d_0 is generated by the binary polynomial $\mathbf{g}(X)$ of minimum degree which has as roots the following consecutive powers of β:

$$\beta^{l_0}, \beta^{l_0+1}, \ldots, \beta^{l_0+d_0-2}$$

For $0 \leq i < d_0 - 1$, let $\psi_i(X)$ and n_i be the minimum polynomial and order of β^{l_0+i}, respectively. Then

$$\mathbf{g}(X) = \text{LCM}\,\{\psi_0(X), \psi_1(X), \ldots, \psi_{d_0-2}(X)\}$$

and the length of the code is

$$n = \text{LCM}\,\{n_0, n_1, \ldots, n_{d_0-2}\}.$$

The BCH code defined above has minimum distance at least d_0 and has no more than $m(d_0 - 1)$ parity-check digits (the proof of these is left as an exercise). Of course, the code is capable of correcting $[(d_0 - 1)/2]$ or fewer errors. If we let $l_0 = 1$, $d_0 = 2t + 1$ and β be a primitive element of GF(2^m), the code becomes a t-error-correcting primitive BCH code of length $2^m - 1$. If $l_0 = 1$, $d_0 = 2t + 1$ and β is not a primitive element of GF(2^m), the code is a nonprimitive t-error-correcting BCH code of length n which is the order of β. We note that, in the definition of a BCH code with designed distance d_0, we require that the generator polynomial $\mathbf{g}(X)$ has $d_0 - 1$ consecutive powers of a field element β as roots. This requirement guarantees that the code has minimum distance at least d_0. This lower bound on the minimum distance is called the *BCH bound.*

In the rest of this chapter, we consider only the primitive BCH codes.

6.2 DECODING OF THE BCH CODES

Suppose that a code word $\mathbf{v}(X) = v_0 + v_1X + v_2X^2 + \cdots + v_{n-1}X^{n-1}$ is transmitted and the transmission errors result in the following received vector:

$$\mathbf{r}(X) = r_0 + r_1X + r_2X^2 + \cdots + r_{n-1}X^{n-1}.$$

Let $\mathbf{e}(X)$ be the error pattern. Then

$$\mathbf{r}(X) = \mathbf{v}(X) + \mathbf{e}(X). \tag{6.11}$$

As usual, the first step of decoding a code is to compute the syndrome from the received vector $\mathbf{r}(X)$. For decoding a t-error-correcting primitive BCH code, the syndrome is a $2t$-tuple,

$$\mathbf{S} = (S_1, S_2, \ldots, S_{2t}) = \mathbf{r} \cdot \mathbf{H}^T, \tag{6.12}$$

where $\mathbf{H}$ is given by (6.6). From (6.6) and (6.12) we find that the ith component of the syndrome is

$$\begin{aligned} S_i &= \mathbf{r}(\alpha^i) \\ &= r_0 + r_1\alpha^i + r_2\alpha^{2i} + \cdots + r_{n-1}\alpha^{(n-1)i} \end{aligned} \tag{6.13}$$

for $1 \leq i \leq 2t$. Note that the syndrome components are elements in the field GF(2^m). These components can be computed from $\mathbf{r}(X)$ as follows. Dividing $\mathbf{r}(X)$ by the minimal polynomial $\phi_i(X)$ of α^i, we obtain

$$\mathbf{r}(X) = \mathbf{a}_i(X)\phi_i(X) + \mathbf{b}_i(X),$$

where $\mathbf{b}_i(X)$ is the remainder with degree less than that of $\phi_i(X)$. Since $\phi_i(\alpha^i) = 0$, we have

$$S_i = \mathbf{r}(\alpha^i) = \mathbf{b}_i(\alpha^i). \tag{6.14}$$

Thus, the syndrome component S_i is obtained by evaluating $\mathbf{b}_i(X)$ with $X = \alpha^i$.

Example 6.4

Consider the double-error-correcting (15, 7) BCH code given in Example 6.1. Suppose that the vector

$$\mathbf{r} = (1\ 0\ 0\ 0\ 0\ 0\ 0\ 0\ 1\ 0\ 0\ 0\ 0\ 0\ 0)$$

is received. The corresponding polynomial is

$$\mathbf{r}(X) = 1 + X^8.$$

The syndrome consists of four components,

$$\mathbf{S} = (S_1, S_2, S_3, S_4).$$

The minimal polynomials for α, α^2, and α^4 are identical and

$$\phi_1(X) = \phi_2(X) = \phi_4(X) = 1 + X + X^4.$$

The minimal polynomial of α^3 is

$$\phi_3(X) = 1 + X + X^2 + X^3 + X^4.$$

Dividing $\mathbf{r}(X) = 1 + X^8$ by $\phi_1(X) = 1 + X + X^4$, the remainder is

$$\mathbf{b}_1(X) = X^2.$$

Dividing $\mathbf{r}(X) = 1 + X^8$ by $\phi_3(X) = 1 + X + X^2 + X^3 + X^4$, the remainder is

$$\mathbf{b}_3(X) = 1 + X^3.$$

Using GF(2^4) given by Table 2.8 and substituting α, α^2, and α^4 into $\mathbf{b}_1(X)$, we obtain

$$S_1 = \alpha^2, \qquad S_2 = \alpha^4, \qquad S_4 = \alpha^8.$$

Substituting α^3 into $\mathbf{b}_3(X)$, we obtain

$$S_3 = 1 + \alpha^9 = 1 + \alpha + \alpha^3 = \alpha^7.$$

Thus,

$$\mathbf{S} = (\alpha^2, \alpha^4, \alpha^7, \alpha^8).$$

Since $\alpha^1, \alpha^2, \ldots, \alpha^{2t}$ are roots of each code polynomial, $\mathbf{v}(\alpha^i) = 0$ for $1 \leq i \leq 2t$. It follows from (6.11) and (6.13) that we obtain the following relationship between the syndrome components and the error pattern:

$$S_i = \mathbf{e}(\alpha^i) \tag{6.15}$$

for $1 \leq i \leq 2t$. From (6.15) we see that the syndrome $\mathbf{S}$ depends on the error pattern $\mathbf{e}$ only. Suppose that the error pattern $\mathbf{e}(X)$ has ν errors at locations $X^{j_1}, X^{j_2}, \ldots, X^{j_\nu}$, that is,

$$\mathbf{e}(X) = X^{j_1} + X^{j_2} + \cdots + X^{j_\nu}, \tag{6.16}$$

where $0 \leq j_1 < j_2 < \cdots < j_\nu < n$. From (6.15) and (6.16) we obtain the following set of equations:

$$\begin{aligned}
S_1 &= \alpha^{j_1} + \alpha^{j_2} + \cdots + \alpha^{j_\nu} \\
S_2 &= (\alpha^{j_1})^2 + (\alpha^{j_2})^2 + \cdots + (\alpha^{j_\nu})^2 \\
S_3 &= (\alpha^{j_1})^3 + (\alpha^{j_2})^3 + \cdots + (\alpha^{j_\nu})^3 \\
&\ \ \vdots \\
S_{2t} &= (\alpha^{j_1})^{2t} + (\alpha^{j_2})^{2t} + \cdots + (\alpha^{j_\nu})^{2t},
\end{aligned} \tag{6.17}$$

where $\alpha^{j_1}, \alpha^{j_2}, \ldots, \alpha^{j_\nu}$ are unknown. *Any method for solving these equations is a decoding algorithm for the BCH codes.* Once $\alpha^{j_1}, \alpha^{j_2}, \ldots, \alpha^{j_\nu}$ have been found, the powers $j_1, j_2, \ldots, j_\nu$ tell us the error locations in $\mathbf{e}(X)$ as in (6.16). In general, the equations of (6.17) have many possible solutions (2^k of them). Each solution yields a different error pattern. If the number of errors in the actual error pattern $\mathbf{e}(X)$ is t or less (i.e., $\nu \leq t$), the solution that yields an error pattern with the *smallest number of errors* is the right solution. That is, the error pattern corresponding to this solution is the most probable error pattern $\mathbf{e}(X)$ caused by the channel noise. For large t, solving the equations of (6.17) directly is difficult and ineffective. In the following, we describe an effective procedure to determine α^{j_l} for $l = 1, 2, \ldots, \nu$ from the syndrome components S_i's.

For convenience, let

$$\beta_l = \alpha^{j_l} \tag{6.18}$$

for $1 \leq l \leq \nu$. We call these elements the *error location numbers* since they tell us the locations of the errors. Now the equations of (6.17) can be expressed in the following form:

$$\begin{aligned}
S_1 &= \beta_1 + \beta_2 + \cdots + \beta_\nu \\
S_2 &= \beta_1^2 + \beta_2^2 + \cdots + \beta_\nu^2 \\
&\ \ \vdots \\
S_{2t} &= \beta_1^{2t} + \beta_2^{2t} + \cdots + \beta_\nu^{2t}.
\end{aligned} \tag{6.19}$$

These $2t$ equations are symmetric functions in $\beta_1, \beta_2, \ldots, \beta_\nu$, which are known as *power-sum symmetric functions.* Now, we define the following polynomial:

$$\begin{aligned}
\sigma(X) &= (1 + \beta_1 X)(1 + \beta_2 X)\cdots(1 + \beta_\nu X) \\
&= \sigma_0 + \sigma_1 X + \sigma_2 X^2 + \cdots + \sigma_\nu X^\nu.
\end{aligned} \tag{6.20}$$

The roots of $\sigma(X)$ are $\beta_1^{-1}, \beta_2^{-1}, \ldots, \beta_\nu^{-1}$, which are the inverses of the error location numbers. For this reason, $\sigma(X)$ is called the *error-location polynomial.* Note that $\sigma(X)$ is an unknown polynomial whose coefficients must be determined. The coefficients of $\sigma(X)$ and the error-location numbers are related by the following equations:

$$
\begin{aligned}
\sigma_0 &= 1 \\
\sigma_1 &= \beta_1 + \beta_2 + \cdots + \beta_\nu \\
\sigma_2 &= \beta_1\beta_2 + \beta_2\beta_3 + \cdots + \beta_{\nu-1}\beta_\nu \\
&\vdots \\
\sigma_\nu &= \beta_1\beta_2\cdots\beta_\nu.
\end{aligned}
\tag{6.21}
$$

The σ_i's are known as *elementary symmetric functions* of β_l's. From (6.19) and (6.21), we see that the σ_i's are related to the syndrome components S_j's. In fact, they are related to the syndrome components by the following *Newton's identities:*

$$
\begin{aligned}
S_1 + \sigma_1 &= 0 \\
S_2 + \sigma_1 S_1 + 2\sigma_2 &= 0 \\
S_3 + \sigma_1 S_2 + \sigma_2 S_1 + 3\sigma_3 &= 0 \\
&\vdots \\
S_\nu + \sigma_1 S_{\nu-1} + \cdots + \sigma_{\nu-1} S_1 + \nu\sigma_\nu &= 0 \\
S_{\nu+1} + \sigma_1 S_\nu + \cdots + \sigma_{\nu-1} S_2 + \sigma_\nu S_1 &= 0 \\
&\vdots
\end{aligned}
\tag{6.22}
$$

For binary case, since $1 + 1 = 2 = 0$, we have

$$
i\sigma_i = \begin{cases} \sigma_i & \text{for odd } i \\ 0 & \text{for even } i. \end{cases}
$$

If it is possible to determine the elementary functions $\sigma_1, \sigma_2, \ldots, \sigma_\nu$ from the equations of (6.22), the error-location numbers $\beta_1, \beta_2, \ldots, \beta_\nu$ can be found by determining the roots of the error-location polynomial $\sigma(X)$. Again, the equations of (6.22) may have many solutions. However, we want to find the solution that yields a $\sigma(X)$ of minimal degree. This $\sigma(X)$ would produce an error pattern with minimum number of errors. If $\nu \leq t$, this $\sigma(X)$ will give the actual error pattern $\mathbf{e}(X)$. In the following, we describe a procedure to determine the polynomial $\sigma(X)$ of minimum degree which satisfies the first $2t$ equations of (6.22) (since we only know S_1 through S_{2t}).

At this point, it would be appropriate to outline the error-correcting procedure for the BCH codes. The procedure consists of three major steps:

1. Compute the syndrome $\mathbf{S} = (S_1, S_2, \ldots, S_{2t})$ from the received polynomial $\mathbf{r}(X)$.
2. Determine the error-location polynomial $\sigma(X)$ from the syndrome components $S_1, S_2, \ldots, S_{2t}$.

3. Determine the error-location numbers $\beta_1, \beta_2, \ldots, \beta_\nu$ by finding the roots of $\sigma(X)$ and correct the errors in $\mathbf{r}(X)$.

The first decoding algorithm that carries out these three steps was devised by Peterson [3]. Steps 1 and 3 are quite simple; step 2 is the most complicated part of decoding a BCH code.

Iterative Algorithm for Finding the Error-Location Polynomial $\sigma(X)$

In the following, we present Berlekamp's iterative algorithm for finding the error-location polynomial. We only describe the algorithm, without giving any proof. The reader who is interested in details of this algorithm is referred to Berlekamp [9], Peterson and Weldon [13], Kasami et al. [14], and MacWilliams and Sloane [15].

The first step of iteration is to find a minimum-degree polynomial $\sigma^{(1)}(X)$ whose coefficients satisfy the first Newton's identity of (6.22). The next step is to test whether the coefficients of $\sigma^{(1)}(X)$ also satisfy the second Newton's identity of (6.22). If the coefficients of $\sigma^{(1)}(X)$ do satisfy the second Newton's identity of (6.22), we set

$$\sigma^{(2)}(X) = \sigma^{(1)}(X).$$

If the coefficients of $\sigma^{(1)}(X)$ do not satisfy the second Newton's identity of (6.22), a correction term is added to $\sigma^{(1)}(X)$ to form $\sigma^{(2)}(X)$ such that $\sigma^{(2)}(X)$ has minimum degree and its coefficients satisfy the first two Newton's identities of (6.22). Therefore, at the end of the second step of iteration, we obtain a minimum-degree polynomial $\sigma^{(2)}(X)$ whose coefficients satisfy the first two Newton's identities of (6.22). The third step of iteration is to find a minimum-degree polynomial $\sigma^{(3)}(X)$ from $\sigma^{(2)}(X)$ such that the coefficients of $\sigma^{(3)}(X)$ satisfy the first three Newton's identities of (6.22). Again, we test whether the coefficients of $\sigma^{(2)}(X)$ satisfy the third Newton's identity of (6.22). If they do, we set $\sigma^{(3)}(X) = \sigma^{(2)}(X)$. If they do not, a correction term is added to $\sigma^{(2)}(X)$ to form $\sigma^{(3)}(X)$. Iteration continues until $\sigma^{(2t)}(X)$ is obtained. Then $\sigma^{(2t)}(X)$ is taken to be the error-location polynomial $\sigma(X)$, that is,

$$\sigma(X) = \sigma^{(2t)}(X).$$

This $\sigma(X)$ will yield an error pattern $\mathbf{e}(X)$ of minimum weight that satisfies the equations of (6.17). If the number of errors in the received polynomial $\mathbf{r}(X)$ is t or less, then $\sigma(X)$ produces the true error pattern.

Let

$$\sigma^{(\mu)}(X) = 1 + \sigma_1{}^{(\mu)}X + \sigma_2{}^{(\mu)}X^2 + \cdots + \sigma_{l_\mu}^{(\mu)}X^{l_\mu} \tag{6.23}$$

be the minimum-degree polynomial determined at the μth step of iteration whose coefficients satisfy the first μ Newton's identities of (6.22). To determine $\sigma^{(\mu+1)}(X)$, we compute the following quantity:

$$d_\mu = S_{\mu+1} + \sigma_1^{(\mu)}S_\mu + \sigma_2^{(\mu)}S_{\mu-1} + \cdots + \sigma_{l_\mu}^{(\mu)}S_{\mu+1-l_\mu}. \tag{6.24}$$

This quantity d_μ is called the μth *discrepancy*. If $d_\mu = 0$, the coefficients of $\sigma^{(\mu)}(X)$ satisfy the $(\mu + 1)$th Newton's identity. In this event, we set

$$\sigma^{(\mu+1)}(X) = \sigma^{(\mu)}(X).$$

If $d_\mu \neq 0$, the coefficients of $\sigma^{(\mu)}(X)$ do not satisfy the $(\mu + 1)$th Newton's identity and a correction term must be added to $\sigma^{(\mu)}(X)$ to obtain $\sigma^{(\mu+1)}(X)$. To accomplish this correction, we go back to the steps *prior to* the μth step and determine a polynomial $\sigma^{(\rho)}(X)$ such that the ρth discrepancy $d_\rho \neq 0$ and $\rho - l_\rho$ [l_ρ is the degree of $\sigma^{(\rho)}(X)$] has the largest value. Then

$$\sigma^{(\mu+1)}(X) = \sigma^{(\mu)}(X) + d_\mu d_\rho^{-1} X^{(\mu-\rho)} \sigma^{(\rho)}(X), \tag{6.25}$$

which is the minimum-degree polynomial whose coefficients satisfy the first $\mu + 1$ Newton's identities. The proof of this is quite complicated and is omitted from this introductory book.

To carry out the iteration of finding $\sigma(X)$, we begin with Table 6.5 and proceed

TABLE 6.5

μ	$\sigma^{(\mu)}(X)$	d_μ	l_μ	$\mu - l_\mu$
-1	1	1	0	-1
0	1	S_1	0	0
1				
2				
.				
.				
.				
$2t$				

to fill out the table, where l_μ is the degree of $\sigma^{(\mu)}(X)$. Assuming that we have filled out all rows up to and including the μth row, we fill out the $(\mu + 1)$th row as follows:

1. If $d_\mu = 0$, then $\sigma_\mu^{(\mu+1)}(X) = \sigma^{(\mu)}(X)$ and $l_{\mu+1} = l_\mu$.
2. If $d_\mu \neq 0$, find another row ρ prior to the μth row such that $d_\rho \neq 0$ and the number $\rho - l_\rho$ in the last column of the table has the largest value. Then $\sigma^{(\mu+1)}(X)$ is given by (6.25) and

$$l_{\mu+1} = \max(l_\mu, l_\rho + \mu - \rho). \tag{6.26}$$

In either case,

$$d_{\mu+1} = S_{\mu+2} + \sigma_1^{(\mu+1)} S_{\mu+1} + \cdots + \sigma_{l_{\mu+1}}^{(\mu+1)} S_{\mu+2-l_{\mu+1}}, \tag{6.27}$$

where the $\sigma_i^{(\mu+1)}$'s are the coefficients of $\sigma^{(\mu+1)}(X)$. The polynomial $\sigma^{(2t)}(X)$ in the last row should be the required $\sigma(X)$. If it has degree greater than t, there are more than t errors in the received polynomial $\mathbf{r}(X)$, and generally it is not possible to locate them.

Example 6.5

Consider the (15, 5) triple-error-correcting BCH code given in Example 6.1. Assume that the code vector of all zeros,

$$\mathbf{v} = (0, 0, 0, 0, 0, 0, 0, 0, 0, 0, 0, 0, 0, 0, 0)$$

is transmitted and the vector

$$\mathbf{r} = (0\ \ 0\ \ 0\ \ 1\ \ 0\ \ 1\ \ 0\ \ 0\ \ 0\ \ 0\ \ 0\ \ 0\ \ 1\ \ 0\ \ 0)$$

is received. Then $\mathbf{r}(X) = X^3 + X^5 + X^{12}$. The minimal polynomials for α, α^2, and α^4 are identical and

$$\phi_1(X) = \phi_2(X) = \phi_4(X) = 1 + X + X^4.$$

The elements α^3 and α^6 have the same minimal polynomial,

$$\phi_3(X) = \phi_6(X) = 1 + X + X^2 + X^3 + X^4.$$

The minimal polynomial for α^5 is

$$\phi_5(X) = 1 + X + X^2.$$

Dividing $\mathbf{r}(X)$ by $\phi_1(X), \phi_3(X)$ and $\phi_5(X)$, respectively, we obtain the following remainders:

$$\mathbf{b}_1(X) = 1,$$

$$\mathbf{b}_3(X) = 1 + X^2 + X^3,$$

$$\mathbf{b}_5(X) = X^2.$$

Using Table 2.8 and substituting α, α^2 and α^4 into $\mathbf{b}_1(X)$, we obtain the following syndrome components:

$$S_1 = S_2 = S_4 = 1.$$

Substituting α^3 and α^6 into $\mathbf{b}_3(X)$, we obtain

$$S_3 = 1 + \alpha^6 + \alpha^9 = \alpha^{10},$$

$$S_6 = 1 + \alpha^{12} + \alpha^{18} = \alpha^5.$$

Substituting α^5 into $\mathbf{b}_5(X)$, we have

$$S_5 = \alpha^{10}.$$

Using the iterative procedure described above, we obtain Table 6.6. Thus, the error-location polynomial is

$$\sigma(X) = \sigma^{(6)}(X) = 1 + X + \alpha^5 X^3.$$

We can easily check that α^3, α^{10}, and α^{12} are the roots of $\sigma(X)$. Their inverses are α^{12}, α^5, and α^3 which are the error-location numbers. Therefore, the error pattern is

$$\mathbf{e}(X) = X^3 + X^5 + X^{12}.$$

Adding $\mathbf{e}(X)$ to the received polynomial $\mathbf{r}(X)$, we obtain the all-zero code vector.

TABLE 6.6

μ	$\sigma^{(\mu)}(X)$	d_μ	l_μ	$\mu - l_\mu$
-1	1	1	0	-1
0	1	1	0	0
1	$1 + X$	0	1	0 (take $\rho = -1$)
2	$1 + X$	α^5	1	1
3	$1 + X + \alpha^5 X^2$	0	2	1 (take $\rho = 0$)
4	$1 + X + \alpha^5 X^2$	α^{10}	2	2
5	$1 + X + \alpha^5 X^3$	0	3	2 (take $\rho = 2$)
6	$1 + X + \alpha^5 X^3$	—	—	—

If the number of errors in the received polynomial $\mathbf{r}(X)$ is less than the designed error-correcting capability t of the code, it is not necessary to carry out the $2t$ steps of iteration to find the error-location polynomial $\sigma(X)$. Let $\sigma^{(\mu)}(X)$ and d_μ be the solution and discrepancy obtained at the μth step of iteration. Let l_μ be the degree of $\sigma^{(\mu)}(X)$. Chen [17] has shown that if d_μ and the discrepancies at the next $t - l_\mu - 1$ steps are all zero, $\sigma^{(\mu)}(X)$ is the error-location polynomial. Based on this fact, if the number of errors in the received polynomial $\mathbf{r}(X)$ is ν ($\nu \leq t$), only $t + \nu$ steps of iteration is needed to determine the error-location polynomial $\sigma(X)$. If ν is small (this is often the case), the reduction in the number of iteration steps results in an increase of decoding speed.

The iterative algorithm for finding $\sigma(X)$ described above not only applies to binary BCH codes but also to nonbinary BCH codes.

Simplified Algorithm for Finding $\sigma(X)$

For binary BCH codes, it is not necessary to fill out the empty $2t$ rows of Table 6.5 for finding $\sigma(X)$. A simplified algorithm [9, 13] can be obtained that requires filling out a table with only t empty rows. Such a table is presented as Table 6.7. Assuming

TABLE 6.7

μ	$\sigma^{(\mu)}(X)$	d_μ	l_μ	$2\mu - l_\mu$
$-\frac{1}{2}$	1	1	0	-1
0	1	S_1	0	0
1				
2				
.				
.				
.				
t				

that we have filled out all rows up to and including the μth row, we fill out the $(\mu + 1)$th row as follows:

1. If $d_\mu = 0$, then $\sigma^{(\mu+1)}(X) = \sigma^{(\mu)}(X)$.
2. If $d_\mu \neq 0$, find another row preceding the μth row, say the ρth, such that the number $2\rho - l_\rho$ in the last column is as large as possible and $d_\rho \neq 0$. Then

$$\sigma^{(\mu+1)}(X) = \sigma^{(\mu)}(X) + d_\mu d_\rho^{-1} X^{2(\mu-\rho)} \sigma^{(\rho)}(X). \tag{6.28}$$

In either case, $l_{\mu+1}$ is exactly the degree of $\sigma^{(\mu+1)}(X)$, and the discrepancy at the $(\mu + 1)$th step is

$$d_{\mu+1} = S_{2\mu+3} + \sigma_1^{(\mu+1)} S_{2\mu+2} + \sigma_2^{(\mu+1)} S_{2\mu+1} + \cdots + \sigma_{l_{\mu+1}}^{(\mu+1)} S_{2\mu+3-l_{\mu+1}}. \tag{6.29}$$

The polynomial $\sigma^{(t)}(X)$ in the last row should be the required $\sigma(X)$. If it has degree greater than t, there were more than t errors, and generally it is not possible to locate them.

The computation required in this simplified algorithm is one-half of the computation required in the general algorithm. However, we must remember that the simplified algorithm applies only to binary BCH codes. Again, if the number of errors in the received polynomial $\mathbf{r}(X)$ is less than t, it is not necessary to carry out the t steps of iteration to determine $\sigma(X)$ for a t-error-correcting binary BCH code. Based on Chen's result, if, for some μ, d_μ and the discrepancies at the next $\lceil(t - l_\mu - 1)/2\rceil$ steps of iteration are zero, $\sigma^{(\mu)}(X)$ is the error-location polynomial [17]. If the number of errors in the received polynomial is ν ($\nu \leq t$), only $\lceil(t + \nu)/2\rceil$ steps of iteration are needed to determine the error-location polynomial $\sigma(X)$.

Example 6.6

The simplified table for finding $\sigma(X)$ for the code considered in Example 6.5 is given in Table 6.8. Thus, $\sigma(X) = \sigma^{(3)}(X) = 1 + X + \alpha^5 X^3$, which is identical to the solution found in Example 6.5.

TABLE 6.8

μ	$\sigma^{(\mu)}(X)$	d_μ	l_μ	$2\mu - l_\mu$
$-\frac{1}{2}$	1	1	0	-1
0	1	$S_1 = 1$	0	0
1	$1 + S_1 X = 1 + X$	$S_3 + S_2 S_1 = \alpha^5$	1	1 (take $\rho = -\frac{1}{2}$)
2	$1 + X + \alpha^5 X^2$	α^{10}	2	2 (take $\rho = 0$)
3	$1 + X + \alpha^5 X^3$	—	3	3 (take $\rho = 1$)

Finding the Error-Location Numbers and Error Correction

The last step in decoding a BCH code is to find the error-location numbers that are the reciprocals of the roots of $\sigma(X)$. The roots of $\sigma(X)$ can be found simply by substituting $1, \alpha, \alpha^2, \ldots, \alpha^{n-1}$ ($n = 2^m - 1$) into $\sigma(X)$. Since $\alpha^n = 1$, $\alpha^{-l} = \alpha^{n-l}$. Therefore, if α^l is a root of $\sigma(X)$, α^{n-l} is an error-location number and the received digit r_{n-l} is an erroneous digit. Consider Example 6.6. The error-location polynomial has been found to be

$$\sigma(X) = 1 + X + \alpha^5 X^3.$$

By substituting $1, \alpha, \alpha^2, \ldots, \alpha^{14}$ into $\sigma(X)$, we find that α^3, α^{10}, and α^{12} are roots of $\sigma(X)$. Therefore, the error-location numbers are α^{12}, α^5, and α^3. The error pattern is

$$\mathbf{e}(X) = X^3 + X^5 + X^{12},$$

which is exactly the assumed error pattern. The decoding of the code is completed by adding (modulo-2) $\mathbf{e}(X)$ to the received vector $\mathbf{r}(X)$.

The substitution method described above for finding the roots of the error location polynomial was first used by Peterson in his algorithm for decoding BCH codes [3]. Later, Chien [6] formulated a procedure to carry out the substitution and error correction. Chien's procedure for searching error-location numbers is described next. The received vector

$$\mathbf{r}(X) = r_0 + r_1 X + r_2 X^2 + \cdots + r_{n-1} X^{n-1}$$

is decoded on a bit-by-bit basis. The high-order bits are decoded first. To decode r_{n-1}, the decoder tests whether α^{n-1} is an error-location number; this is equivalent to testing whether its inverse, α, is a root of $\sigma(X)$. If α is a root, then

$$1 + \sigma_1\alpha + \sigma_2\alpha^2 + \cdots + \sigma_v\alpha^v = 0.$$

Therefore, to decode r_{n-1}, the decoder forms $\sigma_1\alpha, \sigma_2\alpha^2, \ldots, \sigma_v\alpha^v$. If the sum $1 + \sigma_1\alpha + \sigma_2\alpha^2 + \cdots + \sigma_v\alpha^v = 0$, then α^{n-1} is an error-location number and r_{n-1} is an erroneous digit; otherwise, r_{n-1} is a correct digit. To decode r_{n-l}, the decoder forms $\sigma_1\alpha^l, \sigma_2\alpha^{2l}, \ldots, \sigma_v\alpha^{vl}$ and tests the sum

$$1 + \sigma_1\alpha^l + \sigma_2\alpha^{2l} + \cdots + \sigma_v\alpha^{vl}.$$

If this sum is 0, then α^l is a root of $\sigma(X)$ and r_{n-l} is an erroneous digit; otherwise, r_{n-l} is a correct digit.

The testing procedure for error locations described above can be implemented in a straightforward manner by a circuit such as that shown in Figure 6.1 [6]. The t

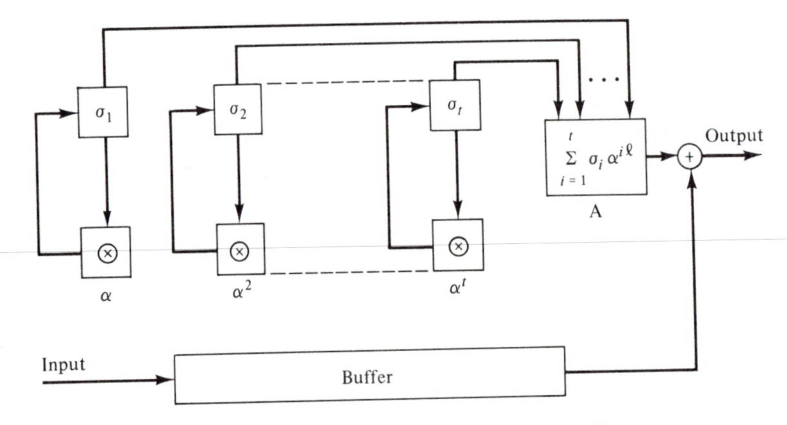

Figure 6.1 Cyclic error location search unit.

σ-registers are initially stored with $\sigma_1, \sigma_2, \ldots, \sigma_t$ calculated in step 2 of the decoding ($\sigma_{v+1} = \sigma_{v+2} = \cdots = \sigma_t = 0$ for $v < t$). Immediately before r_{n-1} is read out of the buffer, the t multipliers $\otimes$ are pulsed once. The multiplications are performed and $\sigma_1\alpha, \sigma_2\alpha^2, \ldots, \sigma_v\alpha^v$ are stored in the σ-registers. The output of the logic circuit A is 1 if and only if the sum $1 + \sigma_1\alpha + \sigma_2\alpha^2 + \cdots + \sigma_v\alpha^v = 0$; otherwise, the output of A is 0. The digit r_{n-1} is read out of the buffer and corrected by the output of A. Having decoded r_{n-1}, the t multipliers are pulsed again. Now $\sigma_1\alpha^2, \sigma_2\alpha^4$, and $\sigma_v\alpha^{2v}$ are stored in the σ-registers. The sum

$$1 + \sigma_1\alpha^2 + \sigma_2\alpha^4 + \cdots + \sigma_v\alpha^{2v}$$

is tested for 0. The digit r_{n-2} is read out of the buffer and corrected in the same manner as r_{n-1} was corrected. This process continues until the whole received vector is read out of the buffer.

The decoding algorithm described above also applies to nonprimitive BCH code. The $2t$ syndrome components are given by

$$S_i = \mathbf{r}(\beta^i)$$

for $1 \leq i \leq 2t$.

6.3 IMPLEMENTATION OF GALOIS FIELD ARITHMETIC

From the discussion above, we see that the decoding of BCH codes requires computations using Galois field arithmetic. Galois field arithmetic can be implemented more easily than ordinary arithmetic because there are no carries. In this section we discuss circuits that perform addition and multiplication over a Galois field. For simplicity, we consider the arithmetic over the Galois field GF(2^4) given by Table 2.8.

To add two field elements, we simply add their vector representations. The resultant vector is then the vector representation of the sum of the two field elements. For example, we want to add α^7 and α^{13} of GF(2^4). From Table 2.8 we find that their vector representations are (1 1 0 1) and (1 0 1 1), respectively. Their vector sum is (1 1 0 1) + (1 0 1 1) = (0 1 1 0), which is the vector representation of α^5. Addition of two field elements can be accomplished with the circuit shown in Figure 6.2. First, the vector representations of the two elements to be added are loaded into the two registers A and B. Their vector sum then appears at the inputs of register A. When register A is pulsed (or clocked), the sum is loaded into register A (register A serves as an accumulator).

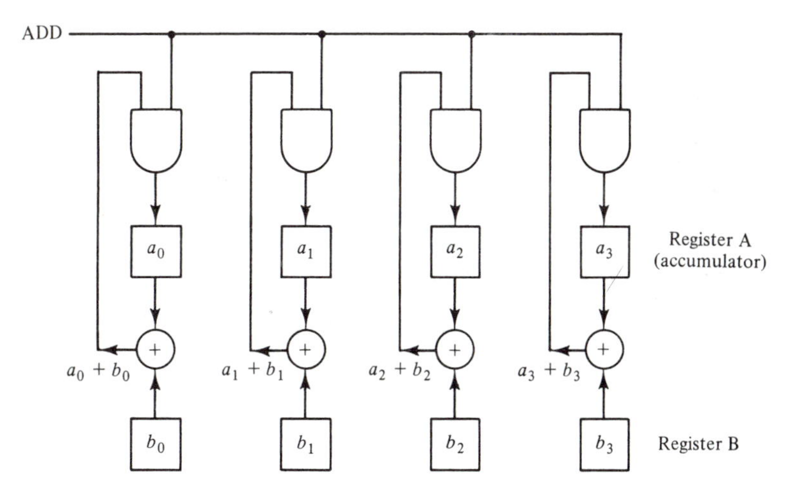

Figure 6.2 Galois field adder.

For multiplication, we first consider multiplying a field element by a fixed element from the same field. Suppose that we want to multiply a field element β in GF(2^4) by the primitive element α whose minimal polynomial is $\phi(X) = 1 + X + X^4$. The element β can be expressed as a polynomial in α as follows:

$$\beta = b_0 + b_1\alpha + b_2\alpha^2 + b_3\alpha^3.$$

Multiplying both sides of the equality above by α and using the fact $\alpha^4 = 1 + \alpha$, we obtain the following equality:

$$\alpha\beta = b_3 + (b_0 + b_3)\alpha + b_1\alpha^2 + b_2\alpha^3.$$

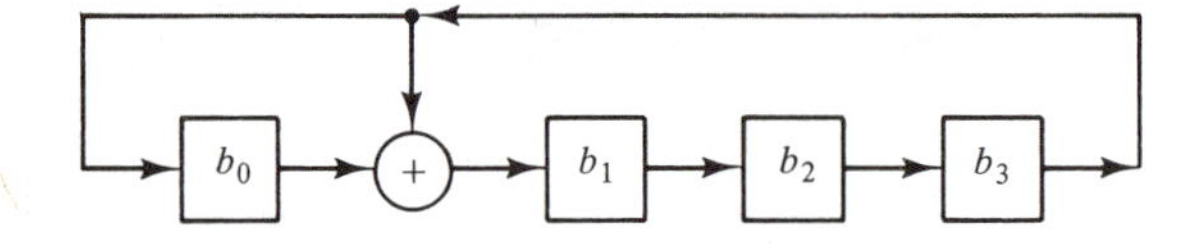

Figure 6.3 Circuit for multiplying arbitrary element in GF(2^4) by α.

This multiplication can be carried out by the feedback shift register shown in Figure 6.3. First, the vector representation (b_0, b_1, b_2, b_3) of β is loaded into the register. Then the register is pulsed. The new contents in the register form the vector representation of $\alpha\beta$. For example, let $\beta = \alpha^7 = 1 + \alpha + \alpha^3$. The vector representation of β is (1 1 0 1). Load this vector into the register of the circuit shown in Figure 6.3. After the register is pulsed, the new contents in the register will be (1 0 1 0), which represents α^8, the product of α^7 and α. The circuit shown in Figure 6.3 can be used to generate (or count) all the nonzero elements of GF(2^4). First, we load (1 0 0 0) (vector representation of $\alpha^0 = 1$) into the register. Successive shifts of the register will generate vector representations of successive powers of α, in exactly the same order as they appear in Table 2.8. At the end of the fifteenth shift, the register will contain (1 0 0 0) again.

As another example, suppose that we want to devise a circuit to multiply an arbitrary element β of GF(2^4) by the element α^3. Again, we express β in polynomial form,

$$\beta = b_0 + b_1\alpha + b_2\alpha^2 + b_3\alpha^3.$$

Multiplying both sides of the above equation by α^3, we have

$$\begin{aligned}\alpha^3\beta &= b_0\alpha^3 + b_1\alpha^4 + b_2\alpha^5 + b_3\alpha^6 \\ &= b_0\alpha^3 + b_1(1 + \alpha) + b_2(\alpha + \alpha^2) + b_3(\alpha^2 + \alpha^3) \\ &= b_1 + (b_1 + b_2)\alpha + (b_2 + b_3)\alpha^2 + (b_0 + b_3)\alpha^3.\end{aligned}$$

Based on the expression above, we obtain a circuit as shown in Figure 6.4, which is capable of multiplying any element β in GF(2^4) by α^3. To multiply, we first load the vector representation (b_0, b_1, b_2, b_3) of β into the register. Then we pulse the register. The new contents in the register will be the vector representation of $\alpha^3\beta$.

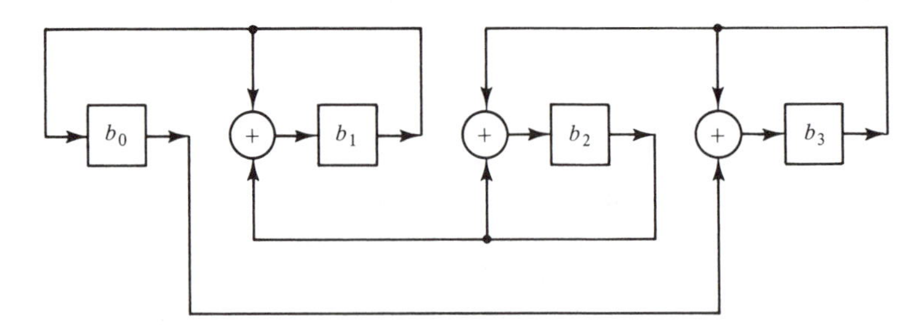

Figure 6.4 Circuit for multiplying arbitrary element in GF(2^4) by α^3.

Next, we consider multiplying two arbitrary field elements. Again, we use GF(2^4) for illustration. Let β and γ be two elements in GF(2^4). Express these two elements in polynomial form:

$$\beta = b_0 + b_1\alpha + b_2\alpha^2 + b_3\alpha^3,$$
$$\gamma = c_0 + c_1\alpha + c_2\alpha^2 + c_3\alpha^3.$$

Then the product $\beta\gamma$ can be expressed in the following form:

$$\beta\gamma = (((c_3\beta)\alpha + c_2\beta)\alpha + c_1\beta)\alpha + c_0\beta \tag{6.30}$$

This product can be carried out with the following steps:

1. Multiply $c_3\beta$ by α and add the product to $c_2\beta$.
2. Multiply $(c_3\beta)\alpha + c_2\beta$ by α and add the product to $c_1\beta$.
3. Multiply $((c_3\beta)\alpha + c_2\beta)\alpha + c_1\beta$ by α and add the product to $c_0\beta$.

Multiplication by α can be carried out by the circuit shown in Figure 6.3. This circuit can be modified to carry out the computation given by (6.30). The resultant circuit is shown in Figure 6.5. In operation of this circuit, the feedback shift register A is initially empty and (b_0, b_1, b_2, b_3) and (c_0, c_1, c_2, c_3), the vector representations of β and γ, are loaded into registers B and C, respectively. Then registers A and C are shifted four times. At the end of the first shift, register A contains $(c_3b_0, c_3b_1, c_3b_2, c_3b_3)$, the vector representation of $c_3\beta$. At the end of the second shift, register A contains the vector representation of $(c_3\beta)\alpha + c_2\beta$. At the end of the third shift, the contents of register A form the vector representation of $((c_3\beta)\alpha + c_2\beta)\alpha + c_1\beta$. At the end of the fourth shift, register A contains the product $\beta\gamma$ in vector form. If we express the product $\beta\gamma$ in the form

$$\beta\gamma = (((c_0\beta) + c_1\beta\alpha) + c_2\beta\alpha^2) + c_3\beta\alpha^3,$$

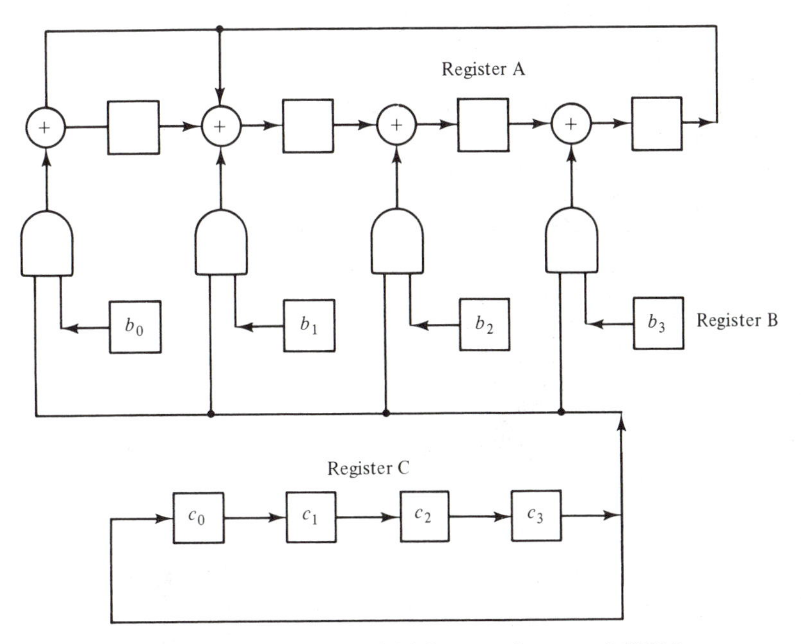

Figure 6.5 Circuit for multiplying two elements of GF(2^4).

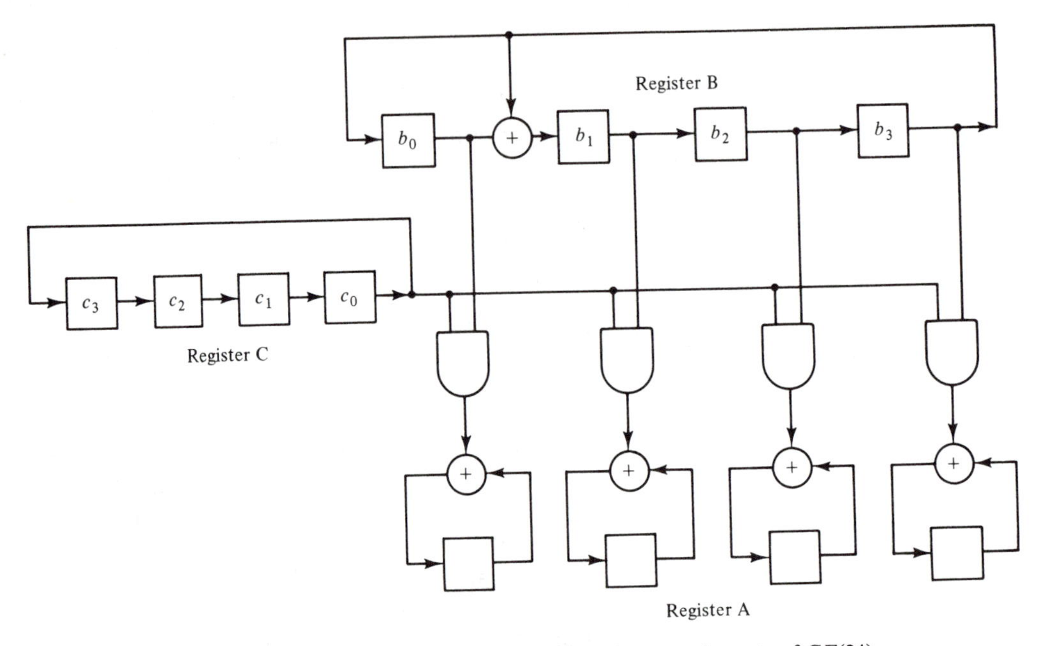

Figure 6.6 Another circuit for multiplying two elements of GF(2^4).

we obtain a different multiplication circuit as shown in Figure 6.6. To multiply, β and γ are loaded into registers B and C, respectively, and register A is initially empty. Then registers A, B, and C are shifted four times. At the end of the fourth shift, register A holds the product $\beta\gamma$. Both multiplication circuits shown in Figures 6.5 and 6.6 are of the same complexity and require the same amount of computation time.

Multiplication of two field elements from GF(2^m) can be implemented in a combinational logic circuit with $2m$ inputs and m outputs. The advantage of this implementation is its speed. However, for $m > 7$, it becomes prohibitively complex and costly. Multiplication can also be programmed in a general-purpose computer; it would require roughly $5m$ instruction executions.

Let $\mathbf{r}(X)$ be a polynomial over GF(2). Next we consider how to compute $\mathbf{r}(\alpha^i)$. This type of computation is required in the first step of decoding of a BCH code. It can be done with a circuit for multiplying a field element by α^i in GF(2^m). Again, we use computation over GF(2^4) for illustration. Suppose that we want to compute

$$\mathbf{r}(\alpha) = r_0 + r_1\alpha + r_2\alpha^2 + \cdots + r_{14}\alpha^{14}, \tag{6.31}$$

where α is a primitive element in GF(2^4) given by Table 2.8. The right-hand side of (6.31) can be expressed in the form

$$\mathbf{r}(\alpha) = (\cdots(((r_{14})\alpha + r_{13})\alpha + r_{12})\alpha + \cdots)\alpha + r_0.$$

Then computation of $\mathbf{r}(\alpha)$ can be accomplished by adding an input to the circuit for multiplying by α shown in Figure 6.3. The resultant circuit for computing $\mathbf{r}(\alpha)$ is shown in Figure 6.7. In operation of this circuit, the register is initially empty. The vector $(r_0, r_1, \ldots, r_{14})$ is shifted into the circuit one digit at a time. After the first

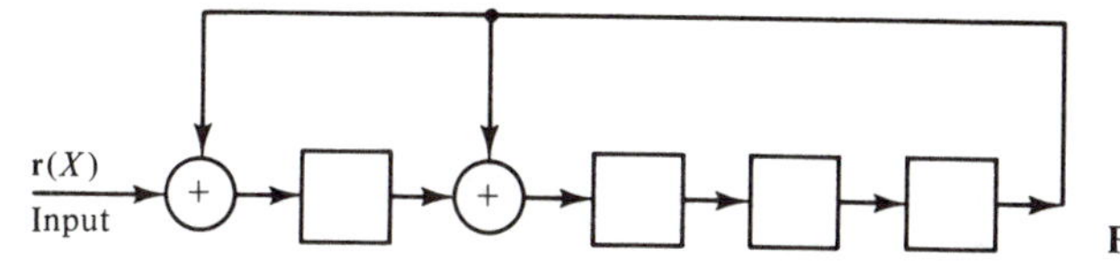

Figure 6.7 Circuit for computing $\mathbf{r}(\alpha)$.

shift, the register contains $(r_{14}, 0, 0, 0)$. At the end of the second shift, the register contains the vector representation of $r_{14}\alpha + r_{13}$. At the completion of the third shift, the register contains the vector representation of $(r_{14}\alpha + r_{13})\alpha + r_{12}$. When the last digit r_0 is shifted into the circuit, the register contains $\mathbf{r}(\alpha)$ in vector form.

Similarly, we can compute $\mathbf{r}(\alpha^3)$ by adding an input to the circuit for multiplying by α^3 of Figure 6.4. The resultant circuit for computing $\mathbf{r}(\alpha^3)$ is shown in Figure 6.8.

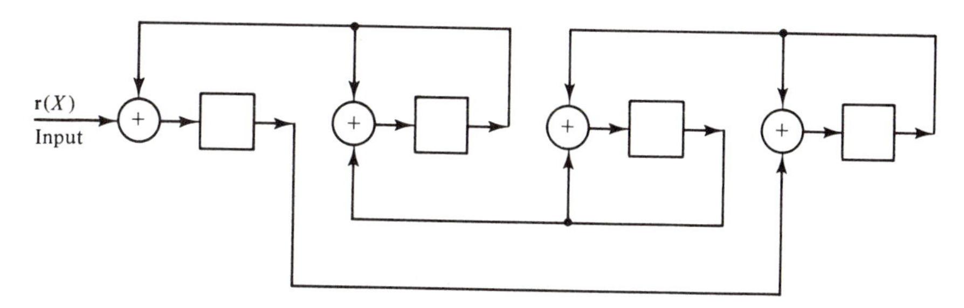

Figure 6.8 Circuit for computing $\mathbf{r}(\alpha^3)$.

There is another way of computing $\mathbf{r}(\alpha^i)$. Let $\phi_i(X)$ be the minimal polynomial of α^i. Let $\mathbf{b}(X)$ be the remainder resulting from dividing $\mathbf{r}(X)$ by $\phi_i(X)$. Then

$$\mathbf{r}(\alpha^i) = \mathbf{b}(\alpha^i).$$

Thus, computing $\mathbf{r}(\alpha^i)$ is equivalent to computing $\mathbf{b}(\alpha^i)$. A circuit can be devised to compute $\mathbf{b}(\alpha^i)$. For illustration, we again consider computation over GF(2^4). Suppose that we want to compute $\mathbf{r}(\alpha^3)$. The minimal polynomial of α^3 is $\phi_3(X) = 1 + X + X^2 + X^3 + X^4$. The remainder resulting from dividing $\mathbf{r}(X)$ by $\phi_3(X)$ has the form

$$\mathbf{b}(X) = b_0 + b_1X + b_2X^2 + b_3X^3.$$

Then

$$\begin{aligned}\mathbf{r}(\alpha^3) &= \mathbf{b}(\alpha^3)\\ &= b_0 + b_1\alpha^3 + b_2\alpha^6 + b_3\alpha^9\\ &= b_0 + b_1\alpha^3 + b_2(\alpha^2 + \alpha^3) + b_3(\alpha + \alpha^3)\\ &= b_0 + b_3\alpha + b_2\alpha^2 + (b_1 + b_2 + b_3)\alpha^3.\end{aligned} \tag{6.32}$$

From the expression above we see that $\mathbf{r}(\alpha^3)$ can be computed by using a circuit that divides $\mathbf{r}(X)$ by $\phi_3(X) = 1 + X + X^2 + X^3 + X^4$ and then combining the coefficients of the remainder $\mathbf{b}(X)$ as given by (6.32). Such a circuit is shown in Figure 6.9, where the feedback connection of the shift register is based on $\phi_3(X) = 1 + X + X^2 + X^3 + X^4$. Since α^6 is a conjugate of α^3, it has the same minimal polynomial as α^3 and therefore $\mathbf{r}(\alpha^6)$ can be computed from the same remainder $\mathbf{b}(X)$ resulting from dividing $\mathbf{r}(X)$ by $\phi_3(X)$. To form $\mathbf{r}(\alpha^6)$, the coefficients of $\mathbf{b}(X)$ are combined in the

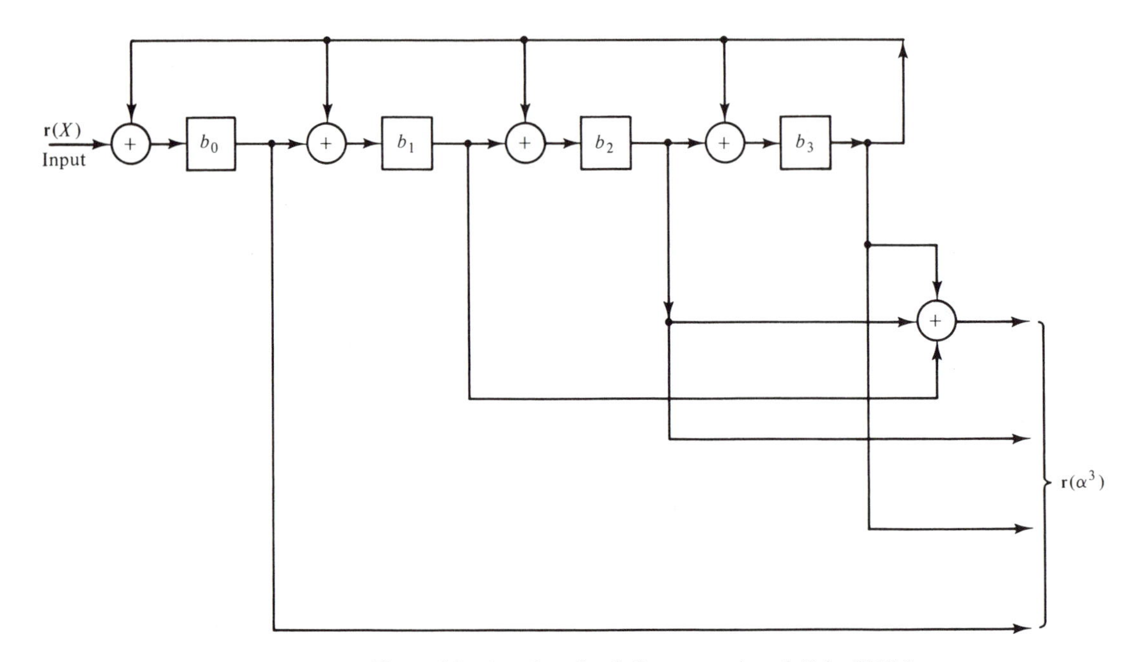

Figure 6.9 Another circuit for computing $r(\alpha^3)$ in GF(2^4).

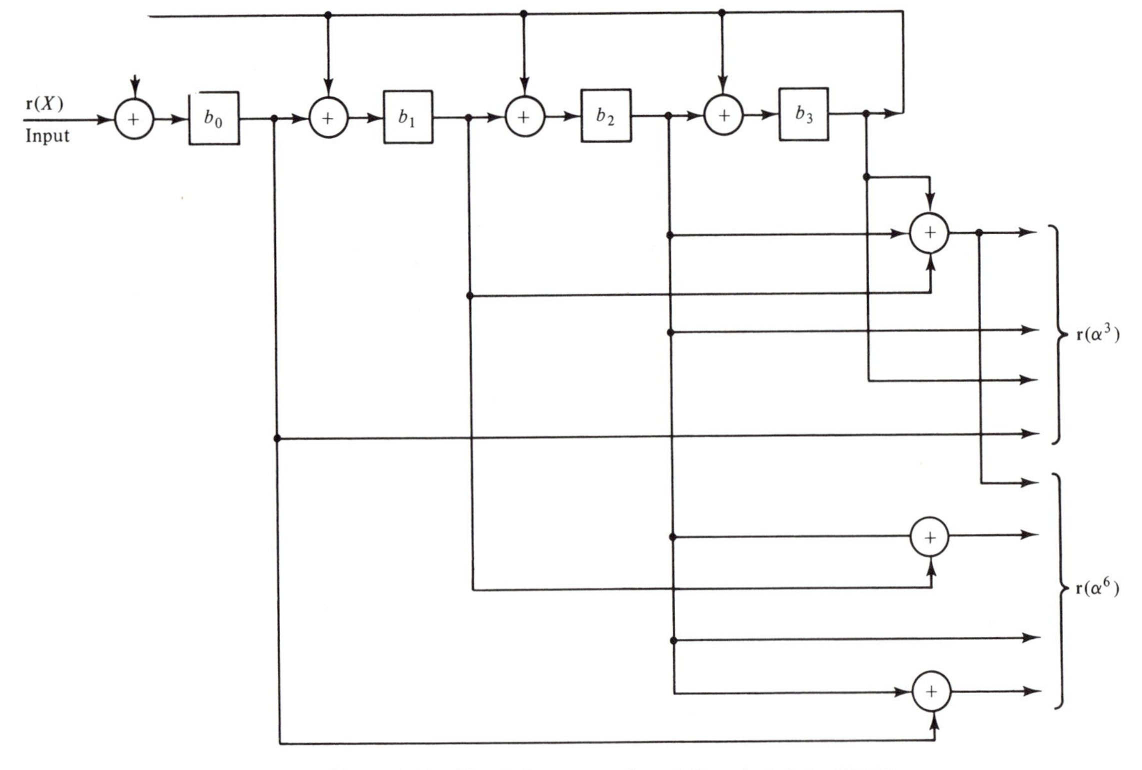

Figure 6.10 Circuit for computing $\mathbf{r}(\alpha^3)$ and $\mathbf{r}(\alpha^6)$ in GF(2^4).

following manner:

$$\begin{aligned}\mathbf{r}(\alpha^6) &= \mathbf{b}(\alpha^6)\\ &= b_0 + b_1\alpha^6 + b_2\alpha^{12} + b_3\alpha^{18}\\ &= b_0 + b_1(\alpha^2 + \alpha^3) + b_2(1 + \alpha + \alpha^2 + \alpha^3) + b_3\alpha^3\\ &= (b_0 + b_2) + b_2\alpha + (b_1 + b_2)\alpha^2 + (b_1 + b_2 + b_3)\alpha^3.\end{aligned}$$

The combined circuit for computing $\mathbf{r}(\alpha^3)$ and $\mathbf{r}(\alpha^6)$ is shown in Figure 6.10.

The arithmetic operation of division over $GF(2^m)$ can be performed by first forming the multiplicative inverse of the divisor β and then multiplying this inverse β^{-1} by the dividend, thus forming the quotient. The multiplicative inverse of β can be found by using the fact $\beta^{2^m-1} = 1$. Thus,

$$\beta^{-1} = \beta^{2^m-2}.$$

6.4 IMPLEMENTATION OF ERROR CORRECTION

Each step in the decoding of a BCH code can be implemented either by digital hardware or by software (i.e., programmed on a general-purpose computer). Each implementation has certain advantages. We consider these implementations next.

Syndrome Computations

The first step in decoding a t-error-correction BCH code is to compute the $2t$ syndrome components $S_1, S_2, \ldots, S_{2t}$. These syndrome components may be obtained by substituting the field elements $\alpha, \alpha^2, \ldots, \alpha^{2t}$ into the received polynomial $\mathbf{r}(X)$. For software implementation, substituting α^i into $\mathbf{r}(X)$ is best accomplished in the following manner:

$$\begin{aligned}S_i = \mathbf{r}(\alpha^i) &= r_{n-1}(\alpha^i)^{n-1} + r_{n-2}(\alpha^i)^{n-2} + \cdots + r_1\alpha^i + r_0\\ &= (\cdots((r_{n-1}\alpha^i + r_{n-2})\alpha^i + r_{n-3})\alpha^i + \cdots + r_1)\alpha^i + r_0.\end{aligned}$$

This computation takes $n - 1$ additions and $n - 1$ multiplications. For binary BCH codes, it can be shown that $S_{2i} = S_i^2$ (see Problem 6.5). With this equality, the $2t$ syndrome components can be computed with $(n - 1)t$ additions and nt multiplications.

For hardware implementation, the syndrome components may be computed with feedback shift registers as described in Section 6.3. We may use either the type of circuits shown in Figures 6.7 and 6.8 or the type of circuit shown in Figure 6.10. The second type of circuit is simpler. From the expression of (6.3), we see that the generator polynomial is a product of at most t minimal polynomials. Therefore, at most t feedback shift registers, each consisting of at most m stages, are needed to form the $2t$ syndrome components. The computation is performed as the received polynomial $\mathbf{r}(X)$ enters the decoder. As soon as the entire $\mathbf{r}(X)$ has entered the decoder, the $2t$ syndrome components are formed. It takes n clock cycles to complete the computation. A syndrome computation circuit for the double-error-correcting (15, 7) BCH code is shown in Figure 6.11, where two feedback shift registers, each with four stages, are employed.

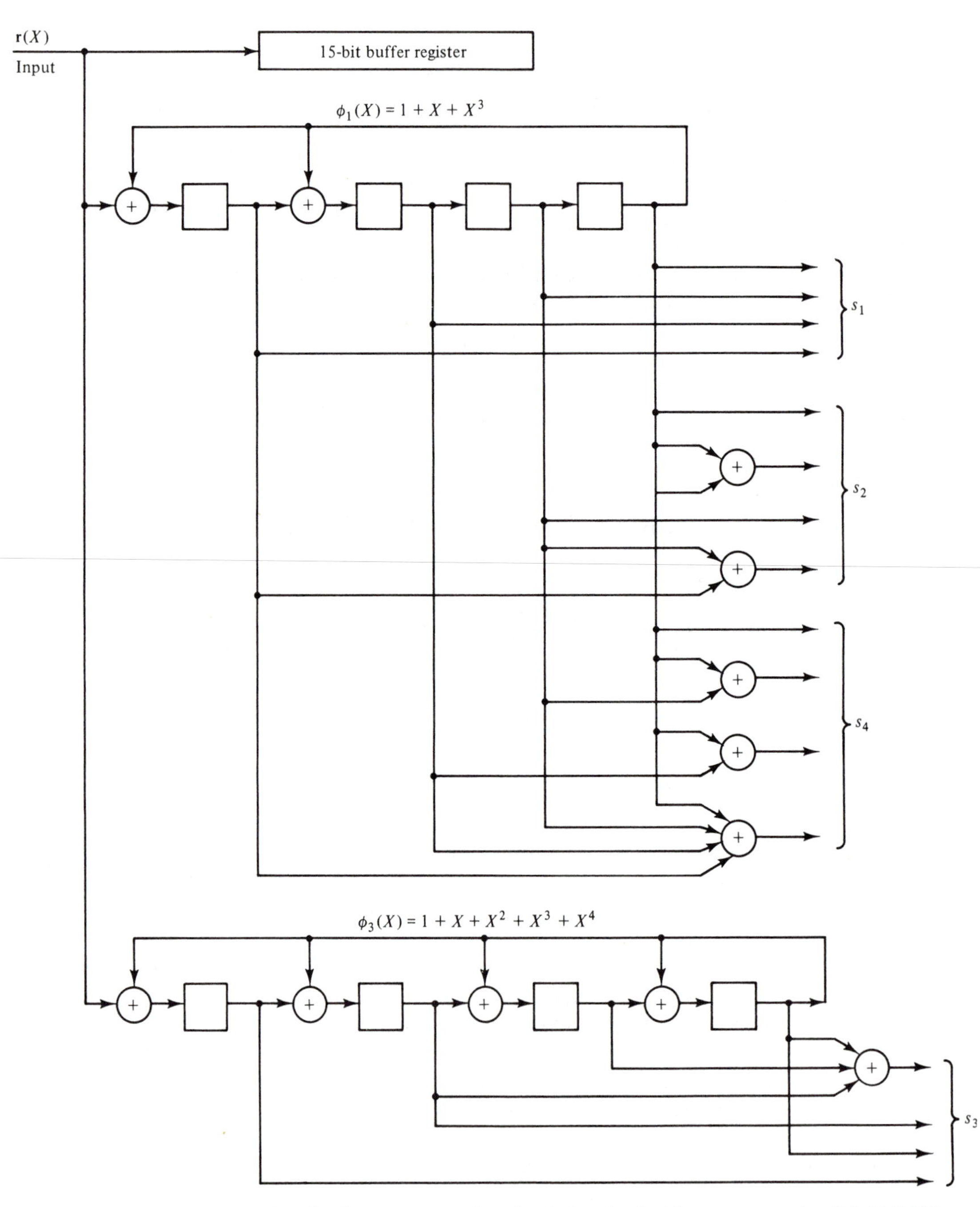

Figure 6.11 Syndrome computation circuit for the double-error-correcting (15, 7) BCH code.

The advantage of hardware implementation of syndrome computation is speed; however, software implementation is less expensive.

Finding the Error-Location Polynomial σ(X)

For this step the software computation requires somewhat less than t additions and t multiplications to compute each $\sigma^{(\mu)}(X)$ and each d_μ, and since there are t of each, the total is roughly $2t^2$ additions and $2t^2$ multiplications. A pure hardware implementation requires the same total, and the speed would depend on how much is done in parallel. The type of circuit shown in Figure 6.2 may be used for addition, and the type of circuits shown in Figures 6.5 and 6.6 may be used for multiplication. A very fast hardware implementation of finding $\sigma(X)$ would probably be very expensive, whereas a simple hardware implementation would probably be organized much like a general-purpose computer, except with a wired rather than a stored program.

Computation of Error-Location Numbers and Error Correction

In the worst case, this step requires substituting n field elements into an error-location polynomial $\sigma(X)$ of degree t to determine its roots. In software this requires nt multiplications and nt additions. This can also be done in hardware using Chien's searching circuit shown in Figure 6.1. Chien's searching circuit requires t multipliers for multiplying by $\alpha, \alpha^2, \ldots, \alpha^t$, respectively. These multipliers may be the type of circuits shown in Figures 6.3 and 6.4. Initially, $\sigma_1, \sigma_2, \ldots, \sigma_t$ found in step 2 are loaded into the registers of the t multipliers. Then these multipliers are shifted n times. At the end the lth shift, the t registers contain $\sigma_1\alpha^l, \sigma_2\alpha^{2l}, \ldots, \sigma_t\alpha^{tl}$. Then the sum

$$1 + \sigma_1\alpha^l + \sigma_2\alpha^{2l} + \cdots + \sigma_t\alpha^{tl}$$

is tested. If the sum is zero, α^{n-l} is an error-location number; otherwise, α^{n-l} is not an error-location number. This sum can be formed by using t m-input modulo-2 adders. A m-input OR gate is used to test whether the sum is zero. It takes n clock cycles to complete this step. If we only want to correct the message digits, only k clock cycles are needed. A Chien's searching circuit for the double-error-correcting (15, 7) BCH code is shown in Figure 6.12.

For large t and m, the cost for building t wired multipliers for multiplying α, $\alpha^2, \ldots, \alpha^t$ in one clock cycle becomes substantial. For more economical but slower multipliers, we may use the type of circuit shown in Figure 6.5 (or shown in Fig. 6.6). Initially, σ_i is loaded into register B and α^i is stored in register C. After m clock cycles, the product $\sigma_i\alpha^i$ is in register A. To form $\sigma_i\alpha^{2i}$, $\sigma_i\alpha^i$ is loaded into register B. After another m clock cycles, $\sigma_i\alpha^{2i}$ will be in register A. Using this type of multipliers, nm clock cycles are needed to complete the third step of decoding a binary BCH code.

Steps 1 and 3 involve roughly the same amount of computation. Since n is generally much larger than t, $4nt$ is much larger than $4t^2$, and steps 1 and 3 involve most of the computation. Thus, the hardware implementation of these steps is essential if high decoding speed is needed. With hardware implementation, step 1

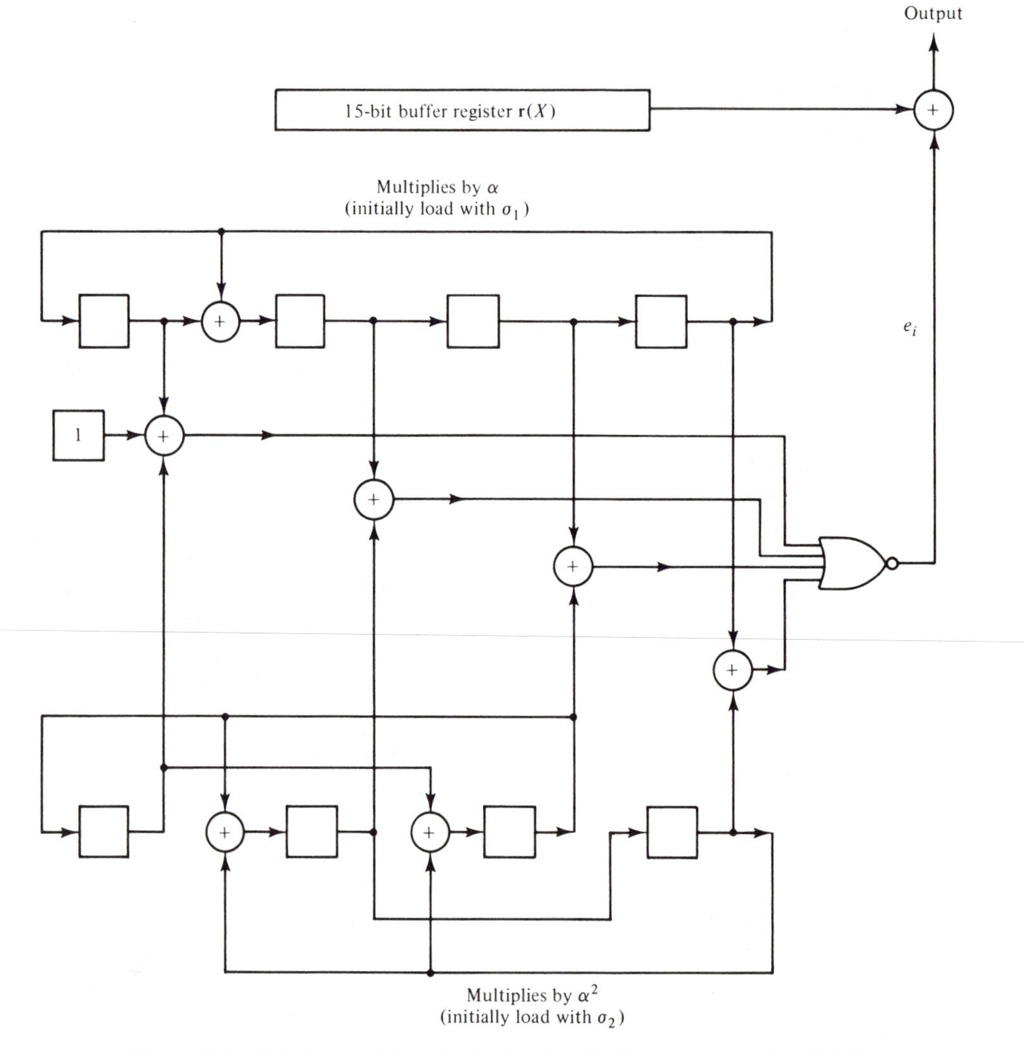

Figure 6.12 Chien's searching circuit for the double-error-correcting (15, 7) BCH code.

can be done as the received polynomial $\mathbf{r}(X)$ is read in and step 3 can be accomplished as $\mathbf{r}(X)$ is read out, and in this case the computation time required in steps 1 and 3 is essentially negligible.

6.5 NONBINARY BCH CODES AND REED–SOLOMON CODES

In addition to the binary codes, there are nonbinary codes. In fact, if p is a prime number and q is any power of p, there are codes with symbols from the Galois field GF(q). These codes are called *q-ary codes*. The concepts and properties developed for the binary codes in the previous chapters apply to q-ary codes with little modification. An (n, k) linear code with symbols from GF(q) is a k-dimensional subspace of the vector space of all n-tuples over GF(q). A q-ary (n, k) cyclic code is generated by a polynomial of degree $n - k$ with coefficients from GF(q), which is a factor of $X^n - 1$. Encoding and decoding of q-ary codes are similar to that of binary codes.

The binary BCH codes defined in Section 6.1 can be generalized to nonbinary codes in a straightforward manner. For any choice of positive integers s and t, there exists a q-ary BCH code of length $n = q^s - 1$, which is capable of correcting any combination of t or fewer errors and requires no more than $2st$ parity-check digits. Let α be a primitive element in the Galois field GF(q^s). The generator polynomial $\mathbf{g}(X)$ of a t-error-correcting q-ary BCH is the polynomial of lowest degree with coefficients from GF(q) for which $\alpha, \alpha^2, \ldots, \alpha^{2t}$ are roots. Let $\phi_i(X)$ be the minimal polynomial of α^i. Then

$$\mathbf{g}(X) = \text{LCM}\,\{\phi_1(X), \phi_2(X), \ldots, \phi_{2t}(X)\}.$$

The degree of each minimal polynomial is s or less. Therefore, the degree of $\mathbf{g}(X)$ is at most $2st$, and hence the number of parity-check digits of the code generated by $\mathbf{g}(X)$ is no more than $2st$. For $q = 2$, we obtain the binary BCH codes. In this section we study only a special subclass of q-ary BCH codes. For details of q-ary BCH codes, the reader is referred to References 4, 7, 9, and 13 to 15.

The special subclass of q-ary BCH codes for which $s = 1$ is the most important subclass of q-ary BCH codes. The codes of this subclass are usually called the *Reed–Solomon* codes in honor of their discoverers [5]. A t-error-correcting Reed–Solomon code with symbols from GF(q) has the following parameters:

Block length:	$n = q - 1$
Number of parity-check digits:	$n - k = 2t,$
Minimum distance:	$d_{\min} = 2t + 1.$

We see that the length of the code is one less than the size of code symbols and the minimum distance is one greater than the number of parity-check digits. In what follows we consider Reed–Solomon codes with code symbols from the Galois field GF(2^m) (i.e., $q = 2^m$). Let α be a primitive element in GF(2^m). The generator polynomial of a primitive t-error-correcting Reed–Solomon code of length $2^m - 1$ is

$$\begin{aligned}\mathbf{g}(X) &= (X + \alpha)(X + \alpha^2)\cdots(X + \alpha^{2t})\\ &= g_0 + g_1X + g_2X^2 + \cdots + g_{2t-1}X^{2t-1} + X^{2t}.\end{aligned}$$

Clearly, $\mathbf{g}(X)$ has $\alpha, \alpha^2, \ldots, \alpha^{2t}$ as all its roots and has coefficients from GF(2^m). The code generated by $\mathbf{g}(X)$ is an $(n, n - 2t)$ cyclic code which consists of those polynomials of degree $n - 1$ or less with coefficients from GF(2^m) that are multiples of $\mathbf{g}(X)$. Encoding of this code is similar to the binary case. Let

$$\mathbf{a}(X) = a_0 + a_1X + a_2X^2 + \cdots + a_{k-1}X^{k-1}$$

be the message to be encoded where $k = n - 2t$. In systematic form, the $2t$ parity-check digits are the coefficients of the remainder $\mathbf{b}(X) = b_0 + b_1X + \cdots + b_{2t-1}X^{2t-1}$ resulting from dividing the message polynomial $X^{2t}\mathbf{a}(X)$ by the generator polynomial $\mathbf{g}(X)$. In hardware implementation, this is accomplished by using a

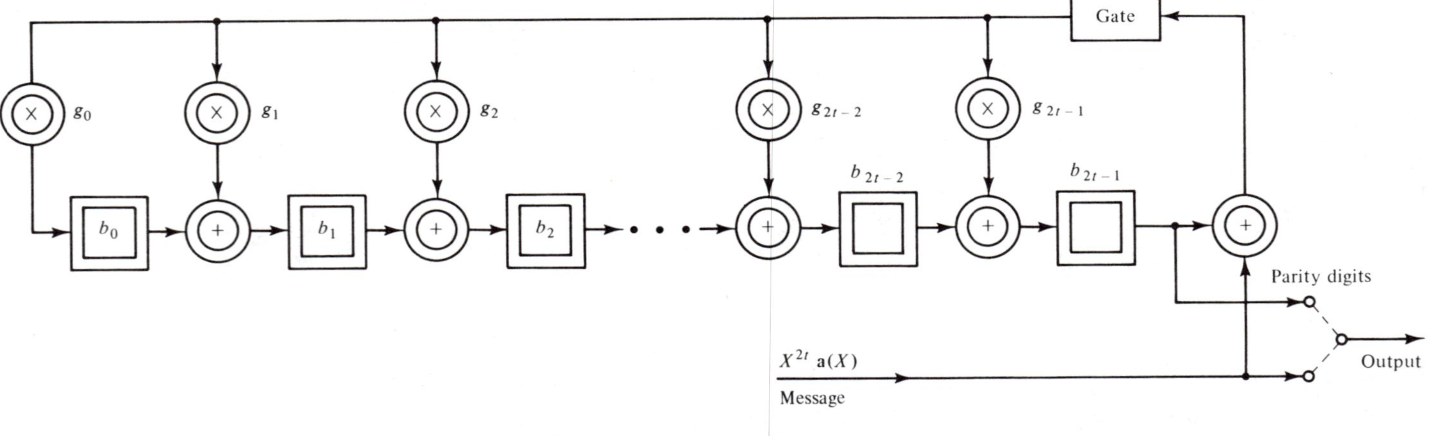

Figure 6.13 Encoding circuit for a nonbinary cyclic code.

division circuit as shown in Figure 6.13, where

1. $\longrightarrow\oplus\longrightarrow$ denotes an adder that adds two elements from GF(2^m).
2. $\longrightarrow\otimes^{g_i}\longrightarrow$ denotes a multiplier that multiplies a field element from GF(2^m) by a fixed element g_i from the same field.
3. $\longrightarrow\boxed{b_i}\longrightarrow$ denotes a storage device that is capable of storing a field element b_i from GF(2^m).

As soon as the message $\mathbf{a}(X)$ has entered the channel and the circuit, the parity-check digits are in the register.

Using the same argument as for the binary BCH codes, we can show that the t-error-correcting Reed–Solomon code has minimum distance at least $2t + 1$. In fact we can prove that the minimum distance is exactly $2t + 1$ (see Problem 6.7).

Let

$$\mathbf{v}(X) = v_0 + v_1 X + \cdots + v_{n-1} X^{n-1}$$

be the transmitted code vector and let

$$\mathbf{r}(X) = r_0 + r_1 X + \cdots + r_{n-1} X^{n-1}$$

be the corresponding received vector. Then the error pattern added by the channel is

$$\begin{aligned}\mathbf{e}(X) &= \mathbf{r}(X) - \mathbf{v}(X) \\ &= e_0 + e_1 X + \cdots + e_{n-1} X^{n-1},\end{aligned}$$

where $e_i = r_i - v_i$ is a symbol from GF(2^m). Suppose that the error pattern $\mathbf{e}(X)$ contains ν errors (nonzero components) at location $X^{j_1}, X^{j_2}, \ldots, X^{j_\nu}$ where $0 \leq j_1 < j_2 < \cdots < j_\nu \leq n - 1$. Then

$$\mathbf{e}(X) = e_{j_1} X^{j_1} + e_{j_2} X^{j_2} + \cdots + e_{j_\nu} X^{j_\nu}.$$

Hence, to determine $\mathbf{e}(X)$, we need to know the error locations X^{j_l}'s and the error values e_{j_l}'s [i.e., we need to know the ν pairs (X^{j_l}, e_{j_l})'s].

As with binary BCH codes, we define

$$\beta_l = \alpha^{j_l} \qquad \text{for } l = 1, 2, \ldots, \nu$$

as error-location numbers. In decoding a Reed–Solomon code (or any q-ary BCH code), the same three steps used for decoding a binary BCH code are required; in addition, a fourth step involving calculation of the error values is required. The $2t$ syndrome components are obtained by substituting α^i into the received polynomial $\mathbf{r}(X)$ for $i = 1, 2, \ldots, 2t$. Thus, we have

$$\begin{aligned}S_1 &= \mathbf{r}(\alpha) = e_{j_1}\beta_1 + e_{j_2}\beta_2 + \cdots + e_{j_\nu}\beta_\nu \\ S_2 &= \mathbf{r}(\alpha^2) = e_{j_1}\beta_1^2 + e_{j_2}\beta_2^2 + \cdots + e_{j_\nu}\beta_\nu^2 \\ &\;\;\vdots \\ S_{2t} &= \mathbf{r}(\alpha^{2t}) = e_{j_1}\beta_1^{2t} + e_{j_2}\beta_2^{2t} + \cdots + e_{j_\nu}\beta_\nu^{2t}.\end{aligned}$$

The syndrome component S_i can also be computed by dividing $\mathbf{r}(X)$ by $X + \alpha^i$. The division results in the equality

$$\mathbf{r}(X) = \mathbf{c}_i(X)(X + \alpha^i) + b_i, \tag{6.33}$$

where the remainder b_i is a constant in GF(2^m). Substituting α^i in both sides of (6.33), we have

$$S_i = b_i.$$

This computation can be accomplished with a division circuit shown in Figure 6.14.

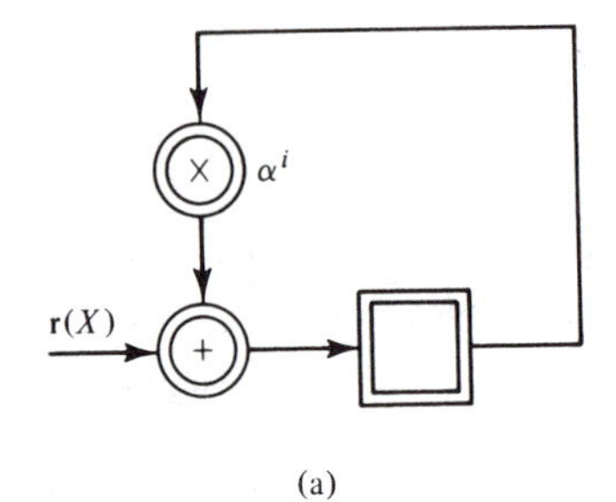

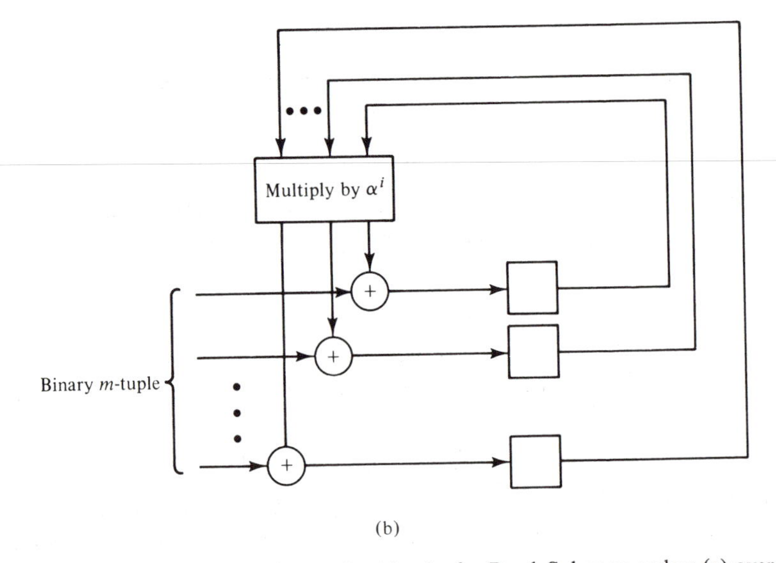

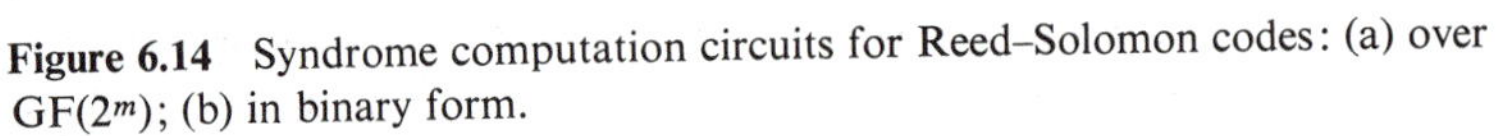

Figure 6.14 Syndrome computation circuits for Reed–Solomon codes: (a) over GF(2^m); (b) in binary form.

To find the error-location polynomial

$$\begin{aligned}\sigma(X) &= (1 + \beta_1 X)(1 + \beta_2 X)\cdots(1 + \beta_\nu X) \\ &= 1 + \sigma_1 X + \cdots + \sigma_\nu X^\nu\end{aligned}$$

with Berlekamp's iterative algorithm, we fill out Table 6.5. Once $\sigma(X)$ is found, we can determine the error values. Let

$$\begin{aligned}Z(X) = 1 &+ (S_1 + \sigma_1)X + (S_2 + \sigma_1 S_1 + \sigma_2)X^2 + \cdots \\ &+ (S_\nu + \sigma_1 S_{\nu-1} + \sigma_2 S_{\nu-2} + \cdots + \sigma_\nu)X^\nu\end{aligned} \tag{6.34}$$

Then the error value at location $\beta_l = \alpha^{j_l}$ is given by [9]

$$e_{j_l} = \frac{Z(\beta_l^{-1})}{\prod_{\substack{i=1 \\ i \neq l}}^{v} (1 + \beta_i \beta_l^{-1})}. \tag{6.35}$$

The decoding computation for a Reed–Solomon code is best explained by an example. Consider a triple-error-correcting Reed–Solomon code with symbols from GF(2^4). The generator polynomial of this code is

$$\begin{aligned} \mathbf{g}(X) &= (X + \alpha)(X + \alpha^2)(X + \alpha^3)(X + \alpha^4)(X + \alpha^5)(X + \alpha^6) \\ &= \alpha^6 + \alpha^9 X + \alpha^6 X^2 + \alpha^4 X^3 + \alpha^{14} X^4 + \alpha^{10} X^5 + X^6. \end{aligned}$$

Let the all-zero vector be the transmitted code vector and let $\mathbf{r} = (0\ 0\ 0\ \alpha^7\ 0\ 0\ \alpha^3\ 0\ 0\ 0\ 0\ 0\ \alpha^4\ 0\ 0)$ be the received vector. Thus, $\mathbf{r}(X) = \alpha^7 X^3 + \alpha^3 X^6 + \alpha^4 X^{12}$.

Step 1. The syndrome components are computed as follows (use Table 2.8):

$$\begin{aligned} S_1 &= \mathbf{r}(\alpha) = \alpha^{10} + \alpha^9 + \alpha = \alpha^{12} \\ S_2 &= \mathbf{r}(\alpha^2) = \alpha^{13} + 1 + \alpha^{13} = 1 \\ S_3 &= \mathbf{r}(\alpha^3) = \alpha + \alpha^6 + \alpha^{10} = \alpha^{14} \\ S_4 &= \mathbf{r}(\alpha^4) = \alpha^4 + \alpha^{12} + \alpha^7 = \alpha^{10} \\ S_5 &= \mathbf{r}(\alpha^5) = \alpha^7 + \alpha^3 + \alpha^4 = 0 \\ S_6 &= \mathbf{r}(\alpha^6) = \alpha^{10} + \alpha^9 + \alpha = \alpha^{12}. \end{aligned}$$

Step 2. To find the error-location polynomial $\sigma(X)$, we fill out Table 6.5. Thus, $\sigma(X) = 1 + \alpha^7 X + \alpha^4 X^2 + \alpha^6 X^3$.

μ	$\sigma^{(\mu)}(X)$	d_μ	l_μ	$\mu - l_\mu$
-1	1	1	0	-1
0	1	α^{12}	0	0
1	$1 + \alpha^{12} X$	α^7	1	0 (take $\rho = -1$)
2	$1 + \alpha^3 X$	1	1	1 (take $\rho = 0$)
3	$1 + \alpha^3 X + \alpha^3 X^2$	α^7	2	1 (take $\rho = 0$)
4	$1 + \alpha^4 X + \alpha^{12} X^2$	α^{10}	2	2 (take $\rho = 2$)
5	$1 + \alpha^7 X + \alpha^4 X^2 + \alpha^6 X^3$	0	3	2 (take $\rho = 3$)
6	$1 + \alpha^7 X + \alpha^4 X^2 + \alpha^6 X^3$	—	—	—

Step 3. By substituting $1, \alpha, \alpha^2, \ldots, \alpha^{14}$ into $\sigma(X)$, we find that α^3, α^9, and α^{12} are roots of $\sigma(X)$. The reciprocals of these roots are α^{12}, α^6, and α^3, which are the error-location numbers of the error pattern $\mathbf{e}(X)$. Thus, errors occur at positions X^3, X^6, and X^{12}.

Step 4. From (6.34) we find that

$$Z(X) = 1 + \alpha^2 X + X^2 + \alpha^6 X^3.$$

Using (6.35), we obtain the error values at locations X^3, X^6, and X^{12}:

$$e_3 = \frac{1 + \alpha^2\alpha^{-3} + \alpha^{-6} + \alpha^6\alpha^{-9}}{(1 + \alpha^6\alpha^{-3})(1 + \alpha^{12}\alpha^{-3})} = \frac{1 + \alpha^{14} + \alpha^9 + \alpha^{12}}{\alpha^{14}\alpha^7} = \frac{\alpha^{13}}{\alpha^6} = \alpha^7,$$

$$e_6 = \frac{1 + \alpha^2\alpha^{-6} + \alpha^{-12} + \alpha^6\alpha^{-18}}{(1 + \alpha^3\alpha^{-6})(1 + \alpha^{12}\alpha^{-6})} = \frac{1 + \alpha^{11} + \alpha^3 + \alpha^3}{\alpha^9} = \frac{\alpha^{12}}{\alpha^9} = \alpha^3,$$

$$e_{12} = \frac{1 + \alpha^2\alpha^{-12} + \alpha^{-24} + \alpha^6\alpha^{-36}}{(1 + \alpha^3\alpha^{-12})(1 + \alpha^6\alpha^{-12})} = \frac{1 + \alpha^5 + \alpha^6 + 1}{\alpha^5} = \frac{\alpha^9}{\alpha^5} = \alpha^4.$$

Thus, the error pattern is

$$\mathbf{e}(X) = \alpha^7 X^3 + \alpha^3 X^6 + \alpha^4 X^{12},$$

which is exactly the difference between the received vector and the transmitted vector. The decoding is completed by taking $\mathbf{r}(X) - \mathbf{e}(X)$.

If β is not a primitive element of GF(2^m), then the 2^m-ary code generated by

$$\mathbf{g}(X) = (X + \beta)(X + \beta^2)\cdots(X + \beta^{2t})$$

is a nonprimitive t-error-correcting Reed–Solomon code. The length n of this code is simply the order of β. Decoding of a nonprimitive Reed–Solomon code is identical to the decoding of a primitive Reed–Solomon code.

Reed–Solomon codes are very effective for correcting multiple bursts of errors. This is discussed in Chapter 9.

Two information symbols can be added to a Reed–Solomon code of length n *without reducing its minimum distance.* The extended Reed–Solomon code has length $n + 2$ and the same number of parity-check symbols as the original code. For a t-error-correcting Reed–Solomon, the parity-check matrix may take the form

$$\mathbf{H} = \begin{bmatrix} 1 & \alpha & \alpha^2 & \cdots & \alpha^{n-1} \\ 1 & \alpha^2 & (\alpha^2)^2 & \cdots & (\alpha^2)^{n-1} \\ \cdot & & & & \\ \cdot & & & & \\ \cdot & & & & \\ 1 & \alpha^{2t} & (\alpha^{2t})^2 & \cdots & (\alpha^{2t})^{n-1} \end{bmatrix}.$$

Then the parity-check matrix of the extended Reed–Solomon code is

$$\mathbf{H}_1 = \left[\begin{array}{cc|c} 0 & 1 & \\ 0 & 0 & \\ \cdot & \cdot & \\ \cdot & \cdot & \mathbf{H} \\ \cdot & \cdot & \\ 0 & 0 & \\ 1 & 0 & \end{array}\right].$$

The result above was first obtained by Kasami, Lin, and Peterson [18,19] and later independently by Wolf [20].

BCH codes form a subclass of a very special class of linear codes, known as Goppa codes [21,22]. It has been proved that the class of Goppa codes contains good codes; however, these good codes have not been explicitly identified. Goppa codes are in general noncyclic (except the BCH codes), and they can be decoded in a manner very similar to the decoding of BCH codes. The decoding also consists of four steps: (1) compute the syndromes, (2) determine the error-location polynomial $\sigma(X)$, (3) find the error-location numbers, and (4) evaluate the error values (this step is not needed for binary Goppa codes). Berlekamp's iterative algorithm for finding the error-location polynomial for a BCH code can be modified for finding the error-location polynomial for Goppa codes [20]. Discussion of Goppa codes is beyond the scope of this introductory book. Moreover, implementation of BCH codes is simpler than that of Goppa codes, and no Goppa codes better than BCH codes have been found. For detail of Goppa codes, the reader is referred to References 14, 15, and 21 to 25.

Our presentation of BCH codes and their implementation is given in time domain. BCH codes can also be defined and implemented in frequency domain using Fourier transforms over Galois fields. Decoding BCH codes in frequency domain sometimes offer computational or implementation advantages. Reference 26 is an excellent source to find transform techniques for decoding BCH codes.

6.6 WEIGHT DISTRIBUTION AND ERROR DETECTION OF BINARY BCH CODES

The weight distributions of double-error-correcting, triple-error-correcting, and some low-rate primitive BCH codes have been completely determined. However, for the other BCH codes, their weight distributions are still unknown. Computation of the weight distribution of a double-error-correcting or a triple-error-correcting primitive BCH code can be achieved by first computing the weight distribution of its dual code and then applying the MacWilliams identity of (3.32). The weight distribution of the dual of a double-error-correcting primitive BCH code of length $2^m - 1$ is given in Tables 6.9 and 6.10. The weight distribution of the dual of a triple-error-correcting primitive BCH code is given in Tables 6.11 and 6.12. Results presented in Tables 6.9 to 6.11 were mainly derived by Kasami [27]. For more on the weight distribution of primitive binary BCH codes, the reader is referred to References 9 and 27.

If a double-error-correcting or a triple-error-correcting primitive BCH code is used for error detection on a BSC with transition probability p, its probability of an undetected error can be computed from (3.36) and one of the weight distribution

TABLE 6.9 WEIGHT DISTRIBUTION OF THE DUAL OF A DOUBLE-ERROR-CORRECTING PRIMITIVE BINARY BCH CODE OF LENGTH $2^m - 1$

Odd $m \geq 3$	
Weight, i	Number of vectors with weight i, B_i
0	1
$2^{m-1} - 2^{(m+1)/2-1}$	$[2^{m-2} + 2^{(m-1)/2-1}](2^m - 1)$
2^{m-1}	$(2^m - 2^{m-1} + 1)(2^m - 1)$
$2^{m-1} + 2^{(m+1)/2-1}$	$[2^{m-2} - 2^{(m-1)/2-1}](2^m - 1)$

TABLE 6.10 WEIGHT DISTRIBUTION OF THE DUAL OF A DOUBLE-ERROR-CORRECTING PRIMITIVE BINARY BCH CODE OF LENGTH $2^m - 1$

Even $m \geq 4$	
Weight, i	Number of vectors with weight i, B_i
0	1
$2^{m-1} - 2^{(m+2)/2-1}$	$2^{(m-2)/2-1}[2^{(m-2)/2} + 1](2^m - 1)/3$
$2^{m-1} - 2^{m/2-1}$	$2^{(m+2)/2-1}(2^{m/2} + 1)(2^m - 1)/3$
2^{m-1}	$(2^{m-2} + 1)(2^m - 1)$
$2^{m-1} + 2^{m/2-1}$	$2^{(m+2)/2-1}(2^{m/2} - 1)(2^m - 1)/3$
$2^{m-1} + 2^{(m+2)/2-1}$	$2^{(m-2)/2-1}[2^{(m-2)/2} - 1](2^m - 1)/3$

TABLE 6.11 WEIGHT DISTRIBUTION OF THE DUAL OF A TRIPLE-ERROR-CORRECTING PRIMITIVE BINARY BCH CODE OF LENGTH $2^m - 1$

Odd $m \geq 5$	
Weight, i	Number of vectors with weight i, B_i
0	1
$2^{m-1} - 2^{(m+1)/2}$	$2^{(m-5)/2}[2^{(m-3)/2} + 1](2^{m-1} - 1)(2^m - 1)/3$
$2^{m-1} - 2^{(m-1)/2}$	$2^{(m-3)/2}[2^{(m-1)/2} + 1](5 \cdot 2^{m-1} + 4)(2^m - 1)/3$
2^{m-1}	$(9 \cdot 2^{2m-4} + 3 \cdot 2^{m-3} + 1)(2^m - 1)$
$2^{m-1} + 2^{(m-1)/2}$	$2^{(m-3)/2}[2^{(m-1)/2} - 1](5 \cdot 2^{m-1} + 4)(2^m - 1)/3$
$2^{m-1} + 2^{(m+1)/2}$	$2^{(m-5)/2}[2^{(m-3)/2} - 1](2^{m-1} - 1)(2^m - 1)/3$

TABLE 6.12 WEIGHT DISTRIBUTION OF THE DUAL OF A TRIPLE-ERROR-CORRECTING PRIMITIVE BINARY BCH CODE OF LENGTH $2^m - 1$

Even $m \geq 6$	
Weight, i	Number of vectors with weight i, B_i
0	1
$2^{m-1} - 2^{(m+4)/2-1}$	$[2^{m-1} + 2^{(m+4)/2-1}](2^m - 4)(2^m - 1)/960$
$2^{m-1} - 2^{(m+2)/2-1}$	$7[2^{m-1} + 2^{(m+2)/2-1}]2^m(2^m - 1)/48$
$2^{m-1} - 2^{m/2-1}$	$2(2^{m-1} + 2^{m/2-1})(3 \cdot 2^m + 8)(2^m - 1)/15$
2^{m-1}	$(29 \cdot 2^{2m} - 4 \cdot 2^m + 64)(2^m - 1)/64$
$2^{m-1} + 2^{m/2-1}$	$2(2^{m-1} - 2^{m/2-1})(3 \cdot 2^m + 8)(2^m - 1)/15$
$2^{m-1} + 2^{(m+2)/2-1}$	$7[2^{m-1} - 2^{(m+2)/2-1}]2^m(2^m - 1)/48$
$2^{m-1} + 2^{(m+4)/2-1}$	$[2^{m-1} - 2^{(m+4)/2-1}](2^m - 4)(2^m - 1)/960$

tables, Tables 6.9 to 6.12. It has been proved [28] that the probability of an undetected error $P_u(E)$ for a double-error-correcting primitive BCH code of length $2^m - 1$ is upper bounded by 2^{-2m} for $p \leq \frac{1}{2}$, where $2m$ is the number of parity-check digits of the code. It is unknown whether the probability of an undetected error for a triple-error-correcting primitive BCH code of length $2^m - 1$ satisfies the upper bound 2^{-3m}, where $3m$ is the number of parity-check digits of the code.

It would be interesting to know how a general t-error-correcting primitive BCH code performs when it is used for error detection on a BSC with transition probability p. It has been proved [29] that, for a t-error-correcting primitive BCH of length $2^m - 1$, if the number of parity-check digits is equal to mt and m is greater than certain constant $m_0(t)$, the number of code vectors of weight i satisfies the following equalities:

$$A_i = \begin{cases} 0 & \text{for } 0 < i \leq 2t \\ (1 + \lambda_0 \cdot n^{-1/10})\binom{n}{i}2^{-(n-k)} & \text{for } i > 2t, \end{cases} \tag{6.36}$$

where $n = 2^m - 1$ and λ_0 is upper bounded by a constant. From (3.19) and (6.36), we obtain the following expression for the probability of an undetected error:

$$P_u(E) = (1 + \lambda_0 \cdot n^{-1/10})2^{-(n-k)} \sum_{i=2t+1}^{n} \binom{n}{i} p^i(1-p)^{n-i}. \tag{6.37}$$

Let $\epsilon = (2t + 1)/n$. Then the summation of (6.37) can be upper bounded as follows [13]:

$$\sum_{i=n\epsilon}^{n} \binom{n}{i} p^i(1-p)^{n-i} \leq 2^{-nE(\epsilon,p)} \tag{6.38}$$

provided that $p < \epsilon$, where

$$E(\epsilon, p) = H(p) + (\epsilon - p)H'(p) - H(\epsilon)$$
$$H(x) = -x \log_2 x - (1 - x) \log_2 (1 - x),$$

and

$$H'(x) = \log_2 \frac{1-x}{x}.$$

$E(\epsilon, p)$ is positive for $\epsilon > p$. Combining (6.37) and (6.38), we obtain the following upper bound on $P_u(E)$ for $\epsilon > p$:

$$P_u(E) \leq (1 + \lambda_0 \cdot n^{-1/10})2^{-nE(\epsilon,p)}2^{-(n-k)} \tag{6.39}$$

For $p < \epsilon$ and sufficient large n, $P_u(E)$ can be made very small. For $p \geq \epsilon$, we can also derive a bound on $P_u(E)$. It is clear from (6.37) that

$$P_u(E) \leq (1 + \lambda_0 \cdot n^{-1/10})2^{-(n-k)} \sum_{i=0}^{n} \binom{n}{i} p^i(1-p)^{n-i}.$$

Since

$$\sum_{i=0}^{n} \binom{n}{i} p^i(1-p)^i = 1,$$

we obtain the following upper bound on $P_u(E)$:

$$P_u(E) \leq (1 + \lambda_0 \cdot n^{-1/10})2^{-(n-k)}. \tag{6.40}$$

We see that, for $p \geq \epsilon$, $P_u(E)$ still decreases exponentially with the number of parity-check digits, $n - k$. If we use sufficient large number of parity-check digits, the probability of an undetected error $P_u(E)$ will become very small. Now, we may summarize the results above as follows: For a t-error-correcting primitive BCH code of length $n = 2^m - 1$ with number of parity-check digits $n - k = mt$ and $m \geq m_0(t)$, its probability of an undetected error on a BSC with transition probability p satisfies the following bounds:

$$P_u(E) \leq \begin{cases} (1 + \lambda_0 \cdot n^{-1/10}) 2^{-n[1-R+E(\epsilon,p)]} & \text{for } p < \epsilon \\ (1 + \lambda_0 \cdot n^{-1/10}) 2^{-n(1-R)} & \text{for } p \geq \epsilon \end{cases} \tag{6.41}$$

where $\epsilon = (2t + 1)/n$, $R = k/n$, and λ_0 is a constant.

The analysis above indicates that primitive BCH codes are very effective for error detection on a BSC. Even though it has not been proved, we believe that the probability of an undetected error for any primitive BCH code satisfies the upper bound $2^{-(n-k)}$.

For the nonbinary case, the weight distribution of Reed–Solomon codes has been completely determined [18,30,31]. For a t-error-correcting Reed–Solomon code of length $q - 1$ with symbols from GF(q), the number of code vectors of weight j is

$$A_j = \binom{q-1}{j} \sum_{i=0}^{j-2t-1} (-1)^i \binom{j}{i} (q^{j-2t-i} - 1).$$

PROBLEMS

6.1. Consider the Galois field GF(2^4) given by Table 2.8. The element $\beta = \alpha^7$ is also a primitive element. Let $\mathbf{g}_0(X)$ be the lowest-degree polynomial over GF(2) which has

$$\beta, \beta^2, \beta^3, \beta^4$$

as its roots. This polynomial also generates a double-error-correcting primitive BCH code of length 15.

(a) Determine $\mathbf{g}_0(X)$.

(b) Find the parity-check matrix for this code.

(c) Show that $\mathbf{g}_0(X)$ is the reciprocal polynomial of the polynomial $\mathbf{g}(X)$ which generates the (15, 7) double-error-correcting BCH code given in Example 6.1.

6.2. Determine the generator polynomials of all the primitive BCH codes of length 31. Use the Galois field GF(2^5) generated by $\mathbf{p}(X) = 1 + X^2 + X^5$.

6.3. Suppose that the double-error-correcting BCH code of length 31 found in Problem 6.2 is used for error correction on a BSC. Decode the received polynomials $\mathbf{r}_1(X) = X^7 + X^{30}$ and $\mathbf{r}_2(X) = 1 + X^{17} + X^{28}$.

6.4. Find the generator polynomial of the double-error-correcting Reed–Solomon code of length $2^4 - 1$ with symbols from GF(2^4). Let α be a primitive element in GF(2^4) given by Table 2.8. Decode the received polynomial $\mathbf{r}(X) = \alpha X^3 + \alpha^{11} X^7$.

6.5. Prove that the syndrome components S_i and S_{2i} are related by $S_{2i} = S_i^2$.

6.6. Consider a t-error-correcting primitive binary BCH code of length $n = 2^m - 1$. If $2t + 1$ is a factor of n, prove that the minimum distance of the code is exactly $2t + 1$.

[*Hint:* Let $n = l(2t + 1)$. Show that $(X^n + 1)/(X^l + 1)$ is a code polynomial of weight $2t + 1$.]

6.7. Show that the t-error-correcting Reed–Solomon code with symbols from GF(2^m) generated by

$$\mathbf{g}(X) = (X + \alpha)(X + \alpha^2) \cdots (X + \alpha^{2t})$$

has minimum distance exactly $2t + 1$, where α is a primitive element in GF(2^m).

6.8. Is there a binary t-error-correcting BCH code of length $2^m + 1$ for $m \geq 3$ and $t < 2^{m-1}$? If there is such a code, determine its generator polynomial.

6.9. Consider the field GF(2^4) generated by $\mathbf{p}(X) = 1 + X + X^4$ (see Table 2.8). Let α be a primitive element in GF(2^4) such that $\mathbf{p}(\alpha) = 0$. Devise a circuit that is capable of multiplying any element in GF(2^4) by α^7.

6.10. Devise a circuit that is capable of multiplying any two elements in GF(2^5). Use $\mathbf{p}(X) = 1 + X^2 + X^5$ to generate GF(2^5).

6.11. Devise a syndrome computation circuit for the binary double-error-correcting (31, 21) BCH code.

6.12. Devise a Chien's searching circuit for the binary double-error-correcting (31, 21) BCH code.

6.13. Devise an encoding circuit for the single-error-correcting Reed–Solomon code of length 15 with symbols from GF(2^4). Use Table 2.8.

6.14. Devise a syndrome computation circuit for the single-error-correcting Reed–Solomon code of length 15 with symbols from GF(2^4).

6.15. Let β be any nonzero element in GF(2^m). Let l_0 be any integer. Let $\mathbf{g}(X)$ be the lowest-degree polynomial over GF(2) which has

$$\beta^{l_0}, \beta^{l_0+1}, \ldots, \beta^{l_0+d-2}$$

as its roots.

(a) Express $\mathbf{g}(X)$ in terms of the minimal polynomials of its roots.

(b) Determine the smallest integer n such that $\mathbf{g}(X)$ divides $X^n + 1$.

(c) Show that the cyclic code of length n generated by $\mathbf{g}(X)$ has minimum distance at least d.

Remark. The code defined above is a binary BCH code in general form. The code is called a BCH code with designed distance d. If $l_0 = 1$, $d = 2t + 1$, and β is a primitive element in GF(2^m), the code is a primitive binary t-error-correcting BCH code.

6.16. Consider the Galois field GF(2^6) given by Table 6.2. Let $\beta = \alpha^3$, $l_0 = 2$, and $d = 5$. Use the results of Problem 6.15 to determine the generator polynomial of the BCH code which has

$$\beta^2, \beta^3, \beta^4, \beta^5$$

as its roots. What is the length of this code?

6.17. In Problem 6.15, let $l_0 = -t$ and $d = 2t + 2$. Then we obtain a BCH code of designed distance $2t + 2$ whose generator polynomial has

$$\beta^{-t}, \ldots, \beta^{-1}, \beta^0, \beta^1, \ldots, \beta^t$$

and their conjugates as all its roots.

(a) Show that this code is a reversible cyclic code.

(b) Show that if t is odd, the minimum distance of this code is at least $2t + 4$. [*Hint:* Show that $\beta^{-(t+1)}$ and β^{t+1} are also roots of the generator polynomial.]

REFERENCES

1. A. Hocquenghem, "Codes corecteurs d'erreurs," *Chiffres*, 2, pp. 147–156, 1959.
2. R. C. Bose and D. K. Ray-Chaudhuri, "On a Class of Error Correcting Binary Group Codes," *Inf. Control*, 3, pp. 68–79, March 1960.
3. W. W. Peterson, "Encoding and Error-Correction Procedures for the Bose-Chaudhuri Codes," *IRE Trans. Inf. Theory*, IT-6, pp. 459–470, September 1960.
4. D. Gorenstein and N. Zierler, "A Class of Cyclic Linear Error-Correcting Codes in p^m Symbols," *J. Soc. Ind. Appl. Math.*, 9, pp. 107–214, June 1961.
5. I. S. Reed and G. Solomon, "Polynomial Codes over Certain Finite Fields," *J. Soc. Ind. Appl. Math.*, 8, pp. 300–304, June 1960.
6. R. T. Chien, "Cyclic Decoding Procedure for the Bose-Chaudhuri-Hocquenghem Codes," *IEEE Trans. Inf. Theory*, IT-10, pp. 357–363, October 1964.
7. G. D. Forney, "On Decoding BCH Codes," *IEEE Trans. Inf. Theory*, IT-11, pp. 549–557, October 1965.
8. E. R. Berlekamp, "On Decoding Binary Bose-Chaudhuri-Hocquenghem Codes," *IEEE Trans. Inf. Theory*, IT-11, pp. 577–580, October 1965.
9. E. R. Berlekamp, *Algebraic Coding Theory*, McGraw-Hill, New York, 1968.
10. J. L. Massey, "Step-by-Step Decoding of the Bose-Chaudhuri-Hocquenghem Codes," *IEEE Trans. Inf. Theory*, IT-11, pp. 580–585, October 1965.
11. J. L. Massey, "Shift-Register Synthesis and BCH Decoding," *IEEE Trans. Inf. Theory*, IT-15, pp. 122–127, January 1969.
12. H. O. Burton, "Inversionless Decoding of Binary BCH Codes," *IEEE Trans. Inf. Theory*, IT-17, pp. 464–466, July 1971.
13. W. W. Peterson and E. J. Weldon, Jr., *Error-Correcting Codes*, 2nd ed., MIT Press, Cambridge, Mass., 1970.
14. T. Kasami, N. Tokura, Y. Iwadare, and Y. Inagaki, *Coding Theory*, Corona, Tokyo, 1974 (in Japanese).
15. F. J. MacWilliams and N. J. A. Sloane, *The Theory of Error-Correcting Codes*, North-Holland, Amsterdam, 1977.
16. H. B. Mann, "On the Number of Information Symbols in Bose-Chaudhuri Codes," *Inf. Control*, 5, pp. 153–162, June 1962.
17. C. L. Chen, "High-Speed Decoding of BCH Codes," *IEEE Trans. Inf. Theory*, IT-27(2), pp. 254–256, March 1981.
18. T. Kasami, S. Lin, and W. W. Peterson, "Some Results on Weight Distributions of BCH Codes," *IEEE Trans. Inf. Theory*, IT-12(2), p. 274, April 1966.
19. T. Kasami, S. Lin, and W. W. Peterson, "Some Results on Cyclic Codes Which Are Invariant under the Affine Group," Scientific Report AFCRL-66-622, Air Force Cambridge Research Labs., Bedford, Mass., 1966.
20. J. K. Wolf, "Adding Two Information Symbols to Certain Nonbinary BCH Codes and Some Applications," *Bell Syst. Tech. J.*, 48, pp. 2405–2424, 1969.
21. V. D. Goppa, "A New Class of Linear Codes," *Probl. Peredachi Inf.*, 6(3), pp. 24–30, September 1970.
22. V. D. Goppa, "Rational Representation of Codes and (L, g) Codes," *Probl. Peredachi Inf.*, 7(3), pp. 41–49, September 1971.

23. N. J. Patterson, "The Algebraic Decoding of Goppa Codes," *IEEE Trans. Inf. Theory*, IT-21, pp. 203–207, March 1975.

24. E. R. Berlekamp, "Goppa Codes," *IEEE Trans. Inf. Theory*, IT-19(5), pp. 590–592, September 1973.

25. Y. Sugiyama, M. Kasahara, S. Hirasawa, and T. Namekawa, "A Method for Solving Key Equation for Decoding Goppa Codes," *Inf. Control*, 27, pp. 87–99, January 1975.

26. R. E. Blahut, "Transform Techniques for Error Control Codes," *IBM J. Res. Dev.*, 23(3), May 1979.

27. T. Kasami, "Weight Distributions of Bose-Chaudhuri-Hocquenghem Codes," *Proc. Conf. Combinatorial Mathematics and Its Applications*, R. C. Bose and T. A. Dowling, eds., University of North Carolina Press, Chapel Hill, N.C., 1968.

28. S. K. Leung-Yan-Cheong, E. R. Barnes, and D. U. Friedman, "On Some Properties of the Undetected Error Probability of Linear Codes," *IEEE Trans. Inf. Theory*, IT-25(1), pp. 110–112, January 1979.

29. V. M. Sidelńikov, "Weight Spectrum of Binary Bose-Chaudhuri-Hocquenghem Code," *Probl. Inf. Transm.*, 7(1), pp. 11–17, 1971.

30. E. F. Assmus, Jr., H. F. Mattson, Jr., and R. J. Turyn, "Cyclic Codes," Report No. AFCRL-65-332, Air Force Cambridge Research Labs., Bedford, Mass, April 1965.

31. G. D. Forney, Jr., *Concatenated Codes*, MIT Press, Cambridge, Mass, 1966.

7

Majority-Logic Decoding for Cyclic Codes

The majority-logic decoding presented in this and the next chapters is another effective scheme for decoding certain classes of block codes, especially for decoding certain classes of cyclic codes. The first majority-logic decoding algorithm was devised in 1954 by Reed [1] for a class of multiple-error-correcting codes discovered by Muller [2]. Reed's algorithm was later extended and generalized by many coding investigators. The first unified formulation of majority-logic decoding algorithms was due to Massey [3].

Most majority-logic decodable codes found so far are cyclic codes. Important cyclic codes of this category are presented in this and the next chapters.

7.1 ONE-STEP MAJORITY-LOGIC DECODING

Consider an (n, k) cyclic code C with parity-check matrix $\mathbf{H}$. The row space of $\mathbf{H}$ is an $(n, n-k)$ cyclic code, denoted by C_d, which is the dual code of C, or the null space of C. For any vector $\mathbf{v}$ in C and any vector $\mathbf{w}$ in C_d, the inner product of $\mathbf{v}$ and $\mathbf{w}$ is zero, that is,

$$\mathbf{w} \cdot \mathbf{v} = w_0 v_0 + w_1 v_1 + \cdots + w_{n-1} v_{n-1} = 0. \tag{7.1}$$

In fact, an n-tuple $\mathbf{v}$ is a code vector in C if and only if, for any vector $\mathbf{w}$ in C_d, $\mathbf{w} \cdot \mathbf{v} = 0$. The equality of (7.1) is called a *parity-check* equation. Clearly, there are $2^{(n-k)}$ such parity-check equations.

Now suppose that a code vector $\mathbf{v}$ in C is trasmitted. Let $\mathbf{e} = (e_0, e_1, \ldots, e_{n-1})$ and $\mathbf{r} = (r_0, r_1, \ldots, r_{n-1})$ be the error vector and the received vector respectively. Then

$$\mathbf{r} = \mathbf{v} + \mathbf{e}. \tag{7.2}$$

For any vector $\mathbf{w}$ in the dual code C_d, we can form the following linear sum of the received digits:

$$A = \mathbf{w} \cdot \mathbf{r} = w_0 r_0 + w_1 r_1 + \cdots + w_{n-1} r_{n-1}, \tag{7.3}$$

which is called a *parity-check sum* or simply *check sum*. If the received vector $\mathbf{r}$ is a code vector in C, this parity-check sum, A, *must be* zero; however, if $\mathbf{r}$ is not a code vector in C, then A *may not* be zero. Combining (7.2) and (7.3) and using the fact that $\mathbf{w} \cdot \mathbf{v} = 0$, we obtain the following relationship between the check sum A and error digits in $\mathbf{e}$:

$$A = w_0 e_0 + w_1 e_1 + \cdots + w_{n-1} e_{n-1}. \tag{7.4}$$

An error digit e_l is said to be *checked* by the check sum A if the coefficient $w_l = 1$. In the following, we show that certain properly formed check sums can be used for estimating the error digits in $\mathbf{e}$.

Suppose that there exist J vectors in the dual code C_d,

$$\begin{aligned}
\mathbf{w}_1 &= (w_{10}, w_{11}, \ldots, w_{1,n-1}), \\
\mathbf{w}_2 &= (w_{20}, w_{21}, \ldots, w_{2,n-1}), \\
&\vdots \\
\mathbf{w}_J &= (w_{J0}, w_{J1}, \ldots, w_{J,n-1}),
\end{aligned}$$

which have the following properties:

1. The $(n-1)$th component of each vector is a "1," that is,
$$w_{1,n-1} = w_{2,n-1} = \cdots = w_{J,n-1} = 1.$$
2. For $i \neq n-1$, there is *at most* one vector whose ith component is a "1"; for example, if $w_{1,i} = 1$, then $w_{2,i} = w_{3,i} = \cdots = w_{J,i} = 0$.

These J vectors are said to be orthogonal on the $(n-1)$th digit position. We call them *orthogonal vectors*. Now, let us form J parity-check sums from these J orthogonal vectors,

$$\begin{aligned}
A_1 &= \mathbf{w}_1 \cdot \mathbf{r} = w_{10} r_0 + w_{11} r_1 + \cdots + w_{1,n-1} r_{n-1} \\
A_2 &= \mathbf{w}_2 \cdot \mathbf{r} = w_{20} r_0 + w_{21} r_1 + \cdots + w_{2,n-1} r_{n-1} \\
&\vdots \\
A_J &= \mathbf{w}_J \cdot \mathbf{r} = w_{J0} r_0 + w_{J1} r_1 + \cdots + w_{J,n-1} r_{n-1}.
\end{aligned} \tag{7.5}$$

Since $w_{1,n-1} = w_{2,n-1} = \cdots = w_{J,n-1} = 1$, these J check sums are related to the error digits in the following manner:

$$\begin{aligned}
A_1 &= w_{10} e_0 + w_{11} e_1 + \cdots + w_{1,n-2} e_{n-2} + e_{n-1} \\
A_2 &= w_{20} e_0 + w_{21} e_1 + \cdots + w_{2,n-2} e_{n-2} + e_{n-1} \\
&\vdots \\
A_J &= w_{J,0} e_0 + w_{J,1} e_1 + \cdots + w_{J,n-2} e_{n-2} + e_{n-1}.
\end{aligned} \tag{7.6}$$

We see that the error digit e_{n-1} is checked by all the check sums above. Because of the second property of the orthogonal vectors, $\mathbf{w}_1, \mathbf{w}_2, \ldots, \mathbf{w}_J$, any error digit other than e_{n-1} is checked by at most one check sum. *These J check sums are said to be orthogonal on the error digit* e_{n-1}. Since $w_{i,j} = 0$ or 1, each of the foregoing check sums orthogonal on e_{n-1} is of the form:

$$A_j = e_{n-1} + \sum_{i \neq n-1} e_i.$$

If all the error digits in the sum A_j are zero for $i \neq n - 1$, *the value of* e_{n-1} *is equal to* A_j (i.e., $e_{n-1} = A_j$). Based on this fact, the parity-check sums orthogonal on e_{n-1} can be used to estimate e_{n-1}, or to decode the received digit r_{n-1}.

Suppose that there are $\lfloor J/2 \rfloor$ or fewer errors in the error vector $\mathbf{e} = (e_0, e_1, \ldots, e_{n-1})$ (i.e., $\lfloor J/2 \rfloor$ or fewer components of $\mathbf{e}$ are 1). If $e_{n-1} = 1$, the other nonzero error digits can distribute among at most $\lfloor J/2 \rfloor - 1$ check sums orthogonal on e_{n-1}. Hence, at least $J - \lfloor J/2 \rfloor + 1$, or *more than one-half* of the check sums orthogonal on e_{n-1}, are equal to $e_{n-1} = 1$. However, if $e_{n-1} = 0$, the nonzero error digits can distribute among at most $\lfloor J/2 \rfloor$ check sums. Hence, at least $J - \lfloor J/2 \rfloor$ or *at least one-half* of the check sums orthogonal on e_{n-1}, are equal to $e_{n-1} = 0$. Thus, the value of e_{n-1} is equal to the value assumed by a *clear majority* of the parity-check sums orthogonal on e_{n-1}; if no value is assumed by a clear majority of the parity-check sums (i.e., there is a *tie*), the error digit e_{n-1} is zero. Based on the facts above, an algorithm for decoding e_{n-1} can be formulated as follows:

> The error digit e_{n-1} is decoded as 1 if a clear majority of the parity-check sums orthogonal on e_{n-1} is 1; otherwise, e_{n-1} is decoded as 0.

Correct decoding of e_{n-1} is guaranteed if there are $\lfloor J/2 \rfloor$ or fewer errors in the error vector $\mathbf{e}$. If it is possible to form J parity-check sums orthogonal on e_{n-1}, it is possible to form J parity-check sums orthogonal on any error digit because of the cyclic symmetry of the code. The decoding of other error digits is identical to the decoding of e_{n-1}. The decoding algorithm described above is called *one-step mojority-logic decoding* [3]. If J is the maximum number of parity-check sums orthogonal on e_{n-1} (or any error digit) that can be formed, then, by one-step majority-logic decoding, any error pattern of $\lfloor J/2 \rfloor$ or fewer errors can be corrected. The parameter $t_{\text{ML}} = \lfloor J/2 \rfloor$ is called the *majority-logic error-correcting capability* of the code. Let $d_{\min}$ be the minimum distance of the code. Clearly, the one-step majority-logic decoding is effective for this code only if $t_{\text{ML}} = \lfloor J/2 \rfloor$ is equal to or close to the error-correcting capability $t = \lfloor (d_{\min} - 1)/2 \rfloor$ of the code; in other words, J should be equal to or close to $d_{\min} - 1$.

Definition: A cyclic code with minimum distance $d_{\min}$ is said to be *completely orthogonalizable* in one step if and only if it is possible to form $J = d_{\min} - 1$ parity-check sums orthogonal on an error digit.

At this point, an example will be helpful in clarifying the notions developed above.

Example 7.1

Consider a (15, 7) cyclic code generated by the polynomial

$$\mathbf{g}(X) = 1 + X^4 + X^6 + X^7 + X^8.$$

The parity-check matrix of this code (in systematic form) is found as follows:

$$\mathbf{H} = \begin{bmatrix} \mathbf{h}_0 \\ \mathbf{h}_1 \\ \mathbf{h}_2 \\ \mathbf{h}_3 \\ \mathbf{h}_4 \\ \mathbf{h}_5 \\ \mathbf{h}_6 \\ \mathbf{h}_7 \end{bmatrix} = \begin{bmatrix} 1 & 0 & 0 & 0 & 0 & 0 & 0 & 0 & 1 & 1 & 0 & 1 & 0 & 0 & 0 \\ 0 & 1 & 0 & 0 & 0 & 0 & 0 & 0 & 0 & 1 & 1 & 0 & 1 & 0 & 0 \\ 0 & 0 & 1 & 0 & 0 & 0 & 0 & 0 & 0 & 0 & 1 & 1 & 0 & 1 & 0 \\ 0 & 0 & 0 & 1 & 0 & 0 & 0 & 0 & 0 & 0 & 0 & 1 & 1 & 0 & 1 \\ 0 & 0 & 0 & 0 & 1 & 0 & 0 & 0 & 1 & 1 & 0 & 1 & 1 & 1 & 0 \\ 0 & 0 & 0 & 0 & 0 & 1 & 0 & 0 & 0 & 1 & 1 & 0 & 1 & 1 & 1 \\ 0 & 0 & 0 & 0 & 0 & 0 & 1 & 0 & 1 & 1 & 1 & 0 & 0 & 1 & 1 \\ 0 & 0 & 0 & 0 & 0 & 0 & 0 & 1 & 1 & 0 & 1 & 0 & 0 & 0 & 1 \end{bmatrix}.$$

Consider the following linear combinations of the rows of $\mathbf{H}$:

$$\begin{array}{lr} \textit{Digit positions:} & 0\ \ 1\ \ 2\ \ 3\ \ 4\ \ 5\ \ 6\ \ 7\ \ 8\ \ 9\ \ 10\ \ 11\ \ 12\ \ 13\ \ 14 \\ \mathbf{w}_1 = \qquad\qquad\quad \mathbf{h}_3 = & (0\ \ 0\ \ 0\ \ 1\ \ 0\ \ 0\ \ 0\ \ 0\ \ 0\ \ 0\ \ 0\ \ 1\ \ 1\ \ 0\ \ 1), \\ \mathbf{w}_2 = \qquad \mathbf{h}_1 + \mathbf{h}_5 = & (0\ \ 1\ \ 0\ \ 0\ \ 0\ \ 1\ \ 0\ \ 0\ \ 0\ \ 0\ \ 0\ \ 0\ \ 0\ \ 1\ \ 1), \\ \mathbf{w}_3 = \mathbf{h}_0 + \mathbf{h}_2 + \mathbf{h}_6 = & (1\ \ 0\ \ 1\ \ 0\ \ 0\ \ 0\ \ 1\ \ 0\ \ 0\ \ 0\ \ 0\ \ 0\ \ 0\ \ 0\ \ 1), \\ \mathbf{w}_4 = \qquad\qquad\quad \mathbf{h}_7 = & (0\ \ 0\ \ 0\ \ 0\ \ 0\ \ 0\ \ 0\ \ 1\ \ 1\ \ 0\ \ 1\ \ 0\ \ 0\ \ 0\ \ 1). \end{array}$$

We see that all these four vectors have a 1 at the digit position 14 (or X^{14}) and, at any other digit position, no more than one vector has a 1. Therefore, these four vectors are orthogonal on the digit position 14. Let $\mathbf{r}$ be the received vector. The four parity-check sums formed from these orthogonal vectors are related to the error digits as follows:

$$\begin{aligned} A_1 &= \mathbf{w}_1 \cdot \mathbf{r} = e_3 + e_{11} + e_{12} + e_{14} \\ A_2 &= \mathbf{w}_2 \cdot \mathbf{r} = e_1 + e_5 + e_{13} + e_{14} \\ A_3 &= \mathbf{w}_3 \cdot \mathbf{r} = e_0 + e_2 + e_6 + e_{14} \\ A_4 &= \mathbf{w}_4 \cdot \mathbf{r} = e_7 + e_8 + e_{10} + e_{14}. \end{aligned}$$

We see that e_{14} is checked by all four check sums and no other error digit is checked by more than one check sum. If $e_{14} = 1$ and if there is one or no error occurring among the other 14 digit positions, then at least three (majority) of the four sums A_1, A_2, A_3, and A_4 are equal to $e_{14} = 1$. If $e_{14} = 0$ and if there are two or fewer errors occurring among the other 14 digit positions, then at least two of the four check sums are equal to $e_{14} = 0$. Hence, if there are two or fewer errors in $\mathbf{e}$, the one-step majority-logic decoding always results in correct decoding of e_{14}. Since the code is cyclic, four parity-check sums orthogonal on any error digit can be formed. It can be checked that four is the maximum number of parity-check sums orthogonal on any error digit that can be formed. Thus, by one-step majority-logic decoding, the code is capable of correcting any error pattern with two or fewer errors. It can be shown that there exists at least one error pattern with three errors which cannot be corrected. Consider an error pattern $\mathbf{e}$ with three errors which are e_0, e_3, and e_8 (i.e., $e_0 = e_3 = e_8 = 1$). From the four parity-check sums orthogonal on e_{14}, we have $A_1 = 1$, $A_2 = 0$, $A_3 = 1$, and $A_4 = 1$. Since the majority of the four sums is 1, according to the decoding rule, e_{14} is decoded as 1. This results in an incorrect decoding. The code given in this example is actually a BCH code with minimum distance exactly 5. Therefore, it is completely orthogonalizable.

Given an (n, k) cyclic code C for which J parity-check sums orthogonal on an error digit can be formed, the one-step majority-logic decoding of the code can be easily implemented. First, from the null space C_d of the code, we determine a set of J vectors $\mathbf{w}_1, \mathbf{w}_2, \ldots, \mathbf{w}_J$ that are orthogonal on the highest-order digit position, X^{n-1}. Then, J parity-check sums $A_1, A_2, \ldots, A_J$ orthogonal on the error digit e_{n-1} are formed from these J orthogonal vectors and the received vector $\mathbf{r}$. From (7.5), we see that the vector $\mathbf{w}_j$ tells what received digits should be summed up to from the check sum A_j. The J check sums can be formed by using J multi-input modulo-2 adders. Once these J check sums are formed, they are used as inputs to a J-input majority-logic gate. The output of a majority-logic gate is "1" if and only if more than one-half of its inputs are 1; otherwise, the output is 0. The output is the estimated value of e_{n-1}. A general one-step majority-logic decoder is shown in Figure 7.1. The error correction procedure can be described as follows:

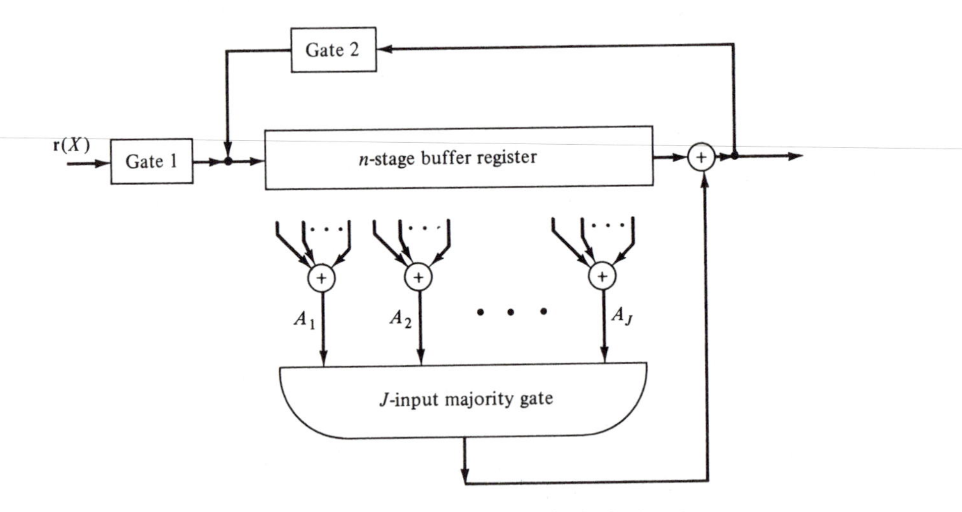

Figure 7.1 General type II one-step majority-logic decoder.

Step 1. With gate 1 turned on and gate 2 turned off, the received vector $\mathbf{r}$ is read into the buffer register.

Step 2. The J parity-check sums orthogonal on e_{n-1} are formed by summing the appropriate received digits.

Step 3. The J orthogonal check sums are fed into a majority-logic gate. The first received digit r_{n-1} is read out of the buffer and is corrected by the output of the majority-logic gate.

Step 4. At the end of step 3, the buffer register has been shifted one place to the right with gate 2 on. Now the second received digit is in the rightmost stage of the buffer register and is corrected in exactly the same manner as the first received digit was. The decoder repeats steps 2 and 3.

Step 5. The received vector is decoded digit by digit in the manner above until a total of n shifts.

If the received vector $\mathbf{r}$ contains $\lfloor J/2 \rfloor$ or fewer errors, the buffer register should contain the transmitted code vector and the inputs to the majority gate should be all zero at the completion of the decoding operation. If not all the inputs to the majority gate are zero, an *uncorrectable error pattern* has been detected. The decoder shown in Figure 7.1 is called the type II one-step majority-logic decoder [3].

The type II one-step majority-logic decoder for the (15, 7) BCH code considered in Example 7.1 is shown in Figure 7.2.

The parity-check sums orthogonal on an error digit can also be formed from the syndrome digits. Let

$$\mathbf{H} = \begin{bmatrix} \mathbf{h}_0 \\ \mathbf{h}_1 \\ \mathbf{h}_2 \\ \cdot \\ \cdot \\ \cdot \\ \mathbf{h}_{n-k-1} \end{bmatrix} = \begin{bmatrix} 1 & 0 & 0 & 0 & \cdots & 0 & p_{00} & p_{01} & & p_{0,k-1} \\ 0 & 1 & 0 & 0 & \cdots & 0 & p_{10} & p_{11} & & p_{1,k-1} \\ 0 & 0 & 1 & 0 & \cdots & 0 & p_{20} & p_{21} & & p_{2,k-1} \\ \cdot & & & & & & \cdot & \cdot & & \cdot \\ \cdot & & & & & & \cdot & \cdot & & \cdot \\ \cdot & & & & & & \cdot & \cdot & & \cdot \\ 0 & 0 & 0 & 0 & \cdots & 1 & p_{n-k-1,0} & p_{n-k-1,1} & \cdots & p_{n-k-1,k-1} \end{bmatrix}$$

be the parity-check matrix for an (n, k) cyclic code C in systematic form. Since the orthogonal vectors $\mathbf{w}_1, \mathbf{w}_2, \ldots, \mathbf{w}_J$ are vectors in the row space of $\mathbf{H}$, they are linear combinations of rows of $\mathbf{H}$. Let

$$\begin{aligned} \mathbf{w}_j &= (w_{j0}, w_{j1}, \ldots, w_{j,n-1}) \\ &= a_{j0}\mathbf{h}_0 + a_{j1}\mathbf{h}_1 + \cdots + a_{j,n-k-1}\mathbf{h}_{n-k-1}. \end{aligned}$$

Because of the systematic structure of $\mathbf{H}$, we see that

$$w_{j0} = a_{j0}, \quad w_{j1} = a_{j1}, \quad \ldots, \quad w_{j,n-k-1} = a_{j,n-k-1}. \tag{7.7}$$

Let $\mathbf{r} = (r_0, r_1, \ldots, r_{n-1})$ be the received vector. Then the syndrome of $\mathbf{r}$ is

$$\mathbf{s} = (s_0, s_1, \ldots, s_{n-k-1}) = \mathbf{r} \cdot \mathbf{H}^T,$$

where the ith syndrome digit is

$$s_i = \mathbf{r} \cdot \mathbf{h}_i \tag{7.8}$$

for $0 \leq i < n - k$. Now, consider the parity-check sum

$$\begin{aligned} A_j &= \mathbf{w}_j \cdot \mathbf{r} \\ &= (a_{j0}\mathbf{h}_0 + a_{j1}\mathbf{h}_1 + \cdots + a_{j,n-k-1}\mathbf{h}_{n-k-1}) \cdot \mathbf{r} \\ &= a_{j0}\mathbf{r} \cdot \mathbf{h}_0 + a_{j1}\mathbf{r} \cdot \mathbf{h}_1 + \cdots + a_{j,n-k-1}\mathbf{r} \cdot \mathbf{h}_{n-k-1}. \end{aligned} \tag{7.9}$$

From (7.7), (7.8), and (7.9), we obtain

$$A_j = w_{j0}s_0 + w_{j1}s_1 + \cdots + w_{j,n-k-1}s_{n-k-1}. \tag{7.10}$$

Thus, the check sum A_j is simply a *linear sum of the syndrome digits with coefficients being the first $n - k$ digits of the orthogonal vector* $\mathbf{w}_j$. Based on (7.10), we obtain a different implementation of the one-step majority-logic decoding as shown in Figure 7.3 (the received vector can be shifted into the syndrome register from the *right*

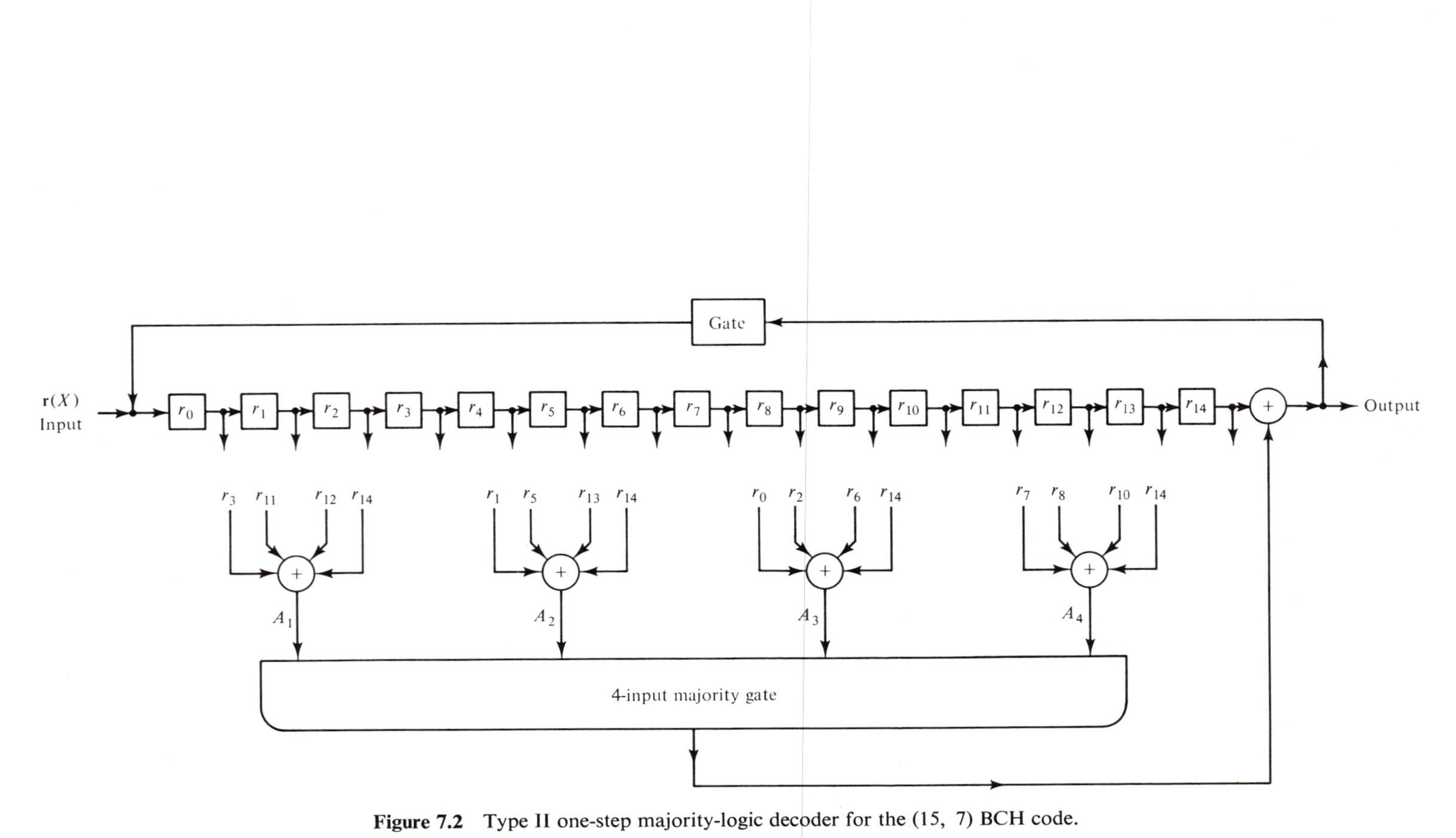

Figure 7.2 Type II one-step majority-logic decoder for the (15, 7) BCH code.

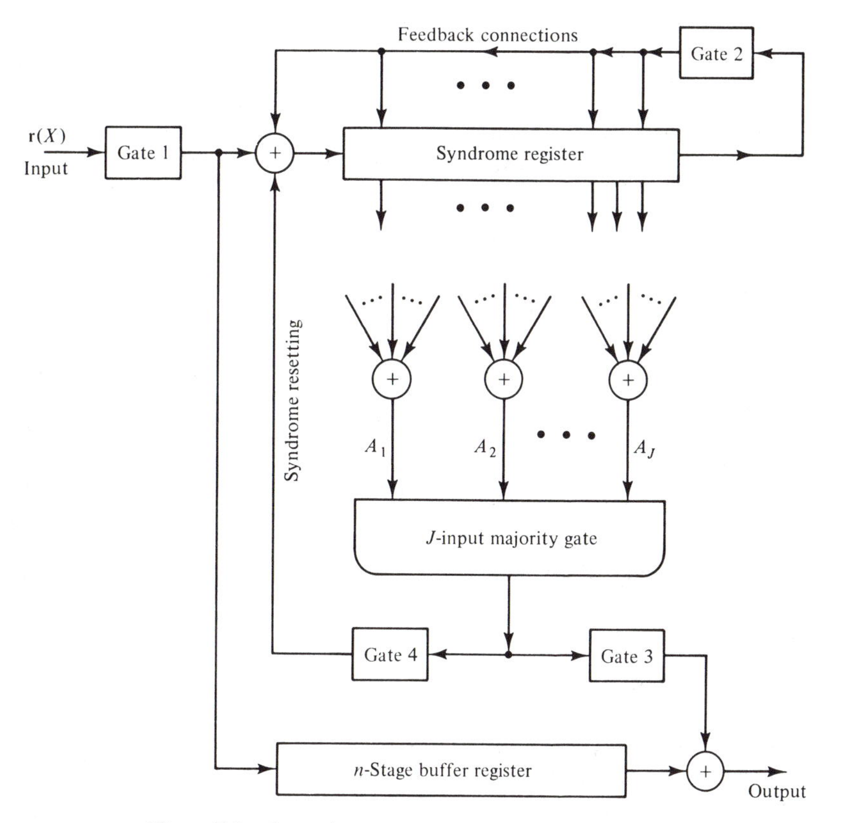

Figure 7.3 General type-I one-step majority-logic decoder.

end). This decoder is called the type I one-step majority-logic decoder [3]. The error correction procedure is described as follows:

Step 1. The syndrome is computed as usual by shifting the received polynomial $\mathbf{r}(X)$ into the syndrome register.

Step 2. The J parity-check sums orthogonal on e_{n-1} are formed by taking proper sums of the syndrome digits. These J check sums are fed into a J-input majority-logic gate.

Step 3. The first received digit is read out of the buffer register and is corrected by the output of the majority gate. At the same time the syndrome register is also shifted once (with gate 2 on) and the effect of e_{n-1} on the syndrome is removed (with gate 4 on). The new contents in the syndrome register form the syndrome of the altered received vector cyclically shifted one place to the right.

Step 4. The new syndrome formed in step 3 is used to decode the next received digit r_{n-2}. The decoder repeats steps 2 and 3. The received digit r_{n-2} is corrected in exactly the same manner as the first received digit r_{n-1} was corrected.

Step 5. The decoder decodes the received vector $\mathbf{r}$ digit by digit in the manner above until a total of n shifts of the buffer and the syndrome registers.

At the completion of the decoding operation, the syndrome register should contain only zeros if the decoder ouput is a code vector. If the syndrome register does not contain all zeros at the end of the decoding, an uncorrectable error pattern has been detected. If we are interested only in decoding the received message digits but not the received parity digits, the buffer register needs only to store the k received message digits and it consists of only k stages. In this case, both type I and type II decoders require roughly the same amount of complexity.

Example 7.2

Consider the (15, 7) BCH code given in Example 7.1. From the vectors $\mathbf{w}_1$, $\mathbf{w}_2$, $\mathbf{w}_3$, and $\mathbf{w}_4$ that are orthogonal on the digit position 14, we find that the parity-check sums orthogonal on e_{14} are equal to the following sums of syndrome digits:

$$A_1 = s_3, \qquad A_2 = s_1 + s_5, \qquad A_3 = s_0 + s_2 + s_6, \qquad A_4 = s_7.$$

Based on these sums we construct the type I one-step majority-logic decoder for the (15, 7) BCH code as shown in Figure 7.4. Suppose that the all-zero code vector (0, 0, . . . , 0) is transmitted and $\mathbf{r}(X) = X^{13} + X^{14}$ is received. Clearly, there are two errors at locations X^{13} and X^{14}. After the entire received polynomial has entered the syndrome register, the syndrome register contains (0 0 1 1 1 0 0 1). The four parity-check sums

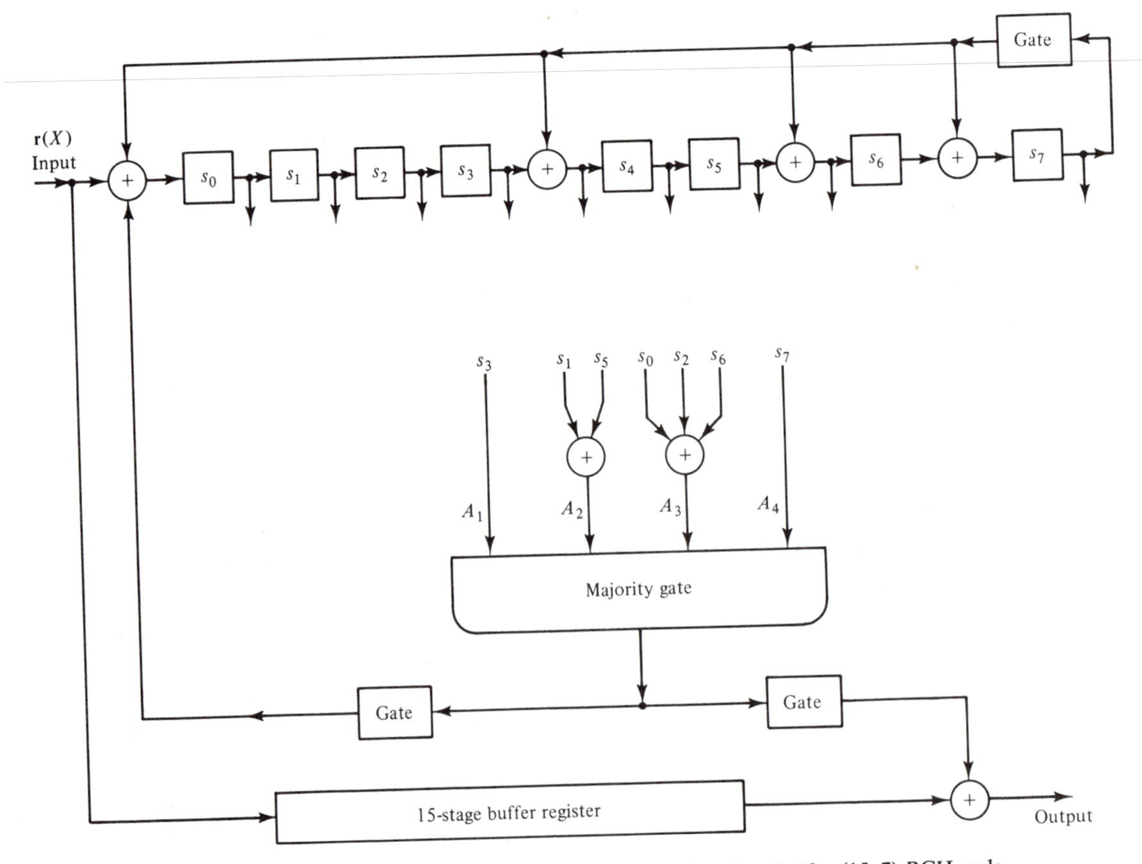

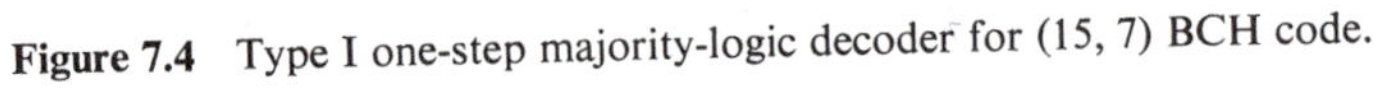
Figure 7.4 Type I one-step majority-logic decoder for (15, 7) BCH code.

orthogonal on e_{14} are

$$A_1 = 1, \qquad A_2 = 0, \qquad A_3 = 1, \qquad A_4 = 1.$$

Since the majority of these four sums is 1, the output of the majority-logic gate is 1, which is the value of e_{14}. Simultaneously shift the buffer and syndrome registers once; the highest-order received digit $r_{14} = 1$ is then corrected by the output of the majority-logic gate and the new contents in the syndrome register are (0 0 0 1 0 1 1 1). The new parity-check sums are now

$$A_1^{(1)} = 1, \qquad A_2^{(1)} = 1, \qquad A_3^{(1)} = 1, \qquad A_4^{(1)} = 1.$$

Again the output of the majority-logic gate is 1, which is the value of e_{13}. Shift both the buffer and syndrome registers once more; the received digit r_{13} would be corrected and the syndrome register would contain only zeros. At this point, both errors have been corrected and the next 13 received digits are error-free.

One-step majority-logic decoding is most efficient for codes that are completely orthogonalizable, or for codes with larger J compared to $d_{\min} - 1$. When J is small compared to $d_{\min} - 1$, one-step majority-logic decoding becomes very inefficient, and much of the error-correcting capability of the code is sacrificed. Given a code C, one would like to know the maximum number of parity-check sums orthogonal on an error digit that can be formed. This is answered by Theorem 7.1.

Theorem 7.1. Let C be an (n, k) cyclic code whose dual code C_d has minimum distance δ. Then the number of parity-check sums orthogonal on an error digit that can be formed, J, is upper bounded by

$$J \leq \left\lfloor \frac{n-1}{\delta - 1} \right\rfloor. \tag{7.11}$$

Proof. Suppose that there exist J vectors $\mathbf{w}_1, \mathbf{w}_2, \ldots, \mathbf{w}_J$ in the dual code of C which are orthogonal on the highest-order digit position, X^{n-1}. Since each of these J vectors has weight at least δ, the total number of 1's in these J vectors is at least $J\delta$. However, because of the orthogonal structure of these J vectors, the total number of 1's in them cannot exceed $J + (n-1)$. Therefore, we have $J\delta \leq J + (n-1)$. This implies that $J \leq (n-1)/(\delta-1)$. Since J is an integer, we must have $J \leq \lfloor (n - 1/(\delta - 1) \rfloor$. Q.E.D.

The dual code of the (15, 7) BCH code has minimum distance 4. Therefore, the maximum number of parity-check sums orthogonal on an error digit is upper bounded by $\lfloor 14/3 \rfloor = 4$. This proves our claim in Example 7.1 that $J = 4$ is the maximum number of parity-check sums orthogonal on an error digit that can be formed for the (15, 7) BCH code.

If it is possible to form J parity-check sums orthogonal on an error digit for a cyclic code, the code has minimum distance at least $J + 1$. The proof of this is left as a problem.

As we pointed out earlier in this section, one-step majority-logic decoding is most effective for cyclic codes which are completely orthogonalizable. Unfortunately, there exist very few good cyclic codes in this category. The double-error-correcting (15, 7) code considered in Example 7.1 is the only known BCH code that is completely orthogonalizable in one step. In the next two sections, several small classes of one-step majority-logic decodable cyclic codes will be presented. Two of the classes are proved to be completely orthogonalizable.

7.2 CLASS OF ONE-STEP MAJORITY-LOGIC DECODABLE CODES

In this section we present a class of one-step majority-logic decodable cyclic codes whose construction is based on certain symmetry property.

Let C be an (n, k) cyclic code generated by $\mathbf{g}(X)$, where $n = 2^m - 1$. We may extend each vector $\mathbf{v} = (v_0, v_1, \ldots, v_{n-1})$ in C by adding an *overall parity-check digit*, denoted by v_∞, to its left. The overall parity-check digit v_∞ is defined as the modulo-2 sum of all the digits of $\mathbf{v}$ (i.e., $v_\infty = v_0 + v_1 + \cdots + v_{n-1}$). Adding v_∞ to $\mathbf{v}$ results in the following vector of $n + 1 = 2^m$ components:

$$\mathbf{v}_e = (v_\infty, v_0, v_1, \ldots, v_{n-1}).$$

The overall parity-check digit is 1 if the weight of $\mathbf{v}$ is odd, and it is 0 if the weight of $\mathbf{v}$ is even. The 2^k extended vectors form an $(n + 1, k)$ linear code, denoted by C_e, which is called an *extension* of C. Clearly, the code vectors of C_e have even weight.

Let α be a primitive element in the Galois field GF(2^m). We may number the components of a vector $\mathbf{v}_e = (v_\infty, v_0, v_1, \ldots, v_{2^m-2})$ in C_e by the elements of GF(2^m) as follows: The component v_∞ is numbered $\alpha^\infty = 0$, the component v_0 is numbered $\alpha = 1$ and, for $1 \leq i < 2^m - 1$, the component v_i is numbered α^i. We call these numbers the *location numbers*. Let Y denote the location of a component of $\mathbf{v}_e$. Consider a permutation that carries the component of $\mathbf{v}_e$ at the location Y to the location $Z = aY + b$, where a and b are elements from the field GF(2^m) and $a \neq 0$. This permutation is called an *affine permutation*. Application of an affine permutation to a vector of 2^m components results in another vector of 2^m components.

Example 7.3

Consider the following vector of 16 components, which are numbered with the elements of GF(2^4) (using Table 2.8):

α^∞	α^0	α^1	α^2	α^3	α^4	α^5	α^6	α^7	α^8	α^9	α^{10}	α^{11}	α^{12}	α^{13}	α^{14}
(1	1	0	0	1	0	0	1	0	0	1	0	0	1	0	0).

Now, we apply the affine permutation

$$Z = \alpha Y + \alpha^{14}$$

to the components of the vector above. The resultant vector is

α^∞	α^0	α^1	α^2	α^3	α^4	α^5	α^6	α^7	α^8	α^9	α^{10}	α^{11}	α^{12}	α^{13}	α^{14}
(0	0	1	1	0	0	0	0	1	0	1	0	1	0	0	1).

For example, the component at the location $Y = \alpha^8$ is carried to the location

$$Z = \alpha \cdot \alpha^8 + \alpha^{14} = \alpha^9 + \alpha^{14} = \alpha^4.$$

An extended cyclic code C_e of length 2^m is said to be *invariant* under the group of affine permutations if every affine permutation carries every code vector in C_e

into another code vector in C_e. In the following we state a necessary and sufficient condition for an extended cyclic code of length 2^m to be invariant under the affine permutations.

Let h be a nonegative integer less than 2^m. The *radix* 2 (binary) expansion of h is

$$h = \delta_0 + \delta_1 2 + \delta_2 2^2 + \cdots + \delta_{m-1} 2^{m-1},$$

where $\delta_i = 0$ or 1 for $0 \leq i < m$. Let h' be another nonnegative integer less than 2^m whose radix 2 expansion is

$$h' = \delta'_0 + \delta'_1 2 + \delta'_2 2^2 + \cdots + \delta'_{m-1} 2^{m-1}.$$

The integer h' is said to be a *descendant* of h if $\delta'_i \leq \delta_i$ for $0 \leq i < m$.

Example 7.4

Let $m = 5$. The integer 21 has the following radix 2 expansion

$$21 = 1 + 0 \cdot 2 + 1 \cdot 2^2 + 0 \cdot 2^3 + 1 \cdot 2^4.$$

The following integers are proper descendants of 21:

$$20 = 0 + 0 \cdot 2 + 1 \cdot 2^2 + 0 \cdot 2^3 + 1 \cdot 2^4,$$
$$17 = 1 + 0 \cdot 2 + 0 \cdot 2^2 + 0 \cdot 2^3 + 1 \cdot 2^4,$$
$$16 = 0 + 0 \cdot 2 + 0 \cdot 2^2 + 0 \cdot 2^3 + 1 \cdot 2^4,$$
$$5 = 1 + 0 \cdot 2 + 1 \cdot 2^2 + 0 \cdot 2^3 + 0 \cdot 2^4,$$
$$4 = 0 + 0 \cdot 2 + 1 \cdot 2^2 + 0 \cdot 2^3 + 0 \cdot 2^4,$$
$$1 = 1 + 0 \cdot 2 + 0 \cdot 2^2 + 0 \cdot 2^3 + 0 \cdot 2^4,$$
$$0 = 0 + 0 \cdot 2 + 0 \cdot 2^2 + 0 \cdot 2^3 + 0 \cdot 2^4.$$

Let $\Delta(h)$ denote the set of all nonzero proper descendants of h. The following theorem characterizes a necessary and sufficient condition for the extension C_e of a cyclic code C of length $2^m - 1$ to be invariant under the affine group of permutations.

Theorem 7.2. Let C be a cyclic code of length $n = 2^m - 1$ generated by $\mathbf{g}(X)$. Let C_e be the extended code obtained from C by appending an overall parity-check digit. Let α be a primitive element of the Galois field GF(2^m). Then the extended code C_e is invariant under the affine permutations if and only if for every α^h that is a root of the generator polynomial $\mathbf{g}(X)$ of C and for every h' in $\Delta(h)$, $\alpha^{h'}$ is also a root of $\mathbf{g}(X)$ and $\alpha^0 = 1$ is not a root of $\mathbf{g}(X)$.

The proof of this theorem is omitted here. For a proof, the reader is referred to References 4 and 5. A cyclic code of length $2^m - 1$ whose generator polynomial satisfies the conditions given in Theorem 7.2 is said to have the *doubly transitive invariant* (*DTI*) *property*.

Given a code C_e of length $n = 2^m$ which is invariant under the affine permutations, the code C obtained by deleting the first digit from each vector of C_e is cyclic. To see this, we apply the permuation $Z = \alpha Y$ to a vector $(v_\infty, v_0, v_1, \ldots, v_{2^m-2})$ in C_e. This permutation keeps the component v_∞ at the same location α^∞ but *cyclically shifts* the other $2^m - 1$ components one place to the right. The resultant vector is

$$(v_\infty, v_{2^m-2}, v_0, v_1, \ldots, v_{2^m-3}),$$

which is also in C_e. Clearly, if we delete v_∞ from each vector of C_e, we obtain a cyclic code of length $2^m - 1$.

Now, we are ready to present a class of one-step majority-logic decodable codes whose dual codes have the DTI property. Let J and L be two factors of $2^m - 1$ such that $J \cdot L = 2^m - 1$. Clearly, both J and L are odd. The polynomial $X^{2^m-1} + 1$ can be factored as follows:

$$X^{2^m-1} + 1 = (1 + X^J)(1 + X^J + X^{2J} + \cdots + X^{(L-1)J}).$$

Let

$$\pi(X) = 1 + X^J + X^{2J} + \cdots + X^{(L-1)J}. \tag{7.12}$$

From Theorem 2.8 we know that the $2^m - 1$ nonzero elements of GF(2^m) form the $2^m - 1$ roots of $X^{2^m-1} + 1$. Let α be a primitive element of GF(2^m). Since $(\alpha^L)^J = \alpha^{2^m-1} = 1$, the polynomial $X^J + 1$ has $\alpha^0 = 1, \alpha^L, \alpha^{2L}, \ldots, \alpha^{(J-1)L}$ as all its roots. Therefore, the polynomial $\pi(X)$ has α^h as a root if and only if h is not a multiple of L and $0 < h < 2^m - 1$.

Now, we form a polynomial $H(X)$ over GF(2) as following: $H(X)$ has α^h as a root if and only if (1) α^h is a root of $\pi(X)$ and (2) for every h' in $\Delta(h)$, $\alpha^{h'}$ is also a root of $\pi(X)$. Let α^i be a root of $H(X)$. Let $\phi_i(X)$ be the minimal polynomial of α^i. Then

$$H(X) = \text{LCM}\,\{\text{minimal polynomials } \phi_i(X) \text{ of the roots of } H(X)\}. \tag{7.13}$$

It is clear that $H(X)$ divides $\pi(X)$ and is a factor of $X^{2^m-1} + 1$. Let C' be the cyclic code of length $2^m - 1$ generated by $H(X)$. It follows from Theorem 7.2 that C' has the doubly transitive invariant property. Thus, the extended code C'_e of C' is invariant under the group of affine permutations. Let C be the dual code of C'. Then C is also cyclic. Since $H(X)$ divides $X^{2^m-1} + 1$, we have

$$X^{2^m-1} + 1 = G(X)H(X).$$

Let k be the degree of $H(X)$. Then the degree of $G(X)$ is $2^m - 1 - k$. The generator polynomial of C is

$$\mathbf{g}(X) = X^{2^m-k-1}G(X^{-1}), \tag{7.14}$$

which is the reciprocal of $G(X)$. Next we will show that the code C is one-step majority-logic decodable and is capable of correcting $t_{ML} = \lfloor J/2 \rfloor$ or fewer errors where $J = (2^m - 1)/L$.

First, we need to determine J vectors from C' (the dual of C) that are orthogonal on the digit at location α^{2^m-2}. Since $\pi(X)$ is a multiple of $H(X)$ and has degree less than $2^m - 1$, it is a code polynomial in C' generated by $H(X)$. Clearly, the polynomials $X\pi(X), X^2\pi(X), \ldots, X^{J-1}\pi(X)$ are also code polynomials in C'. From (7.12), we see that, for $i \neq j$, $X^i\pi(X)$ and $X^j\pi(X)$ do not have any common term. Let $\mathbf{v}_0, \mathbf{v}_1, \ldots, \mathbf{v}_{J-1}$ be the J corresponding code vectors of $\pi(X), X\pi(X), \ldots, X^{J-1}\pi(X)$. The weight of each of these vectors is L. Adding an overall parity-check digit to each of these vectors, we obtain J vectors $\mathbf{u}_0, \mathbf{u}_1, \ldots, \mathbf{u}_{J-1}$ of length 2^m which are code vectors in the extension C'_e of C'. Since L is odd, the overall parity-check digit of each $\mathbf{u}_i$ is 1. Thus, the J vectors $\mathbf{u}_0, \mathbf{u}_1, \ldots, \mathbf{u}_{J-1}$ have the following properties:

1. They all have 1 at location α^∞ (the overall parity-check digit position).
2. One and only one vector has a 1 at the location α^j for $0 \leq j < 2^m - 1$.

Therefore, they form J vectors orthogonal on the digit at location α^∞. Now, we apply the affine permutation

$$Z = \alpha Y + \alpha^{2^m-2}$$

to $\mathbf{u}_0, \mathbf{u}_1, \ldots, \mathbf{u}_{J-1}$. This permutation carries $\mathbf{u}_0, \mathbf{u}_1, \ldots, \mathbf{u}_{J-1}$ into J vectors $\mathbf{z}_0, \mathbf{z}_1, \ldots, \mathbf{z}_{J-1}$ which are also in C'_e (since C'_e is invariant under the group of affine permutations). Note that the permutation $Z = \alpha Y + \alpha^{2^m-2}$ carries the component of $\mathbf{u}_i$ at location α^∞ to the location α^{2^m-2}. Thus, the vectors $\mathbf{z}_0, \mathbf{z}_1, \ldots, \mathbf{z}_{J-1}$ are orthogonal on the digit at location α^{2^m-2}. Deleting the digit at location α^∞ from $\mathbf{z}_0, \mathbf{z}_1, \ldots, \mathbf{z}_{J-1}$, we obtain J vectors $\mathbf{w}_0, \mathbf{w}_1, \ldots, \mathbf{w}_{J-1}$ of length $2^m - 1$, which are vectors in C' and are orthogonal on the digit at the location α^{2^m-2}. From these J vectors, we can form J parity-check sums orthogonal on the error digit e_{2^m-2}. Therefore, the cyclic code C generated by

$$\mathbf{g}(X) = X^{2^m-k-1}G(X^{-1})$$

is one-step majority-logic decodable and is capable of correcting $t_{\text{ML}} = \lfloor J/2 \rfloor$ of fewer errors. For convenience, we call this code a *type* 0 one-step majority-logic decodable DTI code.

Example 7.5

Let $m = 5$. The polynomial $X^{2^4-1} + 1 = X^{15} + 1$ can be factored as

$$X^{15} + 1 = (1 + X^5)(1 + X^5 + X^{10}).$$

Thus, $J = 5, L = 3$, and $\pi(X) = 1 + X^5 + X^{10}$. Let α be a primitive element in GF(2^4) (use Table 2.8) whose minimal polynomial is $\phi_1(X) = 1 + X + X^4$. Since $\alpha^{15} = 1$, the polynomial $1 + X^5$ has $1, \alpha^3, \alpha^6, \alpha^9$, and α^{12} as all its roots. The polynomial $\pi(X)$ has $\alpha, \alpha^2, \alpha^4, \alpha^5, \alpha^7, \alpha^8, \alpha^{10}, \alpha^{11}, \alpha^{13}$, and α^{14} as roots. Next, we determine the polynomial $H(X)$. From the conditions on the roots of $H(X)$, we find that $H(X)$ has $\alpha, \alpha^2, \alpha^4, \alpha^5, \alpha^8$, and α^{10} as its roots. The roots $\alpha, \alpha^2, \alpha^4$, and α^8 are conjugates and they have the same minimum polynomials $\phi_1(X) = 1 + X + X^4$. The roots α^5 and α^{10} are conjugates and they have $\phi_5(X) = 1 + X + X^2$ as their minimal polynomial. Hence,

$$\begin{aligned} H(X) &= \phi_1(X)\phi_5(X) = (1 + X + X^4)(1 + X + X^2) \\ &= 1 + X^3 + X^4 + X^5 + X^6. \end{aligned}$$

We can easily check that $H(X)$ divides $\pi(X)$ and, in fact, $\pi(X) = (1 + X^3 + X^4)H(X)$. Also, $H(X)$ divides $X^{15} + 1$ and $X^{15} + 1 = (1 + X^3 + X^4 + X^5 + X^8 + X^9)H(X)$. Thus, $G(X) = 1 + X^3 + X^4 + X^5 + X^8 + X^9$. The polynomial $H(X)$ generates a (15, 9) cyclic code C' which has the DTI property. The polynomials $\pi(X)$, $X\pi(X)$, $X^2\pi(X)$, $X^3\pi(X)$, and $X^4\pi(X)$ are code polynomials in C'. The dual code of C', C, is generated by

$$\mathbf{g}(X) = X^9G(X^{-1}) = 1 + X + X^4 + X^5 + X^6 + X^9.$$

Thus, C is a (15, 6) cyclic code.

To decode C, we need to determine parity-check sums orthogonal on e_{14}. The vectors corresponding to $\pi(X)$, $X\pi(X)$, $X^2\pi(X)$, $X^3\pi(X)$, and $X^4\pi(X)$ are

Location Numbers

	α^0	α^1	α^2	α^3	α^4	α^5	α^6	α^7	α^8	α^9	α^{10}	α^{11}	α^{12}	α^{13}	α^{14}
$\mathbf{v}_0 =$	(1	0	0	0	0	1	0	0	0	0	1	0	0	0	0)
$\mathbf{v}_1 =$	(0	1	0	0	0	0	1	0	0	0	0	1	0	0	0)
$\mathbf{v}_2 =$	(0	0	1	0	0	0	0	1	0	0	0	0	1	0	0)
$\mathbf{v}_3 =$	(0	0	0	1	0	0	0	0	1	0	0	0	0	1	0)
$\mathbf{v}_4 =$	(0	0	0	0	1	0	0	0	0	1	0	0	0	0	1)

which are code vectors in C'. Adding an overall parity-check digit to these vectors, we obtain the following vectors:

Location Numbers

	α^∞	α^0	α^1	α^2	α^3	α^4	α^5	α^6	α^7	α^8	α^9	α^{10}	α^{11}	α^{12}	α^{13}	α^{14}
$\mathbf{u}_0 =$	(1	1	0	0	0	0	1	0	0	0	0	1	0	0	0	0)
$\mathbf{u}_1 =$	(1	0	1	0	0	0	0	1	0	0	0	0	1	0	0	0)
$\mathbf{u}_2 =$	(1	0	0	1	0	0	0	0	1	0	0	0	0	1	0	0)
$\mathbf{u}_3 =$	(1	0	0	0	1	0	0	0	0	1	0	0	0	0	1	0)
$\mathbf{u}_4 =$	(1	0	0	0	0	1	0	0	0	0	1	0	0	0	0	1)

which are vectors in C'_e (the extension of C'). Now, we apply the affine permutation $Z = \alpha Y + \alpha^{14}$ to permute the components of $\mathbf{u}_0$, $\mathbf{u}_1$, $\mathbf{u}_2$, $\mathbf{u}_3$, and $\mathbf{u}_4$. The permutation results in the following vectors:

Location Numbers

	α^∞	α^0	α^1	α^2	α^3	α^4	α^5	α^6	α^7	α^8	α^9	α^{10}	α^{11}	α^{12}	α^{13}	α^{14}
$\mathbf{z}_0 =$	(0	0	0	0	0	0	0	0	1	1	0	1	0	0	0	1)
$\mathbf{z}_1 =$	(0	0	1	0	0	0	1	0	0	0	0	0	0	0	1	1)
$\mathbf{z}_2 =$	(0	1	0	1	0	0	0	1	0	0	0	0	0	0	0	1)
$\mathbf{z}_3 =$	(1	0	0	0	0	1	0	0	0	0	1	0	0	0	0	1)
$\mathbf{z}_4 =$	(0	0	0	0	1	0	0	0	0	0	0	0	1	1	0	1)

which are also in C'_e. Deleting the overall parity-check digits from these vectors, we obtain the following vectors in C':

Location Numbers

	α^0	α^1	α^2	α^3	α^4	α^5	α^6	α^7	α^8	α^9	α^{10}	α^{11}	α^{12}	α^{13}	α^{14}
$\mathbf{w}_0 =$	(0	0	0	0	0	0	0	1	1	0	1	0	0	0	1)
$\mathbf{w}_1 =$	(0	1	0	0	0	1	0	0	0	0	0	0	0	1	1)
$\mathbf{w}_2 =$	(1	0	1	0	0	0	1	0	0	0	0	0	0	0	1)
$\mathbf{w}_3 =$	(0	0	0	0	1	0	0	0	0	1	0	0	0	0	1)
$\mathbf{w}_4 =$	(0	0	0	1	0	0	0	0	0	0	0	1	1	0	1)

We see that these vectors are orthogonal on the digit at location α^{14}.

Let $\mathbf{r} = (r_0, r_1, r_2, r_3, r_4, r_5, r_6, r_7, r_8, r_9, r_{10}, r_{11}, r_{12}, r_{13}, r_{14})$ be the received vector. Then the parity-check sums orthogonal on e_{14} are

$$
\begin{aligned}
A_0 &= \mathbf{r} \cdot \mathbf{w}_0 = r_7 + r_8 + r_{10} + r_{14}, \\
A_1 &= \mathbf{r} \cdot \mathbf{w}_1 = r_1 + r_5 + r_{13} + r_{14}, \\
A_2 &= \mathbf{r} \cdot \mathbf{w}_2 = r_0 + r_2 + r_6 + r_{14}, \\
A_3 &= \mathbf{r} \cdot \mathbf{w}_3 = r_4 + r_9 + r_{14}, \\
A_4 &= \mathbf{r} \cdot \mathbf{w}_4 = r_3 + r_{11} + r_{12} + r_{14}.
\end{aligned}
$$

Therefore, the (15, 6) cyclic code C generated by $\mathbf{g}(X) = 1 + X + X^4 + X^5 + X^6 + X^9$ is one-step majority-logic decodable and is capable of correcting $t_{ML} = \lfloor 5/2 \rfloor = 2$ or fewer errors. The code has minimum distance at least $J + 1 = 5 + 1 = 6$. However, the generator polynomial has weight exactly 6. Thus, the minimum distance of the code is exactly 6. Hence, the code is completely orthogonalizable.

Recall that $\pi(X)$ has α^h as a root if and only if h is not a multiple of L and $0 < h < 2^m - 1$. Therefore, $\pi(X)$ has the following consecutive powers of α as roots: $\alpha, \alpha^2, \ldots, \alpha^{L-1}$. Since any descendant h' of an integer h is less than h, if $h < L$ and h' in $\Delta(h)$, both α^h and $\alpha^{h'}$ are roots of $\pi(X)$. Consequently, the polynomial $H(X)$ also has $\alpha, \alpha^2, \ldots, \alpha^{L-1}$ as roots. Using the argument that proves the minimum distance of a BCH code, we can show that the minimum distance of C' generated by $H(X)$ is at least L. However, since $\pi(X)$ is a code polynomial of weight L in C', the minimum distance of C' is exactly L. It follows from Theorem 7.1 that the number of parity-check sums orthogonal on an error digit that can be formed for C is upper bounded by

$$\left\lfloor \frac{2^m - 2}{L - 1} \right\rfloor. \tag{7.15}$$

However, $J = (2^m - 1)/L$. Therefore, for large L, J is either equal to or close to the upper bound of (7.15).

In general, it is not known whether the type 0 DTI codes are completely orthogonalizable. There are a number of special cases for which we can prove that the codes are completely orthogonalizable.

The type 0 DTI codes may be modified so that the resultant codes are also one-step majority-logic decodable and $J - 1$ parity-check sums orthogonal on an error digit can be formed. Recall that the polynomial $H(X)$ does not have $(X + 1)$ as a factor (i.e., it does not have $\alpha^0 = 1$ as a root). Let

$$H_1(X) = (X + 1)H(X). \tag{7.16}$$

The cyclic code C_1' generated by $H_1(X)$ is a subcode of C' generated by $H(X)$. In fact, C_1' consists of the *even-weight* vectors of C' as code vectors. Recall that the J orthogonal vectors $\mathbf{w}_0, \mathbf{w}_1, \ldots, \mathbf{w}_{J-1}$ in C' are obtained from the vectors $\mathbf{z}_0, \mathbf{z}_1, \ldots, \mathbf{z}_{J-1}$ by deleting the digit at the location α^∞. Since $\mathbf{z}_0, \mathbf{z}_1, \ldots, \mathbf{z}_{J-1}$ are orthogonal on the digit at the location α^{2^m-2}, there is one and only one vector $\mathbf{z}_i$ that has a "1" at the location α^∞. Since $\mathbf{z}_0, \mathbf{z}_1, \ldots, \mathbf{z}_{J-1}$ all have weight $L + 1$ which is even, all but one of the orthogonal vectors $\mathbf{w}_0, \mathbf{w}_1, \ldots, \mathbf{w}_{J-1}$ have weight $L + 1$. These $J - 1$ even-weight orthogonal vectors are in C_1'. Therefore, the dual code of C_1', denoted by C_1, is one-step majority-logic decodable and $J - 1$ parity-check sums orthogonal on an error digit can be formed. Let

$$G_1(X) = \frac{G(X)}{X+1}. \tag{7.17}$$

Then the generator polynomial for C_1 is

$$\mathbf{g}_1(X) = X^{2^m-k-2}G_1(X^{-1}) = \frac{\mathbf{g}(X)}{X+1}, \tag{7.18}$$

where $\mathbf{g}(X)$ is given by (7.14). C_1 is called a *type* 1 DTI code and its dimension is one greater than that of its corresponding type 0 DTI code.

Example 7.6

For $m = 4$ and $J = 5$, the type 1 DTI code C_1 that corresponds to the type 0 DTI code given in Example 7.5 is generated by

$$\begin{aligned}\mathbf{g}_1(X) &= \frac{1 + X + X^4 + X^5 + X^6 + X^9}{1 + X} \\ &= 1 + X^4 + X^6 + X^7 + X^8.\end{aligned}$$

It is interesting to note that this code is the (15, 7) BCH code. From Example 7.5 we see that $\mathbf{w}_3$ has odd weight, and therefore it is not a vector in C_1'(the dual of C_1). Hence, the four orthogonal vectors in C_1' are

$$\begin{aligned}\mathbf{w}_0 &= (0\ 0\ 0\ 0\ 0\ 0\ 0\ 1\ 1\ 0\ 1\ 0\ 0\ 0\ 1),\\ \mathbf{w}_1 &= (0\ 1\ 0\ 0\ 0\ 1\ 0\ 0\ 0\ 0\ 0\ 0\ 0\ 1\ 1),\\ \mathbf{w}_2 &= (1\ 0\ 1\ 0\ 0\ 0\ 1\ 0\ 0\ 0\ 0\ 0\ 0\ 0\ 1),\\ \mathbf{w}_4 &= (0\ 0\ 0\ 1\ 0\ 0\ 0\ 0\ 0\ 0\ 0\ 1\ 1\ 0\ 1),\end{aligned}$$

which are the same four orthogonal vectors given in Example 7.1.

Since the dual code of type 1 DTI code C_1 has minimum distance $L + 1$, the number of parity-check sums orthogonal on an error digit that can be formed is upper bounded by

$$\left\lfloor \frac{2^m - 2}{L} \right\rfloor = \left\lfloor \frac{2^m - 1}{L} - \frac{1}{L} \right\rfloor = \left\lfloor J - \frac{1}{L} \right\rfloor = J - 1.$$

Therefore, the number of parity-check sums orthogonal on an error digit that can be formed for a type 1 DTI code is equal to its upper bound. Since J is odd, $\lfloor J/2 \rfloor = \lfloor (J-1)/2 \rfloor$. Thus, both type 0 and type 1 DTI codes have the same majority-logic error-correcting capability.

In general, there is no simple formula for enumerating the number of parity-check digits of the one-step majority-logic decodable DTI codes (type 0 or type 1). However, for two special cases, exact formulas for $n - k$ can be obtained [6]:

Case I. For $m = 2sl$ and $J = 2^l + 1$, the number of parity-check digits of the type 1 DTI code of length $2^m - 1$ is

$$n - k = (2^{s+1} - 1)^l - 1.$$

Case II. For $m = \lambda l$ and $J = 2^l - 1$, the number of parity-check digits of the type 1 DTI code of length $2^m - 1$ is

$$n - k = 2^m - (2^\lambda - 1)^l - 1.$$

TABLE 7.1 SOME ONE-STEP MAJORITY-LOGIC DECODABLE TYPE 1 DTI CODES

n	k	t_{ML}	n	k	t_{ML}
15	9	1	2047	1211	11
	7	2		573	44
63	49	1	4095	3969	1
	37	4		3871	2
	13	10		3753	4
255	225	1		3611	6
	207	2		3367	32
	175	8		2707	17
	37	25		2262	19
	21	42		2074	22
511	343	3		1649	45
	139	36		1393	52
1023	961	1		1377	136
	833	5		406	292
	781	16		101	409
	151	46		43	682
	30	170			

A list of one-step majority-logic decodable type 1 DTI codes is given in Table 7.1.

For short length, DTI codes are comparable with BCH codes in efficiency. For example, there exists a (63, 37) one-step majority-logic decodable type 1 DTI code which is capable of correcting four or fewer errors. The corresponding four-error-correcting BCH code of the same length is a (63, 39) code that has two information digits more than the (63, 37) type-1 DTI code. However, the decoding circuit for the (63, 39) BCH code is much more complex than the (63, 37) DTI code. For large block length, the DTI codes are much less efficient than the BCH codes of the same length and the same error-correcting capability.

7.3 OTHER ONE-STEP MAJORITY-LOGIC DECODABLE CODES

There are two other small classes of one-step majority-logic decodable cyclic codes: the maximum-length codes and the difference-set codes. Both classes have been proved to be completely orthogonalizable.

Maximum-Length Codes

For any integer $m \geq 3$, there exists a nontrivial maximum-length code with the following parameters:

Block length:	$n = 2^m - 1$
Number of information digits:	$k = m$
Minimum distance:	$d = 2^{m-1}$.

The generator polynomial of this code is

$$\mathbf{g}(X) = \frac{X^n + 1}{\mathbf{p}(X)} \tag{7.19}$$

where $\mathbf{p}(X)$ is a primitive polynomial of degree m. This code consists of the all-zero code vector and $2^m - 1$ code vectors of weight 2^{m-1} (see Problem 7.11). Maximum-length codes were first shown to be majority-logic decodable by Yale [7] and Zierler [8] independently.

The dual code of the maximum-length code is a $(2^m - 1, 2^m - m - 1)$ cyclic code generated by the reciprocal of the parity polynomial $\mathbf{p}(X)$,

$$\mathbf{p}^*(X) = X^m\mathbf{p}(X^{-1}).$$

Since $\mathbf{p}^*(X)$ is also a primitive polynomial of degree m, the dual code is thus a Hamming code. Therefore, the null space of the maximum-length code contains vectors of weight 3 (this is the minimum weight). Now, consider the following set of distinct code polynomials:

$$Q = \{\mathbf{w}(X) = X^i + X^j + X^{n-1} \mid 0 \leq i < j \leq n - 1\} \tag{7.20}$$

in the Hamming code generated by $\mathbf{p}^*(X)$. No two polynomials in Q can have any common terms except the term X^{n-1}. Otherwise, the sum of these two polynomials would be a code polynomial of only two terms in the Hamming code. This is impossible since the minimum weight of a Hamming code is 3. Therefore, the set Q contains polynomials orthogonal on the highest-order digit position X^{n-1}. To find $\mathbf{w}(X)$, we start with a polynomial $X^{n-1} + X^j$ for $0 \leq j < n - 1$, and then determine X^i such that $X^{n-1} + X^j + X^i$ is divisible by $\mathbf{p}^*(X)$. This can be carried out as follows. Divide $X^{n-1} + X^j$ by $\mathbf{p}^*(X)$ step by step with long division until a single term X^i appears at the end of a certain step. Then $\mathbf{w}(X) = X^{n-1} + X^j + X^i$ is a polynomial orthogonal on digit position X^{n-1}. Clearly, if we start with $X^{n-1} + X^i$, we would obtain the same polynomial $\mathbf{w}(X)$. Thus, we can find $(n - 1)/2 = 2^{m-1} - 1$ polynomials orthogonal on digit position X^{n-1}. That is, $J = 2^{m-1} - 1$ parity-check sums orthogonal on e_{n-1} can be formed. Since the maximum-length code generated by $\mathbf{g}(X)$ of (7.19) has minimum distance exactly 2^{m-1}, it is completely orthogonalizable. The code is capable of correcting $t_{\text{ML}} = 2^{m-2} - 1$ or fewer errors with one-step majority-logic decoding.

Example 7.7

Consider the maximum-length code with $m = 4$ and parity polynomial $\mathbf{p}(X) = 1 + X + X^4$. This code has block length $n = 15$ and minimum distance $d = 8$. The generator polynomial of this code is

$$\begin{aligned}\mathbf{g}(X) &= \frac{X^{15} + 1}{\mathbf{p}(X)} \\ &= 1 + X + X^2 + X^3 + X^5 + X^7 + X^8 + X^{11}.\end{aligned}$$

The null space of this code is generated by

$$\mathbf{p}^*(X) = X^4\mathbf{p}(X^{-1}) = X^4 + X^3 + 1.$$

Divide $X^{14} + X^{13}$ by $\mathbf{p}^*(X) = X^4 + X^3 + 1$ with long division as shown in the following:

$$\begin{array}{r}
X^{10} \qquad\qquad\quad \\
X^4 + X^3 + 1 \overline{) X^{14} + X^{13} \qquad\quad} \\
X^{14} + X^{13} + X^{10} \\
\hline
X^{10} \text{ (stop).}
\end{array}$$

A single term X^{10} appears at the end of the first step of the long division. Then $\mathbf{w}_1(X) = X^{14} + X^{13} + X^{10}$ is a polynomial orthogonal on X^{14}. Now divide $X^{14} + X^{12}$ by $\mathbf{p}^*(X)$:

$$\begin{array}{l}
\qquad\qquad\qquad X^{10} + X^9 + X^6 \\
X^4 + X^3 + 1 \overline{) X^{14} \qquad\quad + X^{12}} \\
\qquad\qquad\qquad X^{14} + X^{13} \qquad + X^{10} \\
\qquad\qquad\qquad \overline{\qquad X^{13} + X^{12} + X^{10}} \\
\qquad\qquad\qquad\qquad X^{13} + X^{12} \qquad + X^9 \\
\qquad\qquad\qquad\qquad \overline{\qquad\qquad X^{10} + X^9} \\
\qquad\qquad\qquad\qquad\qquad\quad X^{10} + X^9 + X^6 \\
\qquad\qquad\qquad\qquad\qquad\quad \overline{\qquad\qquad X^6 \text{ (stop).}}
\end{array}$$

Then $\mathbf{w}_2(X) = X^{14} + X^{12} + X^6$ is another polynomial orthogonal on X^{14}. The rest of the polynomials orthogonal on X^{14} can be found in the same manner; they are

$$\mathbf{w}_3(X) = 1 + X^{11} + X^{14}, \qquad \mathbf{w}_4(X) = X^4 + X^9 + X^{14},$$

$$\mathbf{w}_5(X) = X + X^8 + X^{14}, \qquad \mathbf{w}_6(X) = X^5 + X^7 + X^{14},$$

$$\mathbf{w}_7(X) = X^2 + X^3 + X^{14}.$$

From the set of polynomials orthogonal on X^{14}, we obtain the following seven parity-check sums orthogonal on e_{14}:

$$\begin{aligned}
A_1 &= e_{10} + e_{13} + e_{14}, \\
A_2 &= e_6 \;\; + e_{12} + e_{14}, \\
A_3 &= e_0 \;\; + e_{11} + e_{14}, \\
A_4 &= e_4 \;\; + e_9 \;\; + e_{14}, \\
A_5 &= e_1 \;\; + e_8 \;\; + e_{14}, \\
A_6 &= e_5 \;\; + e_7 \;\; + e_{14}, \\
A_7 &= e_2 \;\; + e_3 \;\; + e_{14}.
\end{aligned}$$

In terms of syndrome bits, we have $A_1 = s_{10}$, $A_2 = s_6$, $A_3 = s_0$, $A_4 = s_4 + s_9$, $A_5 = s_1 + s_8$, $A_6 = s_5 + s_7$, and $A_7 = s_2 + s_3$. The code is capable of correcting three or fewer errors by one-step majority-logic decoding. The type I and type II one-step majority-logic decoders for this code are shown in Figures 7.5 and 7.6, respectively.

Difference-Set Codes

The formulation of difference-set codes is based on the construction of a *perfect difference set*. Let $P = \{l_0, l_1, l_2, \ldots, l_q\}$ be a set of $q + 1$ nonnegative integers such that

$$0 \leq l_0 < l_1 < l_2 < \cdots < l_q \leq q(q + 1).$$

From this set of integers, it is possible to form $q(q + 1)$ *ordered differences* as follows:

$$D = \{l_j - l_i \mid j \neq i\}.$$

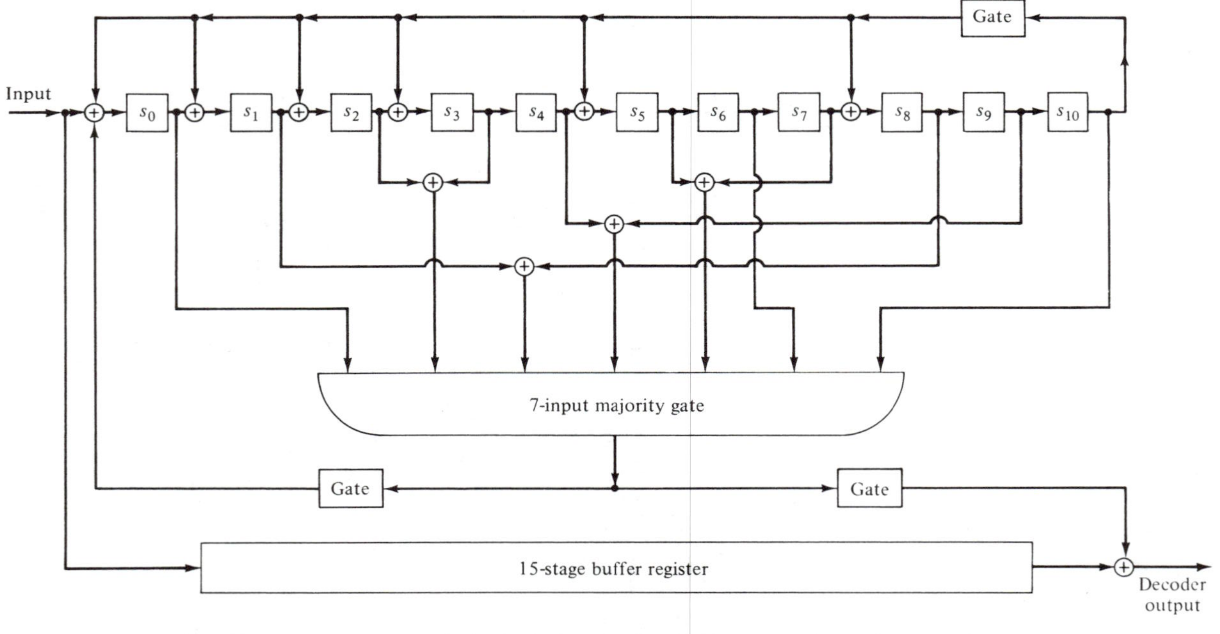

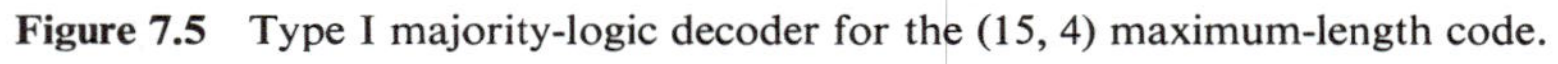
Figure 7.5 Type I majority-logic decoder for the (15, 4) maximum-length code.

$$
\begin{array}{rl}
 & X^{10} \\
X^4 + X^3 + 1 \,) & X^{14} + X^{13} \\
 & X^{14} + X^{13} + X^{10} \\ \hline
 & \qquad\qquad X^{10} \text{ (stop)}.
\end{array}
$$

A single term X^{10} appears at the end of the first step of the long division. Then $\mathbf{w}_1(X) = X^{14} + X^{13} + X^{10}$ is a polynomial orthogonal on X^{14}. Now divide $X^{14} + X^{12}$ by $\mathbf{p}^*(X)$:

$$
\begin{array}{rl}
 & X^{10} + X^9 + X^6 \\
X^4 + X^3 + 1 \,) & X^{14} + X^{12} \\
 & X^{14} + X^{13} + X^{10} \\ \hline
 & X^{13} + X^{12} + X^{10} \\
 & X^{13} + X^{12} + X^9 \\ \hline
 & X^{10} + X^9 \\
 & X^{10} + X^9 + X^6 \\ \hline
 & \qquad\qquad X^6 \text{ (stop)}.
\end{array}
$$

Then $\mathbf{w}_2(X) = X^{14} + X^{12} + X^6$ is another polynomial orthogonal on X^{14}. The rest of the polynomials orthogonal on X^{14} can be found in the same manner; they are

$$
\begin{aligned}
\mathbf{w}_3(X) &= 1 + X^{11} + X^{14}, & \mathbf{w}_4(X) &= X^4 + X^9 + X^{14}, \\
\mathbf{w}_5(X) &= X + X^8 + X^{14}, & \mathbf{w}_6(X) &= X^5 + X^7 + X^{14}, \\
\mathbf{w}_7(X) &= X^2 + X^3 + X^{14}.
\end{aligned}
$$

From the set of polynomials orthogonal on X^{14}, we obtain the following seven parity-check sums orthogonal on e_{14}:

$$
\begin{aligned}
A_1 &= e_{10} + e_{13} + e_{14}, \\
A_2 &= e_6 + e_{12} + e_{14}, \\
A_3 &= e_0 + e_{11} + e_{14}, \\
A_4 &= e_4 + e_9 + e_{14}, \\
A_5 &= e_1 + e_8 + e_{14}, \\
A_6 &= e_5 + e_7 + e_{14}, \\
A_7 &= e_2 + e_3 + e_{14}.
\end{aligned}
$$

In terms of syndrome bits, we have $A_1 = s_{10}$, $A_2 = s_6$, $A_3 = s_0$, $A_4 = s_4 + s_9$, $A_5 = s_1 + s_8$, $A_6 = s_5 + s_7$, and $A_7 = s_2 + s_3$. The code is capable of correcting three or fewer errors by one-step majority-logic decoding. The type I and type II one-step majority-logic decoders for this code are shown in Figures 7.5 and 7.6, respectively.

Difference-Set Codes

The formulation of difference-set codes is based on the construction of a *perfect difference set*. Let $P = \{l_0, l_1, l_2, \ldots, l_q\}$ be a set of $q + 1$ nonnegative integers such that

$$0 \le l_0 < l_1 < l_2 < \cdots < l_q \le q(q + 1).$$

From this set of integers, it is possible to form $q(q + 1)$ *ordered differences* as follows:

$$D = \{l_j - l_i \mid j \neq i\}.$$

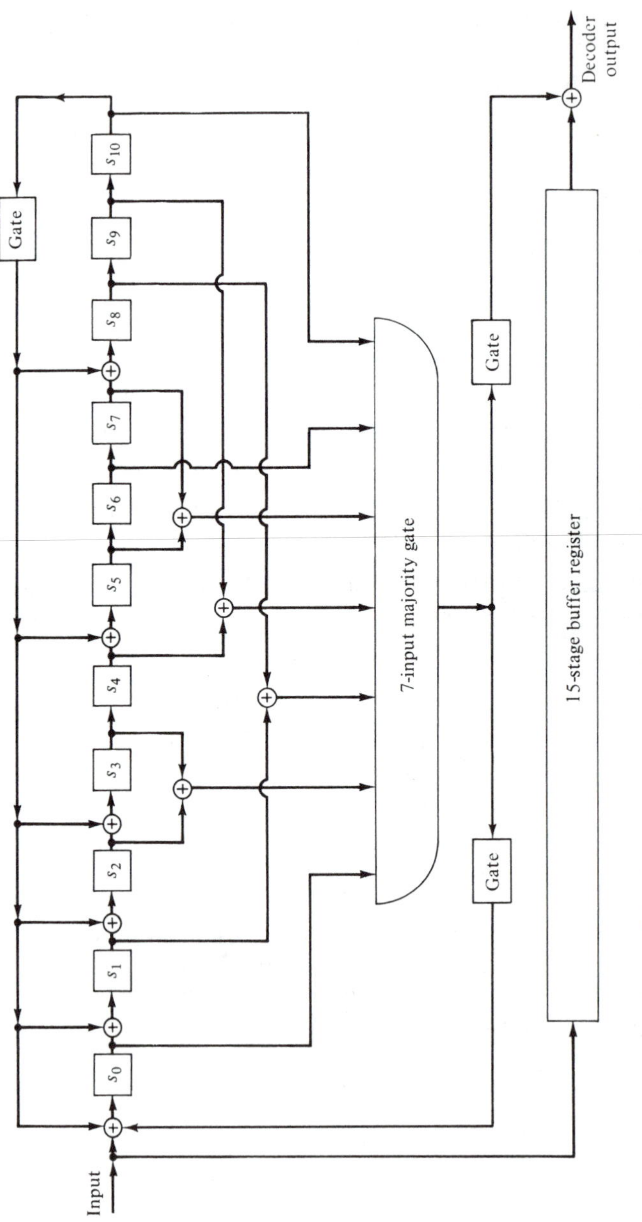

Figure 7.5 Type I majority-logic decoder for the (15, 4) maximum-length code.

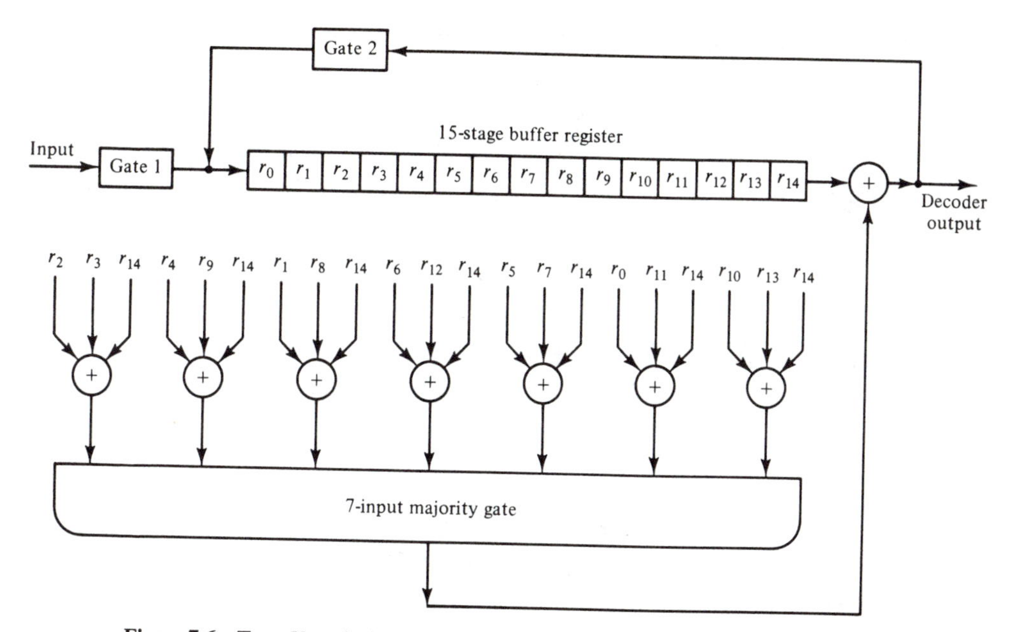

Figure 7.6 Type II majority-logic decoder for the (15, 4) maximum-length code.

Obviously, half of the differences in D are positive and the other half are negative. The set P is said to be a *perfect simple difference set of order q* if and only if it has the following properties:

1. All the positive differences in D are distinct.
2. All the negative differences in D are distinct.
3. If $l_j - l_i$ is a negative difference in D, then $q(q + 1) + 1 + (l_j - l_i)$ is not equal to any positive difference in D.

Clearly, it follows from the definition that $P' = \{0, l_1 - l_0, l_2 - l_0, \ldots, l_q - l_0\}$ is also a perfect simple difference set.

Example 7.8

Consider the set $P = \{0, 2, 7, 8, 11\}$ with $q = 4$. The $4 \cdot 5 = 20$ ordered differences are

$$D = \{2, 7, 8, 11, 5, 6, 9, 1, 4, 3, -2, -7, -8, -11, -5, -6, -9, -1, -4, -3\}.$$

It can be checked easily that P satisfies all three properties of a perfect simple difference set.

Singer [9] has constructed perfect difference sets for order $q = p^s$, where p is a prime and s is any positive integer (see also Reference 10). In what follows we shall only be concerned with $q = 2^s$.

Let $P = \{l_0 = 0, l_1, l_2, \ldots, l_{2^s}\}$ be a perfect simple difference set of order 2^s. Define the polynomial

$$\mathbf{z}(X) = 1 + X^{l_1} + X^{l_2} + \cdots + X^{l_{2^s}}. \tag{7.21}$$

Let $n = 2^s(2^s + 1) + 1$ and $\mathbf{h}(X)$ be the greatest common divisor of $\mathbf{z}(X)$ and $X^n + 1$, that is,

$$\begin{aligned}\mathbf{h}(X) &= \text{GCD}\{\mathbf{z}(X), X^n + 1\} \\ &= 1 + h_1X + h_2X^2 + \cdots + h_{k-1}X^{k-1} + X^k.\end{aligned} \tag{7.22}$$

Then a difference-set code of length n is defined as the cyclic code generated by

$$\begin{aligned}\mathbf{g}(X) &= \frac{X^n + 1}{\mathbf{h}(X)} \\ &= 1 + g_1X + g_2X^2 + \cdots + X^{n-k}.\end{aligned} \tag{7.23}$$

This code has the following parameters:

Code length:	$n = 2^{2s} + 2^s + 1$
Number of parity-check digits:	$n - k = 3^s + 1$
Minimum distance:	$d = 2^s + 2.$

Difference-set codes were discovered by Rudolph [11] and Weldon [12] independently. The formula for the number of parity-check digits was derived by Graham and MacWilliams [13].

Example 7.9

In Example 7.8 we have shown that the set $P = \{0, 2, 7, 8, 11\}$ is a perfect simple difference set of order $q = 2^2$. Let $\mathbf{z}(X) = 1 + X^2 + X^7 + X^8 + X^{11}$. Then

$$\begin{aligned}\mathbf{h}(X) &= \text{GCD}\{1 + X^2 + X^7 + X^8 + X^{11}, 1 + X^{21}\} \\ &= 1 + X^2 + X^7 + X^8 + X^{11}.\end{aligned}$$

The generator polynomial of the difference-set code of length $n = 21$ is

$$\begin{aligned}\mathbf{g}(X) &= \frac{X^{21} + 1}{\mathbf{h}(X)} \\ &= 1 + X^2 + X^4 + X^6 + X^7 + X^{10}.\end{aligned}$$

Thus, the code is a (21, 11) cyclic code.

Let $\mathbf{h}^*(X) = X^k\mathbf{h}(X^{-1})$ be the reciprocal polynomial of $\mathbf{h}(X)$. Then the $(n, n - k)$ cyclic code generated by $\mathbf{h}^*(X)$ is the null space of the difference-set code generated by $\mathbf{g}(X)$ of (7.23). Let

$$\begin{aligned}\mathbf{z}^*(X) &= X^{l_{2s}}\mathbf{z}(X^{-1}) \\ &= 1 + \cdots + X^{l_{2s}-l_2} + X^{l_{2s}-l_1} + X^{l_{2s}}.\end{aligned} \tag{7.24}$$

Since $\mathbf{z}(X)$ is divisible by $\mathbf{h}(X)$, $\mathbf{z}^*(X)$ is divisible by $\mathbf{h}^*(X)$. Thus, $\mathbf{z}^*(X)$ is in the null space of the difference-set code generated by $\mathbf{g}(X)$ of (7.23). Let

$$\begin{aligned}\mathbf{w}_0(X) &= X^{n-1-l_{2s}}\mathbf{z}^*(X) \\ &= X^{n-1-l_{2s}} + \cdots + X^{n-1-l_2} + X^{n-1-l_1} + X^{n-1}.\end{aligned}$$

Obviously, $\mathbf{w}_0(X)$ is divisible by $\mathbf{h}^*(X)$ and is also in the null space of the difference-set code generated by $\mathbf{g}(X)$ of (7.23). Now let

$$\mathbf{w}_i(X) = X^{l_i - l_{i-1} - 1} + X^{l_i - l_{i-2} - 1} + \cdots + X^{l_i - l_1 - 1} + X^{l_i - 1}$$
$$+ X^{n-1-l_{2^s}+l_i} + X^{n-1-l_{2^s-1}+l_i} + \cdots + X^{n-1} \tag{7.25}$$

be the vector obtained by shifting $\mathbf{w}_0(X)$ cyclically to the right l_i times. Since $\{l_0 = 0, l_1, l_2, \ldots, l_{2^s}\}$ is a perfect difference set, no two polynomials $\mathbf{w}_i(X)$ and $\mathbf{w}_j(X)$ for $i \neq j$ can have any common term except X^{n-1}. Thus, $\mathbf{w}_0(X), \mathbf{w}_1(X), \ldots, \mathbf{w}_{2^s}(X)$ form a set of $J = 2^s + 1$ polynomials orthogonal on the digit at the position X^{n-1}. Since the code generated by $\mathbf{g}(X)$ of (7.23) is proved to have minimum distance $2^s + 2$, it is completely orthogonalizable and is capable of correcting $t_{ML} = 2^{s-1}$ or fewer errors.

Example 7.10

Consider the code given in Example 7.9, which is specified by the perfect difference set $P = \{0, 2, 7, 8, 11\}$ of order 2^2. Thus, we have

$$\mathbf{z}^*(X) = X^{11}\mathbf{z}(X^{-1}) = 1 + X^3 + X^4 + X^9 + X^{11}$$

and

$$\mathbf{w}_0(X) = X^9\mathbf{z}^*(X) = X^9 + X^{12} + X^{13} + X^{18} + X^{20}.$$

By shifting $\mathbf{w}_0(X)$ cyclically to the right 2 times, 7 times, 8 times, and 11 times, we obtain

$$\begin{aligned}
\mathbf{w}_1(X) &= X + X^{11} + X^{14} + X^{15} + X^{20}, \\
\mathbf{w}_2(X) &= X^4 + X^6 + X^{16} + X^{19} + X^{20}, \\
\mathbf{w}_3(X) &= 1 + X^5 + X^7 + X^{17} + X^{20}, \\
\mathbf{w}_4(X) &= X^2 + X^3 + X^8 + X^{10} + X^{20}.
\end{aligned}$$

Clearly, $\mathbf{w}_0(X), \mathbf{w}_1(X), \mathbf{w}_2(X), \mathbf{w}_3(X)$, and $\mathbf{w}_4(X)$ are five polynomials orthogonal on X^{20}. From these five orthogonal polynomials, we can form the following five parity-check sums orthogonal on e_{20}:

$$\begin{aligned}
A_1 &= s_9 &&= e_9 + e_{12} + e_{13} + e_{18} + e_{20}, \\
A_2 &= s_1 &&= e_1 + e_{11} + e_{14} + e_{15} + e_{20}, \\
A_3 &= s_4 + s_6 &&= e_4 + e_6 + e_{16} + e_{19} + e_{20}, \\
A_4 &= s_0 + s_5 + s_7 &&= e_0 + e_5 + e_7 + e_{17} + e_{20}, \\
A_5 &= s_2 + s_3 + s_8 &&= e_2 + e_3 + e_8 + e_{10} + e_{20}.
\end{aligned}$$

A type II majority-logic decoder for this code is shown in Figure 7.7. The construction of a type I decoder for this code is left as an exercise.

Difference-set codes are nearly as powerful as the best known cyclic codes in the range of practical interest. Unfortunately, there are relatively few codes with useful parameters in this class. A list of the first few codes with their generator polynomials and their corresponding perfect simple difference sets is given in Table 7.2.

There are other one-step majority-logic decodable cyclic codes which we present in Chapter 8.

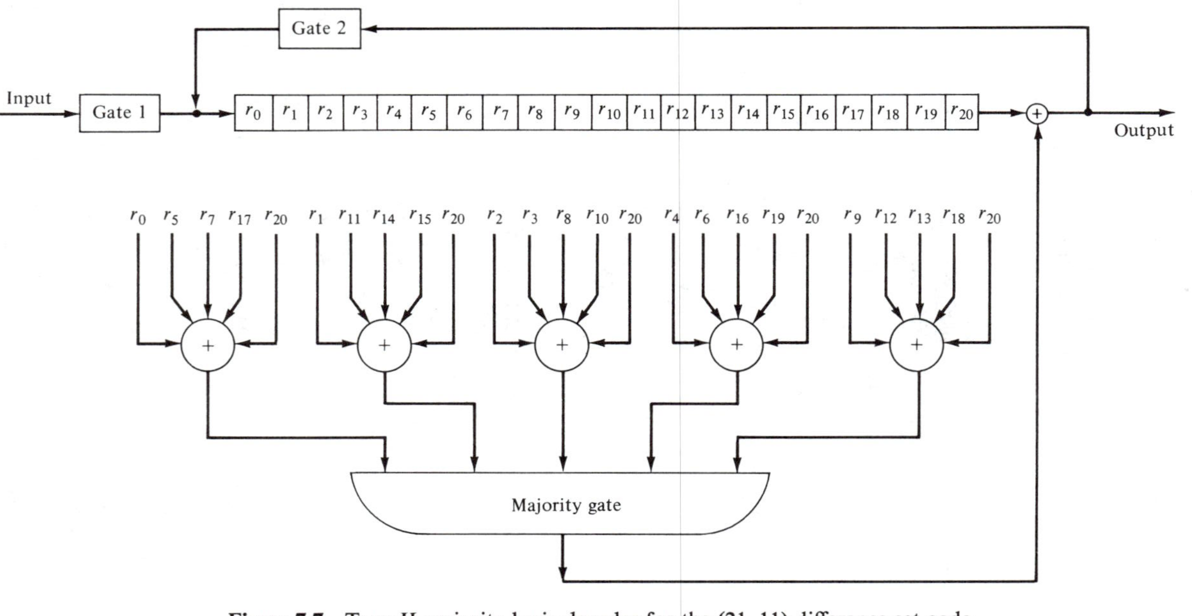

Figure 7.7 Type II majority-logic decoder for the (21, 11) difference-set code.

TABLE 7.2 LIST OF BINARY DIFFERENCE-SET CYCLIC CODES

s	n	k	d	t	Generator polynomial, $g(X)$*	Associated difference set
1	7	3	4	1	0, 2, 3, 4	0, 2, 3
2	21	11	6	2	0, 2, 4, 6, 7, 10	0, 2, 7, 8, 11
3	73	45	10	4	0, 2, 4, 6, 8, 12, 16, 22, 25, 28	0, 2, 10, 24, 25, 29, 36, 42, 45
4	273	191	18	8	0, 4, 10, 18, 22, 24, 34, 36, 40, 48, 52, 56, 66, 67, 71, 76, 77, 82	0, 18, 24, 46, 50, 67, 103, 112, 115, 126, 128, 159, 166, 167, 186, 196, 201
5	1057	813	34	16	0, 1, 3, 4, 5, 11, 14, 17, 18, 22, 23, 26, 27, 28, 32, 33, 35, 37, 39, 41, 43, 45, 47, 48, 51, 52, 55, 59, 62, 68, 70, 71, 72, 74, 75, 76, 79, 81, 83, 88, 95, 96, 98, 101, 103, 105, 106, 108, 111, 114, 115, 116, 120, 121, 122, 123, 124, 126, 129, 131, 132, 135, 137, 138, 141, 142, 146, 147, 149, 150, 151, 153, 154, 155, 158, 160, 161, 164, 165, 166, 167, 169, 174, 175, 176, 177, 178, 179, 180, 181, 182, 183, 184, 186, 188, 189, 191, 193, 194, 195, 198, 199, 200, 201, 202, 203, 208, 209, 210, 211, 212, 214, 216, 222, 224, 226, 228, 232, 234, 236, 242, 244	0, 1, 3, 7, 15, 31, 54, 63, 109, 127, 138, 219, 255, 277, 298, 338, 348, 439, 452, 511, 528, 555, 597, 677, 697, 702, 792, 897, 905, 924, 990, 1023

*Each generator polynomial is represented by the exponents of its nonzero terms. For example, {0, 2, 3, 4} represents $g(X) = 1 + X^2 + X^3 + X^4$.

7.4 MULTIPLE-STEP MAJORITY-LOGIC DECODING

The one-step majority-logic decoding for a cyclic code is based on the condition that a set of J parity-check sums orthogonal on a single error digit can be formed. This decoding method is effective for codes that are completely orthogonalizable or for codes with large J compared to their minimum distance $d_{\min}$. Unfortunately, there are only several small classes of cyclic codes known to be in this category. However, the concept of parity-check sums orthogonal on a single error digit can be generalized in such a way that many cyclic codes can be decoded by employing several *levels* of majority-logic gates.

Let $E = \{e_{i_1}, e_{i_2}, \ldots, e_{i_M}\}$ be a set of M error digits where $0 \leq i_1 < i_2 < \cdots < i_M < n$. The integer M is called the *size* of E.

Definition 7.2. A set of J parity-check sums $A_1, A_2, \ldots, A_J$ is said to be orthogonal on the set E if and only if (1) every error digit e_{i_l} in E is checked by every check sum A_j for $1 \leq j \leq J$, and (2) no other error digit is checked by more than one check sum.

For example, the following four parity-check sums are orthogonal on the set $E = \{e_6, e_8\}$:

$$\begin{aligned} A_1 &= e_0 + e_2 + e_6 + e_8, \\ A_2 &= e_3 + e_4 + e_6 + e_8, \\ A_3 &= e_1 + e_6 + e_7 + e_8, \\ A_4 &= e_5 + e_6 + e_8. \end{aligned}$$

Following the same argument employed for one-step majority-logic decoding, the sum of error digits in E, $e_{i_1} + e_{i_2} + \cdots + e_{j_M}$, can be determined correctly from the check sums, $A_1, A_2, \ldots, A_J$, orthogonal on E provided that there are $\lfloor J/2 \rfloor$ or fewer errors in the error pattern $\mathbf{e}$. This sum of error digits in E may be regarded as an *additional* check sum and so can be used for decoding.

Consider an (n, k) cyclic code C which is used for error control purpose in a communication system. Let $\mathbf{e} = (e_0, e_1, \ldots, e_{n-1})$ denote the error vector that occurs during the transmission of a code vector $\mathbf{v}$ in C. Let $E_1^1, E_2^1, \ldots, E_i^1, \ldots$ be some properly selected sets of error digits of $\mathbf{e}$. Let $S(E_i^1)$ denote the modulo-2 sum of the error digits in E_i^1. Suppose that, for each set E_i^1, it is possible to form at least J parity-check sums orthogonal on it. Then the sum $S(E_i^1)$ can be estimated from these J orthogonal check sums. The estimation can be done by a J-input majority-logic gate with the J orthogonal check sums as inputs. The estimated value of $S(E_i^1)$ is the output of a majority-logic gate, which is 1 if and only if more than one-half of the inputs are 1; otherwise, it is 0. The estimation is correct provided that there are $\lfloor J/2 \rfloor$ or fewer errors in the error vector $\mathbf{e}$. The sums, $S(E_1^1), S(E_2^1), \ldots, S(E_i^1), \ldots$ (possibly together with other check sums) are then used to estimate the sums of error digits in the second selected sets, $E_1^2, E_2^2, \ldots, E_i^2, \ldots$ with size smaller than that of the first selected sets. Suppose that, for each set E_i^2, it is possible to form J or more check sums orthogonal on it. Then the sum $S(E_i^2)$ can be determined correctly from the check sums orthogonal on E_i^2 provided that there are no more than $\lfloor J/2 \rfloor$ errors in $\mathbf{e}$. Once the sums, $S(E_1^2), S(E_2^2), \ldots, S(E_i^2), \ldots$, are determined, they (may be together with other check sums) are used to estimate the sums of error digits in the third selected sets, $E_1^3, E_2^3, \ldots, E_i^3, \ldots$, with size smaller than that of the second selected sets. The process of estimating check sums from known check sums is called *orthogonalization* [3]. The orthogonalization process continues until a set of J or more check sums orthogonal on only a single error digit, say e_{n-1}, is obtained. Then the value of e_{n-1} can be estimated from these orthogonal check sums. Because of the cyclic structure of the code, other error digits can be estimated in the same manner and by the same circuitry. A code is said to be *L-step orthogonalizable* (or *L-step majority-logic decodable*) if L steps of orthogonalization are required to make a decoding decision on an error digit. The decoding process is called *L-step majority-logic decoding*. A code is said to be *completely L-step orthogonalizable* if J is one less than the minimum distance of the code (i.e., $J = d_{\min} - 1$). Since majority-logic gates are used to estimate selected sums of error digits at each step of orthogonalization, a total of L levels of majority-logic gates are required for decoding. The number of gates required at each level depends on the structure of the code.

In the following two examples are used to illustrate the notions of multiple-step majority-logic decoding.

Example 7.11

Consider the (7, 4) cyclic code generated by $\mathbf{g}(X) = 1 + X + X^3$. This is a Hamming code. The parity-check matrix (in systematic form) is found as follows:

$$\mathbf{H} = \begin{bmatrix} \mathbf{h}_0 \\ \mathbf{h}_1 \\ \mathbf{h}_2 \end{bmatrix} = \begin{bmatrix} 1 & 0 & 0 & 1 & 0 & 1 & 1 \\ 0 & 1 & 0 & 1 & 1 & 1 & 0 \\ 0 & 0 & 1 & 0 & 1 & 1 & 1 \end{bmatrix}.$$

We see that the vectors, $\mathbf{h}_0$ and $\mathbf{h}_2$ are orthogonal on digit positions 5 and 6 (or X^5 and X^6). We also see that the vectors, $\mathbf{h}_0 + \mathbf{h}_1$ and $\mathbf{h}_2$, are orthogonal on digit positions 4 and 6. Let $E_1^1 = \{e_5, e_6\}$ and $E_2^1 = \{e_4, e_6\}$ be two selected sets. Let $\mathbf{r} = (r_0, r_1, r_2, r_3, r_4, r_5, r_6)$ be the received vector. Then the parity-check sums formed from $\mathbf{h}_0$ and $\mathbf{h}_2$ are

$$A_1 = \mathbf{r} \cdot \mathbf{h}_0 = e_0 + e_3 + e_5 + e_6$$
$$A_2 = \mathbf{r} \cdot \mathbf{h}_2 = e_2 + e_4 + e_5 + e_6$$

and the parity-check sums formed from $\mathbf{h}_0 + \mathbf{h}_1$ and $\mathbf{h}_2$ are

$$B_1 = \mathbf{r} \cdot (\mathbf{h}_0 + \mathbf{h}_1) = e_0 + e_1 + e_4 + e_6$$
$$B_2 = \mathbf{r} \cdot \mathbf{h}_2 = e_2 + e_4 + e_5 + e_6.$$

The parity-check sums, A_1 and A_2, are orthogonal on the set $E_1^1 = \{e_5, e_6\}$ and the parity-check sums, B_1 and B_2, are orthogonal on the set $E_2^1 = \{e_4, e_6\}$. Therefore, the sum $S(E_1^1) = e_5 + e_6$ can be estimated from A_1 and A_2, and the sum $S(E_2^1) = e_4 + e_6$ can be estimated from B_1 and B_2. The sums $S(E_1^1)$ and $S(E_2^1)$ would be correctly estimated provided that there is no more than one error in the error vector $\mathbf{e}$. Now let $E_1^2 = \{e_6\}$. We see that $S(E_1^1)$ and $S(E_2^1)$ are orthogonal on e_6. Hence, e_6 can be estimated from $S(E_1^1)$ and $S(E_2^1)$. The value of e_6 will be estimated correctly provided that there are no more than one error in $\mathbf{e}$. Therefore, the (7, 4) Hamming code can be decoded with two steps of orthogonalization and it is two-step majority-logic decodable. Since its minimum distance is 3 and $J = 2$, it is two-step completely orthogonalizable. A type II decoder for this code is shown in Figure 7.8.

Figure 7.8 Type II two-step majority-logic decoder for the (7, 4) Hamming code.

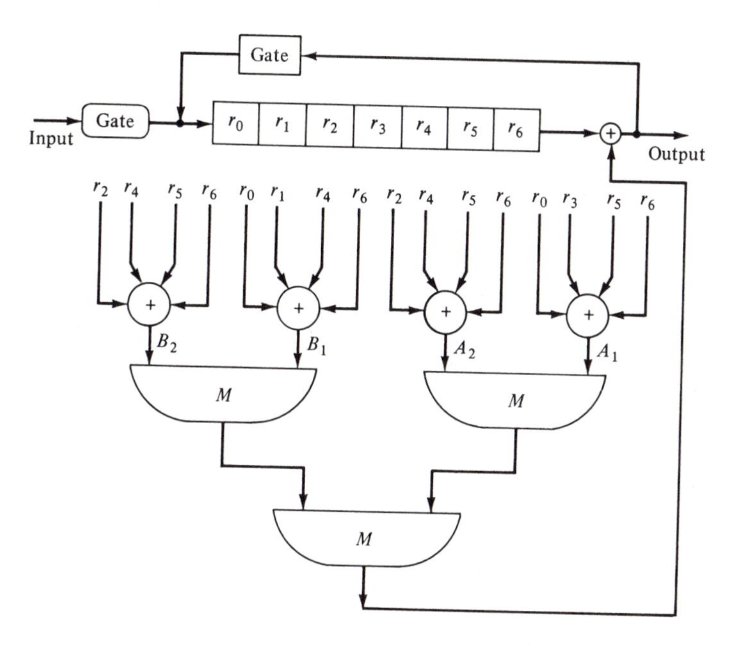

Let $\mathbf{s} = (s_0, s_1, s_2) = \mathbf{r} \cdot \mathbf{H}^T$ be the syndrome of the received vector $\mathbf{r}$. Then we can form the parity-check sums A_1, A_2, B_1, and B_2 from the syndrome digits as follows:

$$A_1 = s_0, \qquad A_2 = s_2,$$
$$B_1 = s_0 + s_1, \qquad B_2 = s_2.$$

Based on these check sums, one may construct a type I majority-logic decoder for the (7, 4) Hamming code.

Example 7.12

Consider the triple-error-correcting (15, 5) BCH code whose generator polynomial is

$$\mathbf{g}(X) = 1 + X + X^2 + X^4 + X^5 + X^8 + X^{10}.$$

The parity-check matrix (in systematic form) is

$$\mathbf{H} = \begin{bmatrix} \mathbf{h}_0 \\ \mathbf{h}_1 \\ \mathbf{h}_2 \\ \mathbf{h}_3 \\ \mathbf{h}_4 \\ \mathbf{h}_5 \\ \mathbf{h}_6 \\ \mathbf{h}_7 \\ \mathbf{h}_8 \\ \mathbf{h}_9 \end{bmatrix} = \begin{bmatrix} 1&0&0&0&0&0&0&0&0&0&1&0&1&0&1 \\ 0&1&0&0&0&0&0&0&0&0&1&1&1&1&1 \\ 0&0&1&0&0&0&0&0&0&0&1&1&0&1&0 \\ 0&0&0&1&0&0&0&0&0&0&0&1&1&0&1 \\ 0&0&0&0&1&0&0&0&0&0&1&0&0&1&1 \\ 0&0&0&0&0&1&0&0&0&0&1&1&1&0&0 \\ 0&0&0&0&0&0&1&0&0&0&0&1&1&1&0 \\ 0&0&0&0&0&0&0&1&0&0&0&0&1&1&1 \\ 0&0&0&0&0&0&0&0&1&0&1&0&1&1&0 \\ 0&0&0&0&0&0&0&0&0&1&0&1&0&1&1 \end{bmatrix}.$$

Let

$$E_1^1 = \{e_{13}, e_{14}\}, \qquad E_2^1 = \{e_{12}, e_{14}\},$$
$$E_3^1 = \{e_{11}, e_{14}\}, \qquad E_4^1 = \{e_{10}, e_{14}\},$$
$$E_5^1 = \{e_5, e_{14}\}, \qquad E_6^1 = \{e_2, e_{14}\}$$

be six selected sets of error digits. For each of the sets above, it is possible to find six parity-check sums orthogonal on it. Let $\mathbf{r} = (r_0, r_1, r_2, r_3, r_4, r_5, r_6, r_7, r_8, r_9, r_{10}, r_{11}, r_{12}, r_{13}, r_{14})$ be the received vector. By taking proper combinations of the rows of $\mathbf{H}$, we find the following parity-check sums orthogonal on E_1^1, E_2^1, E_3^1, E_4^1, E_5^1, and E_6^1:

1. Check sums orthogonal on $E_1^1 = \{e_{13}, e_{14}\}$:

$$\begin{aligned}
A_{11} &= \mathbf{r} \cdot \mathbf{h}_4 &&= e_4 + e_{10} + e_{13} + e_{14} \\
A_{12} &= \mathbf{r} \cdot \mathbf{h}_7 &&= e_7 + e_{12} + e_{13} + e_{14} \\
A_{13} &= \mathbf{r} \cdot \mathbf{h}_9 &&= e_9 + e_{11} + e_{13} + e_{14} \\
A_{14} &= \mathbf{r} \cdot (\mathbf{h}_0 + \mathbf{h}_8) &&= e_0 + e_8 + e_{13} + e_{14} \\
A_{15} &= \mathbf{r} \cdot (\mathbf{h}_1 + \mathbf{h}_5) &&= e_1 + e_5 + e_{13} + e_{14} \\
A_{16} &= \mathbf{r} \cdot (\mathbf{h}_3 + \mathbf{h}_6) &&= e_3 + e_6 + e_{13} + e_{14}.
\end{aligned}$$

2. Check sums orthogonal on $E_2^1 = \{e_{12}, e_{14}\}$:

$$\begin{aligned}
A_{21} &= \mathbf{r} \cdot \mathbf{h}_0 &&= e_0 + e_{10} + e_{12} + e_{14} \\
A_{22} &= \mathbf{r} \cdot \mathbf{h}_3 &&= e_3 + e_{11} + e_{12} + e_{14} \\
A_{23} &= \mathbf{r} \cdot \mathbf{h}_7 &&= e_7 + e_{13} + e_{12} + e_{14}
\end{aligned}$$

$$A_{24} = \mathbf{r} \cdot (\mathbf{h}_1 + \mathbf{h}_2) = e_1 + e_2 \quad + e_{12} + e_{14}$$
$$A_{25} = \mathbf{r} \cdot (\mathbf{h}_4 + \mathbf{h}_8) = e_4 + e_8 \quad + e_{12} + e_{14}$$
$$A_{26} = \mathbf{r} \cdot (\mathbf{h}_6 + \mathbf{h}_9) = e_6 + e_9 \quad + e_{12} + e_{14}.$$

3. Check sums orthogonal on $E_3^1 = \{e_{11}, e_{14}\}$:

$$A_{31} = \mathbf{r} \cdot \mathbf{h}_3 \qquad = e_3 + e_{12} + e_{11} + e_{14}$$
$$A_{32} = \mathbf{r} \cdot \mathbf{h}_9 \qquad = e_9 + e_{13} + e_{11} + e_{14}$$
$$A_{33} = \mathbf{r} \cdot (\mathbf{h}_0 + \mathbf{h}_5) = e_0 + e_5 \quad + e_{11} + e_{14}$$
$$A_{34} = \mathbf{r} \cdot (\mathbf{h}_1 + \mathbf{h}_8) = e_1 + e_8 \quad + e_{11} + e_{14}$$
$$A_{35} = \mathbf{r} \cdot (\mathbf{h}_2 + \mathbf{h}_4) = e_2 + e_4 \quad + e_{11} + e_{14}$$
$$A_{36} = \mathbf{r} \cdot (\mathbf{h}_6 + \mathbf{h}_7) = e_6 + e_7 \quad + e_{11} + e_{14}.$$

4. Check sums orthogonal on $E_4^1 = \{e_{10}, e_{14}\}$:

$$A_{41} = \mathbf{r} \cdot \mathbf{h}_0 \qquad = e_0 + e_{12} + e_{10} + e_{14}$$
$$A_{42} = \mathbf{r} \cdot \mathbf{h}_4 \qquad = e_4 + e_{13} + e_{10} + e_{14}$$
$$A_{43} = \mathbf{r} \cdot (\mathbf{h}_1 + \mathbf{h}_6) = e_1 + e_6 \quad + e_{10} + e_{14}$$
$$A_{44} = \mathbf{r} \cdot (\mathbf{h}_3 + \mathbf{h}_5) = e_3 + e_5 \quad + e_{10} + e_{14}$$
$$A_{45} = \mathbf{r} \cdot (\mathbf{h}_7 + \mathbf{h}_8) = e_7 + e_8 \quad + e_{10} + e_{14}$$
$$A_{46} = \mathbf{r} \cdot (\mathbf{h}_2 + \mathbf{h}_9) = e_2 + e_9 \quad + e_{10} + e_{14}.$$

5. Check sums orthogonal on $E_5^1 = \{e_5, e_{14}\}$:

$$A_{51} = \mathbf{r} \cdot (\mathbf{h}_0 + \mathbf{h}_5) \qquad = e_0 + e_{11} + e_5 + e_{14}$$
$$A_{52} = \mathbf{r} \cdot (\mathbf{h}_1 + \mathbf{h}_5) \qquad = e_1 + e_{13} + e_5 + e_{14}$$
$$A_{53} = \mathbf{r} \cdot (\mathbf{h}_3 + \mathbf{h}_5) \qquad = e_3 + e_{10} + e_5 + e_{14}$$
$$A_{54} = \mathbf{r} \cdot (\mathbf{h}_4 + \mathbf{h}_5 + \mathbf{h}_6) = e_4 + e_6 \quad + e_5 + e_{14}$$
$$A_{55} = \mathbf{r} \cdot (\mathbf{h}_2 + \mathbf{h}_5 + \mathbf{h}_7) = e_2 + e_7 \quad + e_5 + e_{14}$$
$$A_{56} = \mathbf{r} \cdot (\mathbf{h}_5 + \mathbf{h}_8 + \mathbf{h}_9) = e_8 + e_9 \quad + e_5 + e_{14}.$$

6. Check sums orthogonal on $E_6^1 = \{e_2, e_{14}\}$:

$$A_{61} = \mathbf{r} \cdot (\mathbf{h}_1 + \mathbf{h}_2) \qquad = e_1 + e_{12} + e_2 + e_{14}$$
$$A_{62} = \mathbf{r} \cdot (\mathbf{h}_2 + \mathbf{h}_4) \qquad = e_4 + e_{11} + e_2 + e_{14}$$
$$A_{63} = \mathbf{r} \cdot (\mathbf{h}_0 + \mathbf{h}_2 + \mathbf{h}_6) = e_0 + e_6 \quad + e_2 + e_{14}$$
$$A_{64} = \mathbf{r} \cdot (\mathbf{h}_2 + \mathbf{h}_3 + \mathbf{h}_8) = e_3 + e_8 \quad + e_2 + e_{14}$$
$$A_{65} = \mathbf{r} \cdot (\mathbf{h}_2 + \mathbf{h}_5 + \mathbf{h}_7) = e_5 + e_7 \quad + e_2 + e_{14}$$
$$A_{66} = \mathbf{r} \cdot (\mathbf{h}_2 + \mathbf{h}_9) \qquad = e_9 + e_{10} + e_2 + e_{14}.$$

From the orthogonal check sums above, the sums, $S(E_1^1) = e_{13} + e_{14}$, $S(E_2^1) = e_{12} + e_{14}$, $S(E_3^1) = e_{11} + e_{14}$, $S(E_4^1) = e_{10} + e_{14}$, $S(E_5^1) = e_5 + e_{14}$, and $S(E_6^1) = e_2 + e_{14}$, can be correctly estimated provided that there are no more than three errors in the error vector $\mathbf{e}$. Let $E_1^2 = \{e_{14}\}$. We see that the error sums, $S(E_1^1)$, $S(E_2^1)$, $S(E_3^1)$, $S(E_4^1)$, $S(E_5^1)$ and $S(E_6^1)$, are orthogonal on e_{14}. Hence, e_{14} can be estimated from these sums. Therefore, the (15, 5) BCH code is two-step orthogonalizable. Since $J = 6$, it is capable of correcting three or fewer errors with two-step majority-logic decoding.

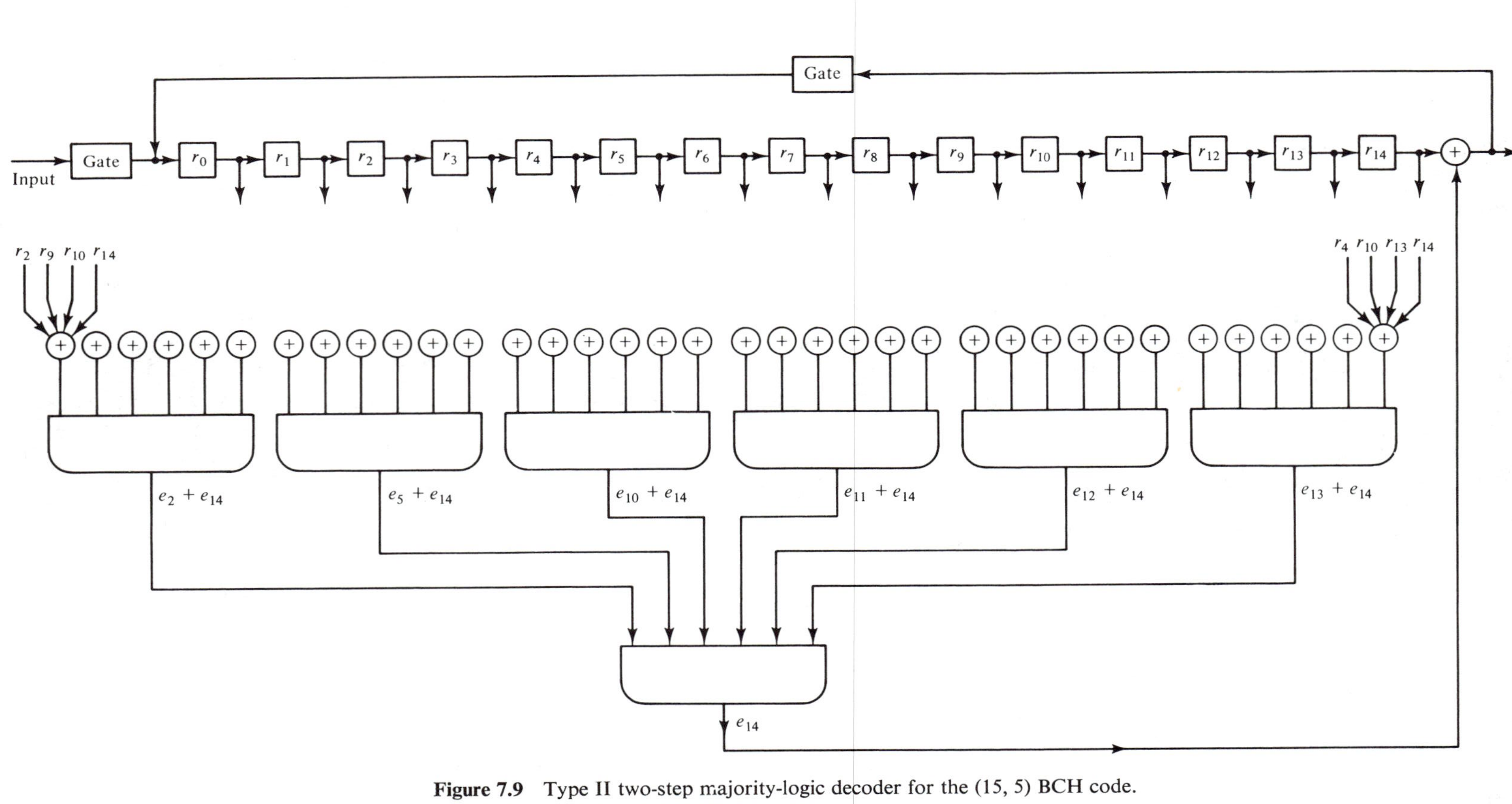

Figure 7.9 Type II two-step majority-logic decoder for the (15, 5) BCH code.

It is known that the code has minimum distance exactly 7. Hence, it is two-step completely orthogonalizable.

The type II decoder for the (15, 5) BCH code is shown in Figure 7.9, where seven six-input majority-logic gates (connected in a tree form) are used. If we examine the parity-check matrix carefully and are willing to trade decoding speed with decoding complexity, we will be able to reduce the number of majority-logic gates. Suppose that we choose the following sets:

$$Q_1^1 = \{e_{12}, e_{13}\}, \qquad Q_2^1 = \{e_{10}, e_{13}\}, \qquad Q_3^1 = \{e_{11}, e_{12}\}.$$

For each of these sets, we are able to form six parity-check sums orthogonal on it as follows:

1. Check sums orthogonal on $Q_1^1 = \{e_{12}, e_{13}\}$:

$$\begin{aligned}
B_{11} &= \mathbf{r} \cdot (\mathbf{h}_3 + \mathbf{h}_9) = e_3 + e_9 + e_{12} + e_{13} \\
B_{12} &= \mathbf{r} \cdot \mathbf{h}_6 = e_6 + e_{11} + e_{12} + e_{13} \\
B_{13} &= \mathbf{r} \cdot \mathbf{h}_8 = e_8 + e_{10} + e_{12} + e_{13} \\
B_{14} &= \mathbf{r} \cdot \mathbf{h}_7 = e_7 + e_{14} + e_{12} + e_{13} \\
B_{15} &= \mathbf{r} \cdot (\mathbf{h}_0 + \mathbf{h}_4) = e_0 + e_4 + e_{12} + e_{13} \\
B_{16} &= \mathbf{r} \cdot (\mathbf{h}_2 + \mathbf{h}_5) = e_2 + e_5 + e_{12} + e_{13}.
\end{aligned}$$

2. Check sums orthogonal on $Q_2^1 = \{e_{10}, e_{13}\}$:

$$\begin{aligned}
B_{21} &= \mathbf{r} \cdot \mathbf{h}_2 = e_2 + e_{11} + e_{10} + e_{13} \\
B_{22} &= \mathbf{r} \cdot \mathbf{h}_8 = e_8 + e_{12} + e_{10} + e_{13} \\
B_{23} &= \mathbf{r} \cdot \mathbf{h}_4 = e_4 + e_{14} + e_{10} + e_{13} \\
B_{24} &= \mathbf{r} \cdot (\mathbf{h}_0 + \mathbf{h}_7) = e_0 + e_7 + e_{10} + e_{13} \\
B_{25} &= \mathbf{r} \cdot (\mathbf{h}_1 + \mathbf{h}_3) = e_1 + e_3 + e_{10} + e_{13} \\
B_{26} &= \mathbf{r} \cdot (\mathbf{h}_5 + \mathbf{h}_6) = e_5 + e_6 + e_{10} + e_{13}.
\end{aligned}$$

3. Check sums orthogonal on $Q_3^1 = \{e_{11}, e_{12}\}$:

$$\begin{aligned}
B_{31} &= \mathbf{r} \cdot (\mathbf{h}_2 + \mathbf{h}_8) = e_2 + e_8 + e_{11} + e_{12} \\
B_{32} &= \mathbf{r} \cdot \mathbf{h}_5 = e_5 + e_{10} + e_{11} + e_{12} \\
B_{33} &= \mathbf{r} \cdot (\mathbf{h}_7 + \mathbf{h}_9) = e_7 + e_9 + e_{11} + e_{12} \\
B_{34} &= \mathbf{r} \cdot \mathbf{h}_6 = e_6 + e_{13} + e_{11} + e_{12} \\
B_{35} &= \mathbf{r} \cdot \mathbf{h}_3 = e_3 + e_{14} + e_{11} + e_{12} \\
B_{36} &= \mathbf{r} \cdot (\mathbf{h}_1 + \mathbf{h}_4) = e_1 + e_4 + e_{11} + e_{12}.
\end{aligned}$$

Based on the orthogonal sums above, we can estimate the sums, $S(Q_1^1) = e_{12} + e_{13}$, $S(Q_2^1) = e_{10} + e_{13}$, and $S(Q_3^1) = e_{11} + e_{12}$. Next, we combine the sums $S(Q_1^1)$, $S(Q_2^1)$, and $S(Q_3^1)$ with *other check sums* to form six check sums orthogonal on $Q_1^2 = \{e_{14}\}$. By examining the rows of $\mathbf{H}$ carefully, we obtain the following check sums orthogonal on e_{14}:

$$\begin{aligned}
D_1 &= S(Q_1^1) + \mathbf{r} \cdot \mathbf{h}_7 = e_7 + e_{14}, \\
D_2 &= S(Q_2^1) + \mathbf{r} \cdot \mathbf{h}_4 = e_4 + e_{14}, \\
D_3 &= S(Q_3^1) + \mathbf{r} \cdot \mathbf{h}_3 = e_3 + e_{14},
\end{aligned}$$

$$D_4 = S(Q_1^1) + S(Q_2^1) + \mathbf{r} \cdot \mathbf{h}_0 = e_0 + e_{14},$$
$$D_5 = S(Q_1^1) + S(Q_3^1) + \mathbf{r} \cdot \mathbf{h}_9 = e_9 + e_{14},$$
$$D_6 = S(Q_2^1) + S(Q_3^1) + \mathbf{r} \cdot \mathbf{h}_1 = e_1 + e_{14}.$$

From these check sums, we can determine e_{14}. At the first level of decoding, we need three six-input majority-logic gates to determine the sums $S(Q_1^1)$, $S(Q_2^1)$, and $S(Q_3^1)$. Once these sums are formed, we can form the sums D_1 to D_6 by using six modulo-2 adders. Then it comes to the last level of decoding, where another six-input majority-logic gate is needed to determine e_{14} from the sums D_1 to D_6. Forming orthogonal parity-check sums as described above, we obtain a simpler type II majority-logic decoder for the (15, 5) BCH code as shown in Figure 7.10. Comparing the two decoding circuits for the (15, 5) BCH code, we see that the first circuit (Figure 7.9) requires 7 six-input majority-logic gates and 26 modulo-2 adders to form the orthogonal sums and to make the decoding decision, whereas the second circuit (Figure 7.10) requires only 4 six-input majority-logic gates and 22 modulo-2 adders. Therefore, there is a reduction in decoding complexity by using the second method of forming the orthogonal check sums. However, this reduction in complexity is achieved at the expense of extra decoding delay caused by the layer of modulo-2 adders between the two levels of majority-logic gates. Construction of a type I majority-logic decoder for the (15, 5) BCH code is left as an exercise (see Problem 7.12).

A general type II L-step majority-logic decoder is shown in Figure 7.11. The error correction procedure is described as follows:

Step 1. The received vector $\mathbf{r}(X)$ is read into the buffer register.

Step 2. Parity-check sums [no more than $(J)^L$ of them] orthogonal on certain properly selected sets of error digits are formed by summing appropriate sets of received digits. These check sums are then fed into the first-level majority-logic gates, at most $(J)^{L-1}$ of them. The outputs of the first-level majority-logic gates are used to form inputs to the second-level majority-logic gates [there are at most $(J)^{L-2}$ of them]. The outputs of the second-level majority-logic gates are then used to form inputs to third-level majority-logic gates [there are at most $(J)^{L-3}$ of them]. This continues until the last level is reached; there is only one gate at the last level. The J inputs to this gate are check sums orthogonal on the highest-order error digit e_{n-1}. The output of this gate is used to correct the received digit r_{n-1}.

Step 3. The received digit r_{n-1} is read out of the buffer and is corrected by the last-level majority-logic gate.

Step 4. At the end of step 3, the buffer register has been shifted one place to the right. Now the second-highest-order received digit r_{n-2} is in the rightmost stage of the buffer register, and it will be corrected in exactly the same manner as the highest-order received digit r_{n-1} was. The decoder repeats steps 2 and 3.

Step 5. The received vector is decoded digit by digit in the manner described above until a total of n shifts.

A general type I decoder for a L-step majority-logic decoder code is shown in Figure 7.12. Its decoding operation is identical to that of the type I decoder for a

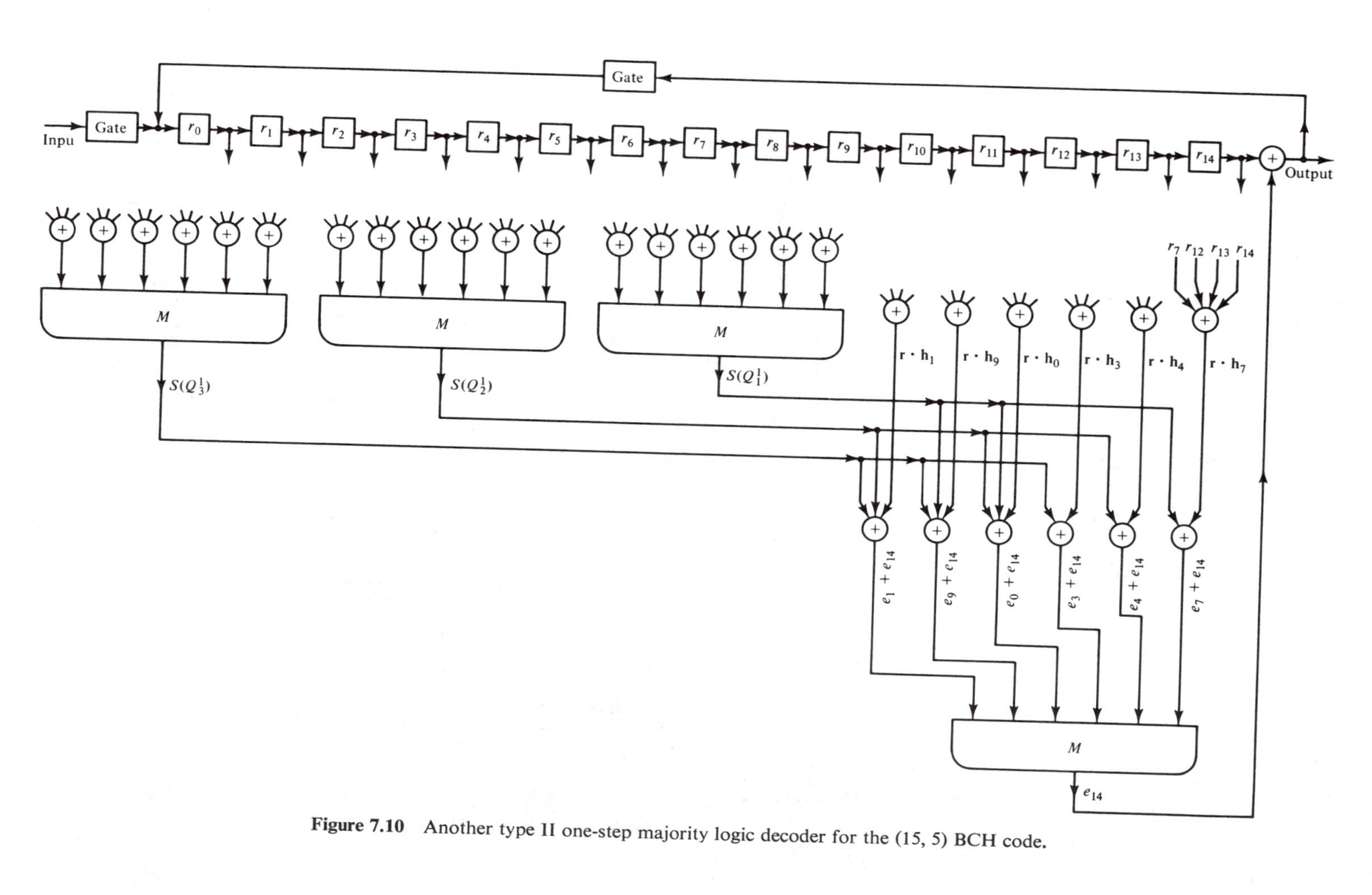

Figure 7.10 Another type II one-step majority logic decoder for the (15, 5) BCH code.

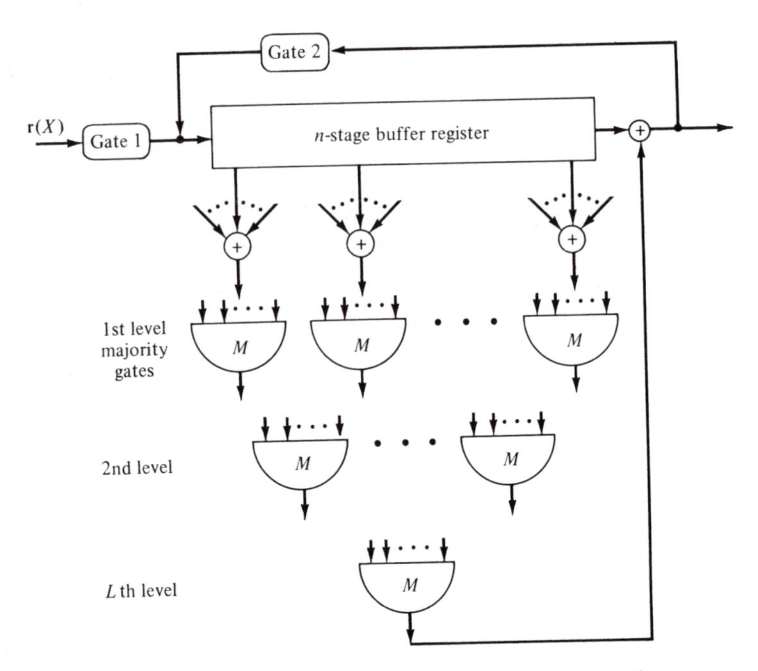

Figure 7.11 General type II L-step majority-logic decoder.

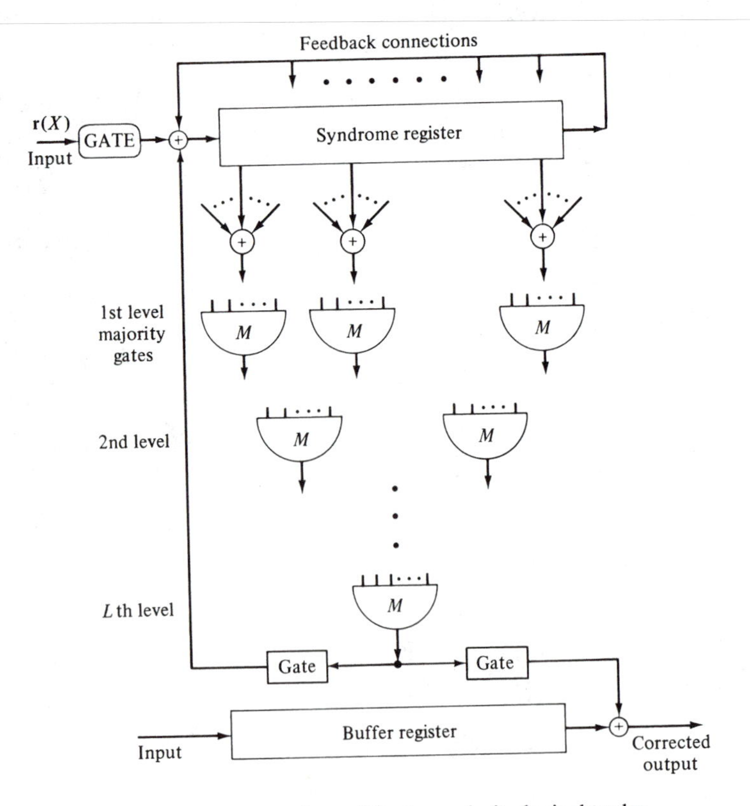

Figure 7.12 General type I L-step majority-logic decoder.

one-step majority-logic decodable code except that L levels of orthogonalization are required.

An L-step majority-logic decoder requires L levels of majority-logic gates. At the ith level, no more than $(J)^{L-i}$ gates are required. Thus, the total number of majority-logic gates needed is upper bounded by $1 + J + J^2 + \cdots + J^{L-1}$. In fact, Massey [3] has proved that, for an (n, k) L-step majority-logic decodable code, no more than k majority-logic gates are ever required. Unfortunately, for a given L-step majority-logic decodable cyclic code, there is no known systematic method for minimizing the number of majority-logic gates except the trial-and-error method. For almost all the known classes of L-step majority-logic decodable codes, the rules for forming orthogonal parity-check sums requires a total of $1 + J + J^2 + \cdots + J^{L-1}$ majority-logic gates. Thus, the complexity is an exponential function of L. For large L, the decoder is likely to be impractical. Fortunately, there are many cyclic codes with useful parameters that can be decoded with a reasonably small L.

It has been shown [3] that the $(2^m - 1, 2^m - m - 1)$ Hamming code is completely orthogonalizable in $m - 1$ steps. Thus, the Hamming codes can be decoded by the majority-logic decoding described above. However, the error-trapping decoding for Hamming codes described in Section 4.6 can be much more simply implemented than the majority-logic decoding. The BCH codes which are known to be completely orthogonalizable are: (1) the subclass of $(2^{m-2} - 1)$-error correcting $(2^m - 1, m + 1)$ codes with $m \geq 3$, which is two-step orthogonalizable [3]; (2) the double-error-correcting (15, 7), code which is one-step orthogonalizable [3]; (3) the triple-error-correcting (31, 16) code, which can be completely orthogonalized in two steps [14–16]; and (4) the triple-error-correcting (63, 45) code and the seven-error-correcting (63, 24) code, both are two-step orthogonalizable [17]. Besides the codes above, several large classes of cyclic codes have been found to be L-step majority-logic decodable. The construction and the rules for orthogonalization of these codes are based on the properties of finite geometries, which are the subjects of Chapter 8.

PROBLEMS

7.1. Consider the (31, 5) maximum-length code whose parity-check polynomial is $\mathbf{p}(X) = 1 + X^2 + X^5$. Find all the polynomials orthogonal on the digit position X^{30}. Devise both type I and type II majority-logic decoders for this code.

7.2. $P = \{0, 2, 3\}$ is a perfect simple difference set. Construct a difference-set code based on this set.

(a) What is the length n of this code?

(b) Determine its generator polynomial.

(c) Find all the polynomials orthogonal on the highest-order digit position X^{n-1}.

(d) Construct a type I majority-logic decoder for this code.

7.3. In Example 7.1 we showed that the (15, 7) BCH code is one-step majority-logic decodable and is capable of correcting any combination of two or fewer errors. Show that the code is also capable of correcting some error patterns of three errors and some error patterns of four errors. List some of these error patterns.

7.4. Consider an (11, 6) linear code whose parity-check matrix is

$$\mathbf{H} = \begin{bmatrix} 1 & 0 & 0 & 0 & 0 & 1 & 1 & 1 & 1 & 1 & 1 \\ 0 & 1 & 0 & 0 & 0 & 1 & 1 & 0 & 1 & 0 & 0 \\ 0 & 0 & 1 & 0 & 0 & 1 & 0 & 1 & 0 & 1 & 0 \\ 0 & 0 & 0 & 1 & 0 & 0 & 1 & 1 & 0 & 0 & 1 \\ 0 & 0 & 0 & 0 & 1 & 0 & 0 & 0 & 1 & 1 & 1 \end{bmatrix}.$$

(This code is not cyclic.)

(a) Show that the minimum distance of this code is exactly 4.

(b) Let $\mathbf{e} = (e_0, e_1, e_2, e_3, e_4, e_5, e_6, e_7, e_8, e_9, e_{10})$ be an error vector. Find the syndrome bits in terms of error digits.

(c) Construct all possible parity-check sums orthogonal on each message error digit e_i for $i = 5, 6, 7, 8, 9, 10$.

(d) Is this code completely orthogonalizable in one step?

7.5. Let $m = 6$. Express the integer 43 in radix 2 form. Find all the nonzero proper descendants of 43.

7.6. Let α be a primitive element of GF(2^4) given by Table 2.8. Apply the affine permutation $Z = \alpha^3 Y + \alpha^{11}$ to the following vector of 16 components:

Location Numbers

	α^∞	α^0	α^1	α^2	α^3	α^4	α^5	α^6	α^7	α^8	α^9	α^{10}	α^{11}	α^{12}	α^{13}	α^{14}
$\mathbf{u} =$	(1	1	1	0	0	1	0	1	0	1	1	0	0	0	0	1)

What is the resultant vector?

7.7. Let $m = 6$. Then $2^6 - 1$ can be factored as follows: $2^6 - 1 = 7 \times 9$. Let $J = 9$ and $L = 7$. Find the generator polynomial of the type I DTI code of length 63 and $J = 9$ (use Table 6.2). Find all the polynomials (or vectors) orthogonal on the digit position X^{62} (or α^{62}).

7.8. Find the generator polynomial of the type I DTI code of length 63 and $J = 7$. Find all the polynomials orthogonal on the digit position X^{62}.

7.9. Show that the all-one vector is not a code vector in a maximum-length code.

7.10. Let $\mathbf{v}(X) = v_0 + v_1 X + \cdots + v_{2^m-2} X^{2^m-2}$ be a nonzero code polynomial in the $(2^m - 1, m)$ maximum-length code whose parity-check polynomial is $\mathbf{p}(X)$. Show that the other $2^m - 2$ nonzero code polynomials are cyclic shifts of $\mathbf{v}(X)$. [*Hint:* Let $\mathbf{v}^{(i)}(X)$ and $\mathbf{v}^{(j)}(X)$ be the ith and jth cyclic shifts of $\mathbf{v}(X)$, respectively, with $0 \leq i < j < 2^m - 2$. Show that $\mathbf{v}^{(i)}(X) \neq \mathbf{v}^{(j)}(X)$.]

7.11. Arrange the 2^m code vectors of a maximum-length code as rows of a $2^m \times (2^m - 1)$ array.

(a) Show that each column of this array has 2^{m-1} ones and 2^{m-1} zeros.

(b) Show that the weight of each nonzero code vector is exactly 2^{m-1}.

7.12. In Example 7.12, we showed that the (15, 5) BCH code is two-step majority-logic decodable and is capable of correcting any combination of three or fewer errors. Devise a type I majority-logic decoder for this code.

7.13. Show that the extended cyclic Hamming code is invariant under the affine permutations.

7.14. Show that the extended primitive BCH code is invariant under the affine permutations.

7.15. Let $P = \{l_0, l_1, l_2, \ldots, l_{2^s}\}$ be a perfect simple difference set of order 2^s such that

$$0 \leq l_0 < l_1 < l_2 < \cdots < l_{2^s} \leq 2^s(2^s + 1).$$

Construct a vector of $n = 2^{2s} + 2^s + 1$ components,

$$\mathbf{v} = (v_0, v_1, \ldots, v_{n-1})$$

whose nonzero components are $v_{l_0}, v_{l_1}, \ldots, v_{l_{2^s}}$, that is,

$$v_{l_0} = v_{l_1} = \cdots = v_{l_{2^s}} = 1.$$

Consider the following $n \times 2n$ matrix:

$$\mathbf{G} = [\mathbf{Q} \quad \mathbf{I}_n],$$

where (1) I_n is an $n \times n$ identity matrix, and (2) Q is an $n \times n$ matrix whose n rows are $\mathbf{v}$ and $n - 1$ cyclic shifts of $\mathbf{v}$. The code generated by $\mathbf{G}$ is a $(2n, n)$ linear code (not cyclic) whose parity-check matrix is

$$\mathbf{H} = [\mathbf{I}_n \quad \mathbf{Q}^T].$$

(a) Show that $J = 2^s + 1$ parity-check sums orthogonal on any message error digit can be formed.

(b) Show that the minimum distance of this code is $d = J + 1 = 2^s + 2$. (This code is referred to as a half-rate *quasi-cyclic code* [18].)

7.16. Prove that if J parity-check sums orthogonal on any digit position can be formed for a linear code (cyclic or noncyclic), the minimum distance of the code is at least $J + 1$.

7.17. Let $\mathbf{H}$ be the parity-check matrix of an (n, k) linear code C. Show that the matrix

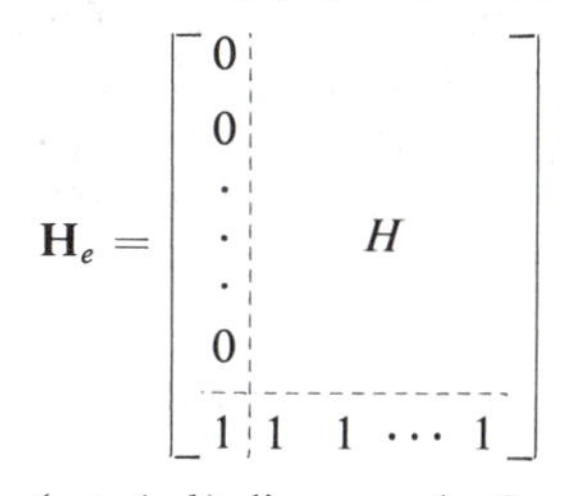

$$\mathbf{H}_e = \left[\begin{array}{c|cccc} 0 & & & & \\ 0 & & & & \\ \cdot & & & & \\ \cdot & & H & & \\ \cdot & & & & \\ 0 & & & & \\ \hline 1 & 1 & 1 & \cdots & 1 \end{array}\right]$$

is a parity-check matrix of the $(n+1, k)$ linear code C_e obtained by adding an overall parity-check digit to each code vector in C.

7.18. Determine the weight enumerator for the extended Hamming code of length 2^m.

7.19. Determine the weight enumerator for the dual of the extended Hamming code.

7.20. Show that the probability of an undetected error $P_u(E)$ of the extended Hamming code of length 2^m satisfies the upper bound $2^{-(m+1)}$.

REFERENCES

1. I. S. Reed, "A Class of Multiple-Error-Correcting Codes and the Decoding Scheme," *IRE Trans.*, IT-4, pp. 38–49, September 1954.
2. D. E. Muller, "Applications of Boolean Algebra to Switching Circuit Design and to Error Detection," *IRE Trans.*, EC-3, pp. 6–12, September 1954.
3. J. L. Massey, *Threshold Decoding*, MIT Press, Cambridge, Mass. 1963.
4. T. Kasami, L. Lin, and W. W. Peterson, "Some Results on Cyclic Codes Which Are Invariant under the Affine Group and Their Applications," *Inf. Control*, 2(5 and 6), pp. 475–496, November 1968.
5. W. W. Peterson and E. J. Weldon, Jr., *Error-Correcting Codes*, 2nd ed. MIT Press, Cambridge, Mass., 1972.

6. S. Lin and G. Markowsky, "On a Class of One-Step Majority-Logic Decodable Cyclic Codes," *IBM J. Res. Dev.*, January 1980.
7. R. B. Yale, "Error-Correcting Codes and Linear Recurring Sequences," Lincoln Laboratory Report, pp. 33–77, Lincoln Labs., MIT, Cambridge, Mass., 1958.
8. N. Zierler, "On a Variation of the First Order Reed-Muller Codes," Lincoln Laboratory Report, pp. 38–80, Lincoln Labs., MIT, Cambridge, Mass., 1958.
9. J. Singer, "A Theorem in Finite Projective Geometry and Some Applications to Number Theory," *AMS Trans.*, 43, pp. 377–385, 1938.
10. T. A. Evans and H. B. Mann, "On Simple Difference Sets," *Sankhya*, 11, pp. 464–481, 1955.
11. L. D. Rudolph, "Geometric Configuration and Majority Logic Decodable Codes," M.E.E. thesis, University of Oklahoma, Norman, Okla., 1964.
12. E. J. Weldon, Jr., "Difference-Set Cyclic Codes," *Bell Syst. Tech. J.*, 45, pp. 1045–1055, September 1966.
13. F. L. Graham and J. MacWilliams, "On the Number of Parity Checks in Difference-Set Cyclic Codes," *Bell Syst. Tech. J.*, 45, pp. 1046–1070, September 1966.
14. L. D. Rudolph, "A Class of Majority Logic Decodable Codes," *IEEE Trans. Inf. Theory*, IT-13, pp. 305–307, April 1967.
15. E. J. Weldon, Jr., "New Generations of the Reed-Muller Codes-Part II: Non-Primitive Codes," *IEEE Trans. Inf. Theory*, IT-14, pp. 199–205, March 1968.
16. J. M. Coethals and P. Delsarte, "On a Class of Majority-Logic Decodable Codes," *IEEE Trans. Inf. Theory*, IT-14(2), pp. 182–189, March 1968.
17. S. Lin, "Multifold Euclidean Geometry Codes," *IEEE Trans. Inf. Theory*, IT-19(4), pp. 537–548, July 1973.
18. E. J. Weldon, Jr., "Quasi-Cyclic Codes," *IEEE Trans. Inf. Theory*, IT-12, pp. 183–195, April 1967.

8

Finite Geometry Codes

The construction and rules of orthogonalization for most of the multiple-step majority-logic decodable cyclic codes are based on the structure of finite geometries, namely *Euclidean* and *projective* geometries. In this chapter a brief discussion of these geometries is given. We introduce some basic concepts and state some useful properties without proofs. For details of finite geometries, the reader is referred to References 1 and 2.

Codes constructed based on the structure of finite geometries are called *finite geometry codes*. Finite geometry codes were first investigated by Rudolph [3]. Rudolph's work was later extended and generalized by many coding investigators. In this chapter several classes of finite geometry codes are presented.

8.1 EUCLIDEAN GEOMETRY

Consider all the m-tuples $(a_0, a_1, \ldots, a_{m-1})$, with components a_i's from the Galois field $\mathrm{GF}(2^s)$. There are $(2^s)^m = 2^{ms}$ such m-tuples. These 2^{ms} m-tuples form a vector space over $\mathrm{GF}(2^s)$. The vector addition and scalar multiplication are defined in the usual way,

$$(a_0, a_1, \ldots, a_{m-1}) + (b_0, b_1, \ldots, b_{m-1}) = (a_0 + b_0, a_1 + b_1, \ldots, a_{m-1} + b_{m-1}),$$
$$\beta(a_0, a_1, \ldots, a_{m-1}) = (\beta a_0, \beta a_1, \ldots, \beta a_{m-1}),$$

where addition $a_i + b_i$ and multiplication βa_i are carried out in $\mathrm{GF}(2^s)$. In combinatorial mathematics, the 2^{ms} m-tuples over $\mathrm{GF}(2^s)$ are also known to form an m-dimensional *Euclidean geometry* over $\mathrm{GF}(2^s)$, denoted $\mathrm{EG}(m, 2^s)$. Each m-tuple $\mathbf{a} =$

$(a_0, a_1, \ldots, a_{m-1})$ is called a *point* in EG(m, 2^s). The all-zero m-tuple, $\mathbf{0} = (0, 0, \ldots, 0)$, is called the *origin* of the geometry EG(m, 2^s).

Let $\mathbf{a}$ be a nonorigin point in EG(m, 2^s) (i.e., $\mathbf{a} \neq \mathbf{0}$). Then the 2^s points, $\{\beta\mathbf{a}: \beta \in \text{GF}(2^s)\}$, constitute a *line* (or 1-*flat*) in EG(m, 2^s). For convenience, we use the notation $\{\beta\mathbf{a}\}$ to represent this line. Since this line contains the origin (with $\beta = 0$), we say that $\{\beta\mathbf{a}\}$ passes through the origin. Let $\mathbf{a}_0$ and $\mathbf{a}$ be two linearly independent points in EG(m, 2^s) (i.e., $\beta_0\mathbf{a}_0 + \beta\mathbf{a} \neq \mathbf{0}$ unless $\beta_0 = \beta = 0$). Then the collection of the following 2^s points,

$$\{\mathbf{a}_0 + \beta\mathbf{a}\}$$

with $\beta \in \text{GF}(2^s)$, constitute a line in EG(m, 2^s) that passes through the point $\mathbf{a}_0$. Line $\{\beta\mathbf{a}\}$ and the line $\{\mathbf{a}_0 + \beta\mathbf{a}\}$ do not have any point in common. Suppose that they have a common point. Then, for some β' and β'' in GF(2^s),

$$\beta'\mathbf{a} = \mathbf{a}_0 + \beta''\mathbf{a}.$$

As a result, $\mathbf{a}_0 + (\beta'' - \beta')\mathbf{a} = \mathbf{0}$. This implies that $\mathbf{a}_0$ and $\mathbf{a}$ are linearly dependent, which is a contradiction to our assumption that $\mathbf{a}_0$ and $\mathbf{a}$ are two linearly independent points in EG(m, 2^s). Therefore, $\{\beta\mathbf{a}\}$ and $\{\mathbf{a}_0 + \beta\mathbf{a}\}$ do not have any common points. We say that $\{\beta\mathbf{a}\}$ and $\{\mathbf{a}_0 + \beta\mathbf{a}\}$ are *parallel* lines. Let $\mathbf{b}_0$ be a point not on line $\{\beta\mathbf{a}\}$ or on line $\{\mathbf{a}_0 + \beta\mathbf{a}\}$. The line $\{\mathbf{b}_0 + \beta\mathbf{a}\}$ passes through the point $\mathbf{b}_0$ and is parallel to both $\{\beta\mathbf{a}\}$ and $\{\mathbf{a}_0 + \beta\mathbf{a}\}$. In EG($m$, 2^s), for every line passing through the origin, there are $2^{(m-1)s} - 1$ lines parallel to it.

Let $\mathbf{a}_1$ and $\mathbf{a}_2$ be two linearly independent points in EG(m, 2^s). The lines $\{\mathbf{a}_0 + \beta\mathbf{a}_1\}$ and $\{\mathbf{a}_0 + \beta\mathbf{a}_2\}$ have only one point, $\mathbf{a}_0$, in common. Suppose that they have another point besides $\mathbf{a}_0$ in common. Then, for some $\beta' \neq 0$ and $\beta'' \neq 0$, we have

$$\mathbf{a}_0 + \beta'\mathbf{a}_1 = \mathbf{a}_0 + \beta''\mathbf{a}_2.$$

This equality implies that $\beta'\mathbf{a}_1 - \beta''\mathbf{a}_2 = \mathbf{0}$ and that $\mathbf{a}_1$ and $\mathbf{a}_2$ are linearly dependent. This is a contradiction to the hypothesis that $\mathbf{a}_1$ and $\mathbf{a}_2$ are linearly independent points in EG(m, 2^s). Therefore, $\{\mathbf{a}_0 + \beta\mathbf{a}_1\}$ and $\{\mathbf{a}_0 + \beta\mathbf{a}_2\}$ have only one point in common, and they both pass through the point $\mathbf{a}_0$. We say that $\{\mathbf{a}_0 + \beta\mathbf{a}_1\}$ and $\{\mathbf{a}_0 + \beta\mathbf{a}_2\}$ intersect at the point $\mathbf{a}_0$. Given a point $\mathbf{a}_0$ in EG(m, 2^s), there are

$$\frac{2^{ms} - 1}{2^s - 1} \tag{8.1}$$

lines in EG(m, 2^s) that intersect at $\mathbf{a}_0$. (This is an important property that will be used to form orthogonal parity-check sums for the codes presented in the next section.) The total number of lines in EG(m, 2^s) is

$$2^{(m-1)s}(2^{ms} - 1).$$

Example 8.1

Let $m = 3$ and $s = 1$. Consider the Euclidean geometry EG(3, 2) over GF(2). There are eight points and 28 lines. Each point $\mathbf{a}_i$ is 3-tuple over GF(2). Each line consists of two points $\{\mathbf{a}_i, \mathbf{a}_j\}$. The points and the lines are given in Table 8.1.

TABLE 8.1

(a) Points in EG(3, 2)

$\mathbf{a}_0 = (0\ 0\ 0)$, $\mathbf{a}_1 = (0\ 0\ 1)$, $\mathbf{a}_2 = (0\ 1\ 0)$, $\mathbf{a}_3 = (0\ 1\ 1)$,
$\mathbf{a}_4 = (1\ 0\ 0)$, $\mathbf{a}_5 = (1\ 0\ 1)$, $\mathbf{a}_6 = (1\ 1\ 0)$, $\mathbf{a}_7 = (1\ 1\ 1)$.

(b) Lines in EG(3, 2)

$\{\mathbf{a}_0, \mathbf{a}_1\}$	$\{\mathbf{a}_1, \mathbf{a}_2\}$	$\{\mathbf{a}_2, \mathbf{a}_4\}$	$\{\mathbf{a}_3, \mathbf{a}_7\}$
$\{\mathbf{a}_0, \mathbf{a}_2\}$	$\{\mathbf{a}_1, \mathbf{a}_3\}$	$\{\mathbf{a}_2, \mathbf{a}_5\}$	$\{\mathbf{a}_4, \mathbf{a}_5\}$
$\{\mathbf{a}_0, \mathbf{a}_3\}$	$\{\mathbf{a}_1, \mathbf{a}_4\}$	$\{\mathbf{a}_2, \mathbf{a}_6\}$	$\{\mathbf{a}_4, \mathbf{a}_6\}$
$\{\mathbf{a}_0, \mathbf{a}_4\}$	$\{\mathbf{a}_1, \mathbf{a}_5\}$	$\{\mathbf{a}_2, \mathbf{a}_7\}$	$\{\mathbf{a}_4, \mathbf{a}_7\}$
$\{\mathbf{a}_0, \mathbf{a}_5\}$	$\{\mathbf{a}_1, \mathbf{a}_6\}$	$\{\mathbf{a}_3, \mathbf{a}_4\}$	$\{\mathbf{a}_5, \mathbf{a}_6\}$
$\{\mathbf{a}_0, \mathbf{a}_6\}$	$\{\mathbf{a}_1, \mathbf{a}_7\}$	$\{\mathbf{a}_3, \mathbf{a}_5\}$	$\{\mathbf{a}_5, \mathbf{a}_7\}$
$\{\mathbf{a}_0, \mathbf{a}_7\}$	$\{\mathbf{a}_2, \mathbf{a}_3\}$	$\{\mathbf{a}_3, \mathbf{a}_6\}$	$\{\mathbf{a}_6, \mathbf{a}_7\}$

The lines, $\{\mathbf{a}_0, \mathbf{a}_1\}$, $\{\mathbf{a}_2, \mathbf{a}_3\}$, $\{\mathbf{a}_4, \mathbf{a}_5\}$, and $\{\mathbf{a}_6, \mathbf{a}_7\}$ are parallel. The lines that intersect at the point $\mathbf{a}_2$ are $\{\mathbf{a}_0, \mathbf{a}_2\}$, $\{\mathbf{a}_1, \mathbf{a}_2\}$, $\{\mathbf{a}_2, \mathbf{a}_3\}$, $\{\mathbf{a}_2, \mathbf{a}_4\}$, $\{\mathbf{a}_2, \mathbf{a}_5\}$, $\{\mathbf{a}_2, \mathbf{a}_6\}$, and $\{\mathbf{a}_2, \mathbf{a}_7\}$.

Now, we extend the concepts of lines to planes in EG(m, 2^s). Let $\mathbf{a}_0, \mathbf{a}_1, \ldots, \mathbf{a}_\mu$ be $\mu + 1$ linearly independent points in EG(m, 2^s), where $\mu < m$. The $2^{\mu s}$ points of the form

$$\mathbf{a}_0 + \beta_1\mathbf{a}_1 + \beta_2\mathbf{a}_2 + \cdots + \beta_\mu\mathbf{a}_\mu,$$

with $\beta_i \in \mathrm{GF}(2^s)$ for $1 \leq i \leq \mu$, constitute a *μ-flat* (or a μ-dimensional *hyperplane*) in EG(m, 2^s) which passes through the point $\mathbf{a}_0$. We denote this μ-flat by $\{\mathbf{a}_0 + \beta_1\mathbf{a}_1 + \cdots + \beta_\mu\mathbf{a}_\mu\}$. The μ-flat that consists of the $2^{\mu s}$ points

$$\beta_1\mathbf{a}_1 + \beta_2\mathbf{a}_2 + \cdots + \beta_\mu\mathbf{a}_\mu$$

passes through the origin. We can readily prove that the μ-flats $\{\beta_1\mathbf{a}_1 + \beta_2\mathbf{a}_2 + \beta_3\mathbf{a}_3 + \cdots + \beta_\mu\mathbf{a}_\mu\}$ and $\{\mathbf{a}_0 + \beta_1\mathbf{a}_1 + \beta_2\mathbf{a}_2 + \cdots + \beta_\mu\mathbf{a}_\mu\}$ do not have any point in common. We say that these two μ-flats are parallel. For any μ-flat passing through the origin, there are $2^{(m-\mu)s} - 1$ μ-flat in EG(m, 2^s) parallel to it.

If $\mathbf{a}_{\mu+1}$ is not a point in the μ-flat $\{\mathbf{a}_0 + \beta_1\mathbf{a}_1 + \cdots + \beta_\mu\mathbf{a}_\mu\}$, then the $(\mu + 1)$-flat $\{\mathbf{a}_0 + \beta_1\mathbf{a}_1 + \cdots + \beta_\mu\mathbf{a}_\mu + \beta_{\mu+1}\mathbf{a}_{\mu+1}\}$ contains the μ-flat $\{\mathbf{a}_0 + \beta_1\mathbf{a}_1 + \cdots + \beta_\mu\mathbf{a}_\mu\}$. Let $\mathbf{b}_{\mu+1}$ be a point not in $\{\mathbf{a}_0 + \beta_1\mathbf{a}_1 + \cdots + \beta_{\mu+1}\mathbf{a}_{\mu+1}\}$. Then the two $(\mu + 1)$-flats $\{\mathbf{a}_0 + \beta_1\mathbf{a}_1 + \cdots + \beta_\mu\mathbf{a}_\mu + \beta_{\mu+1}\mathbf{a}_{\mu+1}\}$ and $\{\mathbf{a}_0 + \beta_1\mathbf{a}_1 + \cdots + \beta_\mu\mathbf{a}_\mu + \beta_{\mu+1}\mathbf{b}_{\mu+1}\}$ intersect on the μ-flat $\{\mathbf{a}_0 + \beta_1\mathbf{a}_1 + \cdots + \beta_\mu\mathbf{a}_\mu\}$ (i.e., they have the points in $\{\mathbf{a}_0 + \beta_1\mathbf{a}_1 + \cdots + \beta_\mu\mathbf{a}_\mu\}$ as all their common points). Given a μ-flat F in EG(m, 2^s), the number of $(\mu + 1)$-flats in EG(m, 2^s) that intersect on F is

$$\frac{2^{(m-\mu)s} - 1}{2^s - 1}. \tag{8.2}$$

Any point outside the μ-flat F is in one and only one of the $(\mu + 1)$-flats that intersect on F. The number of μ-flats in EG(m, 2^s) is

$$2^{(m-\mu)s} \prod_{i=1}^{\mu} \frac{2^{(m-i+1)s} - 1}{2^{(\mu-i+1)s} - 1}.$$

Example 8.2

Consider the geometry EG(3, 2) over GF(2) given in Example 8.1. There are fourteen 2-flats, which are given in Table 8.2. The 2-flats that intersect on the line $\{\mathbf{a}_1, \mathbf{a}_3\}$ are $\{\mathbf{a}_0, \mathbf{a}_1, \mathbf{a}_2, \mathbf{a}_3\}$, $\{\mathbf{a}_1, \mathbf{a}_3, \mathbf{a}_5, \mathbf{a}_7\}$, and $\{\mathbf{a}_1, \mathbf{a}_3, \mathbf{a}_4, \mathbf{a}_6\}$. The 2-flats, $\{\mathbf{a}_0, \mathbf{a}_1, \mathbf{a}_2, \mathbf{a}_3\}$ and $\{\mathbf{a}_4, \mathbf{a}_5, \mathbf{a}_6, \mathbf{a}_7\}$ are parallel.

TABLE 8.2 2-FLATS IN EG(3, 2)

$\{\mathbf{a}_0, \mathbf{a}_1, \mathbf{a}_2, \mathbf{a}_3\}$	$\{\mathbf{a}_4, \mathbf{a}_5, \mathbf{a}_6, \mathbf{a}_7\}$	$\{\mathbf{a}_0, \mathbf{a}_1, \mathbf{a}_4, \mathbf{a}_5\}$	$\{\mathbf{a}_2, \mathbf{a}_3, \mathbf{a}_6, \mathbf{a}_7\}$
$\{\mathbf{a}_0, \mathbf{a}_2, \mathbf{a}_4, \mathbf{a}_6\}$	$\{\mathbf{a}_1, \mathbf{a}_3, \mathbf{a}_5, \mathbf{a}_7\}$	$\{\mathbf{a}_0, \mathbf{a}_1, \mathbf{a}_6, \mathbf{a}_7\}$	$\{\mathbf{a}_2, \mathbf{a}_3, \mathbf{a}_4, \mathbf{a}_5\}$
$\{\mathbf{a}_0, \mathbf{a}_2, \mathbf{a}_5, \mathbf{a}_7\}$	$\{\mathbf{a}_1, \mathbf{a}_3, \mathbf{a}_4, \mathbf{a}_6\}$	$\{\mathbf{a}_0, \mathbf{a}_4, \mathbf{a}_3, \mathbf{a}_7\}$	$\{\mathbf{a}_1, \mathbf{a}_5, \mathbf{a}_3, \mathbf{a}_6\}$
$\{\mathbf{a}_0, \mathbf{a}_3, \mathbf{a}_5, \mathbf{a}_6\}$	$\{\mathbf{a}_1, \mathbf{a}_2, \mathbf{a}_4, \mathbf{a}_7\}$		

Next, we show that the elements in the Galois field GE(2^{ms}) actually forms an m-dimensional Euclidean geometry EG(m, 2^s). Let α be a primitive element of GF(2^{ms}). Then the 2^{ms} elements in GF(2^{ms}) can be expressed as powers of α as follows: $\alpha^\infty = 0$, $\alpha^0 = 1, \alpha^1, \alpha^2, \ldots, \alpha^{2^{ms}-2}$. It is known that GF($2^{ms}$) contains GF($2^s$) as a subfield. Every element α^i in GF(2^{ms}) can be expressed as

$$\alpha^i = a_{i0} + a_{i1}\alpha + a_{i2}\alpha^2 + \cdots + a_{i,m-1}\alpha^{m-1},$$

where $a_{ij} \in$ GF(2^s) for $0 \leq j < m$. There is a *one-to-one correspondence* between the element α^i and the m-tuple $(a_{i0}, a_{i1}, \ldots, a_{i,m-1})$ over GF(2^s). Therefore, the 2^{ms} elements in GF(2^{ms}) may be regarded as the 2^{ms} points in EG(m, 2^s), and GF(2^{ms}) as the geometry EG(m, 2^s). In this case, a μ-flat passing through the point α^{l_0} consists of the following $2^{\mu s}$ points:

$$\alpha^{l_0} + \beta_1\alpha^{l_1} + \cdots + \beta_\mu\alpha^{l_\mu},$$

where $\alpha^{l_0}, \alpha^{l_1}, \ldots, \alpha^{l_\mu}$ are $\mu + 1$ linearly independent elements in GF(2^{ms}) and $\beta_i \in$ GF(2^s).

Example 8.3

Consider the Galois field GF(2^4) given by Table 2.8. Let $m = 2$. Let α be a primitive element whose minimal polynomial is $\phi(X) = 1 + X + X^4$. Let $\beta = \alpha^5$. We see that $\beta^0 = 1$, $\beta^1 = \alpha^5$, $\beta^2 = \alpha^{10}$, and $\beta^3 = \alpha^{15} = 1$. Therefore, the order of β is 3. We can readily check that the elements

$$0, 1, \beta, \beta^2$$

form a field of four elements, GF(2^2). Therefore, GF(2^2) is a subfield of GF(2^4). Table 8.3 shows that every element α^i in GF(2^4) is expressed in the form

$$\alpha^i = a_{i0} + a_{i1}\alpha,$$

with a_{i0} and a_{i1} in GF(2^2) $= \{0, 1, \beta, \beta^2\}$. We may regard GF($2^4$) as the Euclidean geometry EG(2, 2^2) over GF(2^2). Then the points

$$\alpha^{14} + 0\cdot\alpha = \alpha^{14}, \qquad \alpha^{14} + 1\cdot\alpha = \alpha^7,$$
$$\alpha^{14} + \beta\cdot\alpha = \alpha^8, \qquad \alpha^{14} + \beta^2\cdot\alpha = \alpha^{10},$$

TABLE 8.3 ELEMENTS IN GF(2^4)*

	2-Tuples over GF(2^2)
$0 = 0$	$(0, 0)$
$1 = 1$	$(1, 0)$
$\alpha = \alpha$	$(0, 1)$
$\alpha^2 = \beta + \alpha$	$(\beta, 1)$
$\alpha^3 = \beta + \beta^2\alpha$	(β, β^2)
$\alpha^4 = 1 + \alpha$	$(1, 1)$
$\alpha^5 = \beta$	$(\beta, 0)$
$\alpha^6 = \beta\alpha$	$(0, \beta)$
$\alpha^7 = \beta^2 + \beta\alpha$	(β^2, β)
$\alpha^8 = \beta^2 + \alpha$	$(\beta^2, 1)$
$\alpha^9 = \beta + \beta\alpha$	(β, β)
$\alpha^{10} = \beta^2$	$(\beta^2, 0)$
$\alpha^{11} = \beta^2\alpha$	$(0, \beta^2)$
$\alpha^{12} = 1 + \beta^2\alpha$	$(1, \beta^2)$
$\alpha^{13} = 1 + \beta\alpha$	$(1, \beta)$
$\alpha^{14} = \beta^2 + \beta^2\alpha$	(β^2, β^2)

*Elements in GF(2^4) are expressed in the form $a_{i0} + a_{i1}\alpha$, where α is a primitive element in GF(2^4) and a_{ij} is an element in GF(2^2) $=$ $\{0, 1, \beta, \beta^2\}$ with $\beta = \alpha^5$.

form a line passing through the point α^{14}. The other four lines in EG(2, 2^2) passing through α^{14} are

$$\{\alpha^{14}, \alpha^{13}, \alpha, \alpha^5\}, \quad \{\alpha^{14}, \alpha^0, \alpha^6, \alpha^2\},$$

$$\{\alpha^{14}, \alpha^9, \alpha^4, 0\}, \quad \{\alpha^{14}, \alpha^{12}, \alpha^{11}, \alpha^3\}.$$

8.2 MAJORITY-LOGIC DECODABLE CYCLIC CODES BASED ON EUCLIDEAN GEOMETRY

Let

$$\mathbf{v} = (v_0, v_1, \ldots, v_{2^{ms}-2})$$

be a $(2^{ms} - 1)$-tuple over the binary field GF(2). Let α be a primitive element of the Galois field GF(2^{ms}). We may number the components of $\mathbf{v}$ with the nonzero elements of GF(2^{ms}) as follows: the component v_i is numbered α^i for $0 \leq i \leq 2^m - 2$. Hence, α^i is the location number of v_i. Now we regard GF(2^{ms}) as the m-dimensional Euclidean geometry over GF(2^s), EG(m, 2^s). Let F be a μ-flat in EG(m, 2^s) that does not pass through the origin $\alpha^\infty = 0$. Based on this flat F, we may form a vector over GF(2) as follows:

$$\mathbf{v}_F = (v_0, v_1, \ldots, v_{2^{ms}-2}),$$

whose ith component v_i is 1 if its location number α^i is a point in F; otherwise, v_i is 0. In other words, the location numbers for the nonzero components of $\mathbf{v}_F$ form the points of the μ-flat F. The vector $\mathbf{v}_F$ is called the *incidence vector* of the μ-flat F.

Example 8.4

Let $m = 2$ and $s = 2$. Consider the field GF(2^4), which is regarded as the Euclidean geometry over GF(2^2), EG(2, 2^2). From Example 8.3, the four 1-flats passing through the point α^{14} but not the origin are

$$L_1 = \{\alpha^{14}, \alpha^7, \alpha^8, \alpha^{10}\}, \qquad L_2 = \{\alpha^{14}, \alpha^{13}, \alpha, \alpha^5\},$$
$$L_3 = \{\alpha^{14}, \alpha^0, \alpha^6, \alpha^2\}, \qquad L_4 = \{\alpha^{14}, \alpha^{12}, \alpha^{11}, \alpha^3\}.$$

The incidence vectors for these four 1-flats are:

Location Numbers

	α^0	α^1	α^2	α^3	α^4	α^5	α^6	α^7	α^8	α^9	α^{10}	α^{11}	α^{12}	α^{13}	α^{14}
$\mathbf{v}_{L_1} =$	(0	0	0	0	0	0	0	1	1	0	1	0	0	0	1)
$\mathbf{v}_{L_2} =$	(0	1	0	0	0	1	0	0	0	0	0	0	0	1	1)
$\mathbf{v}_{L_3} =$	(1	0	1	0	0	0	1	0	0	0	0	0	0	0	1)
$\mathbf{v}_{L_4} =$	(0	0	0	1	0	0	0	0	0	0	0	1	1	0	1)

Definition 8.1. A (μ, s)th-order binary Euclidean geometry (EG) code of length $2^{ms} - 1$ is the largest cyclic code whose null space contains the incidence vectors of all the $(\mu + 1)$-flats of EG(m, 2^s) that do not pass through the origin.

The generator polynomial of a (μ, s)th-order EG code will be given in terms of its roots in GF(2^{ms}). Let h be a nonnegative integer less than 2^{ms}. Then we can express h in radix 2^s form as follows:

$$h = \delta_0 + \delta_1 2^s + \delta_2 2^{2s} + \cdots + \delta_{m-1} 2^{(m-1)s},$$

where $0 \leq \delta_i < 2^s$ for $0 \leq i < m$. The 2^s-weight of h, denoted $W_{2^s}(h)$, is defined as the real sum of the coefficients in the radix 2^s expansion of h, that is,

$$W_{2^s}(h) = \sum_{i=0}^{m-1} \delta_i. \tag{8.3}$$

As an example, let $m = 3$ and $s = 2$. Then the integer $h = 45$ can be expanded in radix 2^2 form as follows:

$$45 = 1 + 3 \cdot 2^2 + 2 \cdot 2^{2 \times 2},$$

with $\delta_0 = 1$, $\delta_1 = 3$, and $\delta_2 = 2$. The 2^2-weight of 45 is then

$$W_{2^2}(45) = 1 + 3 + 2 = 6.$$

Consider the difference, $h - W_{2^s}(h)$, which can be expressed as follows:

$$h - W_{2^s}(h) = \delta_1(2^s - 1) + \delta_2(2^{2s} - 1) + \cdots + \delta_{m-1}(2^{(m-1)s} - 1).$$

It is clear from this difference that h is divisible by $2^s - 1$ if and only if its 2^s-weight, $W_{2^s}(h)$, is divisible by $2^s - 1$. Let $h^{(l)}$ be the remainder resulting from dividing $2^l h$ by $2^{ms} - 1$, that is,

$$2^l h = q(2^{ms} - 1) + h^{(l)},$$

with $0 \leq h^{(l)} < 2^{ms} - 1$. Clearly, $h^{(l)}$ is divisible by $2^s - 1$ if and only if h is divisible by $2^s - 1$. Note that $h^{(0)} = h$.

Now we state a theorem (without proof) that characterizes the roots of the generator polynomial of a (μ, s)th-order EG code.

Theorem 8.1. Let α be a primitive element of the Galois field $GF(2^{ms})$. Let h be a nonnegative integer less than $2^{ms} - 1$. The generator polynomial $\mathbf{g}(X)$ of the (μ, s)th-order EG code of length $2^{ms} - 1$ has α^h as a root if and only if

$$0 < \max_{0 \le l < s} W_{2^s}(h^{(l)}) \le (m - \mu - 1)(2^s - 1). \tag{8.4}$$

Example 8.5

Let $m = 2$, $s = 2$, and $\mu = 0$. Then the Galois field $GF(2^4)$ may be regarded as the Euclidean geometry $EG(2, 2^2)$ over $GF(2^2)$. Let α be a primitive element in $GF(2^4)$ (use Table 2.8). Let h be a nonnegative integer less than 15. It follows from Theorem 8.1 that the generator polynomial $\mathbf{g}(X)$ of the (0, 2)th-order EG code of length 15 has α^h as a root if and only if

$$0 < \max_{0 \le l < 2} W_{2^2}(h^{(l)}) \le 3.$$

The nonnegative integers less than 15 that satisfy the condition above are 1, 2, 3, 4, 6, 8, 9, and 12. Therefore, $\mathbf{g}(X)$ has $\alpha, \alpha^2, \alpha^3, \alpha^4, \alpha^6, \alpha^8, \alpha^9$, and α^{12} as all its roots. The elements $\alpha, \alpha^2, \alpha^4$, and α^8 have the same minimal polynomial $\phi_1(X) = 1 + X + X^4$, and the elements $\alpha^3, \alpha^6, \alpha^9$, and α^{12} have the same minimal polynomial $\phi_3(X) = 1 + X + X^2 + X^3 + X^4$. Thus, the generator polynomial of the (0, 2)th-order EG code of length 15 is

$$\begin{aligned}\mathbf{g}(X) &= (1 + X + X^4)(1 + X + X^2 + X^3 + X^4) \\ &= 1 + X^4 + X^6 + X^7 + X^8.\end{aligned}$$

It is interesting to note that the (0, 2)th-order EG code is the (15, 7) BCH code considered in Example 7.1. It is one-step majority-logic decodable.

Example 8.6

Let $m = 3$, $s = 2$, and $\mu = 1$. Then the Galois field $GF(2^6)$ may be regarded as the Euclidean geometry $EG(3, 2^2)$ over $GF(2^2)$. Let α be a primitive element in $GF(2^6)$ (use Table 6.2). Let h be a nonnegative integer less than 63. If follows from Theorem 8.1 that the generator polynomial $\mathbf{g}(X)$ of the (1, 2)th-order EG code of length 63 has α^h as a root if and only if

$$0 < \max_{0 \le l < 2} W_{2^2}(h^{(l)}) \le 3.$$

The nonnegative integers less than 63 that satisfy the condition above are

$$1, 2, 3, 4, 6, 8, 9, 12, 16, 18, 24, 32, 33, 48.$$

Thus, $\mathbf{g}(X)$ has the following roots:

$$\alpha^1, \alpha^2, \alpha^3, \alpha^4, \alpha^6, \alpha^8, \alpha^9, \alpha^{12}, \alpha^{16}, \alpha^{18}, \alpha^{24}, \alpha^{32}, \alpha^{33}, \alpha^{48}.$$

From Table 6.3 we find that:

1. $\alpha, \alpha^2, \alpha^4, \alpha^8, \alpha^{16}$, and α^{32} have $\phi_1(X) = 1 + X + X^6$ as their minimal polynomial.
2. $\alpha^3, \alpha^6, \alpha^{12}, \alpha^{24}, \alpha^{33}$, and α^{48} have $\phi_3(X) = 1 + X + X^2 + X^4 + X^6$ as their minimal polynomial.
3. α^9, α^{18}, and α^{36} have the same minimal polynomial $\phi_9(X) = 1 + X^2 + X^3$.

Therefore, the generator polynomial of the (1, 2)th-order EG code of length 63 is

$$\begin{aligned}\mathbf{g}(X) &= (1 + X + X^6)(1 + X + X^2 + X^4 + X^6)(1 + X^2 + X^3)\\ &= 1 + X^2 + X^4 + X^{11} + X^{13} + X^{14} + X^{15}.\end{aligned}$$

Hence, the (1, 2)th-order EG code of length 63 is a (63, 48) cyclic code. Later we will show that this code is two-step orthogonalizable and is capable of correcting any combination of two or fewer errors.

Decoding of the (μ, s)th-order EG code of length $2^{ms} - 1$ is based on the structural properties of the Euclidean geometry EG$(m, 2^s)$. From Definition 8.1, we know that the null space of the code contains the incidence vectors of all the $(\mu + 1)$-flats of EG$(m, 2^s)$ that do not pass through the origin. Let $F^{(\mu)}$ be a μ-flat passing through the point $\alpha^{2^{ms}-2}$. From (8.2) we know that there are

$$J = \frac{2^{(m-\mu)s} - 1}{2^s - 1} \tag{8.5}$$

$(\mu + 1)$-flats not passing through the origin which intersect on $F^{(\mu)}$. The incidence vectors of these J $(\mu + 1)$-flats are orthogonal on the digits at the locations that correspond to the points in $F^{(\mu)}$. Therefore, the parity-check sums formed from these J incidence vectors are orthogonal on the error digits at the locations corresponding to the points in $F^{(\mu)}$. If there are $\lfloor J/2 \rfloor$ or fewer errors in the received vector, the sum of errors at the locations corresponding to the points in $F^{(\mu)}$ can be determined correctly. Let us denote this error sum with $S(F^{(\mu)})$. In this manner, for every μ-flat $F^{(\mu)}$ passing through the point $\alpha^{2^{ms}-2}$, the error sum $S(F^{(\mu)})$ can be determined. This forms the first step of orthogonalization.

The error sums $S(F^{(\mu)})$'s corresponding to all the μ-flats $F^{(\mu)}$ that pass through the point $\alpha^{2^{ms}-2}$ are then used for the second step of orthogonalization. Let $F^{(\mu-1)}$ be a $(\mu - 1)$-flat passing through the point $\alpha^{2^{ms}-2}$. From (8.2) we see that there are

$$J_1 = \frac{2^{(m-\mu+1)s} - 1}{2^s - 1} - 1 > J$$

μ-flats not passing through the origin which intersect on $F^{(\mu-1)}$. The error sums corresponding to these J_1 μ-flats are orthogonal on the error digits at the locations corresponding to the points in $F^{(\mu-1)}$. Let $S(F^{(\mu-1)})$ denote the sum of error digits at the locations corresponding to the points in $F^{(\mu-1)}$. Then $S(F^{(\mu-1)})$ can be determined from the J_1 error sums $S(F^{(\mu)})$'s that are orthogonal on $S(F^{(\mu-1)})$. Since $J_1 > J$, if there are no more than $\lfloor J/2 \rfloor$ errors in the received vector, the error sum $S(F^{(\mu-1)})$ can be determined correctly. In this manner for every $(\mu - 1)$-flat $F^{(\mu-1)}$ passing through the point $\alpha^{2^{ms}-2}$ but not the origin, the error sum $S(F^{(\mu-1)})$ can be determined. This completes the second step of orthogonalization.

The error sums $S(F^{(\mu-1)})$'s now are used for the third step of orthogonalization. Let $F^{(\mu-2)}$ be a $(\mu - 2)$-flat passing through the point $\alpha^{2^{ms}-2}$ but not the origin. From (8.2) we see that there are

$$J_2 = \frac{2^{(m-\mu+2)s} - 1}{2^s - 1} - 1 > J_1 > J$$

error sums $S(F^{(\mu-1)})$'s orthogonal on the error sum $S(F^{(\mu-2)})$. Hence, $S(F^{(\mu-2)})$ can be determined correctly. The error sums $S(F^{(\mu-2)})$'s are then used for the next step of orthogonalization. This process continues until the error sums corresponding to all the 1-flats (lines) passing through the point $\alpha^{2^{ms}-2}$ but not the origin are determined. There are

$$J_\mu = \frac{2^{ms}-1}{2^s-1} > J_{\mu-1} > \cdots > J_1 > J$$

such error sums orthogonal on the error digit $e_{2^{ms}-2}$ at the location $\alpha^{2^{ms}-2}$. Thus, $e_{2^{ms}-2}$ can be determined correctly from these orthogonal error sums provided that there are no more than $\lfloor J/2 \rfloor$ errors in the received vector. Since the code is cyclic, other error digits can be decoded in the same manner successively.

Since the decoding of each error digit requires $\mu + 1$ steps of orthogonalization, the (μ, s)th-order EG code of length $2^{ms} - 1$ is therefore $(\mu + 1)$-step majority-logic decodable. The code is capable of correcting

$$t_{\mathrm{ML}} = \left\lfloor \frac{2^{(m-\mu)s}-1}{2(2^s-1)} - \frac{1}{2} \right\rfloor \tag{8.6}$$

or fewer errors. Note that, at each step of orthogonalization, we need only J orthogonal error sums to determine an error sum for the next step. For $\mu = 0$, a $(0, s)$th-order EG code is one-step majority-logic decodable.

Example 8.7

Let $m = 2$, $s = 2$ and $\mu = 0$. Consider the (0, 2)th-order EG code of length 15. From Example 8.5 we know that this code is the (15, 7) BCH code (also a type 1 DTI code). The null space of this code contains the incidence vectors of all the 1-flats in EG(2, 2^2) that do not pass through the origin. To decode e_{14}, we need to determine the incidence vectors of the 1-flat passing through the point α^{14}, where α is a primitive element in GF(2^4). There are

$$J = \frac{2^{2\cdot 2}-1}{2^2-1} - 1 = 4$$

such incidence vectors, which are given in Example 8.4. These four vectors are orthogonal on the digit position α^{14}. In fact, these are exactly the four orthogonal vectors $\mathbf{w}_1$, $\mathbf{w}_2$, $\mathbf{w}_3$, and $\mathbf{w}_4$ given in Example 7.1.

Example 8.8

Let $m = 4$, $s = 1$, and $\mu = 1$. Consider the (1, 1)th-order EG code of length $2^4 - 1 = 15$. Let α be a primitive element of GF(2^4) given by Table 2.8. Let h be a nonnegative integer less than 15. It follows from Theorem 8.1 that the generator polynomial $\mathbf{g}(X)$ of this code has α^h as a root if and only if

$$0 < W_2(h^{(0)}) \leq 2.$$

Note that $h^{(0)} = h$. From the condition above, we find that $\mathbf{g}(X)$ has the following roots: $\alpha, \alpha^2, \alpha^3, \alpha^4, \alpha^5, \alpha^6, \alpha^8, \alpha^9, \alpha^{10}$, and α^{12}. From Table 2.9 we find that

$$\begin{aligned}\mathbf{g}(X) &= (1 + X + X^4)(1 + X + X^2 + X^3 + X^4)(1 + X + X^2)\\ &= 1 + X + X^2 + X^4 + X^5 + X^8 + X^{10}.\end{aligned}$$

It is interesting to note that this EG code is actually the (15, 5) BCH code studied in Example 7.12.

The null space of this code contains the incidence vectors of all the 2-flats of the EG(4, 2) that do not pass through the origin. Now, we will show how to form orthogonal check sums based on the structure of EG(4, 2). First, we treat GF(2^4) as the geometry EG(4, 2). A 1-flat passing through the point α^{14} consists of the points of the form $\alpha^{14} + a\alpha^i$ with $a \in$ GF(2). There are thirteen 1-flats passing through α^{14} but not the origin $\alpha^\infty = 0$; they are

$$\{\alpha^{13}, \alpha^{14}\}, \quad \{\alpha^{12}, \alpha^{14}\}, \quad \{\alpha^{11}, \alpha^{14}\}, \quad \{\alpha^{10}, \alpha^{14}\}, \quad \{\alpha^{9}, \alpha^{14}\}, \quad \{\alpha^{8}, \alpha^{14}\},$$
$$\{\alpha^{7}, \alpha^{14}\}, \quad \{\alpha^{6}, \alpha^{14}\}, \quad \{\alpha^{5}, \alpha^{14}\}, \quad \{\alpha^{4}, \alpha^{14}\}, \quad \{\alpha^{3}, \alpha^{14}\}, \quad \{\alpha^{2}, \alpha^{14}\}, \quad \{\alpha, \alpha^{14}\}.$$

For each of these 1-flats, there are

$$J = \frac{2^{(4-1)\cdot 1} - 1}{2^1 - 1} - 1 = 6$$

2-flats not passing through the origin that intersect on it. Each of these 2-flats consists of the points of the form $\alpha^{14} + a\alpha^i + b\alpha^j$, with a and b in GF(2). The six 2-flats that intersect on the 1-flat, $\{\alpha^{13}, \alpha^{14}\}$, are

$$\{\alpha^4, \alpha^{10}, \alpha^{13}, \alpha^{14}\}, \quad \{\alpha^7, \alpha^{12}, \alpha^{13}, \alpha^{14}\}, \quad \{\alpha^9, \alpha^{11}, \alpha^{13}, \alpha^{14}\},$$
$$\{\alpha^0, \alpha^8, \alpha^{13}, \alpha^{14}\}, \quad \{\alpha^1, \alpha^5, \alpha^{13}, \alpha^{14}\}, \quad \{\alpha^3, \alpha^6, \alpha^{13}, \alpha^{14}\}.$$

The incidence vectors for these six 2-flats are:

	α^0	α^1	α^2	α^3	α^4	α^5	α^6	α^7	α^8	α^9	α^{10}	α^{11}	α^{12}	α^{13}	α^{14}
$\mathbf{w}_{11} =$	(0	0	0	0	1	0	0	0	0	0	1	0	0	1	1)
$\mathbf{w}_{12} =$	(0	0	0	0	0	0	0	1	0	0	0	0	1	1	1)
$\mathbf{w}_{13} =$	(0	0	0	0	0	0	0	0	0	1	0	1	0	1	1)
$\mathbf{w}_{14} =$	(1	0	0	0	0	0	0	0	1	0	0	0	0	1	1)
$\mathbf{w}_{15} =$	(0	1	0	0	0	1	0	0	0	0	0	0	0	1	1)
$\mathbf{w}_{16} =$	(0	0	0	1	0	0	1	0	0	0	0	0	0	1	1).

Clearly, these six vectors are orthogonal on digits at locations α^{13} and α^{14}. Let $\mathbf{r}$ be the received vector. The parity-check sums formed from these six orthogonal vectors are

$$A_{11} = \mathbf{w}_{11}\cdot\mathbf{r} = e_4 + e_{10} + e_{13} + e_{14}$$
$$A_{12} = \mathbf{w}_{12}\cdot\mathbf{r} = e_7 + e_{12} + e_{13} + e_{14}$$
$$A_{13} = \mathbf{w}_{13}\cdot\mathbf{r} = e_9 + e_{11} + e_{13} + e_{14}$$
$$A_{14} = \mathbf{w}_{14}\cdot\mathbf{r} = e_0 + e_8 + e_{13} + e_{14}$$
$$A_{15} = \mathbf{w}_{15}\cdot\mathbf{r} = e_1 + e_5 + e_{13} + e_{14}$$
$$A_{16} = \mathbf{w}_{16}\cdot\mathbf{r} = e_3 + e_6 + e_{13} + e_{14}.$$

We see that these six check sums orthogonal on $\{e_{13}, e_{14}\}$ are exactly the same check sums given in Example 7.12. Thus, the error sum $e_{13} + e_{14}$ corresponding to the 1-flat $\{\alpha^{13}, \alpha^{14}\}$ can be determined from these six check sums.

In the same manner, we can determine the error sums corresponding to the other twelve 1-flats passing through α^{14}. Since $J = 6$, we only need to determine six error sums corresponding to any six 1-flats passing through α^{14}. These error sums are then used to determine e_{14}. Thus, the (1, 1)th-order EG code of length 15 is a two-step majority-logic decodable code.

Except for certain special cases, there is no simple formula for enumerating the number of parity-check digits of EG codes. Complicate combinatorial expressions for the number of parity-check digits of EG codes can be found in References 4 and 5.

One special case is $\mu = m - 2$. The number of parity-check digits for a $(m - 2, s)$th-order EG code of length $2^{ms} - 1$ is

$$n - k = \binom{m+1}{m}^s - 1.$$

This result was obtained independently by Smith [6] and by MacWilliams and Mann [7].

For $s = 1$, we obtain another special subclass of EG codes, which is known as the class *Reed–Muller* (RM) codes. A $(\mu, 1)$th-order EG code is called a μth-order RM code. Let α be a primitive element of the Galois field GF(2^m). Let h be a nonnegative integer less than 2^m. It follows from Theorem 8.1 that the generator polynomial $\mathbf{g}(X)$ of a μth-order RM code of length $2^m - 1$ has α^h as a root if and only if

$$0 < W_2(h) \leq m - \mu - 1. \tag{8.7}$$

It has been proved that a μth-order RM code of length $n = 2^m - 1$ has the following parameters:

$$k = \sum_{i=0}^{\mu} \binom{m}{i},$$

$$d_{\min} = 2^{m-\mu} - 1,$$

$$J = 2^{m-\mu} - 2.$$

Since $J = d_{\min} - 1$, RM codes are completely orthogonalizable. Reed–Muller codes in noncyclic form were first constructed by Muller in 1954 [8]. In the same year, Reed [9] devised a majority-logic decoding algorithm for these codes. Cyclic structure of RM codes was proved independently by Kasami et al. [10, 11] and Kolesnik and Mironchikov [12].

Except for RM codes and other special cases, EG codes in general are not completely orthogonalizable. For moderate length n, the error-correcting capability of an EG code is slightly inferior to that of a comparable BCH code. However, the majority-logic decoding for EG codes is more simply implemented than the decoding for BCH codes. Thus, for moderate n, EG codes provide rather effective error control. For large length n, EG codes become much more inferior to the comparable BCH codes, and the number of majority-logic gates required for decoding becomes prohibitively large. In this case, BCH codes are definitely superior to the EG codes in error-correcting capability and decoding complexity. A list of EG codes with $n \leq 1023$ is given in Table 8.4. See Reference 13 for a more extensive list.

EG codes were first studied by Rudolph [3]. Rudolph's work was later extended and generalized by other coding theorists [13–18]. Improvements for decoding EG codes were suggested by Weldon [19] and by Chen [20]. Chen proved that any EG code can be decoded in *no more* than three steps. Chen's decoding algorithm is based on further structure of the Euclidean geometry, which is not covered in this introductory book.

There are several classes of generalized EG codes [13, 16–18, 21] which all contain EG codes as subclasses. These generalizations will not be covered here. However, a simple generalization using parallel flats is presented next.

TABLE 8.4 LIST OF EG CODES

m	s	μ	n	k	J	t_{ML}
3	1	1	7	4	2	1
4	1	2	15	11	2	1
4	1	1	15	5	6	3
2	2	0	15	7	4	2
5	1	3	31	26	2	1
5	1	2	31	16	6	3
5	1	1	31	6	14	7
6	1	4	63	57	2	1
6	1	3	63	42	6	3
6	1	2	63	22	14	7
6	1	1	63	7	30	15
3	2	1	63	48	4	2
3	2	0	63	13	20	10
2	3	0	63	37	8	4
7	1	5	127	120	2	1
7	1	4	127	99	6	3
7	1	3	127	64	14	7
7	1	2	127	29	30	15
7	1	1	127	8	62	31
8	1	6	255	247	2	1
8	1	5	255	219	6	3
8	1	4	255	163	14	7
8	1	3	255	93	30	15
8	1	2	255	37	62	31
8	1	1	255	9	126	63
4	2	2	255	231	4	2
4	2	1	255	127	20	10
4	2	0	255	21	84	42
2	4	0	255	175	16	8
9	1	7	511	502	2	1
9	1	6	511	466	6	3
9	1	5	511	382	14	7
9	1	4	511	256	30	15
9	1	3	511	130	62	31
9	1	2	511	46	126	63
9	1	1	511	10	254	127
3	3	1	511	448	8	4
3	3	0	511	139	72	36
10	1	8	1023	1013	2	1
10	1	7	1023	968	6	3
10	1	6	1023	848	14	7
10	1	5	1023	638	30	15
10	1	4	1023	386	62	31
10	1	3	1023	176	126	63
10	1	2	1023	56	254	127
10	1	1	1023	11	510	255
5	2	3	1023	988	4	2
5	2	2	1023	748	20	10
5	2	1	1023	288	84	42
5	2	0	1023	31	340	170
2	5	0	1023	781	32	16

Twofold EG Codes

Let F and F_1 be any two parallel μ-flats in EG$(m, 2^s)$. We say that F and F_1 form a $(\mu, 2)$-*frame* in EG$(m, 2^s)$, denoted by $\{F, F_1\}$. Since F and F_1 do not have any point in common, the $(\mu, 2)$-frame $\{F, F_1\}$ consists of $2^{\mu s+1}$ points. Let F_2 be another μ-flat parallel to F and F_1. Then the two $(\mu, 2)$-frames $\{F, F_1\}$ and $\{F, F_2\}$ intersect on F. Let L be a $(\mu + 1)$-flat that contains the μ-flat F. Then L contains $2^s - 1$ other μ-flats that are parallel to F. Each of these $2^s - 1$ μ-flats together with F form a $(\mu, 2)$-frame. There are $2^s - 1$ such $(\mu, 2)$-frames which intersect on F. Clearly, these $2^s - 1$ $(\mu, 2)$-frames are all contained in the $(\mu + 1)$-flat L. Any point in L but outside F is *in one and only one of these* $2^s - 1$ $(\mu, 2)$-*frames*. Since there are

$$\frac{2^{(m-\mu)s} - 1}{2^s - 1}$$

$(\mu + 1)$-flats that intersect on F, there are

$$(2^s - 1) \cdot \frac{2^{(m-\mu)s} - 1}{2^s - 1} = 2^{(m-\mu)s} - 1 \tag{8.8}$$

$(\mu, 2)$-frames that intersect on F. Any point outside F is in one and only one of these $(\mu, 2)$-frames. We say that these $(\mu, 2)$-frames are orthogonal on the μ-flat F. If F does not pass through the origin, there are

$$2^{(m-\mu)s} - 2 \tag{8.9}$$

$(\mu, 2)$-frames that are orthogonal on F and do not pass through the origin.

Again, we regard the Galois field GF(2^{ms}) as the geometry EG$(m, 2^s)$. Let α be a primitive element of GF(2^{ms}). For any $(2^{ms} - 1)$-tuple

$$\mathbf{v} = (v_0, v_1, \ldots, v_{2^{ms}-2})$$

over GF(2), we again number its components with the nonzero elements of GF(2^{ms}) as usual (i.e., v_i is numbered with α^i for $0 \leq i < 2^{ms} - 1$). For each $(\mu, 2)$-frame Q in EG$(m, 2^s)$, we define its incidence vector as follows:

$$\mathbf{v}_Q = (v_0, v_1, \ldots, v_{2^{ms}-2}),$$

where the ith component

$$v_i = \begin{cases} 1 & \text{if } \alpha^i \text{ is a point in } Q \\ 0 & \text{otherwise.} \end{cases}$$

Definition 8.2. A (μ, s)th-order twofold EG code of length $2^{ms} - 1$ is the largest cyclic code whose null space contains the incidence vectors of all the $(\mu, 2)$-frames in EG$(m, 2^s)$ that do not pass through the origin.

We now state a theorem (without proof) [22] which characterizes the roots of the generator polynomial of a (μ, s)th-order twofold EG code.

Theorem 8.2. Let α be a primitive element of the Galois field GF(2^{ms}). Let h be a nonnegative integer less than $2^{ms} - 1$. The generator polynomial $\mathbf{g}(X)$ of the (μ, s)th-order twofold EG code of length $2^{ms} - 1$ has α^h as a root if and only if

$$0 < \max_{0 \leq l < s} W_{2^s}(h^{(l)}) < (m - \mu)(2^s - 1). \tag{8.10}$$

Example 8.9

Let $m = 2$, $s = 3$, and $\mu = 1$. Consider the (1, 3)th-order twofold EG code of length 63. Let α be a primitive element of GF(2^6) given by Table 6.2. Let h be a nonnegative integer less than 63. It follows from (8.10) that the generator polynomial $\mathbf{g}(X)$ of the (1, 3)th-order twofold EG code of length 63 has α^h as a root if and only if

$$0 < \max_{0 \le l < 3} W_{2^3}(h^{(l)}) < 7.$$

The nonnegative integers less than 63 that satisfy the conditions above are

1, 2, 3, 4, 5, 6, 8, 9, 10, 12, 16, 17, 20, 24, 32, 33, 34, 40, 48.

Thus, the generator polynomial $\mathbf{g}(X)$ has the following roots:

$$\alpha^1,\ \alpha^2,\ \alpha^3,\ \alpha^4,\ \alpha^5,\ \alpha^6,\ \alpha^8,\ \alpha^9,\ \alpha^{10},\ \alpha^{12},$$
$$\alpha^{16},\ \alpha^{17},\ \alpha^{20},\ \alpha^{24},\ \alpha^{32},\ \alpha^{33},\ \alpha^{34},\ \alpha^{40},\ \alpha^{48}.$$

From Table 6.3 we find that:

1. The roots $\alpha, \alpha^2, \alpha^4, \alpha^8, \alpha^{16}$, and α^{32} have the same minimal polynomial, $\phi_1(X) = 1 + X + X^6$.
2. The roots $\alpha^3, \alpha^6, \alpha^{12}, \alpha^{24}, \alpha^{48}$, and α^{33} have the same minimal polynomial, $\phi_3(X) = 1 + X + X^2 + X^4 + X^6$.
3. The roots $\alpha^5, \alpha^{10}, \alpha^{20}, \alpha^{40}, \alpha^{17}$, and α^{34} have the same minimal polynomial, $\phi_5(X) = 1 + X + X^2 + X^5 + X^6$.

Therefore,

$$\begin{aligned}\mathbf{g}(X) &= \phi_1(X)\cdot\phi_3(X)\cdot\phi_5(X)\\ &= 1 + X + X^2 + X^3 + X^6 + X^7 + X^9 + X^{15} + X^{16} + X^{17} + X^{18}.\end{aligned}$$

Therefore, the (1, 3)th-order twofold EG code of length 63 with $m = 3$ is a (63, 45) cyclic code. In fact, it is the (63, 45) BCH code with minimum distance equal to 7.

To decode the (μ, s)th-order twofold EG code of length $2^{ms} - 1$, we first form the parity-check sums from the incidence vectors of all the $(\mu, 2)$-frames in EG($m, 2^s$) that do not pass the origin (note that these incidence vectors are in the null space of the code). Let $F^{(\mu)}$ be a μ-flat that passes through the point $\alpha^{2^{ms}-2}$. From (8.9) we see that there are

$$J = 2^{(m-\mu)s} - 2 \tag{8.11}$$

$(\mu, 2)$-frames not passing through the origin which are orthogonal on $F^{(\mu)}$. The incidence vectors of these $(\mu, 2)$-frames are orthogonal on the digits at the locations that correspond to the points in $F^{(\mu)}$. Therefore, the parity-check sums formed from these J incidence vectors are orthogonal on the error digits at the locations that correspond to the points in $F^{(\mu)}$. Let $S(F^{(\mu)})$ denote the sum of error digits at the locations corresponding to the points in $F^{(\mu)}$. Then this error sum, $S(F^{(\mu)})$, can be determined correctly from the J check sums orthogonal on it provided that there are no more than

$$\left\lfloor \frac{J}{2} \right\rfloor = 2^{(m-\mu)s-1} - 1 \tag{8.12}$$

errors in the received vector. In this manner we can determine the error sums, $S(F^{(\mu)})$'s, that correspond to all the μ-flats passing through the point $\alpha^{2^{ms}-2}$. This completes the first step of orthogonalization. After this step, the rest of orthogonalization steps are the same as those for a μth-order EG code. Therefore, a total of $\mu + 1$ steps of orthogonalization are needed to decode a (μ, s)th-order twofold EG code.

We can easily check that, at each decoding step, there are at least $J = 2^{(m-\mu)s} - 2$ error sums orthogonal on an error sum for the next step. Thus, the (μ, s)th-order twofold EG code of length $2^{ms} - 1$ is capable of correcting

$$t_{\text{ML}} = \left\lfloor \frac{J}{2} \right\rfloor = 2^{(m-\mu)s-1} - 1 \tag{8.13}$$

or fewer errors with majority-logic decoding. It has been proved [22] that the minimum distance of the (μ, s)th-order twofold EG code of length $2^{ms} - 1$ is exactly $2^{(m-\mu)s} - 1$. Therefore, the class of twofold EG codes is completely orthogonalizable.

Example 8.10

Consider the decoding of the (1, 3)th-order twofold EG of length 63 with $m = 2$ and $s = 3$. In Example 8.9 we showed that this code is a (63, 45) cyclic code (also a BCH code). The null space of this code contains the incidence vectors of all the (1, 2)-frames in EG(2, 2^3) that do not pass through the origin. Regard GF(2^6) as the geometry EG(2, 2^3). Let α be a primitive element of GF(2^6) (use Table 6.2). From (8.1) we see that there are nine lines in EG(2, 2^3) that intersect at the point α^{62}. Eight of these lines do not pass the origin. From (8.9) we see that, for each of these eight lines, there are six (1, 2)-frames intersecting on it. The incidence vectors of these six (1, 2)-frames are in the null space of the code and they will be used to form parity-check sums for decoding the error digit e_{62} at the location α^{62}. Since $J = 6$, we only need to find six lines in EG(2, 2^3) that intersect at the point α^{62} and do not pass the origin.

Let $\beta = \alpha^9$. Then $0, 1, \beta, \beta^2, \beta^3, \beta^4, \beta^5$, and β^6 ($\beta^7 = 1$) form a subfield, GF(2^3), of the field GF(2^6) (use Table 6.2). Then each line in EG(2, 2^3) that passes through α^{62} consists of the following points:

$$\alpha^{62} + \eta\alpha^j$$

where $\eta \in \{0, 1, \beta, \beta^2, \beta^3, \beta^4, \beta^5, \beta^6\}$. Six lines passing through α^{62} are as follows:

$$L_1 = \{\alpha^{11}, \alpha^{16}, \alpha^{18}, \alpha^{24}, \alpha^{48}, \alpha^{58}, \alpha^{59}, \alpha^{62}\},$$

$$L_2 = \{\alpha^{1}, \alpha^{7}, \alpha^{31}, \alpha^{41}, \alpha^{42}, \alpha^{45}, \alpha^{57}, \alpha^{62}\},$$

$$L_3 = \{\alpha^{23}, \alpha^{33}, \alpha^{34}, \alpha^{37}, \alpha^{49}, \alpha^{54}, \alpha^{56}, \alpha^{62}\},$$

$$L_4 = \{\alpha^{2}, \alpha^{12}, \alpha^{19}, \alpha^{21}, \alpha^{27}, \alpha^{51}, \alpha^{61}, \alpha^{62}\},$$

$$L_5 = \{\alpha^{0}, \alpha^{3}, \alpha^{15}, \alpha^{20}, \alpha^{22}, \alpha^{28}, \alpha^{52}, \alpha^{62}\},$$

$$L_6 = \{\alpha^{9}, \alpha^{10}, \alpha^{13}, \alpha^{25}, \alpha^{30}, \alpha^{32}, \alpha^{38}, \alpha^{62}\}.$$

For each of the lines above, we want to form six (1, 2)-frames intersecting on it. A (1, 2)-frame that contains the line $\{\alpha^{62} + \eta\alpha^j\}$ is of the form

$$(\{\alpha^{62} + \eta\alpha^j\}, \{\alpha^{62} + \alpha^i + \eta\alpha^j\}),$$

where α^i is not in $\{\alpha^{62} + \eta\alpha^j\}$. Line L_1 given above consists of the points $\{\alpha^{62} + \eta\alpha\}$. The point α^9 is not in L_1. Then the line $\{\alpha^{62} + \alpha^9 + \eta\alpha\}$ is parallel to $\{\alpha^{62} + \eta\alpha\}$. Thus,

$$(\{\alpha^{62} + \eta\alpha\}, \{\alpha^{62} + \alpha^9 + \eta\alpha\})$$

form a (1, 2)-frame containing the line L_1; this (1, 2)-frame consists of the following points:

$$\{\alpha^{11}, \alpha^{16}, , \alpha^{18}, \alpha^{21}, \alpha^{24}, \alpha^{31}, \alpha^{32}, \alpha^{35}, \alpha^{47}, \alpha^{48}, \alpha^{52}, \alpha^{54}, \alpha^{58}, \alpha^{59}, \alpha^{60}, \alpha^{62}\}.$$

In this manner, for each line L_i we can form six (1, 2)-frames orthogonal on it. The incidence vectors of these 36 (1, 2)-frames are given in Tables 8.5A to 8.5F.

To decode the code, the incidence vectors of the 36 (1, 2)-frames given in Tables 8.5A to 8.5F are used to form parity-check sums. Let $S(L_i)$ denote the sum of error digits at the locations corresponding to the points on line L_i for $1 \leq i \leq 6$. Then, for each error sum $S(L_i)$, there are six parity-check sums orthogonal on it. Thus, $S(L_i)$ can be determined correctly provided that there are three or fewer errors in the error

TABLE 8.5A POLYNOMIALS ORTHOGONAL ON $\{e_{11}, e_{16}, e_{18}, e_{24}, e_{48}, e_{58}, e_{59}, e_{62}\}$

$\mathbf{w}_{11}(X)^* = (11, 16, 18, 21, 24, 31, 32, 35, 47, 48, 52, 54, 58, 59, 60, 62)$
$\mathbf{w}_{12}(X) = (11, 12, 16, 18, 22, 23, 24, 26, 38, 43, 45, 48, 51, 58, 59, 62)$
$\mathbf{w}_{13}(X) = (0, 6, 11, 16, 18, 24, 30, 40, 41, 44, 48, 56, 58, 59, 61, 62)$
$\mathbf{w}_{14}(X) = (4, 5, 8, 11, 16, 18, 20, 24, 25, 27, 33, 48, 57, 58, 59, 62)$
$\mathbf{w}_{15}(X) = (3, 11, 13, 14, 16, 17, 18, 24, 29, 34, 36, 42, 48, 58, 59, 62)$
$\mathbf{w}_{16}(X) = (2, 7, 9, 11, 15, 16, 18, 24, 39, 48, 49, 50, 53, 58, 59, 62)$

TABLE 8.5B POLYNOMIALS ORTHOGONAL ON $\{e_1, e_7, e_{31}, e_{41}, e_{42}, e_{45}, e_{57}, e_{62}\}$

$\mathbf{w}_{21}(X) = (1, 7, 13, 23, 24, 27, 31, 39, 41, 42, 44, 45, 46, 52, 57, 62)$
$\mathbf{w}_{22}(X) = (1, 7, 22, 31, 32, 33, 36, 41, 42, 45, 48, 53, 55, 57, 61, 62)$
$\mathbf{w}_{23}(X) = (0, 1, 7, 12, 17, 19, 25, 31, 41, 42, 45, 49, 57, 59, 60, 62)$
$\mathbf{w}_{24}(X) = (1, 5, 6, 7, 9, 21, 26, 28, 31, 34, 41, 42, 45, 57, 58, 62)$
$\mathbf{w}_{25}(X) = (1, 3, 7, 8, 10, 16, 31, 40, 41, 42, 45, 50, 51, 54, 57, 62)$
$\mathbf{w}_{26}(X) = (1, 4, 7, 14, 15, 18, 30, 31, 35, 37, 41, 42, 43, 45, 57, 62)$

*In Table 8.5A to 8.5F, the integers inside the parentheses are powers of X.

TABLE 8.5C POLYNOMIALS ORTHOGONAL ON $\{e_{23}, e_{33}, e_{34}, e_{37}, e_{49}, e_{54}, e_{56}, e_{62}\}$

$\mathbf{w}_{31}(X) = (4, 9, 11, 17, 23, 33, 34, 37, 41, 49, 51, 52, 54, 55, 56, 62)$
$\mathbf{w}_{32}(X) = (5, 15, 16, 19, 23, 31, 33, 34, 36, 37, 38, 44, 49, 54, 56, 62)$
$\mathbf{w}_{33}(X) = (1, 13, 18, 20, 23, 26, 33, 34, 37, 42, 43, 46, 49, 54, 56, 58, 62)$
$\mathbf{w}_{34}(X) = (0, 2, 8, 23, 32, 33, 34, 37, 42, 43, 46, 49, 54, 56, 58, 62)$
$\mathbf{w}_{35}(X) = (14, 23, 24, 25, 28, 33, 34, 37, 40, 45, 47, 49, 53, 54, 56, 62)$
$\mathbf{w}_{36}(X) = (6, 7, 10, 22, 23, 27, 29, 33, 34, 35, 37, 49, 54, 56, 59, 62)$

TABLE 8.5D POLYNOMIALS ORTHOGONAL ON $\{e_2, e_{14}, e_{19}, e_{21}, e_{27}, e_{51}, e_{61}, e_{62}\}$

$\mathbf{w}_{41}(X) = (2, 7, 8, 11, 14, 19, 21, 23, 27, 28, 30, 36, 51, 60, 61, 62)$
$\mathbf{w}_{42}(X) = (0, 2, 14, 19, 21, 24, 27, 34, 35, 38, 50, 51, 55, 57, 61, 62)$
$\mathbf{w}_{43}(X) = (2, 14, 15, 19, 21, 25, 26, 27, 29, 41, 46, 48, 51, 54, 61, 62)$
$\mathbf{w}_{44}(X) = (1, 2, 3, 9, 14, 19, 21, 27, 33, 43, 44, 47, 51, 59, 61, 62)$
$\mathbf{w}_{45}(X) = (2, 6, 14, 16, 17, 19, 20, 21, 27, 32, 37, 39, 45, 51, 61, 62)$
$\mathbf{w}_{46}(X) = (2, 5, 10, 12, 14, 18, 19, 21, 27, 42, 51, 52, 53, 56, 61, 62)$

TABLE 8.5E POLYNOMIALS ORTHOGONAL ON $\{e_0, e_3, e_{15}, e_{20}, e_{22}, e_{28}, e_{52}, e_{62}\}$

$\mathbf{w}_{51}(X) = (0, 3, 6, 11, 13, 15, 19, 20, 22, 28, 43, 52, 53, 54, 57, 62)$
$\mathbf{w}_{52}(X) = (0, 3, 8, 9, 12, 15, 20, 22, 24, 28, 29, 31, 37, 52, 61, 62)$
$\mathbf{w}_{53}(X) = (0, 1, 3, 15, 20, 22, 25, 28, 35, 36, 39, 51, 52, 56, 58, 62)$
$\mathbf{w}_{54}(X) = (0, 3, 15, 16, 20, 22, 26, 27, 28, 30, 42, 47, 49, 52, 55, 62)$
$\mathbf{w}_{55}(X) = (0, 2, 3, 4, 10, 15, 20, 22, 28, 34, 44, 45, 48, 52, 60, 62)$
$\mathbf{w}_{56}(X) = (0, 3, 7, 15, 17, 18, 20, 21, 22, 28, 33, 38, 40, 46, 52, 62)$

TABLE 8.5F POLYNOMIALS ORTHOGONAL ON $\{e_9, e_{10}, e_{13}, e_{25}, e_{30}, e_{32}, e_{38}, e_{62}\}$

$\mathbf{w}_{61}(X) = (3, 5, 9, 10, 11, 13, 25, 30, 32, 35, 38, 45, 46, 49, 61, 62)$
$\mathbf{w}_{62}(X) = (9, 10, 13, 17, 25, 27, 28, 30, 31, 32, 38, 43, 48, 50, 56, 62)$
$\mathbf{w}_{63}(X) = (0, 1, 4, 9, 10, 13, 16, 21, 23, 25, 29, 30, 32, 38, 53, 62)$
$\mathbf{w}_{64}(X) = (8, 9, 10, 13, 18, 19, 22, 25, 30, 32, 34, 38, 39, 41, 47, 62)$
$\mathbf{w}_{65}(X) = (7, 9, 10, 12, 13, 14, 20, 25, 30, 32, 38, 44, 54, 55, 58, 62)$
$\mathbf{w}_{66}(X) = (2, 9, 10, 13, 25, 26, 30, 32, 36, 37, 38, 40, 52, 57, 59, 62)$

vector. The error sums $S(L_1)$, $S(L_2)$, $S(L_3)$, $S(L_4)$, $S(L_5)$, and $S(L_6)$ are orthogonal on e_{62}. Consequently, e_{62} can be determined from these error sums. Thus, the (1, 3)th-order twofold (63, 45) EG code is two-step majority-logic decodable. Since its minimum distance $d_{\min} = 7$ and $J = 6$, it is completely orthogonalizable.

There is no simple formula for enumerating the number of parity-check digits for a general twofold EG code. However, for $\mu = m - 1$, the number of parity-check digits for the $(m - 1, s)$th-order twofold EG code of length $2^{ms} - 1$ is [22]

$$n - k = \binom{m+1}{m}^s - \binom{m}{m-1}^s. \tag{8.14}$$

A list of twofold EG codes is given in Table 8.6. We see that the twofold EG codes are more efficient than their corresponding RM codes and are comparable to their corresponding BCH codes. For example, for error-correcting capability $t = 7$, there

TABLE 8.6 LIST OF TWOFOLD EG CODES*

m	s	μ	n	k	J	t_{ML}
3	2	1	63	24	14	7
2	3	1	63	45	6	3
4	2	1	255	45	62	31
4	2	2	255	171	14	7
2	4	1	255	191	14	7
3	3	1	511	184	62	31
3	3	2	511	475	6	3
5	2	1	1023	76	254	127
5	2	2	1023	438	62	31
5	2	3	1023	868	14	7
2	5	1	1023	813	30	15

*The (63, 24) and (63, 45) codes are BCH codes.

is a twostep majority-logic decodable (255, 191) twofold EG code; the corresponding RM code is a (255, 163) code which is five-step majority-logic decodable (using Chen's decoding algorithm [20], it may be decoded in two steps); the corresponding BCH code is a (255, 199) code.

The concept of twofold EG codes can be generalized to form a large class of multifold EG codes [18, 22]. The class of twofold EG codes is a proper subclass of several large classes of majority-logic decodable codes which are constructed based on Euclidean geometry [16–18, 22].

8.3 PROJECTIVE GEOMETRY AND PROJECTIVE GEOMETRY CODES

Like Euclidean geometry, a projective geometry may be constructed from the elements of a Galois field. Consider the Galois field $GF(2^{(m+1)s})$ which contains $GF(2^s)$ as a subfield. Let α be a primitive element in $GF(2^{(m+1)s})$. Then the powers of α, $\alpha^0, \alpha^1, \ldots, \alpha^{2^{(m+1)s}-2}$, form all the nonzero elements of $GF(2^{(m+1)s})$. Let

$$n = \frac{2^{(m+1)s} - 1}{2^s - 1} = 2^{ms} + 2^{(m-1)s} + \cdots + 2^s + 1. \tag{8.15}$$

Then the order of $\beta = \alpha^n$ is $2^s - 1$. The 2^s elements $0, 1, \beta, \beta^2, \ldots, \beta^{2^s-2}$ form the Galois field $GF(2^s)$.

Consider the first n powers of α,

$$\Gamma = \{\alpha^0, \alpha^1, \alpha^2, \ldots, \alpha^{n-1}\}.$$

No element α^i in Γ can be a product of an element in $GF(2^s)$ and another element α^j in Γ [i.e., $\alpha^i \neq \eta \cdot \alpha^j$ for $\eta \in GF(2^s)$]. Suppose that $\alpha^i = \eta\alpha^j$. Then $\alpha^{i-j} = \eta$. Since $\eta^{2^s-1} = 1$, we obtain $\alpha^{(i-j)(2^s-1)} = 1$. This is impossible since $(i-j)(2^s-1) < 2^{(m+1)s} - 1$ and the order of α is $2^{(m+1)s} - 1$. Therefore, we conclude that, for α^i and α^j in Γ, $\alpha^i \neq \eta\alpha^j$ for any $\eta \in GF(2^s)$. Now, we partition the nonzero elements of $GF(2^{(m+1)s})$ into n disjoint subsets as follows:

$$\begin{gathered}
\{\alpha^0, \beta\alpha^0, \beta^2\alpha^0, \ldots, \alpha^{2^s-2}\alpha^0\}, \\
\{\alpha^1, \beta\alpha^1, \beta^2\alpha^1, \ldots, \beta^{2^s-2}\alpha^1\}, \\
\{\alpha^2, \beta\alpha^2, \beta^2\alpha^2, \ldots, \beta^{2^s-2}\alpha^2\}, \\
\vdots \\
\{\alpha^{n-1}, \beta\alpha^{n-1}, \beta^2\alpha^{n-1}, \ldots, \beta^{2^s-2}\alpha^{n-1}\},
\end{gathered}$$

where $\beta = \alpha^n$, a primitive element in $GF(2^s)$. Each set consists of $2^s - 1$ elements and each element is a multiple of the first element in the set. No element in one set can be a product of an element of $GF(2^s)$ and an element from a different set. Now, *we represent each set by its first element* as follows:

$$(\alpha^i) = \{\alpha^i, \beta\alpha^i, \ldots, \beta^{2^s-2}\alpha^i\},$$

with $0 \leq i < n$. For any α^j in $GF(2^{(m+1)s})$, if $\alpha^j = \beta^l \cdot \alpha^i$ with $0 \leq i < n$, then α^j is represented by (α^i). The n elements

$$(\alpha^0), (\alpha^1), (\alpha^2), \ldots, (\alpha^{n-1})$$

are said to form an *m-dimensional projective geometry over* GF(2^s), denoted by PG(m, 2^s). The elements $(\alpha^0), (\alpha^1), \ldots, (\alpha^{n-1})$ are called the *points* of PG(m, 2^s). Note that the $2^s - 1$ elements in $\{\alpha^i, \beta\alpha^i, \ldots, \beta^{2^s-2}\alpha^i\}$ are considered to be the *same* point. This is a *major difference* between a projective geometry and an Euclidean geometry.

Let (α^i) and (α^j) be any two *distinct* points in PG(m, 2^s). Then the *line* (1-flat) passing through (α^i) and (α^j) consists of points of the following form:

$$(\eta_1\alpha^i + \eta_2\alpha^j), \tag{8.16}$$

where η_1 and η_2 are from GF(2^s) and are not both equal to zero. There are $(2^s)^2 - 1$ possible choices of η_1 and η_2 from GF(2^s) (excluding $\eta_1 = \eta_2 = 0$). However, there are always $2^s - 1$ choices of η_1 and η_2 that result in the same point. For example:

$$\eta_1\alpha^i + \eta_2\alpha^j,\ \beta\eta_1\alpha^i + \beta\eta_2\alpha^j, \ldots, \beta^{2^s-2}\eta_1\alpha^i + \beta^{2^s-2}\eta_2\alpha^j$$

represent the same point in PG($2^{(m+1)s}$). Therefore, a line in PG(m, 2^s) consists of

$$\frac{(2^s)^2 - 1}{2^s - 1} = 2^s + 1$$

points. To generate the $2^s + 1$ distinct points on the line $\{(\eta_1\alpha^i + \eta_2\alpha^j)\}$, we simply choose η_1 and η_2 such that no choice (η_1, η_2) is a multiple of another choice (η_1', η_2') [i.e., $(\eta_1, \eta_2) \neq (\delta\eta_1', \delta\eta_2')$ for any $\delta \in$ GF(2^s)].

Example 8.11

Let $m = 2$ and $s = 2$. Consider the projective geometry PG(2, 2^2). This geometry can be constructed from the field GF(2^6) which contains GF(2^2) as a subfield. Let

$$n = \frac{2^6 - 1}{2^2 - 1} = 2^{2\cdot 2} + 2^2 + 1 = 21.$$

Let α be a primitive of GF(2^6) (use Table 6.2). Let $\beta = \alpha^{21}$. Then 0, 1, β, and β^2 form the field GF(2^2). The geometry PG(2, 2^2) consists of the following 21 points:

$$\begin{array}{lllllll}(\alpha^0), & (\alpha^1), & (\alpha^2), & (\alpha^3), & (\alpha^4), & (\alpha^5), & (\alpha^6), \\ (\alpha^7), & (\alpha^8), & (\alpha^9), & (\alpha^{10}), & (\alpha^{11}), & (\alpha^{12}), & (\alpha^{13}), \\ (\alpha^{14}), & (\alpha^{15}), & (\alpha^{16}), & (\alpha^{17}), & (\alpha^{18}), & (\alpha^{19}), & (\alpha^{20}),\end{array}$$

Consider the line passing through the point (α) and (α^{20}) which consists of five points of the form $(\eta_1\alpha + \eta_2\alpha^{20})$, with η_1 and η_2 from GF(2^2) $= \{0, 1, \beta, \beta^2\}$. The five distinct points are

$$\begin{aligned}&(\alpha),\\ &(\alpha^{20}),\\ &(\alpha + \alpha^{20}) = (\alpha^{57}) = (\beta^2\alpha^{15}) = (\alpha^{15}),\\ &(\alpha + \beta\alpha^{20}) = (\alpha + \alpha^{41}) = (\alpha^{56}) = (\beta^2\alpha^{14}) = (\alpha^{14}),\\ &(\alpha + \beta^2\alpha^{20}) = (\alpha + \alpha^{62}) = (\alpha^{11}).\end{aligned}$$

Thus, $\{(\alpha), (\alpha^{11}), (\alpha^{14}), (\alpha^{15}), (\alpha^{20})\}$ is the line in PG(2, 2^s) that passes through the points (α) and (α^{20}).

Let (α^l) be a point not on the line $\{(\eta_1\alpha^i + \eta_2\alpha^j)\}$. Then the line $\{(\eta_1\alpha^i + \eta_2\alpha^j)\}$ and the line $\{(\eta_1\alpha^l + \eta_2\alpha^j)\}$ have (α^j) as a common point (the only common point).

We say that they intersect at (α^j). The number of lines in PG$(m, 2^s)$ that intersect at a given point is

$$\frac{2^{ms} - 1}{2^s - 1} = 1 + 2^s + \cdots + 2^{(m-1)s}. \tag{8.17}$$

Let $(\alpha^{l_1}), (\alpha^{l_2}), \ldots, (\alpha^{l_{\mu+1}})$ be $\mu + 1$ linearly independent points (i.e., $\eta_1\alpha^{l_1} + \eta_2\alpha^{l_2} + \cdots + \eta_{\mu+1}\alpha^{l_{\mu+1}} = 0$ if and only if $\eta_1 = \eta_2 = \cdots = \eta_{\mu+1} = 0$). Then, a μ-flat in PG$(m, 2^s)$ consists of points of the form

$$(\eta_1\alpha^{l_1} + \eta_2\alpha^{l_2} + \cdots + \eta_{\mu+1}\alpha^{l_{\mu+1}}), \tag{8.18}$$

where $\eta_i \in$ GF(2^s) and not all $\eta_1, \eta_2, \ldots, \eta_{\mu+1}$ are zero. There are $2^{(\mu+1)s} - 1$ choices for $\eta_1, \eta_2, \ldots, \eta_{\mu+1}$ ($\eta_1 = \eta_2 = \cdots = \eta_{\mu+1} = 0$ is not allowed). Since there are always $2^s - 1$ choices of η_1 to $\eta_{\mu+1}$ resulting in the same point in PG$(m, 2^s)$, there are

$$\frac{2^{(\mu+1)s} - 1}{2^s - 1} = 1 + 2^s + \cdots + 2^{\mu s} \tag{8.19}$$

points in a μ-flat in PG$(m, 2^s)$. Let $\alpha^{l'_{\mu+1}}$ be a point not in the μ-flat

$$\{(\eta_1\alpha^{l_1} + \eta_1\alpha^{l_2} + \cdots + \eta_{\mu+1}\alpha^{l_{\mu+1}})\}.$$

Then the μ-flat $\{(\eta_1\alpha^{l_1} + \eta_2\alpha^{l_2} + \cdots + \eta_\mu\alpha^{l_\mu} + \eta_{\mu+1}\alpha^{l_{\mu+1}})\}$ and the μ-flat $\{(\eta_1\alpha^{l_1} + \eta_2\alpha^{l_2} + \cdots + \eta_\mu\alpha^{l_\mu} + \eta_{\mu+1}\alpha^{l'_{\mu+1}})\}$ intersect on the $(\mu - 1)$-flat $\{(\eta_1\alpha^{l_1} + \eta_2\alpha^{l_2} + \cdots + \eta_\mu\alpha^{l_\mu})\}$. The number of μ-flats in PG$(m, 2^s)$ that intersect on a given $(\mu - 1)$-flat in PG$(m, 2^s)$ is

$$\frac{2^{(m-\mu+1)s} - 1}{2^s - 1} = 1 + 2^s + \cdots + 2^{(m-\mu)s}. \tag{8.20}$$

Every point outside a $(\mu - 1)$-flat F is in one and only one of the μ-flats intersecting on F.

Let $\mathbf{v} = (v_0, v_1, \ldots, v_{n-1})$ be a n-tuple over GF(2), where

$$n = \frac{2^{(m+1)s} - 1}{2^s - 1} = 1 + 2^s + \cdots + 2^{ms}.$$

Let α be a primitive element in GF$(2^{(m+1)s})$. We may number the components of $\mathbf{v}$ with the first n powers of α as follows: v_i is numbered α^i for $0 \leq i < n$. As usual, α^i is called the location number of v_i. Let F be a μ-flat in PG$(m, 2^s)$. The incidence vector for F is an n-tuple over GF(2),

$$\mathbf{v}_F = (v_0, v_1, \ldots, v_{n-1}),$$

whose ith component

$$v_i = \begin{cases} 1 & \text{if } (\alpha^i) \text{ is a point in } F \\ 0 & \text{otherwise.} \end{cases}$$

Definition 8.3. A (μ, s)th-order binary projective geometry (PG) code of length $n = (2^{(m+1)s} - 1)/(2^s - 1)$ is defined as the largest cyclic code whose null space contains the incidence vectors of all the μ-flats in PG$(m, 2^s)$.

Let h be a nonnegative integer less than $2^{(m+1)s} - 1$ and $h^{(l)}$ be the remainder resulting from dividing $2^l h$ by $2^{(m+1)s} - 1$. Clearly, $h^{(0)} = h$. The 2^s-weight of h, $W_{2^s}(h)$, is defined by (8.3). The following theorem characterizes the roots of the generator polynomial of a (μ, s)th-order PG code (the proof is omitted).

Theorem 8.3. Let α be a primitive element $GF(\alpha^{(m+1)s})$. Let h be a nonnegative integer less than $2^{(m+1)s} - 1$. Then the generator polynomial $\mathbf{g}(X)$ of a (μ, s)th-order PG code of length $n = (2^{(m+1)s} - 1)/(2^s - 1)$ has α^h as a root if and only if h is divisible by $2^s - 1$ and

$$0 \le \max_{0 \le l < s} W_{2^s}(h^{(l)}) = j(2^s - 1) \tag{8.21}$$

with $0 \le j \le m - \mu$.

Example 8.12

Let $m = 2$, $s = 2$, and $\mu = 1$. Consider the (1, 2)th-order PG code of length

$$n = \frac{2^{(2+1)\cdot 2} - 1}{2^2 - 1} = 21.$$

Let α be a primitive element of $GF(2^6)$. Let h be a nonnegative integer less than 63. It follows form Theorem 8.3 that the generator polynomial $\mathbf{g}(X)$ of the (1, 2)th-order PG code of length 21 has α^h as a root if and only if h is divisible by 3 and

$$0 \le \max_{0 \le l < 2} W_{2^2}(h^{(l)}) = 3j$$

with $0 \le j \le 1$. The integers that are divisible by 3 and satisfy the condition above are 0, 3, 6, 9, 12, 18, 24, 33, 36, and 48. Thus, $\mathbf{g}(X)$ has $\alpha^0 = 1, \alpha^3, \alpha^6, \alpha^9, \alpha^{12}, \alpha^{18}, \alpha^{24}, \alpha^{33}, \alpha^{36}$, and α^{48} as roots. From Table 6.3 we find that (1) $\alpha^3, \alpha^6, \alpha^{12}, \alpha^{24}, \alpha^{33}$, and α^{48} have the same minimal polynomial $\phi_3(X) = 1 + X + X^2 + X^4 + X^6$; and (2) α^9, α^{18}, and α^{36} have $\phi_9(X) = 1 + X^2 + X^3$ as their minimal polynomial. Thus,

$$\begin{aligned}\mathbf{g}(X) &= (1 + X)\phi_3(X)\phi_9(X) \\ &= 1 + X^2 + X^4 + X^6 + X^7 + X^{10}.\end{aligned}$$

Hence, the (1, 2)th-order PG code of length 21 is a (21, 11) cyclic code. It is interesting to note that this code is the (21, 11) difference-set code considered in Example 7.9.

Decoding PG codes is similar to decoding EG codes. Consider the decoding of a (μ, s)th-order PG code of length $n = (2^{(m+1)s} - 1)/(2^s - 1)$. The null space of this code contains the incidence vectors of all the μ-flats of $PG(m, 2^s)$. Let $F^{(\mu-1)}$ be a $(\mu - 1)$-flat in $PG(m, 2^s)$ that contains the point (α^{n-1}). From (8.20) we see that there are

$$J = \frac{2^{(m-\mu+1)s} - 1}{2^s - 1}$$

μ-flats intersecting on the $(\mu - 1)$-flat $F^{(\mu-1)}$. The incidence vectors of these J μ-flats are orthogonal on the digits at the locations corresponding to the points in $F^{(\mu-1)}$. Therefore, the parity-check sums formed from these J incidence vectors are orthogonal on the error digits at the locations corresponding to the points in $F^{(\mu-1)}$. Let $S(F^{(\mu-1)})$ denote the sum of error digits at the locations corresponding to the points in $F^{(\mu-1)}$.

Then this error sum, $S(F^{(\mu-1)})$, can be determined correctly from the J check sums orthogonal on it provided that there are no more than

$$\left\lfloor \frac{J}{2} \right\rfloor = \left\lfloor \frac{2^{(m-\mu+1)s} - 1}{2(2^s - 1)} \right\rfloor$$

errors in the received vector. In this manner we can determine the error sums, $S(F^{(\mu-1)})$'s, corresponding to all the $(\mu - 1)$-flats that contain the point (α^{n-1}). These error sums are then used to determine the error sums, $S(F^{(\mu-2)})$'s, corresponding to all the $(\mu - 2)$-flats that contain (α^{n-1}). This process continues until the error sums, $S(F^{(1)})$'s, corresponding to all the 1-flats that intersect on (α^{n-1}) are formed. These error sums, $S(F^{(1)})$'s, are orthogonal on the error digit e_{n-1} at the location α^{n-1}. Thus, the value of e_{n-1} can be determined. A total of μ steps of orthogonalization are required to decode e_{n-1}. Since the code is cyclic, other error digits can be decoded in the same manner. Thus, the code is μ-step decodable. At the rth step of orthogonalization with $1 \leq r \leq \mu$, the number of error sums, $S(F^{(\mu-r+1)})$'s, that are orthogonal in the error sum corresponding to a given $(\mu - r)$-flat $F^{(\mu-r)}$ is

$$J_{\mu-r+1} = \frac{2^{(m-\mu+r)s} - 1}{2^s - 1} \geq J.$$

Therefore, at each step of orthogonalization, the error sums needed for the next step can always be determined correctly provided that there are no more than $\lfloor J/2 \rfloor$ errors in the received vector. Thus, the μth-order PG code of length $n = (2^{(m+1)s} - 1)/(2^s - 1)$ is capable of correcting

$$t_{\text{ML}} = \left\lfloor \frac{J}{2} \right\rfloor = \left\lfloor \frac{2^{(m-\mu+1)s} - 1}{2(2^s - 1)} \right\rfloor \tag{8.22}$$

or fewer errors with majority-logic decoding.

Example 8.13

Consider the decoding of the (1, 2)th-order (21, 11) PG code with $m = 2$ and $s = 2$. The null space of this code contains the incidence vectors of all the 1-flats (lines) in PG(2, 2^2). Let α be a primitive element in GF(2^6). The geometry PG(2, 2^2) consists of 21 points, (α^0) to (α^{20}), as given in Example 8.11. Let $\beta = \alpha^{21}$. Then, 0, 1, β, and β^2 form the field GF(2^2).

There are $2^2 + 1 = 5$ lines passing through the point (α^{20}), which are

$$\{(\eta_1\alpha^0 + \eta_2\alpha^{20})\} = \{(\alpha^0), (\alpha^5), (\alpha^7), (\alpha^{17}), (\alpha^{20})\},$$
$$\{(\eta_1\alpha^1 + \eta_2\alpha^{20})\} = \{(\alpha^1), (\alpha^{11}), (\alpha^{14}), (\alpha^{15}), (\alpha^{20})\},$$
$$\{(\eta_1\alpha^2 + \eta_2\alpha^{20})\} = \{(\alpha^2), (\alpha^3), (\alpha^8), (\alpha^{10}), (\alpha^{20})\},$$
$$\{(\eta_1\alpha^4 + \eta_2\alpha^{20})\} = \{(\alpha^4), (\alpha^6), (\alpha^{16}), (\alpha^{19}), (\alpha^{20})\},$$
$$\{(\eta_1\alpha^9 + \eta_2\alpha^{20})\} = \{(\alpha^9), (\alpha^{12}), (\alpha^{13}), (\alpha^{18}), (\alpha^{20})\}.$$

The incidence vectors of these lines (in polynomial form) are

$$\mathbf{w}_1(X) = 1 + X^5 + X^7 + X^{17} + X^{20},$$
$$\mathbf{w}_2(X) = X + X^{11} + X^{14} + X^{15} + X^{20},$$
$$\mathbf{w}_3(X) = X^2 + X^3 + X^8 + X^{10} + X^{20},$$
$$\mathbf{w}_4(X) = X^4 + X^6 + X^{16} + X^{19} + X^{20},$$
$$\mathbf{w}_5(X) = X^9 + X^{12} + X^{13} + X^{18} + X^{20}.$$

These vectors are orthogonal on digit position 20. They are exactly the orthogonal vectors for the (21, 11) difference-set code given in Example 7.10.

For $\mu = 1$, we obtain a class of one-step majority-logic decodable PG codes. For $m = 2$, a $(1, s)$th-order PG code becomes a difference-set code. Thus, the difference-set codes form a subclass of the class of $(1, s)$th-order PG codes. For $s = 1$, a (1, 1)th-order PG code becomes a maximum-length code.

There is no simple formula for enumerating the number of parity-check digits for general (μ, s)th-order PG code. Rather complicated combinatorial expressions for the number of parity-check digits of PG codes can be found in References 4 and 5. However for $\mu = m - 1$, the number of parity-check digits for the $(m - 1, s)$th-order of PG code of length $n = (2^{(m+1)s} - 1)/(2^s - 1)$ is

$$n - k = 1 + \binom{m+1}{m}^s. \tag{8.23}$$

This expression was obtained independently by Goethals and Delsarte [23], Smith [6], and MacWilliams and Mann [7]. A list of PG codes is given in Table 8.7.

PG codes were first studied by Rudolph [3] and were later extended and generalized by many others [6, 13–16, 23, 24]. Extensive treatment for EG and PG codes can be found in References 25 to 28.

TABLE 8.7 LIST OF PG CODES

m	s	μ	n	k	J	t_{ML}
2	2	1	21	11	5	2
2	3	1	73	45	9	4
3	2	2	85	68	5	2
3	2	1	85	24	21	10
2	4	1	273	191	17	8
4	2	3	341	315	5	2
4	2	2	341	195	21	10
4	2	1	341	45	85	42
3	3	2	585	520	9	4
3	3	1	585	184	73	36
2	5	1	1057	813	33	16
5	2	4	1365	1328	5	2
5	2	3	1365	1063	21	10
5	2	2	1365	483	85	21
5	2	1	1365	76	341	170
6	2	5	5461	5411	5	2
6	2	4	5461	4900	21	10
6	2	3	5461	3185	85	42
6	2	2	5461	1064	341	170
6	2	1	5461	119	1365	682

8.4 MODIFICATIONS OF MAJORITY-LOGIC DECODING

For large J and μ, the conventional rule of orthogonalization for decoding long finite geometry codes described in Sections 8.2 and 8.3 requires a very large number of majority-logic gates, and thus the decoding complexity becomes prohibitive.

However, if we are willing to sacrifice the decoding speed, the decoding complexity may be reduced. This trade-off between the decoding complexity and the decoding speed is best illustrated by an example.

Consider the (1, 1)th-order (15, 5) EG code. This code is two-step majority-logic decodable and is capable of correcting three or fewer errors (see Examples 7.12 and 8.8). If the conventional rule of orthogonalization for decoding EG codes is used, the decoder would consist of seven six-input majority-logic gates; six gates are used at the first level to determine the check sums corresponding to six 1-flats in EG(4, 2) that pass through the point α^{14} and one gate is used at the second level to determine the value of the error digit e_{14}, as shown in Figure 7.9. We see that the outputs of the *rightmost* first-level majority-logic gate at the *successive times* are

$$S_1 = e_{13} + e_{14}, \quad S_2 = e_{12} + e_{13}, \quad S_3 = e_{11} + e_{12},$$
$$S_4 = e_{10} + e_{11}, \quad S_5 = e_9 + e_{10}, \quad S_6 = e_8 + e_9, \ldots .$$

If we add these sums successively as follows:

$$\begin{aligned}
S_1 &= e_{13} + e_{14}, \\
S_1 + S_2 &= e_{12} + e_{14}, \\
S_1 + S_2 + S_3 &= e_{11} + e_{14}, \\
S_1 + S_2 + S_3 + S_4 &= e_{10} + e_{14}, \\
S_1 + S_2 + S_3 + S_4 + S_5 &= e_9 + e_{14}, \\
S_1 + S_2 + S_3 + S_4 + S_5 + S_6 &= e_8 + e_{14}, \\
&\vdots
\end{aligned}$$

we obtain check sums orthogonal on e_{14}. Hence, e_{14} can be determined from these orthogonal check sums. Forming check sums orthogonal on e_{14} in this manner, only one majority-logic gate is needed at the first level of decoding. Since $S_1, S_2, \ldots, S_6, \ldots$ are formed at successive times, a buffer register is needed to store them. Since $J = 6$, to determine the value of e_{14} we only need to form the check sums $S_1, S_1 + S_2, \ldots, S_1 + S_2 + S_3 + S_4 + S_5 + S_6$. As a result, we obtain a decoding circuit for the (15, 5) EG codes as shown in Figure 8.1. Since five clock times are needed to form the check sums S_1 to S_6, the received vector must be delayed five units of time before it can be decoded. This delay is achieved by a buffer register of five stages.

The implementation described above for decoding the (1, 1)th-order (15, 5) EG code requires only two majority-logic gates, one at each level of orthogonalization. Therefore, a reduction of five majority-logic gates is obtained. This reduction is achieved at the cost of an increase of five units of time for decoding the received vector and 10 extra memory elements for buffering. The memory elements are much cheaper than the majority-logic gates. As a result, a reduction in decoding complexity is obtained.

Rudolph and Hartmann [29] have shown that finite geometry codes of length up to several thousand digits can be decoded in a manner similar to the example given above. For a L-step decodable geometry (EG or PG) code, only one J-input majority-

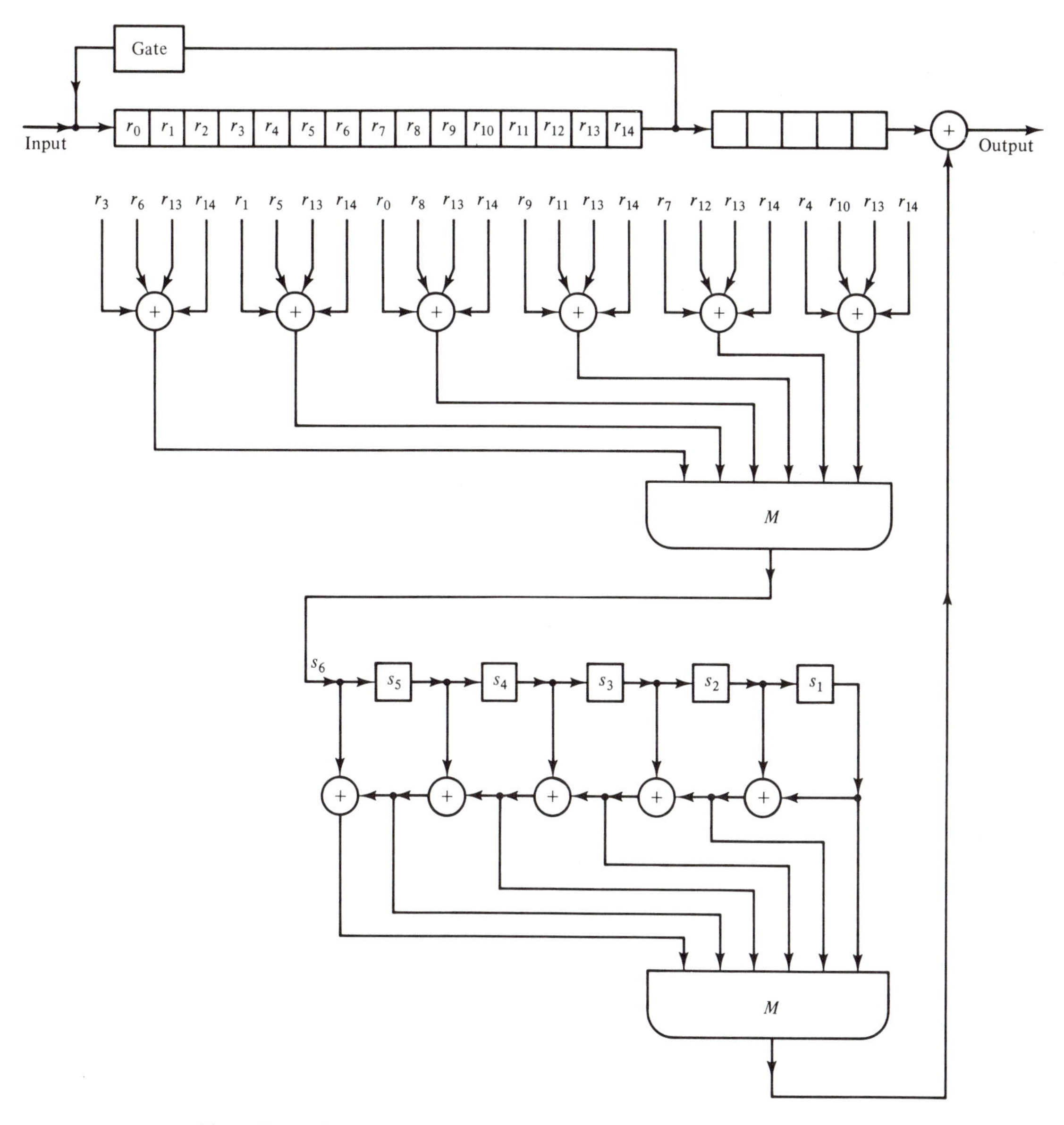

Figure 8.1 SCR majority-logic decoder for the (1, 1)th order (15, 5) EG code.

logic gate is needed at each level of orthogonalization. The successive check sums formed at the output of the ith-level majority-logic gate are first stored in a buffer register and are then *combined* to form J check sums that are orthogonal on a properly selected set of error digits. These J orthogonal check sums are inputs to the $(i + 1)$th-level majority-logic gate. The successive check sums formed at the output of this $(i + 1)$th-level gate are again stored and combined to form J check sums orthogonal on another properly selected set of error digits. These J new orthogonal check sums form the inputs to the $(i + 2)$th-level majority-logic gate. This process continues until the last level of orthogonalization is reached; the output of the last-level majority-logic gate is the estimated value of a specific error digit. The majority-logic decoding algorithm described above is called *sequential-code-reduction* (SCR) decoding. Using this decoding technique, only L majority-logic gates are required for a L-step SCR majority-logic decodable code, one majority-logic gate at each level of orthogonalization. A buffer register of no more than n stages is needed between every two consecutive levels of orthogonalization.

Using SCR technique to decode finite geometry codes with large L and J greatly reduces the number of majority-logic gates, from $1 + J + J^2 + \cdots + J^{L-1}$ to L. Furthermore, the number of modulo-2 adders needed to form check sums are also reduced significantly. There is an increase of buffer storage, from one buffer register of n stages (for storing the received vector) to L buffer registers. The major disadvantage of SCR majority-logic decoding is the increase in decoding time. The conventional majority-logic decoding requires n clock times to decode the received vector; however, the SCR majority-logic decoding requires nL clock times to decode the received vector. In communication systems where decoding speed is critical, the SCR decoding algorithm may not be suitable.

Consider the decoding of the highest-order error digit e_{n-1} with SCR technique. Let

$$S_0 = e_{j_1} + e_{j_2} + \cdots + e_{j_m} + e_{n-1}$$

be the first check sum formed at the output of the $(i-1)$th-level majority-logic gate where $0 \leq j_1 < j_2 < \cdots < j_m < n-1$. Let

$$\mathbf{w}(X) = X^{j_1} + X^{j_2} + \cdots + X^{j_m} + X^{n-1}$$

be the orthogonal polynomial associated to S_0. At the successive clock times, the outputs of the $(i-1)$th-level majority-logic gate are the following check sums:

$$\begin{aligned}
S_1 &= e_{j_1-1} + e_{j_2-1} + \cdots + e_{j_m-1} + e_{n-2}, \\
S_2 &= e_{j_1-2} + e_{j_2-2} + \cdots + e_{j_m-2} + e_{n-3}, \\
&\vdots \\
S_l &= e_{j_1-l} + e_{j_2-l} + \cdots + e_{j_m-l} + e_{n-1-l} \\
&\vdots
\end{aligned}$$

with $1 \leq l < n$. Note that the indices of the error digits are cyclically shifted to the left one place every clock time (if $j_t - l$ is negative, it is replaced by $n + j_t - l$). The polynomial associated to the check sum S_l is

$$\mathbf{w}^{(-l)}(X) = X^{j_1-l} + X^{j_2-l} + \cdots + X^{j_m-l} + X^{n-1-l},$$

which is obtained by cyclically shifting $\mathbf{w}(X)$ to the left l times (or to the right $n-l$ times). In fact, $\mathbf{w}^{(-l)}(X)$ is the remainder resulting from dividing $X^{n-l}\mathbf{w}(X)$ by $X^n + 1$ (see Chapter 4). Therefore, we have the following relationship:

$$X^{n-l}\mathbf{w}(X) = q_l(X)(X^n + 1) + \mathbf{w}^{(-l)}(X). \tag{8.24}$$

Let

$$Q = e_{t_1} + e_{t_2} + \cdots + e_{t_p} + e_{n-1}$$

be the first check sum formed at the output of the ith-level majority-logic gate. Using SCR majority-logic decoding, the check sums orthogonal on Q are linear combinations of $S_0, S_1, S_2, \ldots$. Let ψ_l be a check sum orthogonal on Q. Then

$$\psi_l = b_0 S_0 + b_{n-1} S_1 + b_{n-2} S_2 + \cdots + b_1 S_{n-1}. \tag{8.25}$$

Let $\mathbf{f}_l(X)$ be the orthogonal polynomial associated with the error digits in ψ_l. Then, from (8.25), we have

$$\mathbf{f}_l(X) = b_0 \mathbf{w}(X) + b_{n-1}\mathbf{w}^{(-1)}(X) + b_{n-2}\mathbf{w}^{(-2)}(X) + \cdots + b_1 \mathbf{w}^{(-n+1)}(X). \tag{8.26}$$

Combining (8.24) and (8.26), we obtain

$$\mathbf{f}_l(X) = (b_0 + b_1 X + \cdots + b_{n-1}X^{n-1})\mathbf{w}(X) + \mathbf{a}_l(X)(X^n + 1). \tag{8.27}$$

From (8.27) we see that each orthogonal polynomial $\mathbf{f}_l(X)$ at the ith level of orthogonalization is a multiple of $\mathbf{w}(X)$ plus a term $\mathbf{a}_l(X)(X^n + 1)$. The polynomial $\mathbf{w}(X)$ is called the *generating orthogonal polynomial.* At each level of SCR majority-logic decoding, if we know $\mathbf{w}(X)$ and $\mathbf{f}_l(X)$ and if we are able to express $\mathbf{f}_l(X)$ in the form of (8.27), we can form check sums orthogonal on Q from (8.25). Therefore, a cyclic code is SCR majority-logic decodable if and only if there exists a generating orthogonal polynomial at each level of orthogonalization.

Example 8.14

Let $m = 5$, $s = 1$, and $\mu = 2$. Consider the second-order RM code of length $n = 2^5 - 1 = 31$. Let α be a primitive element of GF(2^5) (see Table 8.8). Let h be a nonnegative integer less than 31. It follows from (8.7) that the generator polynomial $\mathbf{g}(X)$ of this RM code has α^h as a root if and only if

$$0 < W_2(h) \leq 2.$$

The nonnegative integers less than 31 that satisfy the condition above are 1, 2, 3, 4, 5, 6, 8, 9, 10, 12, 16, 17, 18, 20, and 24. Therefore, $\mathbf{g}(X)$ has $\alpha^1, \alpha^2, \alpha^3, \alpha^4, \alpha^5, \alpha^6, \alpha^8, \alpha^9, \alpha^{10}, \alpha^{12}, \alpha^{16}, \alpha^{17}, \alpha^{18}, \alpha^{20}$, and α^{24} as roots. The minimal polynomial of $\alpha, \alpha^2, \alpha^4, \alpha^8$, and α^{16} is $\phi_1(X) = 1 + X^2 + X^5$. The minimal polynomial of $\alpha^3, \alpha^6, \alpha^{12}, \alpha^{17}$, and α^{24} is $\phi_3(X) = 1 + X^2 + X^3 + X^4 + X^5$. The minimal polynomial of $\alpha^5, \alpha^9, \alpha^{10}, \alpha^{18}$, and α^{20} is $\phi_5(X) = 1 + X + X^2 + X^4 + X^5$. Thus,

$$\begin{aligned}\mathbf{g}(X) &= \phi_1(X)\phi_3(X)\phi_5(X)\\ &= 1 + X + X^2 + X^3 + X^5 + X^7 + X^8 + X^9 + X^{10} + X^{11} + X^{15}.\end{aligned}$$

Hence, the code is a (31, 16) cyclic code. Since

$$J = 2^{(5-2)} - 1 - 1 = 6,$$

it is capable of correcting three or fewer errors with majority-logic decoding.

The null space of this code contains incidence vectors of all the 3-flats in EG(5, 2) that do not pass the origin. These incidence vectors can be used to form parity-check sums for decoding. If the conventional rule of orthogonalization is used for decoding this code, three levels of orthogonalization and a total of 43 six-input majority-logic gates are needed. However, if we examine the geometry EG(5, 2) carefully, this code can be decoded in two steps using SCR majority-logic decoding.

TABLE 8.8 GALOIS FIELD GF(2^5) WITH $p(X) = 1 + \alpha^2 + \alpha^5 = 0$

Power representation	Vector
0	(0 0 0 0 0)
1	(1 0 0 0 0)
α	(0 1 0 0 0)
α^2	(0 0 1 0 0)
α^3	(0 0 0 1 0)
α^4	(0 0 0 0 1)
$\alpha^5 = 1 + \alpha^2$	(1 0 1 0 0)
$\alpha^6 = \alpha + \alpha^3$	(0 1 0 1 0)
$\alpha^7 = \alpha^2 + \alpha^4$	(0 0 1 0 1)
$\alpha^8 = 1 + \alpha^2 + \alpha^3$	(1 0 1 1 0)
$\alpha^9 = \alpha + \alpha^3 + \alpha^4$	(0 1 0 1 1)
$\alpha^{10} = 1 + \alpha^4$	(1 0 0 0 1)
$\alpha^{11} = 1 + \alpha + \alpha^2$	(1 1 1 0 0)
$\alpha^{12} = \alpha + \alpha^2 + \alpha^3$	(0 1 1 1 0)
$\alpha^{13} = \alpha^2 + \alpha^3 + \alpha^4$	(0 0 1 1 1)
$\alpha^{14} = 1 + \alpha^2 + \alpha^3 + \alpha^4$	(1 0 1 1 1)
$\alpha^{15} = 1 + \alpha + \alpha^2 + \alpha^3 + \alpha^4$	(1 1 1 1 1)
$\alpha^{16} = 1 + \alpha + \alpha^3 + \alpha^4$	(1 1 0 1 1)
$\alpha^{17} = 1 + \alpha + \alpha^4$	(1 1 0 0 1)
$\alpha^{18} = 1 + \alpha$	(1 1 0 0 0)
$\alpha^{19} = \alpha + \alpha^2$	(0 1 1 0 0)
$\alpha^{20} = \alpha^2 + \alpha^3$	(0 0 1 1 0)
$\alpha^{21} = \alpha^3 + \alpha^4$	(0 0 0 1 1)
$\alpha^{22} = 1 + \alpha^2 + \alpha^4$	(1 0 1 0 1)
$\alpha^{23} = 1 + \alpha + \alpha^2 + \alpha^3$	(1 1 1 1 0)
$\alpha^{24} = \alpha + \alpha^2 + \alpha^3 + \alpha^4$	(0 1 1 1 1)
$\alpha^{25} = 1 + \alpha^3 + \alpha^4$	(1 0 0 1 1)
$\alpha^{26} = 1 + \alpha + \alpha^2 + \alpha^4$	(1 1 1 0 1)
$\alpha^{27} = 1 + \alpha + \alpha^3$	(1 1 0 1 0)
$\alpha^{28} = \alpha + \alpha^2 + \alpha^4$	(0 1 1 0 1)
$\alpha^{29} = 1 + \alpha^3$	(1 0 0 1 0)
$\alpha^{30} = \alpha + \alpha^4$	(0 1 0 0 1)

Consider the following six 2-flats

$$\{\alpha^3, \alpha^{11}, \alpha^{14}, \alpha^{30}\}, \quad \{\alpha^7, \alpha^{23}, \alpha^{29}, \alpha^{30}\}, \quad \{\alpha^{15}, \alpha^{16}, \alpha^{28}, \alpha^{30}\},$$

$$\{\alpha^6, \alpha^{17}, \alpha^{27}, \alpha^{30}\}, \quad \{\alpha^{19}, \alpha^{20}, \alpha^{21}, \alpha^{30}\}, \quad \{\alpha^8, \alpha^{10}, \alpha^{12}, \alpha^{30}\}.$$

For each of these 2-flats, there exist six 3-flats intersecting (orthogonal) on it. For example, the six 3-flats intersecting on $\{\alpha^3, \alpha^{11}, \alpha^{14}, \alpha^{30}\}$ are

$$\{\alpha^0, \alpha^3, \alpha^7, \alpha^{11}, \alpha^{12}, \alpha^{14}, \alpha^{16}, \alpha^{30}\},$$

$$\{\alpha^1, \alpha^3, \alpha^8, \alpha^{11}, \alpha^{14}, \alpha^{21}, \alpha^{26}, \alpha^{30}\},$$

$$\{\alpha^2, \alpha^3, \alpha^{10}, \alpha^{11}, \alpha^{14}, \alpha^{24}, \alpha^{27}, \alpha^{30}\},$$

$$\{\alpha^3, \alpha^{11}, \alpha^{13}, \alpha^{14}, \alpha^{17}, \alpha^{19}, \alpha^{29}, \alpha^{30}\},$$

$$\{\alpha^3, \alpha^4, \alpha^5, \alpha^6, \alpha^{11}, \alpha^{14}, \alpha^{15}, \alpha^{30}\},$$

$$\{\alpha^3, \alpha^{11}, \alpha^{14}, \alpha^{18}, \alpha^{20}, \alpha^{25}, \alpha^{28}, \alpha^{30}\}.$$

The error sums corresponding to the 2-flats above are

$$\psi_0 = e_3 + e_{11} + e_{14} + e_{30}, \qquad \psi_1 = e_7 + e_{23} + e_{29} + e_{30},$$

$$\psi_2 = e_{15} + e_{16} + e_{28} + e_{30}, \qquad \psi_3 = e_6 + e_{17} + e_{27} + e_{30},$$
$$\psi_4 = e_{19} + e_{20} + e_{21} + e_{30}, \qquad \psi_5 = e_8 + e_{10} + e_{12} + e_{30}.$$

Each of these error sums can be estimated from six parity-check sums orthogonal on it which are formed based on the incidence vectors of six 3-flats (in the null space of the code). We see that the error sums, ψ_0 to ψ_5, are orthogonal on the error digit e_{30}. Therefore, once ψ_0 to ψ_5 are formed, the error digit e_{30} can be estimated. Consequently, one level of orthogonalization (estimating error sums corresponding to 1-flats from error sums corresponding to 2-flats) is removed, and the code can be decoded in two steps. If we form ψ_0 to ψ_5 in parallel, the decoding circuit requires seven six-input majority-logic gates.

Next we show that the (31, 16) RM code is SCR majority-logic decodable. The incidence polynomials corresponding to the six 2-flats that intersect on α^{30} are

$$\mathbf{f}_0(X) = X^3 + X^{11} + X^{14} + X^{30}, \qquad \mathbf{f}_1(X) = X^7 + X^{23} + X^{29} + X^{30},$$
$$\mathbf{f}_2(X) = X^{15} + X^{16} + X^{28} + X^{30}, \qquad \mathbf{f}_3(X) = X^6 + X^{17} + X^{27} + X^{30},$$
$$\mathbf{f}_4(X) = X^{19} + X^{20} + X^{21} + X^{30}, \qquad \mathbf{f}_5(X) = X^8 + X^{10} + X^{12} + X^{30}.$$

Let $\mathbf{w}(X) = \mathbf{f}_0(X) = X^3 + X^{11} + X^{14} + X^{30}$ be the polynomial associated to the first parity-check sum S_0 formed at the output of the first-level majority-logic gate, (i.e., $S_0 = e_3 + e_{11} + e_{14} + e_{30}$). Then each of the polynomials $\mathbf{f}_i(X)$'s above can be expressed as a multiple of $\mathbf{w}(X)$ plus $\mathbf{a}_i(X)(X^{31} + 1)$ as follows:

$$\begin{aligned}
\mathbf{f}_0(X) &= \mathbf{w}(X),\\
\mathbf{f}_1(X) &= (1 + X^4 + X^{12} + X^{15})\mathbf{w}(X) + \mathbf{a}_1(X)(X^{31} + 1),\\
\mathbf{f}_2(X) &= (1 + X + X^2 + X^4 + X^6 + X^{10} + X^{12} + X^{13} + X^{15} + X^{17} + X^{18}\\
&\quad + X^{24})\mathbf{w}(X) + \mathbf{a}_2(X)(X^{31} + 1),\\
\mathbf{f}_3(X) &= (1 + X^3 + X^4 + X^8 + X^{16} + X^{19})\mathbf{w}(X) + \mathbf{a}_3(X)(X^{31} + 1),\\
\mathbf{f}_4(X) &= (1 + X + X^2 + X^3 + X^4 + X^5 + X^6 + X^9 + X^{10} + X^{12} + X^{16} + X^{19}\\
&\quad + X^{20} + X^{21} + X^{23} + X^{24})\mathbf{w}(X) + \mathbf{a}_4(X)(X^{31} + 1),\\
\mathbf{f}_5(X) &= (1 + X + X^2 + X^3 + X^5 + X^6 + X^7 + X^8 + X^{10} + X^{16} + X^{19} + X^{20}\\
&\quad + X^{21} + X^{24})\mathbf{w}(X) + \mathbf{a}_5(X)(X^{31} + 1).
\end{aligned} \tag{8.28}$$

Therefore, $\mathbf{w}(X) = \mathbf{f}_0(X) = X^3 + X^{11} + X^{14} + X^{30}$ is the generating orthogonal polynomial for the first-level orthogonalization. At the first level, only one majority-logic gate is needed. The first check sum to be formed at the output of this gate is $S_0 = \psi_0 = e_3 + e_{11} + e_{14} + e_{30}$ [corresponding to the generating orthogonal polynomial $\mathbf{w}(X)$]. Then, at the successive clock times, the outputs of this gate will be

$$\begin{aligned}
S_1 &= e_2 + e_{10} + e_{13} + e_{29},\\
S_2 &= e_1 + e_9 + e_{12} + e_{28},\\
S_3 &= e_0 + e_8 + e_{11} + e_{27},\\
S_4 &= e_7 + e_{10} + e_{26} + e_{30},\\
&\ \ \vdots\\
S_{30} &= e_0 + e_4 + e_{12} + e_{15}.
\end{aligned}$$

From these successive sums and (8.26) [using (8.25) and (8.27)], we obtain the following check sums orthogonal on e_{30}:

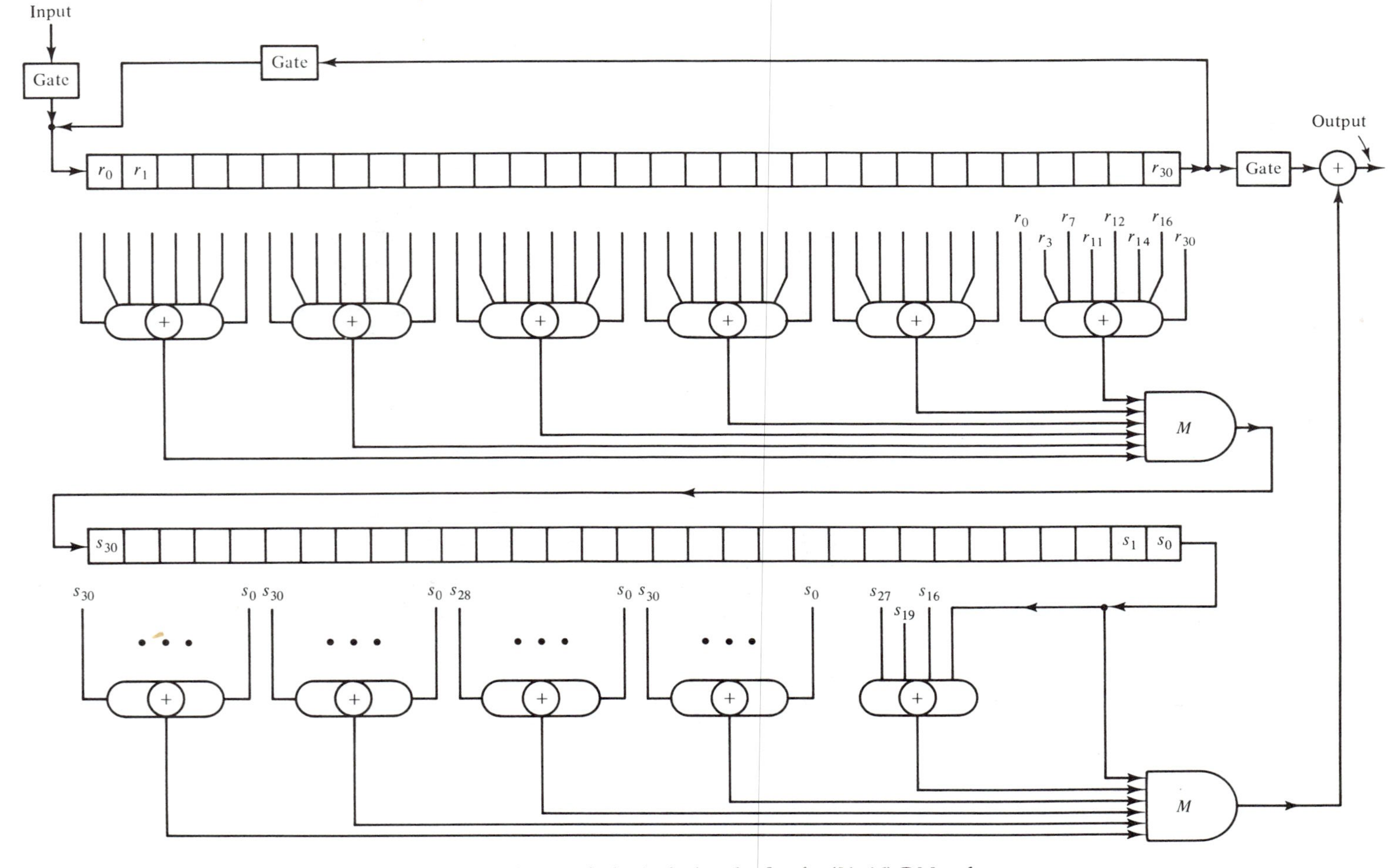

Figure 8.2 SCR majority-logic decoder for the (31, 16) RM code.

$$\psi_0 = S_0,$$

$$\psi_1 = S_0 + S_{16} + S_{19} + S_{27},$$

$$\psi_2 = S_0 + S_7 + S_{13} + S_{14} + S_{16} + S_{18} + S_{19} + S_{21} + S_{25} + S_{27} + S_{29} + S_{30},$$

$$\psi_3 = S_0 + S_{12} + S_{15} + S_{23} + S_{27} + S_{28},$$

$$\psi_4 = S_0 + S_7 + S_8 + S_{10} + S_{11} + S_{12} + S_{15} + S_{19} + S_{21} + S_{22} + S_{25} + S_{26} + S_{27} + S_{28} + S_{29} + S_{30},$$

$$\psi_5 = S_0 + S_7 + S_{10} + S_{11} + S_{12} + S_{15} + S_{21} + S_{23} + S_{24} + S_{25} + S_{26} + S_{28} + S_{29} + S_{30}.$$

Forming ψ_0 to ψ_5 in this manner, we obtain the SCR majority-logic decoding circuit as shown in Figure 8.2, where only two majority-logic gates are used.

The difficulty with SCR majority-logic decoding is that it is hard to find the generating orthogonal polynomial at each level of orthogonalization. For more on SCR majority-logic decoding, readers are referred to Reference 29. Besides EG and PG codes, there are many other generalized finite geometry codes which are SCR majority-logic decodable [13].

Majority-logic decoding using orthogonal parity-check sums can be generalized to decoding with nonorthogonal parity-check sums. Suppose that it is possible to form N parity-check sums such that:

1. Each sum contains all the error digits in $E = \{e_{j_1}, e_{j_2}, \ldots, e_{j_m}\}$.
2. Any error digit not in E is checked by at most λ sums.

Then we can readily see that if there are no more than $\lfloor N/2\lambda \rfloor$ errors in the received vector, the error sum

$$S(E) = e_{j_1} + e_{j_2} + \cdots + e_{j_m}$$

can be correctly determined from the parity-check sums above using majority-logic decision algorithm. These sums are called nonorthogonal parity-check sums. Using nonorthogonal parity-check sums, many more cyclic codes can be decoded with majority-logic decoding, such as the duals of primitive polynomial codes [21], generalized finite geometry codes [13, 16], and multifold EG codes [18, 22].

PROBLEMS

8.1. Consider the Galois field GF(2^4) given by Table 2.8. Let $\beta = \alpha^5$. Then $\{0, 1, \beta, \beta^2\}$ form the subfield GF(2^2) of GF(2^4). Regard GF(2^4) as the two-dimensional Euclidean geometry over GF(2^2), EG(2, 2^2). Find all the 1-flats that pass through the point α^7.

8.2. Consider the Galois field GF(2^6) given by Table 6.2. Let $\beta = \alpha^{21}$. Then $\{0, 1, \beta, \beta^2\}$ form the subfield GF(2^2) of GF(2^6). Regard GF(2^6) as the three-dimensional Euclidean geometry, EG(3, 2^2).

(a) Find all the 1-flats that pass through the point α^{63}.
(b) Find all the 2-flats that intersect on the 1-flat, $\{\alpha^{63} + \eta\alpha\}$, where $\eta \in GF(2^2)$.

8.3. Regard $GF(2^6)$ as the two-dimensional Euclidean geometry $EG(2, 2^3)$. Let $\beta = \alpha^9$. Then $\{0, 1, \beta, \beta^2, \beta^3, \beta^4, \beta^5, \beta^6\}$ form the subfield $GF(2^3)$ of $GF(2^6)$. Determine all the 1-flats that pass through the point α^{21}.

8.4. Consider the two-dimensional projective $PG(2, 2^2)$ constructed from $GF(2^{3\times 2})$. Construct all the 1-flats that pass through the point (α^7).

8.5. Let $m = 2$ and $s = 3$.
(a) Determine the 2^3-weight of 47.
(b) Determine $\max_{0 \leq l < 3} W_{2^3}(47^{(l)})$.
(c) Determine all the positive integers h less than 63 such that

$$0 < \max_{0 \leq l < 3} W_{2^3}(h^{(l)}) \leq 2^3 - 1.$$

8.6. Find the generator polynomial of the first-order RM code of length $2^5 - 1$. Describe how to decode this code.

8.7. Find the generator polynomial of the third-order RM code of length $2^6 - 1$. Describe how to decode this code.

8.8. Let $m = 2$ and $s = 3$. Find the generator polynomial of the (0, 3)th-order EG code of length $2^{2\times 3} - 1$. This code is one-step majority-logic decodable. Find all the polynomials orthogonal on the digit location α^{63} where α is a primitive element in $GF(2^{2\times 3})$. Design a type I majority-logic decoder for this code.

8.9. Let $m = 3$ and $s = 2$. Find the generator polynomial of the (1, 2)th-order twofold EG code of length $2^{3\times 2} - 1$. Describe how to decode this code.

8.10. Let $m = 3$ and $s = 2$. Find the generator polynomial of the (1, 2)th-order PG code of length $(2^{4\times 2} - 1)/(2^2 - 1) = 85$. This code is two-step majority-logic decodable with $J = 4$. Find all the orthogonal polynomials at each step of orthogonalization.

8.11. The (7, 4) Hamming code is two-step majority-logic decodable (see Example 7.11). Show that this code is SCR decodable. Construct such a decoder.

8.12. Prove that the $(m - 2)$th-order RM code of length $2^m - 1$ is a Hamming code. (*Hint:* Show that its generator polynomial is a primitive polynomial of degree m.)

8.13. The first-order RM code of length $2^5 - 1$ is two-step majority-logic decodable.
(a) Find its generator polynomial.
(b) Show that it is SCR decodable.

8.14. Prove that the even-weight code vectors of the first-order RM code of length $2^m - 1$ form the maximum-length code of length $2^m - 1$.

8.15. Let $0 < \mu < m - 1$. Prove that the even-weight code vectors of the $(m - \mu - 1)$th-order RM code of length $2^m - 1$ form the dual of the μth-order RM code of length $2^m - 1$. [*Hint:* Let $\mathbf{g}(X)$ be the generator polynomial of the $(m - \mu - 1)$th-order RM code C. Show that the even-weight code vector of C is a cyclic code generated by $(X + 1)\mathbf{g}(X)$. Show that the dual of the μth-order RM code is also generated by $(X + 1)\mathbf{g}(X)$.]

8.16. The μth-order RM code of length $2^m - 1$ has minimum distance $d_{\min} = 2^{m-\mu} - 1$. Prove that this RM code is a subcode of the primitive BCH code of length $2^m - 1$ and designed distance $2^{m-\mu} - 1$. [*Hint:* Let $\mathbf{g}(X)_{\mathrm{RM}}$ be the generator polynomial of the RM code and let $\mathbf{g}(X)_{\mathrm{BCH}}$ be the generator polynomial of the BCH code. Prove that $\mathbf{g}(X)_{\mathrm{BCH}}$ is a factor of $\mathbf{g}(X)_{\mathrm{RM}}$.]

8.17. Show that extended RM codes are invariant under the affine permutations.

REFERENCES

1. R. D. Carmichael, *Introduction to the Theory of Groups of Finite Order*, Dover, New York, 1956.
2. H. B. Mann, *Analysis and Design of Experiments*, Dover, New York, 1949.
3. L. D. Rudolph, "A Class of Majority Logic Decodable Codes," *IEEE Trans. Inf. Theory*, IT-13, pp. 305–307, April 1967.
4. N. Hamada, "On the *p*-rank of the Incidence Matrix of a Balanced or Partially Balanced Incomplete Block Design and Its Applications to Error-Correcting Codes," *Hiroshima Math. J.*, 3, pp. 153–226, 1973.
5. S. Lin, "On the Number of Information Symbols in Polynomial Codes," *IEEE Trans. Inf. Theory*, IT-18, pp. 785–794, November 1972.
6. K. J. C. Smith, "Majority Decodable Codes Derived from Finite Geometries," Institute of Statistics Mimeo Series No. 561, University of North Carolina, Chapel Hill, N.C., 1967.
7. F. J. MacWilliams and H. B. Mann, "On the *p*-rank of the Design Matrix of a Different Set," *Inf. Control*, 12, pp. 474–488, 1968.
8. D. E. Muller, "Applications of Boolean Algebra to Switching Circuit Design and to Error Detection," *IRE Trans.*, EC-2, pp. 6–12, September 1954.
9. I. S. Reed, "A Class of Multiple-Error-Correcting Codes and the Decoding Scheme," *IRE Trans.*, IT-4, pp. 38–49, September 1954.
10. T. Kasami, S. Lin, and W. W. Peterson, "Linear Codes Which Are Invariant under the Affine Group and Some Results on Minimum Weights in BCH Codes," *Electron. Commun. Jap.*, 50(9), pp. 100–106, September 1967.
11. T. Kasami, S. Lin, and W. W. Peterson, "New Generalizations of the Reed-Muller Codes, Part I: Primitive Codes," *IEEE Trans. Inf. Theory*, IT-14, pp. 189–199, March 1968.
12. V. D. Kolesnik and E. T. Mironchikov, "Cyclic Reed-Muller Codes and Their Decoding," *Probl. Inf. Transm.*, No. 4, pp. 15–19, 1968.
13. C. R. P. Hartmann, J. B. Ducey, and L. D. Rudolph, "On the Structure of Generalized Finite Geometry Codes," *IEEE Trans. Inf. Theory*, IT-20(2), pp. 240–252, March 1974.
14. D. K. Chow, "A Geometric Approach to Coding Theory with Applications to Information Retrieval," CSL Report No. R-368, University of Illinois, Urbana, Ill., 1967.
15. E. J. Weldon, Jr. "Euclidean Geometry Cyclic Codes," Proc. Symp. Combinatorial Math., University of North Carolina, Chapel Hill, N.C., April 1967.
16. P. Delsarte, "A Geometric Approach to a Class of Cyclic Codes," *J. Combinatorial Theory*, 6, pp. 340–358, 1969.
17. P. Delsarte, J. M. Goethals, and J. MacWilliams, "On GRM and Related Codes," *Inf. Control*, 16, pp. 403–442, July 1970.
18. S. Lin and K. P. Yiu, "An Improvement to Multifold Euclidean Geometry Codes," *Inf. Control*, 28(3), July 1975.
19. E. J. Weldon, Jr., "Some Results on Majority-Logic Decoding," Chap. 8, in *Error-Correcting Codes*, H. Mann, ed., Wiley, New York, 1968.
20. C. L. Chen, "On Majority-Logic Decoding of Finite Geometry Codes," *IEEE Trans. Inf. Theory*, IT-17(3), pp. 332–336, May 1971.
21. T. Kasami and S. Lin, "On Majority-Logic Decoding for Duals of Primitive Polynomial Codes," *IEEE Trans. Inf. Theory*, IT-17(3), pp. 322–331, May 1971.

22. S. Lin, "Multifold Euclidean Geometry Codes," *IEEE Trans. Inf. Theory*, IT-19(4), pp. 537–548, July 1973.
23. J. M. Goethals and P. Delsarte, "On a Class of Majority-Logic Decodable Codes," *IEEE Trans. Inf. Theory*, IT-14, pp. 182–189, March 1968.
24. E. J. Weldon, Jr., "New Generations of the Reed-Muller Codes, Part II: Non-primitive Codes," *IEEE Trans. Inf. Theory*, IT-14, pp. 199–205, March 1968.
25. E. R. Berlekamp, *Algebraic Coding Theory*, McGraw-Hill, New York, 1968.
26. W. W. Peterson and E. J. Weldon, Jr., *Error-Correcting Codes*, 2nd ed. MIT Press, Cambridge, Mass., 1972.
27. I. F. Blake and R. C. Mullin, *The Mathematical Theory of Coding*, Academic Press, New York, 1975.
28. F. J. MacWilliams and N. J. A. Sloane, *The Theory of Error-Correcting Codes*, North-Holland, Amsterdam, 1977.
29. L. D. Rudolph and C. R. P. Hartmann, "Decoding by Sequential Code Reduction," *IEEE Trans. Inf. Theory*, IT-19(4), pp. 549–555, July 1973.

9

Burst-Error-Correcting Codes

So far, we have been concerned primarily with coding techniques for channels on which transmission errors occur independently in digit positions (i.e., each transmitted digit is affected independently by noise). However, there are communication channels which are affected by disturbances that cause transmission errors to cluster into bursts. For example, on telephone lines, a stroke of lightening or a human-made electrical disturbance frequently affects many adjacent transmitted digits. On magnetic storage systems, magnetic tape defects may last up to several mils and cause clusters of errors. In general, codes for correcting random errors are not efficient for correcting burst errors. Therefore, it is desirable to design codes specifically for correcting burst errors. Codes of this kind are called burst-error-correcting codes.

Cyclic codes are effective not only for burst-error detection as discussed in Chapter 4; they are also very effective for burst-error correction. Many effective cyclic codes for correcting burst errors have been discovered for the past 20 years. Cyclic codes for single-burst-error correction were first studied by Abramson [1,2]. In an effort to generalize Abramson's results, Fire discovered a large class of burst-error-correcting cyclic codes [3]. Fire codes can be decoded with very simple circuitry. Besides the Fire codes, many other effective burst-error-correcting cyclic codes have been constructed both analytically and with the aid of a computer [4–22].

9.1 INTRODUCTION

A burst of length l is defined as a vector whose nonzero components are confined to l consecutive digit positions, the first and last of which are nonzero. For example, the error vector $\mathbf{e} = (0\ 0\ 0\ 0\ 1\ 0\ 1\ 1\ 0\ 1\ 0\ 0\ 0\ 0\ 0)$ is a burst of length 6. A linear code that is capable of correcting all error bursts of length l or less but not all error bursts of length $l + 1$ is called an *l-burst-error-correcting code*, or the code is said to have *burst-error-correcting capability l*.

It is clear that for given code length n and burst-error-correcting capability l, we desire to construct an (n, k) code with as small a redundancy $n - k$ as possible. We establish next certain restrictions on $n - k$ for given l, or restrictions on l for given $n - k$.

Theorem 9.1. A necessary condition for an (n, k) linear code to be able to correct all burst errors of length l or less is that no burst of length $2l$ or less can be a code vector.

Proof. Suppose that there exists a burst $\mathbf{v}$ of length $2l$ or less as a code vector. This code vector $\mathbf{v}$ can be expressed as a vector sum of two bursts $\mathbf{u}$ and $\mathbf{w}$ of length l or less (except the degenerate case, in which $\mathbf{v}$ is a burst of length 1). Then $\mathbf{u}$ and $\mathbf{w}$ must be in the same coset of a standard array for this code. If one of these two vectors is used as a coset leader (correctable error pattern), the other will be an uncorrectable error burst. As a result, this code would not be able to correct all burst errors of length l or less. Therefore, in order to correct all burst errors of length l or less, no burst of length $2l$ or less can be a code vector. Q.E.D.

Theorem 9.2. The number of parity-check digits of an (n, k) linear code that has no burst of length b or less as a code vector is at least b (i.e., $n - k \geq b$).

Proof. Consider the vectors whose nonzero components are confined to the first b digit positions. There are a toal of 2^b of them. No two such vectors can be in the same coset of a standard array for this code; otherwise, their vector sum, which is a burst of length b or less, would be a code vector. Therefore, these 2^b vectors must be in 2^b distinct cosets. There are a total of 2^{n-k} cosets for an (n, k) code. Thus, $n - k$ must be at least equal to b (i.e., $n - k \geq b$). Q.E.D.

It follows from Theorems 9.1 and 9.2 that we obtain a restriction on the number of parity-check digits of an l-burst-error-correcting code.

Theorem 9.3. The number of parity-check digits of an l-burst-error-correcting code must be at least $2l$, that is,

$$n - k \geq 2l. \tag{9.1}$$

For a given n and k, Theorem 9.3 implies that the burst-error-correcting capability of an (n, k) code is at most $\lfloor (n - k)/2 \rfloor$, that is,

$$l \leq \left\lfloor \frac{n - k}{2} \right\rfloor. \tag{9.2}$$

This is an upper bound on the burst-error-correcting capability of an (n, k) code and is called the Reiger bound [5]. Codes that meet the Reiger bound are said to be *optimal*. The ratio

$$z = \frac{2l}{n - k} \tag{9.3}$$

is used as a measure of the *burst-correcting efficiency* of a code. An optimal code has burst-correcting efficiency equal to 1.

It is possible to show that if an (n, k) code is designed to correct all burst errors of length l or less and simultaneously detect all burst errors of length $d \geq l$ or less, the number of parity-check digits of the code must be at least $l + d$ (see Problem 9.1).

9.2 DECODING OF SINGLE-BURST-ERROR-CORRECTING CYCLIC CODES

An l-burst-error-correcting cyclic code can be most easily decoded by the error-trapping technique, with a slight variation. Suppose that a code word $\mathbf{v}(X)$ from an l-burst-error-correcting (n, k) cyclic code is transmitted. Let $\mathbf{r}(X)$ and $\mathbf{e}(X)$ be the received and error vectors, respectively. Let

$$\mathbf{s}(X) = s_0 + s_1 X + \cdots + s_{n-k-1} X^{n-k-1}$$

be the syndrome of $\mathbf{r}(X)$. If the errors in $\mathbf{e}(X)$ are confined to the l high-order parity-check digit positions, $X^{n-k-l}, \ldots X^{n-k-2}, X^{n-k-1}$, of $\mathbf{r}(X)$, then the l high-order syndrome digits, $s_{n-k-l}, \ldots, s_{n-k-2}, s_{n-k-1}$, match the errors of $\mathbf{e}(X)$ and the $n - k - l$ low-order syndrome digits, $s_0, s_1, \ldots, s_{n-k-l-1}$, are zeros. Suppose that the errors in $\mathbf{e}(X)$ are not confined to the positions $X^{n-k-l}, \ldots, X^{n-k-2}, X^{n-k-1}$, of $\mathbf{r}(X)$ but are confined to l consecutive positions of $\mathbf{r}(X)$ (including the end-around case). Then, after a certain number of cyclic shifts of $\mathbf{r}(X)$, say i cyclic shifts, the errors will be shifted to the positions $X^{n-k-l}, \ldots, X^{n-k-2}, X^{n-k-1}$ of $\mathbf{r}^{(i)}(X)$, the ith shift of $\mathbf{r}(X)$. Let $\mathbf{s}^{(i)}(X)$ be the syndrome of $\mathbf{r}^{(i)}(X)$. Then the first l high-order digits of $\mathbf{s}^{(i)}(X)$ match the errors at the positions $X^{n-k-l}, \ldots, X^{n-k-2}, X^{n-k-1}$ of $\mathbf{r}^{(i)}(X)$, and the $n - k - l$ low-order digits of $\mathbf{s}^{(i)}(X)$ are zeros. Using these facts, we may trap the errors in the syndrome register by cyclic shifting $\mathbf{r}(X)$.

An error-trapping decoder for an l-burst-correcting cyclic code is shown in Figure 9.1, where the received vector is shifted into the syndrome register from the left end. The decoding procedure is described in the following steps:

Step 1. The received vector $\mathbf{r}(X)$ is shifted into the syndrome and buffer registers simultaneously. (If we do not want to decode the received parity-check digits, the buffer register needs only k stages.) As soon as $\mathbf{r}(X)$ has been shifted into the syndrome register, the syndrome $\mathbf{s}(X)$ is formed.

Step 2. The syndrome register starts to shift with gate 2 on. As soon as its $n - k - l$ leftmost stages contain only zeros, its l rightmost stages contain the burst-error pattern. The error correction begins. There are three cases to be considered.

Step 3. If the $n - k - l$ leftmost stages of the syndrome register contain all zeros after the ith shift for $0 \leq i \leq n - k - l$, the errors of the burst $\mathbf{e}(X)$ are confined to the parity-check positions of $\mathbf{r}(X)$. In this event, the k received information digits in the buffer register are error-free. Gate 4 is then activated and the k error-free information digits in the buffer are shifted out to the data sink. If the $n - k - l$ leftmost stages of the syndrome register never contain

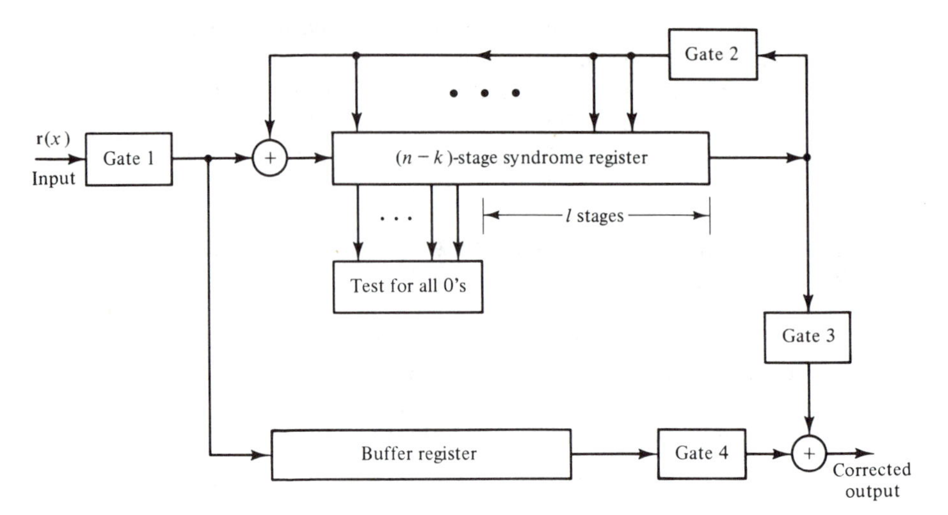

Figure 9.1 Error-trapping decoder for burst-error-correcting codes.

all zeros during the first $n - k - l$ shifts of the syndrome register, the error burst is not confined to the $n - k$ parity-check positions of $\mathbf{r}(X)$.

Step 4. If the $n - k - l$ leftmost stages of the syndrome register contain all zeros after the $(n - k - l + i)$th shift of the syndrome register for $1 \leq i \leq l$, the error burst is confined to positions $X^{n-i}, \ldots, X^{n-1}, X^0, \ldots, X^{l-i-1}$ of $\mathbf{r}(X)$. (This is an end-around burst.) In this event, the $l - i$ digits contained in the $l - i$ rightmost stages of the syndrome register match the errors at the parity-check positions, $X^0, X^1, \ldots, X^{l-i-1}$ of $\mathbf{r}(X)$, and the i digits contained in the next i stages of the syndrome register match the errors at the positions X^{n-i}, $\ldots, X^{n-2}, X^{n-1}$ of $\mathbf{r}(X)$. At this instant, a clock starts to count from $(n - k - l + i + 1)$. The syndrome register is then shifted (in step with the clock) with gate 2 turned off. As soon as the clock has counted up to $n - k$, the i rightmost digits in the syndrome register match the errors at the positions $X^{n-i}, \ldots, X^{n-2}$, X^{n-1} of $\mathbf{r}(X)$. Gates 3 and 4 are then activated. The received information digits are read out of the buffer register and corrected by the error digits shifted out from the syndrome register.

Step 5. If the $n - k - l$ leftmost stages of the syndrome register never contains all zeros by the time that the syndrome register has been shifted $n - k$ times, the received information digits are read out of the buffer register one at a time with gate 4 activated. At the same time, the syndrome register is shifted with gate 2 activated. As soon as the $n - k - l$ leftmost stages of the syndrome register contain all zeros, the digits in the l rightmost stages of the syndrome register match the errors in the next l received information digits to come out of the buffer register. Gate 3 is then activated and the erroneous information digits are corrected by the digits coming out from the syndrome register with gate 2 disabled.

If the $n - k - l$ stages of the syndrome register never contain all zeros by the time the k information digits have been read out of the buffer, an uncorrectable burst of errors has been detected. With the decoder described above, the decoding process takes $2n$ clock cycles; the first n clock cycles are required for syndrome computation and the next n clock cycles are needed for error trapping and error correction. The n clock cycles for syndrome computation are concurrent with the reception of the received vector from the channel; no time delay occurs in this operation. The second n clock cycles for error trapping and correction represent decoding delay.

In this decoder, the received vector is shifted into the syndrome register from the left end. If the received vector is shifted into the syndrome register from the right end, the decoding operation would be slightly different (see Problem 9.2).

This decoder corrects only burst errors of length l or less. The number of these burst error patterns is $n2^{l-1}$, which, for large n, is only a small fraction of 2^{n-k} correctable error patterns (coset leaders). It is possible to modify the decoder in such a way that it corrects all the correctable burst errors of length $n - k$ or less. That is, besides correcting all the bursts of length l or less, the decoder also corrects those bursts of length $l + 1$ to $n - k$ which are used as coset leaders. This modified decoder operates as follows. The entire received vector is first shifted into the syndrome register. Before performing the error correction, the syndrome register is cyclically shifted n times (with feedback connections operative). During this cycling, the length b of the shortest burst that appears in the b rightmost stages of the syndrome register is recorded by a counter. This burst is assumed to be the error burst added by the channel. Having completed these precorrection shifts, the decoder begins its correction process. The syndrome register starts to shift again. As soon as the shortest burst reappears in the b rightmost stages of the syndrome register, the decoder starts to make corrections as described earlier. This decoding is an optimum decoding for burst-error-correcting codes which was proposed by Gallager [23].

9.3 SINGLE-BURST-ERROR-CORRECTING CODES

Fire Codes

Fire codes are the first class of cyclic codes constructed systematically for correcting burst errors. Let $\mathbf{p}(X)$ be an irreducible polynomial of degree m over GF(2). Let ρ be the smallest integer such that $\mathbf{p}(X)$ divides $X^{\rho} + 1$. The integer ρ is called the *period* of $\mathbf{p}(X)$. Let l be a positive integer such that $l \leq m$ and $2l - 1$ is not divisible by ρ. An l-burst-error-correcting Fire code is generated by the following polynomial:

$$\mathbf{g}(X) = (X^{2l-1} + 1)\mathbf{p}(X). \tag{9.4}$$

The length n of this code is the least common multiple of $2l - 1$ and the period ρ of $\mathbf{p}(X)$, that is,

$$n = \text{LCM}\,(2l - 1, \rho). \tag{9.5}$$

The number of parity-check digits of this code is $m + 2l - 1$. Note that the two factors $X^{2l-1} + 1$ and $\mathbf{p}(X)$ of $\mathbf{g}(X)$ are relatively prime.

Example 9.1

Consider the irreducible polynomial $\mathbf{p}(X) = 1 + X^2 + X^5$. Since $\mathbf{p}(X)$ is a primitive polynomial, its period is $\rho = 2^5 - 1 = 31$. Let $l = 5$. Clearly, 31 does not divide $2l - 1 = 9$. The Fire code generated by

$$\mathbf{g}(X) = (X^9 + 1)(1 + X^2 + X^5)$$
$$= 1 + X^2 + X^5 + X^9 + X^{11} + X^{14}$$

has length $n = \text{LCM}(9, 31) = 279$. Therefore, it is a (279, 265) cyclic code that is capable of correcting any burst error of length 5 or less.

Next we prove that the Fire code generated by the polynomial of (9.4) is indeed capable of correcting any burst error of length l or less. To prove this it is sufficient to show that all the bursts of length l or less are in different cosets of the code; so they can be used as coset leaders and form correctable error patterns. Let $X^iA(X)$ and $X^jB(X)$ be the polynomial representations of two bursts of length l_1 and l_2, respectively, with $l_1 \leq l$ and $l_2 \leq l$, where

$$A(X) = 1 + a_1X + a_2X^2 + \cdots + a_{l_1-2}X^{l_1-2} + X^{l_1-1}$$

and

$$B(X) = 1 + b_1X + b_2X^2 + \cdots + b_{l_1-2}X^{l_2-2} + X^{l_2-1}.$$

Suppose that $X^iA(X)$ and $X^jB(X)$ are in the same coset of the code. Then the polynomial

$$\mathbf{v}(X) = X^iA(X) + X^jB(X) \tag{9.6}$$

must be a code polynomial in the code. Without loss of generality, we assume that $i \leq j$. Dividing $j - i$ by $2l - 1$, we obtain

$$j - i = q(2l - 1) + b \tag{9.7}$$

where $0 \leq b < 2l - 1$. Substituting (9.7) into (9.6), the polynomial $\mathbf{v}(X)$ can be expressed in the form

$$\mathbf{v}(X) = X^i[A(X) + X^bB(X)] + X^{i+b}B(X)[X^{q(2l-1)} + 1]. \tag{9.8}$$

Since $\mathbf{v}(X)$ is a code polynomial based on the assumption that $X^iA(X)$ and $X^jB(X)$ are in the same coset, $\mathbf{v}(X)$ must be divisible by $X^{2l-1} + 1$ (a factor of the generator polynomial). Since $X^{q(2l-1)} + 1$ is divisible by $X^{2l-1} + 1$, it follows from (9.8) that $A(X) + X^bB(X)$ is either divisible by $X^{2l-1} + 1$ or equal to zero. Suppose that

$$A(X) + X^bB(X) = D(X)(X^{2l-1} + 1). \tag{9.9}$$

Let d be the degree of $D(X)$. Then the degree of $D(X)(X^{2l-1} + 1)$ is $2l - 1 + d$. Since the degree of $A(X)$ is $l_1 - 1$, which is less than $2l - 1$, the degree of $A(X) + X^bB(X)$ must be the degree of $X^bB(X)$, which is $b + l_2 - 1$. From (9.9) we obtain

$$b + l_2 - 1 = 2l - 1 + d. \tag{9.10}$$

Since $l_1 \leq l$ and $l_2 \leq l$, it follows from (9.10) that

$$b \geq l_1 + d. \tag{9.11}$$

Since $l_1 - 1 \geq 0$, it follows from the equality above that

$$b > l_1 - 1,$$
$$b > d. \tag{9.12}$$

From the equalities of (9.12), we see that $A(X) + X^b B(X)$ has the term X^b. Since $d < b < 2l - 1$, $D(X)(X^{2l-1} + 1)$ does not have the term X^b. This contradicts the hypothesis that $A(X) + X^b B(X) = D(X)(X^{2l-1} + 1)$. Therefore, we must have $D(X) = 0$ and $A(X) + X^b B(X) = 0$. This requires that $b = 0$ and

$$A(X) = B(X). \tag{9.13}$$

Since $b = 0$, it follows from (9.7) that

$$j - i = q(2l - 1). \tag{9.14}$$

Substituting (9.13) and (9.14) into (9.8), we obtain

$$\mathbf{v}(X) = X^i B(X)(X^{j-i} + 1). \tag{9.15}$$

Note that the degree of $B(X)$ is $l_2 - 1$, which is less than l. Therefore, the degree of $B(X)$ is less than the degree m of $\mathbf{p}(X)$, and $B(X)$ and $\mathbf{p}(X)$ are relatively prime. Since $\mathbf{v}(X)$ is assumed to be a code polynomial, $X^{j-i} + 1$ must be divisible by $\mathbf{p}(X)$. As a result, $j - i$ must be a multiple of ρ, the period of $\mathbf{p}(X)$. From (9.14) we see that $j - i$ is also a multiple of $2l - 1$. Therefore, $j - i$ must be a multiple of $n =$ LCM $(2l - 1, \rho)$. This is impossible since both j and i are less than n and $j - i$ cannot be a multiple of n. Therefore, our hypothesis that two bursts, $X^i A(X)$ and $X^j B(X)$, of length l or less are in the same coset is invalid. As a result, all the bursts of length l or less are in different cosets of the Fire code generated by $\mathbf{g}(X)$ of (9.4) and they are correctable error patterns. Since the code is cyclic, it also corrects the end-around bursts of length l or less.

Fire codes can be decoded with the error-trapping circuit shown in Figure 9.1. The error-trapping decoder for the (279, 265) Fire code considered in Example 9.1 is shown in Figure 9.2.

In a data transmission (or storage) system, if the receiver has some computation capability, a fast decoder for Fire codes may be implemented. Consider a Fire code with generator polynomial $\mathbf{g}(X) = (X^{2l-1} + 1)\mathbf{p}(X)$, where $2l - 1$ and the period ρ of $\mathbf{p}(X)$ are relatively prime. Let $\mathbf{r}(X)$ be the received polynomial. Let $\mathbf{s}_1(X)$ and $\mathbf{s}_2(X)$ be the remainders resulting from dividing $\mathbf{r}(X)$ by $X^{2l-1} + 1$ and $\mathbf{p}(X)$, respectively. Then we may take

$$[\mathbf{s}_1(X), \mathbf{s}_2(X)]$$

as a syndrome of $\mathbf{r}(X)$. We can readily see that $\mathbf{s}_1(X) = \mathbf{s}_2(X) = 0$ if and only if $\mathbf{r}(X)$ is a code polynomial. If $\mathbf{r}(X)$ contains a nonzero error burst of length l or less, we must have $\mathbf{s}_1(X) \neq 0$ and $\mathbf{s}_2(X) \neq 0$. If $\mathbf{s}_1(X) = 0$ and $\mathbf{s}_2(X) \neq 0$ [or $\mathbf{s}_1(X) \neq 0$ and $\mathbf{s}_2(X) = 0$], then $\mathbf{r}(X)$ must contain a detectable but uncorrectable burst of length greater than l.

Now, consider an error-trapping decoder as shown in Figure 9.3. This decoder consists of two syndrome registers; one is called the *error-pattern register* and the other is called the *error-location register*. The feedback connections of the error-pattern register are based on the factor $X^{2l-1} + 1$, and the feedback connections of the error-location register are based on the factor $\mathbf{p}(X)$. The received polynomial $\mathbf{r}(X)$ is first read into the two syndrome registers and the buffer register. As soon as the entire $\mathbf{r}(X)$ has been shifted into the two syndrome registers, $\mathbf{s}_1(X)$ and $\mathbf{s}_2(X)$ are formed. The decoder tests $\mathbf{s}_1(X)$ and $\mathbf{s}_2(X)$. If $\mathbf{s}_1(X) = \mathbf{s}_2(X) = 0$, the received poly-

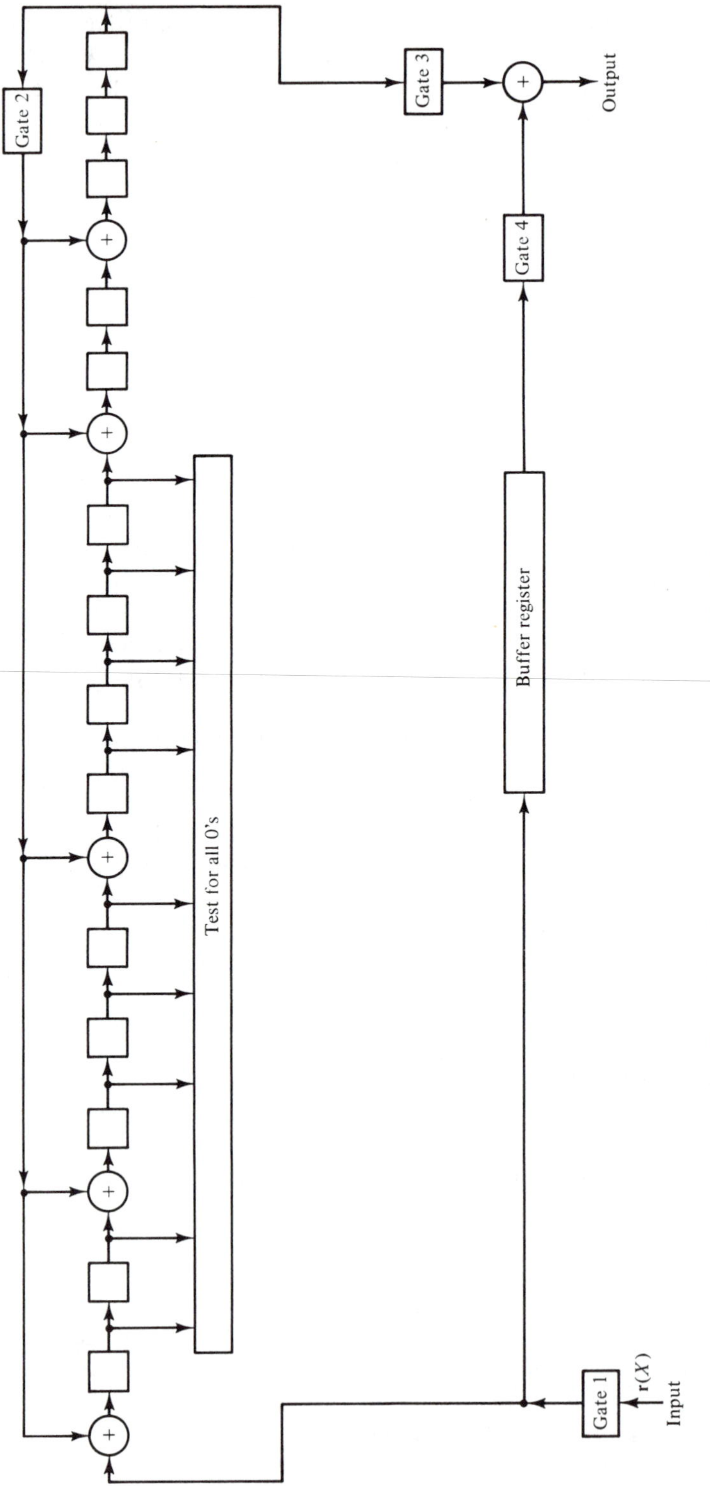

Figure 9.2 Error-trapping decoder for the (279, 265) Fire code.

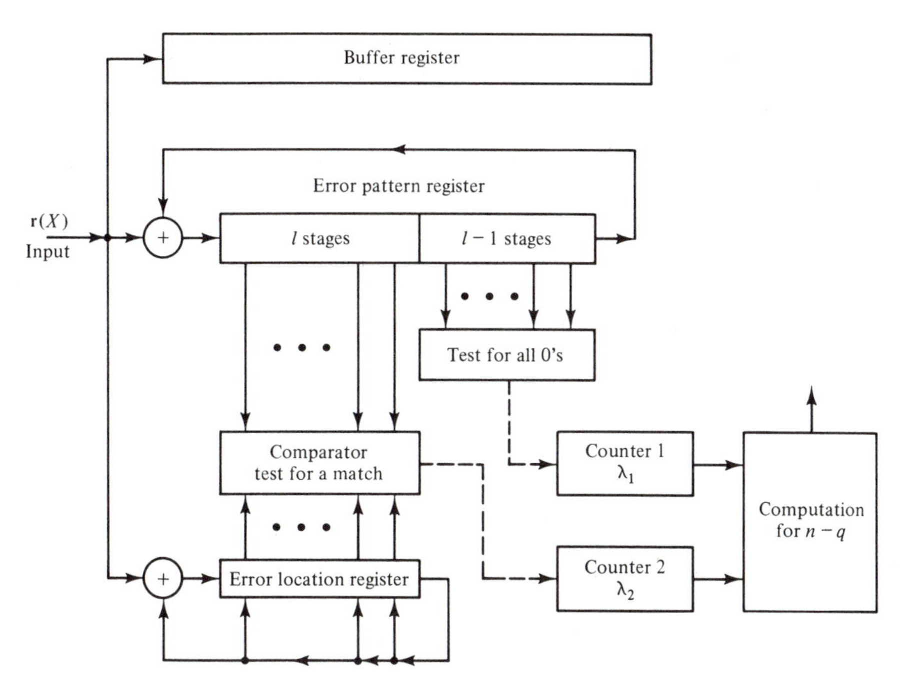

Figure 9.3 High-speed error-trapping decoder for Fire codes.

nomial $\mathbf{r}(X)$ is assumed to be error-free and is then delivered to the user. If $\mathbf{s}_1(X) = 0$ and $\mathbf{s}_2(X) \neq 0$ [or $\mathbf{s}_1(X) \neq 0$ and $\mathbf{s}_2(X) = 0$], then $\mathbf{r}(X)$ contains a detectable but uncorrectable error burst and is therefore discarded. If $\mathbf{s}_1(X) \neq 0$ and $\mathbf{s}_2(X) \neq 0$, then $\mathbf{r}(X)$ is assumed to contain a correctable error burst and the decoder starts the error correction process. The error correction process is discribed as follows:

Step 1. Shift the error-pattern register and test for zeros at the $l - 1$ high-order stages. Stop shifting as soon as the $l - 1$ high-order stages contain all zeros. The error burst is then trapped in the l low-order stages of the error-pattern register. Let λ_1 be the number of shifts performed (in counter 1). Note that no more than $2l - 2$ shifts are needed to trap the error-burst.

Step 2. Shift the error-location register until the contents in its l low-order stages match the burst pattern in the l low-order stages of the error-pattern register. Let the number of shifts be λ_2 (in counter 2). In this step, no more than $\rho - 1$ shifts are required.

Step 3. Since $2l - 1$ and ρ are relatively prime, there exists a unique non-negative integer q less than n (code length) such that the remainders resulting from dividing q by $2l - 1$ and ρ are λ_1 and λ_2, respectively. Determine the integer q by computation. Then the error burst begins at position X^{n-q} and ends at position $X^{n-q+l-1}$ of $\mathbf{r}(X)$. In the case that $q = 0$, the burst begins at position X^0 and ends at position X^{l-1} of $\mathbf{r}(X)$.

Step 4. Let $B(X)$ be the burst pattern trapped in the error-pattern register. Add $X^{n-q}B(X)$ to $\mathbf{r}(X)$ in the buffer register. This completes the error correction process.

If, in step 1, the $l - 1$ high-order stages of the error-pattern register never contain all zeros by the time the register has been shifted $2l - 2$ times, an uncorrectable error burst has been detected. In this event, the decoder stops the error-correction process.

The error-location number $n - q$ can be computed easily. Since $2l - 1$ and ρ are relatively prime, there exist two integers, A_1 and A_2, such that

$$A_1(2l - 1) + A_2\rho = 1.$$

The q is simply the remainder resulting from dividing

$$A_1(2l - 1)\lambda_2 + A_2\rho\lambda_1$$

by n. Once A_1 and A_2 are determined, the numbers $A_1(2l - 1)$ and $A_2\rho$ can be stored in the receiver permanently for use in each decoding. Therefore, computing $n - q$ needs two multiplications, one addition, one division, and one subtraction.

We note that the error-trapping decoder for Fire codes described above requires at most $2l + \rho - 3$ shifts of the two syndrome registers and five arithmetic operations to carry out the error correction process. However, the error-trapping decoder described in Section 9.2 takes n shifts (cycle times) to complete the error-correction process. Since $n = \text{LCM}(2l - 1, \rho)$, it is much greater than $2l + \rho - 3$. Therefore, decoding speed is improved. This improvement in decoding speed is possible only when the receiver has some computation capability or computation facility is available at the receiver. Furthermore, the fast error-trapping decoder requires more logic.

Example 9.2

Consider the (279, 265) Fire code considered in Example 9.1. This code is capable of correcting any burst of length $l = 5$ or less. The fast error-trapping decoder for this code is shown in Figure 9.4. Suppose that the error burst

$$\mathbf{e}(X) = X^2 + X^3 + X^4 + X^5 + X^6$$

has occurred. It is a solid burst of length 5 starting at position $n - q = 2$. The syndrome $\mathbf{s}_1(X)$ and $\mathbf{s}_2(X)$ are remainders resulting from dividing $\mathbf{e}(X)$ by $X^{11} + 1$ and $\mathbf{p}(X) = 1 + X^2 + X^5$, respectively. They are

$$\mathbf{s}_1(X) = X^2 + X^3 + X^4 + X^5 + X^6,$$
$$\mathbf{s}_2(X) = 1 + X + X^4.$$

As soon as the entire received polynomial $\mathbf{r}(X)$ has been shifted into the error-pattern and error-location registers, the contents in the two registers are $\mathbf{s}_1(X)$ and $\mathbf{s}_2(X)$. Since the four high-order stages of the error-pattern register do not contain all zeros, the error burst is not trapped in the five low-order stages. The error-pattern register starts to shift. Table 9.1 shows the contents in the error-pattern register after each shift. We see that the error burst is trapped in the five low-order stages after $\lambda_1 = 7$ shifts. Now, the error-location register begins to shift. Table 9.2 displays the contents in the error-location register after each shift. At the twenty-ninth shift, the contents in the error-location register match the contents in the five low-order stages of the error-

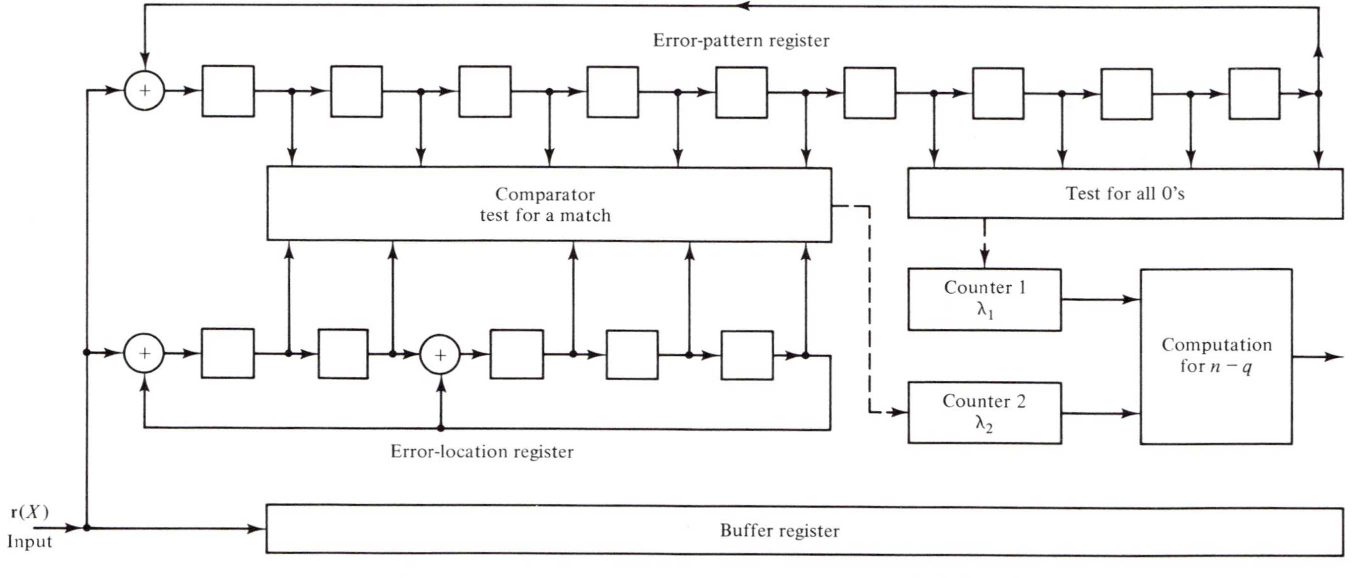

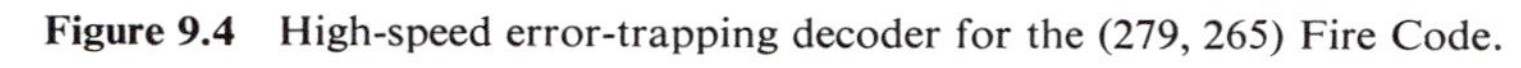

Figure 9.4 High-speed error-trapping decoder for the (279, 265) Fire Code.

TABLE 9.1 CONTENTS IN THE ERROR-PATTERN REGISTER OF THE DECODER SHOWN IN FIGURE 9.4 AFTER EACH SHIFT

Shift	Content
0	0 0 1 1 1 1 1 0 0
1	0 0 0 1 1 1 1 1 0
2	0 0 0 0 1 1 1 1 1
3	1 0 0 0 0 1 1 1 1
4	1 1 0 0 0 0 1 1 1
5	1 1 1 0 0 0 0 1 1
6	1 1 1 1 0 0 0 0 1
7*	1 1 1 1 1 0 0 0 0

*At the seventh shift, the contents in the four high-order stages are all zeros.

TABLE 9.2 CONTENTS IN THE ERROR-LOCATION REGISTER OF THE DECODER SHOWN IN FIGURE 9.4 AFTER EACH SHIFT

Shift	Contents	Shift	Contents
0	1 1 0 0 1	15	0 1 0 0 0
1	1 1 0 0 0	16	0 0 1 0 0
2	0 1 1 0 0	17	0 0 0 1 0
3	0 0 1 1 0	18	0 0 0 0 1
4	0 0 0 0 1	19	1 0 1 0 0
5	1 0 1 0 1	20	0 1 0 1 0
6	1 1 1 1 0	21	0 0 1 0 1
7	0 1 1 1 1	22	1 0 1 1 0
8	1 0 0 1 1	23	0 1 0 1 1
9	1 1 1 0 1	24	1 0 0 0 1
10	1 1 0 1 0	25	1 1 1 0 0
11	0 1 1 0 1	26	0 1 1 1 0
12	1 0 0 1 0	27	0 0 1 1 1
13	0 1 0 0 1	28	1 0 1 1 1
14	1 0 0 0 0	29*	1 1 1 1 1

*At the twenty-ninth shift, the contents match the burst pattern in the error-pattern register.

pattern register. Therefore, $\lambda_2 = 29$. Next we need to compute the error-location number $n - q$. First we find that

$$7 \times 9 + (-2) \times 31 = 1$$

($A_1 = 7$ and $A_2 = -2$). Then we compute

$$7 \times 9 \times 29 + (-2) \times 31 \times 7 = 1393.$$

Dividing 1393 by $n = 279$, we obtain $q = 277$. Consequently, $n - q = 2$, which is exactly the error-location number. Error correction is achieved by adding the error burst $X^2 + X^3 + X^4 + X^5 + X^6$ to the received polynomial $\mathbf{r}(X)$ in the buffer register. Error-correction process takes at most $8 + 30 = 38$ cycle times. Using the decoder shown in Figure 9.2, the error-correction process takes $n = 279$ cycle times.

The fast error-trapping decoder for Fire codes was first devised by Peterson [24] and then refined by Chien [20].

The burst-correcting efficiency of a Fire code is $z = 2l/(m + 2l - 1)$. If l is chosen to be equal to m, then $z = 2m/(3m - 1)$. For large m, z is approximately 2/3. Thus, Fire codes are not very efficient with respect to the Reiger bound. However, they can be simply implemented.

A Fire code that is capable of correcting any burst of length l or less and simultaneously detecting any burst of length $d \geq l$ is generated by

$$\mathbf{g}(X) = (X^c + 1)\mathbf{p}(X),$$

where $c \geq l + d - 1$ and c is not divisible by the period ρ of $\mathbf{p}(X)$. The length of this code is the least common multiple of c and ρ.

Other Codes

Besides Fire codes, some very efficient cyclic codes and shortened cyclic codes for correcting short single bursts have been found either analytically or with the aid of a computer [7,11,14,15]. A list of these codes with their generator polynomials is given in Table 9.3. These codes and the codes derived from them by interleaving, described in the following section, are the most efficient single-burst-error-correction codes known.

TABLE 9.3 SOME BURST-CORRECTING CYCLIC AND SHORTENED CYCLIC CODES

$n - k - 2l$	Code (n, k)	Burst-correcting capability l	Generator polynomial $\mathbf{g}(X)$*
0	(7, 3)	2	35
	(15, 9)	3	171
	(15, 7)	4	721
	(15, 5)	5	2467
	(19, 11)	4	1151
	(21, 9)	6	14515
	(21, 7)	7	47343
	(21, 5)	8	214537
	(21, 3)	9	1647235
	(27, 17)	5	2671

*Generator polynomials are given in an octal representation. Each digit represents three binary digits according to the following code:

$0 \leftrightarrow 0\ 0\ 0$	$2 \leftrightarrow 0\ 1\ 0$	$4 \leftrightarrow 1\ 0\ 0$	$6 \leftrightarrow 1\ 1\ 0$
$1 \leftrightarrow 0\ 0\ 1$	$3 \leftrightarrow 0\ 1\ 1$	$5 \leftrightarrow 1\ 0\ 1$	$7 \leftrightarrow 1\ 1\ 1$

The binary digits are then the coefficients of the polynomial, with the high-order coefficients at the left. For example, the binary representation of 171 is 0 0 1 1 1 1 0 0 1, and the corresponding polynomial is $\mathbf{g}(X) = X^6 + X^5 + X^4 + X^3 + 1$.

TABLE 9.3 CONTINUED

$n - k - 2l$	Code (n, k)	Burst-correcting capability l	Generator polynomial $g(X)$
	(34, 22)	6	15173
	(38, 24)	7	114361
	(50, 34)	8	224531
	(56, 38)	9	1505773
	(59, 39)	10	4003351
1	(15, 10)	2	65
	(21, 14)	3	171
	(21, 12)	4	1663
	(21, 10)	5	7707
	(23, 12)	5	5343
	(27, 20)	3	311
	(31, 20)	5	4673
	(38, 29)	4	1151
	(48, 37)	5	4501
	(63, 50)	6	22377
	(63, 48)	7	105437
	(63, 46)	8	730535
	(63, 44)	9	2002353
	(67, 54)	6	36365
	(96, 79)	7	114361
	(103, 88)	8	501001
2	(17, 9)	3	471
	(21, 15)	2	123
	(31, 25)	2	161
	(31, 21)	4	3551
	(35, 23)	5	13627
	(39, 27)	5	13617
	(41, 21)	9	6647133
	(51, 41)	4	3501
	(51, 35)	7	304251
	(55, 35)	9	7164555
	(57, 39)	8	1341035
	(63, 55)	3	711
	(63, 53)	4	2263
	(63, 51)	5	16447
	(63, 49)	6	61303
	(73, 63)	4	2343
	(85, 75)	4	2651
	(85, 73)	5	10131
	(105, 91)	6	70521
	(131, 119)	5	15163
	(169, 155)	6	55725
3	(51, 42)	3	1455
	(63, 56)	2	305
	(85, 76)	3	1501
	(89, 78)	4	4303
	(93, 82)	4	6137
	(121, 112)	3	1411
	(151, 136)	6	114371

TABLE 9.3 CONTINUED

$n - k - 2l$	Code (n, k)	Burst-correcting capability l	Generator polynomial $g(X)$
	(164, 153)	4	6255
	(195, 182)	5	22475
	(217, 202)	6	120247
	(290, 277)	5	24711
4	(43, 29)	5	52225
	(91, 79)	4	10571
	(93, 83)	3	2065
	(117, 105)	4	13413
	(133, 115)	7	1254355
	(255, 245)	3	3523
	(255, 243)	4	17667
	(255, 241)	5	76305
	(255, 239)	6	301565
	(273, 261)	4	10743
	(511, 499)	4	10451
	(595, 581)	5	64655
5	(465, 454)	3	7275
	(1023, 1010)	4	22365

9.4 INTERLEAVED CODES

Given an (n, k) cyclic code, it is possible to construct a $(\lambda n, \lambda k)$ cyclic code (i.e., a code λ times as long with λ times as many information digits) by *interleaving*. This is done simply by arranging λ code vectors in the original code into λ rows of a rectangular array and then transmitting them column by column as shown in Figure 9.5. The resulting code is called an *interleaved code*. The parameter λ is referred to as the *interleaving degree*.

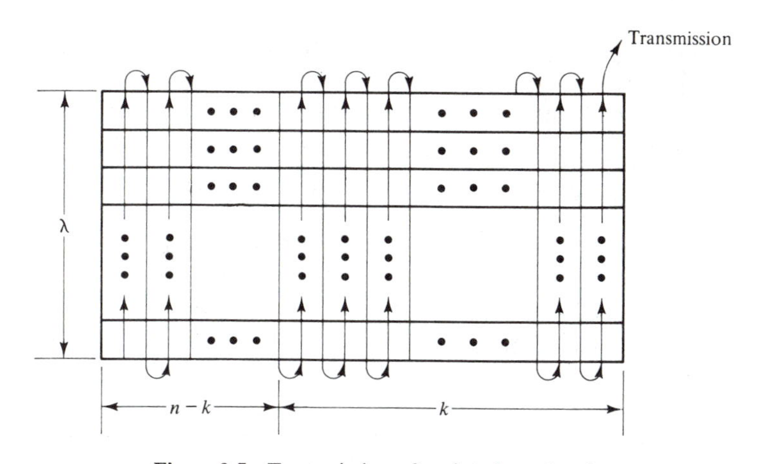

Figure 9.5 Transmission of an interleaved code.

Obviously, a pattern of errors can be corrected for the whole array if and only if the pattern of errors in each row is a correctable pattern for the original code. No matter where it starts, a burst of length λ will affect no more than one digit in each row. Thus, if the original code corrects single errors, the interleaved code corrects single bursts of length λ or less. If the original code corrects any single burst of length l or less, the interleaved code will correct any single burst of length λl or less. If an (n, k) code has maximum possible burst-error-correcting capability (i.e., $n - k - 2l = 0$), the interleaved $(\lambda n, \lambda k)$ code also has maximum possible burst-error-correcting capability. By interleaving short codes with maximum possible burst-error-correcting capability, it is possible to construct codes of practically any length with maximum burst-error-correcting ability. Therefore, the interleaving technique reduces the problem of searching long efficient burst-error-correcting codes to search good short codes.

The obvious way to implement an interleaved code is to set up the array and operate on rows in encoding and decoding. This is generally not the simplest implementation. The simplest implementation results from the observation that *if the original code is cyclic, the interleaved code is also cyclic.* If the generator polynomial of the original code is $\mathbf{g}(X)$, the generator polynomial for the interleaved code is $\mathbf{g}(X^\lambda)$ (see Problem 9.6). Thus, encoding and syndrome computation can be accomplished by using shift registers. It turns out that the decoder for the interleaved code can be derived from the decoder of the original code simply by replacing each register stage of the original decoder by λ stages without changing the other connections. This essentially allows the decoder circuitry to look at successive rows of the code array on successive decoder cycles. Therefore, if the decoder of the original code is simple (this is usually true for short codes), so is the decoder for the interleaved code.

The interleaving technique described above is effective not only for deriving long powerful single-burst-error-correcting codes from short optimal single-burst-error-correcting codes, but also for deriving long powerful burst-and-random-error-correcting codes from short codes.

9.5 PHASED-BURST-ERROR-CORRECTING CODES

Consider an (n, k) code whose length n is a multiple of m, say $n = \sigma m$. The σm digits of each code vector may be grouped into σ subblocks; each subblock consists of m consecutive code digits. For example, let

$$\mathbf{v} = (v_0, v_1, v_2, \ldots, v_{\sigma m-1})$$

be a code vector. Then the ith subblock consists of the following consecutive code digits:

$$v_{im}, v_{im+1}, \ldots, v_{(i+1)m-1},$$

with $0 \leq i < \sigma$. A burst of length λm or less is called a *phased burst* if and only if it is confined to λ consecutive subblocks, where λ is a positive integer less than σ. A linear code of length $n = \sigma m$ that is capable of correcting all phased error bursts confined to λ or fewer subblocks is called a *λm-phased-burst-error-correcting code.* Since a burst of length $(\lambda - 1)m + 1$, no matter where it starts, can affect at most λ subblocks, it is clear that a λm-phased-burst-error-correcting code is capable of correcting any single burst of length $(\lambda - 1)m + 1$ or less. Thus, a λm-phased-burst-error-correcting code can be used as a $[(\lambda - 1)m + 1]$-single-burst-error-correcting code.

Burton Codes

In the following, a class of phased-burst-error-correcting cyclic codes is presented. This class of codes is similar to the class of Fire codes and was discovered by Burton [21]. Let $\mathbf{p}(X)$ be an irreducible polynomial of degree m and period ρ. Let n be the least common multiple of m and ρ. Then $n = \sigma m$. For any positive integer m, there exists an m-phased burst-error-correcting Burton code of length $n = \sigma m$ which is generated by

$$\mathbf{g}(X) = (X^m + 1)\mathbf{p}(X). \tag{9.16}$$

The number of parity-check digits of this code is $2m$. Thus, it is a $[\sigma m, (\sigma - 2)m]$ cyclic code. Each code vector consists of σ subblocks. In order to show that the Burton code generated by $\mathbf{g}(X) = (X^m + 1)\mathbf{p}(X)$ is capable of correcting all phased bursts confined to a single subblock of m digits, it is necessary and sufficient to prove that no two such bursts are in the same coset of a standard array for the code. The proof is similar to the proof given for the Fire codes and is left as an exercise (see Problem 9.7).

The decoding of a Burton code can be accomplished with the error trapping decoder as described in Section 9.2 except that the contents of the m leftmost stages of the syndrome register are tested for zero at every mth shift. If m and the period ρ of $\mathbf{p}(X)$ are relatively prime and if the receiver has some computation power, the Burton codes can be decoded with the fast error-trapping algorithm described in Section 9.3.

It is possible to interleave an m-phased-burst-error-correcting Burton code in such a way that the interleaved $(\lambda n, \lambda k)$ code is capable of correcting any phased burst that is confined to λ consecutive subblocks. To accomplish this, we arrange λ code vectors in the m-phased-burst-error-correcting code into λ rows of a rectangular array as usual. We regard a subblock of each row as a single element. Then the array consists of σ columns, each column consists of λ subblocks. The array is transmitted column by column, one subblock at a time from each row. Therefore, a code vector in the interleaved code consists of $\lambda\sigma$ subblocks. No matter where it starts, any phased-error burst confined to λ or fewer subblocks will affect no more than one subblock in each row. Thus, a phased burst of length λm will be corrected if the array is decoded on a row-by-row basis. If the interleaved code is used as a $[(\lambda - 1)m + 1]$-burst-error-correcting code, its burst-error-correcting efficiency is

$$z = \frac{2[(\lambda - 1)m + 1]}{2\lambda m} = 1 - \frac{1}{\lambda}\left(\frac{m - 1}{m}\right).$$

As the interleaving degree λ becomes large, the burst-error-correcting efficiency of a Burton code approached 1. Thus, by interleaving the Burton codes, we obtain a class of asymptotically optimal burst-error-correcting codes.

The obvious way to implement an interleaved Burton code is to set up the code array and operate on rows in encoding and decoding. Thus, the encoder of the interleaved code consists of the encoder of the original code and a buffer for the storage of the row vectors of the code array; the decoder consists of the decoder of the original code and a buffer for the storage of the received code array. Of course, the interleaved code can be decoded with the error-trapping decoder of Figure 9.1, where the contents of the λm leftmost stages are tested for zeros at every mth shift.

9.6 BURST-AND-RANDOM-ERROR-CORRECTING CODES

On many communication channels, errors occur neither independently at random nor in well-defined single bursts, but occur in a mixed manner. To combat these mixed errors, the random-error-correcting codes or single-burst-error-correcting codes will be either inefficient or inadequate. Consequently, it is desirable to design codes that are capable of correcting random errors and/or single or multiple bursts. There are several methods of constructing codes for the correction of random errors as well as burst errors. The most effective method is the interleaving technique as described in Section 9.4. By interleaving a t-random-error-correcting (n, k) code to degree λ, we obtain a $(\lambda n, \lambda k)$ code which is capable of correcting any combination of t bursts of length λ or less. Several other methods of deriving new codes from known codes for simultaneously correcting random errors and burst errors are discussed in this section.

Product Codes

Let C_1 be an (n_1, k_1) linear code and let C_2 be an (n_2, k_2) linear code. Then an (n_1n_2, k_1k_2) linear code can be formed such that each code word is a rectangular array of n_1 columns and n_2 rows in which every row is a code vector in C_1 and every column is a code vector in C_2, as shown in Figure 9.6. This two-dimensional code is called the *direct product* (or simply the *product*) of C_1 and C_2 [25]. The k_1k_2 digits in the upper right corner of the array are information symbols. The digits in the upper left corner of this array are computed from the parity-check rules for C_1 on rows, and the digits in the lower right corner are computed from the parity-check rules for C_2 on columns. Now, should we compute the check digits in the lower left corner by using

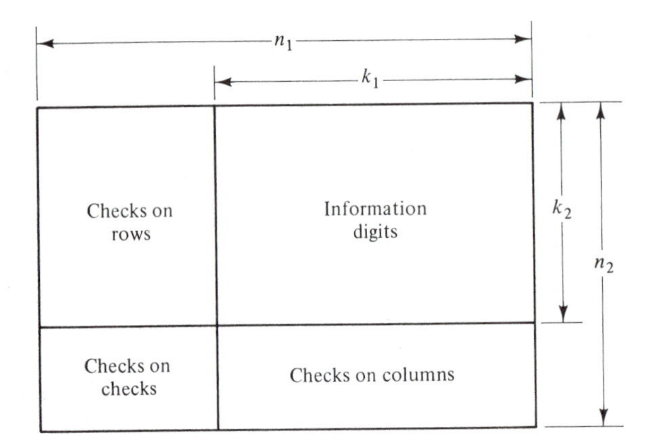

Figure 9.6 Code array for the product code $C_1 \times C_2$.

the parity-check rules for C_2 on columns or the parity-check rules for C_1 on rows? It turns out that either way would yield the same $(n_1 - k_1) \times (n_2 - k_2)$ check digits (see Problem 9.17), and it is possible to have all row code vectors in C_1 and all column code vectors in C_2 simultaneously. Therefore, for encoding the product code $C_1 \times C_2$, we may first encode the k_2 rows of the information array based on the parity-check rules for C_1 and then encode the n_1 resulting columns based on the rules for C_2, or vice versa.

If the code C_1 has minimum weight d_1 and the code C_2 has minimum weight d_2, the minimum weight of the product code is exactly $d_1 d_2$. A minimum-weight code vector in the product code is formed by (1) choosing a minimum-weight code vector in C_1 and a minimum-weight code vector in C_2; and (2) forming an array in which all columns corresponding to zeros in the code vector from C_1 are zeros and all columns corresponding to ones in the code vector C_1 are the minimum-weight code vector chosen from C_2.

It is not easy to characterize the correctable error patterns for the product code; this depends upon how the correction is actually done. One method involves using the correction first on rows and then on columns. In this case, a pattern will be correctable if and only if the uncorrectable patterns on rows after row correction leave correctable patterns on the columns. It generally improves the correction by decoding rows, columns, then columns and rows again. This, of course, increases the decoding delay.

The product code is capable of correcting any combination of $\lfloor (d_1 d_2 - 1)/2 \rfloor$ errors, but the method described above will not achieve this. For example, consider the product code of two Hamming single-error-correcting codes. The minimum distance of each is 3, so the minimum distance of the product is 9. A pattern of four errors at the corners of a rectangle gives two errors in each of the two rows and two columns and is therefore not correctable by simple correction on rows and columns. Nevertheless, simple correction on rows and columns, although nonoptimum, can be very effective.

Let l_1 and l_2 be the burst-error-correcting capabilities of code C_1 and the code C_2, respectively. The burst-error-correcting capability of the product code of C_1 and C_2 can be analyzed as follows. Suppose that a code array is transmitted row by row and that, at the output of the channel, the received digits are rearranged back into an array row by row. No matter where it starts, any existing error burst of length $n_1 l_2$ or less will affect no more than $l_2 + 1$ consecutive rows; when the received digits are rearranged back into an array, each column is at most affected by a burst of length l_2. Now if the array is decoded on a column-by-column basis, the burst will be corrected. Therefore, the burst-error-correcting capability of the product code is at least $n_1 l_2$. Suppose that a code array is transmitted on a column-by-column basis and decoded on a row-by-row basis. By a similar argument, it is possible to show that any error burst of length $n_2 l_1$ or less can be corrected. Thus, the burst-correcting capability of the product code is at least $n_2 l_1$. Consequently, we may conclude that the burst-error-correcting capability l of the product code is at least $\max\{n_1 l_2, n_2 l_1\}$.

So far, we have considered a product code for either random-error correction or burst-error correction. However, a product code can be used for simultaneous random-error correction and burst-error correction. Let d_1 and d_2 be the minimum distances of codes C_1 and C_2, respectively. Then it is possible to show that the product code of C_1 and C_2 is capable of correcting any combination of $t = \lfloor (d_1 d_2 - 1)/2 \rfloor$ or fewer random errors and simultaneously correcting any error burst of length $l = \max(n_1 t_2, n_2 t_1)$ or less, where $t_1 = \lfloor (d_1 - 1)/2 \rfloor$ and $t_2 = \lfloor (d_2 - 1)/2 \rfloor$ [26,27]. To prove this assertion, it is sufficient to show that an error burst of length l or less and a random-error pattern of t or fewer errors cannot be in the same coset of a standard array for the product code. Suppose that $n_1 t_2 \geq n_2 t_1$. Then $l = n_1 t_2$. Consider a burst of length $n_1 t_2$ or less. When this vector is arranged as an array of n_2 rows and n_1 columns, each column contains at most t_2 errors. Suppose that this burst and some random-error pattern of t or fewer errors in the same coset of the product code. Then the sum of these two error patterns (in array form) is a code array in the product code. As a result, each column of the sum array must either have no nonzero components or have at least d_2 nonzero components. Each nonzero column of the sum array must be composed of at least $d_2 - t_2$ errors from the random-error pattern and at most t_2 errors from the burst-error pattern. Since there are at most t random errors, these errors can be distributed among at most $\lfloor t/(d_2 - t_2) \rfloor$ columns. Thus, the sum array contains at most $\lfloor t/(d_2 - t_2) \rfloor t_2 + t$ nonzero components. However,

$$\left\lfloor \frac{t}{d_2 - t_2} \right\rfloor t_2 + t \leq t\left(\frac{t_2}{d_2 - t_2} - 1\right) < 2t.$$

Hence, the sum array contains fewer than $2t < d_1 d_2$ nonzero components and cannot be a code array in the product code. This contradiction implies that a burst of length $l = n_1 t_2$ or less and a random-error pattern of t or fewer errors cannot be in the same coset of a standard array for the product code. Therefore, they can both be used as coset leaders and are correctable error patterns. If $n_2 t_1 > n_1 t_2$, then $l = n_2 t_1$. The same argument can be applied to rows instead of columns of the sum array.

The obvious way to implement a product code is to set up the code array and operate on rows and then columns (or columns and then rows) in encoding and decoding, but there is an alternative that can be extremely attractive. In many cases the product code of cyclic codes is cyclic, and cyclic code implementation is much simpler.

If the component codes, C_1 and C_2, are cyclic and if their lengths, n_1 and n_2, are relatively prime, the product code $C_1 \times C_2$ is cyclic if the code digits are transmitted in a proper order [26]. Start with the upper right corner and move down and *to the left* on a 45° diagonal as shown in Figure 9.7. When we reach the end of a column, move to the top of the next column. When we reach the end of a row, move to the rightmost digit of the next row.

Since n_1 and n_2 are relatively prime, there exists a pair of integers a and b such that

$$an_1 + bn_2 = 1.$$

Let $\mathbf{g}_1(X)$ and $\mathbf{h}_1(X)$ be the generator and parity polynomials of the (n_1, k_1) cyclic code C_1 and let $\mathbf{g}_2(X)$ and $\mathbf{h}_2(X)$ be the generator and parity polynomials of the (n_2, k_2) cyclic code C_2. Then it is possible to show [26,28] that the generator polynomial $\mathbf{g}(X)$ of the cyclic product code of C_1 and C_2 is the greatest common divisor of $X^{n_1 n_2} - 1$ and $\mathbf{g}_1(X^{bn_2})\mathbf{g}_2(X^{an_1})$, that is,

$$\mathbf{g}(X) = \text{GCD}\left[X^{n_1 n_2} - 1, \mathbf{g}_1(X^{bn_2})\mathbf{g}_2(X^{an_1})\right] \tag{9.17}$$

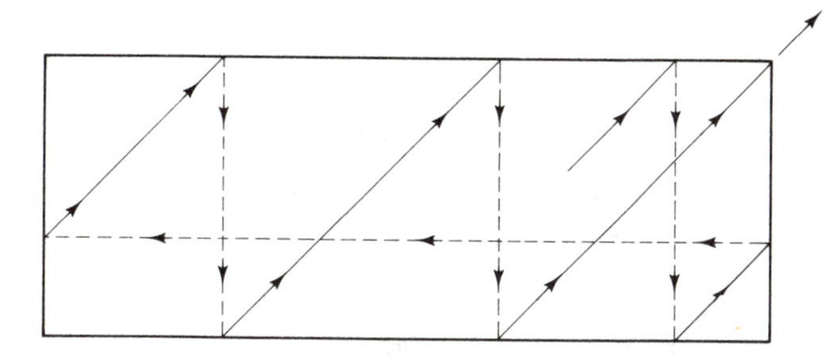

Figure 9.7 Transmission of a cyclic product code.

and the parity polynomial $\mathbf{h}(X)$ of the cyclic product code is the greatest common divisor of $\mathbf{h}_1(X^{bn_2})$ and $\mathbf{h}_2(X^{an_1})$, that is,

$$\mathbf{h}(X) = \text{GCD}\,[\mathbf{h}_1(X^{bn_2}), \mathbf{h}_2(X^{an_1})]. \tag{9.18}$$

Suppose that the cyclic code C_1 has random-error-correcting capability t_1 and burst-error-correcting capability l_1 and the cyclic code C_2 has random-error-correcting capability t_2 and burst-error-correcting capability l_2. Then the burst-error-correcting capability l of the cyclic product code of C_1 and C_2 is at least equal to max $(n_1t_2 + l_1, n_2t_1 + l_2)$, [26], that is,

$$l \geq \max\,(n_1t_2 + l_1, n_2t_1 + l_2). \tag{9.19}$$

This can be shown as follows. Suppose that an error burst of length $n_2t_1 + l_2$ or less occurred during the transmission of a code array. When the received digits are rearranged back into an array, all except l_2 adjacent rows will contain t_1 or fewer errors. Each of these l_2 adjacent rows will contain $t_1 + 1$ or fewer errors. If the rows are decoded first, these l_2 adjacent rows may contain errors after the row decoding. Therefore, after row decoding, each column of the array contains an error burst of length at most l_2. Since the column code C_2 is capable of correcting any error burst of length l_2 or less, all the remaining errors in the array will be corrected by column decoding. By a similar argument, any error burst of length $n_1t_2 + l_1$ or less will be corrected if the column decoding is performed before the row decoding. Therefore, we obtain the result as stated by (9.19).

The complexity of the decoder for cyclic product codes is comparable to the complexity of the decoders for both the (n_1, k_1) code and the (n_2, k_2) code. At the receiving end of the channel, the received vector may again be rearranged as a retangular array. Thus, the decoder can decode each of the row (or column) code vectors separately and then decode each of the column (or row) code vectors. Alternatively, in the transmitted code vector, the set of n_1 digits formed by selecting every (n_2)th digit are the n_1 digits of a code vector of C_1 permuted in a fixed way. They can be permuted back to their original form and corrected in a Meggitt-type decoder. The digits in the permuted form are a code vector in a related code and can be decoded directly in this form in a Meggitt-type decoder. Similarly, correction for the column code C_2 can be done by selecting every (n_1)th digit from the large code vector. Thus, the total equipment required is roughly that required to decode the two individual codes.

There is room for a good deal of engineering ingenuity in divising decoding procedures for cyclic product codes [28–35].

Code Derived from Reed–Solomon Codes

In Chapter 2 it was pointed out that any element β in the Galois field $GF(2^m)$ can be expressed uniquely as a sum of $1, \alpha, \alpha^2, \ldots, \alpha^{m-1}$ in the following form:

$$\beta = a_0 + a_1\alpha + a_2\alpha^2 + \cdots + a_{m-1}\alpha^{m-1},$$

where α is a primitive element in $GF(2^m)$ and $a_i = 0$ or 1. Thus, the correspondence between β and $(a_0, a_1, \ldots, a_{m-1})$ is one-to-one. We shall call the m-tuple $(a_0, a_1, \ldots, a_{m-1})$ an *m-bit byte* representation of β.

Consider a t-error-correcting Reed–Solomon code with code symbols from $GF(2^m)$. If each symbol is represented by its corresponding m-bit byte, we obtain a binary linear code with the following parameters:

$$n = m(2^m - 1),$$
$$n - k = 2mt.$$

This code is capable of correcting any error pattern that affects t or fewer m-bit bytes. It is immaterial whether a byte has one error or whether all the m bits are in error; it is counted as one byte error. This can be seen as follows. At the channel output, the binary received vector is divided into $2^m - 1$ bytes; each byte is transformed back into a symbol in $GF(2^m)$. Thus, if an error pattern affects t or fewer bytes, it affects t or fewer symbols in a Reed–Solomon code. Obviously, the error pattern can be corrected by the decoding method described in Sections 6.2 and 6.5. We shall call this binary code a *t-byte-correcting code*. Actually, this code is a multiple-phased-burst-error-correcting code.

Binary codes derived from Reed–Solomon codes are more effective against clustered errors than random errors since clustered errors usually involve several errors per byte and thus relatively few byte errors. For example, since a burst of length $3m + 1$ cannot affect more than 4 bytes, a 4-byte-correcting code can correct any single burst of length $3m + 1$ or less. It can also simultaneously correct any combination of two bursts of length $m + 1$ or less because each such burst can affect no more than 2 bytes. At the same time, it can correct any combination of four or fewer random errors. In general, a t-byte-correcting binary Reed–Solomon code is capable of correcting any combination of

$$\lambda = \frac{t}{1 + \lfloor (l + m - 2)/m \rfloor}$$

or fewer bursts of length l [or correcting any single burst of length $(t - 1)m + 1$ or less]. Simultaneously, it corrects any combination of t or fewer random errors.

Concatenated Codes

Concatenation is a specific method of constructing long codes from shorter codes. This method was first proposed by Forney [36] as a means of constructing long block codes which can be decoded without the equipment complexity usually required in the use of long codes. A simple concatenated code is formed from two codes: an

(n_1, k_1) binary code C_1 and an (n_2, k_2) nonbinary code C_2 with symbols form GF(2^{k_1}). The symbols of C_2 are represented by their corresponding bytes of k_1 binary symbols. Usually, a Reed–Solomon code is used for C_2. Encoding consists of two steps, as shown in Figure 9.8. First, the k_1k_2 binary information digits are divided into k_2 bytes of k_1 information digits each. These k_2 bytes are encoded according to the rules for C_2 to form an n_2-byte code vector. Second, each k_1-digit byte is encoded into a code vector in C_1, resulting in a string of n_2 code vectors of C_1, a total of n_2n_1 digits. These digits are then transmitted, one C_1 code vector at a time, in succession. Thus, the resultant code is an (n_1n_2, k_1k_2) binary linear code. The component codes C_1 and C_2 are called the *inner* and the *outer* codes, respectively. If the minimum distances of the inner and outer codes are d_1 and d_2, respectively, the minimum distance of their concatenation is at least d_1d_2 (see Problem 9.18).

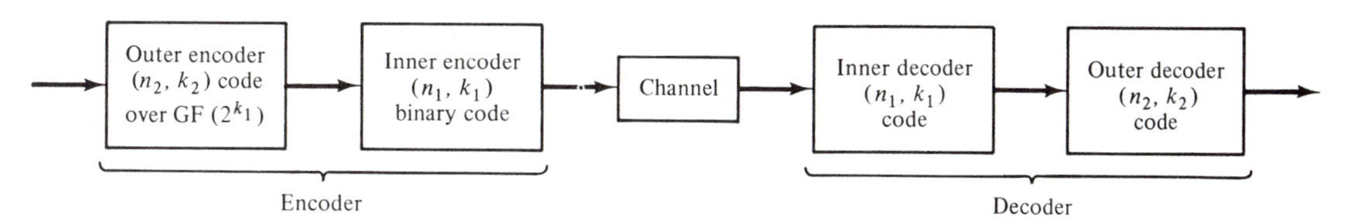

Figure 9.8 Communication system using a concatenated code.

Decoding of the concatenated code of C_1 and C_2 is also done in two steps, as shown in Figure 9.8. First, decoding is done for each C_1 code vector as it arrives, and the check digits are removed, leaving a sequence of n_2k_1-digit bytes. These bytes are then decoded according to the method for C_2, to leave the final corrected message. Decoding implementation is the straightforward combination of the implementations for codes C_1 and C_2, and the amount of hardware required is roughly that required by both codes.

Concatenated codes are effective against a mixture of random errors and bursts, and the pattern of bytes not correctable by the C_1 code must form a correctable error pattern for C_2 if the concatenated code is to correct the error pattern. Scattered random errors are corrected by C_1. Bursts may affect relatively few bytes, but probably so badly that C_1 cannot correct them. These few bytes can then be corrected by C_2.

Example 9.3

Consider the concatenation of the (15, 11) Reed–Solomon code with symbols from GF(2^4) and the (7, 4) binary Hamming code. Each code symbol of the Reed–Solomon code is represented by a byte of four binary digits, as in Table 2.8. Then each 4-bit byte is encoded into a code vector in the (7, 4) Hamming code. The resultant concatenated code is a (105, 44) binary code. Since the minimum distance of the (7, 4) Hamming code is 3 and the minimum distance of the (15, 11) Reed–Solomon code is 5, the concatenated code has a minimum distance at least 15. If the code is decoded in two steps, first the inner code and then the outer code, the decoder is capable of correcting any error pattern such that the number of inner code vectors with more than a single error is less than 3.

It is possible to construct a concatenated code from a single outer code and *multiple* inner codes. For example, one may use an (n_2, k_2) code C_2 with symbols from GF(2^{k_1}) as the outer code and n_2 (n_1, k_1) binary codes, $C_1^{(1)}, C_1^{(2)}, \ldots, C_1^{(n_2)}$, as inner codes. Again, the encoding consists of two steps. First the k_1k_2 information digits are divided into k_2 bytes of k_1 digits each. These k_2 bytes are encoded according to the rules for C_2 to form an n_2-byte code word,

$$(a_0, a_1, \ldots, a_{n_2-1}),$$

where each byte a_i is regarded as an element in GF(2^{k_1}). Second, the ith byte a_i for $0 \leq i < n_2$ is encoded into a code vector in the ith inner code $C_1^{(i)}$. The overall encoding again results in an (n_1n_2, k_1k_2) concatenated code. By concatenating one outer code with multiple inner codes, Justesen [37] was able to construct a class of asymptotically good concatenated codes.

Another interesting generalization of concatenated codes is due to Hirasawa et al. [38]. The outer code of this generalization is formed by interleaving λ codes with symbols from GF(2^k). The jth component of the outer code is an (n, k_j) code with minimum distance d_j and $n \leq 2^k$. Only one inner code is used; it is an $(N, \lambda k)$ binary code. The resultant concatenated code is an (n_0, k_0) code with $n_0 = nN$ and $k_0 = k(k_1 + k_2 + \cdots + k_\lambda)$.

9.7 MODIFIED FIRE CODES FOR SIMULTANEOUS CORRECTION OF BURST AND RANDOM ERRORS

Let β be an element of order n in the Galois field GF(2^m). It follows from Theorem 2.5 that n is a factor of $2^m - 1$. Let $\phi(X)$ be the minimal polynomial of β. The period of $\phi(X)$ is n. The degree of $\phi(X)$, m_0, is either equal to m or a factor of m. Suppose that n has a proper factor b such that

$$\frac{b+1}{2} \leq m_0.$$

Let $n = a \cdot b$. Then

$$X^n + 1 = (X^b + 1)(1 + X^b + X^{2b} + \cdots + X^{(a-1)b}).$$

Since the order of β is n, and $b < n$, β cannot be a root of $X^b + 1$ and must be a root of $1 + X^b + X^{2b} + \cdots + X^{(a-1)b}$. Therefore, $\phi(X)$ divides $1 + X^b + X^{2b} + \cdots + X^{(a-1)b}$. The code generated by

$$\mathbf{g}_1(X) = (X^b + 1)\phi(X) \tag{9.20}$$

is a Fire code C_1 of length n which is capable of correcting any error-burst of length $(b + 1)/2$ or less.

Let $\mathbf{g}_2(X)$ be the generator polynomial of a cyclic code C_2 of length n which is capable of correcting t or fewer random errors. Clearly, $\mathbf{g}_2(X)$ is a factor of $X^n + 1$. Let $\mathbf{g}(X)$ be the least common multiple of $\mathbf{g}_1(X)$ and $\mathbf{g}_2(X)$:

$$\mathbf{g}(X) = \text{LCM}\,\{\mathbf{g}_1(X), \mathbf{g}_2(X)\}. \tag{9.21}$$

Clearly $\mathbf{g}(X)$ divides $X^n + 1$ and can be expressed in the form

$$\mathbf{g}(X) = (X^b + 1)\mathbf{g}_0(X), \tag{9.22}$$

where $\mathbf{g}_0(X)$ is a factor of $1 + X^b + X^{2b} + \cdots + X^{(a-1)b}$. Now, we consider the cyclic code C of length n generated by $\mathbf{g}(X)$. This code C is a *subcode* of both the Fire code C_1 generated by $\mathbf{g}_1(X) = (X^b + 1)\phi(X)$ and the t-error-correcting code C_2 generated by $\mathbf{g}_2(X)$. Since C is a subcode of the Fire code C_1, C is capable of correcting any single error burst of length $(b + 1)/2$ or less. Since C is a subcode of the t-error-correcting code C_2, it is capable of correcting any combination of t or fewer random errors. Since $\mathbf{g}(X)$ has $(X + 1)$ as a factor, the minimum distance of C is even and is at least $2t + 2$. It is possible to show that C is capable of correcting any single error burst of length $(b + 1)/2$ or less *as well as* any combination of t or fewer random errors. To show this, it is necessary and sufficient to prove that a burst of length $(b + 1)/2$ or less and an error pattern of weight t or less cannot be in the same coset of C except that they are identical (see Problem 9.19).

Example 9.4

Let α be a primitive element of the Galois field GF(2^6). The order of α is $2^6 - 1 = 63$ and the minimal polynomial of α is

$$\phi(X) = 1 + X + X^6.$$

The integer 63 can be factored as follows: $63 = 7 \cdot 9$. Thus, we have

$$X^{63} + 1 = (X^9 + 1)(1 + X^9 + X^{18} + X^{27} + X^{36} + X^{45} + X^{54}).$$

The code generated by the polynomial

$$\mathbf{g}_1(X) = (X^9 + 1)(1 + X + X^6)$$

is a Fire code of length 63 which is capable of correcting any single error burst of length 5 or less. Let $\mathbf{g}_2(X)$ be the generator polynomial of the double-error-correcting BCH code of length 63. From Table 6.4 we find that

$$\mathbf{g}_2(X) = (1 + X + X^6)(1 + X + X^2 + X^4 + X^6).$$

Note that both factors of $\mathbf{g}_2(X)$ are factors of $1 + X^9 + X^{18} + X^{27} + X^{36} + X^{45} + X^{54}$. The least common multiple of $\mathbf{g}_1(X)$ and $\mathbf{g}_2(X)$ is

$$\mathbf{g}(X) = (X^9 + 1)(1 + X + X^6)(1 + X + X^2 + X^4 + X^6).$$

Hence, $\mathbf{g}(X)$ generates a (63, 42) cyclic code which is a subcode of both the Fire code generated by $\mathbf{g}_1(X) = (X^9 + 1)(1 + X + X^6)$ and the double-error-correcting BCH code generated by $\mathbf{g}_2(X) = (1 + X + X^6)(1 + X + X^2 + X^4 + X^6)$. Therefore, it is capable of correcting any single error burst of length 5 or less as well as any combination of two or fewer random errors.

Decoding of the codes defined by (9.21) can be implemented in a straightforward manner as shown in Figure 9.9, where the two decoders operate in parallel. Decoder I is an error-trapping decoder which is implemented based on the Fire code generated by $\mathbf{g}_1(X) = (X^b + 1)\phi(X)$; and decoder II is implemented based on the random-error-correcting code generated by $\mathbf{g}_2(X)$. The received polynomial $\mathbf{r}(X)$ is shifted into both decoders simultaneously. Both decoders attempt to decode $\mathbf{r}(X)$. The error-trapping decoder gives a decoded message only if the error pattern is either a burst of length $(b + 1)/2$ or less or an undetectable burst. The random-error decoder gives a decoded message only if the error pattern either contains t or fewer errors or is an undetectable error pattern. The only time when both decoders will provide decoded

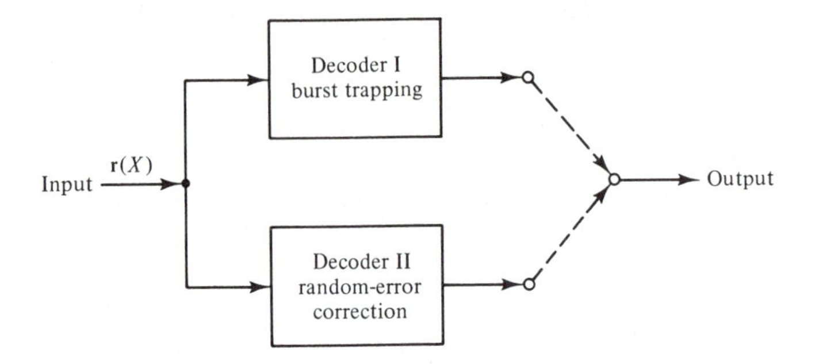

Figure 9.9 Parallel decoding for simultaneous correction of burst and random errors.

messages simultaneously is when the error pattern is in a coset with a coset leader that is a burst of length $\leq (b+1)/2$ and of weight $\leq t$. In this case, the decoded messages from the two decoders are identical. If both decoders fail to decode $\mathbf{r}(X)$, errors are detected.

Example 9.5

In Example 9.4, let us choose $\mathbf{g}_2(X)$ as the generator polynomial of the (1, 3)th-order twofold (63, 45) EG code. From Example 8.9 we find that

$$\mathbf{g}_2(X) = (1 + X + X^6)(1 + X + X^2 + X^4 + X^6)(1 + X + X^2 + X^5 + X^6).$$

Then the least common multiple of $\mathbf{g}_1(X) = (X^9 + 1)(1 + X + X^6)$ and $\mathbf{g}_2(X)$ is

$$\mathbf{g}(X) = (X^9 + 1)(1 + X + X^6)(1 + X + X^2 + X^4 + X^6)(1 + X + X^2 + X^5 + X^6).$$

Hence, $\mathbf{g}(X)$ generates a (63, 36) cyclic code which is capable of correcting any single error burst of length 5 or less as well as any three or fewer random errors with majority-logic decoding.

There is a (63, 36) BCH code which is capable of correcting any combination of five or fewer errors. Clearly, this BCH code is more powerful than the (63, 36) code above. However, decoding for the (63, 36) BCH code is more complex than the decoding of the (63, 36) code above.

By combining Fire codes and BCH codes and with the aid of a computer, Hsu et al. have constructed several classes of shortened cyclic codes which are capable of correcting burst errors as well as random errors [39]. Other works on constructing burst-and-random error-correcting block codes can be found in References 11, 22, and 39 to 42.

PROBLEMS

9.1. Show that if an (n, k) cyclic code is designed to correct all burst errors of length l or less and simultaneously to detect all burst errors of length $d \geq l$ or less, the number of parity-check digits of the code must be at least $l + d$.

9.2. Devise an error-trapping decoder for an l-burst-error-correcting cyclic code. The received polynomial is shifted into the syndrome register from the right end. Describe the decoding operation of your decoder.

9.3. The polynomial $\mathbf{p}(X) = 1 + X + X^4$ is a primitive polynomial over GF(2). Find the generator polynomial of a Fire code that is capable of correcting any single error burst of length 4 or less. What is the length of this code? Devise a simple error-trapping decoder for this code.

9.4. Devise a high-speed error-trapping decoder for the Fire code constructed in Problem 9.3. Describe the decoding operation.

9.5. Use a code from Table 9.3 to derive a new code with burst-error-correcting capability $l = 51$, length $n = 255$, and burst-error-correcting efficiency $z = 1$. Construct a decoder for this new code.

9.6. Let $\mathbf{g}(X)$ be the generator polynomial of an (n, k) cyclic code. Interleave this code to a degree λ. The resultant code is a $(\lambda n, \lambda k)$ linear code. Show that this interleaved code is cyclic and its generator polynomial is $\mathbf{g}(X^\lambda)$.

9.7. Show that the Burton code generated by $\mathbf{g}(X) = (X^m + 1)\mathbf{p}(X)$, where $\mathbf{p}(X)$ is an irreducible polynomial of degree m, is capable of correcting all phased bursts confined to a single subblock of m digits.

9.8. Let $m = 5$. Construct a Burton code that is capable of correcting any phased burst confined to a single subblock of five digits. Suppose that this code is interleaved to a degree $\lambda = 6$. What are the length, the number of parity-check digits, and the burst-error-correcting capability of this interleaved code?

9.9. Interleave the (164, 153) code in Table 9.3 to a degree $\lambda = 6$. Compare this interleaved code with the interleaved Burton code of Problem 9.8. Which code is more efficient?

9.10. Interleave the (15, 7) BCH code to a degree 7. Discuss the error-correcting capability of this interleaved code. Devise a decoder for this code and describe the decoding operation.

9.11. Let C_1 be the (3, 1) cyclic code generated by $\mathbf{g}_1(X) = 1 + X + X^2$ and let C_2 be the (7, 3) maximum-length code generated by $\mathbf{g}_2(X) = 1 + X + X^2 + X^4$. Find the generator and parity-check polynomials of the cyclic product code of C_1 and C_2. What is the minimum distance of this product code? Discuss its error-correcting capability.

9.12. In Problem 9.11, both codes C_1 and C_2 are completely orthogonalizable in one step. Show that the product of these two codes is also completely orthogonalizable in one step.

9.13. Consider the cyclic product code whose component codes are the (3, 2) cyclic code generated by $\mathbf{g}_1(X) = 1 + X$ and the (7, 4) Hamming code generated by $\mathbf{g}_2(X) = 1 + X + X^3$. The component code C_1 is completely orthogonalizable in one step and the component code C_2 is completely orthogonalizable in two steps. Show that the product code is completely orthogonalized in two steps. (In general, if one component code is completely orthogonalizable in one step and the other component code is completely orthogonalizable in L steps, the product code is completely orthogonalizable in L steps [28].)

9.14. Consider the (31, 15) Reed–Solomon code with symbols from GF(2^5). Use Table 8.8 to find its generator polynomial. Convert this RS code to a binary code. Discuss the error-correcting capability of the binary RS code.

9.15. Suppose that the Fire code constructed in Problem 9.3 is shortened by deleting the 15 high-order message digits. Devise a decoder for the shortened code such that the 15 extra shifts of the syndrome register after the received vector has entered can be avoided.

9.16. Find a modified Fire code of length 63 that is capable of correcting any single burst of length 4 or less as well as any combination of two or fewer random errors. Determine its generator polynomial.

9.17. Show that the digits for checking the parity-check digits of a product code array shown in Figure 9.6 are the same no matter whether they are formed by using the parity-check rules for C_2 on columns or the parity-check rules for C_1 on rows.

9.18. Prove that the concatenation of an (n_1, k_1) inner code with minimum distance d_1 and an (n_2, k_2) outer code with minimum distance d_2 has minimum distance at least $d_1 d_2$.

9.19. Consider the modified Fire code C generated by $\mathbf{g}(X)$ of (9.21). Show that a burst of length $(b + 1)/2$ or less and error pattern of weight t or less cannot be in the same coset.

REFERENCES

1. N. Abramson, "A Class of Systematic Codes for Non-independent Errors," *IRE Trans. Inf. Theory*, IT-4(4), pp. 150–157, December 1959.
2. N. Abramson and B. Elspas, "Double-Error-Correcting Coders and Decoders for Non-independent Binary Errors," presented at the UNESCO Inf. Process. Conf., Paris, 1959.
3. P. Fire, "A Class of Multiple-Error-Correcting Binary Codes for Non-independent Errors," Sylvania Report No. RSL-E-2, Sylvania Electronic Defense Laboratory, Reconnaissance Systems Division, Mountain View, Calif., March 1959.
4. C. M. Melas, "A New Group of Codes for Correction of Dependent Errors in Data Transmission," *IBM J. Res. Dev.*, 4, pp. 58–64, January 1960.
5. S. H. Reiger, "Codes for the Correction of 'Clustered' Errors," *IRE Trans. Inf. Theory*, IT-6, pp. 16–21, March 1960.
6. B. Elspas, "A Note on Binary Adjacent-Error-Correcting Codes," *IRE Trans. Inf. Theory*, IT-6, pp. 13–15, March 1960.
7. B. Elspas and R. A. Short, "A Note on Optimum Burst-Error-Correction Codes," *IRE Trans. Inf. Theory*, IT-8, pp. 39–42, January 1962.
8. J. E. Meggitt, "Error Correcting Codes for Correcting Bursts of Errors," *IBM J. Res. Dev.*, 4, pp. 329–334, July 1960.
9. J. E. Meggitt, "Error Correcting Codes for Correcting Bursts of Errors," *Trans. AIEE*, 80, pp. 708–711, January 1961.
10. C. R. Foulk, "Some Properties of Maximally-Efficient Cyclic Burst-Correcting Codes and Results of a Computer Search for Such Codes," File No. 375, Digital Computer Lab., University of Illinois, Urbana, Ill., June 12, 1961.
11. J. J. Stone, "Multiple Burst Error Correction," *Inf. Control*, 4, pp. 324–331, December 1961.
12. L. H. Zetterburg, "Cyclic Codes from Irreducible Polynomials for Correction of Multiple Errors," *IEEE Trans. Inf. Theory*, IT-8, pp. 13–21, January 1962.
13. T. Kasami, "Cyclic Codes for Burst-Error-Correction," *J. Inst. Elec. Comm. Eng. Jap.*, 45, pp. 9–16, January 1962.
14. T. Kasami, "Optimum Shortened Cyclic Codes for Burst-Error-Correction," *IEEE Trans. Inf. Theory*, IT-9, pp. 105–109, April 1963.

15. T. Kasami and S. Matoba, "Some Efficient Shortened Cyclic Codes for Burst-Error-Correction," *IEEE Trans. Inf. Theory*, IT-10, pp. 252–253, July 1964.
16. A. J. Gross, "Binary Group Codes Which Correct in Bursts of Three or Less for Odd Redundancy," *IEEE Trans. Inf. Theory*, IT-8, pp. 356–359, October 1962.
17. A. J. Gross, "A Note on Some Binary Group Codes Which Correct Errors in Bursts of Four or Less," *IRE Trans. Inf. Theory*, IT-8, p. 384, October 1962.
18. A. J. Gross, "Augmented Bose-Chaudhuri Codes Which Correct Single Bursts of Errors," *IEEE Trans. Inf. Theory*, IT-9, p. 121, April 1963.
19. E. Gorog, "Some New Classes of Cyclic Codes Used for Burst Error Correction," *IBM J. Res. Dev.*, 7, pp. 102–111, 1963.
20. R. T. Chien, "Burst-Correction Codes with High-Speed Decoding," *IEEE Trans. Inf. Theory*, IT-1, pp. 109–113, January 1969.
21. H. O. Burton, "A Class of Asymptotically Optimal Burst Correcting Block Codes," presented at the ICCC, Boulder, Colo., June 1969.
22. S. E. Tavares and S. G. S. Shiva, "Detecting and Correction Multiple Bursts for Binary Cyclic Codes," *IEEE Trans. Inf. Theory*, IT-16, pp. 643–644, 1970.
23. R. G. Gallager, *Information Theory and Reliable Communication*, Wiley, New York, 1968.
24. W. W. Peterson, *Error-Correcting Codes*, MIT Press, Cambridge, Mass., 1961.
25. P. Elias, "Error-Free Coding," *IRE Trans. Inf. Theory*, PGIT-4, pp. 29–37, September 1954.
26. H. O. Burton and E. J. Weldon, Jr., "Cyclic Product Codes," *IEEE Trans. Inf. Theory*, IT-11, pp. 433–440, July 1965.
27. W. W. Peterson and E. J. Weldon, Jr., *Error-Correcting Codes*, 2nd ed., MIT Press, Cambridge, Mass., 1970.
28. S. Lin and E. J. Weldon, Jr., "Further Results on Cyclic Product Codes," *IEEE Trans. Inf. Theory*, IT-6(4), pp. 452–459, July 1970.
29. N. M. Abramson, "Cascade Decoding of Cyclic Product Codes," *IEEE Trans. Commun. Technol.*, COM-16, pp. 398–402, 1968.
30. W. C. Gore, "Further Results on Product Codes," *IEEE Trans. Inf. Theory*, IT-16(4), pp. 446–451, July 1970.
31. S. M. Reddy, "On Decoding Iterated Codes," *IEEE Trans. Inf. Theory*, IT-16, pp. 624–627, September 1970.
32. E. J. Weldon, Jr., "Decoding Binary Block Codes on Q-ary Output Channels," *IEEE Trans. Inf. Theory*, IT-17, pp. 713–718, November 1971.
33. S. M. Reddy and J. P. Robinson, "Random Error and Burst Correction by Iterated Codes," *IEEE Trans. Inf. Theory*, IT-18, pp. 182–185, January 1972.
34. N. Q. Duc and L. V. Skattebol, "Further Results on Majority-Logic Decoding of Product Codes," *IEEE Trans. Inf. Theory*, IT-18(2), pp. 308–310, March 1972.
35. S. Wainberg, "Error-Erasure Decoding of Product Codes," *IEEE Trans. Inf. Theory*, IT-8(6), pp. 821–823, November 1972.
36. G. D. Forney, Jr., *Concatenated Codes*, MIT Press, Cambridge, Mass., 1966.
37. J. Justesen, "A Class of Constructive Asymptotically Good Algebraic Codes," *IEEE Trans. Inf. Theory*, IT-18(5), pp. 652–656, September 1972.

38. S. Hirasawa, M. Kasahara, Y. Sugiyama, and T. Namekawa, "Certain Generalization of Concatenated Codes—Exponential Error Bounds and Decoding Complexity," *IEEE Trans. Inf. Theory*, IT-26(5), pp. 527–534, September 1980.
39. H. T. Hsu, T. Kasami, and R. T. Chien, "Error-Correction Codes for a Compound Channel," *IEEE Trans. Inf. Theory*, It-14, pp. 135–139, January 1968.
40. J. K. Wolf, "On Codes Derivable from the Tensor Product of Check Matrices," *IEEE Trans. Inf. Theory*, IT-11, pp. 281–284, April 1965.
41. W. Posner, "Simultaneous Error-Correction and Burst-Error Detection Binary Linear Cyclic Codes," *J. Soc. Ind. Appl. Math.*, 13, pp. 1087–1095, December 1965.
42. L. Bahl and R. T. Chien, "A Class of Multiple-Burst-Error-Correcting Codes," presented at the IEEE Int. Symp. Inf. Theory, Ellenville, N.Y., 1969.

10

Convolutional Codes

As noted in Chapter 1, convolutional codes differ from block codes in that the encoder contains memory and the n encoder outputs at any given time unit depend not only on the k inputs at that time unit but also on m previous input blocks. An (n, k, m) convolutional code can be implemented with a k-input, n-output linear sequential circuit with input memory m. Typically, n and k are small integers with $k < n$, but the memory order m must be made large to achieve low error probabilities. In the important special case when $k = 1$, the information sequence is not divided into blocks and can be processed continuously.

Convolutional codes were first introduced by Elias [1] in 1955 as an alternative to block codes. Shortly thereafter, Wozencraft [2] proposed sequential decoding as an efficient decoding scheme for convolutional codes, and experimental studies soon began to appear. In 1963, Massey [3] proposed a less efficient but simpler-to-implement decoding method called threshold decoding. This advance spawned a number of practical applications of convolutional codes to digital transmission over wire and radio channels. Then in 1967, Viterbi [4] proposed a maximum likelihood decoding scheme that was relatively easy to implement for codes with small memory orders. This scheme, called Viterbi decoding, together with improved versions of sequential decoding, led to the application of convolutional codes to deep-space and satellite communication in the early 1970s. A complete survey of the applications of convolutional codes to practical communication channels is given in Chapter 17.

10.1 ENCODING OF CONVOLUTIONAL CODES

The encoder for a binary (2, 1, 3) code is shown in Figure 10.1. Note that the encoder consists of an $m = 3$-stage shift register together with $n = 2$ modulo-2 adders and a multiplexer for serializing the encoder outputs. The mod-2 adders can be implemented as EXCLUSIVE-OR gates. Since mod-2 addition is a linear operation, the encoder is a linear feedforward shift register. All convolutional encoders can be implemented using a linear feedforward shift register of this type.

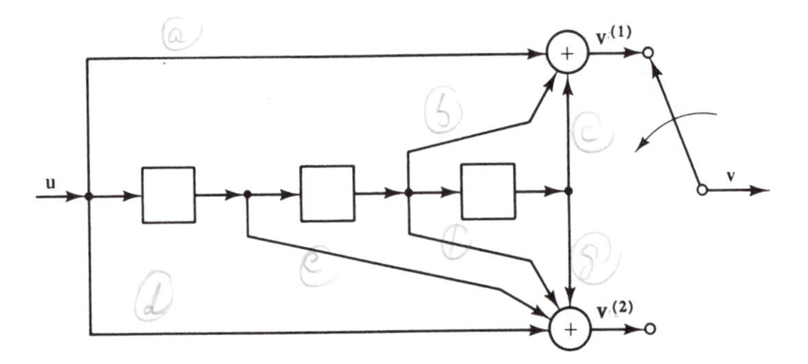

Figure 10.1 A (2, 1, 3) binary convolutional encoder.

The *information sequence* $\mathbf{u} = (u_0, u_1, u_2, \ldots)$ enters the encoder one bit at a time. Since the encoder is a linear system, the two encoder *output sequences* $\mathbf{v}^{(1)} = (v_0^{(1)}, v_1^{(1)}, v_2^{(1)}, \ldots)$ and $\mathbf{v}^{(2)} = (v_0^{(2)}, v_1^{(2)}, v_2^{(2)}, \ldots)$ can be obtained as the convolution of the input sequence $\mathbf{u}$ with the two encoder "impulse responses." The impulse responses are obtained by letting $\mathbf{u} = (1\ 0\ 0\ \cdots)$ and observing the two output sequences. Since the encoder has an m-time unit memory, the impulse responses can last at most $m + 1$ time units, and are written $\mathbf{g}^{(1)} = (g_0^{(1)}, g_1^{(1)}, \ldots, g_m^{(1)})$ and $\mathbf{g}^{(2)} = (g_0^{(2)}, g_1^{(2)}, \ldots, g_m^{(2)})$. For the encoder of Figure 10.1,

$$\mathbf{g}^{(1)} = (1\ \ 0\ \ 1\ \ 1)$$
$$\mathbf{g}^{(2)} = (1\ \ 1\ \ 1\ \ 1).$$

The impulse responses $\mathbf{g}^{(1)}$ and $\mathbf{g}^{(2)}$ are called the *generator sequences* of the code.

The *encoding equations* can now be written as

$$\mathbf{v}^{(1)} = \mathbf{u} * \mathbf{g}^{(1)} \tag{10.1a}$$
$$\mathbf{v}^{(2)} = \mathbf{u} * \mathbf{g}^{(2)}, \tag{10.1b}$$

where $*$ denotes discrete convolution and all operations are modulo-2. The convolution operation implies that for all $l \geq 0$,

$$v_l^{(j)} = \sum_{i=0}^{m} u_{l-i} g_i^{(j)} = u_l g_0^{(j)} + u_{l-1} g_1^{(j)} + \cdots + u_{l-m} g_m^{(j)}, \qquad j = 1, 2,. \tag{10.2}$$

where $u_{l-i} \triangleq 0$ for all $l < i$. Hence, for the encoder of Figure 10.1,

$$v_l^{(1)} = u_l \qquad\qquad + u_{l-2} + u_{l-3}$$
$$v_l^{(2)} = u_l + u_{l-1} + u_{l-2} + u_{l-3},$$

as can easily be verified by direct inspection of the encoding circuit. After encoding, the two output sequences are multiplexed into a single sequence, called the *code word*, for transmission over the channel. The code word is given by

$$\mathbf{v} = (v_0^{(1)} v_0^{(2)}, v_1^{(1)} v_1^{(2)}, v_2^{(1)} v_2^{(2)}, \ldots).$$

Example 10.1

Let the information sequence $\mathbf{u} = (1\ 0\ 1\ 1\ 1)$. Then the output sequences are

$$\mathbf{v}^{(1)} = (1\quad 0\quad 1\quad 1\quad 1) * (1\quad 0\quad 1\quad 1) = (1\quad 0\quad 0\quad 0\quad 0\quad 0\quad 0\quad 1)$$

$$\mathbf{v}^{(2)} = (1\quad 0\quad 1\quad 1\quad 1) * (1\quad 1\quad 1\quad 1) = (1\quad 1\quad 0\quad 1\quad 1\quad 1\quad 0\quad 1)$$

and the code word is

$$\mathbf{v} = (1\ \ 1,\ \ 0\ \ 1,\ \ 0\ \ 0,\ \ 0\ \ 1,\ \ 0\ \ 1,\ \ 0\ \ 1,\ \ 0\ \ 0,\ \ 1\ \ 1).$$

If the generator sequences $\mathbf{g}^{(1)}$ and $\mathbf{g}^{(2)}$ are interlaced and then arranged in the matrix

$$\mathbf{G} = \begin{bmatrix} g_0^{(1)}g_0^{(2)} & g_1^{(1)}g_1^{(2)} & g_2^{(1)}g_2^{(2)} & \cdots & g_m^{(1)}g_m^{(2)} & & \\ & g_0^{(1)}g_0^{(2)} & g_1^{(1)}g_1^{(2)} & \cdots & g_{m-1}^{(1)}g_{m-1}^{(2)} & g_m^{(1)}g_m^{(2)} & \\ & & g_0^{(1)}g_0^{(2)} & \cdots & g_{m-2}^{(1)}g_{m-2}^{(2)} & g_{m-1}^{(1)}g_{m-1}^{(2)} & g_m^{(1)}g_m^{(2)} \\ & & \cdot & & & & \cdot \\ & & \cdot & & & & \cdot \\ & & \cdot & & & & \cdot \end{bmatrix}, \tag{10.3}$$

where the blank areas are all zeros, the encoding equations can be rewritten in matrix form as

$$\mathbf{v} = \mathbf{uG}, \tag{10.4}$$

where all operations are modulo-2. $\mathbf{G}$ is called the *generator matrix* of the code. Note that each row of $\mathbf{G}$ is identical to the preceding row but shifted $n = 2$ places to the right, and that $\mathbf{G}$ is a semi-infinite matrix, corresponding to the fact that the information sequence $\mathbf{u}$ is of arbitrary length. If $\mathbf{u}$ has finite length L, then $\mathbf{G}$ has L rows and $2(m + L)$ columns, and $\mathbf{v}$ has length $2(m + L)$.

Example 10.2

If $\mathbf{u} = (1\ 0\ 1\ 1\ 1)$, then

$$\begin{aligned} \mathbf{v} &= \mathbf{uG} \\ &= (1\quad 0\quad 1\quad 1\quad 1)\begin{bmatrix} 11 & 01 & 11 & 11 & & & & \\ & 11 & 01 & 11 & 11 & & & \\ & & 11 & 01 & 11 & 11 & & \\ & & & 11 & 01 & 11 & 11 & \\ & & & & 11 & 01 & 11 & 11 \end{bmatrix} \\ &= (1\ \ 1,\ \ 0\ \ 1,\ \ 0\ \ 0,\ \ 0\ \ 1,\ \ 0\ \ 1,\ \ 0\ \ 1,\ \ 0\ \ 0,\ \ 1\ \ 1), \end{aligned}$$

agreeing with our previous calculation using discrete convolution.

As a second example of a convolutional encoder, consider the (3, 2, 1) code shown in Figure 10.2. Since $k = 2$, the encoder consists of two $m = 1$-stage shift registers together with $n = 3$ mod-2 adders and two multiplexers. The information sequence enters the encoder $k = 2$ bits at a time, and can be written as $\mathbf{u} = (u_0^{(1)} u_0^{(2)},$

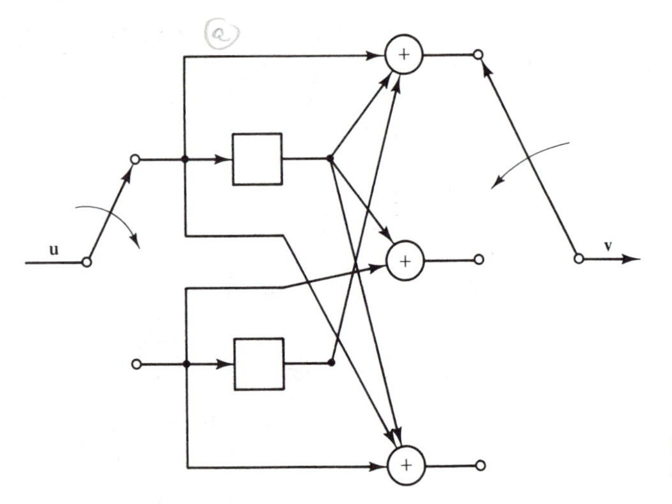

Figure 10.2 A (3, 2, 1) binary convolutional encoder.

$u_1^{(1)}u_1^{(2)}, u_2^{(1)}u_2^{(2)}, \ldots)$ or as the two input sequences $\mathbf{u}^{(1)} = (u_0^{(1)}, u_1^{(1)}, u_2^{(1)}, \ldots)$ and $\mathbf{u}^{(2)} = (u_0^{(2)}, u_1^{(2)}, u_2^{(2)}, \ldots)$. There are three generator sequences corresponding to each input sequence. Letting $\mathbf{g}_i^{(j)} = (g_{i,0}^{(j)}, g_{i,1}^{(j)}, \ldots, g_{i,m}^{(j)})$ represent the generator sequence corresponding to input i and output j, the generator sequences for the encoder of Figure 10.2 are

$$\mathbf{g}_1^{(1)} = (1 \quad 1), \qquad \mathbf{g}_1^{(2)} = (0 \quad 1), \qquad \mathbf{g}_1^{(3)} = (1 \quad 1),$$
$$\mathbf{g}_2^{(1)} = (0 \quad 1), \qquad \mathbf{g}_2^{(2)} = (1 \quad 0), \qquad \mathbf{g}_2^{(3)} = (1 \quad 0),$$

and the encoding equations can be written as

$$\mathbf{v}^{(1)} = \mathbf{u}^{(1)} * \mathbf{g}_1^{(1)} + \mathbf{u}^{(2)} * \mathbf{g}_2^{(1)} \tag{10.5a}$$
$$\mathbf{v}^{(2)} = \mathbf{u}^{(1)} * \mathbf{g}_1^{(2)} + \mathbf{u}^{(2)} * \mathbf{g}_2^{(2)} \tag{10.5b}$$
$$\mathbf{v}^{(3)} = \mathbf{u}^{(1)} * \mathbf{g}_1^{(3)} + \mathbf{u}^{(2)} * \mathbf{g}_2^{(3)}. \tag{10.5c}$$

The convolution operation implies that

$$v_l^{(1)} = u_l^{(1)} + \qquad + u_{l-1}^{(1)} + u_{l-1}^{(2)}$$
$$v_l^{(2)} = \qquad u_l^{(2)} + u_{l-1}^{(1)}$$
$$v_l^{(3)} = u_l^{(1)} + u_l^{(2)} + u_{l-1}^{(1)},$$

as can be seen from the encoding circuit. After multiplexing, the code word is given by

$$\mathbf{v} = (v_0^{(1)}v_0^{(2)}v_0^{(3)}, v_1^{(1)}v_1^{(2)}v_1^{(3)}, v_2^{(1)}v_2^{(2)}v_2^{(3)}, \ldots).$$

Example 10.3

If $\mathbf{u}^{(1)} = (1\ 0\ 1)$ and $\mathbf{u}^{(2)} = (1\ 1\ 0)$, then

$$\mathbf{v}^{(1)} = (1 \quad 0 \quad 1) * (1 \quad 1) + (1 \quad 1 \quad 0) * (0 \quad 1) = (1 \quad 0 \quad 0 \quad 1)$$
$$\mathbf{v}^{(2)} = (1 \quad 0 \quad 1) * (0 \quad 1) + (1 \quad 1 \quad 0) * (1 \quad 0) = (1 \quad 0 \quad 0 \quad 1)$$
$$\mathbf{v}^{(3)} = (1 \quad 0 \quad 1) * (1 \quad 1) + (1 \quad 1 \quad 0) * (1 \quad 0) = (0 \quad 0 \quad 1 \quad 1)$$

and

$$\mathbf{v} = (1 \quad 1 \quad 0, \quad 0 \quad 0 \quad 0, \quad 0 \quad 0 \quad 1, \quad 1 \quad 1 \quad 1).$$

The generator matrix of a (3, 2, m) code is

$$\mathbf{G} = \begin{bmatrix} g_{1,0}^{(1)}g_{1,0}^{(2)}g_{1,0}^{(3)} & g_{1,1}^{(1)}g_{1,1}^{(2)}g_{1,1}^{(3)} & \cdots & g_{1,m}^{(1)} \; g_{1,m}^{(2)} \; g_{1,m}^{(3)} & & \\ g_{2,0}^{(1)}g_{2,0}^{(2)}g_{2,0}^{(3)} & g_{2,1}^{(1)}g_{2,1}^{(2)}g_{2,1}^{(3)} & \cdots & g_{2,m}^{(1)} \; g_{2,m}^{(2)} \; g_{2,m}^{(3)} & & \\ & g_{1,0}^{(1)}g_{1,0}^{(2)}g_{1,0}^{(3)} & \cdots & g_{1,m-1}^{(1)}g_{1,m-1}^{(2)}g_{1,m-1}^{(3)} & g_{1,m}^{(1)}g_{1,m}^{(2)}g_{1,m}^{(3)} & \\ & g_{2,0}^{(1)}g_{2,0}^{(2)}g_{2,0}^{(3)} & \cdots & g_{2,m-1}^{(1)}g_{2,m-1}^{(2)}g_{2,m-1}^{(3)} & g_{2,m}^{(1)}g_{2,m}^{(2)}g_{2,m}^{(3)} & \\ & & \ddots & & & \ddots \end{bmatrix} \tag{10.6}$$

and the encoding equations in matrix form are again given by $\mathbf{v} = \mathbf{uG}$. Note that each set of $k = 2$ rows of $\mathbf{G}$ is identical to the preceding set of rows but shifted $n = 3$ places to the right.

Example 10.4

If $\mathbf{u}^{(1)} = (1 \;\; 0 \;\; 1)$ and $\mathbf{u}^{(2)} = (1 \;\; 1 \;\; 0)$, then $\mathbf{u} = (1 \;\; 1, \;\; 0 \;\; 1, \;\; 1 \;\; 0)$ and

$$\begin{aligned} \mathbf{v} &= \mathbf{uG} \\ &= (1 \;\; 1, \;\; 0 \;\; 1, \;\; 1 \;\; 0) \begin{bmatrix} 1\,0\,1 & 1\,1\,1 & & \\ 0\,1\,1 & 1\,0\,0 & & \\ & 1\,0\,1 & 1\,1\,1 & \\ & 0\,1\,1 & 1\,0\,0 & \\ & & 1\,0\,1 & 1\,1\,1 \\ & & 0\,1\,1 & 1\,0\,0 \end{bmatrix} \\ &= (1 \;\; 1 \;\; 0, \;\; 0 \;\; 0 \;\; 0, \;\; 0 \;\; 0 \;\; 1, \;\; 1 \;\; 1 \;\; 1), \end{aligned}$$

again agreeing with our previous calculation using discrete convolution.

This second example clearly illustrates that encoding, and the notation used to describe it, is somewhat complicated when the number of input sequences $k > 1$. In particular, the encoder now contains k shift registers, not all of which must have the same length. If K_i is the length of the ith shift register, then the encoder *memory order* m is defined as

$$m \triangleq \max_{1 \le i \le k} K_i \tag{10.7}$$

(i.e., the maximum length of all k shift registers). An example of a (4, 3, 2) convolutional encoder in which the shift register lengths are 0, 1, and 2 is shown in Figure 10.3. In Chapter 1 the *constraint length* was defined as

$$n_A \triangleq n(m + 1). \tag{10.8}$$

Since each information bit remains in the encoder for up to $m + 1$ time units, and during each time unit can affect any of the n encoder outputs (depending on the shift register connections), n_A can be interpreted as the maximum number of encoder outputs that can be affected by a single information bit. For example, the constraint lengths of the codes in Figures 10.1, 10.2, and 10.3 are 8, 6, and 12, respectively.

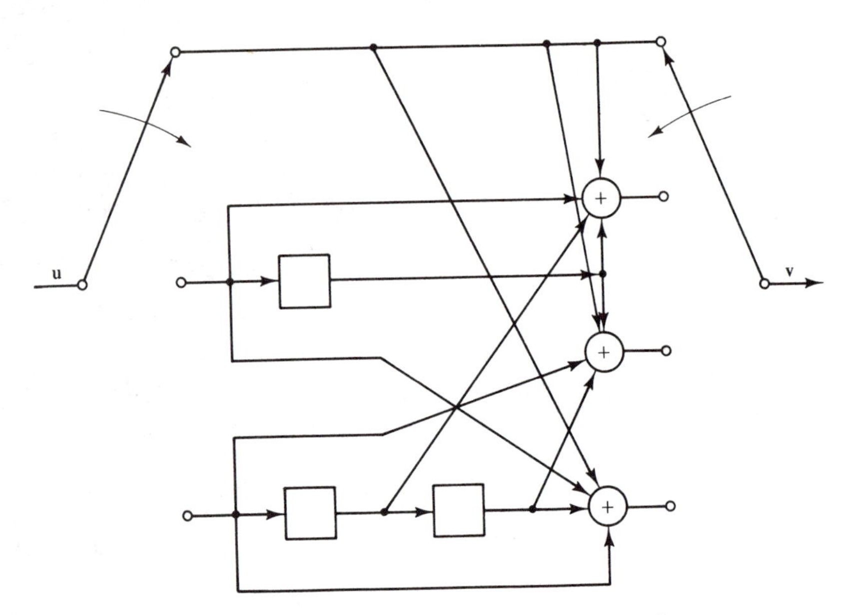

Figure 10.3 A (4, 3, 2) binary convolutional encoder.

In the general case of an (n, k, m) code, the generator matrix is

$$\mathbf{G} = \begin{bmatrix} \mathbf{G}_0 & \mathbf{G}_1 & \mathbf{G}_2 & \cdots & \mathbf{G}_m & & & \\ & \mathbf{G}_0 & \mathbf{G}_1 & \cdots & \mathbf{G}_{m-1} & \mathbf{G}_m & & \\ & & \mathbf{G}_0 & \cdots & \mathbf{G}_{m-2} & \mathbf{G}_{m-1} & \mathbf{G}_m & \\ & & & \ddots & & & & \ddots \end{bmatrix}, \tag{10.9}$$

where each $\mathbf{G}_l$ is a $k \times n$ submatrix whose entries are

$$\mathbf{G}_l = \begin{bmatrix} g_{1,l}^{(1)} & g_{1,l}^{(2)} & \cdots & g_{1,l}^{(n)} \\ g_{2,l}^{(1)} & g_{2,l}^{(2)} & \cdots & g_{2,l}^{(n)} \\ \vdots & \vdots & & \vdots \\ g_{k,l}^{(1)} & g_{k,l}^{(2)} & \cdots & g_{k,l}^{(n)} \end{bmatrix}. \tag{10.10}$$

Again note that each set of k rows of $\mathbf{G}$ is identical to the previous set of rows but shifted n places to the right. For an information sequence $\mathbf{u} = (\mathbf{u}_0, \mathbf{u}_1, \ldots) = (u_0^{(1)}u_0^{(2)} \cdots u_0^{(k)}, u_1^{(1)}u_1^{(2)} \cdots u_1^{(k)}, \ldots)$, the code word $\mathbf{v} = (\mathbf{v}_0, \mathbf{v}_1, \ldots) = (v_0^{(1)}v_0^{(2)} \cdots v_0^{(n)}, v_1^{(1)}v_1^{(2)} \cdots v_1^{(n)}, \ldots)$ is again given by $\mathbf{v} = \mathbf{uG}$. Since the code word $\mathbf{v}$ is a linear combination of rows of the generator matrix $\mathbf{G}$, an (n, k, m) convolutional code is a linear code.

A convolutional encoder generates n encoded bits for each k information bits, and $R = k/n$ is called the *code rate*. Note, however, that for an information sequence of finite length $k \cdot L$, the corresponding code word has length $n(L + m)$, where the final $n \cdot m$ outputs are generated after the last nonzero information block has entered the encoder. In other words, an information sequence is terminated with all-zero blocks in order to allow the encoder memory to clear. Viewing a convolutional code as a linear block code with generator matrix **G**, the block code rate is given by $kL/n(L+m)$, the ratio of the number of information bits to the length of the code word. If $L \gg m$, then $L/(L+m) \approx 1$, and the block code rate and convolutional code rate are approximately equal. This represents the normal mode of operation for convolutional codes, and henceforth we will not distinguish between the rate of a convolutional code and its rate when viewed as a block code. If L were small, however, the ratio $kL/n(L+m)$, which is the effective rate of information transmission, would be reduced below the code rate by a fractional amount

$$\frac{k/n - kL/n(L+m)}{k/n} = \frac{m}{L+m} \tag{10.11}$$

called the *fractional rate loss*. Hence, to keep the fractional rate loss small (i.e., near zero), L is always assumed to be much larger than m.

Example 10.5

For the information sequence in Example 10.1, $L = 5$ and the fractional rate loss is $3/8 = 37.5\%$. However, for an information sequence of length $L = 1000$, the fractional rate loss is only $3/1003 = 0.3\%$.

In any linear system, time-domain operations involving convolution can be replaced by more convenient transform-domain operations involving polynomial multiplication. Since a convolutional encoder is a linear system, each sequence in the encoding equations can be replaced by a corresponding polynomial, and the convolution operation replaced by polynomial multiplication. In the polynomial representation of a binary sequence, the sequence itself is represented by the coefficients of the polynomial. For example, for a $(2, 1, m)$ code, the encoding equations become

$$\mathbf{v}^{(1)}(D) = \mathbf{u}(D)\mathbf{g}^{(1)}(D) \tag{10.12a}$$

$$\mathbf{v}^{(2)}(D) = \mathbf{u}(D)\mathbf{g}^{(2)}(D), \tag{10.12b}$$

where

$$\mathbf{u}(D) = u_0 + u_1 D + u_2 D^2 + \cdots$$

is the information sequence,

$$\mathbf{v}^{(1)}(D) = v_0^{(1)} + v_1^{(1)} D + v_2^{(1)} D^2 + \cdots$$

and

$$\mathbf{v}^{(2)}(D) = v_0^{(2)} + v_1^{(2)} D + v_2^{(2)} D^2 + \cdots$$

are the encoded sequences,

$$\mathbf{g}^{(1)}(D) = g_0^{(1)} + g_1^{(1)} D + \cdots + g_m^{(1)} D^m$$

and

$$\mathbf{g}^{(2)}(D) = g_0^{(2)} + g_1^{(2)} D + \cdots + g_m^{(2)} D^m$$

are the *generator polynomials* of the code, and all operations are modulo-2. After multiplexing, the code word becomes

$$\mathbf{v}(D) = \mathbf{v}^{(1)}(D^2) + D\mathbf{v}^{(2)}(D^2). \tag{10.13}$$

The indeterminate D can be interpreted as a delay operator, the power of D denoting the number of time units a bit is delayed with respect to the initial bit in the sequence.

Example 10.6

For the (2, 1, 3) code of Figure 10.1, the generator polynomials are $\mathbf{g}^{(1)}(D) = 1 + D^2 + D^3$ and $\mathbf{g}^{(2)}(D) = 1 + D + D^2 + D^3$. For the information sequence $\mathbf{u}(D) = 1 + D^2 + D^3 + D^4$, the encoding equations are

$$\mathbf{v}^{(1)}(D) = (1 + D^2 + D^3 + D^4)(1 + D^2 + D^3) = 1 + D^7$$

$$\mathbf{v}^{(2)}(D) = (1 + D^2 + D^3 + D^4)(1 + D + D^2 + D^3)$$
$$= 1 + D + D^3 + D^4 + D^5 + D^7,$$

and the code word is

$$\mathbf{v}(D) = 1 + D + D^3 + D^7 + D^9 + D^{11} + D^{14} + D^{15}.$$

Note that in each case the result is the same as previously computed using convolution and matrix multiplication.

The generator polynomials of an encoder can be determined directly from its circuit diagram. Since each shift register stage represents a one-time-unit delay, the sequence of connections (a 1 representing a connection and a 0 no connection) from a shift register to an output is the sequence of coefficients in the corresponding generator polynomial (i.e., it is the generator sequence). For example, in Figure 10.1, the sequence of connections from the shift register to the first output is $\mathbf{g}^{(1)} = (1\ 0\ 1\ 1)$, and the corresponding generator polynomial is $\mathbf{g}^{(1)}(D) = 1 + D^2 + D^3$.

Since the last stage of the shift register in an $(n, 1)$ code must be connected to at least one output, the degree of at least one generator polynomial must be equal to the shift register length m, that is,

$$m = \max_{1 \leq j \leq n} [\deg \mathbf{g}^{(j)}(D)]. \tag{10.14}$$

In an (n, k) code where $k > 1$, there are n generator polynomials for each of the k inputs. Each set of n generators represents the connections from one of the shift registers to the n outputs. Hence, the length K_i of the ith shift register is given by

$$K_i = \max_{1 \leq j \leq n} [\deg \mathbf{g}_i^{(j)}(D)], \qquad 1 \leq i \leq k, \tag{10.15}$$

where $\mathbf{g}_i^{(j)}(D)$ is the generator polynomial relating the ith input to the jth output, and the encoder memory order m is

$$m = \max_{1 \leq i \leq k} K_i = \max_{\substack{1 \leq j \leq n \\ 1 \leq i \leq k}} [\deg \mathbf{g}_i^{(j)}(D)]. \tag{10.16}$$

Since the encoder is a linear system, and $\mathbf{u}^{(i)}(D)$ represents the ith input sequence and $\mathbf{v}^{(j)}(D)$ represents the jth output sequence, the generator polynomial $\mathbf{g}_i^{(j)}(D)$ can be interpreted as the encoder transfer function relating input i to output j. As with any k-input, n-output linear system, there are a total of $k \cdot n$ transfer functions. These can be represented by the $k \times n$ *transfer function matrix*

$$\mathbf{G}(D) = \begin{bmatrix} \mathbf{g}_1^{(1)}(D) & \mathbf{g}_1^{(2)}(D) & \cdots & \mathbf{g}_1^{(n)}(D) \\ \mathbf{g}_2^{(1)}(D) & \mathbf{g}_2^{(2)}(D) & \cdots & \mathbf{g}_2^{(n)}(D) \\ \vdots & \vdots & & \vdots \\ \mathbf{g}_k^{(1)}(D) & \mathbf{g}_k^{(2)}(D) & \cdots & \mathbf{g}_k^{(n)}(D) \end{bmatrix}. \tag{10.17}$$

Using the transfer function matrix, the encoding equations for an (n, k, m) code can be expressed as

$$\mathbf{V}(D) = \mathbf{U}(D)\mathbf{G}(D), \tag{10.18}$$

where $\mathbf{U}(D) \triangleq [\mathbf{u}^{(1)}(D), \mathbf{u}^{(2)}(D), \ldots, \mathbf{u}^{(k)}(D)]$ is the k-tuple of input sequences and $\mathbf{V}(D) \triangleq [\mathbf{v}^{(1)}(D), \mathbf{v}^{(2)}(D), \ldots, \mathbf{v}^{(n)}(D)]$ is the n-tuple of output sequences. After multiplexing, the code word becomes

$$\mathbf{v}(D) = \mathbf{v}^{(1)}(D^n) + D\mathbf{v}^{(2)}(D^n) + \cdots + D^{n-1}\mathbf{v}^{(n)}(D^n). \tag{10.19}$$

Example 10.7

For the (3, 2, 1) code of Figure 10.2,

$$\mathbf{G}(D) = \begin{bmatrix} 1 + D & D & 1 + D \\ D & 1 & 1 \end{bmatrix}.$$

For the input sequences $\mathbf{u}^{(1)}(D) = 1 + D^2$ and $\mathbf{u}^{(2)}(D) = 1 + D$, the encoding equations are

$$\begin{aligned} \mathbf{V}(D) = [\mathbf{v}^{(1)}(D), \mathbf{v}^{(2)}(D), \mathbf{v}^{(3)}(D)] &= [1 + D^2, 1 + D]\begin{bmatrix} 1 + D & D & 1 + D \\ D & 1 & 1 \end{bmatrix} \\ &= [1 + D^3, 1 + D^3, D^2 + D^3] \end{aligned}$$

and the code word is

$$\mathbf{v}(D) = 1 + D + D^8 + D^9 + D^{10} + D^{11}.$$

Again these results are the same as for those calculated using convolution and matrix multiplication.

Equations (10.17), (10.18), and (10.19) can be rewritten to provide a means of representing the code word $\mathbf{v}(D)$ directly in terms of the input sequences. A little algebraic manipulation yields

$$\mathbf{v}(D) = \sum_{i=1}^{k} \mathbf{u}^{(i)}(D^n)\mathbf{g}_i(D), \tag{10.20}$$

where

$$\mathbf{g}_i(D) \triangleq \mathbf{g}_i^{(1)}(D^n) + D\mathbf{g}_i^{(2)}(D^n) + \cdots + D^{n-1}\mathbf{g}_i^{(n)}(D^n), \qquad 1 \le i \le k, \tag{10.21}$$

is a *composite generator polynomial* relating the ith input sequence to $\mathbf{v}(D)$.

Example 10.8

For the (2, 1, 3) code of Figure 10.1, the composite generator polynomial is

$$\mathbf{g}(D) = \mathbf{g}^{(1)}(D^2) + D\mathbf{g}^{(2)}(D^2) = 1 + D + D^3 + D^4 + D^5 + D^6 + D^7,$$

and for $\mathbf{u}(D) = 1 + D^2 + D^3 + D^4$, the code word is

$$\begin{aligned} \mathbf{v}(D) = \mathbf{u}(D^2)\mathbf{g}(D) &= (1 + D^4 + D^6 + D^8)(1 + D + D^3 + D^4 + D^5 + D^6 + D^7) \\ &= 1 + D + D^3 + D^7 + D^9 + D^{11} + D^{14} + D^{15}, \end{aligned}$$

again agreeing with previous calculations.

10.2 STRUCTURAL PROPERTIES OF CONVOLUTIONAL CODES

Since a convolutional encoder is a sequential circuit, its operation can be described by a state diagram. The state of the encoder is defined as its shift register contents. For an (n, k, m) code with $k > 1$, the ith shift register contains K_i previous information bits. Defining $K \triangleq \sum_{i=1}^{k} K_i$ as the *total encoder memory*,[1] the encoder state at time unit l [when $u_l^{(1)}u_l^{(2)} \cdots u_l^{(k)}$ are the encoder inputs] is the binary K-tuple of inputs

$$(u_{l-1}^{(1)}u_{l-2}^{(1)} \cdots u_{l-K_1}^{(1)} \quad u_{l-1}^{(2)}u_{l-2}^{(2)} \cdots u_{l-K_2}^{(2)} \quad \cdots \quad u_{l-1}^{(k)}u_{l-2}^{(k)} \cdots u_{l-K_k}^{(k)})$$

and there are a total of 2^K different possible states. For an $(n, 1, m)$ code, $K = K_1 = m$ and the encoder state at time unit l is simply

$$(u_{l-1}u_{l-2} \cdots u_{l-m}).$$

Each new block of k inputs causes a transition to a new state. Hence, there are 2^k branches leaving each state, one corresponding to each different input block. For an $(n, 1, m)$ code, therefore, there are only two branches leaving each state. Each branch is labeled with the k inputs causing the transition $(u_l^{(1)}u_l^{(2)} \cdots u_l^{(k)})$ and the n corresponding outputs $(v_l^{(1)}v_l^{(2)} \cdots v_l^{(n)})$. The state diagrams for the (2, 1, 3) code of Figure 10.1 and the (3, 2, 1) code of Figure 10.2 are shown in Figure 10.4(a) and (b). The states are labeled $S_0, S_1, \ldots, S_{2^K-1}$, where by convention S_i represents the state whose binary K-tuple representation $b_0, b_1, \ldots, b_{K-1}$ is equivalent to the integer $i = b_0 2^0 + b_1 2^1 + \cdots + b_{K-1}2^{K-1}$.

Assuming that the encoder is initially in state S_0 (the all-zero state), the code word corresponding to any given information sequence can be obtained by following the path through the state diagram determined by the information sequence and noting the corresponding outputs on the branch labels. Following the last nonzero information block, the encoder is returned to state S_0 by a sequence of m all-zero blocks appended to the information sequence. For example, in Figure 10.4(a), if $\mathbf{u} = (1\ 1\ 1\ 0\ 1)$, the code word $\mathbf{v} = (1\ 1,\ 1\ 0,\ 0\ 1,\ 0\ 1,\ 1\ 1,\ 1\ 0,\ 1\ 1,\ 1\ 1)$.

The state diagram can be modified to provide a complete description of the Hamming weights of all nonzero code words (i.e., a weight distribution function for the code). State S_0 is split into an initial state and a final state, the self-loop around state S_0 is deleted, and each branch is labeled with a *branch gain* X^i, where

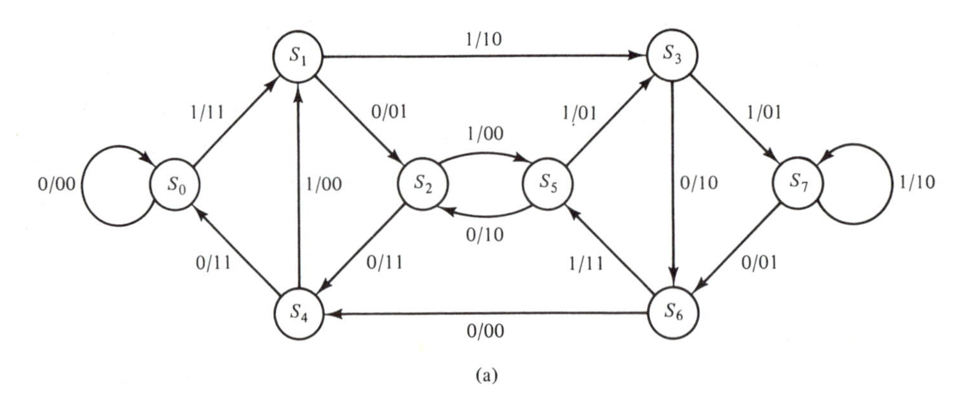

Figure 10.4 Encoder state diagrams: (a) (2, 1, 3) code.

[1]Note the distinction between the total encoder memory K and the encoder memory order m. K is the sum of all the shift register lengths, whereas m is the maximum length of any shift register.

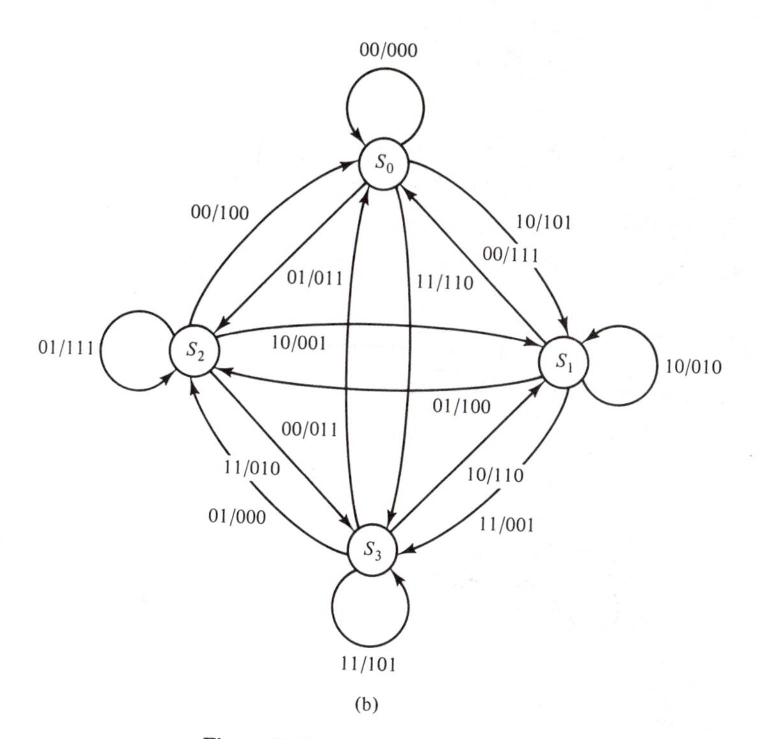

Figure 10.4 (*cont.*) (b) (3, 2, 1) code.

i is the weight of the n encoded bits on that branch. Each path connecting the initial state to the final state represents a nonzero code word that diverges from and remerges with state S_0 exactly once. Code words that diverge from and remerge with state S_0 more than once can be considered as a sequence of shorter code words. The *path gain* is the product of the branch gains along a path, and the weight of the associated code word is the power of X in the path gain. The modified state diagrams for the codes of Figures 10.1 and 10.2 are shown in Figure 10.5(a) and (b).

Example 10.9

(a) The path representing the state sequence $S_0S_1S_3S_7S_6S_5S_2S_4S_0$ in Figure 10.5(a) has path gain $X^2 \cdot X^1 \cdot X^1 \cdot X^1 \cdot X^2 \cdot X^1 \cdot X^2 \cdot X^2 = X^{12}$, and the corresponding code word has weight 12.

(b) The path representing the state sequence $S_0S_1S_3S_2S_0$ in Figure 10.5(b) has path gain $X^2 \cdot X^1 \cdot X^0 \cdot X^1 = X^4$, and the corresponding code word has weight 4.

The weight distribution function of a code can be determined by considering the modified state diagram as a signal flow graph and applying Mason's gain formula [5] to compute its "generating function"

$$T(X) = \sum_i A_i X^i, \tag{10.22}$$

where A_i is the number of code words of weight i. In a signal flow graph, a path connecting the initial state to the final state which does not go through any state twice is called a *forward path*. Let F_i be the gain of the ith forward path. A closed

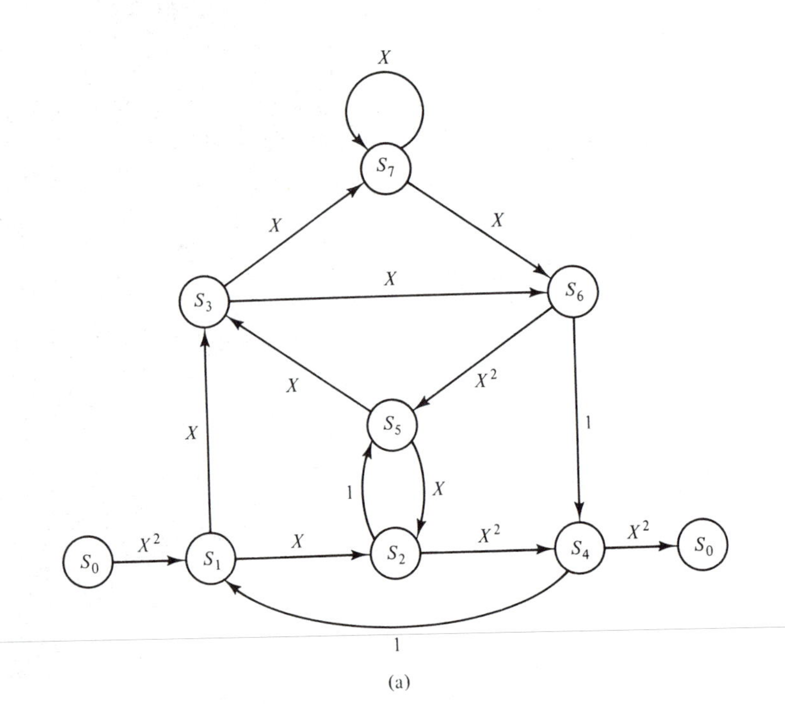

(a)

(b)

Figure 10.5 Modified encoder state diagrams: (a) (2, 1, 3) code; (b) (3, 2, 1) code.

path starting at any state and returning to that state without going through any other state twice is called a *loop*. Let C_i be the gain of the ith loop. A set of loops is *nontouching* if no state belongs to more than one loop in the set. Let $\{i\}$ be the set of all loops, $\{i', j'\}$ be the set of all pairs of nontouching loops, $\{i'', j'', l''\}$ be the set of all triples of nontouching loops, and so on. Then define

$$\Delta = 1 - \sum_i C_i + \sum_{i',j'} C_{i'}C_{j'} - \sum_{i'',j'',l''} C_{i''}C_{j''}C_{l''} + \cdots, \tag{10.23}$$

where $\sum_i C_i$ is the sum of the loop gains, $\sum_{i',j'} C_{i'}C_{j'}$ is the product of the loop gains of two nontouching loops summed over all pairs of nontouching loops, $\sum_{i'',j'',l''} C_{i''}C_{j''}C_{l''}$ is the product of the loop gains of three nontouching loops summed over all triples of nontouching loops, and so on. Finally, Δ_i is defined exactly like Δ, but only for that portion of the graph not touching the ith forward path; that is, all states along the ith forward path, together with all branches connected to these states, are removed from the graph when computing Δ_i. Mason's formula for computing the generating function $T(X)$ of a graph can now be stated as

$$T(X) = \frac{\sum_i F_i \Delta_i}{\Delta}, \tag{10.24}$$

where the sum in the numerator is over all forward paths. Two illustrative examples will clarify the procedure.

Example 10.10

(a) There are 11 loops in the graph of Figure 10.5(a):

Loop 1:	$S_1S_3S_7S_6S_5S_2S_4S_1$	$(C_1 = X^8)$
Loop 2:	$S_1S_3S_7S_6S_4S_1$	$(C_2 = X^3)$
Loop 3:	$S_1S_3S_6S_5S_2S_4S_1$	$(C_3 = X^7)$
Loop 4:	$S_1S_3S_6S_4S_1$	$(C_4 = X^2)$
Loop 5:	$S_1S_2S_5S_3S_7S_6S_4S_1$	$(C_5 = X^9)$
Loop 6:	$S_1S_2S_5S_3S_6S_4S_1$	$(C_6 = X^8)$
Loop 7:	$S_1S_2S_4S_1$	$(C_7 = X^3)$
Loop 8:	$S_2S_5S_2$	$(C_8 = X)$
Loop 9:	$S_3S_7S_6S_5S_3$	$(C_9 = X^5)$
Loop 10:	$S_3S_6S_5S_3$	$(C_{10} = X^4)$
Loop 11:	S_7S_7	$(C_{11} = X)$.

There are 10 pairs of nontouching loops:

Loop pair 1:	(loop 2, loop 8)	$(C_2C_8 = X^4)$
Loop pair 2:	(loop 3, loop 11)	$(C_3C_{11} = X^8)$
Loop pair 3:	(loop 4, loop 8)	$(C_4C_8 = X^3)$
Loop pair 4:	(loop 4, loop 11)	$(C_4C_{11} = X^3)$
Loop pair 5:	(loop 6, loop 11)	$(C_6C_{11} = X^9)$
Loop pair 6:	(loop 7, loop 9)	$(C_7C_9 = X^8)$
Loop pair 7:	(loop 7, loop 10)	$(C_7C_{10} = X^7)$

Loop pair 8: (loop 7, loop 11) $(C_7C_{11} = X^4)$

Loop pair 9: (loop 8, loop 11) $(C_8C_{11} = X^2)$

Loop pair 10: (loop 10, loop 11) $(C_{10}C_{11} = X^5)$.

There are two triples of nontouching loops:

Loop triple 1: (loop 4, loop 8, loop 11) $(C_4C_8C_{11} = X^4)$

Loop triple 2: (loop 7, loop 10, loop 11) $(C_7C_{10}C_{11} = X^8)$.

There are no other sets of nontouching loops. Therefore,

$$\begin{aligned}\Delta = 1 &- (X^8 + X^3 + X^7 + X^2 + X^4 + X^3 + X^3 + X + X^5 + X^4 + X)\\ &+ (X^4 + X^8 + X^3 + X^3 + X^4 + X^8 + X^7 + X^4 + X^2 + X^5)\\ &- (X^4 + X^8)\\ = 1 &- 2X - X^3.\end{aligned}$$

There are seven forward paths in the graph of Figure 10.5(a):

Forward path 1: $S_0S_1S_3S_7S_6S_5S_2S_4S_0$ $(F_1 = X^{12})$

Forward path 2: $S_0S_1S_3S_7S_6S_4S_0$ $(F_2 = X^7)$

Forward path 3: $S_0S_1S_3S_6S_5S_2S_4S_0$ $(F_3 = X^{11})$

Forward path 4: $S_0S_1S_3S_6S_4S_0$ $(F_4 = X^6)$

Forward path 5: $S_0S_1S_2S_5S_3S_7S_6S_4S_0$ $(F_5 = X^8)$

Forward path 6: $S_0S_1S_2S_5S_3S_6S_4S_0$ $(F_6 = X^7)$

Forward path 7: $S_0S_1S_2S_4S_0$ $(F_7 = X^7)$.

Forward paths 1 and 5 touch all states in the graph, and hence the subgraph not touching these paths contains no states. Therefore,

$$\Delta_1 = \Delta_5 = 1.$$

The subgraph not touching forward paths 3 and 6 is shown in Figure 10.6(a), and hence

$$\Delta_3 = \Delta_6 = 1 - X.$$

The subgraph not touching forward path 2 is shown in Figure 10.6(b), and hence

$$\Delta_2 = 1 - X.$$

The subgraph not touching forward path 4 is shown in Figure 10.6(c), and hence

$$\Delta_4 = 1 - (X + X) + (X^2) = 1 - 2X + X^2.$$

The subgraph not touching forward path 7 is shown in Figure 10.6(d), and hence

$$\Delta_7 = 1 - (X + X^4 + X^5) + (X^5) = 1 - X - X^4.$$

The generating function for this graph is then given by

$$\begin{aligned}T(X) &= \frac{\begin{array}{l}X^{12}\cdot 1 + X^7(1 - X) + X^{11}(1 - X)\\ \qquad + X^6(1 - 2X + X^2) + X^8\cdot 1 + X^7(1 - X) + X^7(1 - X - X^4)\end{array}}{1 - 2X - X^3}\\ &= \frac{X^6 + X^7 - X^8}{1 - 2X - X^3}\\ &= X^6 + 3X^7 + 5X^8 + 11X^9 + 25X^{10} + \cdots.\end{aligned}$$

$T(X)$ provides a complete description of the weight distribution of all nonzero code

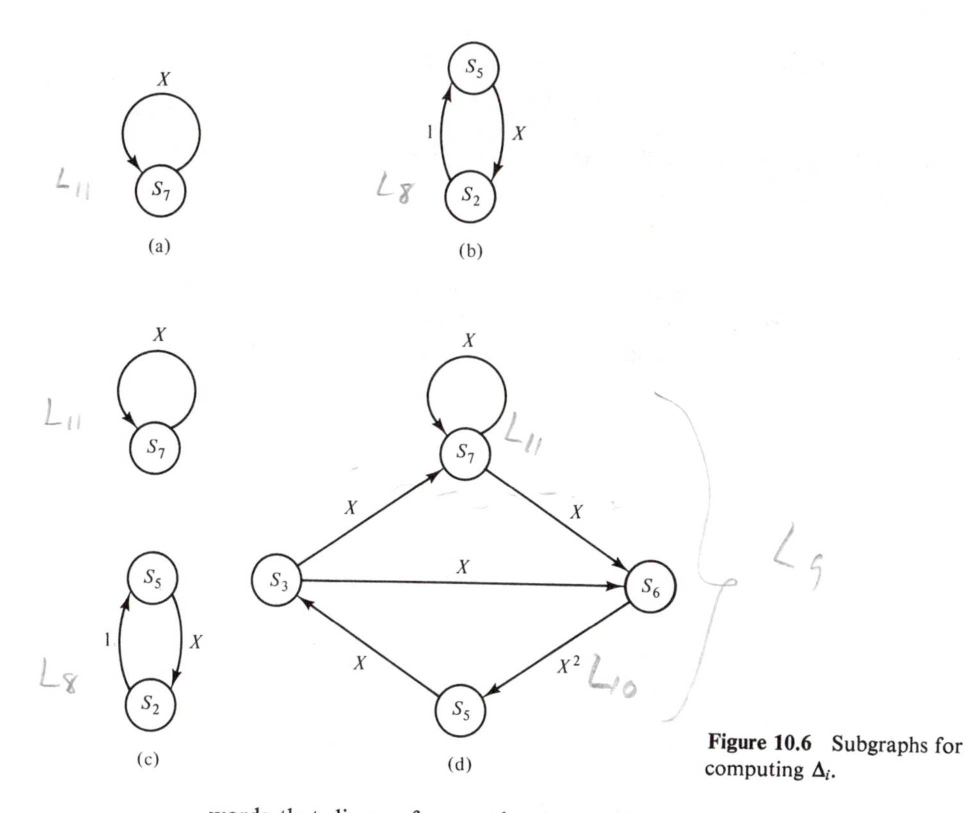

Figure 10.6 Subgraphs for computing Δ_i.

words that diverge from and remerge with state S_0 exactly once. In this case there is one such code word of weight 6, three of weight 7, five of weight 8, and so on.

(b) There are eight loops, six pairs of nontouching loops, and one triple of nontouching loops in the graph of Figure 10.5(b), and

$$\begin{aligned}\Delta &= 1 - (X^2 + X^4 + X^3 + X + X^2 + X + X^2 + X^3) \\ &\quad + (X^6 + X^2 + X^4 + X^3 + X^4 + X^5) - (X^6) \\ &= 1 - 2X - 2X^2 - X^3 + X^4 + X^5.\end{aligned}$$

There are 15 forward paths in the graph of Figure 10.5(b), and

$$\begin{aligned}\sum_i F_i\Delta_i &= X^5(1 - X - X^2 - X^3 + X^5) + X^4(1 - X^2) + X^6 \cdot 1 + X^5(1 - X^3) \\ &\quad + X^4 \cdot 1 + X^3(1 - X - X^2) + X^6(1 - X^2) + X^6 \cdot 1 + X^5(1 - X) \\ &\quad + X^8 \cdot 1 + X^4(1 - X - X^2 - X^3 + X^4) + X^7(1 - X^3) \\ &\quad + X^6 \cdot 1 + X^3(1 - X) + X^6 \cdot 1 \\ &= 2X^3 + X^4 + X^5 + X^6 - X^7 - X^8.\end{aligned}$$

Hence, the generating function is

$$\begin{aligned}T(X) &= \frac{2X^3 + X^4 + X^5 + X^6 - X^7 - X^8}{1 - 2X - 2X^2 - X^3 + X^4 + X^5} \\ &= 2X^3 + 5X^4 + 15X^5 + \cdots,\end{aligned}$$

and this code contains two nonzero code words of weight 3, five of weight 4, fifteen of weight 5, and so on.

Additional information about the structure of a code can be obtained using the same procedure. If the modified state diagram is augmented by labeling each branch corresponding to a nonzero information block with Y^j, where j is the weight of the k information bits on that branch, and labeling every branch with Z, the generating function is given by

$$T(X, Y, Z) = \sum_{i,j,l} A_{i,j,l} X^i Y^j Z^l. \tag{10.25}$$

The coefficient $A_{i,j,l}$ denotes the number of code words with weight i, whose associated information sequence has weight j, and whose length is l branches. The augmented state diagram for the code of Figure 10.1 is shown in Figure 10.7.

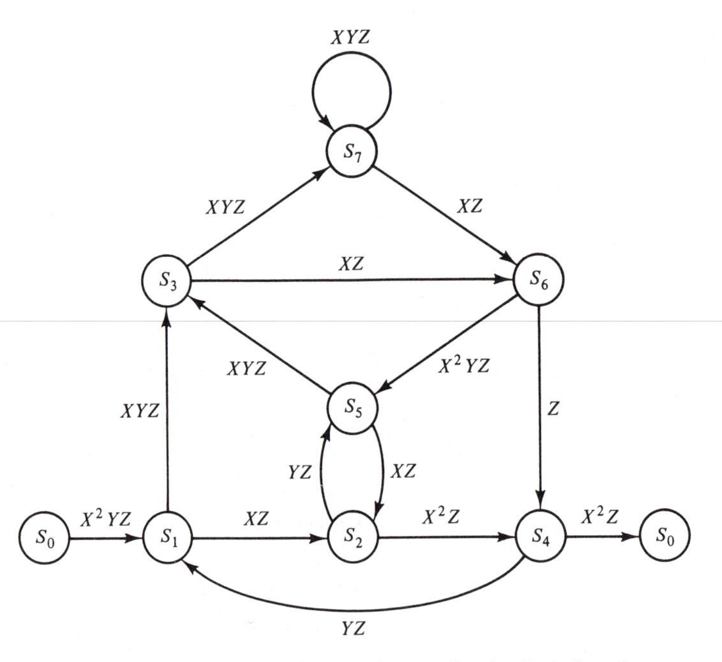

Figure 10.7 Augmented state diagram for the (2, 1, 3) code.

Example 10.11

For the graph of Figure 10.7,

$$\begin{aligned}\Delta = 1 &- (X^8Y^4Z^7 + X^3Y^3Z^5 + X^7Y^3Z^6 + X^2Y^2Z^4 + X^4Y^4Z^7\\ &+ X^3Y^3Z^6 + X^3YZ^3 + XYZ^2 + X^5Y^3Z^4 + X^4Y^2Z^3 + XYZ)\\ &+ (X^4Y^4Z^7 + X^8Y^4Z^7 + X^3Y^3Z^6 + X^3Y^3Z^5 + X^4Y^4Z^7\\ &+ X^8Y^4Z^7 + X^7Y^3Z^6 + X^4Y^2Z^4 + X^2Y^2Z^3 + X^5Y^3Z^4)\\ &- (X^4Y^4Z^7 + X^8Y^4Z^7)\\ = 1 &+ XY(Z + Z^2) - X^2Y^2(Z^4 - Z^3) - X^3(YZ^3 - Y^3Z^6)\\ &- X^4Y^2(Z^3 - Z^4) - X^8(Y^3Z^6 - Y^4Z^7) - X^9Y^4Z^7\end{aligned}$$

and

$$\sum_i F_i\Delta_i = X^{12}Y^4Z^8 \cdot 1 + X^7Y^3Z^6(1 - XYZ^2) + X^{11}Y^3Z^7(1 - XYZ)$$

$$+ X^6Y^2Z^5(1 - XY(Z + Z^2) + X^2Y^2Z^3) + X^8Y^4Z^8 \cdot 1$$
$$+ X^7Y^3Z^7(1 - XYZ) + X^7YZ^4(1 - XYZ - X^4Y^2Z^3)$$
$$= X^6Y^2Z^5 + X^7YZ^4 - X^8Y^2Z^5.$$

Hence, the generating function is

$$T(X, Y, Z) = \frac{X^6Y^2Z^5 + X^7YZ^4 - X^8Y^2Z^5}{\Delta}$$
$$= X^6Y^2Z^5 + X^7(YZ^4 + Y^3Z^6 + Y^3Z^7)$$
$$+ X^8(Y^2Z^6 + Y^4Z^7 + Y^4Z^8 + 2Y^4Z^9) + \cdots.$$

This implies that the code word of weight 6 has length 5 branches and an information sequence of weight 2, one code word of weight 7 has length 4 branches and information sequence weight 1, another has length 6 branches and information sequence weight 3, the third has length 7 branches and information sequence weight 3, and so on.

It should be clear from the above examples that the generating function approach to finding the weight distribution of a convolutional code quickly becomes impractical as the total encoder memory K exceeds 4 or 5. As will be seen in the next section, the important distance properties of a code necessary for estimating its performance can be computed without the aid of the generating function. Nevertheless, the generating function remains an important conceptual tool, and will be used in Section 11.2 to obtain performance bounds for maximum likelihood decoding.

An important subclass of convolutional codes is the class of *systematic codes.* In a systematic code, the first k output sequences are exact replicas of the k input sequences, i.e.,

$$\mathbf{v}^{(i)} = \mathbf{u}^{(i)}, \qquad i = 1, 2, \ldots, k, \tag{10.26}$$

and the generator sequences satisfy

$$\mathbf{g}_i^{(j)} = \begin{cases} 1 & \text{if } j = i \\ 0 & \text{if } j \neq i \end{cases}, \qquad i = 1, 2, \ldots, k. \tag{10.27}$$

The generator matrix is given by

$$\mathbf{G} = \begin{bmatrix} \mathbf{I} & \mathbf{P}_0 & \mathbf{0} & \mathbf{P}_1 & \mathbf{0} & \mathbf{P}_2 & \cdots & \mathbf{0} & \mathbf{P}_m & & & & \\ & & \mathbf{I} & \mathbf{P}_0 & \mathbf{0} & \mathbf{P}_1 & \cdots & \mathbf{0} & \mathbf{P}_{m-1} & \mathbf{0} & \mathbf{P}_m & & \\ & & & & \mathbf{I} & \mathbf{P}_0 & \cdots & \mathbf{0} & \mathbf{P}_{m-2} & \mathbf{0} & \mathbf{P}_{m-1} & \mathbf{0} & \mathbf{P}_m \\ & & & & & & \cdot & & & & & & \cdot \\ & & & & & & \cdot & & & & & & \cdot \\ & & & & & & \cdot & & & & & & \cdot \end{bmatrix}, \tag{10.28}$$

where $\mathbf{I}$ is the $k \times k$ identity matrix, $\mathbf{0}$ is the $k \times k$ all-zero matrix, and $\mathbf{P}_l$ is the $k \times (n - k)$ matrix

$$\mathbf{P}_l = \begin{bmatrix} g_{1,l}^{(k+1)} & g_{1,l}^{(k+2)} & \cdots & g_{1,l}^{(n)} \\ g_{2,l}^{(k+1)} & g_{2,l}^{(k+2)} & \cdots & g_{2,l}^{(n)} \\ \cdot & \cdot & & \cdot \\ \cdot & \cdot & & \cdot \\ \cdot & \cdot & & \cdot \\ g_{k,l}^{(k+1)} & g_{k,l}^{(k+2)} & \cdots & g_{k,l}^{(n)} \end{bmatrix}, \tag{10.29}$$

and the transfer function matrix becomes

$$\mathbf{G}(D) = \begin{bmatrix} 1 & 0 & \cdots & 0 & \mathbf{g}_1^{(k+1)}(D) & \cdots & \mathbf{g}_1^{(n)}(D) \\ 0 & 1 & \cdots & 0 & \mathbf{g}_2^{(k+1)}(D) & \cdots & \mathbf{g}_2^{(n)}(D) \\ \cdot & \cdot & & \cdot & \cdot & & \cdot \\ \cdot & \cdot & & \cdot & \cdot & & \cdot \\ \cdot & \cdot & & \cdot & \cdot & & \cdot \\ 0 & 0 & \cdots & 1 & \mathbf{g}_k^{(k+1)}(D) & \cdots & \mathbf{g}_k^{(n)}(D) \end{bmatrix}. \tag{10.30}$$

Since the first k output sequences equal the input sequences, they are called *information sequences* and the last $n - k$ output sequences are called *parity sequences.* Note that whereas in general $k \cdot n$ generator sequences must be specified to define an (n, k, m) convolutional code, only $k \cdot (n - k)$ sequences must be specified to define a systematic code. Hence, systematic codes represent a subclass of the set of all possible codes. Any code not satisfying (10.26) to (10.30) is said to be *nonsystematic.*

Example 10.12

Consider the (2, 1, 3) systematic code whose encoder is shown in Figure 10.8. The generator sequences are $\mathbf{g}^{(1)} = (1\ 0\ 0\ 0)$ and $\mathbf{g}^{(2)} = (1\ 1\ 0\ 1)$. The generator matrix is

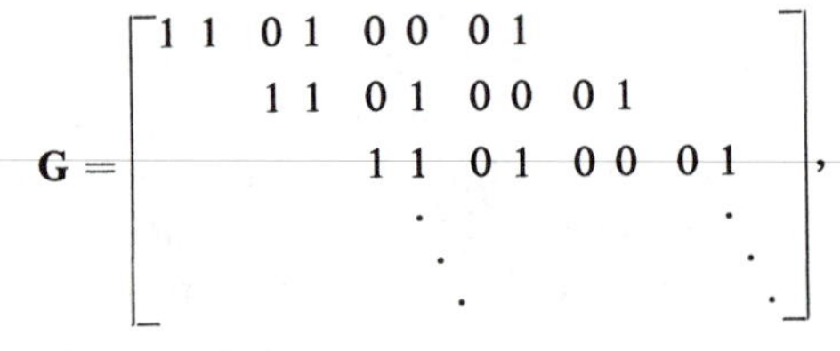

$$\mathbf{G} = \begin{bmatrix} 11 & 01 & 00 & 01 & & & \\ & 11 & 01 & 00 & 01 & & \\ & & 11 & 01 & 00 & 01 & \\ & & & \cdot & & & \cdot \\ & & & & \cdot & & & \cdot \\ & & & & & \cdot & & & \cdot \end{bmatrix},$$

and the transfer function matrix is

$$\mathbf{G}(D) = [1 \quad 1 + D + D^3].$$

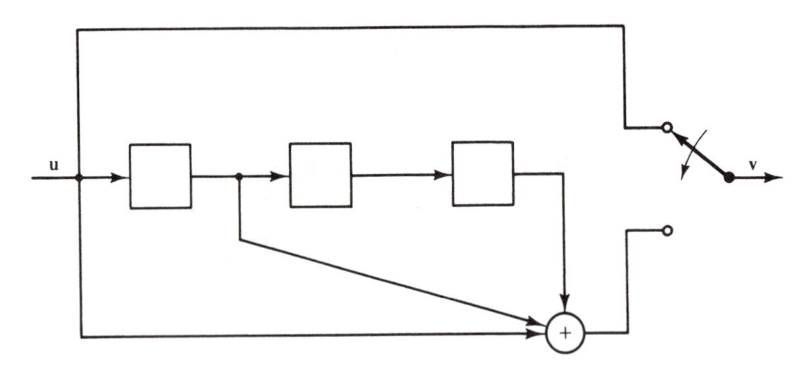

Figure 10.8 A (2, 1, 3) systematic encoder.

For an input sequence $\mathbf{u}(D) = 1 + D^2 + D^3$, the information sequence is

$$\mathbf{v}^{(1)}(D) = \mathbf{u}(D)\mathbf{g}^{(1)}(D) = (1 + D^2 + D^3)\cdot 1 = 1 + D^2 + D^3,$$

and the parity sequence is

$$\begin{aligned} \mathbf{v}^{(2)}(D) &= \mathbf{u}(D)\mathbf{g}^{(2)}(D) = (1 + D^2 + D^3)(1 + D + D^3) \\ &= 1 + D + D^2 + D^3 + D^4 + D^5 + D^6. \end{aligned}$$

One advantage of systematic codes is that encoding is somewhat simpler than for nonsystematic codes because less hardware is required. For example, the (2, 1, 3) nonsystematic code of Figure 10.1 requires two modulo-2 adders with a total of seven inputs, whereas the (2, 1, 3) systematic code of Figure 10.8 requires only one modulo-2 adder with three inputs. Also, for an (n, k, m) systematic code with $k > n - k$, there exists a modified encoding circuit which normally requires fewer than K shift register stages.

Example 10.13

Consider the (3, 2, 2) systematic code with transfer function matrix

$$\mathbf{G}(D) = \begin{bmatrix} 1 & 0 & 1 + D + D^2 \\ 0 & 1 & 1 + D^2 \end{bmatrix}.$$

The straightforward realization of the encoder requires a total of $K = K_1 + K_2 = 4$ shift register stages, and is shown in Figure 10.9(a). On the other hand, since the information sequences are given by $\mathbf{v}^{(1)}(D) = \mathbf{u}^{(1)}(D)$ and $\mathbf{v}^{(2)}(D) = \mathbf{u}^{(2)}(D)$, and the parity sequence is given by

$$\mathbf{v}^{(3)}(D) = \mathbf{u}^{(1)}(D)\mathbf{g}_1^{(3)}(D) + \mathbf{u}^{(2)}(D)\mathbf{g}_2^{(3)}(D),$$

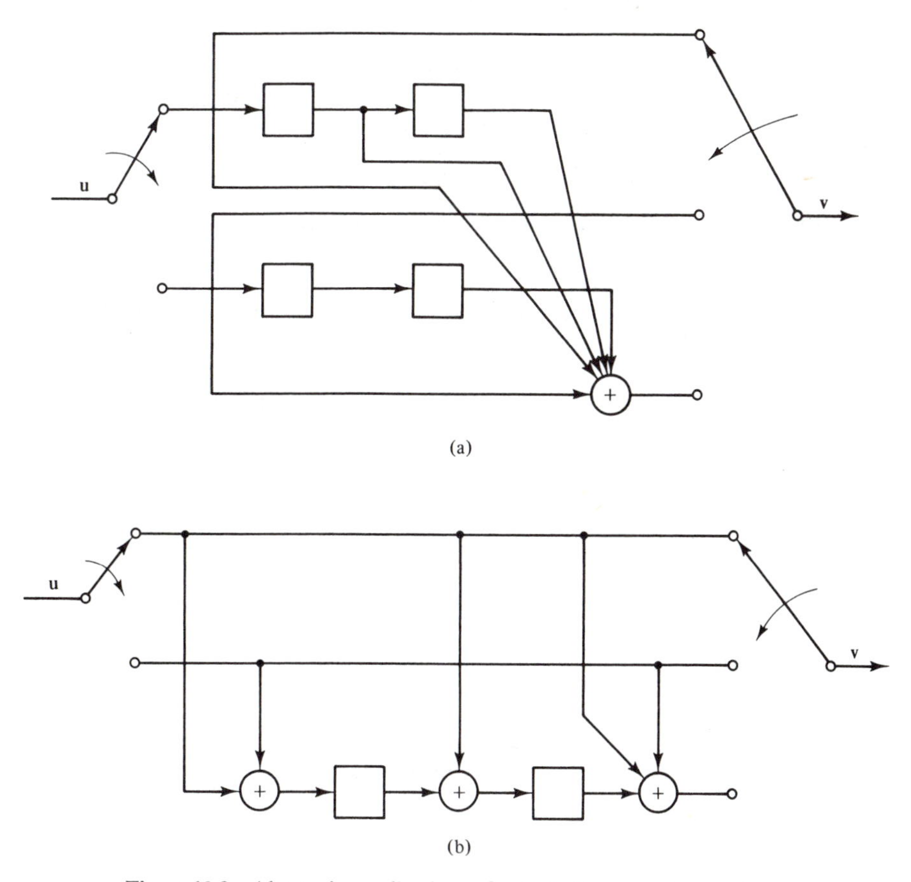

Figure 10.9 Alternative realizations of a (3, 2, 2) systematic encoder.

the encoder can also be realized as shown in Figure 10.9(b). Note that this realization requires only two stages of encoder memory rather than 4.

A complete discussion of the minimal encoder memory required to realize a convolutional code is given by Forney [6]. In most cases the straightforward realization requiring K stages of shift register memory is the most efficient. In the case of an (n, k, m) systematic code with $k > n - k$, however, a simpler realization usually exists, as noted in Example 10.13.

Another advantage of systematic codes is that no inverting circuit is needed for recovering the information sequence from the code word. Nonsystematic codes, on the other hand, require an inverter to recover the information sequence; that is, an $n \times k$ matrix $\mathbf{G}^{-1}(D)$ must exist such that

$$\mathbf{G}(D)\mathbf{G}^{-1}(D) = \mathbf{I}D^l \tag{10.31}$$

for some $l \geq 0$, where $\mathbf{I}$ is the $k \times k$ identity matrix. Equations (10.18) and (10.31) then imply that

$$\mathbf{V}(D)\mathbf{G}^{-1}(D) = \mathbf{U}(D)\mathbf{G}(D)\mathbf{G}^{-1}(D) = \mathbf{U}(D)D^l, \tag{10.32}$$

and the information sequence can be recovered with an l-time-unit delay from the code word by letting $\mathbf{V}(D)$ be the input to the n-input, k-output linear sequential circuit whose transfer function matrix is $\mathbf{G}^{-1}(D)$.

For an $(n, 1, m)$ code, a transfer function matrix $\mathbf{G}(D)$ has a feedforward inverse $\mathbf{G}^{-1}(D)$ of delay l if and only if

$$\text{GCD}\,[\mathbf{g}^{(1)}(D), \mathbf{g}^{(2)}(D), \ldots, \mathbf{g}^{(n)}(D)] = D^l \tag{10.33}$$

for some $l \geq 0$, where GCD denotes the greatest common divisor. For an (n, k, m) code with $k > 1$, let $\Delta_i(D)$, $i = 1, 2, \ldots, \binom{n}{k}$, be the determinants of the $\binom{n}{k}$ distinct $k \times k$ submatrices of the transfer function matrix $\mathbf{G}(D)$. Then a feedforward inverse of delay l exists if and only if

$$\text{GCD}\left[\Delta_i(D) \colon i = 1, 2, \ldots, \binom{n}{k}\right] = D^l \tag{10.34}$$

for some $l \geq 0$. Procedures for constructing $\mathbf{G}^{-1}(D)$ are given in References 6 and 7.

Example 10.14

(a) For the (2, 1, 3) code of Figure 10.1,

$$\text{GCD}\,[1 + D^2 + D^3, 1 + D + D^2 + D^3] = 1,$$

and the transfer function matrix

$$\mathbf{G}^{-1}(D) = \begin{bmatrix} 1 + D + D^2 \\ D + D^2 \end{bmatrix}$$

provides the required inverse of delay 0 [i.e., $\mathbf{G}(D)\mathbf{G}^{-1}(D) = 1$]. The implementation of the inverse is shown in Figure 10.10(a).

(b) For the (3, 2, 1) code of Figure 10.2, the 2×2 submatrices of $\mathbf{G}(D)$ yield determinants $1 + D + D^2$, $1 + D^2$, and 1. Since

$$\text{GCD}\,[1 + D + D^2, 1 + D^2, 1] = 1,$$

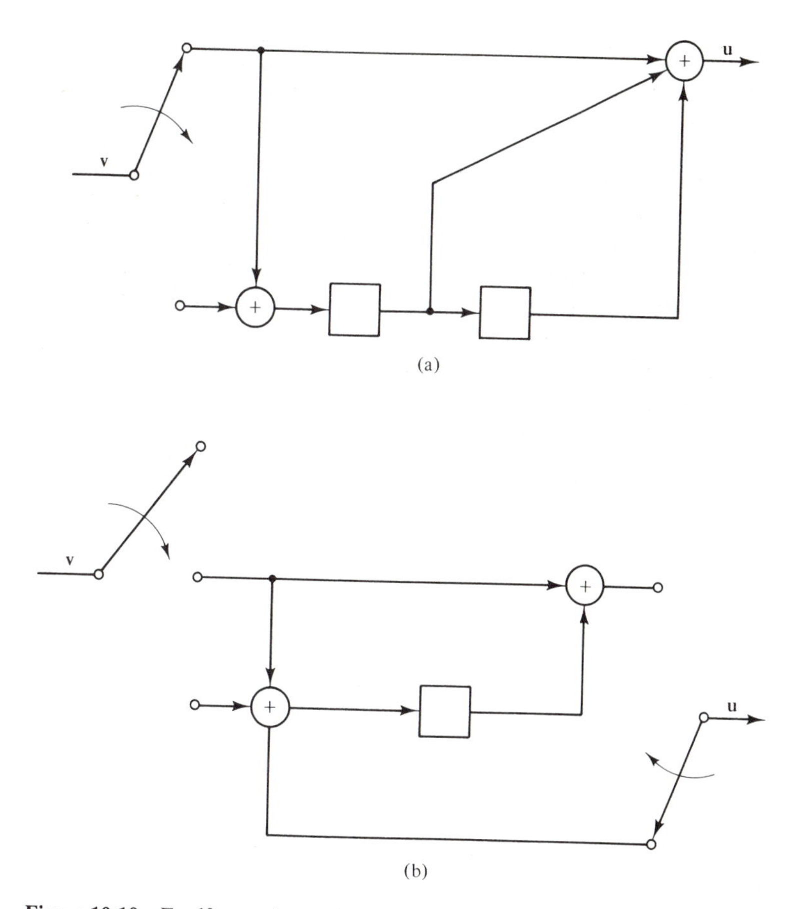

Figure 10.10 Feedforward encoder inverses: (a) (2, 1, 3) code; (b) (3, 2, 1) code.

there exists a feedforward inverse of delay 0. The required transfer function matrix is given by

$$\mathbf{G}^{-1}(D) = \begin{bmatrix} 0 & 0 \\ 1 & 1 + D \\ 1 & D \end{bmatrix},$$

and its implementation is shown in Figure 10.10(b).

To understand what happens when a feedforward inverse does not exist, it is best to consider an example. For the (2, 1, 2) code with $\mathbf{g}^{(1)}(D) = 1 + D$ and $\mathbf{g}^{(2)}(D) = 1 + D^2$,

$$\text{GCD}\,[1 + D, 1 + D^2] = 1 + D,$$

and a feedforward inverse does not exist. If the information sequence is $\mathbf{u}(D) = 1/(1 + D) = 1 + D + D^2 + \cdots$, the output sequences are $\mathbf{v}^{(1)}(D) = 1$ and $\mathbf{v}^{(2)}(D) = 1 + D$; that is, the code word contains only three nonzero bits even though the information sequence has infinite weight. If this code word is transmitted over a BSC, and the three nonzero bits are changed to zeros by the channel noise, the received sequence will be all zeros. A MLD will then produce the all-zero code word as its estimate, since this is a valid code word and it agrees exactly with the received sequence. Thus, the estimated information sequence will be $\hat{\mathbf{u}}(D) = \mathbf{0}$, implying an infinite number of decoding errors caused by a finite number (only three in this case) of channel errors. Clearly, this is a very undesirable circumstance, and the code is said to be subject to catastrophic error propagation, and is called a *catastrophic code*. Massey and Sain [7] have shown that (10.33) and (10.34) are necessary and sufficient conditions for a code to be *noncatastrophic*. Hence, any code for which a feedforward inverse exists is noncatastrophic. Another advantage of systematic codes is that they are always noncatastrophic.

In terms of the state diagram, a code is catastrophic if and only if the state diagram contains a loop of zero weight other than the self-loop around the state S_0. The state diagram for the catastrophic code discussed above is shown in Figure 10.11. Note that the self-loop around the state S_3 has zero weight.

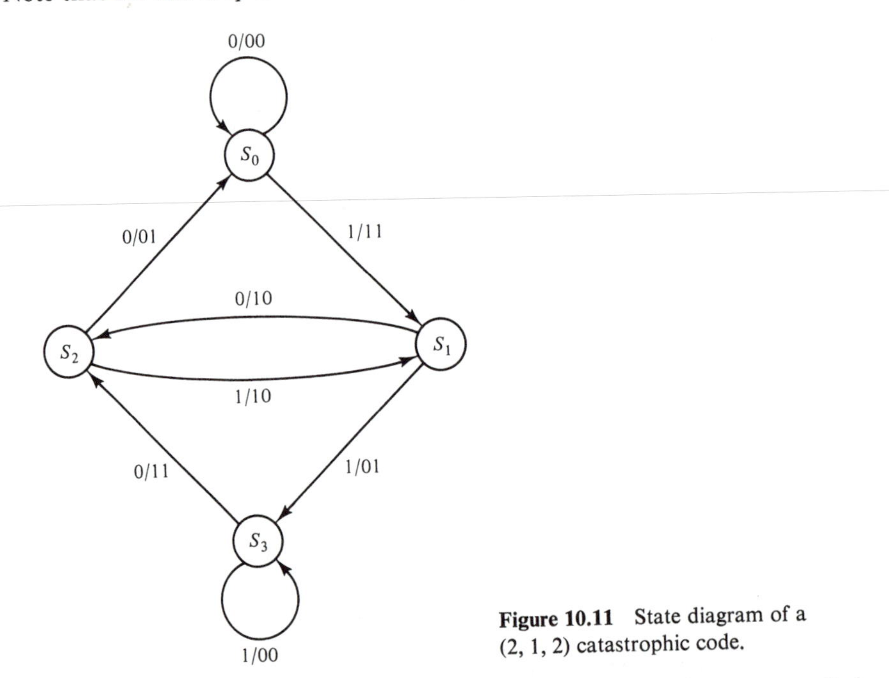

Figure 10.11 State diagram of a (2, 1, 2) catastrophic code.

In choosing nonsystematic codes for use in a communication system, it is important to avoid the selection of catastrophic codes. Rosenberg [8] has shown that only a fraction $1/(2^n - 1)$ of $(n, 1, m)$ nonsystematic codes are catastrophic. A similar result for (n, k, m) codes with $k > 1$ is still lacking.

10.3 DISTANCE PROPERTIES OF CONVOLUTIONAL CODES

The performance of a convolutional code depends on the decoding algorithm employed and the distance properties of the code. In this section several distance measures for convolutional codes are introduced. Their relation to code performance is discussed in subsequent chapters.

The most important distance measure for convolutional codes is the minimum *free distance* d_{free}, defined as

$$d_{\text{free}} \triangleq \min \{d(\mathbf{v}', \mathbf{v}''): \mathbf{u}' \neq \mathbf{u}''\}, \tag{10.35}$$

where $\mathbf{v}'$ and $\mathbf{v}''$ are the code words corresponding to the information sequences $\mathbf{u}'$ and $\mathbf{u}''$, respectively. [In (10.35) it is assumed that if $\mathbf{u}'$ and $\mathbf{u}''$ are of different lengths, zeros are added to the shorter sequence so that their corresponding code words have equal lengths.] Hence, d_{free} is the minimum distance between any two code words in the code. Since a convolutional code is a linear code,

$$\begin{aligned} d_{\text{free}} &= \min \{w(\mathbf{v}' + \mathbf{v}''): \mathbf{u}' \neq \mathbf{u}''\} \\ &= \min \{w(\mathbf{v}): \mathbf{u} \neq \mathbf{0}\} \\ &= \min \{w(\mathbf{uG}): \mathbf{u} \neq \mathbf{0}\}, \end{aligned} \tag{10.36}$$

where $\mathbf{v}$ is the code word corresponding to the information sequence $\mathbf{u}$. Hence, d_{free} is the minimum-weight code word of any length produced by a nonzero information sequence. Also, it is the minimum weight of all paths in the state diagram that diverge from and remerge with the all-zero state S_0, and it is the lowest power of X in the code-generating function $T(X)$. For example, $d_{\text{free}} = 6$ for the (2, 1, 3) code of Figure 10.4(a) and Example 10.10(a), and $d_{\text{free}} = 3$ for the (3, 2, 1) code of Figure 10.4(b) and Example 10.10(b).

As noted earlier, it is not practical to compute $T(X)$ if K exceeds 4 or 5. Hence, other means must be used to find d_{free} for most codes. An efficient algorithm for finding d_{free} based on the state diagram of the code has been developed by Bahl et al. [9] and modified by Larsen [10]. This algorithm is capable of computing d_{free} for values of K up to about 20. For larger values of K, the number of storage locations required by the algorithm, 2^K, becomes unacceptably large, and other means of finding d_{free} must be tried. No general solution to the problem of finding d_{free} for large values of K has yet been discovered. We shall see in Section 12.4, however, that some decoding algorithms for convolutional codes can, with proper modification, be used to compute the free distance of a code.

Another important distance measure for convolutional codes is the *column distance function* (CDF). Letting

$$[\mathbf{v}]_i = (v_0^{(1)}v_0^{(2)} \cdots v_0^{(n)}, v_1^{(1)}v_1^{(2)} \cdots v_1^{(n)}, \ldots, v_i^{(1)}v_i^{(2)} \cdots v_i^{(n)})$$

denote the ith truncation of the code word $\mathbf{v}$, and

$$[\mathbf{u}]_i = (u_0^{(1)}u_0^{(2)} \cdots u_0^{(k)}, u_1^{(1)}u_1^{(2)} \cdots u_1^{(k)}, \ldots, u_i^{(1)}u_i^{(2)} \cdots u_i^{(k)})$$

denote the ith truncation of the information sequence $\mathbf{u}$, the column distance function of order i, d_i, is defined as

$$\begin{aligned} d_i &\triangleq \min \{d([\mathbf{v}']_i, [\mathbf{v}'']_i, : [\mathbf{u}']_0 \neq [\mathbf{u}'']_0\} \\ &= \min \{w[\mathbf{v}]_i : [\mathbf{u}]_0 \neq \mathbf{0}\}, \end{aligned} \tag{10.37}$$

where again $\mathbf{v}$ is the code word corresponding to the information sequence $\mathbf{u}$. Hence, d_i is the minimum-weight code word over the first $(i + 1)$ time units whose initial information block is nonzero. In terms of the generator matrix of the code,

$$[\mathbf{v}]_i = [\mathbf{u}]_i[\mathbf{G}]_i, \tag{10.38}$$

where $[\mathbf{G}]_i$ is a $k(i+1) \times n(i+1)$ submatrix of $\mathbf{G}$ with the form

$$[\mathbf{G}]_i = \begin{bmatrix} \mathbf{G}_0 & \mathbf{G}_1 & \cdots & \mathbf{G}_i \\ & \mathbf{G}_0 & \cdots & \mathbf{G}_{i-1} \\ & & \ddots & \vdots \\ & & & \mathbf{G}_0 \end{bmatrix}, \quad i \leq m, \tag{10.39a}$$

or

$$[\mathbf{G}]_i = \begin{bmatrix} \mathbf{G}_0 & \mathbf{G}_1 & \cdots & \mathbf{G}_{m-1} & \mathbf{G}_m & & & & & \\ & \mathbf{G}_0 & \cdots & \mathbf{G}_{m-2} & \mathbf{G}_{m-1} & \mathbf{G}_m & & & & \\ & & \ddots & \vdots & \vdots & \mathbf{G}_{m-1} & \ddots & & & \\ & & & \mathbf{G}_1 & \vdots & \vdots & & \mathbf{G}_m & & \\ & & & \mathbf{G}_0 & \mathbf{G}_1 & \vdots & & \mathbf{G}_{m-1} & \mathbf{G}_m & \\ & & & & \mathbf{G}_0 & \mathbf{G}_1 & & \vdots & \mathbf{G}_{m-1} & \\ & & & & & \mathbf{G}_0 & & \vdots & \vdots & \\ & & & & & & \ddots & \mathbf{G}_1 & \vdots & \\ & & & & & & & \mathbf{G}_0 & \mathbf{G}_1 & \\ & & & & & & & & \mathbf{G}_0 & \end{bmatrix}, \quad i > m. \tag{10.39b}$$

Then

$$d_i = \min \{w([\mathbf{u}]_i[\mathbf{G}]_i) : [\mathbf{u}]_0 \neq \mathbf{0}\} \tag{10.40}$$

is seen to depend only on the first $n(i+1)$ columns of $\mathbf{G}$. This accounts for the name "column distance function." The definition implies that d_i cannot decrease with increasing i (i.e., it is a monotonically nondecreasing function of i). The most efficient way of computing the CDF of a code is to modify one of the decoding algorithms to be introduced in later chapters. This is discussed in Section 12.4. The complete CDF of the (2, 1, 16) code with

$$\mathbf{G}(D) = [1 + D + D^2 + D^5 + D^6 + D^8 + D^{13} + D^{16},$$
$$1 + D^3 + D^4 + D^7 + D^9 + D^{10} + D^{11} + D^{12} + D^{14} + D^{15} + D^{16}]$$

is shown in Figure 10.12.

Two cases are of specific interest: $i = m$ and $i \longrightarrow \infty$. For $i = m$, d_m is called the *minimum distance* of a convolutional code and will also be denoted $d_{\min}$. From (10.40) we see that it represents the minimum-weight code word over the first constraint length whose initial information block is nonzero. For the code of Figure 10.12, $d_{\min} = d_{16} = 8$. Much of the early work in convolutional codes treated $d_{\min}$ as the distance parameter of most interest. This was due to the fact that the principal decoding techniques at that time had a decoding memory of one constraint length. More recently, as Viterbi decoding and sequential decoding have become more prominent, d_{free} and the CDF have replaced $d_{\min}$ as the distance parameters of primary

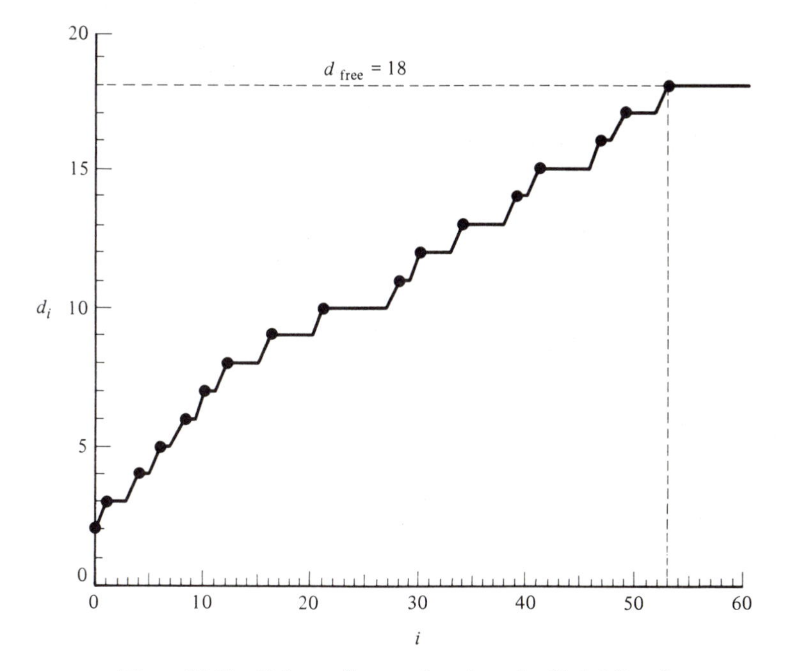

Figure 10.12 Column distance function of a (2, 1, 16) code.

interest, since the decoding memory of these techniques is unlimited. A thorough discussion of the relation between distance measures and decoding algorithms is included in Chapters 11 to 13.

For $i \longrightarrow \infty$, $\lim_{i \to \infty} d_i$ is the minimum-weight code word of any length whose first information block is nonzero. Comparing the definitions of $\lim_{i \to \infty} d_i$ and d_{free}, it can be shown that [11] for noncatastrophic codes

$$\lim_{i \to \infty} d_i = d_{\text{free}}. \tag{10.41}$$

Hence, d_i eventually reaches d_{free}, and then increases no more. This usually happens within three to four constraint lengths (i.e., when i reaches $3m$ or $4m$). For the code of Figure 10.12 with $m = 16$, $d_{\text{free}} = 18$ and $d_i = 18$ for $i \geq 53$. Equation (10.41) is not necessarily true for catastrophic codes. Take as an example the code whose state diagram is shown in Figure 10.11. For this code, $d_0 = 2$ and $d_1 = d_2 = \cdots = \lim_{i \to \infty} d_i = 3$, since the truncated information sequence $[\mathbf{u}]_i = (1, 1, 1, \ldots, 1)$ always produces the truncated code word $[\mathbf{v}]_i = (1\ 1, 0\ 1, 0\ 0, 0\ 0, \ldots, 0\ 0)$, even in the limit as $i \longrightarrow \infty$. Note, however, that all paths in the state diagram that diverge from and remerge with the all-zero state S_0 have weight at least 4, and hence $d_{\text{free}} = 4$. Hence, we have a situation in which $\lim_{i \to \infty} d_i = 3 \neq d_{\text{free}} = 4$.

It is characteristic of catastrophic codes that an infinite-weight information sequence produces a finite-weight code word. In some cases, as in the example above, this code word can have weight less than the free distance of the code. This is due to the zero weight loop in the state diagram. In other words, an information sequence that cycles around this zero weight loop forever will itself pick up infinite weight without adding to the weight of the code word. In a noncatastrophic code, which contains no zero weight loop other than the self-loop around the all-zero state S_0, all infinite-weight information sequences must generate infinite-weight code words, and the minimum-weight code word always has finite length. Unfortunately, the information sequence that produces the minimum-weight code word may be quite long in some cases, thereby causing the calculation of d_{free} to be a rather formidable task.

The best achievable d_{free} for a convolutional code with a given rate and encoder memory has not been determined exactly. However, upper and lower bounds on d_{free} for the best code have been obtained using a random coding approach. These bounds are thoroughly discussed in References 12 to 14. A comparison of the bounds for nonsystematic codes with the bounds for systematic codes implies that more free distance is available with nonsystematic codes of a given rate and encoder memory than with systematic codes. This observation is verified by the code construction results presented in succeeding chapters, and has important consequences when a code with large d_{free} must be selected for use with either Viterbi or sequential decoding. This point is elaborated on further in Chapters 11 and 12.

PROBLEMS

10.1. Consider the (3, 1, 2) convolutional code with

$$\mathbf{g}^{(1)} = (1 \quad 1 \quad 0)$$
$$\mathbf{g}^{(2)} = (1 \quad 0 \quad 1)$$
$$\mathbf{g}^{(3)} = (1 \quad 1 \quad 1).$$

(a) Draw the encoder block diagram.
(b) Find the generator matrix **G**.
(c) Find the code word **v** corresponding to the information sequence $\mathbf{u} = (1\ 1\ 1\ 0\ 1)$.

10.2. Consider the (4, 3, 2) convolutional code of Figure 10.3.
(a) Find the generator sequences of the code.
(b) Find the generator matrix **G**.
(c) Find the code word **v** corresponding to the information sequence $\mathbf{u} = (1\ 1\ 0,\ \ 0\ 1\ 1,\ \ 1\ 0\ 1)$.

10.3. Consider the (3, 1, 2) code of Problem 10.1.
(a) Find the transfer function matrix $\mathbf{G}(D)$.
(b) Find the set of output sequences $\mathbf{V}(D)$, and the code word $\mathbf{v}(D)$, corresponding to the information sequence $\mathbf{u}(D) = 1 + D^2 + D^3 + D^4$.

10.4. Consider the (3, 2, 1) code of Figure 10.2.
(a) Find the composite generator polynomials $\mathbf{g}_1(D)$ and $\mathbf{g}_2(D)$.
(b) Find the code word $\mathbf{v}(D)$ corresponding to the set of input sequences $\mathbf{U}(D) = [1 + D + D^3, 1 + D^2 + D^3]$.

10.5. Consider the (3, 1, 2) code of Problem 10.1.
(a) Draw the state diagram of the encoder.
(b) Draw the modified state diagram of the encoder.
(c) Find the code-generating function $T(X)$.

(d) Draw the augmented state diagram of the encoder.
(e) Find the generating function $T(X, Y, Z)$.

10.6. Repeat Problem 10.5 for the (4, 3, 2) code of Figure 10.3.

10.7. Consider the (3, 1, 5) systematic code with

$$\mathbf{g}^{(2)} = (1 \quad 0 \quad 1 \quad 1 \quad 0 \quad 1)$$
$$\mathbf{g}^{(3)} = (1 \quad 1 \quad 0 \quad 0 \quad 1 \quad 1).$$

(a) Find the generator matrix $\mathbf{G}$.
(b) Find the parity sequences corresponding to the information sequence $\mathbf{u} = (1\ 1\ 0\ 1)$.

10.8. Consider the (3, 2, 3) systematic code with

$$\mathbf{g}_1^{(3)}(D) = 1 + D^2 + D^3$$
$$\mathbf{g}_2^{(3)}(D) = 1 + D + D^3.$$

(a) Draw the straightforward realization of the encoder.
(b) Draw a simpler encoder realization which requires only three shift register stages.

10.9. Consider the (2, 1, 2) code with $\mathbf{G}(D) = [1 + D^2, 1 + D + D^2]$.
(a) Find the GCD of its generator polynomials.
(b) Find the transfer function matrix $\mathbf{G}^{-1}(D)$ of its minimum-delay feedforward inverse.

10.10. Consider the (2, 1, 3) code with $\mathbf{G}(D) = [1 + D^2, 1 + D + D^2 + D^3]$.
(a) Find the GCD of its generator polynomials.
(b) Draw the encoder state diagram.
(c) Find an infinite-weight information sequence that generates a code word of finite weight.
(d) Is this code catastrophic or noncatastrophic?

10.11. Consider the (3, 1, 2) code of Problem 10.1.
(a) Find its free distance d_{free}.
(b) Plot its complete CDF.
(c) Find its minimum distance $d_{\min}$.

10.12. Repeat Problem 10.11 for the code of Problem 10.10.

10.13. Show that:
(a) $T(X, Y) = T(X, Y, Z)|_{Z=1}$
(b) $T(X) = T(X, Y)|_{Y=1} = T(X, Y, Z)|_{Y=Z=1}$

10.14. Find the transfer function matrix $\mathbf{G}^{-1}(D)$ of the feedforward inverse of an (n, k, m) systematic code. What is the minimum delay l?

10.15. Prove that for noncatastrophic codes,

$$\lim_{i \to \infty} d_i = d_{\text{free}}.$$

REFERENCES

1. P. Elias, "Coding for Noisy Channels," *IRE Conv. Rec.*, Part 4, pp. 37–47, 1955.
2. J. M. Wozencraft and B. Reiffen, *Sequential Decoding*, MIT Press, Cambridge, Mass., 1961.

3. J. L. Massey, *Threshold Decoding*, MIT Press, Cambridge, Mass., 1963.
4. A. J. Viterbi, "Error Bounds for Convolutional Codes and an Asymptotically Optimum Decoding Algorithm," *IEEE Trans. Inf. Theory*, IT-13, pp. 260–269, April 1967.
5. S. Mason and H. Zimmermann, *Electronic Circuits, Signals, and Systems*, Wiley, New York, 1960.
6. G. D. Forney, Jr., "Convolutional Codes I: Algebraic Structure," *IEEE Trans. Inf. Theory*, IT-16, pp. 720–738, November 1970.
7. J. L. Massey and M. K. Sain, "Inverses of Linear Sequential Circuits," *IEEE Trans. Comput.*, C-17, pp. 330–337, April 1968.
8. W. J. Rosenberg, "Structural Properties of Convolutional Codes," Ph.D. thesis, University of California, Los Angeles, 1971.
9. L. R. Bahl, C. D. Cullum, W. D. Frazer, and F. Jelinek, "An Efficient Algorithm for Computing Free Distance," *IEEE Trans. Inf. Theory*, IT-18, pp. 437–439, May 1972.
10. K. J. Larsen, "Comments on 'An Efficient Algorithm for Computing Free Distance'," *IEEE Trans. Inf. Theory*, IT-19, pp. 577–579, July 1973.
11. D. J. Costello, Jr., "Construction of Convolutional Codes for Sequential Decoding," Ph.D. thesis, University of Notre Dame, Notre Dame, Ind., 1969.
12. J. Layland and R. J. McEliece, "An Upper Bound on the Free Distance of a Tree Code," Jet Propulsion Laboratory, *SPS 37–62*, Vol. 3, pp. 63–64, April 1970.
13. G. D. Forney, Jr., "Convolutional Codes II: Maximum Likelihood Decoding," *Inf. Control*, 25, pp. 222–266, July 1974.
14. D. J. Costello, Jr., "Free Distance Bounds for Convolutional Codes," *IEEE Trans. Inf. Theory*, IT-20, pp. 356–365, May 1974.

11

Maximum Likelihood Decoding of Convolutional Codes

In 1967, Viterbi [1] introduced a decoding algorithm for convolutional codes which has since become known as the *Viterbi algorithm.* Later, Omura [2] showed that the Viterbi algorithm was equivalent to a dynamic programming solution to the problem of finding the shortest path through a weighted graph. Finally, Forney [3,4] recognized that it was in fact a maximum likelihood decoding algorithm for convolutional codes; that is, the decoder output selected is always the code word that gives the largest value of the log-likelihood function.

Forney [5] also was the first to point out that the Viterbi algorithm could be used to produce the maximum likelihood estimate of the transmitted sequence over a bandlimited channel with intersymbol interference. This observation becomes apparent when one recognizes that the channel memory which produces the intersymbol interference is analogous to the encoder memory in a convolutional code. A thorough discussion of the use of the Viterbi algorithm to combat intersymbol interference is given in Viterbi and Omura [6].

11.1 THE VITERBI ALGORITHM

In order to understand Viterbi's decoding algorithm, it is convenient to expand the state diagram of the encoder in time (i.e., to represent each time unit with a separate state diagram). The resulting structure is called a *trellis diagram,* and is shown in Figure 11.1 for the (3, 1, 2) code with

$$\mathbf{G}(D) = [1 + D, 1 + D^2, 1 + D + D^2]$$

and an information sequence of length $L = 5$. The trellis diagram contains $L +$

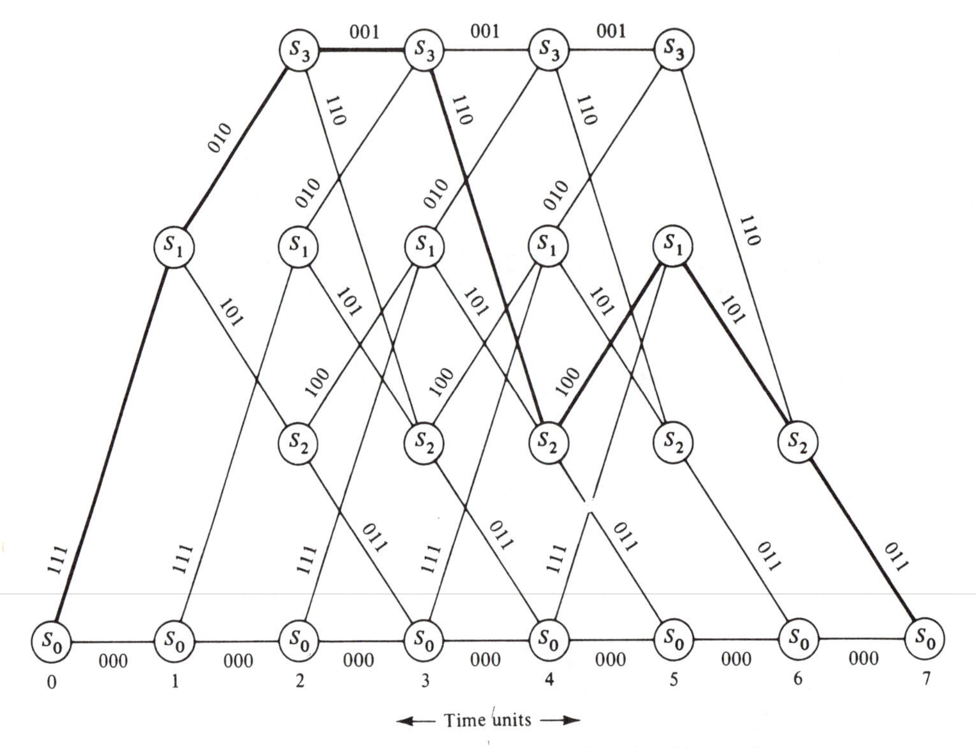

Figure 11.1 Trellis diagram for a (3, 1, 2) code with $L = 5$.

$m + 1$ time units or levels, and these are labeled from 0 to $L + m$ in Figure 11.1. Assuming that the encoder always starts in state S_0 and returns to state S_0, the first m time units correspond to the encoder's departure from state S_0, and the last m time units correspond to the encoder's return to state S_0. It follows that not all states can be reached in the first m or the last m time units. However, in the center portion of the trellis, all states are possible, and each time unit contains a replica of the state diagram. There are two branches leaving and entering each state. The upper branch leaving each state at time unit i represents the input $u_i = 1$, while the lower branch represents $u_i = 0$. Each branch is labeled with the n corresponding outputs $\mathbf{v}_i$, and each of the 2^L code words of length $N = n(L + m)$ is represented by a unique path through the trellis. For example, the code word corresponding to the information sequence $\mathbf{u} = (1\ 1\ 1\ 0\ 1)$ is shown highlighted in Figure 11.1. In the general case of an (n, k, m) code and an information sequence of length kL, there are 2^k branches leaving and entering each state, and 2^{kL} distinct paths through the trellis corresponding to the 2^{kL} code words.

Now assume that an information sequence $\mathbf{u} = (\mathbf{u}_0, \ldots, \mathbf{u}_{L-1})$ of length kL is encoded into a code word $\mathbf{v} = (\mathbf{v}_0, \mathbf{v}_1, \ldots, \mathbf{v}_{L+m-1})$ of length $N = n(L + m)$, and that a Q-ary sequence $\mathbf{r} = (\mathbf{r}_0, \mathbf{r}_1, \ldots, \mathbf{r}_{L+m-1})$ is received over a binary input, Q-ary output *discrete memoryless channel* (DMC). Alternatively, these sequences can be written as $\mathbf{u} = (u_0, u_1, \ldots, u_{kL-1})$, $\mathbf{v} = (v_0, v_1, \ldots, v_{N-1})$, and $\mathbf{r} = (r_0, r_1, \ldots, r_{N-1})$, where the subscripts now simply represent the ordering of the symbols in each sequence. As discussed in Section 1.4, the decoder must produce an estimate $\hat{\mathbf{v}}$ of the code word $\mathbf{v}$ based on the received sequence $\mathbf{r}$. A *maximum likelihood decoder* (MLD) for a DMC chooses $\hat{\mathbf{v}}$ as the code word $\mathbf{v}$ which maximizes the log-likelihood function $\log P(\mathbf{r} \mid \mathbf{v})$. Since for a DMC

$$P(\mathbf{r} \mid \mathbf{v}) = \prod_{i=0}^{L+m-1} P(\mathbf{r}_i \mid \mathbf{v}_i) = \prod_{i=0}^{N-1} P(r_i \mid v_i), \tag{11.1}$$

it follows that

$$\log P(\mathbf{r} \mid \mathbf{v}) = \sum_{i=0}^{L+m-1} \log P(\mathbf{r}_i \mid \mathbf{v}_i) = \sum_{i=0}^{N-1} \log P(r_i \mid v_i), \tag{11.2}$$

where $P(r_i \mid v_i)$ is a channel transition probability. This is a minimum error probability decoding rule when all code words are equally likely.

The log-likelihood function $\log P(\mathbf{r} \mid \mathbf{v})$ is called the *metric* associated with the path $\mathbf{v}$, and is denoted $M(\mathbf{r} \mid \mathbf{v})$. The terms $\log P(\mathbf{r}_i \mid \mathbf{v}_i)$ in the sum of (11.2) are called *branch metrics*, and are denoted $M(\mathbf{r}_i \mid \mathbf{v}_i)$, whereas the terms $\log P(r_i \mid v_i)$ are called *bit metrics*, and are denoted $M(r_i \mid v_i)$. Hence, the path metric $M(\mathbf{r} \mid \mathbf{v})$ can be written as

$$M(\mathbf{r} \mid \mathbf{v}) = \sum_{i=0}^{L+m-1} M(\mathbf{r}_i \mid \mathbf{v}_i) = \sum_{i=0}^{N-1} M(r_i \mid v_i). \tag{11.3}$$

A partial path metric for the first j branches of a path can now be expressed as

$$M([\mathbf{r} \mid \mathbf{v}]_j) = \sum_{i=0}^{j-1} M(\mathbf{r}_i \mid \mathbf{v}_i). \tag{11.4}$$

The following algorithm, when applied to the received sequence $\mathbf{r}$ from a DMC, finds the path through the trellis with the largest metric (i.e., the *maximum likelihood path*). The algorithm processes $\mathbf{r}$ in an iterative manner. At each step, it compares the metrics of all paths entering each state, and stores the path with the largest metric, called the *survivor*, together with its metric.

The Viterbi Algorithm

Step 1. Beginning at time unit $j = m$, compute the partial metric for the single path entering each state. Store the path (the survivor) and its metric for each state.

Step 2. Increase j by 1. Compute the partial metric for all the paths entering a state by adding the branch metric entering that state to the metric of the connecting survivor at the preceding time unit. For each state, store the path with the largest metric (the survivor), together with its metric, and eliminate all other paths.

Step 3. If $j < L + m$, repeat step 2. Otherwise, stop.

There are 2^K survivors from time unit m through time unit L, one for each of the 2^K states. After time unit L there are fewer survivors, since there are fewer states while the encoder is returning to the all-zero state. Finally, at time unit $L + m$, there is only one state, the all-zero state, and hence only one survivor, and the algorithm terminates. We now prove that this final survivor is the maximum likelihood path.

Theorem 11.1. The final survivor $\hat{\mathbf{v}}$ in the Viterbi algorithm is the maximum likelihood path, that is,

$$M(\mathbf{r} \mid \hat{\mathbf{v}}) \geq M(\mathbf{r} \mid \mathbf{v}), \qquad \text{all } \mathbf{v} \neq \hat{\mathbf{v}}.$$

Proof. Assume that the maximum likelihood path is eliminated by the algorithm at time unit j, as illustrated in Figure 11.2. This implies that the partial path metric of the survivor exceeds that of the maximum likelihood path at this point. Now if the remaining portion of the maximum likelihood path is appended onto the survivor at time unit j, the total metric of this path will exceed the total metric of the maximum likelihood path. But this contradicts the definition of the maximum likelihood path as the path with the largest metric. Hence, the maximum likelihood path cannot be eliminated by the algorithm, and must be the final survivor. Q.E.D.

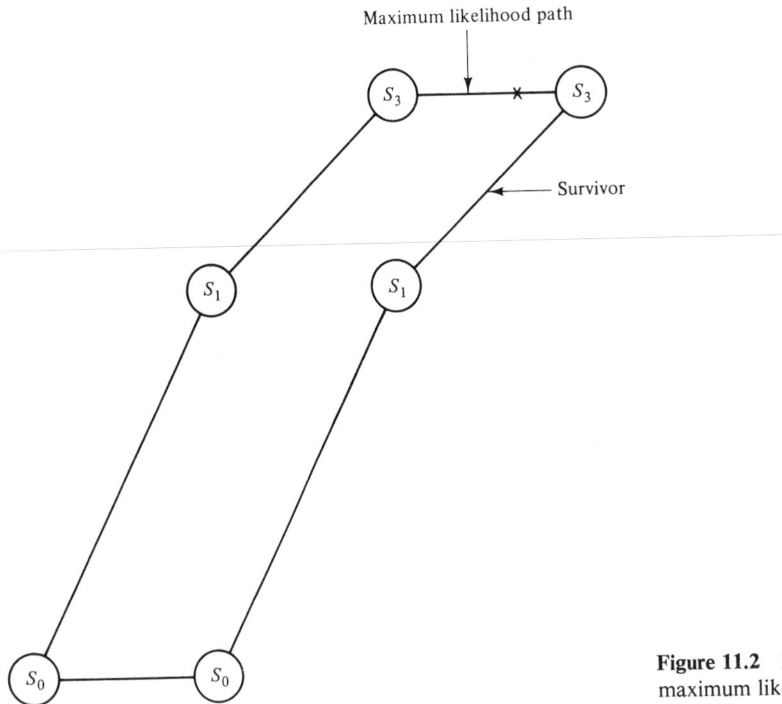

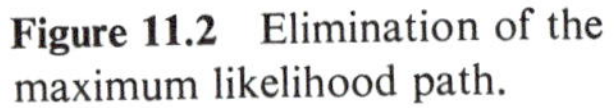
Figure 11.2 Elimination of the maximum likelihood path.

Therefore, the Viterbi algorithm is optimum in the sense that it always finds the maximum likelihood path through the trellis.

From an implementation point of view, it is more convenient to use positive integers as metrics rather than the actual bit metrics. The bit metric $M(r_i \mid v_i) = \log P(r_i \mid v_i)$ can be replaced by $c_2[\log P(r_i \mid v_i) + c_1]$, where c_1 is any real number and c_2 is any positive real number. It can be shown (see Problem 11.2) that a path $\mathbf{v}$ which maximizes

$$M(\mathbf{r} \mid \mathbf{v}) = \sum_{i=0}^{N-1} M(r_i \mid v_i) = \sum_{i=0}^{N-1} \log P(r_i \mid v_i)$$

also maximizes

$$\sum_{i=0}^{N-1} c_2[\log P(r_i \mid v_i) + c_1],$$

and hence the modified metrics can be used without affecting the performance of the Viterbi algorithm. If c_1 is chosen to make the smallest metric 0, c_2 can then be chosen so that all metrics can be approximated by integers. There are many sets of integer metrics possible for a given DMC, depending on the choice of c_2. The performance of the Viterbi algorithm is now slightly suboptimum due to the modified metric approximation by integers, but the degradation is typically very slight.

Example 11.1

As an example, consider the binary input, quaternary output ($Q = 4$) DMC shown in Figure 11.3. Using logarithms to the base 10, the bit metrics for this channel are displayed in a *metric table* in Figure 11.4(a). Choosing $c_1 = 1$ and $c_2 = 17.3$ yields the *integer metric table* shown in Figure 11.4(b). Now assume that a code word from the trellis diagram of the (3, 1, 2) code of Figure 11.1 is transmitted over the DMC of Figure 11.3, and that the quaternary received sequence is

$$\mathbf{r} = (1_1 1_2 0_1,\ 1_1 1_1 0_2,\ 1_1 1_1 0_1,\ 1_1 1_1 1_1,\ 0_1 1_2 0_1,\ 1_2 0_2 1_1,\ 1_2 0_1 1_1).$$

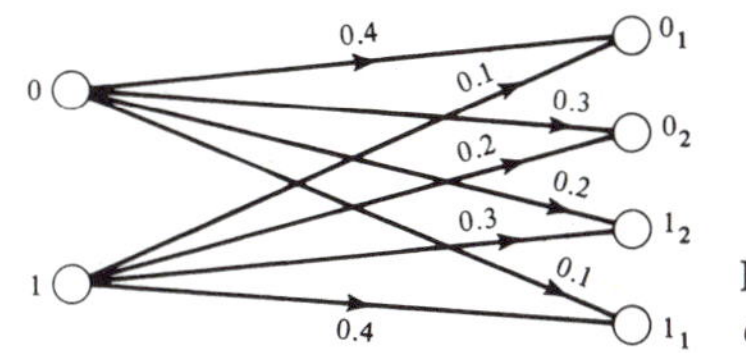

Figure 11.3 Binary-input, quaternary-output DMC.

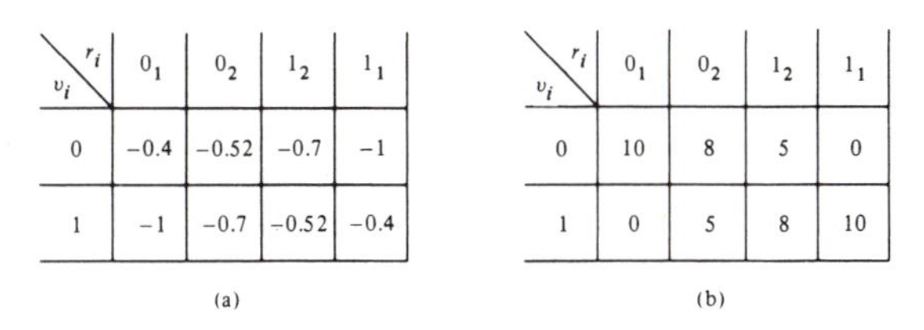

v_i \ r_i	0_1	0_2	1_2	1_1
0	−0.4	−0.52	−0.7	−1
1	−1	−0.7	−0.52	−0.4

(a)

v_i \ r_i	0_1	0_2	1_2	1_1
0	10	8	5	0
1	0	5	8	10

(b)

Figure 11.4 Metric tables for the channel of Figure 11.3.

The application of the Viterbi algorithm to this received sequence is shown in Figure 11.5. The numbers above each state represent the metric of the survivor for that state, and the paths eliminated at each state are shown crossed out on the trellis diagram. The final survivor,

$$\hat{\mathbf{v}} = (1\ 1\ 1,\ 0\ 1\ 0,\ 1\ 1\ 0,\ 0\ 1\ 1,\ 0\ 0\ 0,\ 0\ 0\ 0,\ 0\ 0\ 0),$$

is shown as the highlighted path. This corresponds to the decoded information sequence

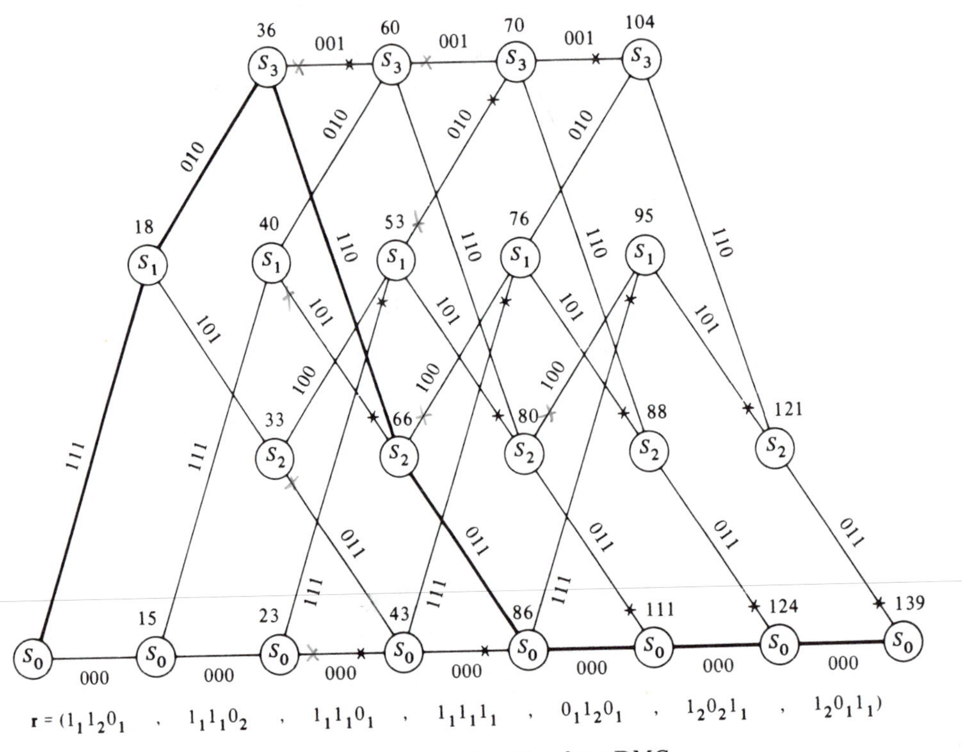

Figure 11.5 Viterbi algorithm for a DMC.

$\hat{\mathbf{u}} = (1\ 1\ 0\ 0\ 0)$. Note that the final m branches in any trellis path always correspond to 0 inputs, and hence are not considered part of the information sequence.

In the special case of a *binary symmetric channel* (BSC) with transition probability $p < \frac{1}{2}$,[1] the received sequence $\mathbf{r}$ is binary ($Q = 2$) and the log-likelihood function becomes [see (1.11)]

$$\log P(\mathbf{r}\,|\,\mathbf{v}) = d(\mathbf{r}, \mathbf{v}) \log \frac{p}{1-p} + N \log (1-p), \tag{11.5}$$

where $d(\mathbf{r}, \mathbf{v})$ is the Hamming distance between $\mathbf{r}$ and $\mathbf{v}$. Since $\log [p/(1-p)] < 0$ and $N \log (1-p)$ is a constant for all $\mathbf{v}$, an MLD for a BSC chooses $\mathbf{v}$ as the code word $\hat{\mathbf{v}}$ that minimizes the Hamming distance

$$d(\mathbf{r}, \mathbf{v}) = \sum_{i=0}^{L+m-1} d(\mathbf{r}_i, \mathbf{v}_i) = \sum_{i=0}^{N-1} d(r_i, v_i). \tag{11.6}$$

Hence, in applying the Viterbi algorithm to the BSC, $d(\mathbf{r}_i, \mathbf{v}_i)$ becomes the branch metric, $d(r_i, v_i)$ becomes the bit metric, and the algorithm must find the path through

[1]A BSC with $p > \frac{1}{2}$ can always be converted to a BSC with $p < \frac{1}{2}$ simply by reversing the output labels.

the trellis with the smallest metric (i.e., the path closest to $\mathbf{r}$ in Hamming distance). The details of the algorithm are exactly the same, except that the Hamming distance replaces the log-likelihood function as the metric and the survivor at each state is the path with the smallest metric.

Example 11.2

An example of the application of the Viterbi algorithm to a BSC is shown in Figure 11.6. Assume that a code word from the trellis diagram of the (3, 1, 2) code of Figure 11.1 is transmitted over a BSC and that the received sequence is

$$\mathbf{r} = (1\ 1\ 0,\ 1\ 1\ 0,\ 1\ 1\ 0,\ 1\ 1\ 1,\ 0\ 1\ 0,\ 1\ 0\ 1,\ 1\ 0\ 1).$$

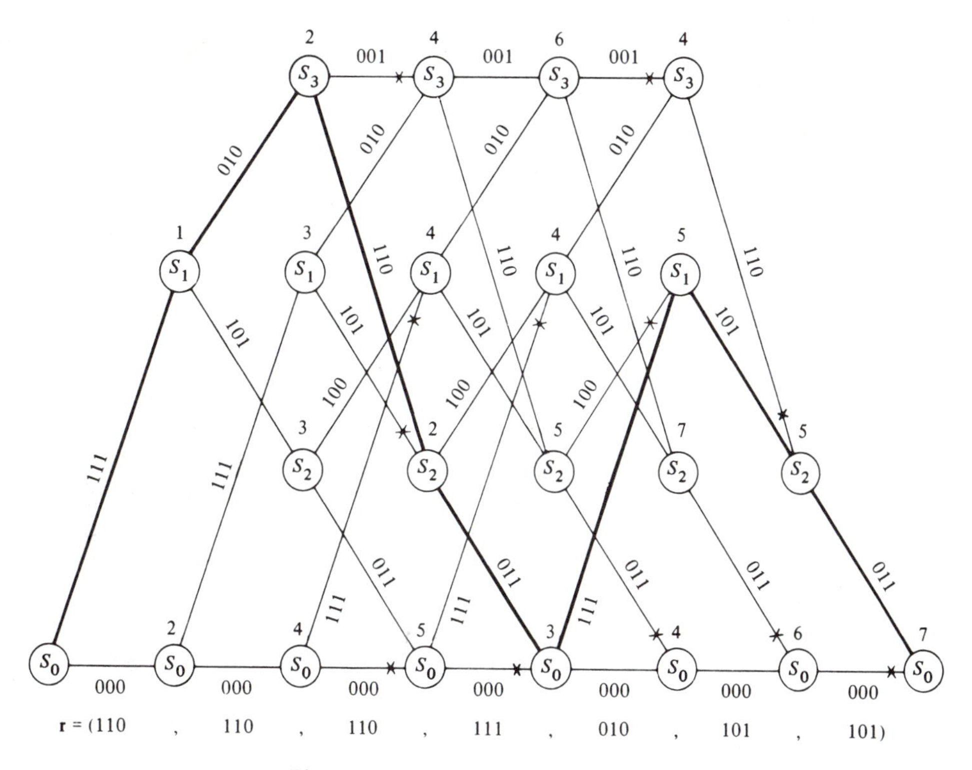

Figure 11.6 Viterbi algorithm for a BSC.

The final survivor

$$\hat{\mathbf{v}} = (1\ 1\ 1,\ 0\ 1\ 0,\ 1\ 1\ 0,\ 0\ 1\ 1,\ 1\ 1\ 1,\ 1\ 0\ 1,\ 0\ 1\ 1)$$

is shown as the highlighted path in the figure, and the decoded information sequence is $\hat{\mathbf{u}} = (1\ 1\ 0\ 0\ 1)$. The fact that the final survivor has a metric of 7 means that no other path through the trellis differs from $\mathbf{r}$ in fewer than seven positions. Note that at some states neither path is crossed out. This indicates a tie in the metric values of the two paths entering that state. If the final survivor goes through any of these states, there is more than one maximum likelihood path (i.e., more than one path whose distance from $\mathbf{r}$ is minimum). From an implementation point of view, whenever a tie in metric values occurs, one path is arbitrarily selected as the survivor, because of the impracticality of storing a variable number of paths. This arbitrary resolution of ties has no effect on the decoding error probability.

In Section 1.3 we noted that making soft demodulator decisions ($Q > 2$) results in a performance advantage over making hard decisions ($Q = 2$). The two examples of the application of the Viterbi algorithm presented above serve to illustrate this point. If the quaternary outputs 0_1 and 0_2 are coalesced into a single output 0 and 1_1 and 1_2 into a single output 1, the soft decision DMC is converted to a hard decision BSC with transition probability $p = 0.3$. In the examples above, the sequence $\mathbf{r}$ in the hard-decision case is the same as in the soft-decision case with 0_1 and 0_2 converted to 0 and 1_1 and 1_2 converted to 1. But the Viterbi algorithm yields different results in the two cases. In the soft-decision case ($Q = 4$), the information sequence $\mathbf{u} = (1\ 1\ 0\ 0\ 0)$ produces the maximum likelihood path, which has a final metric of 139. In the hard-decision case ($Q = 2$), however, the maximum likelihood path is $\mathbf{u} = (1\ 1\ 0\ 0\ 1)$. The metric of this path on the quaternary output channel is 135, and it is clearly not the maximum likelihood path in the soft-decision case. However, since hard decisions mask the distinction between certain soft-decision outputs (e.g., outputs 0_1 and 0_2 are treated as equivalent outputs in the hard-decision case), the hard-decision decoder makes an estimate that would not be made if more information about the channel were available (i.e., if soft decisions were made).

As a final comment, both of the channels described above can be classified as "very noisy" channels. The code rate $R = \frac{1}{2}$ exceeds the channel capacity C in both cases. Hence, we would not expect the performance of this code to be very good with either channel. This is reflected by the relatively low value (139) of the final metric of the maximum likelihood path in the DMC case, as compared with a maximum possible metric of 210 for a path that "agrees" completely with $\mathbf{r}$. Also, in the BSC case, the final Hamming distance of 7 for the maximum likelihood path is large for paths only 21 bits long. Lower code rates would be needed to achieve good performance over these channels. More about code performance is discussed in the next section.

11.2 PERFORMANCE BOUNDS FOR CONVOLUTIONAL CODES

We begin this section by analyzing the performance of the Viterbi algorithm for a specific code on a BSC. More general channel models will be discussed later. First assume, without loss of generality, that the all-zero code word $\mathbf{v} = \mathbf{0}$ is transmitted from the (3, 1, 2) code of Figure 11.1. This code has a generating function (see Problem 10.5)

$$\begin{aligned} T(X, Y, Z) &= \frac{X^7YZ^3}{1 - XYZ(1 + X^2Z)} \\ &= X^7YZ^3[1 + XYZ(1 + X^2Z) + X^2Y^2Z^2(1 + X^2Z^2) + \cdots] \\ &= X^7YZ^3 + X^8Y^2Z^4 + X^9Y^3Z^5 + X^{10}(Y^2Z^5 + Y^4Z^6) + \cdots, \end{aligned} \tag{11.7}$$

that is, it has one code word of weight 7 and length 3 branches generated by an information sequence of weight 1, one code word of weight 8 and length 4 branches generated by an information sequence of weight 2, and so on.

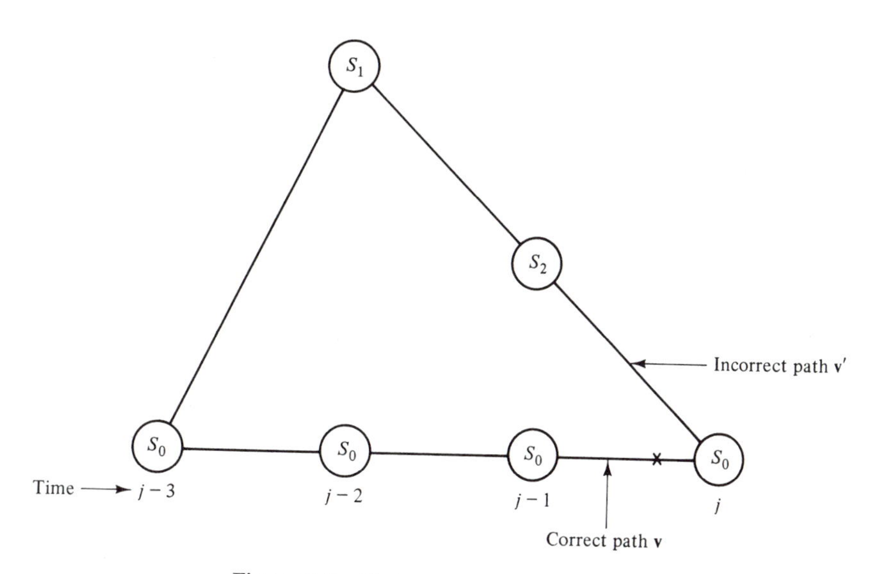

Figure 11.7 First–event error at time unit j.

We say that a *first event error* is made at an arbitrary time unit j if the all-zero path (the *correct path*) is eliminated *for the first time* at time unit j in favor of a competitor (the *incorrect path*). This is illustrated in Figure 11.7, where the correct path $\mathbf{v}$ is eliminated by the incorrect path $\mathbf{v}'$ at time unit j. The incorrect path must be some path that had previously diverged from the all-zero state and is now remerging for the first time at time j (i.e., it must be one of the paths enumerated by the code-generating function). If it is the weight 7 path, a first event error will be made if, in the seven positions in which the correct and incorrect paths differ, the binary received sequence $\mathbf{r}$ agrees with the incorrect path in four or more of these positions (i.e., if $\mathbf{r}$ contains four or more 1's in these seven positions). If the BSC transition probability is p, this probability is

$$\begin{aligned} P_7 &= P[4 \text{ or more 1's in seven positions}] \\ &= \sum_{e=4}^{7} \binom{7}{e} p^e (1-p)^{7-e}. \end{aligned} \tag{11.8}$$

If the weight 8 path is the incorrect path, a first event error is made with probability

$$P_8 = \frac{1}{2}\binom{8}{4} p^4 (1-p)^4 + \sum_{e=5}^{8} \binom{8}{e} p^e (1-p)^{8-e}, \tag{11.9}$$

since if the metrics of the correct and incorrect paths are tied, an error is made with probability $\frac{1}{2}$. In general, if the incorrect path has weight d, a first event error is made with probability

$$P_d = \begin{cases} \displaystyle\sum_{e=(d+1)/2}^{d} \binom{d}{e} p^e (1-p)^{d-e}, & d \text{ odd} \\ \displaystyle\frac{1}{2}\binom{d}{d/2} p^{d/2} (1-p)^{d/2} + \sum_{e=(d/2)+1}^{d} \binom{d}{e} p^e (1-p)^{d-e}, & d \text{ even.} \end{cases} \tag{11.10}$$

Since all incorrect paths of length j branches or less can cause a first event error at time unit j, the *first event error probability at time unit* j, $P_f(E, j)$, can be overbounded, using a union bound, by the sum of the error probabilities of each of these paths. If all incorrect paths of length greater than j branches are also included, $P_f(E, j)$ is overbounded by

$$P_f(E, j) < \sum_{d=d_{free}}^{\infty} A_d P_d, \tag{11.11}$$

where A_d is the number of code words of weight d (i.e., it is the weight enumerator of the code). Since this bound is independent of j, it holds for all time units, and the *first event error probability* at any time unit, $P_f(E)$, is bounded by

$$P_f(E) < \sum_{d=d_{free}}^{\infty} A_d P_d. \tag{11.12}$$

The bound of (11.12) can be further simplified by noting that for d odd,

$$\begin{aligned} P_d &= \sum_{e=(d+1)/2}^{d} \binom{d}{e} p^e (1-p)^{d-e} \\ &< \sum_{e=(d+1)/2}^{d} \binom{d}{e} p^{d/2} (1-p)^{d/2} \\ &= p^{d/2}(1-p)^{d/2} \sum_{e=(d+1)/2}^{d} \binom{d}{e} \\ &< p^{d/2}(1-p)^{d/2} \sum_{e=0}^{d} \binom{d}{e} \\ &= 2^d p^{d/2}(1-p)^{d/2}. \end{aligned} \tag{11.13}$$

It can also be shown (see Problem 11.7) that (11.13) is an upper bound on P_d for d even. Hence,

$$P_f(E) < \sum_{d=d_{free}}^{\infty} A_d [2\sqrt{p(1-p)}]^d, \tag{11.14}$$

and for an arbitrary convolutional code with generating function $T(X) = \sum_{d=d_{free}}^{\infty} A_d X^d$, it follows by comparing (11.14) with the expression for $T(X)$ that

$$P_f(E) < T(X)|_{X=2\sqrt{p(1-p)}}. \tag{11.15}$$

The final decoded path can diverge from and remerge with the correct path any number of times (i.e., it can contain any number of event errors). This is illustrated in Figure 11.8. After one or more event errors have occurred, the two paths compared at the all-zero state will both be incorrect paths, one of which contains at least one previous event error. This is illustrated in Figure 11.9, where it is assumed that the correct path $\mathbf{v}$ has been eliminated for the first time at time unit $j - l$ by incorrect path $\mathbf{v}'$, and that at time unit j incorrect paths $\mathbf{v}'$ and $\mathbf{v}''$ are compared. If the partial metric for $\mathbf{v}''$ exceeds the partial metric for $\mathbf{v}'$ at time unit j, it must also exceed the partial metric for $\mathbf{v}$, since $\mathbf{v}'$ has a better metric than $\mathbf{v}$ at time j due to the first event error at time unit $j - l$. Hence, if $\mathbf{v}''$ were compared to $\mathbf{v}$ at time unit j, a first event error would be made. We say that an event error occurs at time unit j if $\mathbf{v}''$ survives over $\mathbf{v}'$, and the *event error probability at time unit* j, $P(E, j)$ is bounded by

$$P(E, j) \leq P_f(E, j), \tag{11.16}$$

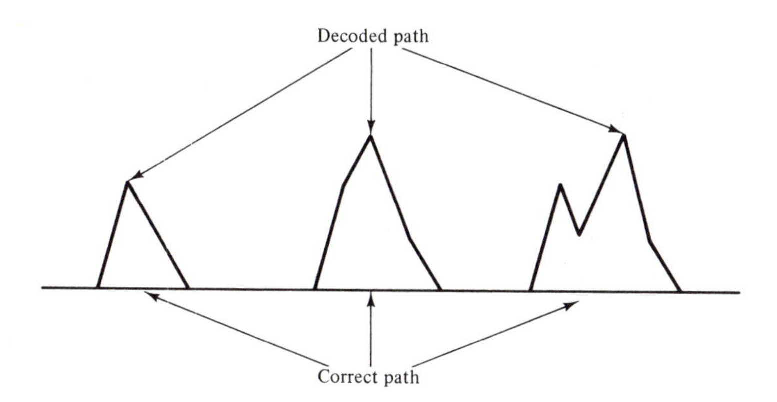

Figure 11.8 Multiple error events.

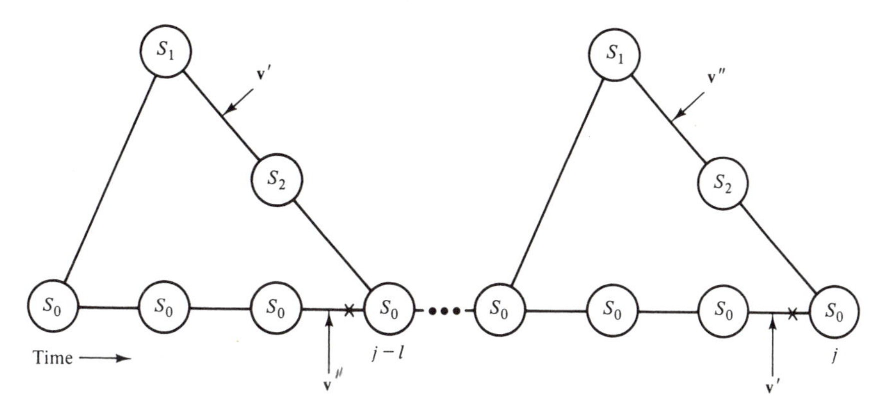

Figure 11.9 Comparison of two incorrect paths.

since if $\mathbf{v}''$ survives over $\mathbf{v}'$ at time unit j, it would also survive if compared to $\mathbf{v}$. The situation illustrated in Figure 11.9 is not the only way in which an event error at time unit j can follow a first event error made earlier. Two other possibilities are shown in Figure 11.10. In these cases, the event error made at time unit j either totally or partially replaces the event error made previously. Using exactly the same arguments as above, it follows that (11.16) holds for these cases also, and hence is a valid bound for any event error occurring at time unit j.[2]

The bound of (11.11) can be applied to (11.16), and since it is independent of j, it holds for all time units, and the *event error probability* at any time unit, $P(E)$, is bounded by

$$P(E) < \sum_{d=d_{free}}^{\infty} A_d P_d < T(X)\big|_{X=2\sqrt{p(1-p)}}, \tag{11.17}$$

just as in (11.15). For small p, the bound is dominated by its first term (i.e., the free

[2]In the two cases shown in Figure 11.10, the event error at time unit j replaces at least a portion of a previous error event. The net effect may be a decrease in the number of decoding errors (i.e., the number of positions in which the decoded path differs from the correct path). Hence, using the first event error probability as a bound may be conservative in these cases.

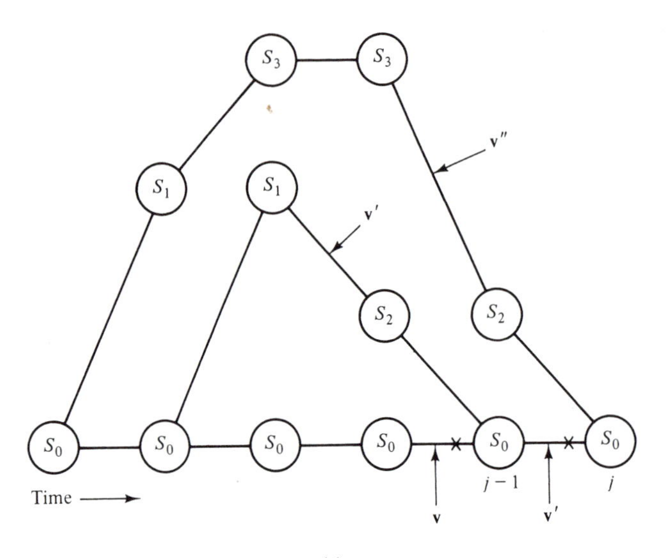

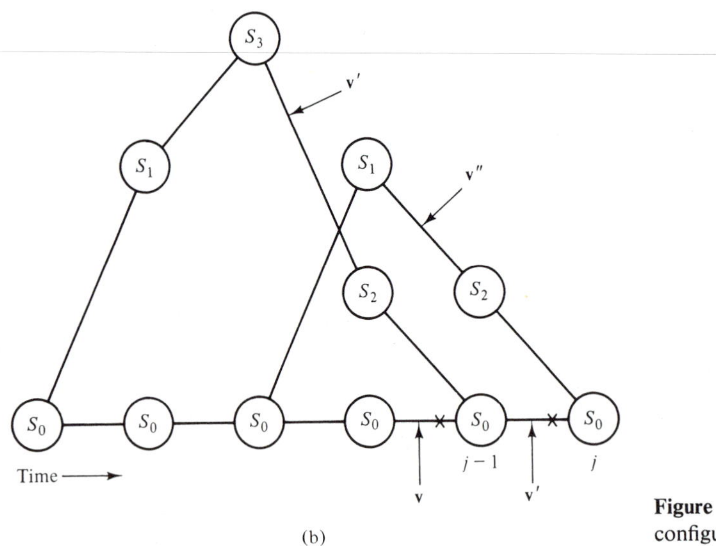

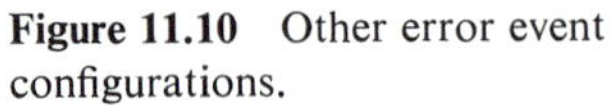
Figure 11.10 Other error event configurations.

distance term), and the event error probability can be approximated as

$$P(E) \approx A_{d_{\text{free}}}[2\sqrt{p(1-p)}]^{d_{\text{free}}} \approx A_{d_{\text{free}}} 2^{d_{\text{free}}} p^{d_{\text{free}}/2}. \tag{11.18}$$

Example 11.3

For the (3, 1, 2) code of Figure 11.1, $d_{\text{free}} = 7$ and $A_{d_{\text{free}}} = 1$, and for $p = 10^{-2}$,

$$P(E) \approx 2^7 p^{7/2} = 1.28 \times 10^{-5}.$$

The event error probability bound of (11.17) can be modified to provide a bound on the *bit error probability*, $P_b(E)$, that is, the expected number of information bit decoding errors per decoded information bit. Each event error causes a number of information bit errors equal to the number of nonzero information bits on the incorrect path. Hence, if each event error probability term P_d is weighted by the number of nonzero information bits on the weight d path, or, if there is more than one weight d path, by the total number of nonzero information bits on all weight d paths, a bound on the expected number of information bit decoding errors made at any time unit results. This can then be divided by k, the number of information bits per unit time, to obtain a bound on $P_b(E)$. In other words, the bit error probability is bounded by

$$P_b(E) < \frac{1}{k} \sum_{d=d_{\text{free}}}^{\infty} B_d P_d, \tag{11.19}$$

where B_d is the total number of nonzero information bits on all weight d paths. Since $T(X, Y) = \sum_{d=d_{\text{free}}}^{\infty} \sum_{b=1}^{\infty} A_{d,b} X^d Y^b$, it follows that

$$\left.\frac{\partial T(X, Y)}{\partial Y}\right|_{Y=1} = \sum_{d=d_{\text{free}}}^{\infty} \sum_{b=1}^{\infty} b A_{d,b} X^d = \sum_{d=d_{\text{free}}}^{\infty} B_d X^d, \tag{11.20}$$

where we have noted that $B_d = \sum_{b=1}^{\infty} b A_{d,b}$. Now using (11.13) in (11.19) and comparing that bound with (11.20), we see that

$$P_b(E) < \frac{1}{k} \sum_{d=d_{\text{free}}}^{\infty} B_d [2\sqrt{p(1-p)}]^d = \frac{1}{k} \left.\frac{\partial T(X, Y)}{\partial Y}\right|_{X=2\sqrt{p(1-p)},\, Y=1} \tag{11.21}$$

for an arbitrary convolutional code with generating function $T(X, Y)$. For small p, the bound is dominated by its first term, so that

$$P_b(E) \approx \frac{1}{k} B_{d_{\text{free}}} [2\sqrt{p(1-p)}]^{d_{\text{free}}} \approx \frac{1}{k} B_{d_{\text{free}}} 2^{d_{\text{free}}} p^{d_{\text{free}}/2}. \tag{11.22}$$

Example 11.4

For the (3, 1, 2) code of Figure 11.1, $B_{d_{\text{free}}} = 1$ and $d_{\text{free}} = 7$, and for $p = 10^{-2}$,

$$P_b(E) \approx 2^7 p^{7/2} = 1.28 \times 10^{-5},$$

the same as for the event error probability. In other words, when p is small, the most likely error event is that the weight 7 path is decoded instead of the all-zero path, thereby causing one information bit error. Typically, then, each event error causes one bit error, and this is reflected in the approximate expressions for $P(E)$ and $P_b(E)$.

If the BSC is derived from an AWGN channel with BPSK modulation, optimum coherent detection, and binary output quantization, then from (1.4)

$$p = Q\left(\sqrt{\frac{2E}{N_0}}\right), \tag{11.23}$$

where E is the energy per transmitted symbol and N_0 is the (one-sided) noise power spectral density. Using the bound of (1.5) as an approximation yields

$$p \approx \tfrac{1}{2} e^{-E/N_0}, \tag{11.24}$$

and for a convolutional code with free distance d_{free},

$$P_b(E) \approx \frac{1}{k} B_{d_{\text{free}}} 2^{d_{\text{free}}/2} e^{-(d_{\text{free}}/2)\cdot(E/N_0)} \tag{11.25}$$

when p is small (i.e., when E/N_0 is large). Defining the *energy per information bit*, E_b, as

$$E_b \triangleq \frac{E}{R}, \tag{11.26}$$

since $1/R$ is the number of transmitted symbols per information bit, we can write

$$P_b(E) \approx \frac{1}{k} B_{d_{\text{free}}} 2^{d_{\text{free}}/2} e^{-(Rd_{\text{free}}/2)\cdot(E_b/N_0)} \qquad \text{(with coding)} \tag{11.27}$$

for large E_b/N_0.

On the other hand, if no coding is used (i.e., $R = 1$), the BSC transition probability p is the bit error probability $P_b(E)$ and

$$P_b(E) \approx \tfrac{1}{2} e^{-E_b/N_0} \qquad \text{(without coding).} \tag{11.28}$$

Comparing (11.27) and (11.28), we see that for a fixed E_b/N_0 ratio, the (negative) exponent with coding is larger by a factor of $Rd_{\text{free}}/2$ than the exponent without coding. Since the exponential term dominates the error probability expressions for large E_b/N_0, the factor $Rd_{\text{free}}/2$, in decibels, is called the *asymptotic coding gain*, γ:

$$\gamma \triangleq 10 \log_{10} \frac{Rd_{\text{free}}}{2} \qquad \text{dB} \tag{11.29}$$

in the hard decision case. It is worth noting that coding gains become smaller as E_b/N_0 becomes smaller. In fact, if E_b/N_0 is reduced to the point where the code rate R is greater than the channel capacity C, reliable communication with coding is no longer possible, and an uncoded system will outperform a coded system. This point is illustrated in Problem 11.10.

Bounds similar to (11.17) and (11.21) can also be obtained for more general channel models than the BSC. For a binary input AWGN channel with no output quantization, the bounds become

$$P(E) < T(X)\big|_{X=e^{-RE_b/N_0}} \tag{11.30}$$

and

$$P_b(E) < \frac{1}{k} \frac{\partial T(X, Y)}{\partial Y}\bigg|_{X=e^{-RE_b/N_0}, Y=1}. \tag{11.31}$$

For a binary input AWGN channel with finite output quantization (a DMC), the bounds become

$$P(E) < T(X)\big|_{X=D_0} \tag{11.32}$$

and

$$P_b(E) < \frac{1}{k} \frac{\partial T(X, Y)}{\partial Y}\bigg|_{X=D_0, Y=1}, \tag{11.33}$$

where $D_0 \triangleq \sum_j \sqrt{P(j|0)P(j|1)}$ is a function of the channel transition probabilities. Note that when $Q = 2$, a BSC results, and $D_0 = 2\sqrt{p(1-p)}$. Complete derivations of slightly tighter versions of these bounds can be found in Viterbi and Omura [6].

It is instructive to compare the approximate expression for $P_b(E)$ given in (11.27) for a BSC with a similar expression obtained for a binary-input AWGN channel from (11.31). For large E_b/N_0, the first term in the generating function dominates the bound of (11.31) and $P_b(E)$ can be approximated as

$$P_b(E) \approx \frac{1}{k} B_{d_{\text{free}}} (e^{-RE_b/N_0})^{d_{\text{free}}} = \frac{1}{k} B_{d_{\text{free}}} e^{-Rd_{\text{free}}E_b/N_0}. \tag{11.34}$$

Comparing the exponent of (11.34) with that of (11.27), we see that the exponent of (11.34) is larger by factor of 2. This is equivalent to a 3-dB energy (or power) advantage for the AWGN channel over the BSC, since to achieve the same error probability on the BSC, the transmitter must generate an additional 3 dB of signal energy (or power). This illustrates the benefits of allowing an unquantized demodulator output instead of making hard decisions. The asymptotic coding gain in this case is increased by 3 dB over the hard-decision case. The decoder complexity increases, however, due to the need to accept analog inputs.

The foregoing analysis is based on performance bounds for specific codes, and is valid for large values of E_b/N_0. A similar comparison of soft decisions (finite output quantization with $Q > 2$) and hard decisions can be made by computing the approximate expression for (11.33) for a particular binary input DMC and comparing with (11.22). Generally, it is found that $Q = 8$ allows one to achieve a performance within about $\frac{1}{4}$ dB of the optimum performance achievable with an unquantized demodulator output, while avoiding the need for an analog decoder [7].

A random coding analysis has also been used to demonstrate the advantages of soft decisions over hard decisions [8]. For small values of E_b/N_0, this analysis shows that there is about a 2-dB penalty in signal power attached to the use of hard decisions; that is, to achieve the same error probability, 2 dB more signal power must be generated at the transmitter when the demodulator output is hard quantized rather than unquantized. Over the entire range of E_b/N_0 ratios, the decibel loss associated with hard decisions runs between 2 and 3 dB. Hence, the use of soft decisions ($Q > 2$ but still finite) has become increasingly popular in recent years as a means of regaining most of the 2 to 3 dB loss due to hard quantization, while retaining the advantages of digital decoding.

11.3 CONSTRUCTION OF GOOD CONVOLUTIONAL CODES

The problem of constructing good codes for use with the Viterbi algorithm can now be addressed. Once the desired code rate has been selected, the performance bounds of (11.18), (11.22), (11.27), and (11.34) suggest constructing codes with as large a free distance as possible. Then $A_{d_{free}}$, the number of code words with weight d_{free}, and $B_{d_{free}}$, the total information sequence weight of all the weight-d_{free} code words, can be used as secondary criteria. The use of catastrophic codes should be avoided under all circumstances.

Most code construction for convolutional codes has been done by computer search. Algebraic structures that guarantee good distance properties, similar to the BCH construction for block codes, have proved difficult to find for convolutional codes. This has prevented the construction of good long codes, since most computer search techniques are time consuming and limited to relatively short constraint lengths. Computer construction techniques for noncatastrophic codes with maximal free distance are described in References 9 to 12. Lists of the best codes for rates $R = \frac{1}{4}, \frac{1}{3}, \frac{1}{2}, \frac{2}{3}$, and $\frac{3}{4}$ are given in Table 11.1. The generator sequences are expressed in octal form. For instance, the best (2, 1, 5) code has

$$\mathbf{G}(D) = [1 + D + D^3 + D^5, 1 + D^2 + D^3 + D^4 + D^5].$$

TABLE 11.1 CODES WITH MAXIMAL d_{free}

(a) $R = \frac{1}{4}$

m	$\mathbf{g}^{(1)}$	$\mathbf{g}^{(2)}$	$\mathbf{g}^{(3)}$	$\mathbf{g}^{(4)}$	d_{free}	γ (dB)
2	5	7	7	7	10	0.97
3	54	64	64	74	13	2.11
4	52	56	66	76	16	3.01
5	53	67	71	75	18	3.52
6	564	564	634	714	20	3.98
7	472	572	626	736	22	4.39
8	463	535	733	745	24	4.77
9	4474	5724	7154	7254	27	5.28
10	4656	4726	5562	6372	29	5.59
11	4767	5723	6265	7455	32	6.02
12	44624	52374	66754	73534	33	6.15
13	42226	46372	73256	73276	36	6.53

(b) $R = \frac{1}{3}$

m	$\mathbf{g}^{(1)}$	$\mathbf{g}^{(2)}$	$\mathbf{g}^{(3)}$	d_{free}	γ (dB)
2	5	7	7	8	1.25
3	54	64	74	10	2.21
4	52	66	76	12	3.01
5	47	53	75	13	3.35
6	554	624	764	15	3.98
7	452	662	756	16	4.26
8	557	663	711	18	4.77
9	4474	5724	7154	20	5.22
10	4726	5562	6372	22	5.64
11	4767	5723	6265	24	6.02
12	42554	43364	77304	24	6.02
13	43512	73542	76266	26	6.36

(c) $R = \frac{1}{2}$

m	$\mathbf{g}^{(1)}$	$\mathbf{g}^{(2)}$	d_{free}	γ (dB)
2	5	7	5	0.97
3	64	74	6	1.76
4	46	72	7	2.43
5	65	57	8	3.01
6	554	744	10	3.98
7	712	476	10	3.98
8	561	753	12	4.77
9	4734	6624	12	4.77
10	4672	7542	14	5.44
11	4335	5723	15	5.74
12	42554	77304	16	6.02
13	43572	56246	16	6.02
14	56721	61713	18	6.53
15	447254	627324	19	6.77
16	716502	514576	20	6.99

TABLE 11.1 *(cont.)*

(d) $R = \frac{2}{3}$

m	K	Generator sequences			d_{free}	γ (dB)
1	2	$\mathbf{g}_1^{(1)} = 6$ $\mathbf{g}_2^{(1)} = 2$	$\mathbf{g}_1^{(2)} = 2$ $\mathbf{g}_2^{(2)} = 4$	$\mathbf{g}_1^{(3)} = 6$ $\mathbf{g}_2^{(3)} = 4$	3	0
2	3	4 1	2 4	6 7	4	1.25
2	4	7 2	1 5	4 7	5	2.21
3	5	60 14	30 40	70 74	6	3.01
3	6	64 30	30 64	64 74	7	3.68
4	7	60 16	34 46	54 74	8	4.26
4	8	64 26	12 66	52 44	8	4.26
5	9	52 05	06 70	74 53	9	4.77
5	10	63 32	15 65	46 61	10	5.22

(e) $R = \frac{3}{4}$

m	K	Generator sequences				d_{free}	γ (dB)
1	3	$\mathbf{g}_1^{(1)} = 4$ $\mathbf{g}_2^{(1)} = 0$ $\mathbf{g}_3^{(1)} = 0$	$\mathbf{g}_1^{(2)} = 4$ $\mathbf{g}_2^{(2)} = 6$ $\mathbf{g}_3^{(2)} = 2$	$\mathbf{g}_1^{(3)} = 4$ $\mathbf{g}_2^{(3)} = 2$ $\mathbf{g}_3^{(3)} = 5$	$\mathbf{g}_1^{(4)} = 4$ $\mathbf{g}_2^{(4)} = 4$ $\mathbf{g}_3^{(4)} = 5$	4	1.76
2	5	6 1 0	2 6 2	2 0 5	6 7 5	5	2.73
2	6	6 3 2	1 4 3	0 1 7	7 6 4	6	3.52
3	8	70 14 04	30 50 10	20 00 74	40 54 40	7	4.19
3	9	40 04 34	14 64 00	34 20 60	60 70 64	8	4.77

Its generator sequences are $\mathbf{g}^{(1)} = (1\ 1\ 0\ 1\ 0\ 1)$ and $\mathbf{g}^{(2)} = (1\ 0\ 1\ 1\ 1\ 1)$. These are listed in Table 11.1 as $\mathbf{g}^{(1)} = (65)$ and $\mathbf{g}^{(2)} = (57)$. Since the search procedures employed are essentially exhaustive, results have been obtained only for short constraint lengths. The asymptotic coding gain of each code is also listed in Table 11.1. These coding gains are for a BSC (i.e., hard-decision decoding). Note that the

(2, 1, 16) code with $d_{\text{free}} = 20$ achieves an asymptotic coding gain of 7 dB with hard-decision decoding. Up to 3 dB of additional coding gain is available with soft-decision or analog decoding.

It is possible to construct codes with longer constraint lengths by defining a reduced ensemble of codes, and confining the search only to this ensemble. Bahl and Jelinek [13] did this by considering only $R = \frac{1}{2}$ codes with complementary generators, that is, codes for which

$$\mathbf{g}^{(1)}(D) + \mathbf{g}^{(2)}(D) = D + D^2 + \cdots + D^{m-1} \tag{11.35}$$

and $g_0^{(1)} = g_0^{(2)} = g_m^{(1)} = g_m^{(2)} = 1$. There are only 2^{m-1} codes in this ensemble, considerably reduced from the 2^{2m} codes in the general ensemble of $R = \frac{1}{2}$ codes with $g_0^{(1)} = g_0^{(2)} = 1$. ($d_{\text{free}}$ cannot be increased by considering codes with $g_0^{(1)} = 0$ or $g_0^{(2)} = 0$.) The free distance of the best noncatastrophic *complementary codes* are given in Table 11.2. The complementary codes are very close to optimal out to $m = 13$. Whether "good" complementary codes can always be found for longer constraint lengths is still an unanswered question, although recent results [14] imply that complementary codes have catastrophic-like properties at long constraint lengths.

Note that all the codes listed in Tables 11.1 and 11.2 are nonsystematic. This is because, for a given rate and encoder memory, more free distance is available with nonsystematic codes than with systematic codes, as pointed out in Chapter 10.

TABLE 11.2 $R = 1/2$ COMPLEMENTARY CODES WITH MAXIMAL d_{free}

m	$\mathbf{g}^{(1)}$	d_{free}
2	5	5
3	54	6
4	62	7
5	61	8
6	504	9
7	422	10
8	503	11
9	4324	12
10	5032	13
11	5121	14
12	50214	15
13	51042	16
14	51303	17
15	503214	18
16	715022	18
17	425551	20
18	6044204	20
19	5671412	20
20	5011303	22
21	44236204	22
22	45236046	24
23	51202215	24

However, systematic codes do possess some advantages over nonsystematic codes. It was also pointed out in Chapter 10 that systematic encoders require fewer connections from the shift registers to the modulo-2 adders than do nonsystematic encoders, and that no inverting circuit is needed to recover the information sequence from the code word in a systematic code. The latter property allows the user to take a "quick look" at the received information sequence prior to any decoding without having to invert the code word. This can be an important advantage in systems where decoding is done off-line, or where the decoder is subject to temporary failures, or where the channel is known to be "noiseless" during certain time intervals and decoding becomes unnecessary.

It is desirable, therefore, to try to combine the quick-look property of systematic codes with the superior free distance available with nonsystematic codes. To this end, Massey and Costello [15] developed a class of $R = \frac{1}{2}$ nonsystematic codes called *quick-look-in codes* which have a property similar to the quick-look capability of systematic codes. They are defined by

$$\mathbf{g}^{(1)}(D) + \mathbf{g}^{(2)}(D) = D \tag{11.36}$$

and $g_0^{(1)} = g_0^{(2)} = g_m^{(1)} = g_m^{(2)} = 1$ (i.e., the two generator sequences differ in only a single position). These codes are always noncatastrophic, and their feedforward inverse has the trivial transfer function matrix

$$\mathbf{G}^{-1}(D) = \begin{bmatrix} 1 \\ 1 \end{bmatrix}. \tag{11.37}$$

Since

$$\mathbf{G}(D)\mathbf{G}^{-1}(D) = [\mathbf{g}^{(1)}(D),\ D + \mathbf{g}^{(1)}(D)]\begin{bmatrix} 1 \\ 1 \end{bmatrix} = D, \tag{11.38}$$

the information sequence $\mathbf{u}(D)$ can be recovered from the code word $\mathbf{v}(D)$ with a one-time-unit delay. The recovery equation is

$$\mathbf{v}(D)\mathbf{G}^{-1}(D) = \mathbf{v}^{(1)}(D) + \mathbf{v}^{(2)}(D) = D\mathbf{u}(D), \tag{11.39}$$

and we see that if p is the probability of error in the code word $\mathbf{v}(D)$, the probability of error in recovering $\mathbf{u}(D)$ is roughly $2p$. This follows since an error in recovering u_i can be caused by an error in either $v_{i+1}^{(1)}$ or $v_{i+1}^{(2)}$.[3] Quick-look-in codes are therefore said to have an error probability amplification factor of 2.

For any noncatastrophic $R = \frac{1}{2}$ code with feedforward inverse

$$\mathbf{G}^{-1}(D) = \begin{bmatrix} \mathbf{g}_1^{-1} & (D) \\ \mathbf{g}_2^{-1} & (D) \end{bmatrix} \tag{11.40}$$

the recovery equation is

$$\mathbf{v}^{(1)}(D)\mathbf{g}_1^{-1}(D) + \mathbf{v}^{(2)}(D)\mathbf{g}_2^{-1}(D) = D^l\mathbf{u}(D), \tag{11.41}$$

for some l, and an error in recovering u_i can be caused by an error in any of $w[\mathbf{g}_1^{-1}(D)]$ positions in $\mathbf{v}^{(1)}(D)$ or any of $w[\mathbf{g}_2^{-1}(D)]$ positions in $\mathbf{v}^{(2)}(D)$. Hence, the probability of error in recovering $\mathbf{u}(D)$ is $A \triangleq w[\mathbf{g}_1^{-1}(D)] + w[\mathbf{g}_2^{-1}(D)]$ times the probability of

[3] We are ignoring here the unlikely event that two errors in $\mathbf{v}(D)$ will cancel, causing no error in recovering $\mathbf{u}(D)$.

error in the code word. A is called the *error probability amplification factor* of the code. $A = 2$ for quick-look-in codes, and this is the minimum value of A for any $R = \frac{1}{2}$ nonsystematic code. For $R = \frac{1}{2}$ systematic codes,

$$\mathbf{G}^{-1}(D) = \begin{bmatrix} 1 \\ 0 \end{bmatrix} \tag{11.42}$$

and $A = 1$ for systematic codes. Hence, quick-look-in codes are "almost systematic," in the sense that they have the minimum value of A for any nonsystematic code. Catastrophic codes, on the other hand, have no feedforward inverse and their error probability amplification factor is infinite.

The capability of quick-look-in codes to provide a quick estimate (prior to decoding) of the information sequence from a noisy version of the code word with an error probability amplification factor of only 2 makes them desirable from a practical standpoint. The free distances of the best quick-look-in codes are given in Table 11.3 [16]. Note that d_{free} for the best quick-look-in codes lags somewhat behind d_{free} for the complementary codes. Thus, the "almost systematic" quick-look-in codes appear to have a reduced d_{free} compared to other nonsystematic codes. Their free distances are superior, however, to what can be achieved with systematic codes.

As noted previously, there are few available algebraic constructions for con-

TABLE 11.3 $R = 1/2$ QUICK-LOOK-IN CODES WITH MAXIMAL d_{free}

m	$\mathbf{g}^{(1)}$	d_{free}
2	5	5
3	54	6
4	46	7
5	55	8
6	454	9
7	542	9
8	551	10
9	4704	11
10	5522	12
11	5503	13
12	56414	14
13	56406	14
14	42651	15
15	523034	16
16	511542	16
17	522415	17
18	—	—
19	4521722	18
20	5404155	18
21	45477404	19
22	50756602	19
23	53615441	20

volutional codes. One exception is the construction of orthogonal codes for use with threshold decoding. These constructions are covered in detail in Chapter 13. Another approach, initiated by Massey et al. [17], uses the minimum-distance properties of a cyclic block code to provide a lower bound on the free distance of an associated noncatastrophic convolutional code. If $\mathbf{g}(X)$ is the generator polynomial of any (n, k) cyclic code of odd length n with minimum distance d_g, and $\mathbf{h}(X) = (X^n - 1)/\mathbf{g}(X)$ is the generator polynomial of the $(n, n-k)$ dual code with minimum distance d_h, the following construction results.

Construction 1. For any positive integer l, the rate $R = 1/2l$ convolutional code with composite generator polynomial $\mathbf{g}(D)$ is noncatastrophic and has $d_{\text{free}} \geq \min(d_g, 2d_h)$.

Since the lower bound on d_{free} is independent of l, the best codes obtained from construction 1 will be for $l = 1$ (i.e., $R = \frac{1}{2}$). The cyclic codes should be selected so that $d_g \approx 2d_h$. This suggests trying cyclic codes with rates in the range $\frac{1}{3} \leq R \leq \frac{1}{2}$. A list of $R = \frac{1}{2}$ convolutional codes formed from cyclic codes using construction 1 is given in Table 11.4A.

TABLE 11.4A $R = 1/2$ CODES FORMED USING CONSTRUCTION 1

Cyclic code	Exponents of the roots of $\mathbf{g}(X)$*	m	d_{free} bound
(7, 4)	1	1	3
(7, 3)	0, 1	2	4
(15, 8)	0, 1, 5	3	4
(17, 8)	0, 1	4	6
(15, 5)	1, 3, 5	5	7
(23, 12)	1	5	7
(23, 11)	0, 1	6	8
(31, 11)	1, 3, 5, 7	10	11
(47, 24)	1	11	11
(47, 23)	0, 1	12	12
(63, 30)	1, 3, 5, 7, 9, 11	16	13
(63, 24)	1, 5, 7, 11, 15, 27, 31	19	15
(79, 40)	1	19	15
(79, 39)	0, 1	20	16
(89, 44)	0, 1, 5, 9, 11	22	18
(103, 52)	1	25	19
(103, 51)	0, 1	26	20

*$\mathbf{g}(X)$ is the product of the minimum polynomials of these roots.

Example 11.5

Consider the (15, 5) BCH code with generator polynomial $\mathbf{g}(X) = 1 + X + X^2 + X^4 + X^5 + X^8 + X^{10}$. This code has minimum distance $d_g = 7$. The generator polynomial of the dual code is $\mathbf{h}(X) = (X^{15} - 1)/\mathbf{g}(X) = X^5 + X^3 + X + 1$, and $d_h = 4$.

The $R = 1/2$ convolutional code with composite generator polynomial $\mathbf{g}(D) = 1 + D + D^2 + D^4 + D^5 + D^8 + D^{10}$ then has $d_{\text{free}} \geq \min(7, 8) = 7$. The transfer function matrix is

$$\mathbf{G}(D) = [1 + D + D^2 + D^4 + D^5, 1 + D^2],$$

and the code has memory order $m = 5$. If $\mathbf{u}(D) = 1$, the code word

$$\mathbf{v}(D) = \mathbf{u}(D^2)\mathbf{g}(D) = 1 + D + D^2 + D^4 + D^5 + D^8 + D^{10}$$

has weight 7, and hence $d_{\text{free}} = 7$ for this code.

A similar construction can be used to produce codes with rate $R = \frac{1}{4}$.

Construction 2. For any positive integer l, the rate $R = 1/4l$ convolutional code with composite generator polynomial $\mathbf{g}(D^2) + D\mathbf{h}(D^2)$ is noncatastrophic and has $d_{\text{free}} \geq \min(d_g + d_h, 3d_g, 3d_h)$.

Since the lower bound on d_{free} is independent of l, the best codes obtained from construction 2 will be for $l = 1$ (i.e., $R = \frac{1}{4}$). The cyclic codes should be selected so that $d_g \approx d_h$. This suggests trying cyclic codes with rates near $R = \frac{1}{2}$. A list of $R = \frac{1}{4}$ convolutional codes formed from cyclic codes using construction 2 is given in Table 11.4B.

TABLE 11.4B $R = 1/4$ CODES FORMED USING CONSTRUCTION 2

Cyclic code	Exponents of the roots of $\mathbf{g}(X)$*	m	d_{free} bound
(7, 3)	0, 1	2	7
(17, 8)	0, 1	4	11
(23, 11)	0, 1	6	15
(41, 20)	0, 1	10	19
(47, 23)	0, 1	12	23
(79, 39)	0, 1	20	31
(89, 44)	0, 1, 5, 9, 11	22	35
(103, 51)	0, 1	26	39

*$\mathbf{g}(X)$ is the product of the minimum polynomials of these roots.

Two difficulties prevent these constructions from yielding good long convolutional codes. The first is the absence of good long cyclic codes. The second is the dependence of the bound on the minimum distance of the duals of cyclic codes. This second difficulty was circumvented in a subsequent paper by Justesen [18]. Justesen's construction yields the bound $d_{\text{free}} \geq d_g$, but involves a rather complicated condition on the roots of $\mathbf{g}(X)$ and, in the binary case, can be used to construct convolutional codes with odd values of n only. Some of the codes constructed by Justesen are given in Table 11.4C.

TABLE 11.4C CODES FORMED USING JUSTESEN'S CONSTRUCTION

Cyclic code	Exponents of the roots of $\mathbf{g}(X)$*	R	m	d_{free} bound
(65, 52)	0, 1	4/5	2	6
(65, 40)	0, 1, 3	3/5	5	10
(65, 28)	0, 1, 3, 5	2/5	7	14
(65, 16)	0, 1, 3, 5, 7	1/5	9	22
(65, 13)	1, 3, 5, 7, 13	1/5	10	25
(129, 100)	0, 1, 3	2/3	9	10
(129, 44)	0, 1, 3, 5, 7, 9, 11	1/3	28	26
(255, 198)	−3, −1, 0, 1, 3, 5, 7, 9	2/3	19	16
(255, 126)	−17, −15, . . . , −1, 0, 1, 3, . . . , 17	1/3	43	38
(1023, 922)	−9, −7, . . . , −1, 0, 1, 3, . . . , 9	2/3	33	22
(1023, 507)	−31, −29, . . . , −1, 0, 1, 3, . . . , 79	1/3	172	116

*$\mathbf{g}(X)$ is the product of the minimum polynomials of these roots.

11.4 IMPLEMENTATION OF THE VITERBI ALGORITHM

The basic workings of the Viterbi algorithm were presented in Section 11.1. In the practical implementation of the algorithm, several other factors must be considered. In this section we discuss some of these factors and how they affect decoder performance.

Decoder memory. Since there are 2^K states in the state diagram of the encoder, the decoder must reserve 2^K words of storage for the survivors. Each word must be capable of storing the surviving path along with its metric. Since the storage requirements increase exponentially with K, in practice it is not feasible to use codes with large K, and a value for K of about 8 is considered the practical limit for the Viterbi algorithm. This also limits the available free distance, and the performance bounds of (11.18), (11.22), (11.27), and (11.35) imply that the achievable error probability cannot be made arbitrarily small with the Viterbi algorithm. Soft decision coding gains of around 7 dB and bit error probabilities of around 10^{-5} are considered the practical limit for the Viterbi algorithm in most cases. The exact error probabilities achieved depend on the code, its rate, its free distance, the available channel SNR, and the demodulator output quantization, as well as other factors.

Path memory. We noted in Section 10.1 that for an encoder of memory order m and an information sequence of length kL, the factor $m/(L + m)$ represents the fractional loss in the effective rate of information transmission compared to the code rate R. Since energy per information bit is inversely proportional to rate [see (11.26)], a lower effective rate means a larger required E_b/N_0 to achieve a given performance. Hence, it is desirable to have L as large as possible so that the effective rate is nearly R. This also reduces the need for inserting m blocks of 0's into the input stream after every L blocks of information bits.

The difficulty with large L is that each of the 2^K words of storage must be capable of storing a kL-bit path plus its metric. For very large L, this is clearly impossible, and some compromises must be made. The approach that is usually taken is to truncate the path memory of the decoder by storing only the most recent τ blocks of information bits for each survivor, where $\tau \ll L$. Hence after the first τ blocks of the received sequence have been processed by the decoder, the decoder memory is full. After the next block is processed, a decoding decision must be made on the first block of k information bits, since they can no longer be stored in the decoder's memory.

There are several possible strategies for making this decision. Among these are:

1. An arbitrary survivor is selected, and the first k bits on this path are chosen as the decoded bits.
2. From among the 2^k possible first information blocks, that one is selected which appears most often in the 2^K survivors.
3. The survivor with the best metric is selected, and the first k bits on this path are chosen as the decoded bits.

After the first decoding decision is made, additional decoding decisions are made in the same way for each new received block processed. Hence, the decoding decisions always lag the progress of the decoder by an amount equal to the path memory (i.e., τ blocks).

Decoding decisions made in this way are no longer maximum likelihood, but can be almost as good if τ is not too small. Experience and analysis have shown that if τ is on the order of five times the encoder memory or more, with probability approaching 1 all 2^K survivors stem from the same information block τ time units back, and there is no ambiguity in making the decoding decision. This situation is illustrated in Figure 11.11. In addition, this must be the maximum likelihood decision, since no matter which survivor eventually becomes the maximum likelihood path, it must contain this decoded information block. Hence, if τ is large enough, almost all decoding decisions will be maximum likelihood, and the final decoded sequence will be close to the maximum likelihood path.

Note that with a truncated decoder memory, the final decoded sequence may not correspond to an actual path through the trellis, since decisions are made one branch at a time and do not necessarily follow a single path. If this is the case, however, some decoding errors must have occurred, since the actual transmitted sequence must follow a single path through the trellis. This point is illustrated in Problem 11.16.

There are two ways in which errors can occur in a truncated decoder. Assume that a branch decision at time unit j is made by selecting the survivor at time unit $j + \tau$ with the best metric, and then decoding the information bits on that path at time unit j. If a standard decoder would contain a decoding error at time unit j, it is reasonable to assume that the maximum likelihood path is the best survivor at time unit $j + \tau$, and hence the truncated decoder will make the same error. An additional source of error with a truncated decoder occurs when an incorrect path, which is unmerged with the correct path from time unit j through time unit $j + \tau$, is the best survivor at time unit $j + \tau$. In this case a decoding error is made at time

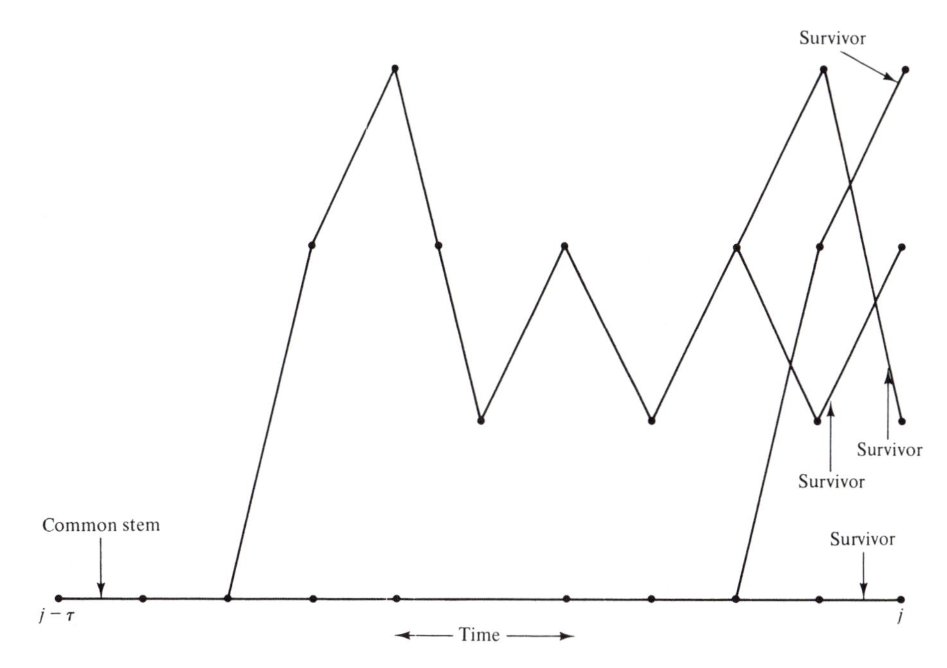

Figure 11.11 Decoding decisions with a finite path memory.

unit j, even though this incorrect path may be eliminated when it later remerges with the correct path, and thus would not cause a decoding error in a standard decoder. Decoding errors of this type are called *decoding errors due to truncation*. The subset of incorrect paths that can cause a decoding error due to truncation is shown in Figure 11.12.

For a convolutional code with generating function $T(X, Y, Z)$, the bit error probability on a BSC of a truncated decoder at any arbitrary time unit is bounded by

$$P_b(E) < \frac{1}{k}\left[\frac{\partial T(X, Y, Z)}{\partial Y} + \sum_{i=1}^{2^K-1} T_i^\tau(X, Y, Z)\right]\Bigg|_{X=2\sqrt{p(1-p)},Y=1,Z=1} \qquad (11.43)$$

where $\sum_{i=1}^{2^K-1} T_i^\tau(X, Y, Z)$ is the generating function for the subset of incorrect paths that can cause decoding errors due to truncation [19]. In other words, $T_i^\tau(X, Y, Z)$ is the generating function for the set of all paths in the modified encoder state diagram connecting the all-zero state with the ith state and expurgated so as to include only paths of length τ branches or greater. The first term in (11.43) represents the bit errors made by a standard decoder, whereas the second term represents the decoding errors due to truncation. Clearly, (11.43) can be generalized to other DMCs and the unquantized AWGN channel in the same way as (11.31) and (11.33).

When p is small (if the BSC is derived from a hard quantized AWGN channel this means large E_b/N_0), (11.43) can be approximated as

$$P_b(E) \approx \frac{1}{k}\left[B_{d_{\text{free}}}(2\sqrt{p(1-p)})^{d_{\text{free}}} + A_{d(\tau)}(2\sqrt{p(1-p)})^{d(\tau)}\right], \qquad (11.44)$$

where $d(\tau)$ is the smallest power of D in the generating function $\sum_{i=1}^{2^K-1} T_i^\tau(X, Y, Z)$,

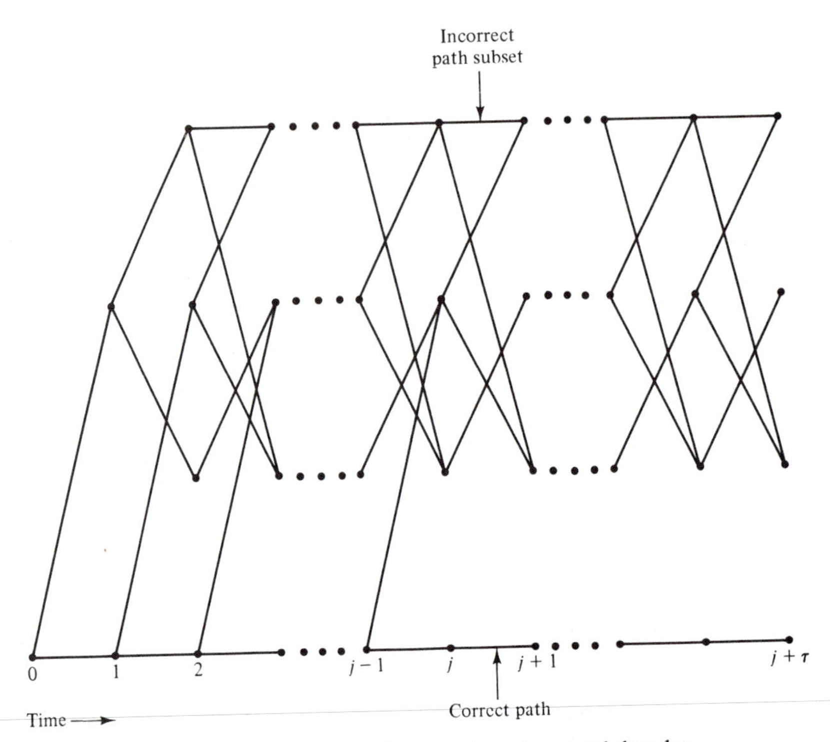

Figure 11.12 Incorrect path subset for a truncated decoder.

and $A_{d(\tau)}$ is the number of terms of length τ and weight $d(\tau)$. Further simplification of (11.44) yields

$$P_b(E) \approx \frac{1}{k}[B_{d_{\text{free}}} 2^{d_{\text{free}}} p^{d_{\text{free}}/2} + A_{d(\tau)} 2^{d(\tau)} p^{d(\tau)/2}]. \tag{11.45}$$

From (11.45) it is clear that if $d(\tau) > d_{\text{free}}$, the second term is negligible compared with the first term, and the additional error probability due to truncation is small compared to the error probability of a standard decoder. Hence, the path memory τ should be chosen large enough so that $d(\tau) > d_{\text{free}}$ in a truncated decoder. The minimum value of τ for which $d(\tau) > d_{\text{free}}$ is called the *minimum truncation length* τ_{min} of the code. The minimum truncation lengths for some of the $R = \frac{1}{2}$ optimum free-distance codes listed in Table 11.1(c) are given in Table 11.5. Note that $\tau_{\text{min}} \approx$

TABLE 11.5 MINIMUM TRUNCATION LENGTHS FOR $R = 1/2$ OPTIMUM FREE-DISTANCE CODES

m	d_{free}	τ_{min}	$d(\tau_{\text{min}})$
2	5	8	6
3	6	10	7
4	7	15	8
5	8	19	9
6	10	27	11
7	10	28	11
8	12	33	13

$4m$ in most cases. A random coding analysis by Forney [4] shows that in the small SNR case, a minimum truncation length of $\tau_{\min} \approx 5.8m$ is sufficient to ensure that the additional error probability due to truncation is negligible. Extensive simulations and actual experience have verified that four to five times the memory order of the encoder is usually an acceptable truncation length to use in practice.

Example 11.6

Consider the (2, 1, 2) code with $\mathbf{G}(D) = [1 + D^2, 1 + D + D^2]$. The augmented state diagram for this code is shown in Figure 11.13, and the generating function is given by

$$T(X, Y, Z) = \frac{X^5 YZ^3}{1 - XYZ(1 + Z)}.$$

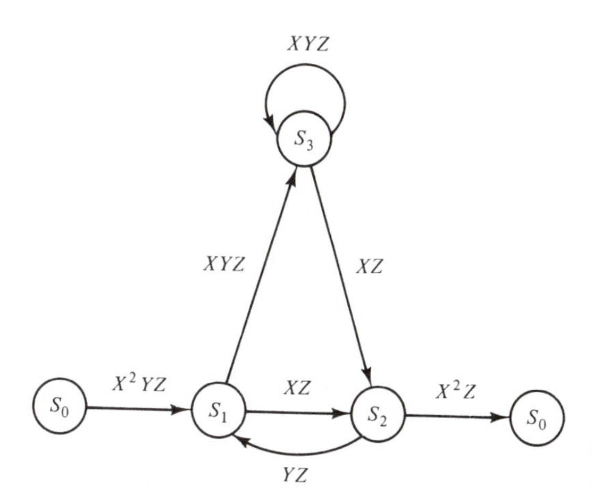

Figure 11.13 Augmented state diagram for a (2, 1, 2) code.

Letting $T_i(X, Y, Z)$ be the generating function for all paths connecting the all-zero state (S_0) with the ith state (S_i), we find that

$$T_1(X, Y, Z) = \frac{X^2 YZ(1 - XYZ)}{1 - XYZ(1 + Z)},$$

$$T_2(X, Y, Z) = \frac{X^3 YZ^2}{1 - XYZ(1 + Z)},$$

$$T_3(X, Y, Z) = \frac{X^3 Y^2 Z^2}{1 - XYZ(1 + Z)}.$$

If we now expurgate each of these generating functions to include only paths of length τ branches or more, $d(\tau)$ is the smallest power of X in any of the three expurgated functions. For example, $d(1) = 2$, $d(2) = 3$, $d(3) = 3$, and so on. Since $d_{\text{free}} = 5$ for this code, $\tau_{\min}$ is the minimum value of τ for which $d(\tau) = d_{\text{free}} + 1 = 6$. $T_1(X, Y, Z)$ contains a term $X^5 Y^4 Z^7$, and hence $d(7) \leq 5$ and $\tau_{\min}$ must be at least 8. The particular path yielding this term is highlighted on the trellis diagram of Figure 11.14. A careful inspection of the trellis diagram shows that there is no path of length 8 branches that terminates on S_1, S_2, or S_3 and has weight less than 6. There are five such paths, however, which have weight 6, and these are also highlighted in Figure 11.14. Hence, $d(8) = 6$, $A_{d(8)} = 5$, and $\tau_{\min} = 8$, which is exactly four times the code memory order. Hence, a path memory of 8 should be sufficient to ensure a negligible increase in bit error probability over a standard decoder for this code.

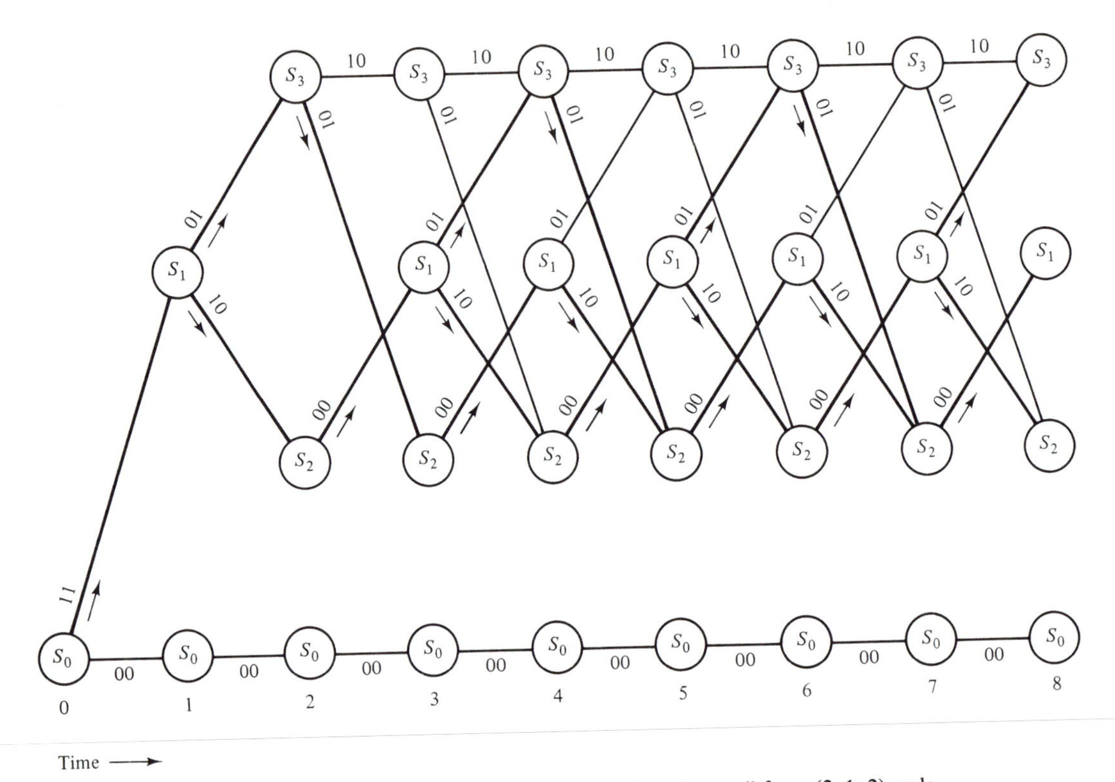

Figure 11.14 Determining "truncation distance" for a (2, 1, 2) code.

As a final comment, it is important to point out the distinction between the column distance d_i defined in Section 10.3 and the "truncation distance" $d(\tau)$ defined above. d_i is the minimum weight of any code word of length $i + 1$ branches. Since this includes code words that have remerged with the all-zero state and hence whose weight has stopped increasing beyond a certain point, d_i reaches a maximum value of d_{free} as i increases. $d(\tau)$, however, is the minimum weight of any code word of length τ branches which has not yet remerged with the all-zero state. Since remergers are not allowed, $d(\tau)$ will continue to increase without bound as τ increases. For example, for the code of Figure 11.13, $d(10) = 7$, $d(20) = 12$, $d(30) = 17, \ldots,$ and in general $d(\tau) = d(\tau + 1) = (\tau/2) + 2$ for even τ.

Catastrophic codes are the only exception to this rule. In a catastrophic code, the zero-weight loop in the state diagram prevents $d(\tau)$ from increasing as τ increases. Hence, catastrophic codes contain very long code words with low weight, which makes them susceptible to high bit error probabilities when used with either a standard or truncated Viterbi decoder. For example, the catastrophic code of Figure 10.11 has $d_{\text{free}} = 4$, but contains an unmerged code word of infinite length with weight 3, and $d(\tau) = 3$ for all $\tau \geq 2$. Hence, the additional error probability due to truncation will dominate the bit error probability expression of (11.45), no matter what truncation length is chosen, and the code will not perform as well as a noncatastrophic code with $d_{\text{free}} = 4$. Further discussion of this performance difference between catastrophic and noncatastrophic codes is given in Reference 14.

Decoder synchronization. In practice, decoding does not always commence with the first branch transmitted after the encoder is set to the all-zero state, but may begin with the encoder in an unknown state, in midstream, so to speak. In this case, all state metrics are initially set to zero, and decoding starts in the middle of the trellis. If path memory truncation is used, the initial decisions taken from the survivor with the best metric are unreliable, causing some initial decoding errors. But Forney [4] has shown, again using random coding arguments, that after about $5m$ branches are decoded, the effect of the initial lack of branch synchronization becomes negligible. Hence, in practice the decoding decisions over the first $5m$ branches are usually discarded, and all later decisions are then treated as reliable.

Bit (or symbol) synchronization is also required by the decoder; that is, the decoder must know which of n consecutive received symbols is the first one on a branch. In this case, the decoder makes an initial assumption. If this assumption is incorrect, the survivor metrics will remain relatively close together. This is usually indicative of a stretch of noisy received data, since if the received sequence is noise-free, the metric of the correct path will clearly dominate the other survivor metrics. This point is illustrated in Problem 11.18. If this condition persists over a long-enough span, it is indicative of incorrect symbol synchronization, since long stretches of noise are very unlikely. In this case, the symbol synchronization assumption is changed until correct synchronization is achieved. Note that at most n attempts are needed to acquire symbol synchronization.

Code performance. Heller and Jacobs [7] have conducted extensive computer simulation studies of the performance of the Viterbi algorithm. Some of their results are presented in Figure 11.15. The bit error probability $P_b(E)$ is plotted as a function of E_b/N_0 (in decibels) for six $R = \frac{1}{2}$ codes with encoder memories $K = 2$ through $K = 7$ and demodulator output quantization $Q = 8$ in Figure 11.15(a). These curves are repeated for $Q = 2$ in Figure 11.15(b). In both cases the path memory $\tau = 32$. Note that there is about a 2-dB improvement in the performance of soft decisions ($Q = 8$) over hard decisions ($Q = 2$). This is illustrated again in Figure 11.15(c), where the performance of a $K = 4$, $R = \frac{1}{2}$ code with $Q = 2, 4, 8$ and $\tau = 32$ is shown. Shown also in Figure 11.15(c) is the no-coding curve of (11.28). Comparing this with the coding curves shows real coding gains of 3 dB in the hard-decision case ($Q = 2$) and 5 dB in the soft-decision case ($Q = 8$) at a bit error probability of 10^{-4}. In Figure 11.15(d), the performance of this same code is shown for $\tau = 8, 16, 32$ and $Q = 2, 8$. Note that path memory $\tau = 8 = 2K$ degrades the performance by about 0.7 dB, but that $\tau = 16 = 4K$ is almost as good as $\tau = 32 = 8K$. The performance of three $R = \frac{1}{3}$ codes with $K = 3, 5, 7$ and $Q = 8$ is shown in Figure 11.15(e). Note that these codes do better than the corresponding $R = \frac{1}{2}$ codes of Figure 11.15(a) by about 0.5 dB. This is because more free distance is available at $R = \frac{1}{3}$ than at $R = \frac{1}{2}$ for the same encoder memory.[4] Finally, Figure 11.15(f) shows the performance of an $R = \frac{2}{3}$ code with $K = 6$ ($m = 3$) for $Q = 2, 8$. Again

[4]Recall from Chapter 1, however, that a rate-1/3 code requires more channel bandwidth than does a rate-1/2 code.

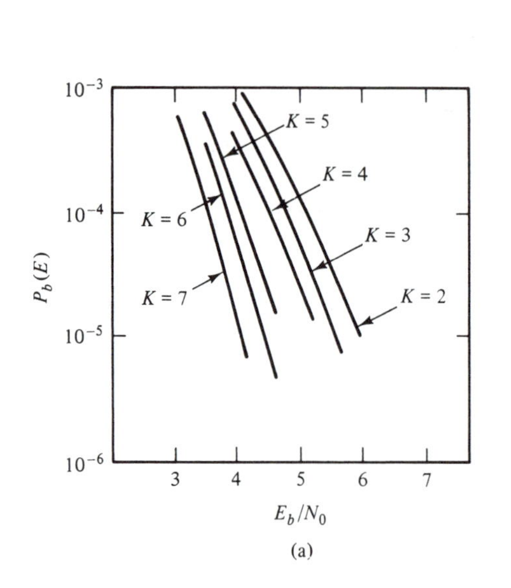

(a)

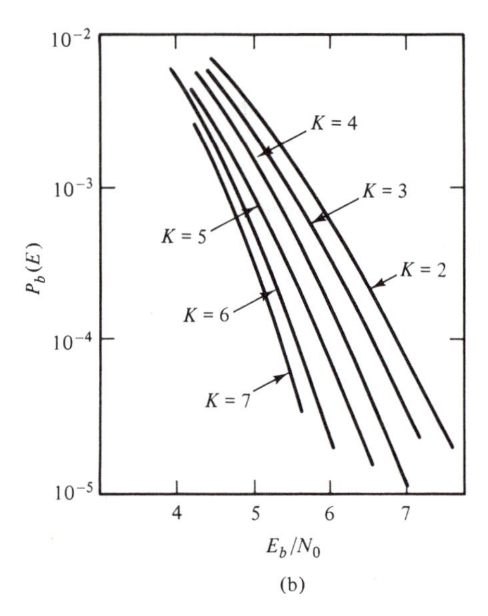

(b)

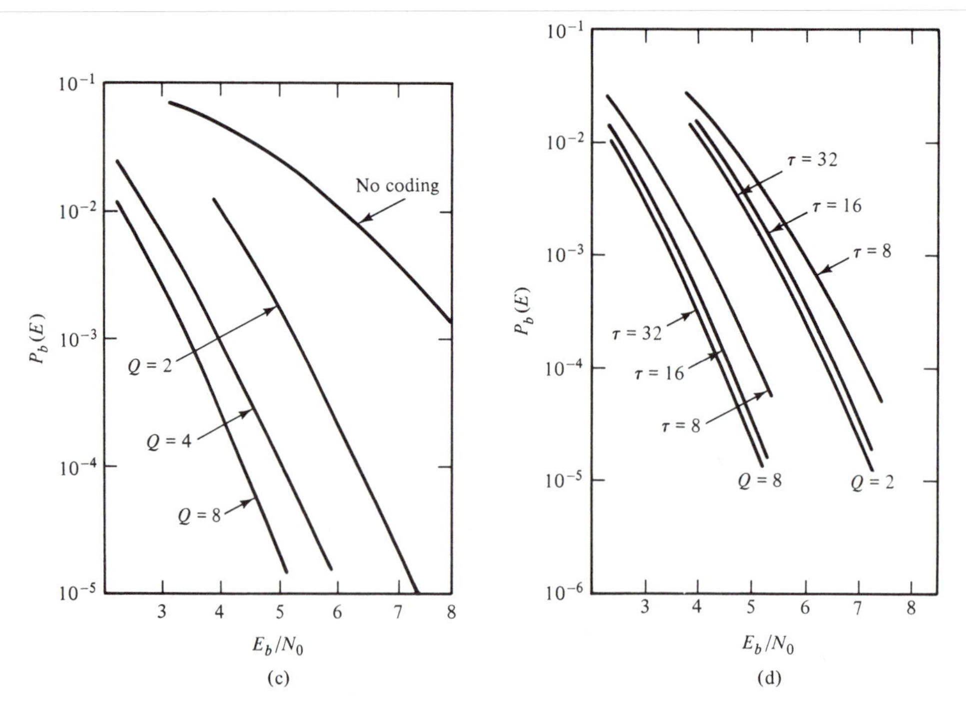

Figure 11.15 Simulation curves for the Viterbi algorithm.

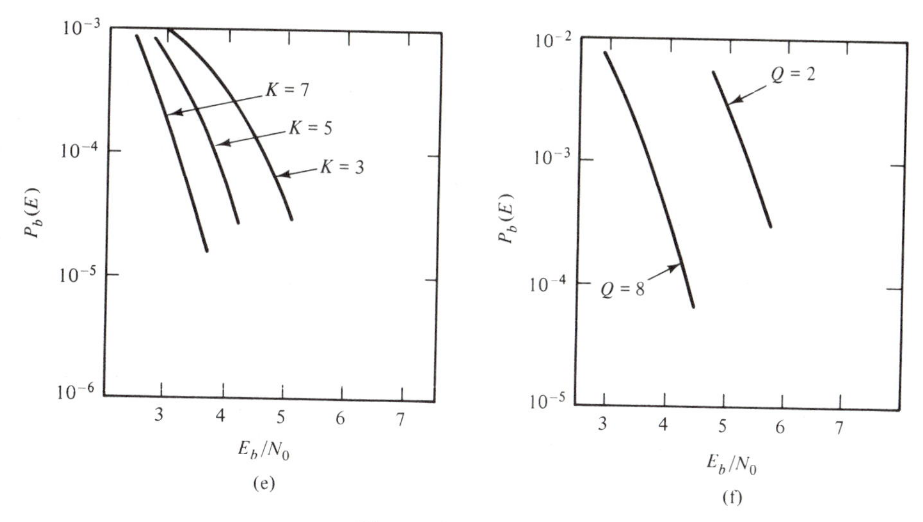

Figrue 11.15 *(cont.)*

note that the corresponding $R = \frac{1}{2}$ code with $K = 6$ of Figure 11.15(a) and (b) does better than the $R = \frac{2}{3}$ code by about 0.5 dB. All of these observations are consistent with the performance analysis presented earlier in this chapter.

11.5 MODIFICATIONS OF THE VITERBI ALGORITHM

The principal limitation on the practical application of the Viterbi algorithm is that the complexity of decoding is proportional to 2^K, the number of encoder states. As noted in Section 11.4, the decoder memory is proportional to 2^K. Also, since 2^K comparisons must be performed per time unit, decoding time is also proportional to 2^K. This exponential dependence on K is what limits use of the Viterbi algorithm to values of K equal to 8 or less.

The speed limitation of the Viterbi algorithm can be alleviated by employing parallel decoding. Since the 2^K comparisons that must be performed at each time unit are identical, 2^K identical processors can be provided to do the comparisons in parallel rather than having one processor do all 2^K comparisons in sequence. Each processor would compute the metric values of the 2^K paths entering some state, select the path with the best metric, and store that path and its metric in the decoder's memory. This parallel implementation of the Viterbi algorithm then requires 2^K processors doing one computation per time unit, as opposed to having one processor do 2^K computations per time unit. Parallel decoding therefore implies a factor-of-2^K speed advantage over a standard decoder, but requires 2^K times as much hardware.

One approach to reducing the complexity of the Viterbi algorithm has been to try to reduce the number of states in the decoder. This approach has had some success in applications of the Viterbi algorithm to reducing intersymbol interference on a bandlimited channel. The basic idea is to design the decoder as if the encoder memory were less than it actually is, thereby reducing the number of states in the decoder. This normally will reduce the "effective" free distance of the code, since code words will remerge with the all-zero state sooner than they would if the decoder used the full memory of the encoder. Improvement can be expected only if the effective free distance of the "reduced code" is larger than the free distance of a standard code with the same encoder memory as the decoder memory of the "reduced code." This problem has been discussed by Conan and Haccoun [20], but no substantial improvement over standard Viterbi decoding has been reported.

Another approach has been to try to identify subclasses of codes that allow a simplified decoding structure. Some successes have been reported here, but in general the subclasses of codes obtained do not contain the optimum free-distance codes of a given rate and encoder memory. Schalkwijk et al. [21] have developed a decoding algorithm similar to the Viterbi algorithm which is based on the state diagram of the syndrome former (see Section 13.1) for the code rather than on the encoder state diagram. Some codes are then found which exploit symmetries in the state diagram of the syndrome former to reduce the amount of decoder hardware required. Cain et al. [22] have found some $R = (n - 1)/n$ codes which are obtained by deleting or puncturing certain bits from the code words of an $R \leq 1/n^*$ code, where $n^* \leq n$. These codes can then be decoded using the Viterbi algorithm with roughly the same decoding complexity as the $R = 1/n^*$ code requires. As a consequence, only binary comparisons are performed at each state, rather than the much more complex 2^{n-1}-ary comparisons required by a standard decoder.

Since the Viterbi algorithm finds the most likely path through the encoder state diagram, it minimizes the sequence or event error probability $P(E)$. The bit error probability $P_b(E)$ is closely related to $P(E)$, and hence the Viterbi algorithm also results in a small (but not necessarily minimum) value of $P_b(E)$. Hartmann and Rudolph [23] have reported a bit-by-bit decoding rule for any linear code (block or convolutional) that minimizes $P_b(E)$. In terms of decoder complexity, the Viterbi algorithm is simpler for low-rate codes, but the bit-by-bit decoding rule is less complex in the high-rate case.

PROBLEMS

11.1. Draw the trellis diagram for the (3, 2, 1) code with $K = 2$ listed in Table 11.1(d) and an information sequence of length $L = 3$ blocks. Find the code word corresponding to the information sequence $\mathbf{u} = (1\ 1,\ \ 0\ 1,\ \ 1\ 0)$. Compare the result with Example 10.3.

11.2. Show that the path $\mathbf{v}$ which maximizes $\sum_{i=0}^{N-1} \log P(r_i | v_i)$ also maximizes $\sum_{i=0}^{N-1} c_2[\log P(r_i | v_i) + c_1]$, where c_1 is any real number and c_2 is any positive real number.

11.3. Find the integer metric table for the DMC of Figure 11.3 when $c_1 = 1$ and $c_2 = 10$. Decode the received sequence $\mathbf{r}$ of Example 11.1 using this integer metric table and the trellis diagram of Figure 11.1. Compare with the result of Example 11.1.

11.4. Consider a binary-input, 8-ary-output DMC with transition probabilities $P(r_i | v_i)$ given by the following table:

v_i \ r_i	0_1	0_2	0_3	0_4	1_4	1_3	1_2	1_1
0	0.434	0.197	0.167	0.111	0.058	0.023	0.008	0.002
1	0.002	0.008	0.023	0.058	0.111	0.167	0.197	0.434

Find the metric table and an integer metric table for this channel.

11.5. Consider the (2, 1, 3) code of Figure 10.1 with

$$\mathbf{G}(D) = [1 + D^2 + D^3, 1 + D + D^2 + D^3].$$

(a) Draw the trellis diagram for an information sequence of length $L = 4$.
(b) Assume that a code word is transmitted over the DMC of Problem 11.4. Decode the received sequence $\mathbf{r} = [1_2 1_1, 1_2 0_1, 0_3 0_1, 0_1 1_3, 1_2 0_2, 0_3 1_1, 0_3 0_2]$.

11.6. The DMC of Problem 11.4 is converted to a BSC by combining the soft-decision outputs 0_1, 0_2, 0_3, and 0_4 into a single hard-decision output 0 and combining the soft-decision outputs 1_1, 1_2, 1_3, and 1_4 into a single hard-decision output 1. A code word from the code of Problem 11.5 is transmitted over this channel. Decode the hard-decision version of the received sequence in Problem 11.5, and compare with the result of Problem 11.5.

11.7. Show that (11.13) is an upper bound on P_d for d even.

11.8. Consider the (2, 1, 3) code of Problem 11.5. Evaluate the upper bounds on event error probability (11.17) and bit error probability (11.21) for a BSC with transition probability
(a) $p = 0.01$
(b) $p = 0.001$
(*Hint:* Use the generating function derived for this code in Examples 10.10 and 10.11.)

11.9. Repeat Problem 11.8 using the approximate expressions for $P(E)$ and $P_b(E)$ given by (11.18) and (11.22).

11.10. Consider the (3, 1, 2) code of Figure 11.1. Plot the approximate expression (11.27) for bit error probability $P_b(E)$ on a BSC as a function of E_b/N_0 in decibels. Also plot on the same set of axes the approximate expression (11.28) for $P_b(E)$ without coding. The *coding gain* (in decibels) is defined as the difference between the E_b/N_0 ratio needed to achieve a given bit error probability with coding and without coding. Plot the coding gain as a function of $P_b(E)$. Find the value of E_b/N_0 for which the coding gain is 0 dB.

11.11. Repeat Problem 11.10 for an AWGN channel with unquantized demodulator output.

11.12. Consider the (3, 1, 2) code of Figure 11.1 and the DMC of Problem 11.4. Calculate an approximate value for the bit error probability $P_b(E)$ using the bound of (11.33). Now convert the DMC to a BSC as described in Problem 11.6, compute an approximate value for $P_b(E)$ on this BSC using (11.22), and compare the two results.

11.13. Consider the following three (2, 1, 3) codes: (1) the optimum code listed in Table 11.1(c), (2) the complementary code listed in Table 11.2, and (3) the quick-look-in code listed in Table 11.3. For each of these codes, find:
(a) the approximate event error probability on a BSC with $p = 10^{-2}$;
(b) the approximate bit error probability on a BSC with $p = 10^{-2}$;
(c) the error probability amplification factor A.

11.14. Using trial-and-error methods, construct a (2, 1, 3) systematic code with maximal d_{free}. Find its asymptotic coding gain γ. Repeat Problem 11.13 for this code.

11.15. Consider the (15, 7) and (31, 16) cyclic BCH codes. For each of these codes find:

(a) the transfer function matrix and a lower bound on d_{free} for the $R = 1/2$ convolutional code derived from the cyclic code using construction 1;

(b) the transfer function matrix and a lower bound on d_{free} for the $R = 1/4$ convolutional code derived from the cyclic code using construction 2.

[*Hint:* d_h is at least one more than the maximum number of consecutive powers of α that are roots of $\mathbf{h}(X)$.]

11.16. **(a)** Repeat Examples 11.1 and 11.2 for a truncated decoder with a path memory of $\tau = 2$ blocks. Assume that at each level the survivor with the best metric is selected, and that the information bit τ time units back on this path is decoded.

(b) Repeat part (a) for a path memory of $\tau = 3$ blocks.

11.17. Consider the (3, 1, 2) code of Figure 11.1 and Problem 10.5.

(a) Find $T_1(X, Y, Z)$, $T_2(X, Y, Z)$, and $T_3(X, Y, Z)$.

(b) Find τ_{min}.

(c) Find $d(\tau)$ and $A_{d(\tau)}$ for $\tau = 1, 2, \ldots, \tau_{min}$.

(d) Find an expression for $\lim_{\tau \to \infty} d(\tau)$.

11.18. A code word from the trellis diagram of Figure 11.1 is transmitted over a BSC. To determine correct bit synchronization, the three 21-bit subsequences of the received sequence

$$\mathbf{r} = 0\ 1\ 1\ 1\ 0\ 0\ 1\ 1\ 0\ 0\ 1\ 0\ 1\ 1\ 0\ 0\ 1\ 0\ 0\ 0\ 1\ 1\ 1$$

must each be decoded. Decode each of these subsequences, and determine which is most likely to be the correctly synchronized received sequence.

REFERENCES

1. A. J. Viterbi, "Error Bounds for Convolutional Codes and an Asymptotically Optimum Decoding Algorithm," *IEEE Trans. Inf. Theory*, IT-13, pp. 260–269, April 1967.
2. J. K. Omura, "On the Viterbi Decoding Algorithm," *IEEE Trans. Inf. Theory*, IT-15, pp. 177–179, January 1969.
3. G. D. Forney, Jr., "The Viterbi Algorithm," *Proc. IEEE*, 61, pp. 268–278, March 1973.
4. G. D. Forney, Jr., "Convolutional Codes II: Maximum Likelihood Decoding," *Inf. Control*, 25, pp. 222–266, July 1974.
5. G. D. Forney, Jr., "Maximum Likelihood Sequence Estimation of Digital Sequences in the Presence of Intersymbol Interference," *IEEE Trans. Inf. Theory*, IT-18, pp. 363–378, May 1972.
6. A. J. Viterbi and J. K. Omura, *Principles of Digital Communication and Coding*, McGraw-Hill, New York, 1979.
7. J. A. Heller and I. M. Jacobs, "Viterbi Decoding for Satellite and Space Communication," *IEEE Trans. Commun. Technol.*, COM-19, pp. 835–848, October 1971.
8. J. M. Wozencraft and I. M. Jacobs, *Principles of Communication Engineering*, Wiley, New York, 1965.
9. J. P. Odenwalder, "Optimal Decoding of Convolutional Codes," Ph.D. thesis, University of California, Los Angeles, 1970.

10. K. J. Larsen, "Short Convolutional Codes with Maximum Free Distance for Rates 1/2, 1/3, and 1/4," *IEEE Trans. Inf. Theory*, IT-19, pp. 371–372, May 1973.
11. E. Paaske, "Short Binary Convolutional Codes with Maximal Free Distance for Rates 2/3 and 3/4," *IEEE Trans. Inf. Theory*, IT-20, pp. 683–689, September 1974.
12. R. Johannesson and E. Paaske, "Further Results on Binary Convolutional Codes with an Optimum Distance Profile," *IEEE Trans. Inf. Theory*, IT-24, pp. 264–268, March 1978.
13. L. R. Bahl, and F. Jelinek, "Rate 1/2 Convolutional Codes with Complementary Generators," *IEEE Trans. Inf. Theory*, IT-17, pp. 718–727, November 1971.
14. F. Hemmati and D. J. Costello, Jr., "Asymptotically Catastrophic Convolutional Codes," *IEEE Trans. Inf. Theory*, IT-26, pp. 298–304, May 1980.
15. J. L. Massey and D. J. Costello, Jr., "Nonsystematic Convolutional Codes for Sequential Decoding in Space Applications," *IEEE Trans. Commun. Technol.*, COM-19, pp. 806–813, October 1971.
16. E. Paaske, unpublished results.
17. J. L. Massey, D. J. Costello, Jr., and J. Justesen, "Polynomial Weights and Code Constructions," *IEEE Trans. Inf. Theory*, IT-19, pp. 101–110, January 1973.
18. J. Justesen "New Convolutional Code Constructions and a Class of Asymptotically Good Time-Varying Codes," *IEEE Trans. Inf. Theory*, IT-19, pp. 220–225, March 1973.
19. F. Hemmati and D. J. Costello, Jr., "Truncation Error Probability in Viterbi Decoding," *IEEE Trans. Commun.*, COM-25, pp. 530–532, May 1977.
20. J. Conan and D. Haccoun, "Reduced-State Viterbi Decoding of Convolutional Codes," Proc. 14th Allerton Conf. Circuit and Syst. Theory, pp. 695–703, September 1976.
21. J. P. M. Schalkwijk, A. J. Vinck, and K. A. Post, "Syndrome Decoding of Binary Rate k/n Convolutional Codes," *IEEE Trans. Inf. Theory*, IT-24, pp. 553–562, September 1978.
22. J. B. Cain, G. C. Clark, Jr., and J. M. Geist, "Punctured Convolutional Codes of Rate $(n-1)/n$ and Simplified Maximum Likelihood Decoding," *IEEE Trans. Inf. Theory*, IT-25, pp. 97–100, January 1979.
23. C. R. P. Hartmann and L. D. Rudolph, "An Optimum Symbol-by-Symbol Decoding Rule for Linear Codes," *IEEE Trans. Inf. Theory*, IT-22, pp. 514–517, September 1976.

12

Sequential Decoding of Convolutional Codes

The primary difficulty with Viterbi decoding of convolutional codes is that the arbitrarily small error probabilities promised by the random coding bound of (1.13) are not achievable in practice. This is because only small constraint lengths can be used due to the limitations on the encoder memory K. Another difficulty with the Viterbi algorithm is the fixed number 2^K of computations that must be performed per decoded information block. Clearly, this amount of computation is not always needed, particularly when the noise is light. For example, assume that the entire encoded sequence of length N is transmitted without error over a BSC (i.e., $\mathbf{r} = \mathbf{v}$). The Viterbi algorithm still performs 2^K computations per decoded information block, all of which is wasted effort in this case. In other words, it is desirable to have a decoding procedure whose decoding effort is adaptable to the noise level. Sequential decoding is such a procedure. As we shall see, the decoding effort of sequential decoding is essentially independent of K, so that large constraint lengths can be used, and hence arbitrarily low error probabilities can be achieved. Its major drawback is that noisy frames take large amounts of computation, and decoding times occasionally exceed some upper limit, causing information to be lost or erased.

Historically, sequential decoding was introduced by Wozencraft [1] as the first practical decoding method for convolutional codes. In 1963, Fano [2] introduced a new version of sequential decoding, subsequently referred to as the Fano algorithm. This approach to sequential decoding is discussed in Section 12.2. A few years later, another version of sequential decoding, called the stack or ZJ algorithm, was discovered independently by Zigangirov [3] and Jelinek [4]. Since this algorithm is the easiest to understand, we begin our discussion of sequential decoding in Section 12.1 with the stack algorithm.

12.1 THE STACK ALGORITHM

In discussing sequential decoding, it is convenient to represent the 2^{kL} code words of length $N = n(L + m)$ for an (n, k, m) code and an information sequence of length kL as paths through a *code tree* containing $L + m + 1$ time units or levels. The code tree is just an expanded version of the trellis diagram in which every path is totally distinct from every other path. The code tree for the (3, 1, 2) code with

$$\mathbf{G}(D) = [1 + D, 1 + D^2, 1 + D + D^2],$$

is shown in Figure 12.1 for an information sequence of length $L = 5$. The trellis diagram for this code was shown in Figure 11.1. The $L + m + 1$ tree levels are labeled from 0 to $L + m$ in Figure 12.1. The leftmost node in the tree is called the *origin node*, and represents the starting state S_0 of the encoder. There are 2^k branches leaving each node in the first L levels for an (n, k, m) code. This is called the *dividing part of the tree*. In Figure 12.1, the upper branch leaving each node in the dividing part of the tree represents the input $u_i = 1$, while the lower branch represents $u_i = 0$. After L time units, there is only one branch leaving each node. This represents the inputs $\mathbf{u}_i = \mathbf{0}$, for $i = L, L + 1, \ldots, L + m - 1$, and corresponds to the encoder's return to the all-zero state S_0. Hence, it is called the *tail of the tree*, and the 2^{kL} rightmost nodes are called *terminal nodes*. Each branch is labeled with the n outputs $\mathbf{v}_i$ corresponding to the input sequence, and each of the 2^{kL} code words of length N is represented by a totally distinct path through the tree. For example, the code word corresponding to the information sequence $\mathbf{u} = (1\ 1\ 1\ 0\ 1)$ is shown highlighted in Figure 12.1. It is important to realize that the code tree contains exactly the same information about the code as the trellis diagram or state diagram. As we shall see, however, the tree is better suited to understanding the operation of a sequential decoder.

There are a variety of tree-searching algorithms which fall under the general heading of sequential decoding. In this chapter we discuss the two most common of these, the stack or ZJ algorithm and the Fano algorithm, in considerable detail. The purpose of a sequential decoding algorithm is to search through the nodes of the code tree in an efficient way (i.e., without having to examine too many nodes) in an attempt to find the maximum likelihood path. Each node examined represents a path through part of the tree. Whether or not a particular path is likely to be part of the maximum likelihood path depends on the metric value associated with that path. The metric is a measure of the "closeness" of a path to the received sequence.

For a binary-input, Q-ary-output DMC, the metrics in the Viterbi algorithm are given by the log-likelihood function of (11.1). This is the optimum metric for the Viterbi algorithm since the paths being compared at any decoding step are all of the same length. In sequential decoding, however, the set of paths that have been examined after any decoding step are generally of many different lengths. If the log-likelihood function metric is used for these paths, a distorted picture of the "closeness" of the paths to the received sequence results.

Example 12.1

Consider the code tree of Figure 12.1. Assume that a code word is transmitted from this code over a BSC, and that the sequence

$$\mathbf{r} = (0\ 1\ 0,\ 0\ 1\ 0,\ 0\ 0\ 1,\ 1\ 1\ 0,\ 1\ 0\ 0,\ 1\ 0\ 1,\ 0\ 1\ 1)$$

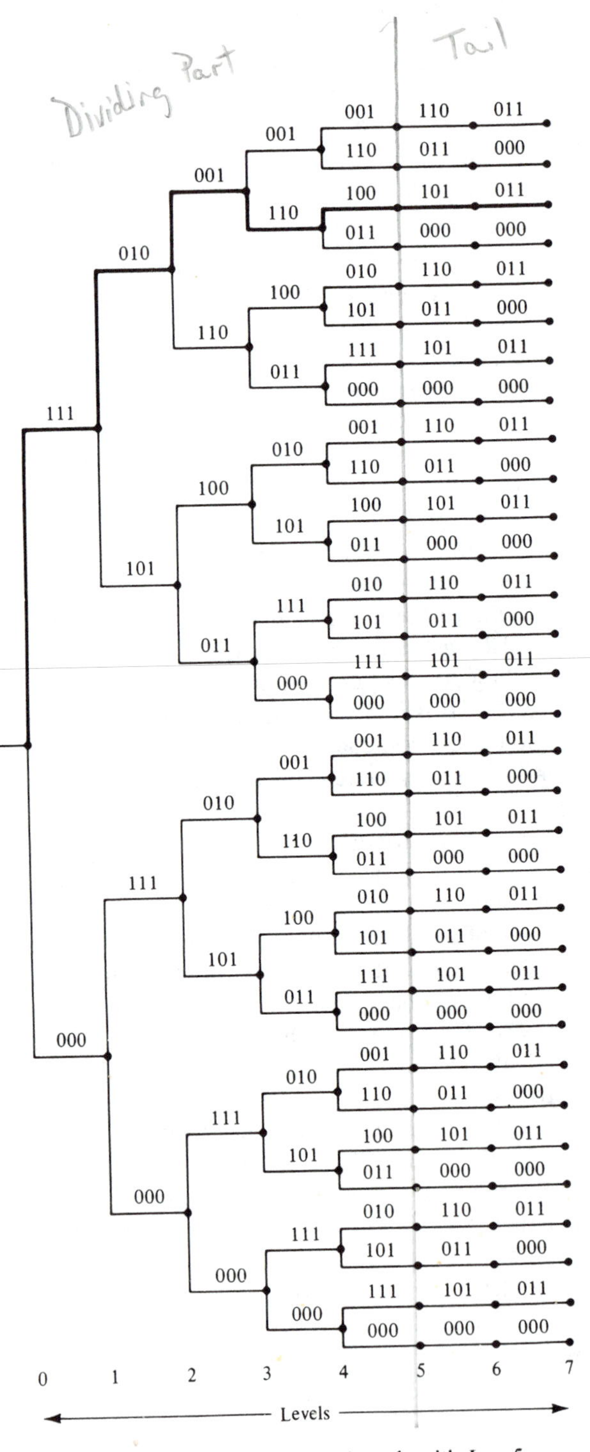

Figure 12.1 Code tree for a (3, 1, 2) code with $L = 5$.

is received. For a BSC, the metric value for a path $\mathbf{v}$ in the Viterbi algorithm is given by $d(\mathbf{r}, \mathbf{v})$, with the maximum likelihood path being the one with the smallest metric. Now consider comparing the partial path metrics for two paths of different lengths [e.g., the truncated code words $[\mathbf{v}]_5 = (1\ 1\ 1,\ 0\ 1\ 0,\ 0\ 0\ 1,\ 1\ 1\ 0,\ 1\ 0\ 0,\ 1\ 0\ 1)$ and $[\mathbf{v}']_0 = (0\ 0\ 0)$]. The partial path metrics are $d([\mathbf{r}]_5, [\mathbf{v}]_5) = 2$ and $d([\mathbf{r}]_0, [\mathbf{v}']_0) = 1$, indicating that $[\mathbf{v}']_0$ is the "better" of the two paths. However, our intuition tells us that the path $[\mathbf{v}]_5$ is more likely to be part of the maximum likelihood path than $[\mathbf{v}']_0$, since to complete a path beginning with $[\mathbf{v}']_0$ requires 18 additional bits, compared to only 3 additional bits required to complete the path beginning with $[\mathbf{v}]_5$. In other words, a path beginning with $[\mathbf{v}']_0$ is much more likely to accumulate additional distance from $\mathbf{r}$ than the path beginning with $[\mathbf{v}]_5$.

It is necessary, therefore, to adjust the metric used in sequential decoding to take into account the lengths of the different paths being compared. For a binary-input, Q-ary-output DMC, the best bit metric to use when comparing paths of different lengths is

$$M(r_i \mid v_i) = \log_2 \frac{P(r_i \mid v_i)}{P(r_i)} - R, \tag{12.1}$$

where $P(r_i \mid v_i)$ is a channel transition probability, $P(r_i)$ is a channel output symbol probability, and R is the code rate [5]. The partial path metric for the first l branches of a path $\mathbf{v}$ is given by

$$M([\mathbf{r} \mid \mathbf{v}]_{l-1}) = \sum_{j=0}^{l-1} M(\mathbf{r}_j \mid \mathbf{v}_j) = \sum_{i=0}^{nl-1} M(r_i \mid v_i), \tag{12.2}$$

where $M(\mathbf{r}_j \mid \mathbf{v}_j)$, the branch metric for the jth branch, is computed by adding the bit metric for the n bits on that branch. Combining (12.1) and (12.2) results in

$$M([\mathbf{r} \mid \mathbf{v}]_{l-1}) = \sum_{i=0}^{nl-1} \log_2 P(r_i \mid v_i) - \sum_{i=0}^{nl-1} \log_2 P(r_i) - nlR. \tag{12.3}$$

A binary-input, Q-ary-output DMC is said to be *symmetric* if

$$P(j \mid 0) = P(Q - 1 - j \mid 1), \qquad j = 0, 1, \ldots, Q - 1. \tag{12.4}$$

For a symmetric channel with equally likely input symbols,[1] the channel output symbols must satisfy $P(r_i = j) = P(r_i = Q - 1 - j) \leq \frac{1}{2}$ for $0 \leq j \leq Q - 1$ and all i (see Problem 12.2), and (12.3) reduces to

$$\begin{aligned} M([\mathbf{r} \mid \mathbf{v}]_{l-1}) &= \sum_{i=0}^{nl-1} \log_2 P(r_i \mid v_i) - \sum_{i=0}^{nl-1} [\log_2 P(r_i) + R] \\ &= \sum_{i=0}^{nl-1} \log_2 P(r_i \mid v_i) + \sum_{i=0}^{nl-1} \left[\log_2 \frac{1}{P(r_i)} - R\right]. \end{aligned} \tag{12.5}$$

The first term in (12.5) is the metric for the Viterbi algorithm. The second term represents a positive (since $1/P(r_i) \geq 2$ and $R \leq 1$) bias which increases linearly with path length. Hence, longer paths have a larger bias than shorter paths, reflecting the fact that they are closer to the end of the tree and hence more likely to be part of the maximum likelihood path. The bit metric of (12.1) was first introduced by Fano

[1] Since convolutional codes are linear, the set of all code words contains an equal number of zeros and ones (see Problem 3.6), and hence the channel input symbols are equally likely.

[2] on intuitive grounds, and hence is called the *Fano metric*. It is the metric most commonly used for sequential decoding, although some other metrics have been proposed. When comparing paths of different lengths, the path with the largest Fano metric is considered the "best" path (i.e., most likely to be part of the maximum likelihood path).

Example 12.2

For a BSC ($Q = 2$) with transition probability p, the Fano metrics for the truncated code words $[\mathbf{v}]_5$ and $[\mathbf{v}']_0$ in Example 12.1 are given by

$$\begin{aligned} M([\mathbf{r}\,|\,\mathbf{v}]_5) &= 16 \log_2 (1 - p) + 2 \log_2 p + 18(1 - 1/3) \\ &= 16 \log_2 (1 - p) + 2 \log_2 p + 12 \end{aligned}$$

and

$$\begin{aligned} M([\mathbf{r}\,|\,\mathbf{v}']_0) &= 2 \log_2 (1 - p) + \log_2 p + 3(1 - 1/3) \\ &= 2 \log_2 (1 - p) + \log_2 p + 2. \end{aligned}$$

For $p = 0.10$,

$$M([\mathbf{r}\,|\,\mathbf{v}]_5) = 2.92 > M([\mathbf{r}\,|\,\mathbf{v}']_0) = -1.63,$$

indicating that $[\mathbf{v}]_5$ is the "better" of the two paths. This differs from the result obtained using the Viterbi algorithm metric, since the bias term in the Fano metric reflects the difference in the path lengths.

In general, for a BSC with transition probability p, the bit metrics are

$$M(r_i\,|\,v_i) = \begin{cases} \log_2 2p - R & \text{if } r_i \neq v_i \\ \log_2 2(1 - p) - R & \text{if } r_i = v_i. \end{cases} \tag{12.6}$$

Example 12.3

If $R = 1/3$ and $p = 0.10$,

$$M(r_i\,|\,v_i) = \begin{cases} -2.65 & \text{if } r_i \neq v_i \\ 0.52 & \text{if } r_i = v_i, \end{cases}$$

and we have the metric table shown in Figure 12.2(a). It is common practice to scale the metrics by a positive constant so that they can be closely approximated as integers for ease of implementation. In this case, the scaling factor of 1/0.52 yields the integer metric table shown in Figure 12.2(b).

In the *stack* or *ZJ algorithm*, an ordered list or stack of previously examined paths of different lengths is kept in storage. Each stack entry contains a path along with its metric, the path with the largest metric is placed on top, and the others are listed in order of decreasing metric. Each decoding step consists of extending the top path in the stack by computing the branch metrics of its 2^k succeeding branches, and then adding these to the metric of the top path to form 2^k new paths, called the *successors* of the top path. The top path is then deleted from the stack, its 2^k successors are inserted, and the stack is rearranged in order of decreasing metric values. When the top path in the stack is at the end of the tree, the algorithm terminates.

v_i \ r_i	0	1
0	0.52	−2.65
1	−2.65	0.52

(a)

v_i \ r_i	0	1
0	1	−5
1	−5	1

(b)

Figure 12.2 Metric tables for an $R = 1/3$ code and a BSC with $p = 0.10$.

The Stack Algorithm

Step 1. Load the stack with the origin node in the tree, whose metric is taken to be zero.

Step 2. Compute the metric of the successors of the top path in the stack.

Step 3. Delete the top path from the stack.

Step 4. Insert the new paths in the stack, and rearrange the stack in order of decreasing metric values.

Step 5. If the top path in the stack ends at a terminal node in the tree, stop. Otherwise, return to step 2.

When the algorithm terminates, the top path in the stack is taken as the decoded path. A complete flowchart for the stack algorithm is shown in Figure 12.3.

In the dividing part of the tree, there are 2^k new metrics to be computed at step 1. In the tail of the tree, only one new metric is computed. Note that for $(n, 1, m)$ codes, the size of the stack increases by one for each decoding step in the dividing part of the tree, but does not increase at all when the decoder is in the tail of the tree. Since the dividing part of the tree is typically much longer than the tail ($L \gg m$), the size

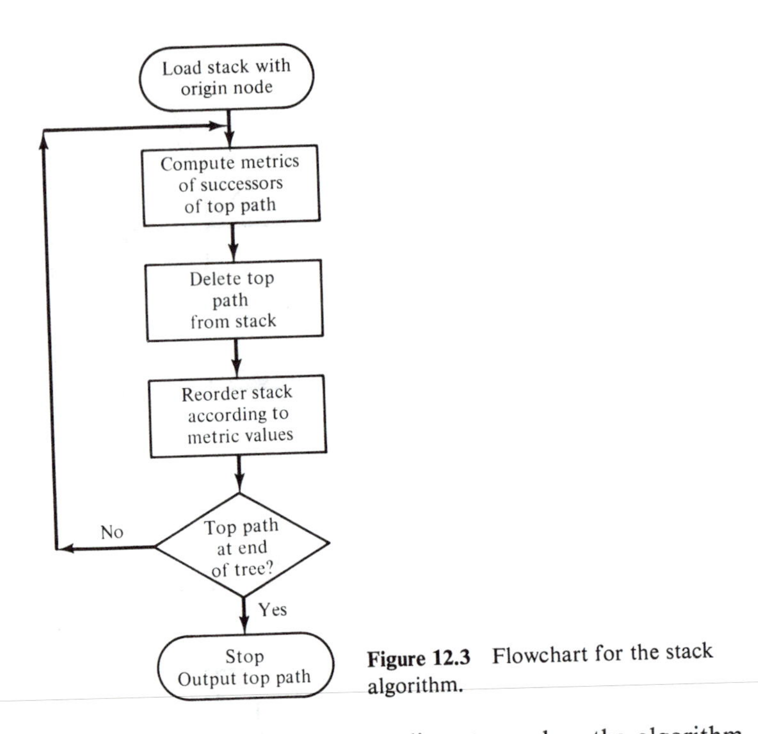

Figure 12.3 Flowchart for the stack algorithm.

of the stack is roughly equal to the number of decoding steps when the algorithm terminates.

Example 12.4

Consider the application of the stack algorithm to the $R = 1/3$ code tree of Figure 12.1. Assume that a code word is transmitted from this code over a BSC with $p = 0.10$, and the sequence

$$\mathbf{r} = (0\ 1\ 0,\ 0\ 1\ 0,\ 0\ 0\ 1,\ 1\ 1\ 0,\ 1\ 0\ 0,\ 1\ 0\ 1,\ 0\ 1\ 1)$$

is received. Using the integer metric table of Figure 12.2(b), the contents of the stack after each step of the algorithm are shown in Figure 12.4. The algorithm terminates after 10 decoding steps, and the final decoded path is

$$\hat{\mathbf{v}} = (1\ 1\ 1,\ 0\ 1\ 0,\ 0\ 0\ 1,\ 1\ 1\ 0,\ 1\ 0\ 0,\ 1\ 0\ 1,\ 0\ 1\ 1),$$

corresponding to the information sequence $\hat{\mathbf{u}} = (1\ 1\ 1\ 0\ 1)$. In this example, ties in the metric values were resolved by placing the longest path on top. This has the effect of slightly reducing the total number of decoding steps. In general, however, the resolution of ties is arbitrary, and does not affect the error probability of the decoder.

It is interesting to compare the number of decoding steps required by the stack algorithm with the number required by the Viterbi algorithm. A decoding step or computation for the Viterbi algorithm is the compare and select operation performed for each state in the trellis diagram beyond time unit m. Hence, the Viterbi algorithm would require 15 computations to decode the received sequence in Example 12.4 (see the trellis diagram of Figure 11.6). Although a decoding step or computation is somewhat more complicated in the stack algorithm since the stack must be reordered

Step 1	Step 2	Step 3	Step 4	Step 5
0 (−3)	00 (−6)	000 (−9)	1 (−9)	11 (−6)
1 (−9)	1 (−9)	1 (−9)	0001 (−12)	0001 (−12)
	01 (−12)	01 (−12)	01 (−12)	01 (−12)
		001 (−15)	001 (−15)	001 (−15)
			0000 (−18)	0000 (−18)
				10 (−24)

Step 6	Step 7	Step 8	Step 9	Step 10
111 (−3)	1110 (0)	11101 (+3)	111010 (+6)	1110100 (+9)
0001 (−12)	0001 (−12)	0001 (−12)	0001 (−12)	0001 (−12)
01 (−12)	01 (−12)	01 (−12)	01 (−12)	01 (−12)
001 (−15)	001 (−15)	11100 (−15)	11100 (−15)	11100 (−15)
0000 (−18)	1111 (−18)	001 (−15)	001 (−15)	001 (−15)
110 (−21)	0000 (−18)	1111 (−18)	1111 (−18)	1111 (−18)
10 (−24)	110 (−21)	0000 (−18)	0000 (−18)	0000 (−18)
	10 (−24)	110 (−21)	110 (−21)	110 (−21)
		10 (−24)	10 (−24)	10 (−24)

Figure 12.4 Stack contents in Example 12.4.

after each path is extended,[2] the stack algorithm requires only 10 computations to decode the received sequence in Example 12.4. This computational advantage of sequential decoding over the Viterbi algorithm is typical when the received sequence is not too noisy, that is, when it contains a fraction of errors not too much greater than the channel transition probability p. Note that $\hat{\mathbf{v}}$ in Example 12.4 disagrees with $\mathbf{r}$ in only 2 positions and agrees in the other 19 positions. Assuming that $\hat{\mathbf{v}}$ was actually transmitted, the fraction of errors in $\mathbf{r}$ is $2/21 = 0.095$, roughly equal to the channel transition probability of $p = 0.10$ in this case.

The situation is somewhat different when the received sequence is very noisy. This is illustrated with another example.

Example 12.5

For the same code, channel, and metric table as in Example 12.4, assume that the sequence

$$\mathbf{r} = (1\ \ 1\ \ 0,\ \ 1\ \ 1\ \ 0,\ \ 1\ \ 1\ \ 0,\ \ 1\ \ 1\ \ 1,\ \ 0\ \ 1\ \ 0,\ \ 1\ \ 0\ \ 1,\ \ 1\ \ 0\ \ 1)$$

is received. The contents of the stack after each step of the algorithm are shown in Figure 12.5. The algorithm terminates after 20 decoding steps and the final decoded path is

$$\hat{\mathbf{v}} = (1\ \ 1\ \ 1,\ \ 0\ \ 1\ \ 0,\ \ 1\ \ 1\ \ 0,\ \ 0\ \ 1\ \ 1,\ \ 1\ \ 1\ \ 1,\ \ 1\ \ 0\ \ 1,\ \ 0\ \ 1\ \ 1),$$

corresponding to the information sequence $\hat{\mathbf{u}} = (1\ 1\ 0\ 0\ 1)$.

In this example the sequential decoder performs 20 computations, whereas the Viterbi algorithm would again require only 15 computations. This points out one of the most important differences between sequential decoding and Viterbi decoding, that the number of computations performed by a sequential decoder is a random

[2]This reordering problem is eliminated, with very little loss in performance, in Jelinek's stack-bucket algorithm [4], discussed later in this section.

Step 1	Step 2	Step 3	Step 4	Step 5	Step 6	Step 7
1 (−3)	11 (−6)	110 (−3)	1100 (−6)	11000 (−9)	0 (−9)	1101 (−12)
0 (−9)	0 (−9)	0 (−9)	0 (−9)	0 (−9)	1101 (−12)	01 (−12)
	10 (−12)	10 (−12)	1101 (−12)	1101 (−12)	10 (−12)	10 (−12)
		111 (−21)	10 (−12)	10 (−12)	11001 (−15)	11001 (−15)
			111 (−21)	11001 (−15)	110000 (−18)	110000 (−18)
				111 (−21)	111 (−21)	00 (−18)
						111 (−21)

Step 8	Step 9	Step 10	Step 11	Step 12	Step 13	Step 14
11011 (−9)	01 (−12)	10 (−12)	11001 (−15)	110010 (−12)	101 (−15)	011 (−15)
01 (−12)	10 (−12)	11001 (−15)	101 (−15)	101 (−15)	011 (−15)	110110 (−18)
10 (−12)	11001 (−15)	011 (−15)	011 (−15)	011 (−15)	110110 (−18)	110000 (−18)
11001 (−15)	110110 (−18)	110110 (−18)	110110 (−18)	110110 (−18)	110000 (−18)	1010 (−18)
110000 (−18)	110000 (−18)	110000 (−18)	110000 (−18)	110000 (−18)	00 (−18)	00 (−18)
00 (−18)	00 (−18)	00 (−18)	00 (−18)	00 (−18)	1100100 (−21)	1100100 (−21)
111 (−21)	111 (−21)	010 (−21)	100 (−21)	100 (−21)	100 (−21)	100 (−21)
11010 (−27)	11010 (−27)	111 (−21)	010 (−21)	010 (−21)	010 (−21)	010 (−21)
		11010 (−27)	111 (−21)	111 (−21)	111 (−21)	111 (−21)
			11010 (−27)	11010 (−27)	11010 (−27)	1011 (−24)
						11010 (−27)

Step 15	Step 16	Step 17	Step 18	Step 19	Step 20
110110 (−18)	110000 (−18)	0110 (−18)	1010 (−18)	00 (−18)	1100100 (−21)
110000 (−18)	0110 (−18)	1010 (−18)	00 (−18)	1100100 (−21)	10100 (−21)
0110 (−18)	1010 (−18)	00 (−18)	1100100 (−21)	10100 (−21)	01100 (−21)
1010 (−18)	00 (−18)	1100100 (−21)	01100 (−21)	01100 (−21)	001 (−21)
00 (−18)	1100100 (−21)	100 (−21)	100 (−21)	100 (−21)	100 (−21)
1100100 (−21)	100 (−21)	010 (−21)	010 (−21)	010 (−21)	010 (−21)
100 (−21)	010 (−21)	111 (−21)	111 (−21)	111 (−21)	111 (−21)
010 (−21)	111 (−21)	0111 (−24)	0111 (−24)	0111 (−24)	0111 (−24)
111 (−21)	0111 (−24)	1011 (−24)	1011 (−24)	1011 (−24)	1011 (−24)
0111 (−24)	1011 (−24)	1100000 (−27)	1100000 (−27)	1100000 (−27)	1100000 (−27)
1011 (−24)	1101100 (−27)	1101100 (−27)	1101100 (−27)	1101100 (−27)	1101100 (−27)
11010 (−27)	11010 (−27)	11010 (−27)	01101 (−27)	10101 (−27)	10101 (−27)
			11010 (−27)	01101 (−27)	01101 (−27)
				11010 (−27)	11010 (−27)
					000 (−27)

Figure 12.5 Stack contents in Example 12.5

variable, whereas the computational load of the Viterbi algorithm is fixed. Very noisy received sequences typically require a large number of computations with a sequential decoder, sometimes more than the fixed number of computations required by the Viterbi algorithm, as in the example above. However, since very noisy received sequences do not occur very often, the *average* number of computations performed by a sequential decoder is normally much less than the fixed number performed by the Viterbi algorithm.

The received sequence in Example 12.5 was also decoded using the Viterbi algorithm in Section 11.1. The final decoded path was the same in both cases. This illustrates the important fact that a sequential decoder almost always produces the maximum likelihood path, even when the received sequence is very noisy. Hence, for a given code, the error probability of sequential decoding is essentially the same as for Viterbi decoding.

However, there are several problems associated with the implementation of the stack sequential decoding algorithm which limit its overall performance. First, since the decoder traces a somewhat random path back and forth through the code tree, jumping from node to node, the decoder must have an input buffer to store incoming blocks of the received sequence while they are waiting to be processed. Depending on the *speed factor* of the decoder, that is, the ratio of the speed at which computations are performed to the speed of the incoming data (in branches received per second), long searches can cause the input buffer to overflow, resulting in a loss of data, or an *erasure*. The buffer accepts data at the fixed rate of $1/nT$ branches per second, where T is the time interval allotted for each transmitted bit. It outputs these branches to the decoder asynchronously, as demanded by the algorithm. Normally, the information is divided into "frames" of L branches, each terminated by a sequence of $m \ll L$ all-zero inputs to return the encoder to the all-zero state. Even if the input buffer is one or two orders of magnitude larger than L, there is some probability that it will fill up during the decoding of a given frame, and that the next branch received from the channel will then cause an undecoded branch from the frame being processed to be shifted out of the buffer. These bits are then lost, and we say that an erasure results. Erasure probabilities of approximately 10^{-3} are common in sequential decoding, where this means that a particular frame has a probability of 10^{-3} of not being decoded due to an overflow of the input buffer.

The average number of computations performed by a sequential decoder, and also its erasure probability, are essentially independent of the encoder memory K. Therefore, codes with large values of K,[3] and hence large free distance, can be selected for use with sequential decoding. Undetected errors (complete but incorrect decoding) then become extremely unlikely, and the major limitation on the code's performance is the erasure probability due to buffer overflow.

Even though an erasure probability of approximately 10^{-3} can be rather troublesome in some systems, it may actually be beneficial in others. Since erasures usually occur when the received sequence is very noisy, if decoding were completed, and even if the maximum likelihood path were obtained, there is a fairly high probability that this estimate will be incorrect. In many systems it would be more desirable to erase such a frame than to decode it incorrectly. In other words, a complete decoder, such as the Viterbi algorithm, would always decode such a frame, even though it is likely to be decoded incorrectly, whereas a sequential decoder trades errors for erasures by "sensing" noisy frames and just erasing them. This property of sequential decoding can be used in an ARQ retransmission scheme as an indicator of when a frame should be retransmitted. This application of sequential decoding is discussed fully in Section 17.5. A second problem in any practical implementation of the stack algorithm is

[3] $K \approx km$ cannot be made too large for fixed L, since then the fractional rate loss [see (10.11)] would be significant. However, values of K up to about 50 are certainly feasible for sequential decoders with $L \approx 1000$.

that the size of the stack must be finite. In other words, there is always some probability that the stack will fill up before decoding is completed (or the buffer overflows) on a given frame. The most common way of handling this problem is simply to allow the path at the bottom of the stack to be pushed out of the stack on the next decoding step. This path is then "lost" and can never return to the stack. For typical stack sizes on the order of 1000 entries, the probability that a path on the bottom of the stack would ever recover to reach the top of the stack and be extended is so small that the loss in performance is negligible.

A related problem has to do with the reordering of the stack after each decoding step. This can become quite time consuming as the number of entries in the stack becomes large, and places severe limitations on the decoding speed that can be achieved with the basic algorithm. Jelinek [4] has proposed a modified algorithm, called the *stack-bucket* algorithm, in which the contents of the stack do not have to be reordered after each decoding step. In the stack-bucket algorithm, the range of possible metric values (from $+21$ to -110 in the foregoing examples) is quantized into a fixed number of intervals. Each metric interval is assigned a certain number of locations in storage, called a bucket. When a path is extended, it is deleted from storage and each new path is inserted as the top entry in the bucket which includes its metric value. No reordering of paths within buckets is required. The top path in the top nonempty bucket is chosen to be extended. A computation now involves only finding the correct bucket in which to place each new path, which is independent of the number of previously extended paths, and is therefore faster than reordering an increasingly larger stack. The disadvantage of the bucket algorithm is that it is not always the best path that gets extended, but only a "very good" path (i.e., a path in the top nonempty bucket, or the best bucket). Typically, though, if the bucket quantization is not too large and the received sequence is not too noisy, the best bucket contains only the best path, and the degradation in performance from the basic algorithm is very slight. The speed savings of the bucket algorithm is considerable, though, and all practical implementations of the stack algorithm have used this approach.

12.2 THE FANO ALGORITHM

Another approach to sequential decoding, the *Fano algorithm*, sacrifices some speed compared to the stack algorithm, but requires essentially no storage. The speed disadvantage of the Fano algorithm is due to the fact that it generally extends more nodes than the stack algorithm, although this is mitigated somewhat by the fact that there is no stack reordering to be done. In the Fano algorithm, the decoder examines a sequence of nodes in the tree, starting with the origin node and ending with one of the terminal nodes. The decoder never jumps from node to node as in the stack algorithm, but always moves to an adjacent node, either forward to one of the 2^k nodes leaving the present node, or backward to the node leading to the present node. The metric of the next node to be examined can then be computed by adding (or subtracting) the metric of the connecting branch to the metric of the present node. This eliminates the need for storing the metrics of previously examined nodes as required by the stack algorithm. However, some nodes are visited more than once,

and in this case their metric values must be recomputed. The decoder will move forward through the tree as long as the metric value along the path being examined continues to increase. When the metric value dips below a threshold, the decoder backs up and begins to examine other paths. If no path can be found whose metric value stays above the threshold, the threshold is then lowered and the decoder attempts to move forward again with a lower threshold. Each time a given node is visited in the forward direction, the threshold is lower than on the previous visit to that node. This prevents looping in the algorithm, and the decoder eventually must reach the end of the tree, at which point the algorithm terminates.

A complete flowchart of the Fano algorithm is shown in Figure 12.6. The decoder starts at the origin node with the *threshold* $T = 0$ and the *metric value* $M = 0$. It then looks forward to the best of the 2^k succeeding nodes (i.e., the one with the

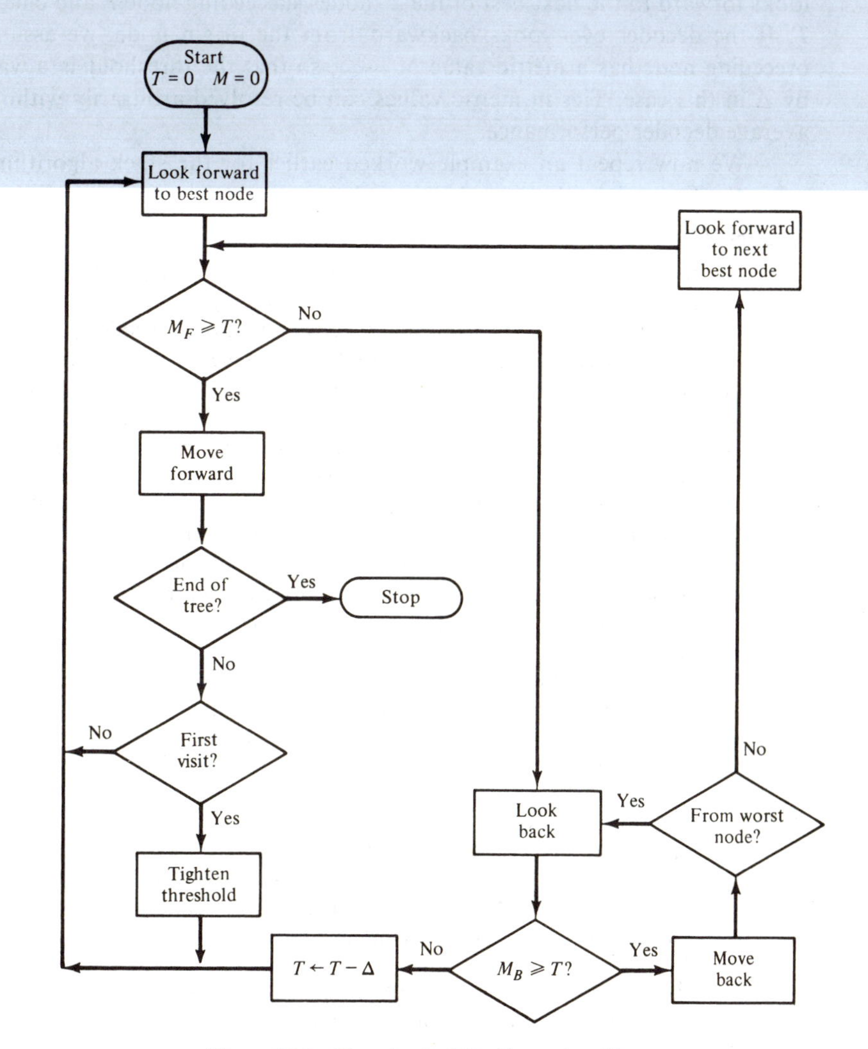

Figure 12.6 Flowchart of the Fano algorithm.

largest metric). If M_F is the *metric of the forward node being examined*, and if $M_F \geq T$, the decoder moves to this node. After checking whether the end of the tree has been reached, a "threshold tightening" is performed if this node is being examined for the first time. This involves increasing T by the largest possible multiple of a *threshold increment* Δ so that the new threshold does not exceed the current metric. If this node has been examined previously, no threshold tightening is performed. Then the decoder again looks forward to the best succeeding node. If $M_F < T$, the decoder then looks backward to the preceding node. If M_B is the *metric of the backward node being examined*, and if $M_B < T$, then T is lowered by Δ and the look forward to the best node step is repeated. If $M_B \geq T$, the decoder moves back to the preceding node. Call this node P. If this backward move was from the worst of the 2^k nodes succeeding node P, the decoder again looks back to the node preceding node P. If not, the decoder looks forward to the next best of the 2^k nodes succeeding node P and checks if $M_F \geq T$. If the decoder ever looks backward from the origin node, we assume that the preceding node has a metric value of $-\infty$, so that the threshold is always lowered by Δ in this case. Ties in metric values can be resolved arbitrarily without affecting average decoder performance.

We now repeat an example worked earlier for the stack algorithm, this time using the Fano algorithm.

Example 12.6

For the same received sequence decoded by the stack algorithm in Example 12.4, the steps of the Fano algorithm are shown in Figure 12.7 for a value of $\Delta = 1$. The algorithm terminates after 40 decoding steps, and the final decoded path is the same one found by the stack algorithm. In Figure 12.7, LFB means "look forward to best node," LFNB means "look forward to next best node," and X denotes the origin node.

A computation in the Fano algorithm is usually counted each time the "look forward" step is performed. Hence, for this example, the Fano algorithm requires 40 computations, compared to only 10 for the stack algorithm. Note that some nodes are visited several times. In fact, the origin node is visited 8 times, the path 0 node 11 times, the path 00 node 5 times, and the nodes representing the paths 000, 1, 11, 111, 1110, 11101, 111010, and 1110100 one time each, for a total of 32 nodes visited. This is less than the 40 computations, since not every forward look results in a move to a different node, but sometimes only in a threshold lowering.

The number of computations performed by the Fano algorithm depends on how the threshold increment Δ is selected. In general, if Δ is too small, a large number of computations results, as in Example 12.6. Making Δ larger will reduce the number of computations.

Example 12.7

The Fano algorithm is repeated for the same received sequence decoded in Example 12.6 and for a value of $\Delta = 3$. The results are shown in Figure 12.8. Twenty-two computations are required, 20 nodes are visited, and the same path is decoded. Hence, in this case raising Δ to 3 reduced the number of computations by almost a factor of 2.

The threshold increment Δ cannot be raised indefinitely, however, without affecting the error probability. For the algorithm to find the maximum likelihood path, T must at some point be lowered below the minimum metric along the maximum

Step	Look	M_F	M_B	Node	Metric	T
0	–	–	–	X	0	0
1	LFB	−3	−∞	X	0	−1
2	LFB	−3	−∞	X	0	−2
3	LFB	−3	−∞	X	0	−3
4	LFB	−3	–	0	−3	−3
5	LFB	−6	0	X	0	−3
6	LFNB	−9	−∞	X	0	−4
7	LFB	−3	–	0	−3	−4
8	LFB	−6	0	X	0	−4
9	LFNB	−9	−∞	X	0	−5
10	LFB	−3	–	0	−3	−5
11	LFB	−6	0	X	0	−5
12	LFNB	−9	−∞	X	0	−6
13	LFB	−3	–	0	−3	−6
14	LFB	−6	–	00	−6	−6
15	LFB	−9	−3	0	−3	−6
16	LFNB	−12	0	X	0	−6
17	LFNB	−9	−∞	X	0	−7
18	LFB	−3	–	0	−3	−7
19	LFB	−6	–	00	−6	−7
20	LFB	−9	−3	0	−3	−7
21	LFNB	−12	0	X	0	−7
22	LFNB	−9	−∞	X	0	−8
23	LFB	−3	–	0	−3	−8
24	LFB	−6	–	00	−6	−8
25	LFB	−9	−3	0	−3	−8
26	LFNB	−12	0	X	0	−8
27	LFNB	−9	−∞	X	0	−9
28	LFB	−3	–	0	−3	−9
29	LFB	−6	–	00	−6	−9
30	LFB	−9	–	000	−9	−9
31	LFB	−12	−6	00	−6	−9
32	LFNB	−15	−3	0	−3	−9
33	LFNB	−12	0	X	0	−9
34	LFNB	−9	–	1	−9	−9
35	LFB	−6	–	11	−6	−6
36	LFB	−3	–	111	−3	−3
37	LFB	0	–	1110	0	0
38	LFB	+3	–	11101	+3	+3
39	LFB	+6	–	111010	+6	+6
40	LFB	+9	–	1110100	+9	Stop

Figure 12.7 Decoding steps for the Fano algorithm with $\Delta = 1$.

likelihood path. If Δ is too large, when T is lowered below the minimum metric of the maximum likelihood path, it may also be lowered below the minimum metric of several other paths, thereby making it possible for any of these to be decoded before the maximum likelihood path. Making Δ too large can also cause the number of computations to increase again, since more "bad" paths can be followed further into the tree if T is lowered too much. Experience has shown that, if unscaled metrics are used, Δ should be chosen between 2 and 8. If the metrics are scaled, Δ should be scaled by the same factor. In Example 12.7, the scaling factor was 1/0.52, indicating that Δ should be chosen between 3.85 and 15.38. A choice of Δ between 6 and 10 would be a good compromise in this case (see Problem 12.7).

Step	Look	M_F	M_B	Node	Metric	T
0			–	X	0	0
1	LFB	−3	−∞	X	0	−3
2	LFB	−3	–	0	−3	−3
3	LFB	−6	0	X	0	−3
4	LFNB	−9	−∞	X	0	−6
5	LFB	−3	–	0	−3	−6
6	LFB	−6	–	00	−6	−6
7	LFB	−9	−3	0	−3	−6
8	LFNB	−12	0	X	0	−6
9	LFNB	−9	−∞	X	0	−9
10	LFB	−3	–	0	−3	−9
11	LFB	−6	–	00	−6	−9
12	LFB	−9	–	000	−9	−9
13	LFB	−12	−6	00	−6	−9
14	LFNB	−15	−3	0	−3	−9
15	LFNB	−12	0	X	0	−9
16	LFNB	−9	–	1	−9	−9
17	LFB	−6	–	11	−6	−6
18	LFB	−3	–	111	−3	−3
19	LFB	0	–	1110	0	0
20	LFB	+3	–	11101	+3	+3
21	LFB	+6	–	111010	+6	+6
22	LFB	+9	–	1110100	+9	Stop

Figure 12.8 Decoding steps for the Fano algorithm with $\Delta = 3$.

In the examples of the application of the Fano algorithm presented earlier, the same path chosen by the stack algorithm was decoded in both cases. The Fano algorithm almost always finds the same path as the stack-bucket algorithm when Δ is chosen equal to the interval of quantization for the metric values in the stack [7]. Also, the Fano algorithm generally decodes faster than the stack-bucket algorithm for moderate rates. This is because the Fano algorithm is not slowed down by the stack control problems of the stack-bucket algorithm. For higher rates, the stack-bucket algorithm is somewhat faster due to the additional computational load of the Fano algorithm [8]. Because it does not require any storage, the Fano algorithm is usually selected in practical implementations of sequential decoding.

12.3 PERFORMANCE CHARACTERISTICS OF SEQUENTIAL DECODING

The performance of sequential decoding is not as easy to characterize as Viterbi decoding because of the interplay between errors and erasures and the need to assess computational behavior. Since the number of computations performed in decoding a frame of data is a random variable, its probability distribution must be computed in order to determine computational performance. A great deal of work has gone into a random coding analysis of this probability distribution. Only a brief summary of these results is given here. Readers interested in more detail are referred to References 9 to 12.

Let the jth *incorrect subset* of the code tree be the set of all nodes branching from the jth node on the correct path, $0 \leq j \leq L-1$, as shown in Figure 12.9,

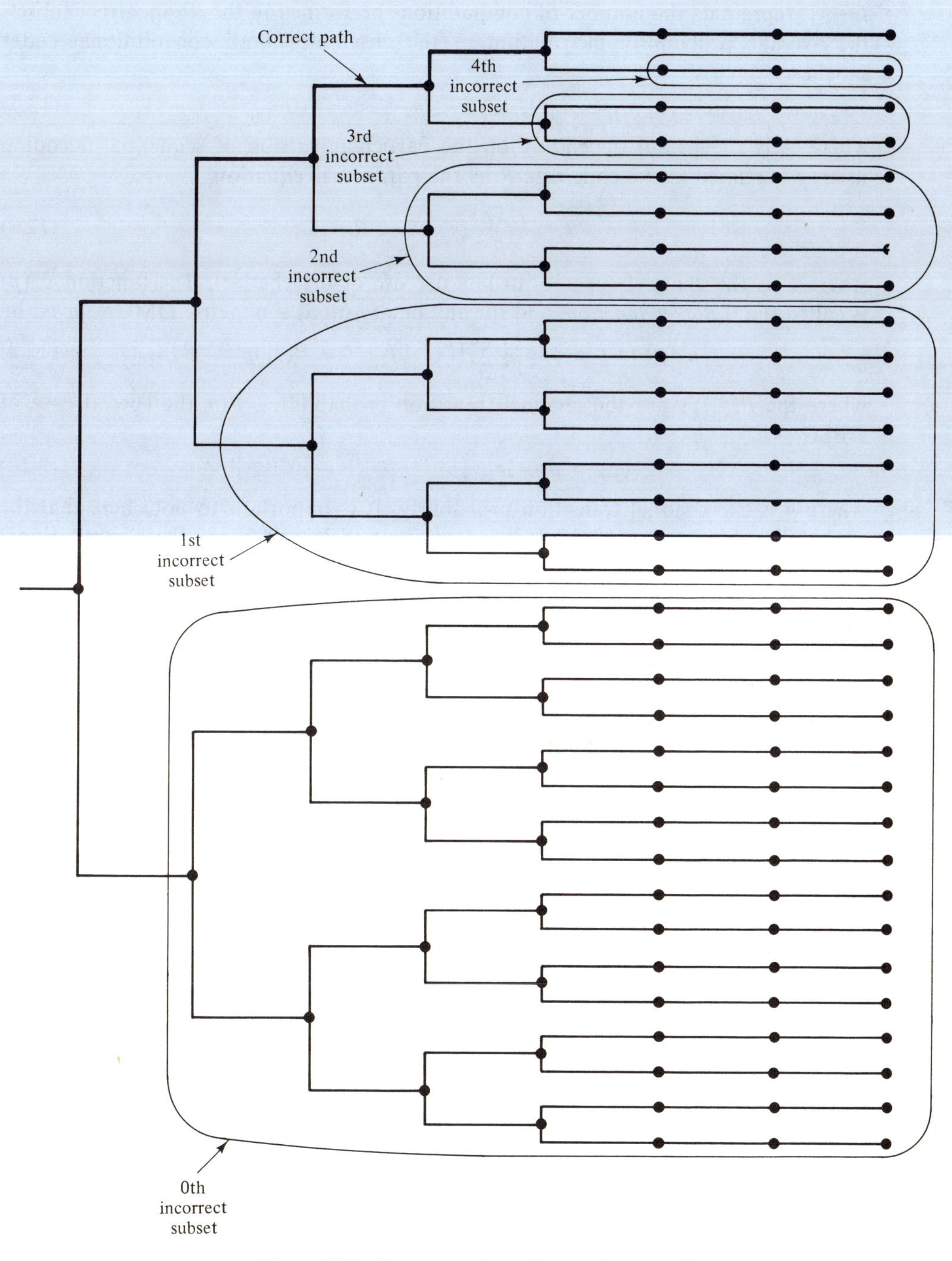

Figure 12.9 Incorrect subsets for a code tree with $L = 5$.

where L represents the length of the information sequence in branches. If C_j, $0 \leq j \leq L - 1$, represents the number of computations performed in the jth incorrect subset, the average probability distribution of the ensemble of all convolutional codes satisfies

$$\Pr[C_j \geq \eta] \approx A\eta^{-\rho}, \qquad 0 < \rho < \infty, \qquad 0 \leq j \leq L - 1, \tag{12.7}$$

where A is a constant depending on the particular version of sequential decoding used. ρ is related to the code rate R by the parametric equation

$$R = \frac{E_0(\rho)}{\rho}, \qquad 0 < R < C, \tag{12.8}$$

where C is the channel capacity in bits per use of the channel. The function $E_0(\rho)$ is called the *Gallager function*, and for any binary-input symmetric DMC is given by

$$E_0(\rho) = \rho - \log_2 \tfrac{1}{2} \sum_j [P(j|0)^{1/1+\rho} + P(j|1)^{1/1+\rho}]^{1+\rho}, \tag{12.9}$$

where the $P(j|i)$'s are the channel transition probabilities. For the special case of a BSC,

$$E_0(\rho) = \rho - \log_2 [p^{1/1+\rho} + (1 - p)^{1/1+\rho}]^{1+\rho}, \tag{12.10}$$

where p is the channel transition probability. It is important to note here that the probability distribution of (12.7) is independent of K, and hence the computational behavior of sequential decoding does not depend on code constraint length.

The distribution of (12.7) is a *Pareto distribution*, and ρ is called the *Pareto exponent*. ρ is an extremely important parameter for sequential decoding, since it determines how rapidly $\Pr[C_j \geq \eta]$ decreases as a function of η. For a fixed rate R, ρ can be calculated for a BSC by solving (12.8) using the expression of (12.10) for $E_0(\rho)$. Results are shown in Figure 12.10 for $R = 1/4$, 1/3, 1/2, 2/3, and 3/4, where ρ is plotted as a function of $E_b/N_0 = E/RN_0$ and $p = Q(\sqrt{2E/N_0}) = Q(\sqrt{2RE_b/N_0})$.

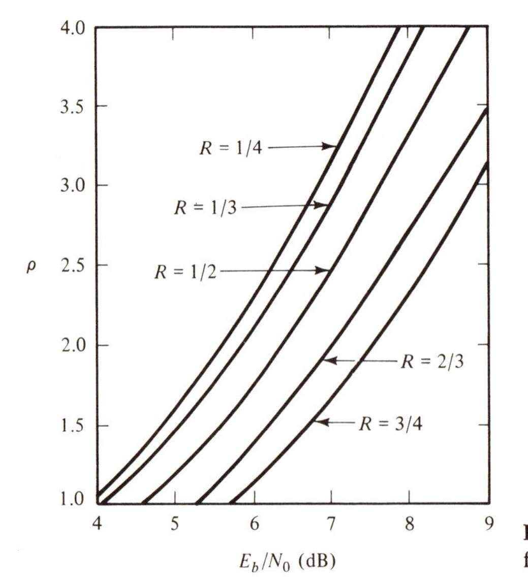

Figure 12.10 Pareto exponent as a function of E_b/N_0 for a BSC.

Note that when $\rho = 1$,

$$R = E_0(1) = 1 - \log_2 \tfrac{1}{2} \sum_j [\sqrt{P(j|0)} + \sqrt{P(j|1)}]^2, \tag{12.11}$$

and for a BSC

$$\begin{aligned} R = E_0(1) &= 1 - \log_2 [\sqrt{p} + \sqrt{1-p}]^2 \\ &= 1 - \log_2 [1 + 2\sqrt{p(1-p)}]. \end{aligned} \tag{12.12}$$

The significance of $E_0(1)$ for sequential decoding is related to the moments of the computational distribution. In order to compute $E[C_j^i]$, the ith moment of C_j, we first form

$$F_{C_j}(X) = \Pr[C_j \leq X] = 1 - \Pr[C_j \geq X] = 1 - AX^{-\rho}, \tag{12.13}$$

the cumulative distribution function of C_j. Differentiating this yields the probability density function

$$f_{C_j}(X) = \frac{dF_{C_j}(X)}{dX} = \rho A X^{-\rho-1}. \tag{12.14}$$

Now the ith moment $E[C_j^i]$ can be computed as

$$\begin{aligned} E[C_j^i] &= \int_1^\infty X^i f_{C_j}(X)\, dX \\ &= \int_1^\infty \rho A X^{i-\rho-1}\, dX \\ &= \frac{\rho A}{i-\rho} X^{i-\rho}\Big|_1^\infty \\ &= \lim_{X\to\infty} \frac{\rho A}{i-\rho}(X^{i-\rho} - 1). \end{aligned} \tag{12.15}$$

For the ith moment to be finite, that is,

$$E[C_j^i] = \lim_{X\to\infty} \frac{\rho A}{i-\rho}(X^{i-\rho} - 1) < \infty, \tag{12.16}$$

we require that $\rho > i$. In other words, if $\rho \leq 1$, the mean number of computations per decoded branch is unbounded. It can be shown from (12.8) and (12.9) that ρ goes to ∞ as $R \longrightarrow 0$ and ρ goes to 0 as $R \longrightarrow C$ [i.e., $R = E_0(\rho)/\rho$ and ρ are inversely related (see Problem 12.9)]. Hence, for a bounded mean number of computations, $\rho > 1$ is equivalent to

$$R = \frac{E_0(\rho)}{\rho} < E_0(1) \triangleq R_0, \tag{12.17}$$

where R_0 is called the *computational cutoff rate* of the channel. In other words, R must be less than R_0 for the average computational load of sequential decoding to be finite, and this is independent of the code constraint length. For example, for a BSC, (12.12) implies that $R_0 = E_0(1) = 1/2$ when $p = 0.045$, and for an $R = 1/2$ code, it is necessary that $p < 0.045$ in order for $R < R_0$, and the average computational load to be finite. This is the reason R_0 is called the computational cutoff rate, because sequential decoding becomes computationally impractical for rates above R_0. Even though this result applies strictly only to sequential decoding, many other decoding methods also become impractical at rates above R_0 [13].

The probability distribution of (12.7) can also be used to calculate the probability of buffer overflow or the *erasure probability* P_{erasure}. Let B be the capacity of the input buffer in branches (nB bits), and μ be the speed factor of the decoder; that is, the decoder can perform μ branch computations in nT seconds, the time required to receive one branch. Then if more than μB computations are required in the jth incorrect subset, the buffer will overflow before the jth received branch can be discarded, even if the buffer was initially empty before the jth branch was received. From (12.7), this probability is approximately $A(\mu B)^{-\rho}$. Since there are L information branches in a frame, the erasure probability for a frame is approximately

$$P_{\text{erasure}} \approx LA(\mu B)^{-\rho}, \tag{12.18}$$

where ρ must satisfy (12.8). Note that (12.18) does not depend on code constraint length. Experimental studies indicate that the constant A is typically between 1 and 10, depending on the particular algorithm employed [14–16].

Example 12.8

Assume that $L = 1000$, $A = 3$, $\mu = 10$, and $B = 10^4$. For a BSC with transition probability $p = 0.08$, $R_0 = 3/8$. Choosing $R = 1/3 < R_0 = 3/8$, and solving for ρ from (12.8) and (12.10) yields $\rho = 1.31$. Substituting these values into (12.18) gives an erasure probability of $P_{\text{erasure}} \approx 0.85 \times 10^{-3}$.

The undetected bit error probability $P_b(E)$ of sequential decoding has also been analyzed using random coding arguments [12-17]. The results are the same as those given in Chapter 1 for maximum likelihood decoding [see (1.13)] for rates $R > R_0$. Thus, for high rates ($R > R_0$), sequential decoding has the same average error probability as maximum likelihood decoding, and is therefore optimum. For low rates ($R < R_0$), the average error probability of sequential decoding is suboptimum. Since maximum likelihood decoding is practical only for small values of K, and the computational behavior of sequential decoding is independent of constraint length, this suboptimum performance of sequential decoding at low rates can be compensated for by simply using codes with larger values of K than can be used with Viterbi decoding. For rates near R_0, only a slightly larger K is needed to provide the same error performance as a Viterbi decoder.

The overall performance of sequential decoding can only be judged by considering the undetected bit error probability, the erasure probability, and the average computational load. It is possible to obtain trade-offs between these three factors by adjusting various parameters. For example, reducing the threshold increment Δ in the Fano algorithm (or the bucket quantization interval in the stack algorithm) will increase $E[C_j]$ and P_{erasure} but reduce $P_b(E)$. On the other hand, increasing the size B of the input buffer will reduce P_{erasure} but increase $E[C_j]$ and $P_b(E)$.

The choice of a metric can also affect the overall performance of sequential decoding. Although the Fano metric of (12.1) is normally selected, it is not always necessary to do so. The bias term $-R$ is chosen to achieve a reasonable balance among $P_b(E)$, P_{erasure}, and $E[C_j]$. For example, using the integer metric table of Figure 12.2(b) for an $R = 1/3$ code and a BSC with $p = 0.10$, a path of length 12 that is distance 2 from the received sequence $\mathbf{r}$ would have a metric value of 0, the same as a path of length 6 that is distance 1 from $\mathbf{r}$. This is intuitively satisfying since both paths contain the same percentage of transmission errors, and leads to a certain balance among errors, erasures, and decoding speed. However, if no bias term were used in the definition of $M(r_i | v_i)$, the integer metric table shown in Figure 12.11(a) would result. In this case the length 12 path would have a metric of $+4$ and the length 6 path a metric of $+2$. This "bias" in favor of the longer path would result in less searching for the maximum likelihood path, and hence faster decoding and fewer erasures at the expense of more errors. On the other hand, if a larger bias term, say $-1/2$, were used, the integer metric table of Figure 12.11(b) would result. The length 12 path would then have a metric of -6 and the length 6 path a metric of -3. This "bias" in favor of the shorter path would result in more searching for the maximum likelihood path, and hence fewer errors at the expense of more erasures and slower decoding. Therefore, although the Fano metric is optimum in the restricted sense of identifying the one path of all those examined up to some point in the algorithm that is most likely to be part of the maximum likelihood path, it is not necessarily the best metric to use in all cases.

v_i \ r_i	0	1
0	1	−3
1	−3	1

(a)

v_i \ r_i	0	1
0	1	−8
1	−8	1

(b)

Figure 12.11 Effect of the bias term on the integer metric table.

The performance results for sequential decoding mentioned so far have all been random coding results; that is, they are representative of the average performance over the ensemble of all codes. Bounds on computational behavior and error probability for specific codes have also been obtained [18]. These bounds relate code performance to specific code parameters, and indicate how codes should be selected to provide the best possible performance. As such, they are analogous to the performance bounds for specific codes given in Chapter 11 for maximum likelihood decoding. For a BSC with transition probability p and an (n, k, m) code with column distance function d_l,

$$\Pr[C_j \geq \eta] < \sigma n_d e^{-\mu d_l + \phi l}, \tag{12.19}$$

where σ, μ, and ϕ are functions of p and R only, $l \triangleq \lfloor \log_{2^k} \eta \rfloor$, $\lfloor x \rfloor$ denotes the integer part of x, n_d is the number of code words of length $l + 1$ branches with weight d_l, and R satisfies

$$R < 1 + 2p \log_2 p + (1 - 2p) \log_2 (1 - p) \triangleq R_{\max}. \tag{12.20}$$

The bound of (12.20) indicates a maximum rate $R_{\max}$ for which (12.19) is known to hold. The significance of (12.19) is that it shows the dependence of the distribution of computation on the code's CDF. Fast decoding speeds and low erasure probabilities require that $\Pr[C_j \geq \eta]$ decrease rapidly as a function of η. From (12.19), this implies that the CDF should increase rapidly. The logarithm in the CDF's index has the effect of enhancing the significance of the initial portion of the CDF. This is illustrated in Figure 12.12. The CDF's of two (2, 1, 16) codes are shown in Figure 12.12(a). Their computational distributions obtained using extensive computer simulations [19] are shown in Figure 12.12(b). Note that the code with the faster column distance growth has a much better computational distribution. In fact, $E[C_j] = 3.14$ for code 1 and $E[C_j] = 7.24$ for code 2, a difference in average decoding speed of more than a factor of 2.[4]

The event error probability of a specific code when used with sequential decoding is bounded by a function that decreases exponentially with d_{free}, the free distance, and increases linearly with $A_{d_{\text{free}}}$, the number of code words with weight d_{free} [18]. This is the same general performance as that established in Section 11.2 for a specific code with maximum likelihood decoding. In other words, the undetected error probability of a code used with sequential decoding will differ little from that obtained with maximum likelihood decoding.

The performance of a (2, 1, 47) systematic code on a prototype hard-decision Fano sequential decoder has been thoroughly tested by Forney and Bower [14]. The results of some of these tests are shown in Figure 12.13. Three of the figures are for an incoming data rate $1/T$ of 1 Mbps, while the other three are for a 5-Mbps data rate. The computational rate of the decoder was 13.3 MHz, corresponding to decoder speed factors of 13.3 and 2.66, respectively. The BSC transition probability was varied between $p = 0.016$ and 0.059. This represents an E_b/N_0 range from 3.9 to 6.55 dB [see (11.22) and (11.25)], and an R_0 range from 0.44 to 0.68 [see (12.12)]. Rather than continuously vary the metric values with p, two sets of bit metrics, $+1/-9$ and $+1/-11$, were chosen. Input buffer sizes ranging from $2^{10} = 1024$ to $2^{16} = 65{,}536$ branches were tested. Long searches were minimized by a technique called backsearch limiting, in which the decoder is never allowed to move more than some maximum number J levels back from its furthest penetration into the tree. Whenever a forward move is made to a new level, the k information bits on that path J branches back are decoded. When the backsearch limit is reached, the decoder is forced forward by lowering the threshold until it falls below the metric of the best forward node. If the buffer overflows, the decoder is resynchronized by jumping forward to the most recently received branch, and accepting the received information bits on the intervening branches as decoded information. This resynchronization of the decoder typically introduces errors into the decoded sequence as the price to be paid for eliminating long searches.

[4]As an interesting comparison, the Viterbi algorithm would require a fixed number $2^K = 65{,}536$ computations per decoded information bit for this code, which is clearly impractical.

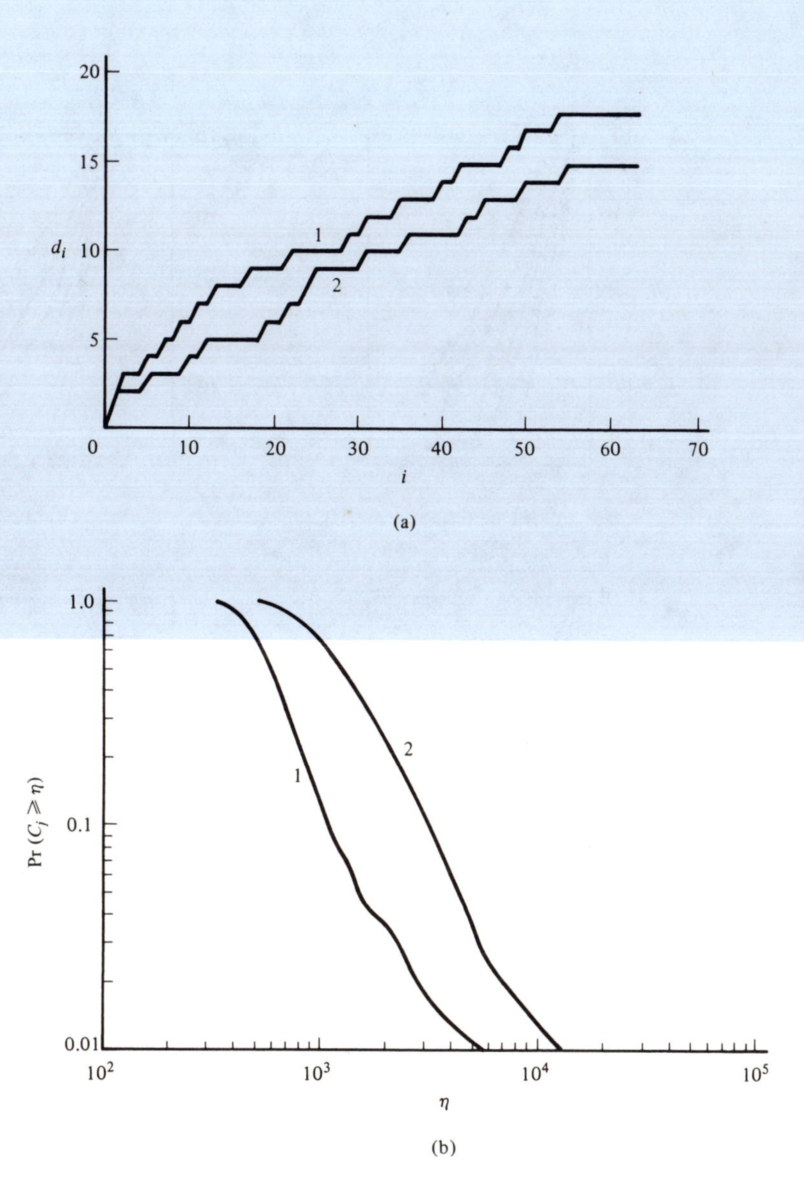

Figure 12.12 (a) The CDF and (b) $\Pr[C_j \geq \eta]$ for two (2, 1, 16) codes.

The performance curves of Figure 12.13(a) plot bit error probability $P_b(E)$ as a function of E_b/N_0, with buffer size as a parameter. The bit metrics were $+1/-11$, the data rate was 1 Mbps, and the backsearch limit was $J = 240$. Identical conditions hold for Figure 12.13(b), except that the data rate was increased to 5 Mbps. Note that performance is up to 0.8 dB worse in this case, reflecting the fact that the higher incoming data rate means that the decoder must jump further ahead during a resynchronization period to catch up with the received sequence, thereby introducing additional errors. In Figure 12.13(c) and (d), the backsearch limit J is a parameter,

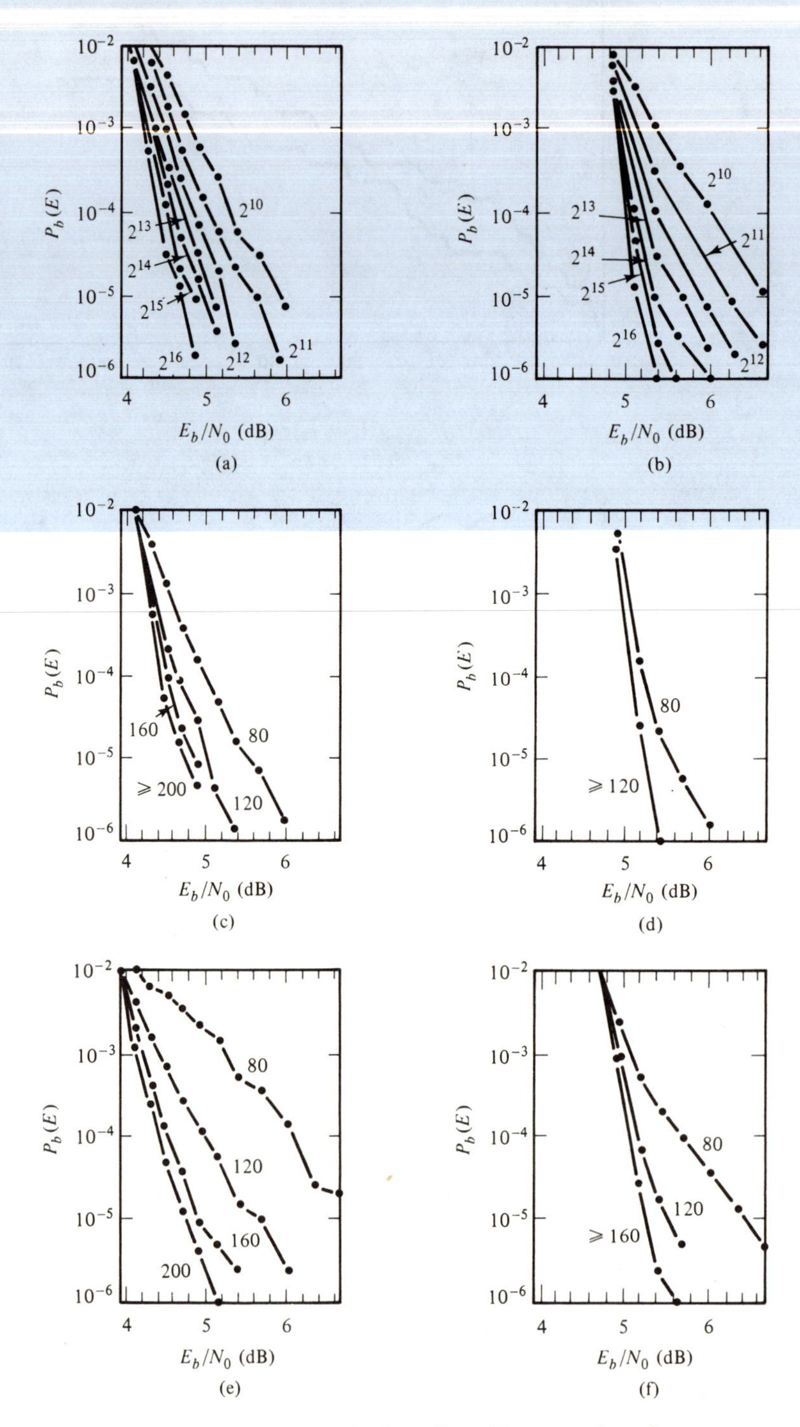

Figure 12.13 Test results for a (2, 1, 47) systematic code.

buffer size is 2^{16} branches, and the data rates are again 1 and 5 Mbps, respectively. Note that in this case the lower data rate gives up to 0.6 dB better performance if J is large enough. However, if J is too small, little improvement is obtained, reflecting the fact that the basic limitation on decoder performance is now J rather than the data rate. Figure 12.13(e) and (f) repeat the previous test conditions with bit metrics of $+1/-9$ instead of $+1/-11$. Performance is about the same for large J, but as much as 1.1 dB worse for small J. This reflects the fact that the $+1/-9$ bit metrics generate more searching before an incorrect path is eliminated, thereby causing increased reliance on the backsearch limit to force the decoder forward. Note that values of E_b/N_0 below 4.6 dB correspond to $R_0 < R = 1/2$, and that performance, although poor, does not degrade catastrophically in this range. In other words, the computational cutoff rate R_0 is obtained from an ensemble average upper bound, and although it generally indicates the maximum rate at which good performance can be obtained, it does not mean that sequential decoders can only be operated at rates $R < R_0$.

It is usually impossible to compare the performance of sequential decoding directly with that of Viterbi decoding because of the interplay between errors and erasures in a sequential decoder. However, the elimination of erasures in sequential decoding by backsearch limiting and decoder resynchronization makes a rudimentary comparison possible. Figure 11.15(b) shows the performance of a hard-decision Viterbi decoder for optimum $R = 1/2$ codes with $K = 2$ through 7. To obtain $P_b(E) = 10^{-4}$ requires E_b/N_0 in the range 5.4 to 7.0 dB. The decoder hardware in this case must perform 2^K computations per received branch. This implies a decoder speed factor ranging from 4 to 128 (in the absence of parallel decoding). The decoder must also be capable of storing 2^K 32-bit paths. Figure 12.13(a) shows the performance of a hard-decision sequential decoder for a suboptimum[5] $R = 1/2$ code with $K = 47$ and buffer sizes from 2^{10} through 2^{16} branches. To obtain $P_b(E) = 10^{-4}$ requires E_b/N_0 in the range 4.4 to 5.3 dB, a 1.0- to 1.7-dB improvement over the Viterbi decoder. The speed factor of the sequential decoder is 13.3, comparable with that of the Viterbi decoder. The sequential decoder memory requirement of from 2^{11} to 2^{17} bits somewhat exceeds the Viterbi decoder's requirement of from 2^7 to 2^{12} bits. To pick a specific point of comparison, consider $K = 5$ for the Viterbi decoder and $B = 2^{11}$ branches for the sequential decoder. At $P_b(E) = 10^{-4}$, the sequential decoder requires $E_b/N_0 = 5.0$ dB compared to the Viterbi algorithm's requirement of $E_b/N_0 = 6.0$ dB, a 1.0-dB advantage for the sequential decoder. The Viterbi decoder requires a speed factor of 32, roughly $2\frac{1}{2}$ times that of the sequential decoder, whereas the sequential decoder's memory requirement of 2^{12} bits is four times that of the Viterbi algorithm. It is well to note here that this comparison ignores the different hardware needed to perform a "computation" in the two algorithms, as well as the differing amounts of control logic required. In addition, when the ability to make soft decisions is considered, the balance tends to shift toward Viterbi decoding. A thorough discussion of the comparison between sequential and Viterbi decoding is given in Reference 20.

[5]Optimum codes of this length are unknown. Since the computational behavior of sequential decoding is independent of K, a systematic code with large K was chosen for convenience of decoder implementation. A nonsystematic code with the same d_{free} and a smaller K would also give approximately the same performance.

12.4 CODE CONSTRUCTION FOR SEQUENTIAL DECODING

The performance results of Section 12.3 can be used to select good codes for use with sequential decoding. In particular, optimum sequential decoding performance requires rapid initial column distance growth for fast decoding and minimum erasure probability, and a large free distance for minimum error probability. The *distance profile* of a code is defined as its CDF over only the first constraint length (i.e., d_0, $d_1, \ldots, d_m$). Since the distance profile determines the initial column distance growth, and is easier to compute than the entire CDF, it is often used instead of the entire CDF as a criterion for selecting codes for use with sequential decoding.

A code is said to have a distance profile $d_0, d_1, \ldots, d_m$ *superior* to the distance profile $d'_0, d'_1, \ldots, d'_m$ of another code of the same memory order m if, for some $l, 0 \leq l \leq m$,

$$d_i \begin{cases} = d'_i, & i = 0, 1, \ldots, l-1 \\ > d'_i, & i = l. \end{cases} \tag{12.21}$$

In other words, the initial portion of the CDF determines which code has the superior distance profile. A code is said to have an *optimum distance profile* (ODP) if its distance profile is superior to that of any other code of the same memory order.

Since an ODP guarantees fast initial column distance growth, ODP codes with large d_{free} make excellent choices for sequential decoding. A list of ODP codes with $R = 1/3$, 1/2, and 2/3, both systematic and nonsystematic, is given in Tables 12.1A

TABLE 12.1A $R = 1/3$ SYSTEMATIC CODES WITH AN OPTIMUM DISTANCE PROFILE

m	$\mathbf{g}^{(2)}$	$\mathbf{g}^{(3)}$	d_{free}
1	6	6	5
2	5	7	6
3	64	74	8
4	56	72	9
5	57	73	10
6	564	754	12
7	626	736	12
8	531	676	13
9	5314	6764	15
10	5312	6766	16
11	5312	6766	16
12	65304	71274	17
13	65306	71276	18
14	65305	71273	19
15	653764	712614	20
16	514112	732374	20
17	653761	712611	22
18	6530574	7127304	24
19	5141132	7323756	24
20	6530547	7127375	26
21	65376164	71261060	26
22	51445036	73251266	26
23	65305477	71273753	28

TABLE 12.1B $R = 1/3$ NONSYSTEMATIC CODES WITH AN OPTIMUM DISTANCE PROFILE

m	$\mathbf{g}^{(1)}$	$\mathbf{g}^{(2)}$	$\mathbf{g}^{(3)}$	d_{free}
1	4	6	6	5
2	5	7	7	8
3	54	64	74	10
4	52	66	76	12
5	47	53	75	13
6	—	—	—	—
7	516	552	656	16

TABLE 12.1C $R = 1/2$ SYSTEMATIC CODES WITH AN OPTIMUM DISTANCE PROFILE

m	$\mathbf{g}^{(2)}$	d_{free}
1	6	3
2	7	4
3	64	4
4	72	5
5	73	6
6	734	6
7	714	6
8	715	7
9	7154	8
10	7152	8
11	7153	9
12	67114	9
13	67114	9
14	67115	10
15	714474	10
16	671166	12
17	671166	12
18	6711454	12
19	7144616	12
20	7144761	12
21	67114544	12
22	71446166	14
23	67114543	14
24	714461654	15
25	671145536	15
26	671151433	16
27	7144760524	16
28	6711454306	16
29	7144760535	18
30	71446162654	≥ 16
31	67114543066	18
32	71447605247	≥ 18
33	714461626554	≥ 18
34	714461625306	≥ 18
35	714461626555	≥ 19

TABLE 12.1D $R = 1/2$ NONSYSTEMATIC CODES WITH AN OPTIMUM DISTANCE PROFILE

m	$\mathbf{g}^{(1)}$	$\mathbf{g}^{(2)}$	d_{free}
1	6	4	3
2	7	5	5
3	74	54	6
4	62	56	7
5	75	55	8
6	634	564	10
7	626	572	10
8	751	557	12
9	7664	5714	12
10	7512	5562	14
11	6643	5175	14
12	63374	47244	15
13	45332	77136	16
14	65231	43677	17
15	517604	664134	18
16	717066	522702	19
17	506477	673711	20
18	5653664	7746714	21
19	4305226	6574374	22
20	6567413	5322305	22
21	67520654	50371444	24
22	67132702	50516146	24
23	55346125	75744143	25

to 12.1F, together with the corresponding values of d_{free}. These codes were constructed by Johannesson and Paaske [21–23]. Note that d_{free} for the systematic codes lags considerably behind d_{free} for the nonsystematic codes, verifying that more free distance is available with nonsystematic codes of a given rate and encoder memory than with systematic codes. Both systematic and nonsystematic codes are listed since in some applications it is desirable to have the quick-look property of systematic codes. This can be obtained with no penalty in sequential decoding, since computational behavior is independent of K and the deficiency in d_{free} can be overcome simply by choosing codes with larger values of K. This is not possible in Viterbi decoding, where a severe penalty in reduced d_{free} or increased K (and hence increased computation) is paid when using a systematic code. Other constructions for codes specifically designed for use with sequential decoding are given in References 19 and 24.

Both the Viterbi algorithm and a sequential decoder can be used to analyze the distance properties of a code. This is done by assuming that the received sequence is all zeros, confining the search through the tree or trellis to only those paths starting with a nonzero information block, using a metric of 0 for an agreement and $+1$ for a disagreement, and searching for the path with the minimum metric. The Viterbi algorithm is best suited for determining the free distance of short-constraint-length codes. As soon as the metric of the survivor at the all-zero state is less than or equal

TABLE 12.1E $R = 2/3$ SYSTEMATIC CODES WITH AN OPTIMUM DISTANCE PROFILE

m	K*	$\mathbf{g}_1^{(3)}$	$\mathbf{g}_2^{(3)}$	d_{free}
1	1	4	6	2
2	2	5	7	3
3	3	54	64	4
4	4	56	62	4
5	5	57	63	5
6	6	554	704	5
7	7	664	742	6
8	8	665	743	6
9	9	5734	6370	6
10	10	5736	6322	7
11	11	5736	6323	8
12	12	66414	74334	8
13	13	57372	63226	8
14	14	57371	63225	8
15	15	664150	743314	8
16	16	664072	743346	10
17	17	573713	632255	10
18	18	6640344	7431024	10
19	19	5514632	7023726	10
20	20	5514633	7023725	11
21	21	57361424	63235074	12
22	22	66415416	74311464	11
23	23	66415417	74311465	12

*Assumes the encoder realization described in Example 10.13.

TABLE 12.1F $R = 2/3$ NONSYSTEMATIC CODES WITH AN OPTIMUM DISTANCE PROFILE

m	K	Generator sequences			d_{free}
2	3	$\mathbf{g}_1^{(1)} = 6$ $\mathbf{g}_2^{(1)} = 1$	$\mathbf{g}_1^{(2)} = 2$ $\mathbf{g}_2^{(2)} = 4$	$\mathbf{g}_1^{(3)} = 4$ $\mathbf{g}_2^{(3)} = 7$	4
2	4	6 1	3 5	7 5	5
3	5	60 34	30 74	70 40	6
3	6	50 24	24 70	54 54	6
4	7	54 00	30 46	64 66	7
4	8	64 26	12 66	52 44	8
5	9	54 25	16 71	66 60	8
5	10	53 36	23 53	51 67	9

TABLE 12.1F CONTINUED

m	K	Generator sequences			d_{free}
6	11	710 320	260 404	670 714	10
8	12	740 367	260 414	520 515	10
8	13	710 140	260 545	670 533	11
7	14	676 256	046 470	704 442	12
8	15	722 302	054 457	642 435	12
9	16	7640 0724	2460 5164	7560 4260	12
9	17	5330 0600	3250 7650	5340 5434	13
9	18	6734 1574	1734 5140	4330 7014	14
10	19	5044 1024	3570 5712	4734 5622	14
10	20	7030 0012	3452 6756	7566 5100	14
11	21	6562 0431	2316 4454	4160 7225	15
12	22	57720 15244	12140 70044	63260 47730	16
12	23	51630 05460	25240 61234	42050 44334	16

to the metric of the other survivors, the algorithm can be terminated. The metric at the all-zero state then equals d_{free}, since none of the other survivors can ever remerge to the all-zero state with a smaller metric. A trellis depth of several constraint lengths is typically required to find d_{free} for a noncatastrophic code. For a catastrophic code, the survivor at the all-zero state may never achieve the minimum metric, because of the zero loop in the state diagram. This can be used as an indicator of when a code is catastrophic.

A sequential decoder is better suited to find the free distance of longer-constraint-length codes and the CDF of a code. Forney [25] describes how the Fano algorithm can be modified to compute the CDF of a code, and Chevillat [26] gives a version of the stack algorithm used for computing the CDF. A flowchart of Chevillat's algorithm is shown in Figure 12.14 for an $R = 1/2$ code. Each stack entry consists of a path, its length (in branches), and its weight. The stack is initially loaded with the path $\mathbf{u} = (1)$ of length 1 and weight 2. The stack is ordered so that the path of greatest length among those of lowest weight is at the top. When a path of weight w reaches

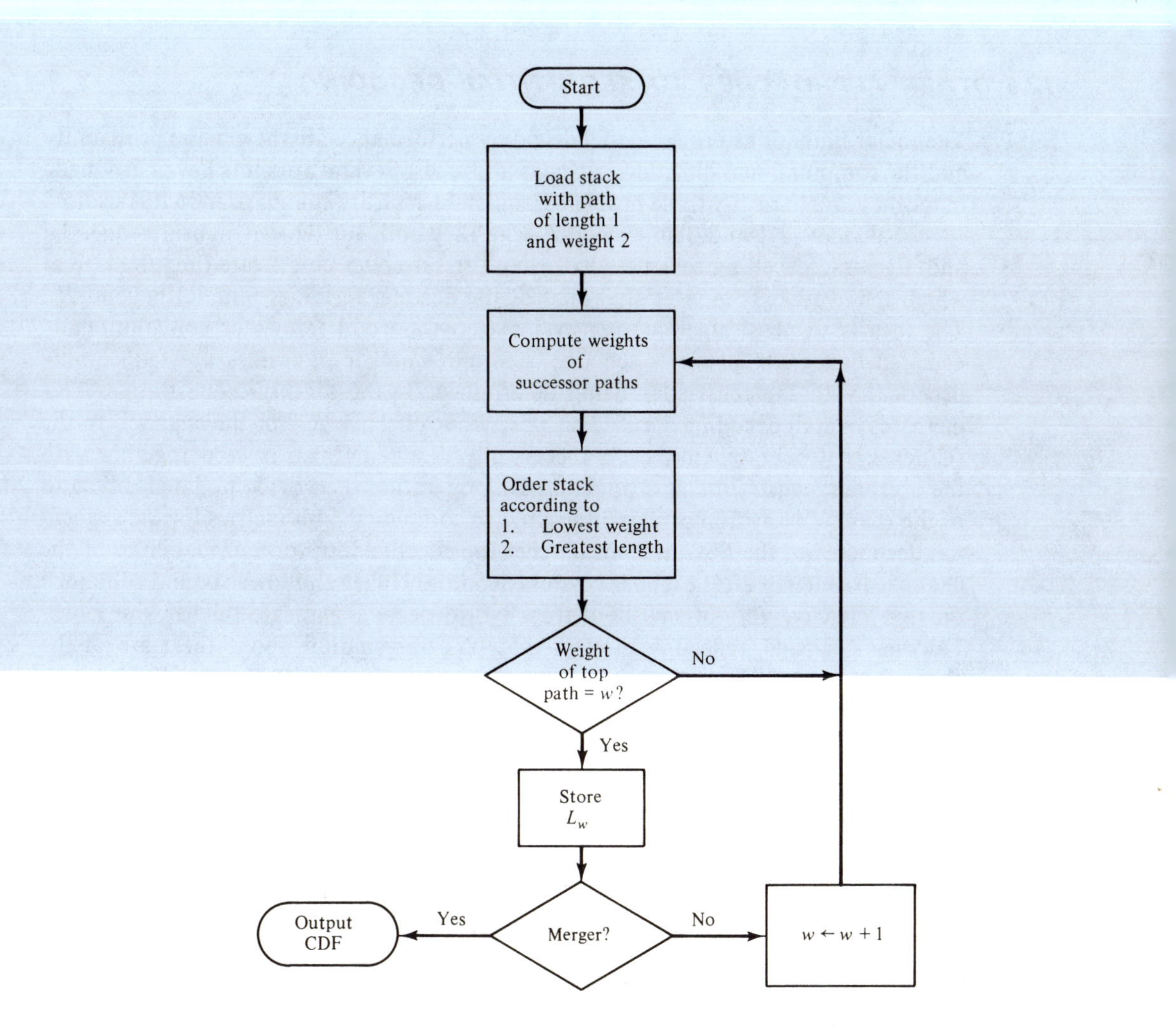

Figure 12.14 Stack algorithm for computing the CDF.

the top of the stack for the first time, its length L_w is the shortest length such that all paths of that length have weight at least w (i.e., $L_w - 1$ is the smallest order for which the code's CDF equals w). Starting with $w = 3$, the algorithm computes the length L_w and continues to increment w and compute the length L_w until the path at the top of the stack is merged with the all-zero state. This merged path's weight equals d_{free}, and the set of lengths L_w for $w = 3, 4, \ldots, d_{\text{free}}$ uniquely determines the code's CDF. For example, for the (2, 1, 16) code with $d_{\text{free}} = 18$ whose CDF is shown in Figure 10.12, the set of lengths put out by the algorithm above is given by $L_w = 2, 5, 7, 9, 11, 13, 17, 22, 29, 31, 35, 40, 42, 48, 50, 54$ for $w = 3, 4, \ldots, 18$, and these lengths uniquely specify the orders $L_w - 1$ at which increases in the code's CDF occur.

These methods are usually feasible to compute the distance properties of codes with encoder memories K on the order of 30 or less. For larger values of K, computationally efficient algorithms for computing the distance properties of codes are not yet known.

12.5 OTHER APPROACHES TO SEQUENTIAL DECODING

The major limitations on sequential decoding performance are the erasure probability and the computational difficulties at rates above R_0. Several attempts have been made in recent years to rectify these problems, some of which are summarized in this section.

Falconer [27] and Jelinek and Cocke [28] both introduced hybrid sequential and algebraic decoding schemes that extend the decoder cutoff rate (i.e., the rate at which the average computational load of the decoder becomes infinite) beyond R_0. The idea is to place algebraic constraints across several frames of convolutionally encoded data. In Falconer's scheme, a certain number of frames are sequentially decoded, with the remainder being determined by the algebraic constraints. This increases overall decoding speed, and raises the "effective R_0" of the channel. In the Jelinek and Cocke scheme, each successfully decoded frame is used together with the algebraic constraints in a bootstrapping operation to provide updated estimates of the channel transition probabilities used to compute the bit metrics for the sequential decoder. For the BSC, this means that the effective transition probabilities of the channel are altered after each successful decoding. Upper and lower bounds obtained on the "effective R_0" of this bootstrap hybrid decoder indicate further gains over Falconer's scheme, reflecting the fact that less information about the state of the channel is wasted. These results are summarized in Figure 12.15 for a BSC, where R_0^F denotes the effective R_0 of Falconer's decoder, and R_0^{BL} and R_0^{BU} denote the lower and upper bounds, respectively, on the effective R_0 of the bootstrap decoder. Note that these schemes offer considerable improvement in practical decoding range at the cost of some increase in the complexity of implementation.

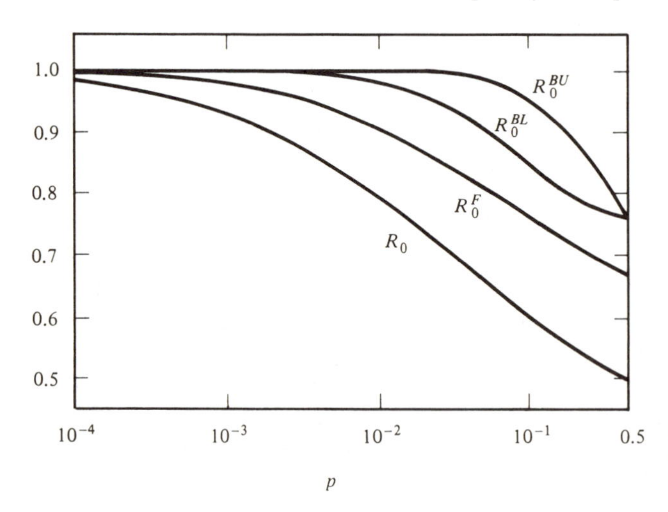

Figure 12.15 Comparison of channel cutoff rates for hybrid decoders on a BSC.

Haccoun and Ferguson [29] have attempted to close the gap between Viterbi and sequential decoding by considering them as special cases of a more general decoding algorithm called the *generalized stack algorithm*. In the generalized stack algorithm, paths in an ordered stack are extended as in the stack algorithm, but more than one path can be extended at the same time. On the other hand, remergers are exploited as in the Viterbi algorithm in order to eliminate redundant paths from the stack.

Analysis and simulation have shown that the variability of the computational distribution is reduced compared to the ordinary stack algorithm, at a cost of a larger average number of computations. The error probability also more closely approaches that of the Viterbi algorithm. The complexity of implementation is somewhat increased, however, over the ordinary stack algorithm.

Chevillat and Costello [30] have introduced a *multiple stack algorithm* which eliminates erasures entirely and whose performance can hence be compared directly to that of the Viterbi algorithm. The multiple stack algorithm begins exactly like the ordinary stack algorithm, except that the stack size is limited to a fixed number of entries Z_1. Starting with the origin node, the top node in the stack is extended. After its elimination from the stack, the successors are inserted and the stack is ordered in the usual way. If decoding is completed before the stack fills up, the multiple stack algorithm outputs exactly the same decoding decision as the ordinary stack algorithm. However, if the received sequence is one of those few that requires extended searches (i.e., a potential erasure), the stack fills up. In this case, the best T paths in the stack are transferred to a smaller second stack with $Z \ll Z_1$ entries. Decoding then proceeds in the second stack using only these T transferred nodes. If the best path in the second stack reaches the end of the tree before the stack fills up, this path is stored as a tentative decision. The decoder then deletes the remaining paths in the second stack and returns to the first stack where decoding continues. If the decoder reaches the end of the tree before the first stack fills up again, the metric of this path is compared to that of the tentative decision. The path with the better metric is retained and becomes the final decoding decision. However, if the first stack fills up again, a new second stack is formed by again transferring the best T paths from the first stack. If the second stack fills up also, a third stack of size Z is formed by transferring the best T paths from the second stack. Additional stacks of the same size are formed in a similar manner until a tentative decision is made. The decoder always compares each new tentative decision with the previous one, and retains the path with the best metric. The rest of the paths in that stack are then deleted and decoding proceeds in the previous stack. The algorithm terminates decoding if it reaches the end of the tree in the first stack. The only other way decoding can be completed is by exceeding a computational limit C_{lim}. In this case the best tentative decision becomes the final decoding decision. A complete flow diagram of the multiple stack algorithm is shown in Figure 12.16.

The general philosophy of the multiple stack algorithm emerges from its use of additional stacks. The first stack is made large enough so that only very noisy code words (i.e., potential erasures) force the use of additional stacks. Rather than follow the strategy of the ordinary stack algorithm, which advances slowly in these noisy cases because it is forced to explore many incorrect subsets before extending the correct path, the mutiple stack algorithm advances quickly through the tree and finds a reasonably good tentative decision. Only then does it explore in detail all the alternatives in search of the maximum likelihood path (i.e., previous stacks are revisited). This change in priorities is achieved by making the additional stacks substantially smaller than the first one ($Z \ll Z_1$). Since the T paths at the top of a full stack are almost always further into the tree than the T paths used to initialize this stack, the creation of each new stack forces the decoder further ahead until the end

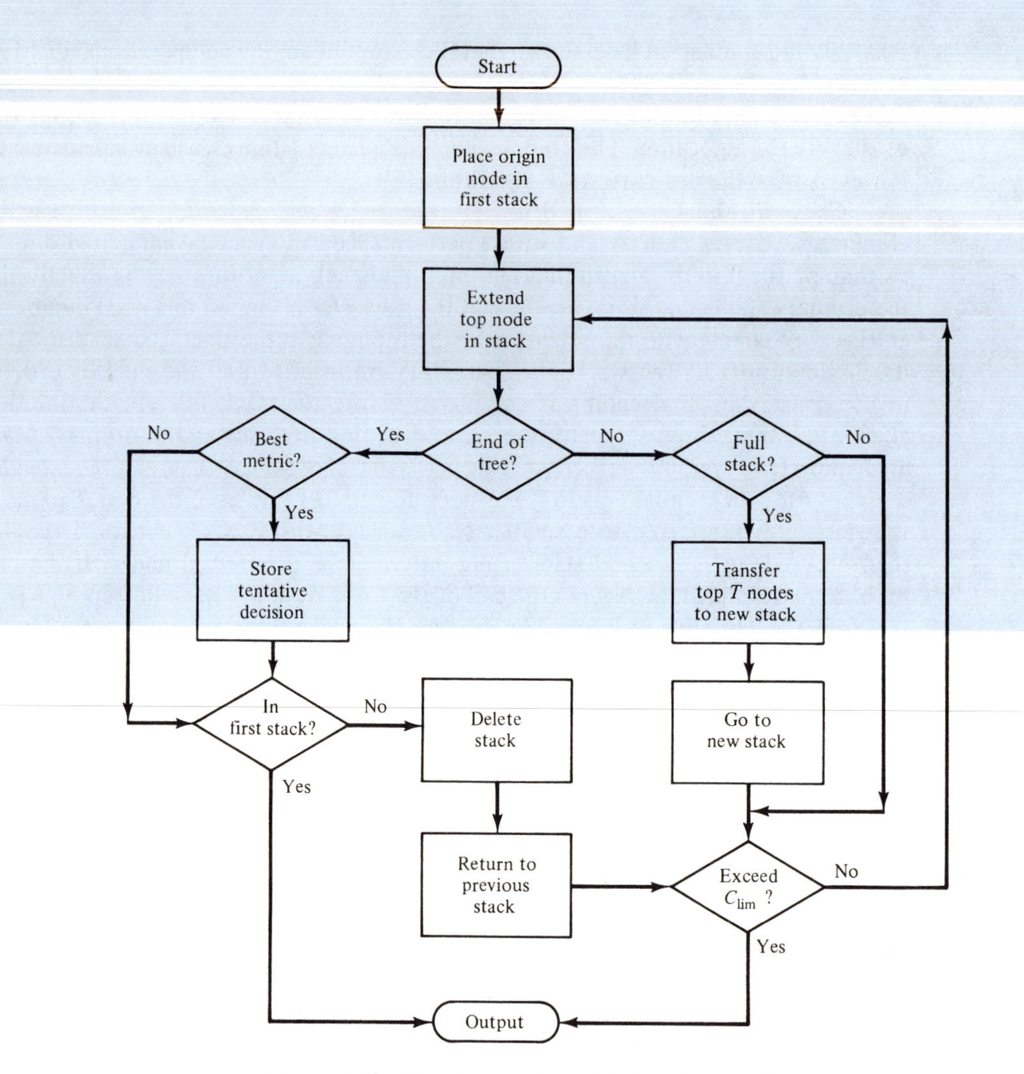

Figure 12.16 Flowchart of the multiple stack algorithm.

of the tree is reached. Hence, if C_{lim} is not too small, at least one tentative decision is always made, and erasures are eliminated. Once a tentative decision is made, the decoder returns to previous stacks trying to improve its final decision.

The performance of the multiple stack algorithm can be compared directly to the Viterbi algorithm since it is erasure-free. As with ordinary sequential decoding, the multiple stack algorithm's computational effort is independent of code constraint length, and hence it can be used with longer codes than the Viterbi algorithm. The performance comparison is thus made by comparing decoding speed and implementation complexity at comparable error probabilities rather than for equal constraint lengths. The performance of the Viterbi algorithm on a BSC with the best (2, 1, 7) code (curve 1) is compared to that of the multiple stack algorithm with four different sets of parameters (curves 2, 3, 4, and 5) in Figure 12.17. These curves were generated

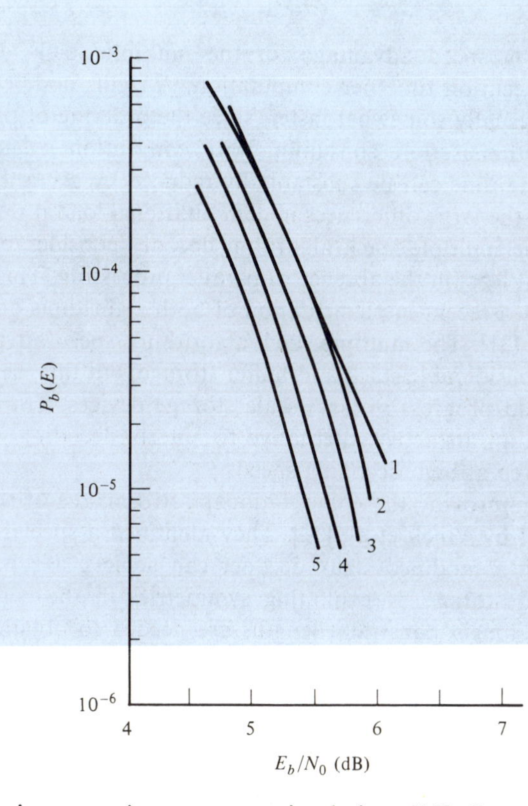

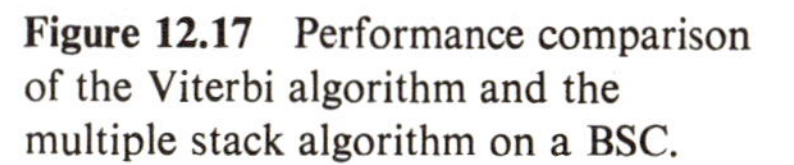
Figure 12.17 Performance comparison of the Viterbi algorithm and the multiple stack algorithm on a BSC.

using extensive computer simulations [30]. Curve 2 is for a (2, 1, 12) code with $Z_1 = 1365$, $Z = 11$, $T = 3$, and $C_{\text{lim}} = 6144$. Note that at $E_b/N_0 = 5.5$ dB both decoders achieve a bit error probability of approximately 7×10^{-5}. However, the Viterbi algorithm requires 128 computations per decoded information bit, almost two orders of magnitude larger than $E[C_j] = 1.37$ for the multiple stack algorithm. Since $L = 60$ bit information sequences were used in the simulations, even the computational limit $C_{\text{lim}} = 6144$ of the multiple stack algorithm is smaller than the constant number of almost $60 \times 128 = 7680$ computations which the Viterbi algorithm executes per received sequence. The multiple stack algorithm's stack requires approximately 1640 entries, compared to 128 for the Viterbi algorithm. Increasing the size of the first stack to $Z_1 = 2048$ (curve 3) lowers the bit error probability to about 3.5×10^{-5} at 5.5 dB without noticeable change in computational effort and a stack storage requirement of about 2400 entries. Using the multiple stack algorithm (curve 4) with a (2, 1, 15) code, $Z_1 = 3413$, $Z = 11$, $T = 3$, and $C_{\text{lim}} = 8192$ achieves $P_b(E) = 1.5 \times 10^{-5}$ at 5.5 dB with $E[C_j] = 1.41$ and a total storage of about 3700. Increasing the first stack size to $Z_1 = 4778$ (curve 5) yields $P_b(E) = 7 \times 10^{-6}$ at 5.5 dB with $E[C_j] = 1.42$ and a total storage of about 5000. In this last example the multiple stack algorithm's bit error probability is 10 times smaller and its average computational load about 90 times smaller than that of the Viterbi decoder for the (2, 1, 7) code. Since the exponentially increasing computational load limits Viterbi decoding to short constraint lengths, it can achieve such low error probabilities only with soft decisions.

Whether the large differences in computational load noted above result in an equally pronounced decoding speed advantage for the multiple stack algorithm depends primarily on the execution time per computation. A trellis node extension for the Viterbi algorithm is usually somewhat faster, since the ordering of the stacks in the multiple stack algorithm is time consuming with conventionally addressed stacks. However, the ordering time can be substantially reduced by using the stack-bucket technique. In view of the large differences in computational load it thus seems justified to conclude that the multiple stack algorithm decodes considerably faster than the Viterbi algorithm, at least in the absence of parallel processing. This conclusion is strongly supported by a recent implementation of both algorithms on a Zilog Z-80 microcomputer system [31]. The multiple stack algorithm's speed advantage is achieved at the expense of considerable stack and buffer storage, but this seems to be tolerable in view of the rapid progress in large-scale storage devices. The multiple stack algorithm is therefore an attractive alternative to Viterbi decoding when low error probabilities are required at high decoding speeds.

Another approach to improving the computational performance of sequential decoding has been proposed by Vinck et al. [32]. They identified a special subclass of $R = 1/2$ codes for which a modified stack decoder can achieve a considerable savings in computation and storage by exploiting symmetries in the code. Since these codes are suboptimal, longer constraint lengths are needed to obtain the free distance required for a certain error probability, but this does not affect the computational behavior of the decoder.

PROBLEMS

12.1. Consider the (2, 1, 3) code with

$$\mathbf{G}(D) = [1 + D^2 + D^3, 1 + D + D^2 + D^3].$$

(a) Draw the code tree for an information sequence of length $L = 4$.

(b) Find the code word corresponding to the information sequence $\mathbf{u} = (1\ 0\ 0\ 1)$.

12.2. For a binary-input, Q-ary-output symmetric DMC with equally likely input symbols, show that $P(r_i = j) = P(r_i = Q - 1 - j) \leqq \frac{1}{2}$ for $0 \leqq j \leqq Q - 1$ and all i.

12.3. Consider the (2, 1, 3) code of Problem 12.1.

(a) For a BSC with $p = 0.045$, find an integer metric table for the Fano metric.

(b) Decode the received sequence

$$\mathbf{r} = [1\ \ 1,\ \ 0\ \ 0,\ \ 1\ \ 1,\ \ 0\ \ 0,\ \ 0\ \ 1,\ \ 1\ \ 0,\ \ 1\ \ 1]$$

using the stack algorithm. Compare the number of decoding steps with the number required by the Viterbi algorithm.

(c) Repeat part (b) for the received sequence

$$\mathbf{r} = [1\ \ 1,\ \ 1\ \ 0,\ \ 0\ \ 0,\ \ 0\ \ 1,\ \ 1\ \ 0,\ \ 0\ \ 1,\ \ 0\ \ 0].$$

Compare the final decoded path with the result of Problem 11.6, where the same received sequence is decoded using the Viterbi algorithm.

12.4. Consider the (2, 1, 3) code of Problem 12.1.

(a) For the binary-input, 8-ary-output DMC of Problem 11.4, find an integer metric table for the Fano metric. (*Hint:* Scale each metric by an appropriate factor and round to the nearest integer.)

(b) Decode the received sequence

$$\mathbf{r} = [1_2 1_1, 1_2 0_1, 0_3 0_1, 0_1 1_3, 1_2 0_2, 0_3 1_1, 0_3 0_2]$$

using the stack algorithm. Compare the final decoded path with the result of Problem 11.5(b), where the same received sequence is decoded using the Viterbi algorithm.

12.5. Repeat parts (b) and (c) of Problem 12.3 with the size of the stack limited to 5 entries. When the stack is full, each additional entry causes the path on the bottom of the stack to be discarded. What is the effect on the final decoded path?

12.6. **(a)** Repeat Example 12.4 using the stack-bucket algorithm with a bucket quantization interval of 5. Assume that the bucket intervals are $\ldots, +4$ to $0, -1$ to -5, -6 to $-10, \ldots$.

(b) Repeat part (a) for a quantization interval of 9, where the bucket intervals are $\ldots, +8$ to $0, -1$ to $-9, -10$ to $-18, \ldots$.

12.7. Repeat Example 12.6 for the Fano algorithm with threshold increments of $\Delta = 5$ and $\Delta = 9$. Compare the final decoded path and the number of computations to the results of Examples 12.6 and 12.7. Also compare the final decoded path to the results of the sfack-bucket algorithm in Problem 12.6.

12.8. Use a programmable calculator to verify the results of Figure 12.10, and to plot ρ as a function of E_b/N_0 (dB) for $R = 1/5$ and $R = 4/5$.

12.9. Show that $\lim_{R\to 0} \rho = \infty$ and $\lim_{R\to C} \rho = 0$ for fixed channel transition probabilities. Also show that $\partial R/\partial \rho < 0$.

12.10. **(a)** Calculate R_0 for the binary-input, 8-ary-output DMC of Problem 11.4.

(b) Repeat Example 12.8 for the DMC of part (a) and a code rate of $R = 1/2$. [*Hint:* Use a programmable calculator to find ρ from (12.8) and (12.9).]

(c) For the value of ρ calculated in part (b), find the buffer size B needed to guarantee an erasure probability of 10^{-3} using the values of L, A, and μ given in Example 12.8.

12.11. **(a)** Sketch P_{erasure} versus E_b/N_0 for an $R = 1/2$ code on a BSC using the values of L, A, μ, and B given in Example 12.8.

(b) Sketch the required buffer size B to guarantee an erasure probability of 10^{-3} as a function of E_b/N_0 for an $R = 1/2$ code on a BSC using the values of L, A, and μ given in Example 12.8.

(*Hint:* ρ can be found as a function of E_b/N_0 using the results of Problem 12.8.)

12.12. Repeat Example 12.4 using the integer metric tables of Figure 12.11. Note any changes in the final decoded path or the number of decoding steps.

12.13. For a BSC with transition probability p, plot both the computational cutoff rate R_0 from (12.12) and $R_{\max}$ from (12.20) as functions of p.

12.14. Find the complete CDFs of the (2, 1, 3) complementary code in Table 11.2 and the (2, 1, 3) quick-look-in code in Table 11.3 using the algorithm shown in Figure 12.14. Which code has the superior distance profile?

12.15. Use the Viterbi algorithm as described in Section 12.4 to find d_{free} for the two codes of Problem 12.14.

REFERENCES

1. J. M. Wozencraft and B. Reiffen, *Sequential Decoding*, MIT Press, Cambridge, Mass., 1961.
2. R. M. Fano, "A Heuristic Discussion of Probabilistic Decoding," *IEEE Trans. Inf. Theory*, IT-9, pp. 64–74, April 1963.
3. K. Zigangirov, "Some Sequential Decoding Procedures," *Probl. Peredachi Inf.*, 2, pp. 13–25, 1966.
4. F. Jelinek, "A Fast Sequential Decoding Algorithm Using a Stack," *IBM J. Res. and Dev.*, 13, pp. 675–685, November 1969.
5. J. L. Massey, "Variable-Length Codes and the Fano Metric," *IEEE Trans. Inf. Theory*, IT-18, pp. 196–198, January 1972.
6. J. L. Massey and D. J. Costello, Jr., "Nonsystematic Convolutional Codes for Sequential Decoding in Space Applications," *IEEE Trans. Commun. Technol.*, COM-19, pp. 806–813, October 1971.
7. J. M. Geist, "Search Properties of Some Sequential Decoding Algorithms," *IEEE Trans. Inf. Theory*, IT-19, pp. 519–526, July 1973.
8. J. M. Geist, "An Empirical Comparison of Two Sequential Decoding Algorithms," *IEEE Trans. Commun. Technol.*, COM-19, pp. 415–419, August 1971.
9. J. E. Savage, "Sequential Decoding—the Computation Problem," *Bell Syst. Tech. J.*, 45, pp. 149–175, January 1966.
10. I. M. Jacobs and E. R. Berlekamp, "A Lower Bound to the Distribution of Computation for Sequential Decoding," *IEEE Trans. Inf. Theory*, IT-13, pp. 167–174, April 1967.
11. F. Jelinek, "An Upper Bound on Moments of Sequential Decoding Effort," *IEEE Trans. Inf. Theory*, IT-15, pp. 140–149, January 1969.
12. G. D. Forney, Jr., "Convolutional Codes III: Sequential Decoding," *Inf. Control*, 25, pp. 267–297, July 1974.
13. A. J. Viterbi and J. K. Omura, *Principles of Digital Communication and Coding*, McGraw-Hill, New York, 1979.
14. G. D. Forney, Jr., and E. K. Bower, "A High-Speed Sequential Decoder: Prototype Design and Test," *IEEE Trans. Commun. Tech.*, COM-19, pp. 821–835, October 1971.
15. K. S. Gilhousen, J. A. Heller, I. M. Jacobs, and A. J. Viterbi, "Coding Study for High Data Rate Telemetry Links," Linkabit Corp. NASA CR-114278 Contract NAS 2-6024, 1971.
16. I. Richer, "Sequential Decoding with a Small Digital Computer," MIT Lincoln Laboratory Tech. Report No. 491, January 1972.
17. H. L. Yudkin, "Channel State Testing in Information Decoding," Sc.D. thesis, M.I.T., Cambridge, Mass., 1964.
18. P. R. Chevillat and D. J. Costello, Jr., "An Analysis of Sequential Decoding for Specific Time-Invariant Convolutional Codes," *IEEE Trans. Inf. Theory*, IT-24, pp. 443–451, July 1978.
19. P. R. Chevillat and D. J. Costello, Jr., "Distance and Computation in Sequential Decoding," *IEEE Trans. Commun.*, COM-24, pp. 440–447, April 1976.
20. J. A. Heller and I. M. Jacobs, "Viterbi Decoding for Satellite and Space Communication," *IEEE Trans. Commun. Technol.*, COM-19, pp. 835–848, October 1971.

21. R. Johannesson, "Robustly-Optimal Rate One-Half Binary Convolutional Codes," *IEEE Trans. Inf. Theory*, IT-21, pp. 464–468, July 1975.

22. R. Johannesson, "Some Rate 1/3 and 1/4 Binary Convolutional Codes with an Optimum Distance Profile," *IEEE Trans. Inf. Theory*, IT-23, pp. 281–283, March 1977.

23. R. Johannesson and E. Paaske, "Further Results on Binary Convolutional Codes with an Optimum Distance Profile," *IEEE Trans. Inf. Theory*, IT-24, pp. 264–268, March 1978.

24. R. Johannesson, "Some Long Rate One-Half Binary Convolutional Codes with an Optimum Distance Profile," *IEEE Trans. Inf. Theory*, IT-22, pp. 629–631, September 1976.

25. G. D. Forney, "Use of a Sequential Decoder to Analyze Convolutional Code Structure," *IEEE Trans. Inf. Theory*, IT-16, pp. 793–795, November 1970.

26. P. R. Chevillat, "Fast Sequential Decoding and a New Complete Decoding Algorithm," Ph.D. thesis, I.I.T., Chicago, 1976.

27. D. D. Falconer, "A Hybrid Sequential and Algebraic Decoding Scheme," Sc.D. thesis, M.I.T., Cambridge, Mass., 1967.

28. F. Jelinek and J. Cocke, "Bootsrap Hybrid Decoding for Symmetrical Binary Input Channels," *Inf. Control*, 18, pp. 261–298, April 1971.

29. D. Haccoun and M. J. Ferguson, "Generalized Stack Algorithms for Decoding Convolutional Codes," *IEEE Trans. Inf. Theory*, IT-21, pp. 638–651, November 1975.

30. P. R. Chevillat and D. J. Costello, Jr., "A Multiple Stack Algorithm for Erasurefree Decoding of Convolutional Codes," *IEEE Trans. Commun.*, COM-25, pp. 1460–1470, December 1977.

31. H. H. Ma, "The Multiple Stack Algorithm Implemented on a Zilog Z-80 Microcomputer," *IEEE Trans. Commun.*, COM-28, pp. 1876–1887, November 1980.

32. A. J. Vinck, A. J. P. de Paepe, and J. P. M. Schalkwijk, "A Class of Binary Rate 1/2 Convolutional Codes That Allows an Improved Stack Decoder," *IEEE Trans. Inf. Theory*, IT-26, pp. 389–392, July 1980.

13

Majority-Logic Decoding of Convolutional Codes

Viterbi decoding is an optimum decoding method for convolutional codes. Its performance depends on the quality of the channel, and the decoding effort is fixed and grows exponentially with code constraint length. Hence, it is only useful for short-constraint-length codes. The performance of sequential decoding is slightly suboptimum, but its decoding effort is independent of code constraint length. This allows sequential decoding to be used with long constraint lengths. The number of computations needed to decode a frame of data is a random variable, however. Although most frames are decoded very quickly, some undergo long searches, causing a few to be incompletely decoded or erased. This variable nature of the decoding effort has led sequential decoding to be referred to as a probabilistic decoding method.

An algebraic approach can also be taken to the decoding of convolutional codes. Majority-logic or threshold decoding, first introduced in Chapter 7 for block codes, was shown by Massey [1] in 1963 also to be applicable to convolutional codes. It differs from Viterbi decoding and sequential decoding in that the final decision made on a given information block is based on only one constraint length of received blocks rather than on the entire received sequence. This results in inferior performance when compared to Viterbi or sequential decoding, but the implementation of the decoder is much simpler. This has led to the use of majority-logic decoding in applications such as telephony and HF radio, where a moderate amount of coding gain is desired at a relatively low cost.

13.1 FEEDBACK DECODING

We begin our discussion of majority-logic decoding by considering an $R = 1/2$ systematic code with generator sequences $\mathbf{g}^{(1)} = (1\ 0\ 0\ \cdots)$ and $\mathbf{g}^{(2)} = (g_0^{(2)}\ g_1^{(2)}\ g_2^{(2)}\ \cdots)$ and generator matrix

$$\mathbf{G} = \begin{bmatrix} 1 & g_0^{(2)} & 0 & g_1^{(2)} & 0 & g_2^{(2)} & \cdots & 0 & g_m^{(2)} & & & & \\ & & 1 & g_0^{(2)} & 0 & g_1^{(2)} & \cdots & 0 & g_{m-1}^{(2)} & 0 & g_m^{(2)} & & \\ & & & & 1 & g_0^{(2)} & \cdots & 0 & g_{m-2}^{(2)} & 0 & g_{m-1}^{(2)} & 0 & g_m^{(2)} \\ & & & & & & \ddots & & & & & & \ddots \end{bmatrix}. \tag{13.1}$$

For an information sequence $\mathbf{u}$, the encoding equations are given by

$$\begin{aligned} \mathbf{v}^{(1)} &= \mathbf{u} * \mathbf{g}^{(1)} = \mathbf{u} \\ \mathbf{v}^{(2)} &= \mathbf{u} * \mathbf{g}^{(2)}, \end{aligned} \tag{13.2}$$

and the transmitted code word is $\mathbf{v} = \mathbf{uG}$. If $\mathbf{v}$ is sent over a BSC, the binary received sequence $\mathbf{r}$ can be written as

$$\mathbf{r} = (r_0^{(1)} r_0^{(2)}, r_1^{(1)} r_1^{(2)}, r_2^{(1)} r_2^{(2)}, \ldots,) = \mathbf{v} + \mathbf{e}, \tag{13.3}$$

where the binary sequence $\mathbf{e} = (e_0^{(1)} e_0^{(2)}, e_1^{(1)} e_1^{(2)}, e_2^{(1)} e_2^{(2)}, \ldots)$ is called the *channel error sequence*. An error bit $e_i^{(j)} = 1$ if and only if $r_i^{(j)} \neq v_i^{(j)}$ (i.e., an error occurred during transmission). This model for the BSC is illustrated in Figure 13.1. The received sequence $\mathbf{r}$ can be divided into a *received information sequence*

$$\mathbf{r}^{(1)} = (r_0^{(1)}, r_1^{(1)}, r_2^{(1)}, \ldots) = \mathbf{v}^{(1)} + \mathbf{e}^{(1)} = \mathbf{u} + \mathbf{e}^{(1)} \tag{13.4a}$$

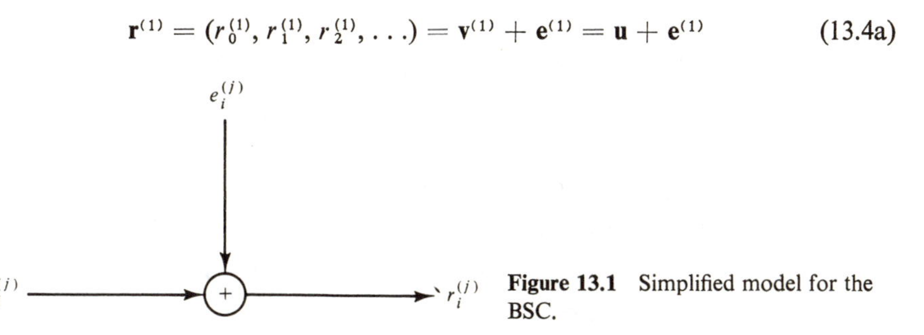

Figure 13.1 Simplified model for the BSC.

and a *received parity sequence*

$$\mathbf{r}^{(2)} = (r_0^{(2)}, r_1^{(2)}, r_2^{(2)}, \ldots) = \mathbf{v}^{(2)} + \mathbf{e}^{(2)} = \mathbf{u} * \mathbf{g}^{(2)} + \mathbf{e}^{(2)}, \tag{13.4b}$$

where $\mathbf{e}^{(1)} = (e_0^{(1)}, e_1^{(1)}, e_2^{(1)}, \ldots)$ is the *information error sequence*, and $\mathbf{e}^{(2)} = (e_0^{(2)}, e_1^{(2)}, e_2^{(2)}, \ldots)$ is the *parity error sequence.*

The *syndrome sequence* $\mathbf{s} = (s_0, s_1, s_2, \ldots)$ is defined as

$$\mathbf{s} \triangleq \mathbf{rH}^T, \tag{13.5}$$

where the *parity-check matrix* $\mathbf{H}$ is given by

$$\mathbf{H} = \begin{bmatrix} g_0^{(2)} & 1 & & & & & & & & & & & & \\ g_1^{(2)} & 0 & g_0^{(2)} & 1 & & & & & & & & & & \\ g_2^{(2)} & 0 & g_1^{(2)} & 0 & g_0^{(2)} & 1 & & & & & & & & \\ \cdot & \cdot & \cdot & \cdot & \cdot & \cdot & \cdot & & & & & & & \\ \cdot & \cdot & \cdot & \cdot & \cdot & \cdot & & \cdot & & & & & & \\ \cdot & \cdot & \cdot & \cdot & \cdot & \cdot & & & \cdot & & & & & \\ g_m^{(2)} & 0 & g_{m-1}^{(2)} & 0 & g_{m-2}^{(2)} & 0 & \cdots & & g_0^{(2)} & 1 & & & & \\ & & g_m^{(2)} & 0 & g_{m-1}^{(2)} & 0 & \cdots & & g_1^{(2)} & 0 & g_0^{(2)} & 1 & & \\ & & & & g_m^{(2)} & 0 & \cdots & & g_2^{(2)} & 0 & g_1^{(2)} & 0 & g_0^{(2)} & 1 \\ & & & & & \cdot & & & & & & & \cdot & \\ & & & & & & \cdot & & & & & & & \cdot \\ & & & & & & & \cdot & & & & & & & \cdot \end{bmatrix}. \tag{13.6}$$

As is the case with block codes, $\mathbf{GH}^T = \mathbf{0}$, and $\mathbf{v}$ is a code word if and only if $\mathbf{vH}^T = \mathbf{0}$. However, unlike block codes, $\mathbf{G}$ and $\mathbf{H}$ are semi-infinite matrices, reflecting the fact that the information sequences and code words are of arbitrary length.

Since $\mathbf{r} = \mathbf{v} + \mathbf{e}$, the syndrome sequence $\mathbf{s}$ can be written as

$$\mathbf{s} = (\mathbf{v} + \mathbf{e})\mathbf{H}^T = \mathbf{vH}^T + \mathbf{eH}^T = \mathbf{eH}^T, \tag{13.7}$$

and we see that $\mathbf{s}$ depends only on the channel error sequence and not on the particular code word transmitted. For decoding purposes, knowing $\mathbf{s}$ is equivalent to knowing $\mathbf{r}$, and hence the decoder can be designed to operate on $\mathbf{s}$ rather than on $\mathbf{r}$. Such a decoder is called a *syndrome decoder*.

Using the polynomial notation introduced in Chapter 10, we can write

$$\mathbf{v}^{(1)}(D) = \mathbf{u}(D)\mathbf{g}^{(1)}(D) = \mathbf{u}(D) \tag{13.8a}$$

$$\mathbf{v}^{(2)}(D) = \mathbf{u}(D)\mathbf{g}^{(2)}(D) \tag{13.8b}$$

and

$$\mathbf{r}^{(1)}(D) = \mathbf{v}^{(1)}(D) + \mathbf{e}^{(1)}(D) = \mathbf{u}(D) + \mathbf{e}^{(1)}(D) \tag{13.9a}$$

$$\mathbf{r}^{(2)}(D) = \mathbf{v}^{(2)}(D) + \mathbf{e}^{(2)}(D) = \mathbf{u}(D)\mathbf{g}^{(2)}(D) + \mathbf{e}^{(2)}(D). \tag{13.9b}$$

At the receiver, the syndrome sequence is formed as

$$\mathbf{s}(D) = \mathbf{r}^{(1)}(D)\mathbf{g}^{(2)}(D) + \mathbf{r}^{(2)}(D). \tag{13.10}$$

Since $\mathbf{v}^{(2)}(D) = \mathbf{u}(D)\mathbf{g}^{(2)}(D) = \mathbf{v}^{(1)}(D)\mathbf{g}^{(2)}(D)$, forming the syndrome is equivalent to "encoding" $\mathbf{r}^{(1)}(D)$ and then adding it to $\mathbf{r}^{(2)}(D)$. A block diagram of a syndrome former for an $R = 1/2$ systematic code is shown in Figure 13.2. Using (13.9) in (13.10) yields

$$\begin{aligned} \mathbf{s}(D) &= [\mathbf{u}(D) + \mathbf{e}^{(1)}(D)]\mathbf{g}^{(2)}(D) + \mathbf{u}(D)\mathbf{g}^{(2)}(D) + \mathbf{e}^{(2)}(D) \\ &= \mathbf{e}^{(1)}(D)\mathbf{g}^{(2)}(D) + \mathbf{e}^{(2)}(D), \end{aligned} \tag{13.11}$$

and we again see that $\mathbf{s}(D)$ depends only on the channel error sequence and not on the code word transmitted.

Majority-logic decoding of convolutional codes is based on the concept of orthogonal parity-check sums. From (13.7) and (13.11), we see that any syndrome

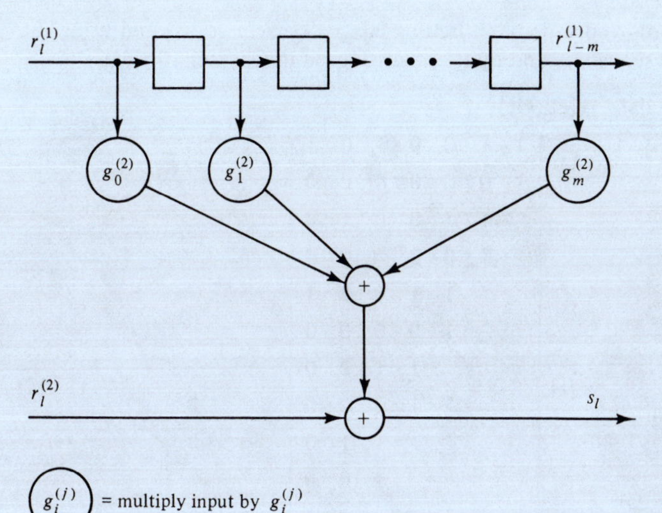

Figure 13.2 Syndrome-forming circuit for an $R = 1/2$ systematic code.

bit, or any sum of syndrome bits, represents a known sum of channel error bits, and is called a *parity-check sum*. If the received sequence is a code word, the channel error sequence is also a code word, and all syndrome bits, and hence all check sums, must be zero. However, if the received sequence is not a code word, some check sums will not be zero. An error bit e_j is said to be *checked* by a check sum if e_j is included in the sum. A set of J check sums is *orthogonal* on e_j if each check sum checks e_j, but no other error bit is checked by more than one check sum. Given a set of J orthogonal check sums on an error bit e_j, the *majority-logic decoding rule* can be used to estimate the value of e_j.

Majority-Logic Decoding Rule: Define $t_{ML} \triangleq \lfloor J/2 \rfloor$. Choose the estimate $\hat{e}_j =$ 1 if and only if more than t_{ML} of the J check sums orthogonal on e_j have value 1.

Theorem 13.1. If the error bits checked by the J orthogonal check sums contain t_{ML} or fewer channel errors (1's), the majority-logic decoding rule correctly estimates e_j.

Proof. If $e_j = 0$, the at most t_{ML} errors can cause at most t_{ML} of the J check sums to have value 1. Hence, $\hat{e}_j = 0$, which is correct. On the other hand, if $e_j = 1$, the at most $t_{ML} - 1$ other errors can cause no more than $t_{ML} - 1$ of the J check sums to have value 0, so that at least $t_{ML} + 1$ will have value 1. Hence, $\hat{e}_j = 1$, which is again correct. Q.E.D.

As a consequence of Theorem 13.1, t_{ML} is called the *majority-logic error correcting capability* of the code.

Example 13.1

Consider finding a set of orthogonal check sums on $e_0^{(1)}$, the first information error bit, for the $R = 1/2$ systematic code with $\mathbf{g}^{(2)}(D) = 1 + D + D^4 + D^6$. First note from (13.7) and (13.11) that $e_0^{(1)}$ can affect only syndrome bits s_0 through s_6 (i.e., the first constraint length of syndrome bits). Letting $[\mathbf{s}]_6 = (s_0, s_1, \ldots, s_6)$ and using the notation for truncated sequences and matrices introduced in Chapter 10 yields

$$
[\mathbf{s}]_6 = [\mathbf{e}]_6[\mathbf{H}^T]_6
$$

$$
= [\mathbf{e}]_6 \begin{bmatrix}
1 & 1 & 0 & 0 & 1 & 0 & 1 \\
1 & 0 & 0 & 0 & 0 & 0 & 0 \\
 & 1 & 1 & 0 & 0 & 1 & 0 \\
 & 1 & 0 & 0 & 0 & 0 & 0 \\
 & & 1 & 1 & 0 & 0 & 1 \\
 & & 1 & 0 & 0 & 0 & 0 \\
 & & & 1 & 1 & 0 & 0 \\
 & & & 1 & 0 & 0 & 0 \\
 & & & & 1 & 1 & 0 \\
 & & & & 1 & 0 & 0 \\
 & & & & & 1 & 1 \\
 & & & & & 1 & 0 \\
 & & & & & & 1 \\
 & & & & & & 1
\end{bmatrix}. \tag{13.12}
$$

Taking the transpose of both sides of (13.12) results in

$$
[\mathbf{s}]_6^T = \begin{bmatrix}
1 & 1 & & & & & & & & & & & & \\
1 & 0 & 1 & 1 & & & & & & & & & & \\
0 & 0 & 1 & 0 & 1 & 1 & & & & & & & & \\
0 & 0 & 0 & 0 & 1 & 0 & 1 & 1 & & & & & & \\
1 & 0 & 0 & 0 & 0 & 0 & 1 & 0 & 1 & 1 & & & & \\
0 & 0 & 1 & 0 & 0 & 0 & 0 & 0 & 1 & 0 & 1 & 1 & & \\
1 & 0 & 0 & 0 & 1 & 0 & 0 & 0 & 0 & 0 & 1 & 0 & 1 & 1
\end{bmatrix} [\mathbf{e}]_6^T. \tag{13.13}
$$

Note that the even-numbered columns of $\mathbf{H}^T$ (i.e., those columns corresponding to the parity error sequence) form an identity matrix. Hence, (13.13) can be rewritten as

$$
[\mathbf{s}]_6^T = \begin{bmatrix} s_0 \\ s_1 \\ s_2 \\ s_3 \\ s_4 \\ s_5 \\ s_6 \end{bmatrix} = \begin{bmatrix}
1 & & & & & & \\
1 & 1 & & & & & \\
0 & 1 & 1 & & & & \\
0 & 0 & 1 & 1 & & & \\
1 & 0 & 0 & 1 & 1 & & \\
0 & 1 & 0 & 0 & 1 & 1 & \\
1 & 0 & 1 & 0 & 0 & 1 & 1
\end{bmatrix} \begin{bmatrix} e_0^{(1)} \\ e_1^{(1)} \\ e_2^{(1)} \\ e_3^{(1)} \\ e_4^{(1)} \\ e_5^{(1)} \\ e_6^{(1)} \end{bmatrix} + \begin{bmatrix} e_0^{(2)} \\ e_1^{(2)} \\ e_2^{(2)} \\ e_3^{(2)} \\ e_4^{(2)} \\ e_5^{(2)} \\ e_6^{(2)} \end{bmatrix}. \tag{13.14}
$$

The matrix in (13.14) which multiplies the information error sequence is called the *parity triangle* of the code. Note that the first column of the parity triangle is the generator sequence $\mathbf{g}^{(2)}$, which is shifted down by one and truncated in each succeeding column.

The parity triangle of the code can now be used to select a set of orthogonal check sums on $e_0^{(1)}$. First note that no syndrome bit can be used in more than one orthogonal check sum, since then a parity error bit would be checked more than once. Since there are only four syndrome bits that check $e_0^{(1)}$, it is not possible to obtain more than four orthogonal check sums on $e_0^{(1)}$ in this example. The orthogonal check sums selected can be illustrated using the parity triangle as follows:

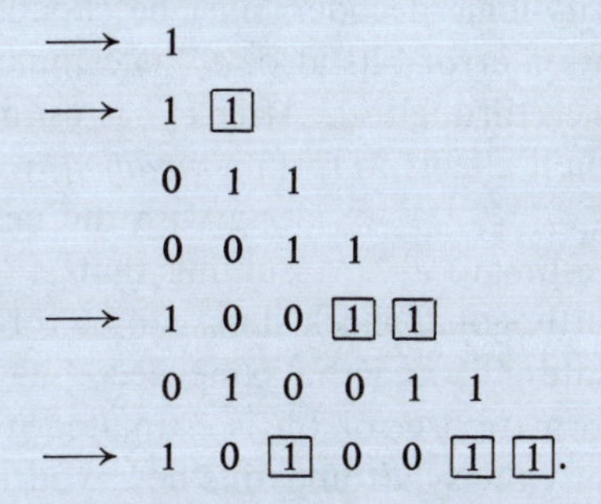

The arrows indicate the syndrome bits, or sums of syndrome bits, which are selected as orthogonal check sums on $e_0^{(1)}$, and the boxes indicate which information error bits, other than $e_0^{(1)}$, are checked. Clearly, at most one arrow can point to each row and at most one box can appear in each column. The equations for the orthogonal check sums are given by

$$\begin{aligned} s_0 &= e_0^{(1)} && && + e_0^{(2)} \\ s_1 &= e_0^{(1)} + e_1^{(1)} && && + e_1^{(2)} \\ s_4 &= e_0^{(1)} && + e_3^{(1)} + e_4^{(1)} && + e_4^{(2)} \\ s_6 &= e_0^{(1)} && + e_2^{(1)} \qquad + e_5^{(1)} + e_6^{(1)} && + e_6^{(2)}. \end{aligned} \tag{13.15}$$

Note that $e_0^{(1)}$ appears in each check sum, but no other error bit appears more than once. Since each check sum is a single syndrome bit, and not a sum of syndrome bits, this is called a *self-orthogonal code*. These codes are discussed in detail in Section 13.4. Since a total of 11 different channel error bits are checked by the four orthogonal check sums of (13.15), the majority-logic decoding rule will correctly estimate $e_0^{(1)}$ whenever $t_{ML} = 2$ or fewer of these 11 error bits are 1's (channel errors). The total number of channel error bits checked by the orthogonal check sum equations is called the *effective constraint length* n_E of the code. Hence, $n_E = 11$ for this code.

Example 13.2

Consider the $R = 1/2$ systematic code with $\mathbf{g}^{(2)}(D) = 1 + D^3 + D^4 + D^5$. A set of four orthogonal check sums on $e_0^{(1)}$ can be constructed from the parity triangle as follows:

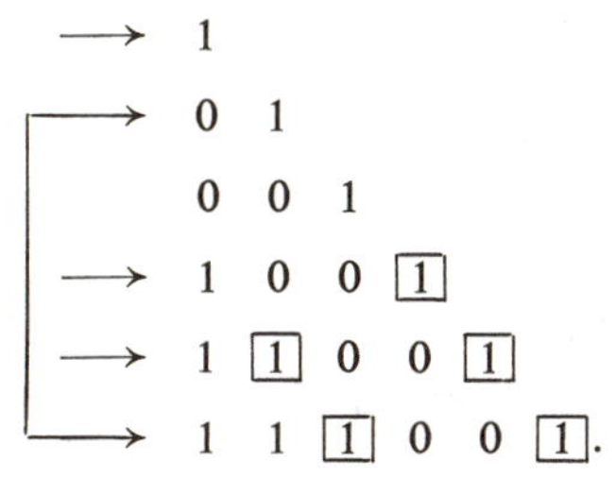

The syndrome bits s_1 and s_5 must be added to form the last check sum in order to eliminate the effect of $e_1^{(1)}$, which is already checked by s_4. The check sums s_0, s_3, s_4, and $s_5 + s_1$ form an orthogonal set on $e_0^{(1)}$. (Methods for constructing sets of orthogonal check sums are discussed in Section 13.4.) The effective constraint length $n_E = 11$ for this code, and the majority-logic decoding rule correctly estimates $e_0^{(1)}$ whenever $t_{ML} = 2$ or fewer of these 11 error bits are 1's.

A majority-logic decoder must be capable of estimating not only $e_0^{(1)}$, but all other information error bits also. $e_0^{(1)}$ is estimated from the first constraint length of syndrome bits s_0 through s_m. After $e_0^{(1)}$ is estimated, it is subtracted from each syndrome equation it affects to form a *modified syndrome* set $s'_0, s'_1, \ldots, s'_m$. The modified syndrome bits $s'_1, s'_2, \ldots, s'_m$ along with the newly calculated syndrome bit s_{m+1} are then used to estimate $e_1^{(1)}$. Assuming that $e_0^{(1)}$ was correctly estimated [i.e., $\hat{e}_0^{(1)} = e_0^{(1)}$], a set of orthogonal check sums can be formed on $e_1^{(1)}$ which is identical to those used to estimate $e_0^{(1)}$, and the same decoding rule can therefore be applied. Each successive information error bit is estimated in the same way. The present estimate is used to modify the syndrome, one new syndrome bit is calculated, the set of orthogonal check sums is formed, and the same decoding rule is applied.

Example 13.3

Assume that $e_0^{(1)}$ has been correctly estimated [i.e., $\hat{e}_0^{(1)} = e_0^{(1)}$]. If it is subtracted from each syndrome equation it affects, the modified syndrome equations for the self-orthogonal code of Example 13.1 become

$$\begin{aligned}
s'_0 &= s_0 - e_0^{(1)} = && && && && && && && && e_0^{(2)} \\
s'_1 &= s_1 - e_0^{(1)} = e_1^{(1)} && && && && && && && &&+ e_1^{(2)} \\
s'_2 &= s_2 \qquad = e_1^{(1)} + e_2^{(1)} && && && && && && && &&+ e_2^{(2)} \\
s'_3 &= s_3 \qquad = \quad e_2^{(1)} + e_3^{(1)} && && && && && && && &&+ e_3^{(2)} \\
s'_4 &= s_4 - e_0^{(1)} = \quad e_3^{(1)} + e_4^{(1)} && && && && && && && &&+ e_4^{(2)} \\
s'_5 &= s_5 \qquad = e_1^{(1)} + e_4^{(1)} + e_5^{(1)} && && && && && && && &&+ e_5^{(2)} \\
s'_6 &= s_6 - e_0^{(1)} = \quad e_2^{(1)} + e_5^{(1)} + e_6^{(1)} && && && && && && && &&+ e_6^{(2)}.
\end{aligned} \tag{13.16}$$

Since s'_0 no longer checks any information error bits, it is of no use in estimating $e_1^{(1)}$. The syndrome bit s_7, however, checks $e_1^{(1)}$. Hence, the modified syndrome bits s'_1 through s'_6 together with s_7 can be used to estimate $e_1^{(1)}$. The equations are

$$\begin{bmatrix} s'_1 \\ s'_2 \\ s'_3 \\ s'_4 \\ s'_5 \\ s'_6 \\ s_7 \end{bmatrix} = \begin{bmatrix} 1 & & & & & & \\ 1 & 1 & & & & & \\ 0 & 1 & 1 & & & & \\ 0 & 0 & 1 & 1 & & & \\ 1 & 0 & 0 & 1 & 1 & & \\ 0 & 1 & 0 & 0 & 1 & 1 & \\ 1 & 0 & 1 & 0 & 0 & 1 & 1 \end{bmatrix} \begin{bmatrix} e_1^{(1)} \\ e_2^{(1)} \\ e_3^{(1)} \\ e_4^{(1)} \\ e_5^{(1)} \\ e_6^{(1)} \\ e_7^{(1)} \end{bmatrix} + \begin{bmatrix} e_1^{(2)} \\ e_2^{(2)} \\ e_3^{(2)} \\ e_4^{(2)} \\ e_5^{(2)} \\ e_6^{(2)} \\ e_7^{(2)} \end{bmatrix}. \tag{13.17}$$

Since the parity triangle is unchanged from (13.14), the syndrome bits s'_1, s'_2, s'_5, and s_7 form a set of four orthogonal check sums on $e_1^{(1)}$ as follows:

$$\begin{aligned}
s'_1 &= e_1^{(1)} && && + e_1^{(2)} \\
s'_2 &= e_1^{(1)} + e_2^{(1)} && && + e_2^{(2)} \\
s'_5 &= e_1^{(1)} + e_4^{(1)} + e_5^{(1)} && && + e_5^{(2)} \\
s_7 &= e_1^{(1)} + e_3^{(1)} + e_6^{(1)} + e_7^{(1)} && && + e_7^{(2)}.
\end{aligned} \tag{13.18}$$

From this orthogonal set, $e_1^{(1)}$ will be correctly estimated if there are $t_{ML} = 2$ or fewer errors among the $n_E = 11$ error bits checked by (13.18). In other words, the majority-logic error correcting capability of the code is the same for $e_1^{(1)}$ as for $e_0^{(1)}$, assuming that $e_0^{(1)}$ was estimated correctly. Moreover, exactly the same decoding rule can be used to estimate $e_1^{(1)}$ as was used for $e_0^{(1)}$, since the check-sum equations (13.18) are identical to (13.15), except that different error bits are checked. This means that after estimating $e_0^{(1)}$, the implementation of the decoding circuitry need not be changed in order to estimate $e_1^{(1)}$.

In general, after each information error bit is estimated, it is subtracted from each syndrome equation it affects. Assuming each estimate to be correct, for the self-orthogonal code of Example 13.1 the syndrome equations

$$\begin{aligned} s_i' &= e_i^{(1)} &&&&+ e_i^{(2)} \\ s_{i+1}' &= e_i^{(1)} + e_{i+1}^{(1)} &&&&+ e_{i+1}^{(2)} \\ s_{i+4}' &= e_i^{(1)} && + e_{i+3}^{(1)} + e_{i+4}^{(1)} &&+ e_{i+4}^{(2)} \\ s_{i+6} &= e_i^{(1)} + e_{i+2}^{(1)} &&& + e_{i+5}^{(1)} + e_{i+6}^{(1)} &+ e_{i+6}^{(2)} \end{aligned} \tag{13.19}$$

form a set of four orthogonal check sums on $e_i^{(1)}$, and the same decoding rule can again be used to estimate $e_i^{(1)}$. Hence, $e_i^{(1)}$ will be correctly estimated if there are $t_{ML} = 2$ or fewer errors among the $n_E = 11$ error bits checked by (13.19).

A complete encoder/decoder block diagram for the self-orthogonal code of Example 13.1 with majority-logic decoding is shown in Figure 13.3. The operation of the decoder can be described as follows:

Step 1. The first constraint length of syndrome bits $s_0, s_1, \ldots, s_6$ is calculated.

Step 2. A set of four orthogonal check sums on $e_0^{(1)}$ is formed from the syndrome bits calculated in step 1.

Step 3. The four check sums are fed into a majority-logic gate which produces an output of "1" if and only if three or four (more than half) of its inputs are "1's". If its output is "1" [$\hat{e}_0^{(1)} = 1$], $r_0^{(1)}$ is assumed to be incorrect, and hence must be corrected. If its output is "0" [$\hat{e}_0^{(1)} = 0$], $r_0^{(1)}$ is assumed to be correct. The correction is performed by adding the output of the threshold gate [$\hat{e}_0^{(1)}$] to $r_0^{(1)}$. The output of the threshold gate [$\hat{e}_0^{(1)}$] is also fed back and subtracted from each syndrome bit it affects. [It is not necessary to subtract $\hat{e}_0^{(1)}$ from s_0 since this syndrome bit is not used in any future estimates.]

Step 4. The estimated information bit $\hat{u}_0 = r_0^{(1)} + \hat{e}_0^{(1)}$ is shifted out of the decoder. The syndrome register is shifted once to the right, the next block of received bits [$r_7^{(1)}$ and $r_7^{(2)}$] is shifted into the decoder, and the next syndrome bit s_7 is calculated and shifted into the leftmost stage of the syndrome register.

Step 5. The syndrome register now contains the modified syndrome bits s_1', $s_2', \ldots, s_6'$ together with s_7. The decoder repeats steps 2, 3, and 4 and estimates $e_1^{(1)}$. All successive information error bits are then estimated in the same way.

Because each estimate must be fed back to modify the syndrome register before the next estimate can be made, this is called a *feedback decoder*. It is analogous to the

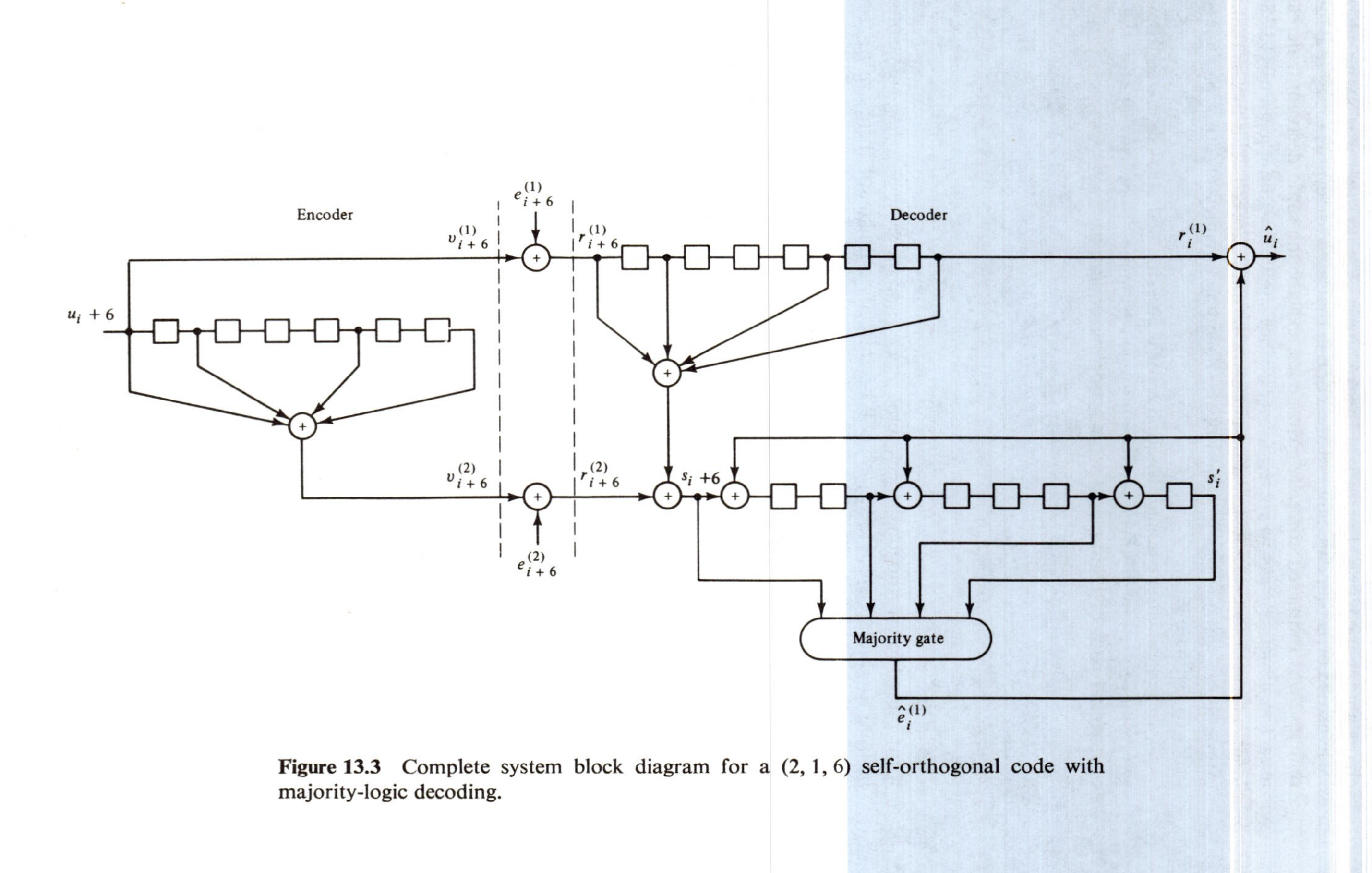

Figure 13.3 Complete system block diagram for a (2, 1, 6) self-orthogonal code with majority-logic decoding.

feedback of each estimate in a Meggitt decoder for cyclic codes. Note that each estimate made by a feedback majority-logic decoder depends on only one constraint length of error bits, the effect of previously estimated error bits having been removed by the feedback. As will be seen in Section 13.5, this leads to suboptimum performance compared to maximum likelihood decoding. The simplicity of the decoder, however, makes majority-logic decoding an attractive alternative to Viterbi or sequential decoding in some applications.

In the general case of an (n, k, m) systematic code with generator matrix

$$\mathbf{G} = \begin{bmatrix} \mathbf{IP}_0 & \mathbf{0P}_1 & \mathbf{0P}_2 & \cdots & \mathbf{0P}_m & & & \\ & \mathbf{IP}_0 & \mathbf{0P}_1 & \cdots & \mathbf{0P}_{m-1} & \mathbf{0P}_m & & \\ & & \mathbf{IP}_0 & \cdots & \mathbf{0P}_{m-2} & \mathbf{0P}_{m-1} & \mathbf{0P}_m & \\ & & & \cdot & & & & \cdot \\ & & & & \cdot & & & & \cdot \\ & & & & & \cdot & & & & \cdot \end{bmatrix}, \tag{13.20}$$

where $\mathbf{I}$ is the $k \times k$ identity matrix, $\mathbf{0}$ is the $k \times k$ all-zero matrix, and $\mathbf{P}_i$ is a $k \times (n-k)$ matrix whose entries are

$$\mathbf{P}_i = \begin{bmatrix} g_{1,i}^{(k+1)} & g_{1,i}^{(k+2)} & \cdots & g_{1,i}^{(n)} \\ g_{2,i}^{(k+1)} & g_{2,i}^{(k+2)} & \cdots & g_{2,i}^{(n)} \\ \cdot & \cdot & & \cdot \\ \cdot & \cdot & & \cdot \\ \cdot & \cdot & & \cdot \\ g_{k,i}^{(k+1)} & g_{k,i}^{(k+2)} & \cdots & g_{k,i}^{(n)} \end{bmatrix}, \tag{13.21}$$

the parity-check matrix is given by

$$\mathbf{H} = \begin{bmatrix} \mathbf{P}_0^T & \mathbf{I} \\ \mathbf{P}_1^T & \mathbf{0} & \mathbf{P}_0^T & \mathbf{I} \\ \mathbf{P}_2^T & \mathbf{0} & \mathbf{P}_1^T & \mathbf{0} & \mathbf{P}_0^T & \mathbf{I} \\ \cdot & \cdot & \cdot & \cdot & \cdot & \cdot & \cdot \\ \cdot & \cdot & \cdot & \cdot & \cdot & \cdot & & \cdot \\ \cdot & \cdot & \cdot & \cdot & \cdot & \cdot & & & \cdot \\ \mathbf{P}_m^T & \mathbf{0} & \mathbf{P}_{m-1}^T & \mathbf{0} & \mathbf{P}_{m-2}^T & \mathbf{0} & \cdots & \mathbf{P}_0^T & \mathbf{I} \\ & & \mathbf{P}_m^T & \mathbf{0} & \mathbf{P}_{m-1}^T & \mathbf{0} & \cdots & \mathbf{P}_1^T & \mathbf{0} & \mathbf{P}_0^T & \mathbf{I} \\ & & & & \mathbf{P}_m^T & \mathbf{0} & \cdots & \mathbf{P}_2^T & \mathbf{0} & \mathbf{P}_1^T & \mathbf{0} & \mathbf{P}_0^T & \mathbf{I} \\ & & & & & \cdot & & & & & & \cdot \\ & & & & & & \cdot & & & & & & \cdot \\ & & & & & & & \cdot & & & & & & \cdot \end{bmatrix}, \tag{13.22}$$

where, in this case, $\mathbf{I}$ is the $(n-k) \times (n-k)$ identity matrix and $\mathbf{0}$ is the $(n-k) \times (n-k)$ all-zero matrix.

The syndrome equations are still given by

$$\mathbf{s} = \mathbf{r}\mathbf{H}^T = \mathbf{e}\mathbf{H}^T, \tag{13.23}$$

but there are now $(n-k)$ syndrome sequences, one corrresponding to each parity error sequence, and $\mathbf{s} = (s_0^{(k+1)} \ldots s_0^{(n)}, s_1^{(k+1)} \ldots s_1^{(n)}, s_2^{(k+1)} \ldots s_2^{(n)}, \ldots)$. In polynomial notation, for $j = k+1, \ldots, n$,

$$\begin{aligned} \mathbf{s}^{(j)}(D) &= \sum_{i=1}^{k} \mathbf{r}^{(i)}(D)\mathbf{g}_i^{(j)}(D) + \mathbf{r}^{(j)}(D) \\ &= \sum_{i=1}^{k} \mathbf{e}^{(i)}(D)\mathbf{g}_i^{(j)}(D) + \mathbf{e}^{(j)}(D), \end{aligned} \tag{13.24}$$

and forming the syndrome vector $\mathbf{s}(D) = [\mathbf{s}^{(k+1)}(D), \ldots, \mathbf{s}^{(n)}(D)]$ is equivalent to "encoding" the received information sequences and then adding each output to the corresponding received parity sequence. A block diagram of a syndrome former for an (n, k, m) systematic code is shown in Figure 13.4. Equation (13.24) can be put in matrix form as follows:

$$\mathbf{s}(D) = \mathbf{r}(D)\mathbf{H}^T(D) = \mathbf{e}(D)\mathbf{H}^T(D), \tag{13.25}$$

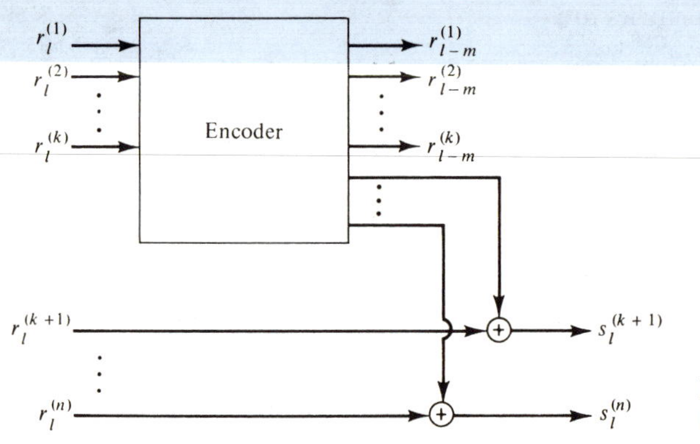

Figure 13.4 Syndrome forming circuit for an (n, k, m) systematic code.

where $\mathbf{r}(D) = [\mathbf{r}^{(1)}(D), \mathbf{r}^{(2)}(D), \ldots, \mathbf{r}^{(n)}(D)]$ is the n-tuple of received sequences, $\mathbf{e}(D) = [\mathbf{e}^{(1)}(D), \mathbf{e}^{(2)}(D), \ldots, \mathbf{e}^{(n)}(D)]$ is the n-tuple of error sequences, and

$$\mathbf{H}(D) = \begin{bmatrix} \mathbf{g}_1^{(k+1)}(D) & \mathbf{g}_2^{(k+1)}(D) & \cdots & \mathbf{g}_k^{(k+1)}(D) & 1 & 0 & \cdots & 0 \\ \mathbf{g}_1^{(k+2)}(D) & \mathbf{g}_2^{(k+2)}(D) & \cdots & \mathbf{g}_k^{(k+2)}(D) & 0 & 1 & \cdots & 0 \\ \cdot & \cdot & & \cdot & \cdot & \cdot & & \cdot \\ \cdot & \cdot & & \cdot & \cdot & \cdot & & \cdot \\ \cdot & \cdot & & \cdot & \cdot & \cdot & & \cdot \\ \mathbf{g}_1^{(n)}(D) & \mathbf{g}_2^{(n)}(D) & \cdots & \mathbf{g}_k^{(n)}(D) & 0 & 0 & \cdots & 1 \end{bmatrix} \tag{13.26}$$

is an $(n-k) \times n$ matrix called the *parity transfer function matrix*.

In general, for an (n, k, m) systematic code there are $n-k$ new syndrome bits to be formed and k information error bits to be estimated at each time unit. There are a total of $k \times (n-k)$ parity triangles, one corresponding to each nontrivial generator sequence. The general structure of the parity triangles used to form orthogonal parity checks is as follows:

$$
\begin{bmatrix}
s_0^{(k+1)} \\ \vdots \\ s_m^{(k+1)} \\ s_0^{(k+2)} \\ \vdots \\ s_m^{(k+2)} \\ \vdots \\ s_0^{(n)} \\ \vdots \\ s_m^{(n)}
\end{bmatrix}
=
\begin{bmatrix}
g_{1,0}^{(k+1)} & & & g_{2,0}^{(k+1)} & & & & g_{k,0}^{(k+1)} & & \\
\vdots & \ddots & & \vdots & \ddots & & & \vdots & \ddots & \\
g_{1,m}^{(k+1)} & \cdots & g_{1,0}^{(k+1)} & g_{2,m}^{(k+1)} & \cdots & g_{2,0}^{(k+1)} & \cdots & g_{k,m}^{(k+1)} & \cdots & g_{k,0}^{(k+1)} \\
g_{1,0}^{(k+2)} & & & g_{2,0}^{(k+2)} & & & & g_{k,0}^{(k+2)} & & \\
\vdots & \ddots & & \vdots & \ddots & & & \vdots & \ddots & \\
g_{1,m}^{(k+2)} & \cdots & g_{1,0}^{(k+2)} & g_{2,m}^{(k+2)} & \cdots & g_{2,0}^{(k+2)} & \cdots & g_{k,m}^{(k+2)} & \cdots & g_{k,0}^{(k+2)} \\
\vdots & & & \vdots & & & & \vdots & & \\
g_{1,0}^{(n)} & & & g_{2,0}^{(n)} & & & & g_{k,0}^{(n)} & & \\
\vdots & \ddots & & \vdots & \ddots & & & \vdots & \ddots & \\
g_{1,m}^{(n)} & \cdots & g_{1,0}^{(n)} & g_{2,m}^{(n)} & \cdots & g_{2,0}^{(n)} & \cdots & g_{k,m}^{(n)} & \cdots & g_{k,0}^{(n)}
\end{bmatrix}
\begin{bmatrix}
e_0^{(1)} \\ \vdots \\ e_m^{(1)} \\ e_0^{(2)} \\ \vdots \\ e_m^{(2)} \\ \vdots \\ e_0^{(k)} \\ \vdots \\ e_m^{(k)}
\end{bmatrix}
+
\begin{bmatrix}
e_0^{(k+1)} \\ \vdots \\ e_m^{(k+1)} \\ e_0^{(k+2)} \\ \vdots \\ e_m^{(k+2)} \\ \vdots \\ e_0^{(n)} \\ \vdots \\ e_m^{(n)}
\end{bmatrix}.
\tag{13.27}
$$

The first set of $m + 1$ rows corresponds to the first constraint length of syndrome sequence $\mathbf{s}^{(k+1)}$, the second set of $m + 1$ rows corresponds to the first constraint length of syndrome sequence $\mathbf{s}^{(k+2)}$, and so on. Syndrome bits, or sums of syndrome bits, are then used to form orthogonal check sums on the information error bits $e_0^{(1)}, e_0^{(2)}, \ldots, e_0^{(k)}$. If at least J orthogonal check sums can be formed on each of these information error bits, then $t_{ML} = \lfloor J/2 \rfloor$ is the majority-logic error-correcting capability of the code; that is, any pattern of t_{ML} or fewer errors within the n_E error bits checked by the k sets of orthogonal check sums will be corrected by the majority-logic decoding rule. The operation of a feedback majority-logic decoder for a t_{ML}-error-correcting (n, k, m) systematic code can be described as follows:

Step 1. The $(m + 1)(n - k)$ syndrome bits in the first constraint length are calculated.

Step 2. A set of J orthogonal check sums is formed on each of the k information error bits from the syndrome bits calculated above.

Step 3. Each set of J check sums is fed into a majority-logic gate which produces an output of 1 if and only if more than half of its inputs are 1's. If the output of the ith gate is 1 $[\hat{e}_0^{(i)} = 1]$, $r_0^{(i)}$ is assumed to be incorrect, and hence must be corrected. If its output is 0 $[\hat{e}_0^{(i)} = 0]$, $r_0^{(i)}$ is assumed to be correct. The corrections are performed by adding the output of each majority-logic gate to the corresponding received bit. The output of each gate is also fed back and subtracted from each syndrome it affects. (It is not necessary to modify the time unit 0 syndrome bits.)

Step 4. The estimated information bits $\hat{u}_0^{(i)} = r_0^{(i)} + \hat{e}_0^{(i)}$, $i = 1, 2, \ldots, k$, are shifted out of the decoder. The syndrome registers are shifted once to the right, the next block of n received bits is shifted into the decoder, and the next set of $n - k$ syndrome bits is calculated and shifted into the leftmost stages of the $n - k$ syndrome registers.

Step 5. The syndrome registers now contain the modified syndrome bits together with the new set of syndrome bits. The decoder repeats steps 2, 3, and 4 and estimates the next block of information error bits $e_1^{(1)}, e_1^{(2)}, \ldots, e_1^{(k)}$. All successive blocks of information error bits are then estimated in the same way.

As noted earlier, if all previous sets of k estimates are correct, their effect is removed by the feedback, and the next set of k estimates depends on only one constraint length of error bits. A block diagram of a general majority-logic feedback decoder for an (n, k, m) systematic code is shown in Figure 13.5. Massey [1] has shown that the L-step orthogonalization process for block codes does not apply to convolutional codes.

Example 13.4

Consider the (3, 2, 13) systematic code with $\mathbf{g}_1^{(3)}(D) = 1 + D^8 + D^9 + D^{12}$ and $\mathbf{g}_2^{(3)}(D) = 1 + D^6 + D^{11} + D^{13}$. Following the procedure of Example 13.1, the first constraint length of syndrome bits can be written as

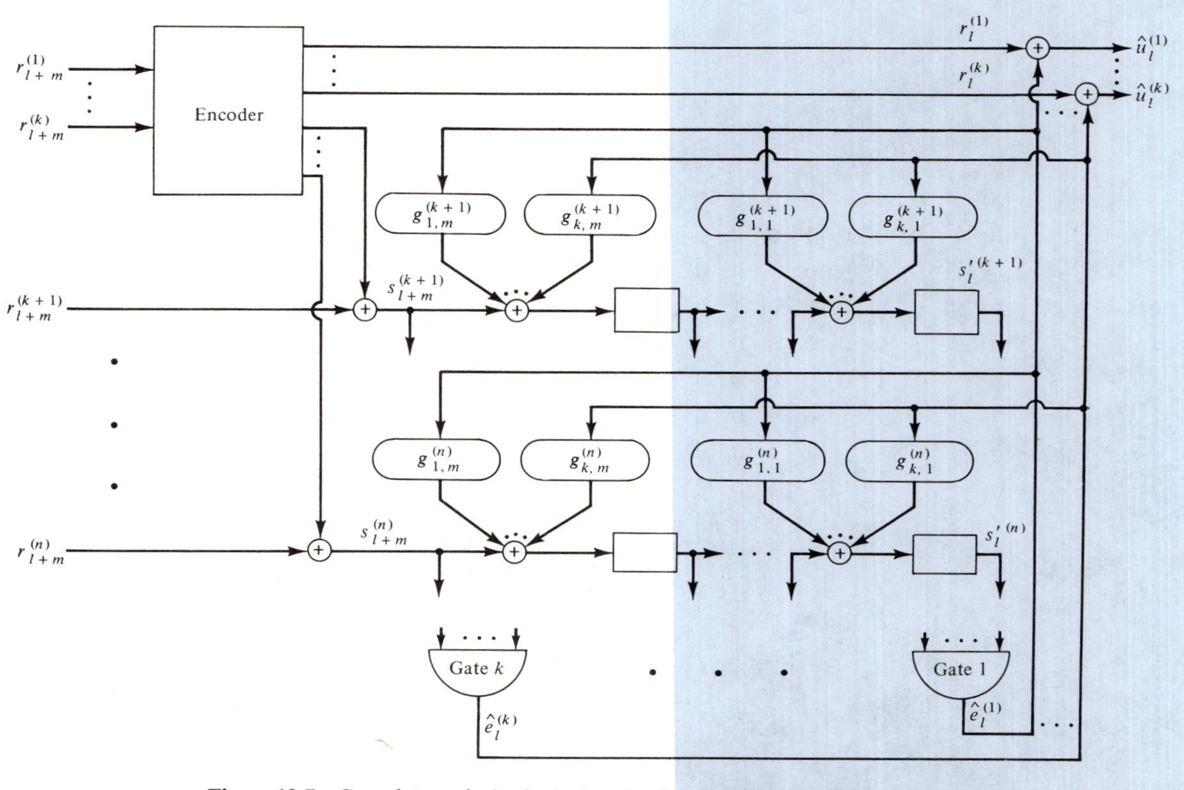

Figure 13.5 Complete majority-logic decoder for an (n, k, m) systematic code.

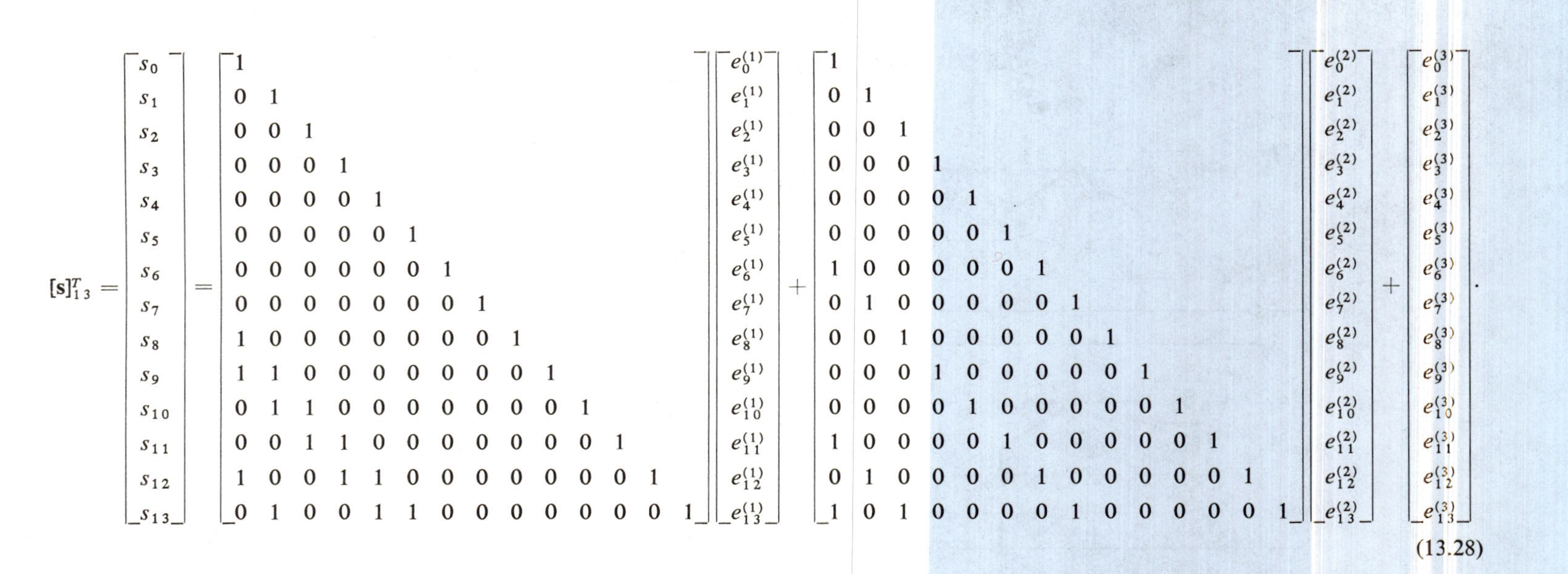

$$[\mathbf{s}]_{13}^{T} = \begin{bmatrix} s_0 \\ s_1 \\ s_2 \\ s_3 \\ s_4 \\ s_5 \\ s_6 \\ s_7 \\ s_8 \\ s_9 \\ s_{10} \\ s_{11} \\ s_{12} \\ s_{13} \end{bmatrix} = \begin{bmatrix} 1 &&&&&&&&&&&&& \\ 0 & 1 &&&&&&&&&&&& \\ 0 & 0 & 1 &&&&&&&&&&& \\ 0 & 0 & 0 & 1 &&&&&&&&&& \\ 0 & 0 & 0 & 0 & 1 &&&&&&&&& \\ 0 & 0 & 0 & 0 & 0 & 1 &&&&&&&& \\ 0 & 0 & 0 & 0 & 0 & 0 & 1 &&&&&&& \\ 0 & 0 & 0 & 0 & 0 & 0 & 0 & 1 &&&&&& \\ 1 & 0 & 0 & 0 & 0 & 0 & 0 & 0 & 1 &&&&& \\ 1 & 1 & 0 & 0 & 0 & 0 & 0 & 0 & 0 & 1 &&&& \\ 0 & 1 & 1 & 0 & 0 & 0 & 0 & 0 & 0 & 0 & 1 &&& \\ 0 & 0 & 1 & 1 & 0 & 0 & 0 & 0 & 0 & 0 & 0 & 1 && \\ 1 & 0 & 0 & 1 & 1 & 0 & 0 & 0 & 0 & 0 & 0 & 0 & 1 & \\ 0 & 1 & 0 & 0 & 1 & 1 & 0 & 0 & 0 & 0 & 0 & 0 & 0 & 1 \end{bmatrix} \begin{bmatrix} e_0^{(1)} \\ e_1^{(1)} \\ e_2^{(1)} \\ e_3^{(1)} \\ e_4^{(1)} \\ e_5^{(1)} \\ e_6^{(1)} \\ e_7^{(1)} \\ e_8^{(1)} \\ e_9^{(1)} \\ e_{10}^{(1)} \\ e_{11}^{(1)} \\ e_{12}^{(1)} \\ e_{13}^{(1)} \end{bmatrix} + \begin{bmatrix} 1 &&&&&&&&&&&&& \\ 0 & 1 &&&&&&&&&&&& \\ 0 & 0 & 1 &&&&&&&&&&& \\ 0 & 0 & 0 & 1 &&&&&&&&&& \\ 0 & 0 & 0 & 0 & 1 &&&&&&&&& \\ 0 & 0 & 0 & 0 & 0 & 1 &&&&&&&& \\ 1 & 0 & 0 & 0 & 0 & 0 & 1 &&&&&&& \\ 0 & 1 & 0 & 0 & 0 & 0 & 0 & 1 &&&&&& \\ 0 & 0 & 1 & 0 & 0 & 0 & 0 & 0 & 1 &&&&& \\ 0 & 0 & 0 & 1 & 0 & 0 & 0 & 0 & 0 & 1 &&&& \\ 0 & 0 & 0 & 0 & 1 & 0 & 0 & 0 & 0 & 0 & 1 &&& \\ 1 & 0 & 0 & 0 & 0 & 1 & 0 & 0 & 0 & 0 & 0 & 1 && \\ 0 & 1 & 0 & 0 & 0 & 0 & 1 & 0 & 0 & 0 & 0 & 0 & 1 & \\ 1 & 0 & 1 & 0 & 0 & 0 & 0 & 1 & 0 & 0 & 0 & 0 & 0 & 1 \end{bmatrix} \begin{bmatrix} e_0^{(2)} \\ e_1^{(2)} \\ e_2^{(2)} \\ e_3^{(2)} \\ e_4^{(2)} \\ e_5^{(2)} \\ e_6^{(2)} \\ e_7^{(2)} \\ e_8^{(2)} \\ e_9^{(2)} \\ e_{10}^{(2)} \\ e_{11}^{(2)} \\ e_{12}^{(2)} \\ e_{13}^{(2)} \end{bmatrix} + \begin{bmatrix} e_0^{(3)} \\ e_1^{(3)} \\ e_2^{(3)} \\ e_3^{(3)} \\ e_4^{(3)} \\ e_5^{(3)} \\ e_6^{(3)} \\ e_7^{(3)} \\ e_8^{(3)} \\ e_9^{(3)} \\ e_{10}^{(3)} \\ e_{11}^{(3)} \\ e_{12}^{(3)} \\ e_{13}^{(3)} \end{bmatrix}. \quad (13.28)$$

There are two parity triangles in this case, one corresponding to each generator sequence. These can be used to form a set of four orthogonal check sums on the information error bit $e_0^{(1)}$ as follows:

$\longrightarrow$	1	$\boxed{1}$
	0 1	0 1
	0 0 1	0 0 1
	0 0 0 1	0 0 0 1
	0 0 0 0 1	0 0 0 0 1
	0 0 0 0 0 1	0 0 0 0 0 1
	0 0 0 0 0 0 1	1 0 0 0 0 0 1
	0 0 0 0 0 0 0 1	0 1 0 0 0 0 0 1
$\longrightarrow$	1 0 0 0 0 0 0 0 $\boxed{1}$	0 0 $\boxed{1}$ 0 0 0 0 0 $\boxed{1}$
$\longrightarrow$	1 $\boxed{1}$ 0 0 0 0 0 0 0 $\boxed{1}$	0 0 0 $\boxed{1}$ 0 0 0 0 0 $\boxed{1}$
	0 1 1 0 0 0 0 0 0 0 1	0 0 0 0 1 0 0 0 0 0 1
	0 0 1 1 0 0 0 0 0 0 0 1	1 0 0 0 0 1 0 0 0 0 0 1
$\longrightarrow$	1 0 0 $\boxed{1}$ $\boxed{1}$ 0 0 0 0 0 0 0 $\boxed{1}$	0 $\boxed{1}$ 0 0 0 0 $\boxed{1}$ 0 0 0 0 0 $\boxed{1}$
	0 1 0 0 1 1 0 0 0 0 0 0 0 1	1 0 1 0 0 0 0 1 0 0 0 0 0 1.

Similarly, a set of four orthogonal check sums on $e_0^{(2)}$ can be formed as follows:

$\longrightarrow$	$\boxed{1}$	1
	0 1	0 1
	0 0 1	0 0 1
	0 0 0 1	0 0 0 1
	0 0 0 0 1	0 0 0 0 1
	0 0 0 0 0 1	0 0 0 0 0 1
$\longrightarrow$	0 0 0 0 0 0 $\boxed{1}$	1 0 0 0 0 0 $\boxed{1}$
	0 0 0 0 0 0 0 1	0 1 0 0 0 0 0 1
	1 0 0 0 0 0 0 0 1	0 0 1 0 0 0 0 0 1
	1 1 0 0 0 0 0 0 0 1	0 0 0 1 0 0 0 0 0 1
	0 1 1 0 0 0 0 0 0 0 1	0 0 0 0 1 0 0 0 0 0 1
$\longrightarrow$	0 0 $\boxed{1}$ $\boxed{1}$ 0 0 0 0 0 0 0 $\boxed{1}$	1 0 0 0 0 $\boxed{1}$ 0 0 0 0 0 $\boxed{1}$
	1 0 0 1 1 0 0 0 0 0 0 1 1	0 1 0 0 0 0 1 0 0 0 0 0 1
$\longrightarrow$	0 $\boxed{1}$ 0 0 $\boxed{1}$ $\boxed{1}$ 0 0 0 0 0 0 0 $\boxed{1}$	1 0 $\boxed{1}$ 0 0 0 0 $\boxed{1}$ 0 0 0 0 0 $\boxed{1}$.

Since syndrome bits alone, not sums of syndrome bits, comprise all check sums, this is a self-orthogonal code. A total of 31 different channel error bits (24 information error bits and 7 parity error bits) are checked by the two orthogonal sets, and the effective constraint length is $n_E = 31$. Hence, the majority-logic decoding rule will correctly estimate both $e_0^{(1)}$ and $e_0^{(2)}$ whenever $t_{ML} = 2$ or fewer of these $n_E = 31$ error bits are 1's. A block diagram of the decoder is shown in Figure 13.6.

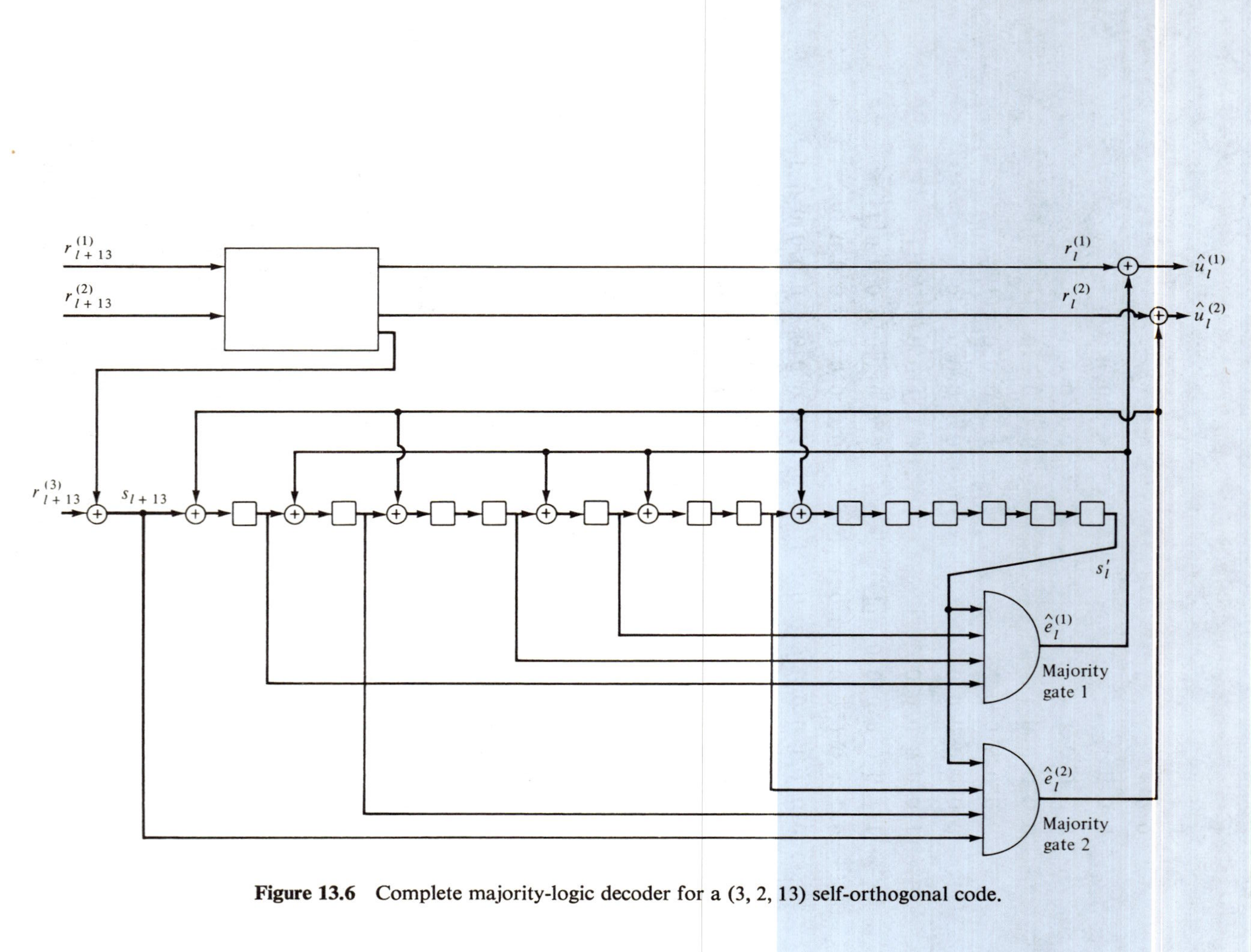

Figure 13.6 Complete majority-logic decoder for a (3, 2, 13) self-orthogonal code.

Note that a total of $n_A = n(m + 1) = 3(14) = 42$ channel error bits are involved in (13.28). Hence, 11 channel error bits have no effect on the estimates of $e_0^{(1)}$ and $e_0^{(2)}$. As decoding proceeds, however, these 11 error bits will affect the estimates of successive information error bits. Also note that there are many patterns of more than $t_{ML} = 2$ channel errors which will be corrected by this code. However, there are some patterns of three channel errors that cannot be corrected.

Example 13.5

Now consider a (3, 1, 4) systematic code with $\mathbf{g}^{(2)}(D) = 1 + D$ and $\mathbf{g}^{(3)}(D) = 1 + D^2 + D^3 + D^4$. In this case, there are two syndrome sequences, $\mathbf{s}^{(2)} = (s_0^{(2)}, s_1^{(2)}, s_2^{(2)}, \ldots)$ and $\mathbf{s}^{(3)} = (s_0^{(3)}, s_1^{(3)}, s_2^{(3)}, \ldots)$, and the first constraint length of syndrome bits can be written as

$$[\mathbf{s}]_4^T = \begin{bmatrix} s_0^{(2)} \\ s_1^{(2)} \\ s_2^{(2)} \\ s_3^{(2)} \\ s_4^{(2)} \\ s_0^{(3)} \\ s_1^{(3)} \\ s_2^{(3)} \\ s_3^{(3)} \\ s_4^{(3)} \end{bmatrix} = \begin{bmatrix} 1 & & & & \\ 1 & 1 & & & \\ 0 & 1 & 1 & & \\ 0 & 0 & 1 & 1 & \\ 0 & 0 & 0 & 1 & 1 \\ 1 & & & & \\ 0 & 1 & & & \\ 1 & 0 & 1 & & \\ 1 & 1 & 0 & 1 & \\ 1 & 1 & 1 & 0 & 1 \end{bmatrix} \begin{bmatrix} e_0^{(1)} \\ e_1^{(1)} \\ e_2^{(1)} \\ e_3^{(1)} \\ e_4^{(1)} \end{bmatrix} + \begin{bmatrix} e_0^{(2)} \\ e_1^{(2)} \\ e_2^{(2)} \\ e_3^{(2)} \\ e_4^{(2)} \\ e_0^{(3)} \\ e_1^{(3)} \\ e_2^{(3)} \\ e_3^{(3)} \\ e_4^{(3)} \end{bmatrix}. \tag{13.29}$$

As in the case of the (3, 2, 13) code of Example 13.4, there are two parity triangles. In Example 13.4, the two parity triangles were used to form two separate sets of four orthogonal check sums on two different information error bits. In this example there is only one information error bit per unit time, and the two parity triangles can be used to form a set of six orthogonal check sums on the information error bit $e_0^{(1)}$ as follows:

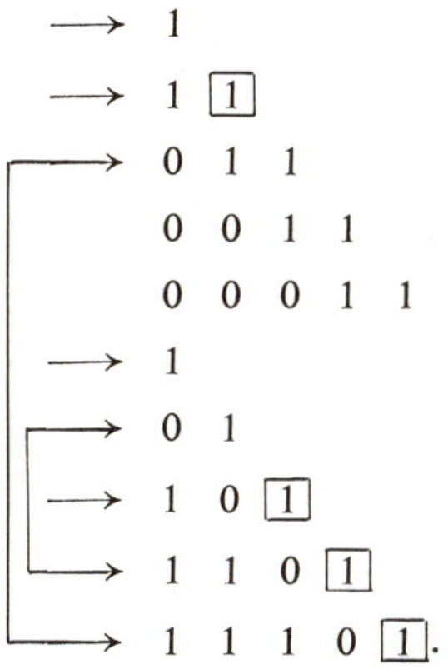

$$\begin{array}{llllll} \longrightarrow & 1 & & & & \\ \longrightarrow & 1 & \boxed{1} & & & \\ \longrightarrow & 0 & 1 & 1 & & \\ & 0 & 0 & 1 & 1 & \\ & 0 & 0 & 0 & 1 & 1 \\ \longrightarrow & 1 & & & & \\ \longrightarrow & 0 & 1 & & & \\ \longrightarrow & 1 & 0 & \boxed{1} & & \\ \longrightarrow & 1 & 1 & 0 & \boxed{1} & \\ \longrightarrow & 1 & 1 & 1 & 0 & \boxed{1}. \end{array}$$

The check sums $s_0^{(2)}$, $s_0^{(3)}$, $s_1^{(2)}$, $s_2^{(3)}$, $s_1^{(3)} + s_3^{(3)}$, and $s_2^{(2)} + s_4^{(3)}$ form a set of six orthogonal check sums on $e_0^{(1)}$. The effective constraint length $n_E = 13$ for this code, and the majority-logic decoding rule correctly estimates $e_0^{(1)}$ whenever $t_{ML} = 3$ or fewer of these 13 error bits are 1's. A block diagram of the decoder is shown in Figure 13.7.

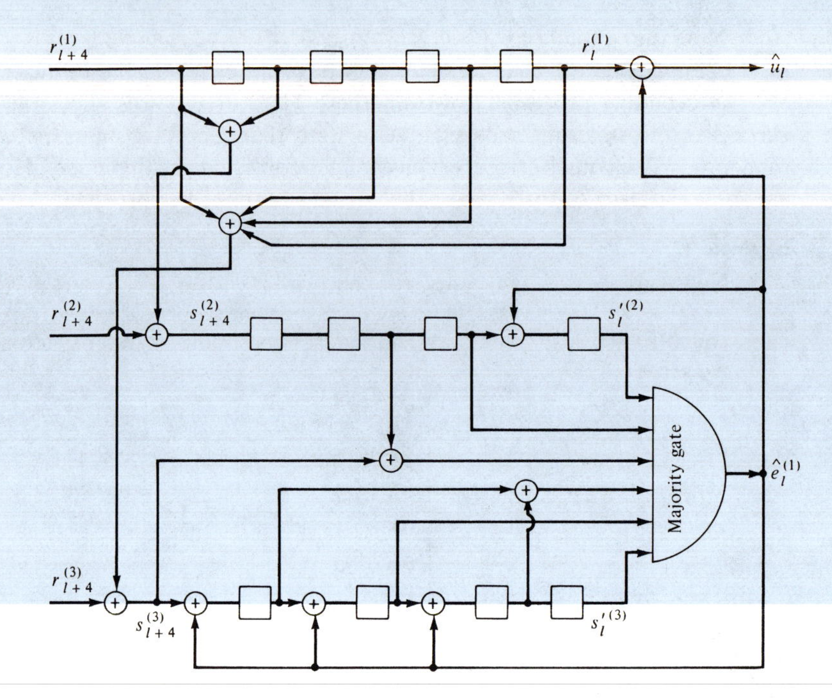

Figure 13.7 Complete majority-logic decoder for a (3, 1, 4) orthogonalizable code.

13.2 ERROR PROPAGATION AND DEFINITE DECODING

In the discussion of feedback decoding above, it was assumed that the past estimates subtracted from the syndrome were all correct. This is, of course, not always true. When an incorrect estimate is fed back to the syndrome, it has the same effect as an additional transmission error, and can cause further decoding errors which would not occur otherwise. This is called the *error propagation effect* of feedback decoders [2].

Example 13.6

For the self-orthogonal code of Example 13.1, let $\tilde{e}_j^{(1)} = e_j^{(1)} + \hat{e}_j^{(1)}$ be the result of adding the estimate $\hat{e}_j^{(1)}$ to a syndrome equation containing $e_j^{(1)}$. Since $\tilde{e}_j^{(1)}$ is formed after $e_j^{(1)}$ has been decoded, it is called a *post-decoding error*. The modified check sums of (13.19) are as follows:

$$\begin{aligned}
s_i' &= \tilde{e}_{i-6}^{(1)} + \tilde{e}_{i-4}^{(1)} + \tilde{e}_{i-1}^{(1)} + e_i^{(1)} + e_i^{(2)} \\
s_{i+1}' &= \tilde{e}_{i-5}^{(1)} + \tilde{e}_{i-3}^{(1)} + e_i^{(1)} + e_{i+1}^{(1)} + e_{i+1}^{(2)} \\
s_{i+4}' &= \tilde{e}_{i-2}^{(1)} + e_i^{(1)} + e_{i+3}^{(1)} + e_{i+4}^{(1)} + e_{i+4}^{(2)} \\
s_{i+6} &= e_i^{(1)} + e_{i+2}^{(1)} + e_{i+5}^{(1)} + e_{i+6}^{(1)} + e_{i+6}^{(2)}.
\end{aligned} \tag{13.30}$$

If all past estimates are correct [i.e., $\tilde{e}_j^{(1)} = 0, j = i - 6, \ldots, i - 1$], then (13.30) reduces to (13.19). However, if any $\tilde{e}_j^{(1)} = 1$, it has the same effect as a transmission error in the equations above, and can cause error propagation.

Several approaches have been taken to reducing the effects of error propagation. One is to resynchronize the decoder periodically by inserting a string of km zeros into the information sequence after every kL information bits. When the resynchronization sequence is received, the decoder is instructed to decode a string of km consecutive zeros. During this m-time-unit span of correct decoding, all post-decoding errors must be zero, and the effect of the past is removed from the syndrome. This periodic resynchronization of the decoder limits error propagation to at most $L + m$ time units. However, for the fractional rate loss $m/(L + m)$ to be small, L must be much larger than m, and the effects of error propagation can still be quite long.

Another approach to limiting error propagation is to select codes with automatic resynchronization properties. Certain codes have the property that if the channel is error-free over a limited span of time units, the effect of past errors on the syndrome is automatically removed, thus halting error propagation. Several classes of codes with automatic resynchronization properties are introduced in Section 13.4.

Error propagation can be completely eliminated simply by not using feedback in the decoder (i.e., past estimates are not fed back to modify the syndrome). This approach, first suggested by Robinson [3], is called *definite decoding*. Since the effects of previously estimated error bits are not removed from the syndrome in a definite decoder, these error bits can continue to influence future decoding estimates, possibly causing decoding errors that would not be made by a feedback decoder.

Example 13.7

For the self-orthogonal code of Example 13.1, if past decoding estimates are not fed back to modify the syndrome, the unmodified check sums of (13.19) become

$$\begin{aligned}
s_i &= e_{i-6}^{(1)} + e_{i-4}^{(1)} + e_{i-1}^{(1)} + e_i^{(1)} + e_i^{(2)} \\
s_{i+1} &= e_{i-5}^{(1)} + e_{i-3}^{(1)} + e_i^{(1)} + e_{i+1}^{(1)} + e_{i+1}^{(2)} \\
s_{i+4} &= e_{i-2}^{(1)} + e_i^{(1)} + e_{i+3}^{(1)} + e_{i+4}^{(1)} + e_{i+4}^{(2)} \\
s_{i+6} &= e_i^{(1)} + e_{i+2}^{(1)} + e_{i+5}^{(1)} + e_{i+6}^{(1)} + e_{i+6}^{(2)}.
\end{aligned} \tag{13.31}$$

Note that (13.31) still forms a set of four orthogonal parity checks on $e_i^{(1)}$, but that the effective constraint length n_E is increased from 11 to 17 since the six past information error bits are now included in the orthogonal check sums. The majority-logic decoding rule will correctly estimate $e_i^{(1)}$ whenever $t_{ML} = 2$ or fewer of these $n_E = 17$ error bits are 1's. Hence, the majority-logic error-correcting capability of the code is weakened by definite decoding, since there are more channel error bits in the check sums which can cause decoding errors. Error propagation has been eliminated, however, by removing the feedback. The definite decoding circuitry for this code is exactly the same as shown in Figure 13.3, except that the feedback connection from the output of the majority-logic gate to the syndrome register is eliminated.

Self-orthogonal codes have the property that the same number of orthogonal check sums can be formed for definite decoding as for feedback decoding, although the effective constraint length is always increased. This is usually not the case with codes that are not self-orthogonal, however.

Example 13.8

Consider the (2, 1, 5) code of Example 13.2. The unmodified syndrome equations are

$$\begin{aligned}
s_i &= e_{i-5}^{(1)} + e_{i-4}^{(1)} + e_{i-3}^{(1)} + e_i^{(1)} + e_i^{(2)} \\
s_{i+1} &= e_{i-4}^{(1)} + e_{i-3}^{(1)} + e_{i-2}^{(1)} + e_{i+1}^{(1)} + e_{i+1}^{(2)} \\
s_{i+2} &= + e_{i-3}^{(1)} + e_{i-2}^{(1)} + e_{i-1}^{(1)} + e_{i+2}^{(1)} + e_{i+2}^{(2)} \\
s_{i+3} &= e_{i-2}^{(1)} + e_{i-1}^{(1)} + e_i^{(1)} + e_{i+3}^{(1)} + e_{i+3}^{(2)} \\
s_{i+4} &= e_{i-1}^{(1)} + e_i^{(1)} + e_{i+1}^{(1)} + e_{i+4}^{(1)} + e_{i+4}^{(2)} \\
s_{i+5} &= e_i^{(1)} + e_{i+1}^{(1)} + e_{i+2}^{(1)} + e_{i+5}^{(1)} + e_{i+5}^{(2)}.
\end{aligned} \tag{13.32}$$

If the effect of past error bits is removed, the check sums s_i, s_{i+3}, s_{i+4}, and $s_{i+1} + s_{i+5}$ form an orthogonal set on $e_i^{(1)}$. With an unmodified syndrome, however, the check sums

$$\begin{aligned}
s_i &= e_{i-5}^{(1)} + e_{i-4}^{(1)} + e_{i-3}^{(1)} + e_i^{(1)} + e_i^{(2)} \\
s_{i+3} &= e_{i-2}^{(1)} + e_{i-1}^{(1)} + e_i^{(1)} + e_{i+3}^{(1)} + e_{i+3}^{(2)} \\
s_{i+4} &= + e_{i-1}^{(1)} + e_i^{(1)} + e_{i+1}^{(1)} + e_{i+4}^{(1)} + e_{i+4}^{(2)} \\
s_{i+1} + s_{i+5} &= e_{i-4}^{(1)} + e_{i-3}^{(1)} + e_{i-2}^{(1)} + e_i^{(1)} + e_{i+2}^{(1)} + e_{i+5}^{(1)} + e_{i+5}^{(2)}
\end{aligned} \tag{13.33}$$

are clearly not orthogonal. In fact, there is no set of four check sums orthogonal on $e_i^{(1)}$. The syndrome bits s_i, s_{i+3}, and s_{i+5} form a set of three orthogonal check sums on $e_i^{(1)}$, and this is the maximum number that can be formed. Hence, the majority-logic error-correcting capability $t_{ML} = \lfloor J/2 \rfloor$ of this code is reduced from 2 to 1 when definite decoding is used.

In Examples 13.7 and 13.8, we have seen that the error-correcting capability of a code is weakened when definite decoding is used, either by an increase in the effective constraint length n_E, or by a reduction in the majority-logic error-correcting capability t_{ML}, or a combination of both. On the other hand, error propagation due to erroneous decoding estimates is eliminated in definite decoding. Comparing equations (13.30) (feedback decoding) with (13.31) (definite decoding) for the self-orthogonal code of Example 13.1 implies that the error probability of a feedback decoder will be less than that of a definite decoder if $\Pr[\tilde{e}_j^{(1)} = 1] < \Pr[e_j^{(1)} = 1]$, that is, if the feedback decoding error probability is less than the channel transition probability. This will normally be the case unless the code is very weak or the channel is very noisy. An analysis by Morrissey [4] comparing the effect of error propagation with feedback to the reduced error-correcting capability without feedback also concludes that feedback decoders will usually outperform definite decoders.

13.3 DISTANCE PROPERTIES AND CODE PERFORMANCE

In Section 10.3 the *minimum distance* of a convolutional code was defined as

$$\begin{aligned}
d_{\min} &= \min \{d([\mathbf{v}']_m, [\mathbf{v}'']_m) : \mathbf{u}'_0 \neq \mathbf{u}''_0\} \\
&= \min \{w[\mathbf{v}]_m : \mathbf{u}_0 \neq \mathbf{0}\},
\end{aligned} \tag{13.34}$$

where $\mathbf{v}$, $\mathbf{v}'$, and $\mathbf{v}''$ are the code words corresponding to the information sequences

$\mathbf{u}$, $\mathbf{u}'$, and $\mathbf{u}''$, respectively. Note that, in computing $d_{\min}$, only the first constraint length is considered, and only code words that differ in the first information block are compared. Consequently, the minimum distance of a convolutional code guarantees a certain error-correcting capability for a decoder that estimates $\mathbf{u}_0$ from only the first constraint length of the received sequence, as demonstrated in the following theorem.

Theorem 13.2. For an (n, k, m) convolutional code with minimum distance $d_{\min}$, $\mathbf{u}_0$ can be correctly decoded given $\lfloor (d_{\min} - 1)/2 \rfloor$ or fewer channel errors in the first constraint length $[\mathbf{r}]_m$ of the received sequence.

Proof. Assume that the code word $\mathbf{v}$ corresponding to the information sequence $\mathbf{u}$ is transmitted, and that $\mathbf{r} = \mathbf{v} + \mathbf{e}$ is received. Then

$$d([\mathbf{r}]_m, [\mathbf{v}]_m) = w([\mathbf{r}]_m - [\mathbf{v}]_m) = w([\mathbf{e}]_m).$$

Now let $\mathbf{v}'$ be the code word corresponding to an information sequence $\mathbf{u}'$ with $\mathbf{u}'_0 \neq \mathbf{u}_0$. Then the triangle inequality (3.14) yields

$$\begin{aligned} d([\mathbf{r}]_m, [\mathbf{v}']_m) &\geq d([\mathbf{v}]_m, [\mathbf{v}']_m) - d([\mathbf{r}]_m, [\mathbf{v}]_m) \\ &= d([\mathbf{v}]_m, [\mathbf{v}']_m) - w([\mathbf{e}]_m) \\ &\geq d_{\min} - w([\mathbf{e}]_m). \end{aligned}$$

Since the number of channel errors $w([\mathbf{e}]_m) \leq (d_{\min} - 1)/2$,

$$d([\mathbf{r}]_m, [\mathbf{v}']_m) \geq \frac{d_{\min} + 1}{2} > \frac{d_{\min} - 1}{2} \geq d([\mathbf{r}]_m, [\mathbf{v}]_m).$$

Hence, no code word $\mathbf{v}'$ with $\mathbf{u}'_0 \neq \mathbf{u}_0$ can be closer to $\mathbf{r}$ over the first constraint length than the transmitted code word $\mathbf{v}$. If $\mathbf{u}''$ is the information sequence corresponding to the code word $\mathbf{v}''$ for which $d([\mathbf{r}]_m, [\mathbf{v}'']_m)$ is a minimum, a decoder that estimates $\hat{\mathbf{u}}_0 = \mathbf{u}''_0$ will correctly decode the first information block $\mathbf{u}_0$. Note that $\mathbf{u}''$ may not equal $\mathbf{u}$, but it must agree with $\mathbf{u}$ in the first block. Q.E.D.

In Theorem 13.2 we see that a decoder which finds the code word that is closest to the received sequence over the first constraint length, and then chooses the first information block in this code word as its estimate of $\mathbf{u}_0$, guarantees correct decoding of $\mathbf{u}_0$ if there are $\lfloor (d_{\min} - 1)/2 \rfloor$ or fewer channel errors in the first constraint length. Theorem 13.2 also has a converse: that a decoder which correctly estimates $\mathbf{u}_0$ from $[\mathbf{r}]_m$ whenever $[\mathbf{r}]_m$ contains $\lfloor (d_{\min} - 1)/2 \rfloor$ or fewer channel errors cannot correctly estimate $\mathbf{u}_0$ from $[\mathbf{r}]_m$ for all $[\mathbf{r}]_m$ containing $\lfloor (d_{\min} - 1)/2 \rfloor + 1$ channel errors. In other words, there are some received sequences containing $\lfloor (d_{\min} - 1)/2 \rfloor + 1$ errors in the first constraint length which will result in $\mathbf{u}_0$ being incorrectly decoded. Hence, $t \triangleq \lfloor (d_{\min} - 1)/2 \rfloor$ is called the *maximum error-correcting capability* of a code when the decoding of $\mathbf{u}_0$ is based only on the first constraint length of the received sequence. If each decoding decision is fed back to remove its effect from the syndrome registers (i.e., feedback decoding is used), and if this does not cause any post-decoding errors, the same maximum error-correcting capability applies to the decoding of each successive information block $\mathbf{u}_l$ based on the constraint length of received blocks $(\mathbf{r}_l,$

$\mathbf{r}_{i+1}, \ldots, \mathbf{r}_{i+m}$). A feedback decoder with this error-correcting capability is called an *optimum feedback decoder*.

Since a feedback majority-logic decoder bases its estimate of $\mathbf{u}_0$ on only the first constraint length of the received sequence, its maximum error-correcting capability is $t = \lfloor (d_{\min} - 1)/2 \rfloor$. Since the J orthogonal parity checks guarantee correct decoding of $\mathbf{u}_0$ whenever $[\mathbf{r}]_m$ contains $t_{ML} = \lfloor J/2 \rfloor$ or fewer channel errors, it follows that

$$J \leq d_{\min} - 1. \tag{13.35}$$

If $d_{\min} - 1$ orthogonal parity checks can be formed on each information error bit, the code is said to be *completely orthogonalizable* (i.e., the maximum error-correcting capability can be achieved with majority-logic decoding). Hence, it is desirable when using majority-logic decoding to select codes that are completely orthogonalizable. This restricts the choice of codes that can be used with majority-logic decoding, since most codes are not completely orthogonalizable. However, as will be seen in Section 13.4, several classes of convolutional codes have been found which are completely orthogonalizable.

Note that the result of Theorem 13.2 does not apply to maximum likelihood or sequential decoding, since these decoding methods process the entire received sequence $\mathbf{r}$ before making a final decision on $\mathbf{u}_0$.[1] This longer decoding delay of maximum likelihood and sequential decoding accounts for their superior performance when compared to majority-logic decoding. However, majority-logic decoders are much simpler to implement since they must store only one constraint length of the received sequence at any time. A complete comparison of the three major decoding methods for convolutional codes is given in Section 13.5.

The minimum distance of a convolutional code can be found by computing the CDF d_i of the code, and then letting $i = m$ (since $d_{\min} = d_m$). A more direct way of finding $d_{\min}$ makes use of the parity-check matrix $\mathbf{H}$. Since $\mathbf{v}$ is a code word if and only if $\mathbf{v}\mathbf{H}^T = \mathbf{0}$, the minimum number of rows of $\mathbf{H}^T$, or columns of $\mathbf{H}$, that add to $\mathbf{0}$ correspond to the minimum-weight nonzero code word. Since $d_{\min}$ is the code word with $\mathbf{u}_0 \neq \mathbf{0}$ which has minimum weight over only one constraint length, it can be computed by finding the minimum number of rows of $\mathbf{H}^T$, or columns of $\mathbf{H}$, including at least one of the first k, which add to $\mathbf{0}$ over only one constraint length. In other words, we must find the minimum-weight code word $[\mathbf{v}]_m$ with $\mathbf{u}_0 \neq \mathbf{0}$ for which $[\mathbf{v}]_m[\mathbf{H}^T]_m = \mathbf{0}$, where $[\mathbf{H}^T]_m$ includes only the first $(m+1)(n-k)$ columns of $\mathbf{H}^T$. This is equivalent to forming the first $(m+1)(n-k)$ rows of $\mathbf{H}$, and then finding the minimum number of columns of this matrix, including at least one of the first k, which add to $\mathbf{0}$.

Example 13.9

Consider the (2, 1, 6) systematic code of Example 13.1. The first $(m+1)(n-k) = 7 \times 1 = 7$ rows of $\mathbf{H}$ are

[1] Even a truncated Viterbi decoder, or a backsearch limited sequential decoder, processes several constraint lengths of the received sequence before making any final decisions.

$$\begin{bmatrix} 1 & 1 & & & & & & & & & & & & \\ 1 & 0 & 1 & 1 & & & & & & & & & & \\ 0 & 0 & 1 & 0 & 1 & 1 & & & & & & & & \\ 0 & 0 & 0 & 0 & 1 & 0 & 1 & 1 & & & & & & \\ 1 & 0 & 0 & 0 & 0 & 0 & 1 & 0 & 1 & 1 & & & & \\ 0 & 0 & 1 & 0 & 0 & 0 & 0 & 0 & 1 & 0 & 1 & 1 & & \\ 1 & 0 & 0 & 0 & 1 & 0 & 0 & 0 & 0 & 0 & 1 & 0 & 1 & 1 \end{bmatrix}.$$

$d_{\min}$ is the minimum number of columns of this matrix, including the first column, which add to zero. Since $J = 4$ orthogonal check sums can be found for this code, $d_{\min} \geq J + 1 = 5$. But columns 1, 2, 4, 10, and 14 add to zero, implying that $d_{min} \leq 5$. Hence, $d_{\min}$ must equal 5 for this code, and the code is completely orthogonalizable. The minimum-weight code word with $\mathbf{u}_0 \neq \mathbf{0}$ in this case is given by

$$[\mathbf{v}]_m = (1\ \ 1,\ \ 0\ \ 1,\ \ 0\ \ 0,\ \ 0\ \ 0,\ \ 0\ \ 1,\ \ 0\ \ 0,\ \ 0\ \ 1).$$

$d_{\min}$ can also be obtained by finding the minimum-weight linear combination of rows of the generator matrix $[\mathbf{G}]_m$ which includes at least one of the first k rows. This corresponds to the code word with $\mathbf{u}_0 \neq \mathbf{0}$ which has minimum weight over the first constraint length.

Example 13.10

Consider the (2, 1, 5) systematic code with $\mathbf{g}^{(2)}(D) = 1 + D + D^3 + D^5$. Then

$$[\mathbf{G}]_m = \begin{bmatrix} 1 & 1 & 0 & 1 & 0 & 0 & 0 & 1 & 0 & 0 & 0 & 1 \\ & & 1 & 1 & 0 & 1 & 0 & 0 & 0 & 1 & 0 & 0 \\ & & & & 1 & 1 & 0 & 1 & 0 & 0 & 0 & 1 \\ & & & & & & 1 & 1 & 0 & 1 & 0 & 0 \\ & & & & & & & & 1 & 1 & 0 & 1 \\ & & & & & & & & & & 1 & 1 \end{bmatrix}.$$

There are several information sequences with $u_0 = 1$ which produce code words of weight 5. For example, the information sequence $[\mathbf{u}]_m = (1\ 1\ 0\ 0\ 0)$ produces the code word $[\mathbf{v}]_m = (1\ 1, 1\ 0, 1\ 0, 0\ 0, 0\ 1, 0\ 0)$. However, there are no linear combinations of rows of $[\mathbf{G}]_m$ including the first row which have weight 4. Hence, $d_{\min} = 5$ for this code. The parity triangle is given by:

$$\begin{matrix} 1 & & & & & \\ 1 & 1 & & & & \\ 0 & 1 & 1 & & & \\ 1 & 0 & 1 & 1 & & \\ 0 & 1 & 0 & 1 & 1 & \\ 1 & 0 & 1 & 0 & 1 & 1. \end{matrix}$$

The maximum number of orthogonal check sums that can be formed in this case is $J = 3$. For example, $\{s_0, s_1, s_3\}$ and $\{s_0, s_2 + s_3, s_5\}$ are both sets of three orthogonal check sums on $e_0^{(1)}$. However, since $J = 3 < d_{\min} - 1 = 4$, this code is not completely orthogonalizable, that is, its maximum error-correcting capability is 2, but its majority-logic error-correcting capability is only 1.

The discussion so far in this chapter has centered strictly on systematic codes. This is due to the fact that any (n, k, m) nonsystematic code, by means of a linear transformation on the rows of its generator matrix, can be converted to an (n, k, m) systematic code with the same $d_{\min}$. Hence, the maximum error-correcting capability over one constraint length cannot be improved by considering nonsystematic codes. This result, first established by Bussgang [5], implies that when selecting codes with large $d_{\min}$ for use with majority-logic decoding, it suffices to consider only systematic codes. This differs markedly from the situation with d_{free}, where nonsystematic codes offer substantially larger values of d_{free} than systematic codes of the same constraint length, as was noted in previous chapters. This advantage in d_{free} of nonsystematic codes over systematic codes accounts for the almost exclusive use of nonsystematic codes in applications involving Viterbi or sequential decoding.

Example 13.11

Consider the (2, 1, 3) nonsystematic code with $\mathbf{G}(D) = [1 + D^2 + D^3, 1 + D + D^2 + D^3]$. Then

$$[\mathbf{G}]_3 = \begin{bmatrix} 1 & 1 & 0 & 1 & 1 & 1 & 1 & 1 \\ & & 1 & 1 & 0 & 1 & 1 & 1 \\ & & & & 1 & 1 & 0 & 1 \\ & & & & & & 1 & 1 \end{bmatrix},$$

and $d_{\min}$ is the minimum-weight linear combination of rows of $[\mathbf{G}]_3$ which includes the first row. Clearly, $d_{\min} = 4$ for this code. Now consider multiplying $[\mathbf{G}]_3$ by the transformation matrix

$$\mathbf{T} = \begin{bmatrix} 1 & 0 & 1 & 1 \\ 0 & 1 & 0 & 1 \\ 0 & 0 & 1 & 0 \\ 0 & 0 & 0 & 1 \end{bmatrix}$$

to obtain

$$[\mathbf{G}']_3 = \mathbf{T}[\mathbf{G}]_3 = \begin{bmatrix} 1 & 1 & 0 & 1 & 0 & 0 & 0 & 1 \\ & & 1 & 1 & 0 & 1 & 0 & 0 \\ & & & & 1 & 1 & 0 & 1 \\ & & & & & & 1 & 1 \end{bmatrix}.$$

$\mathbf{G}'$ is then the generator matrix of a (2, 1, 3) systematic code with $\mathbf{G}'(D) = [1, 1 + D + D^3]$. The minimum distance $d'_{\min}$ of this code is the minimum-weight linear combination of rows of $[\mathbf{G}']_3$ which includes the first row. A quick calculation yields $d'_{\min} = 4 = d_{\min}$. Hence, the transformation matrix $\mathbf{T}$ produces a systematic code $\mathbf{G}'$ with the same minimum distance as the original nonsystematic code $\mathbf{G}$. The free distance of the nonsystematic code is 6, however, whereas the free distance of the systematic code is only 4.

The $(m + 1) \times (m + 1)$ transformation matrix T needed to convert an $(n, 1, m)$ nonsystematic code to an $(n, 1, m)$ systematic code with the same $d_{\min}$ can be specified easily. The first row of $\mathbf{T}$ is chosen to correspond to the unique linear combination of rows of $[\mathbf{G}]_m$ which yields the generator sequence $\mathbf{g}^{(1)'} = (1\ 0 \cdots 0)$ for the systematic code. Each successive row of $\mathbf{T}$ is then obtained by shifting the previous row one column to the right, dropping the last position, and inserting a "0" in the first position. This completely specifies the $\mathbf{T}$ matrix. The procedure is essentially the same but slightly more complicated for an (n, k, m) code with $k > 1$ (see Problem 13.10).

The performance of an (n, k, m) convolutional code with majority-logic decoding on a BSC can be estimated from the distance properties of the code. First for an optimum feedback decoder, it follows from Theorem 13.2 that a decoding error can occur in estimating $\mathbf{u}_0$ only if more than $t = \lfloor (d_{\min} - 1)/2 \rfloor$ channel errors occur in the first constraint length. Hence, the bit error probability in decoding the first information block, $P_{b1}(E)$, can be upper bounded by

$$P_{b1}(E) \leq \frac{1}{k} \sum_{i=t+1}^{n_A} \binom{n_A}{i} p^i (1-p)^{n_A - i}, \tag{13.36}$$

where $n_A = n(m + 1)$ is the code constraint length and p is the channel transition probability. For small p, this bound is dominated by its first term, so that

$$P_{b1}(E) \approx \frac{1}{k} \binom{n_A}{t+1} p^{t+1} (1-p)^{n_A - t - 1} \approx \frac{1}{k} \binom{n_A}{t+1} p^{t+1}. \tag{13.37}$$

For a majority-logic decoder with error-correcting capability $t_{\mathrm{ML}} = \lfloor J/2 \rfloor$ and effective constraint length n_E, an analogous argument yields

$$P_{b1}(E) \leq \frac{1}{k} \sum_{i=t_{\mathrm{ML}}+1}^{n_E} \binom{n_E}{i} p^i (1-p)^{n_E - i} \tag{13.38}$$

and

$$P_{b1}(E) \approx \frac{1}{k} \binom{n_E}{t_{\mathrm{ML}} + 1} p^{t_{\mathrm{ML}}+1}. \tag{13.39}$$

These results strictly apply only to decoding the first information block. However, if each estimated information error block $\hat{\mathbf{e}}_l$ is subtracted from the syndrome equations it affects, and if these estimates are correct, the modified syndrome equations used to estimate $\mathbf{e}_{l+1}$ are identical to those used to estimate $\mathbf{e}_l$, except that different error bits are checked. This point was illustrated in Section 13.1 in connection with Example 13.1. Hence, for a feedback decoder, the bit error probability in decoding any information block, $P_b(E)$, equals $P_{b1}(E)$, assuming that previous estimates have all been correct. Under this assumption, an optimum feedback decoder has bit error probability upper bounded by

$$P_b(E) \leq \frac{1}{k} \sum_{i=t+1}^{n_A} \binom{n_A}{i} p^i (1-p)^{n_A - i}, \tag{13.40}$$

and, for small p, is approximated by

$$P_b(E) \approx \frac{1}{k} \binom{n_A}{t+1} p^{t+1}. \tag{13.41}$$

Similarly, for a feedback majority-logic decoder,

$$P_b(E) \leq \frac{1}{k} \sum_{i=t_{\mathrm{ML}}+1}^{n_E} \binom{n_E}{i} p^i (1-p)^{n_E - i} \tag{13.42}$$

and

$$P_b(E) \approx \frac{1}{k}\binom{n_E}{t_{\mathrm{ML}}+1}p^{t_{\mathrm{ML}}+1}. \tag{13.43}$$

As noted in Section 13.2, if previous decoding estimates are not all correct, post-decoding errors appear in the modified syndrome equations, thereby causing error propagation and degrading the performance of the decoder. If the channel transition probability p is not too large, however, and if the code contains good resynchronization properties, the effect on bit error probability is small and (13.40) through (13.43) remain valid [4]. In the next section we examine the resynchronization properties of several classes of convolutional codes.

13.4 CODE CONSTRUCTION FOR MAJORITY-LOGIC DECODING

Systematic convolutional codes with optimum minimum distance were first constructed by Bussgang [5]. A list of optimum codes for $R = 1/2$ and $R = 1/3$ is given in Table 13.1. Nonsystematic codes are not listed since they cannot achieve larger minimum distances than systematic codes. As a general rule, these codes cannot be completely orthogonalized, and hence cannot be used with majority-logic decoding.

We will now discuss three distinct classes of completely orthogonalizable codes for use with majority-logic decoding: self-orthogonal codes, orthogonalizable codes, and uniform codes. The resynchronization properties of each of these classes when used with feedback decoding is also discussed.

Self-Orthogonal Codes

An (n, k, m) code is said to be *self-orthogonal* if, for each information error bit in block zero, the set of all syndrome bits that check an error bit forms an orthogonal check set on that bit. In other words, sums of syndrome bits are not used to form orthogonal check sets in a self-orthogonal code.

Self-orthogonal codes were first constructed by Massey [1]. A more efficient construction, based on the notion of difference sets, was introduced by Robinson and Bernstein [6]. The *positive difference set* Δ associated with a set of nonnegative integers $\{l_1, l_2, \ldots, l_J\}$, where $l_1 < l_2 < \cdots < l_J$, is defined as the set of $J(J-1)/2$ positive differences $l_b - l_a$, where $l_b > l_a$. A positive difference set Δ is said to be *full* if all the differences in Δ are distinct, and two positive difference sets Δ_i and Δ_j are said to be *disjoint* if they do not contain any differences in common. Now consider an $(n, n-1, m)$ systematic convolutional code with generator polynomials

$$\mathbf{g}_i^{(n)}(D) = g_{i,0}^{(n)} + g_{i,1}^{(n)}D + \cdots + g_{i,K_i}^{(n)}D^{K_i}, \qquad i = 1, 2, \ldots, n-1, \tag{13.44}$$

where the memory order $m = \max_{1 \le i \le n-1} K_i$. Let $g_{i,l_1}^{(n)}, g_{i,l_2}^{(n)}, \ldots, g_{i,l_{J_i}}^{(n)}$ be the nonzero components of $\mathbf{g}_i^{(n)}(D)$, where $l_1 < l_2 < \cdots < l_{J_i}$, $i = 1, 2, \ldots, n-1$, and let Δ_i be the positive difference set associated with the set of integers of $\{l_1, l_2, \ldots, l_{J_i}\}$. The following theorem forms the basis for the construction of self-orthogonal convolutional codes.

TABLE 13.1 SYSTEMATIC CODES WITH OPTIMUM MINIMUM DISTANCE

(a) $R = 1/2$

m	$\mathbf{g}^{(2)}$	d_{min}
1	6	3
2	6*	3
	7	
3	64	4
	70*	
4	64*	4
	72	
5	65	5
	73	
6	650*	5
	730*	
7	670*	6
	714*	
8	670*	6
	715	
9	6710*	6
	7144	
10	6710*	7
	7144*	
11	6517	7
	6703	
	6711	
	6752*	
	7115	
	7144*	
	7154*	
	7306*	
12	67114	8
	71444	
13	67114*	8
	71446	
14	62754*	8
	67027	
	67034*	
	71547	
	71556*	
	75501	
15	653134	9
	732440*	

*The actual value of m for these codes is less than shown, and the optimum value of d_{min} listed is achieved by adding 0's to the generator sequences.

TABLE 13.1 (CONTINUED)

(b) $R = 1/3$ m	$\mathbf{g}^{(2)}$	$\mathbf{g}^{(3)}$	$d_{\min}$
1	4	6	4
	6	6	
2	5	6	5
	5	7	
	6	7	
3	54	60	6
	54	70	
	64	74	
4	54	70*	7
	56	62	
	66	74	
5	55	70	8
	55	71	
	57	62	
	57	63	
	66	74*	
	66	75	
6	550	704	9
	570	634	
	660	744	

Theorem 13.3. An $(n, n-1, m)$ systematic code is self-orthogonal if and only if the positive difference sets $\Delta_1, \Delta_2, \ldots, \Delta_{n-1}$ associated with the code generator polynomials are full and mutually disjoint.

Proof. The proof of this theorem consists of two parts.

(1) Assume that the code is self-orthogonal, and suppose that a positive difference set Δ_i exists which is not full. Then at least two differences in Δ_i are equal, say $l_b - l_a = l_d - l_c$, where $l_b > l_a$ and $l_d > l_c$. From (13.24) the syndrome sequence is given by

$$\mathbf{s}^{(n)}(D) = \sum_{k=1}^{n-1} \mathbf{e}^{(k)}(D)\mathbf{g}_k^{(n)}(D) + \mathbf{e}^{(n)}(D),$$

which can be written as

$$\begin{aligned}\mathbf{s}^{(n)}(D) &= \mathbf{e}^{(i)}(D)\mathbf{g}_i^{(n)}(D) + \sum_{\substack{k=1\\k\neq i}}^{n-1} \mathbf{e}^{(k)}(D)\mathbf{g}_k^{(n)}(D) + \mathbf{e}^{(n)}(D)\\ &= (D^{l_1} + D^{l_2} + \cdots + D^{l_{J_i}})\mathbf{e}^{(i)}(D) + \sum_{\substack{k=1\\k\neq i}}^{n-1} \mathbf{e}^{(k)}(D)\mathbf{g}_k^{(n)}(D) + \mathbf{e}^{(n)}(D).\end{aligned}$$

Since l_a, l_b, l_c, and l_d all belong to the set $\{l_1, l_2, \ldots, l_{J_i}\}$, the syndrome bits $s_{l_a}^{(n)}$, $s_{l_b}^{(n)}$, $s_{l_c}^{(n)}$, and $s_{l_d}^{(n)}$ all check the information error bit $e_0^{(i)}$. In particular,

$$s_{l_b}^{(n)} = e_0^{(i)} + e_{l_b-l_a}^{(i)} + \text{other terms}$$

and

$$s_{l_d}^{(n)} = e_0^{(i)} + e_{l_d-l_c}^{(i)} + \text{other terms}.$$

Since $l_b - l_a = l_d - l_c$, $e_{l_b-l_a}^{(i)} = e_{l_d-l_c}^{(i)}$, and the set of syndrome bits that check the information error bit $e_0^{(i)}$ is not orthogonal, which contradicts the assumption that the code is self-orthogonal.

Now suppose that two difference sets Δ_i and Δ_j exist which are not disjoint. Then they must have at least one difference in common. Let $l_b - l_a$ and $f_d - f_c$ be the common differences in Δ_i and Δ_j, respectively. Then $l_b - l_a = f_d - f_c$, and the syndrome sequence can be written as

$$\begin{aligned}\mathbf{s}^{(n)}(D) &= \mathbf{e}^{(i)}(D)\mathbf{g}_i^{(n)}(D) + \mathbf{e}^{(j)}(D)\mathbf{g}_j^{(n)}(D) + \sum_{\substack{k=1\\k\neq i,j}}^{n-1} \mathbf{e}^{(k)}(D)\mathbf{g}_k^{(n)}(D) + \mathbf{e}^{(n)}(D)\\ &= (D^{l_1} + \cdots + D^{l_{J_i}})\mathbf{e}^{(i)}(D) + (D^{f_1} + \cdots + D^{f_{J_j}})\mathbf{e}^{(j)}(D)\\ &\quad + \sum_{\substack{k=1\\k\neq i,j}}^{n-1} \mathbf{e}^{(k)}(D)\mathbf{g}_k^{(n)}(D) + \mathbf{e}^{(n)}(D).\end{aligned}$$

Assume, without loss of generality, that $l_b = f_d$. Then $l_a = f_c$, and since l_a and l_b belong to the set $\{l_1, l_2, \ldots, l_{J_i}\}$, the syndrome bits $s_{l_a}^{(n)}$ and $s_{l_b}^{(n)}$ both check the information error bit $e_0^{(i)}$, that is,

$$s_{l_a}^{(n)} = e_0^{(i)} + e_{l_a - f_c}^{(j)} + \text{other terms}$$

and

$$s_{l_b}^{(n)} = e_0^{(i)} + e_{l_b - f_d}^{(j)} + \text{other terms.}$$

But $l_b - l_a = f_d - f_c$ implies that $l_b - f_d = l_a - f_c$, and hence $e_{l_a - f_c}^{(j)} = e_{l_b - f_d}^{(j)}$. Therefore, the set of syndrome bits that check the information error bit $e_0^{(i)}$ is not orthogonal, which again contradicts the assumption that the code is self-orthogonal. We conclude that all $n - 1$ positive difference sets must be full and disjoint.

(2) Now assume that the positive difference sets $\Delta_1, \Delta_2, \ldots, \Delta_{n-1}$ are full and disjoint, and suppose that the code is not self-orthogonal. Then there must exist at least one pair of syndrome bits for which

$$s_{l_a}^{(n)} = e_0^{(i)} + e_{l_a - f_c}^{(j)} + \text{other terms}$$

and

$$s_{l_b}^{(n)} = e_0^{(i)} + e_{l_b - f_d}^{(j)} + \text{other terms}$$

and $l_a - f_c = l_b - f_d$. If $i = j$, the differences $l_a - f_c$ and $l_b - f_d$ are both in Δ_i, and hence Δ_i cannot be full. This contradicts the assumption that Δ_i is full. If $i \neq j$, then the difference $l_b - l_a$ is in Δ_i, and the difference $f_d - f_c$ is in Δ_j. Since $l_a - f_c = l_b - f_d$ implies that $l_b - l_a = f_d - f_c$, the positive difference sets Δ_i and Δ_j cannot be disjoint. This contradicts the assumption that Δ_i and Δ_j are disjoint. We conclude that the code must be self-orthogonal. Q.E.D.

Since each of the J_i nonzero components of the generator polynomial $\mathbf{g}_i^{(n)}(D)$ is used to form one of a set of orthogonal check sums on $e_0^{(i)}$, J_i orthogonal check sums are formed on $e_0^{(i)}$. If $J_i = J$, $i = 1, 2, \ldots, n - 1$, then each generator polynomial has the same weight J, and J orthogonal parity checks are formed on each information error bit. Therefore, if $\mathbf{u}^{(1)}(D) = 1$ and $\mathbf{u}^{(2)}(D) = \cdots = \mathbf{u}^{(n-1)}(D) = 0$, the code word $\mathbf{V}(D) = [1, 0, 0, \ldots, 0, \mathbf{g}_1^{(n)}(D)]$ has weight $J + 1$, and $d_{\min} \leq J + 1$. On the other hand, $d_{\min} \geq J + 1$, since otherwise the majority-logic error-correcting capability $t_{\mathrm{ML}} = \lfloor J/2 \rfloor$ would exceed the maximum error-correcting capability $t = \lfloor (d_{\min} - 1)/2 \rfloor$, which is impossible. Hence, $d_{\min} = J + 1$, and the $(n, n - 1, m)$ self-orthogonal code is completely orthogonalizable. If the J_i's are unequal, it can also be shown that $d_{\min} = J + 1$, where $J \triangleq \min_{1 \le i \le n-1} J_i$ (see Problem 13.13). Hence, in this case also, the code is completely orthogonalizable.

Robinson and Bernstein [6] have developed a procedure based on Theorem 13.3 for constructing $(n, n-1, m)$ self-orthogonal convolutional codes. Each of these codes has $J_i = J = d_{\min} - 1$, $i = 1, 2, \ldots, n-1$, and error-correcting capability $t_{ML} = \lfloor J/2 \rfloor = t = \lfloor (d_{\min} - 1)/2 \rfloor$. A list of these codes for $n = 2, 3, 4, 5$ and various values of t and m is given in Table 13.2. In the table, each generator polynomial

TABLE 13.2 SELF-ORTHOGONAL CODES

(a) $R = 1/2$ codes

t_{ML}	m	$\mathbf{g}_1^{(2)}$
1	1	{0, 1}
2	6	{0, 2, 5, 6}
3	17	{0, 2, 7, 13, 16, 17}
4	35	{0, 7, 10, 16, 18, 30, 31, 35}
5	55	{0, 2, 14, 21, 29, 32, 45, 49, 54, 55}
6	85	{0, 2, 6, 24, 29, 40, 43, 55, 68, 75, 76, 85}
7	127	{0, 5, 28, 38, 41, 49, 50, 68, 75, 92, 107, 121, 123, 127}
8	179	{0, 6, 19, 40, 58, 67, 78, 83, 109, 132, 133, 162, 165, 169, 177, 179}
9	216	{0, 2, 10, 22, 53, 56, 82, 83, 89, 98, 130, 148, 153, 167, 188, 192, 205, 216}
10	283	{0, 24, 30, 43, 55, 71, 75, 89, 104, 125, 127, 162, 167, 189, 206, 215, 272, 275, 282, 283}
11	358	{0, 3, 16, 45, 50, 51, 65, 104, 125, 142, 182, 206, 210, 218, 228, 237, 289, 300, 326, 333, 356, 358}
12	425	{0, 22, 41, 57, 72, 93, 99, 139, 147, 153, 197, 200, 214, 253, 263, 265, 276, 283, 308, 367, 368, 372, 396, 425}

(b) $R = 2/3$ codes

t_{ML}	m	$\mathbf{g}_1^{(3)}$	$\mathbf{g}_2^{(3)}$
1	2	{0, 1}	{0, 2}
2	13	{0, 8, 9, 12}	{0, 6, 11, 13}
3	40	{0, 2, 6, 24, 29, 40}	{0, 3, 15, 28, 35, 36}
4	86	{0, 1, 27, 30, 61, 73, 81, 83}	{0, 18, 23, 37, 58, 62, 75, 86}
5	130	{0, 1, 6, 25, 32, 72, 100, 108, 120, 130}	{0, 23, 39, 57, 60, 74, 101, 103, 112, 116}
6	195	{0, 17, 46, 50, 52, 66, 88, 125, 150, 165, 168, 195}	{0, 26, 34, 47, 57, 58, 112, 121, 140, 181, 188, 193}
7	288	{0, 2, 7, 42, 45, 117, 163, 185, 195, 216, 229, 246, 255, 279}	{0, 8, 12, 27, 28, 64, 113, 131, 154, 160, 208, 219, 233, 288}

(c) $R = 3/4$ codes

t_{ML}	m	$\mathbf{g}_1^{(4)}$	$\mathbf{g}_2^{(4)}$	$\mathbf{g}_3^{(4)}$
1	3	{0, 1}	{0, 2}	{0, 3}
2	19	{0, 3, 15, 19}	{0, 8, 17, 18}	{0, 6, 11, 13}
3	67	{0, 5, 15, 34, 35, 42}	{0, 31, 33, 44, 47, 56}	{0, 17, 21, 43, 49, 67}
4	129	{0, 9, 33, 37, 38, 97, 122, 129}	{0, 11, 13, 23, 62, 76, 79, 123}	{0, 19, 35, 50, 71, 77, 117, 125}
5	202	{0, 7, 27, 76, 113, 137, 155, 156, 170, 202}	{0, 8, 38, 48, 59, 82, 111, 146, 150, 152}	{0, 12, 25, 26, 76, 81 98, 107, 143, 197}

TABLE 13.2 (CONTINUED)

(d) $R = 4/5$ codes

t_{ML}	m	$\mathbf{g}_1^{(5)}$	$\mathbf{g}_2^{(5)}$	$\mathbf{g}_3^{(5)}$	$\mathbf{g}_4^{(5)}$
1	4	{0, 1}	{0, 2}	{0, 3}	{0, 4}
2	26	{0, 16, 20, 21}	{0, 2, 10, 25}	{0, 14, 17, 26}	{0, 11, 18, 24}
3	78	{0, 5, 26, 51, 55, 69}	{0, 6, 7, 41, 60, 72}	{0, 8, 11, 24, 44, 78}	{0, 10, 32, 47, 49, 77}
4	178	{0, 19, 59, 68, 85, 88, 103, 141}	{0, 39, 87, 117, 138, 148, 154, 162}	{0, 2, 13, 25, 96, 118, 168, 172}	{0, 7, 65, 70, 97, 98, 144, 178}

$\mathbf{g}_i^{(n)}(D)$ is identified by the set of integers $\{l_1, l_2, \ldots, l_{J_i}\}$, which specify the locations of its nonzero components.

Example 13.12

Consider the (2, 1, 6) self-orthogonal code from Table 13.2(a) whose generator polynomial $\mathbf{g}^{(2)}(D)$ is identified by the set of integers {0, 2, 5, 6} [i.e., $\mathbf{g}^{(2)}(D) = 1 + D^2 + D^5 + D^6$]. This code has $J = 4$ and will correctly estimate the information error bit $e_0^{(1)}$ whenever the first constraint length of received bits contains $t = t_{ML} = \lfloor J/2 \rfloor = 2$ or fewer errors. The positive difference set associated with this generator polynomial is given by $\Delta = \{2, 5, 6, 3, 4, 1\}$. Since this positive difference set contains all positive integers from 1 to $J(J-1)/2 = 6$, it follows that the memory order $m = 6$ of this code is as small as possible for any $R = 1/2$ double-error-correcting self-orthogonal code. Note that the (2, 1, 6) self-orthogonal code of Example 13.1 also has $J = 4$ orthogonal parity checks and is double-error-correcting.

Example 13.13

Consider the (3, 2, 13) self-orthogonal code from Table 13.2(b) whose generator polynomials $\mathbf{g}_1^{(3)}(D)$ and $\mathbf{g}_2^{(3)}(D)$ are identified by the sets of integers {0, 8, 9, 12} and {0, 6, 11, 13}, respectively. This code was previously shown to be self-orthogonal with $J = 4$ in Example 13.4. The positive difference sets associated with $\mathbf{g}_1^{(3)}(D)$ and $\mathbf{g}_2^{(3)}(D)$ are $\Delta_1 = \{8, 9, 12, 1, 4, 3\}$ and $\Delta_2 = \{6, 11, 13, 5, 7, 2\}$. These positive difference sets are full and disjoint, as required by Theorem 13.3 for any self-orthogonal code.

A self-orthogonal $(n, 1, m)$ code can be obtained from a self-orthogonal $(n, n-1, m)$ code in the following way. Consider an $(n, n-1, m)$ self-orthogonal code with generator polynomials $\mathbf{g}_1^{(n)}(D), \mathbf{g}_2^{(n)}(D), \ldots, \mathbf{g}_{n-1}^{(n)}(D)$, and let J_i be the weight of $\mathbf{g}_i^{(n)}(D)$. An $(n, 1, m)$ code with generator polynomials $\mathbf{h}^{(2)}(D), \mathbf{h}^{(3)}(D), \ldots, \mathbf{h}^{(n)}(D)$ is obtained from this $(n, n-1, m)$ self-orthogonal code by setting

$$\mathbf{h}^{(i+1)}(D) = \mathbf{g}_i^{(n)}(D), \qquad i = 1, 2, \ldots, n-1. \tag{13.45}$$

Since the positive difference sets associated with the generator polynomials $\mathbf{g}_i^{(n)}(D)$, $i = 1, 2, \ldots, n-1$, of the $(n, n-1, m)$ self-orthogonal code are full and disjoint, the positive difference sets associated with the generator polynomials $\mathbf{h}^{(j)}(D)$, $i = 2, 3, \ldots, n$, of the $(n, 1, m)$ code are also full and disjoint. It is shown in Problem 13.15 that this condition is necessary and sufficient for the $(n, 1, m)$ code to be self-orthogonal. Since J_i is the weight of $\mathbf{h}^{(i+1)}(D)$, there are a total of $J \triangleq J_1 + J_2 + \cdots +$

J_{n-1} orthogonal check sums on the information error bit $e_0^{(1)}$, and the $(n, 1, m)$ code has a majority-logic error-correcting capability of $t_{ML} = \lfloor J/2 \rfloor$. Since the information sequence $\mathbf{u}(D) = 1$ results in the code word $\mathbf{V}(D) = [1, \mathbf{h}^{(2)}(D), \ldots, \mathbf{h}^{(n)}(D)]$, which has weight $J + 1$, it follows that $d_{min} \leq J + 1$. But $t = \lfloor (d_{min} - 1)/2 \rfloor \geq t_{ML} = \lfloor J/2 \rfloor$ implies that $d_{min} \geq J + 1$. Hence, $d_{min} = J + 1$, and the $(n, 1, m)$ self-orthogonal code must be completely orthogonalizable.

Example 13.14

Consider the (3, 1, 13) self-orthogonal code with generator polynomials $\mathbf{h}^{(2)}(D) = 1 + D^8 + D^9 + D^{12}$ and $\mathbf{h}^{(3)}(D) = 1 + D^6 + D^{11} + D^{13}$ derived from the (3, 2, 13) self-orthogonal code of Example 13.13. This code has $J = J_1 + J_2 = 4 + 4 = 8$ orthogonal check sums on the information error bit $e_0^{(1)}$, and has an error-correcting capability of $t = t_{ML} = \lfloor J/2 \rfloor = 4$; that is, it correctly estimates $e_0^{(1)}$ whenever there are four or fewer errors in the first constraint length of received bits. Each $(n, n - 1, m)$ self-orthogonal code in Table 13.2 can be converted to an $(n, 1, m)$ self-orthogonal code in the same way.

If a self-orthogonal code is used in the feedback decoding mode, each successive block of information error bits is estimated using the same decoding rule, and the error-correcting capability is the same as for block zero, assuming that previous estimates have all been correct. This was illustrated in equations (13.16) to (13.19) for the self-orthogonal code of Example 13.1. On the other hand, if all previous estimates have not been correct, post-decoding errors appear in the syndrome equations, as illustrated in (13.30) for the same self-orthogonal code. In this case, the modified syndrome equations are still orthogonal, but post-decoding errors as well as transmission errors can cause further decoding errors. This is the error propagation effect of feedback decoders discussed in Section 13.2.

We now demostrate that self-orthogonal codes possess automatic resynchronization properties with respect to error propagation. First note that whenever a decoding estimate is 1 [i.e., $\hat{e}_i^{(j)} = 1$], it is fed back and reduces the total number of 1's stored in the syndrome registers. This follows from the fact that more than half the syndrome bits that are used to estimate $e_i^{(j)}$ must be 1's in order to cause an estimate of 1; and hence when $\hat{e}_i^{(j)} = 1$ is fed back it changes more 1's to 0's than 0's to 1's, thereby reducing the total weight in the syndrome registers. Now assume that there are no more transmission errors after time unit i. Then beginning at time unit $i + m$, only 0's can enter the left most stages of the syndrome registers. No matter what the contents of the registers are at time unit $i + m$, they must soon clear to all 0's since only 0's are shifted in and each estimate of 1 reduces the total number of 1's in the registers. (Clearly, estimates of 0 have no effect on the register weights.) Once the syndrome registers have resynchronized to all 0's, no further decoding errors are possible unless there are additional transmission errors (i.e., error propagation has been terminated).

Orthogonalizable Codes

Completely orthogonalizable convolutional codes can also be constructed using a trial-and-error approach. In this case some of the orthogonal parity checks are formed from sums of syndrome bits, and these codes are not self-orthogonal. The

(2, 1, 5) code of Example 13.2 contains $J = 4$ orthogonal parity checks constructed by trial-and-error, and hence $d_{min} \geq 5$. Since the generator polynomials have weight 5, d_{min} must equal 5, and therefore the code is completely orthogonalizable with error-correcting capability $t = t_{ML} = \lfloor J/2 \rfloor = 2$. Similarly the (3, 1, 4) code of Example 13.5 contains $J = 6$ orthogonal parity checks constructed by trial-and-error, is completely orthogonalizable, and has error-correcting capability $t = t_{ML} = \lfloor J/2 \rfloor = 3$. A list of orthogonalizable codes constructed by Massey [1] is given in Table 13.3. The notation used to describe the rules for forming the orthogonal check sums will be explained in an example.

Example 13.15

Consider the (3, 1, 7) code listed in Table 13.3(b) whose generator polynomials $\mathbf{g}^{(2)}(D)$ and $\mathbf{g}^{(3)}(D)$ are specified by the sets of integers {0, 1, 7} and {0, 2, 3, 4, 6} [i.e., $\mathbf{g}^{(2)}(D) = 1 + D + D^7$ and $\mathbf{g}^{(3)}(D) = 1 + D^2 + D^3 + D^4 + D^6$]. The rules for forming

TABLE 13.3 ORTHOGONALIZABLE CODES

(a) $R = 1/2$ codes

t_{ML}	m	$\mathbf{g}_1^{(2)}$	Orthogonalization rules
2	5	{0, 3, 4, 5}	$(0^2)(3^2)(4^2)(1^2 5^2)$
3	11	{0, 6, 7, 9, 10, 11}	$(0^2)(6^2)(7^2)(9^2)(1^2 3^2 10^2)(4^2 8^2 11^2)$
4	21	{0, 11, 13, 16, 17, 19, 20, 21}	$(0^2)(11^2)(13^2)(16^2)(17^2)(2^2 3^2 6^2 19^2)(4^2 14^2 20^2)(1^2 5^2 8^2 15^2 21^2)$
5	35	{0, 18, 19, 27, 28, 29, 30, 32, 33, 35}	$(0^2)(18^2)(19^2)(27^2)(1^2 9^2 28^2)(10^2 20^2 29^2)(11^2 30^2 31^2)(13^2 21^2 23^2 32^2)(14^2 33^2 34^2)(2^2 3^2 16^2 24^2 26^2 35^2)$
6	51	{0, 26, 27, 39, 40, 41, 42, 44, 45, 47, 48, 51}	$(0^2)(26^2)(27^2)(39^2)(1^2 13^2 40^2)(14^2 28^2 41^2)(15^2 42^2 43^2)(17^2 29^2 31^2 44^2)(18^2 45^2 46^2)(2^2 3^2 20^2 32^2 34^2 47^2)(21^2 35^2 48^2 49^2 50^2)(24^2 30^2 33^2 36^2 38^2 51^2)$

(b) $R = 1/3$ codes

t_{ML}	m	$\mathbf{g}_1^{(2)}$	$\mathbf{g}_1^{(3)}$	Orthogonalization rules
3	4	{0, 1}	{0, 2, 3, 4}	$(0^2)(0^3)(1^2)(2^3)(1^3 3^3)(2^2 4^3)$
4	7	{0, 1, 7}	{0, 2, 3, 4, 6}	$(0^2)(0^3)(1^2)(2^3)(1^3 3^3)(2^2 4^3)(7^2)(3^2 5^2 6^2 6^3)$
5	10*	{0, 1, 9}	{0, 1, 2, 3, 5, 8, 9}	$(0^2)(0^3)(1^2)(2^2 2^3)(9^2)(3^3 4^3)(3^2 5^2 5^3)(1^3 4^2 6^2 6^3)(8^2 8^3)(7^3 9^3 10^3)$
6	17*	{0, 4, 5, 6, 7, 9, 12, 13, 16}	{0, 1, 14, 15, 16}	$(0^2)(0^3)(1^2 1^3)(4^2)(5^2)(2^3 6^2)(14^3)(7^2 10^2 11^2 11^3)(3^3 5^3 9^2)(6^3 8^3 12^2)(3^3 16^3 17^3)(4^3 10^3 12^3 16^2)$
7	22	{0, 4, 5, 6, 7, 9, 12, 13, 16, 19, 20, 21}	{0, 1, 20, 22}	$(0^2)(0^3)(1^2 1^3)(4^2)(5^2)(2^3 6^2)(7^2 10^2 11^2 11^3)(3^3 5^3 9^2)(19^3 20^3)(22^3)(6^3 8^3 12^2)(4^3 10^3 12^3 16^2)(3^2 7^3 13^3 15^3 19^2)(9^3 13^2 14^3 18^2 20^2 21^2 21^3)$
8	35	{0, 4, 5, 6, 7, 9, 12, 16, 17, 30, 31}	{0, 1, 22, 25, 35}	$(0^2)(0^3)(1^2 1^3)(4^2)(5^2)(2^3 6^2)(22^3)(7^2 10^2 11^2 11^3)(3^2 25^3)(3^3 5^3 9^2)(6^3 8^3 12^2)(7^3 14^2 17^2 18^2 18^3)(9^3 16^2 19^2 20^2 20^3)(14^3 15^3 35^3)(12^3 21^3 28^2 31^2 32^2)(10^3 13^3 19^3 26^3 29^3 30^2)$

*The actual value of m for these codes is less than shown, and the error-correcting capability t listed is achieved by adding 0's to the generator sequences.

TABLE 13.3 (CONTINUED)

(c) $R = 1/5$ codes

t_{ML}	m	$\mathbf{g}_1^{(2)}$	$\mathbf{g}_1^{(3)}$	$\mathbf{g}_1^{(4)}$	$\mathbf{g}_1^{(5)}$	Orthogonalization rules
3	1	{0, 1}	{0, 1}	{0}	{0}	$(0^2)(0^3)(0^4)(0^5)(1^2 1^4)(1^3 1^5)$
4	2	{0, 1, 2}	{0, 1}	{0, 2}	{0}	$(0^2)(0^3)(0^4)(0^5)(1^2 1^4)(1^3 1^5)(2^2 2^3)$ $(2^4 2^5)$
5	3	{0, 1, 2, 3}	{0, 1}	{0, 2}	{0, 3}	$(0^2)(0^3)(0^4)(0^5)(1^2 1^4)(1^3 1^5)(2^2 2^3)$ $(2^4 2^5)(3^5)(3^2 3^3)$
6	5	{0, 1, 2, 3, 4}	{0, 1}	{0, 2, 5}	{0, 3, 5}	$(0^2)(0^3)(0^4)(0^5)(1^2 1^4)(1^3 1^5)(2^2 2^3)$ $(2^4 2^5)(3^5)(3^2 3^3)(3^4 4^2 4^4)(5^5)$ $(4^3 5^3 5^4)$
7	6	{0, 1, 2, 3, 4}	{0, 1}	{0, 2, 5, 6}	{0, 3, 5}	$(0^2)(0^3)(0^4)(0^5)(1^2 1^4)(1^3 1^5)(2^2 2^3)$ $(2^4 2^5)(3^5)(3^2 3^3)(3^4 4^2 4^4)(5^5)$ $(4^3 5^3 5^4)(4^5 6^4)$
8	8	{0, 1, 2, 3, 4}	{0, 1, 8}	{0, 2, 5, 6, 7}	{0, 3, 5}	$(0^2)(0^3)(0^4)(0^5)(1^2 1^4)(1^3 1^5)(2^2 2^3)$ $(2^4 2^5)(3^5)(3^2 3^3)(3^4 4^2 4^4)(5^5)$ $(4^3 5^3 5^4)(4^5 6^4)(8^3)(5^2 6^3 7^2 7^4)$
9	10	{0, 1, 2, 3, 5, 6, 8, 10}	{0, 3, 5, 6, 8}	{0, 1}	{0, 2, 10}	$(0^2)(0^3)(0^4)(0^5)(1^2)(2^2 2^4)(3^3)(1^4 1^5)$ $(2^3 2^5)(3^2 4^2)(3^4 5^2 5^4)(9^4 10^2 10^3)$ $(5^3)(3^5 6^3)(10^5)(1^3 4^4 6^2 6^4)$ $(7^2 7^4 8^2 9^2)(4^3 5^5 7^3 8^3 8^4)$
10	12	{0, 1, 2, 3, 5, 6, 8, 10}	{0, 3, 5, 6, 8}	{0, 1, 10}	{0, 2, 10, 12}	$(0^2)(0^3)(0^4)(0^5)(1^2)(2^2 2^4)(3^3)(1^4 1^5)$ $(2^3 2^5)(3^2 4^2)(3^4 5^2 5^4)(9^4 10^2 10^3)$ $(5^3)(3^5 6^3)(10^5)(1^3 4^4 6^2 6^4)$ $(7^2 7^4 8^2 9^2)(4^3 5^5 7^3 8^3 8^4)$ $(6^5 9^5 12^3)(10^4 11^5 12^4 12^5)$
11	15	{0, 1, 2, 3, 5, 6, 8, 10, 11, 13, 14}	{0, 3, 5, 6, 8}	{0, 1, 10}	{0, 2, 10, 12, 15}	$(0^2)(0^3)(0^4)(0^5)(1^2)(2^2 2^4)(3^3)(1^4 1^5)$ $(2^3 2^5)(3^2 4^2)(3^4 5^2 5^4)(9^4 10^2 10^3)$ $(5^3)(3^5 6^3)(10^5)(1^3 4^4 6^2 6^4)$ $(7^2 7^4 8^2 9^2)(4^3 5^5 7^3 8^3 8^4)$ $(6^5 9^5 12^3)(10^4 11^5 12^4 12^5)$ $(4^5 13^4 14^2 14^3)(13^5 14^4 15^4 15^5)$

$J = 8$ orthogonal parity checks are given by the set

$$\{(0^2), (0^3), (1^2), (2^3), (1^3 3^3), (2^2 4^3), (7^2), (3^2 5^2 6^2 6^3)\},$$

where the notation k^i indicates that the syndrome bit $s_k^{(i)}$ forms an orthogonal check sum on $e_0^{(1)}$ and $(k^i l^j)$ indicates that the sum $s_k^{(i)} + s_l^{(j)}$ forms an orthogonal check sum on $e_0^{(1)}$. The $J = 8$ orthogonal parity checks on $e_0^{(1)}$ are then given by the set

$$\{s_0^{(2)}, s_0^{(3)}, s_1^{(2)}, s_2^{(3)}, s_1^{(3)} + s_3^{(3)}, s_2^{(2)} + s_4^{(3)}, s_7^{(2)}, s_3^{(2)} + s_5^{(2)} + s_6^{(2)} + s_6^{(3)}\}.$$

This code has $d_{\min} = 9$, is completely orthogonalizable, and has error-correcting capability $t = t_{\text{ML}} = \lfloor J/2 \rfloor = 4$.

Note that (n, k, m) orthogonalizable codes can achieve a given error-correcting capability t with a smaller memory order m than self-orthogonal codes. This is due to the added flexibility available in using sums of syndrome bits to form orthogonal parity checks for orthogonalizable codes. The major disadvantage of orthogonalizable codes is that they do not possess the automatic resynchronization properties that limit error propagation in the feedback decoding mode. In addition, their error-correcting capability is reduced when used with definite decoding, as was demonstrated in Example 13.8.

Uniform Codes

Now consider the class of $(2^m, 1, m)$ codes whose generator polynomials $\mathbf{g}^{(j)}(D) = g_0^{(j)} + g_1^{(j)}D + \cdots + g_m^{(j)}D^m$ are such that the sets $(g_1^{(j)}, g_2^{(j)}, \ldots, g_m^{(j)})$, $j = 1, 2, \ldots, 2^m$, are all 2^m distinct m-tuples, where for convenience we choose $\mathbf{g}^{(1)}(D) = 1$. These codes have the property that the code word $\mathbf{v}$ corresponding to any information sequence $\mathbf{u}$ with $\mathbf{u}_0 \neq \mathbf{0}$ has $w([\mathbf{v}]_m) = (m + 2)2^{m-1}$; that is, all code words with a nonzero first information block have the same weight over the first constraint length. This implies that $d_{\min} = (m + 2)2^{m-1}$ for these codes (see Problem 13.17). Massey [7] was the first to study these codes, and he called them *uniform codes*. They represent a class of low-rate convolutional codes with large error-correcting capability similar to the maximal-length block codes. Like maximal-length block codes, they can be completely orthogonalized and can be decoded by majority-logic decoding. This will be illustrated by means of an example.

Example 13.16

Consider the (4, 1, 2) uniform code with $\mathbf{g}^{(1)}(D) = 1$, $\mathbf{g}^{(2)}(D) = 1 + D$, $\mathbf{g}^{(3)}(D) = 1 + D^2$, and $\mathbf{g}^{(4)}(D) = 1 + D + D^2$. This code has $d_{\min} = (m + 2)2^{m-1} = 8$. The modified syndrome bits from time unit i through time unit $i + m$ can be written as

$$\begin{bmatrix} s_i'^{(2)} \\ s_{i+1}'^{(2)} \\ s_{i+2}^{(2)} \\ s_i'^{(3)} \\ s_{i+1}'^{(3)} \\ s_{i+2}^{(3)} \\ s_i'^{(4)} \\ s_{i+1}'^{(4)} \\ s_{i+2}^{(4)} \end{bmatrix} = \begin{bmatrix} 0 & 1 & 1 & & \\ & 0 & 1 & 1 & \\ & & 0 & 1 & 1 \\ 1 & 0 & 1 & & \\ & 1 & 0 & 1 & \\ & & 1 & 0 & 1 \\ 1 & 1 & 1 & & \\ & 1 & 1 & 1 & \\ & & 1 & 1 & 1 \end{bmatrix} \begin{bmatrix} \tilde{e}_{i-2}^{(1)} \\ \tilde{e}_{i-1}^{(1)} \\ e_i^{(1)} \\ e_{i+1}^{(1)} \\ e_{i+2}^{(1)} \end{bmatrix} + \begin{bmatrix} e_i^{(2)} \\ e_{i+1}^{(2)} \\ e_{i+2}^{(2)} \\ e_i^{(3)} \\ e_{i+1}^{(3)} \\ e_{i+2}^{(3)} \\ e_i^{(4)} \\ e_{i+1}^{(4)} \\ e_{i+2}^{(4)} \end{bmatrix}. \tag{13.46}$$

The parity triangles in this case have been replaced by "parity parallelograms" due to the inclusion of the post-decoding errors in the equations. Now form a set of check sums that check $e_i^{(1)}$ by:

1. Using alone the first row in each parity parallelogram.
2. Using alone any row that checks only one bit in the information error vector $[\tilde{e}_{i-2}^{(1)} \;\; \tilde{e}_{i-1}^{(1)} \;\; e_i^{(1)} \;\; e_{i+1}^{(1)} \;\; e_{i+2}^{(1)}]^T$ besides $e_i^{(1)}$.
3. Adding to any other row that checks $e_i^{(1)}$ and at least two other bits in $[\tilde{e}_{i-2}^{(1)} \;\; \tilde{e}_{i-1}^{(1)} \;\; e_i^{(1)} \;\; e_{i+1}^{(1)} \;\; e_{i+2}^{(1)}]^T$ the unique row that checks the same other bits in $[\tilde{e}_{i-2}^{(1)} \;\; \tilde{e}_{i-1}^{(1)} \;\; e_i^{(1)} \;\; e_{i+1}^{(1)} \;\; e_{i+2}^{(1)}]^T$ but does not check $e_i^{(1)}$.

This results in the check set

$$
\begin{aligned}
s_i'^{(2)} &= \qquad\quad\;\; \tilde{e}_{i-1}^{(1)} + e_i^{(1)} \qquad\qquad + e_i^{(2)} \\
s_i'^{(3)} &= \tilde{e}_{i-2}^{(1)} \qquad\quad\;\; + e_i^{(1)} \qquad\qquad + e_i^{(3)} \\
s_i'^{(4)} &= \tilde{e}_{i-2}^{(1)} + \tilde{e}_{i-1}^{(1)} + e_i^{(1)} \qquad\qquad + e_i^{(4)} \\
s_{i+1}'^{(2)} &= \qquad\qquad\qquad\;\; e_i^{(1)} + e_{i+1}^{(1)} \qquad + e_{i+1}^{(2)} \qquad\qquad (13.47) \\
s_{i+2}^{(3)} &= \qquad\qquad\qquad\;\; e_i^{(1)} \qquad\quad + e_{i+2} + e_{i+2}^{(3)} \\
s_{i+1}'^{(4)} + s_{i+1}'^{(3)} &= \qquad\qquad\qquad\;\; e_i^{(1)} \qquad\qquad\quad + e_{i+1}^{(4)} + e_{i+1}^{(3)} \\
s_{i+2}^{(4)} + s_{i+2}^{(2)} &= \qquad\qquad\qquad\;\; e_i^{(1)} \qquad\qquad\quad + e_{i+2}^{(4)} + e_{i+2}^{(2)}.
\end{aligned}
$$

Assuming that all past decoding decisions affecting these check equations are correct [i.e., $\tilde{e}_{i-1}^{(1)} = \tilde{e}_{i-2}^{(1)} = 0$], (13.47) forms a set of $J = 7 = d_{\min} - 1$ orthogonal parity checks on $e_i^{(1)}$, and the code is completely orthogonalizable. It can be shown that for general m the construction described above always results in a set of $J = d_{\min} - 1$ orthogonal parity checks on $e_i^{(1)}$ (see Problem 13.19).

Sullivan [8] has shown that uniform codes also possess automatic resynchronization properties. From Example 13.16 we see that the post-decoding errors affect only the first three check equations, and that any combination of post-decoding errors will change exactly two parity checks. For example, if $\tilde{e}_{i-1}^{(1)} = 1$, syndrome bits $s_i'^{(2)}$ and $s_i'^{(4)}$ are changed, if $\tilde{e}_{i-2}^{(1)} = 1$, $s_i'^{(3)}$ and $s_i'^{(4)}$ are changed, and if $\tilde{e}_{i-1}^{(1)} = \tilde{e}_{i-2}^{(1)} = 1$, $s_i'^{(2)}$ and $s_i'^{(3)}$ are changed. Hence, any combination of post-decoding errors has the same effect on the orthogonal check equations as two channel errors. But $d_{\min} = 8$, so that at least one actual channel error can be correctly decoded even when post-decoding errors have been made. Thus, there is not only no error propagation after a decoding error, but the code still has the ability to correct some channel errors. In general, post-decoding errors affect only the first $2^m - 1$ check equations, and have the same effect as exactly 2^{m-1} channel errors. Since $d_{\min} = (m + 2)2^{m-1}$, the error-correcting capability in the absence of error propagation is $t = \lfloor (d_{\min} - 1)/2 \rfloor = m2^{m-2} + 2^{m-1} - 1$. When post-decoding errors affect the syndrome equations, the code is still capable of correcting $m2^{m-2} - 1$ channel errors.

13.5 COMPARISON WITH PROBABILISTIC DECODING

The three principal ways of decoding convolutional codes, Viterbi decoding, sequential decoding, and majority-logic decoding, all have different characteristics. They can be compared on the basis of their performance, decoding speed, and implementation complexity.

Performance. Viterbi decoding is an optimum decoding procedure for convolutional codes, and its bit error probability on the BSC was shown in (11.27) to be approximated by

$$P_b(E) \approx C_v e^{-(Rd_{\text{free}}/2)\cdot(E_b/N_0)} \tag{13.48}$$

when E_b/N_0 is large, where C_v is a constant associated with code structure. In Chapter 12 it was shown that sequential decoding achieved nearly the same performance.

For an optimum feedback decoder, it was shown in Section 13.3 that for a BSC,

$$P_b(E) \approx \frac{1}{k}\binom{n_A}{t+1}p^{t+1} \tag{13.49}$$

for small p. Using the approximation

$$p \approx \tfrac{1}{2}e^{-E_bR/N_0} \tag{13.50}$$

yields

$$\begin{aligned} P_b(E) &\approx \frac{1}{k}\binom{n_A}{t+1}\left(\frac{1}{2}\right)^{t+1} e^{-E_bR(t+1)/N_0} \\ &\approx C_t e^{-(Rd_{\min}/2)\cdot(E_b/N_0)} \end{aligned} \tag{13.51}$$

when E_b/N_0 is large, where C_t is a constant associated with code structure. If the code is completely orthogonalizable, (13.51) also represents the performance of majority-logic decoding. Equations (13.48) and (13.51) differ in that $P_b(E)$ decreases exponentially with d_{free} for Viterbi (or sequential) decoding, whereas $P_b(E)$ decreases exponentially with $d_{\min}$ for feedback (or majority-logic) decoding. Since for most nonsystematic codes

$$d_{\min} = d_m < \lim_{i\to\infty} d_i = d_{\text{free}}, \tag{13.52}$$

the asymptotic performance of feedback decoding is inferior to that of Viterbi decoding. This is due to the fact that a Viterbi decoder delays making decisions until the entire received sequence is processed, or, in the case of a truncated decoder, until at least four or five constraint lengths are received. It thus bases its decisions on the total (or free) distance between sequences. A feedback decoder, on the other hand, makes decisions with a delay of only one constraint length. Hence, its decisions are based on the minimum distance over just one constraint length.

Random coding bounds indicate that for a given rate and constraint length, the free distance of the best nonsystematic codes is about twice the minimum distance of the best systematic codes [9]. (Recall that nonsystematic codes cannot achieve larger values of $d_{\min}$ than systematic codes.) This gives Viterbi decoding roughly a 3-dB advantage in performance over feedback decoding for the same rate and constraint length. In other words, the constraint length for feedback decoding must be about twice that for Viterbi decoding in order to achieve comparable performance. If majority-logic decoding is being used, even a longer constraint length must be employed. This is due to the fact that optimum minimum distance codes are usually not completely orthogonalizable; that is, to achieve a given $d_{\min}$ for a completely orthogonalizable code requires a longer constraint length than would otherwise be necessary. If the code must be self-orthogonal (say, to protect against error propagation), the constraint length must be longer yet. For example, the best $R = 1/2$ self-orthogonal code with $d_{\min} = 7$ has constraint length $n_A = 36$, the best $R = 1/2$ orthogonalizable code with $d_{\min} = 7$ has $n_A = 24$, the best $R = 1/2$ systematic code with $d_{\min} = 7$ has $n_A = 22$, and the best $R = 1/2$ nonsystematic code with $d_{\text{free}} = 7$ has $n_A = 14$.

Decoding speed. A Viterbi decoder requires 2^K computations per decoded information bit, a sequential decoder has a variable computational load which typically averages between 1 and 2 computations per bit, depending on E_b/N_0, and a majority-logic decoder requires only one computation (one cycle of the decoder registers) per bit. Although the time required to perform a computation is different in each case, this comparison indicates that majority-logic decoders are capable of higher-speed operation than Viterbi or sequential decoders. As noted in Chapter 11, the speed of Viterbi decoding can be increased by a factor of 2^K, making it attractive for high-speed applications, by using a parallel implementation. This greatly increases the cost of the decoder, however.

Decoding delay. A majority-logic decoder has a decoding delay of exactly one constraint length; that is, bits received at time unit i are decoded at time unit $i + m$. Viterbi decoding and sequential decoding, on the other hand, have a decoding delay equal to the entire frame length $L + m$; that is, no decoding decisions are made until all $L + m$ encoded blocks have been received. Since typically $L \gg m$, the decoding delay is substantial in these cases. If truncated Viterbi decoding, or backsearch limited sequential decoding, is used, the decoding delay is reduced to four or five constraint lengths with a minor penalty in performance.

Implementation complexity. Majority-logic decoders are simpler to implement than either Viterbi decoders or sequential decoders. Besides a replica of the encoder and a buffer register to store the received information bits, all that is needed is a syndrome register, some modulo-2 adders (exclusive-or gates), and a majority-logic gate. These relatively modest implementation requirements make majority-logic decoders particularly attractive in low-cost applications. However, if large minimum distances are needed to achieve high performance, very long constraint lengths are required, and the implementation complexity increases. In these cases, maximum likelihood decoders (Viterbi or sequential) can provide a more efficient trade-off between performance and complexity.

Finally, we have seen in Chapter 11 that Viterbi decoders can easily be adapted to take advantage of soft demodulator decisions. Soft-decision majority-logic decoders have also been developed, but at a considerable increase in implementation complexity. Massey's [1] APP decoding and Rudolph's [10] generalized majority-logic decoding are examples of such schemes.

PROBLEMS

13.1. Show that for the BSC, the received sequence $\mathbf{r}$ is a code word if and only if the error sequence $\mathbf{e}$ is a code word.

13.2. Using the definition of the $\mathbf{H}$ matrix for $R = 1/2$ systematic codes given by (13.6), show that (13.5) and (13.10) are equivalent.

13.3. Draw the complete encoder/decoder block diagram for Example 13.2.

13.4. Consider the (2, 1, 11) systematic code with $\mathbf{g}^{(2)}(D) = 1 + D + D^3 + D^5 + D^8 + D^9 + D^{10} + D^{11}$.

(a) Find the parity-check matrix $\mathbf{H}$.

(b) Write equations for the first constraint length of syndrome bits $s_0, s_1, \ldots, s_{11}$ in terms of the channel error bits.

(c) Write equations for the modified syndrome bits $s'_i, s'_{i+1}, \ldots, s_{i+11}$, assuming that the effect of error bits prior to time unit i has been removed by feedback.

13.5. Consider the (3, 2, 13) code of Example 13.4 and the (3, 1, 4) code of Example 13.5.

(a) Find the transfer function matrix $\mathbf{G}(D)$.

(b) Find the parity transfer function matrix $\mathbf{H}(D)$.

(c) Show that in each case $\mathbf{G}(D)\mathbf{H}^T(D) = \mathbf{0}$.

13.6. Consider the (3, 2, 13) code of Example 13.4.

(a) Write equations for the unmodified syndrome bits $s_i, s_{i+1}, \ldots, s_{i+13}$, which includes the effect of error bits prior to time unit i (assume that $i \geq 13$).

(b) Find a set of orthogonal parity checks for both $e_i^{(1)}$ and $e_i^{(2)}$ from the unmodified syndrome equations.

(c) Determine the majority-logic error-correcting capability t_{ML} and the effective constraint length n_E with definite decoding.

(d) Draw the block diagram of the definite decoder.

13.7. Consider an (n, k, m) convolutional code with minimum distance $d_{\min} = 2t + 1$. Prove that there is at least one error sequence $\mathbf{e}$ with weight $t + 1$ in its first constraint length for which an optimum feedback decoder will decode $\mathbf{u}_0$ incorrectly.

13.8. Consider the (2, 1, 11) code of Problem 13.4.

(a) Find the minimum distance $d_{\min}$.

(b) Is this code self-orthogonal?

(c) Find the maximum number of orthogonal parity checks which can be formed on $e_0^{(1)}$.

(d) Is this code completely orthogonalizable?

13.9. Consider the (3, 1, 3) nonsystematic code with $\mathbf{g}^{(1)}(D) = 1 + D + D^3$, $\mathbf{g}^{(2)}(D) = 1 + D^3$, $\mathbf{g}^{(3)}(D) = 1 + D + D^2$.

(a) Find the transformation matrix $\mathbf{T}$ needed to convert this code to a (3, 1, 3) systematic code with the same $d_{\min}$.

(b) Find the transfer function matrix of the systematic code.

(c) Find the minimum distance $d_{\min}$.

13.10. Describe a general procedure for converting an (n, k, m) nonsystematic code to an (n, k, m) systematic code with the same $d_{\min}$.

13.11. Consider the (2, 1, 5) code of Example 13.10.

(a) Estimate the bit error probability $P_b(E)$ of an optimum feedback decoder on a BSC with small transition probability p.

(b) Repeat part (a) for a feedback threshold decoder.

(c) Compare the results of parts (a) and (b) for $p = 10^{-2}$.

13.12. Repeat Problem 13.11 for the (2, 1, 6) code of Example 13.9.

13.13. Consider an $(n, n - 1, m)$ self-orthogonal code with J_i orthogonal check sums on $e_0^{(i)}$, $i = 1, 2, \ldots, n - 1$. Show that $d_{\min} = J + 1$, where $J \triangleq \max_{1 \leq i \leq n-1} J_i$.

13.14. Consider the (2, 1, 17) self-orthogonal code in Table 13.2(a).

(a) Form the orthogonal check sums on information error bit $e_0^{(1)}$.

(b) Draw the block diagram of the feedback majority-logic decoder for this code.

13.15. Consider an $(n, 1, m)$ systematic code with generator polynomials $\mathbf{g}^{(j)}(D)$, $j = 2, 3, \ldots, n$. Show that the code is self-orthogonal if and only if the positive difference sets associated with each generator polynomial are full and disjoint.

13.16. Consider the (2, 1, 11) orthogonalizable code in Table 13.3(a).

(a) Form the orthogonal check sums on information error bit $e_0^{(1)}$.

(b) Draw the block diagram of the feedback majority-logic decoder for this code.

13.17. For a uniform code, show that the code word $\mathbf{v}$ corresponding to any information sequence $\mathbf{u}$ with $\mathbf{u}_0 \neq \mathbf{0}$ has $w([\mathbf{v}]_m) = (m + 2)2^{m-1}$.

13.18. Consider the (8, 1, 3) uniform code. Form a set of $J = 19$ orthogonal check sums on information error bit $e_0^{(1)}$.

13.19. Show that the procedure described in Example 13.16 for constructing orthogonal check sums for uniform codes always results in a set of $J = d_{min} - 1$ parity checks orthogonal on $e_i^{(1)}$.

13.20. Find and compare the constraint lengths of the following codes:

(a) The best $R = 1/2$ self-orthogonal code with $d_{min} = 9$.

(b) The best $R = 1/2$ orthogonalizable code with $d_{min} = 9$.

(c) The best $R = 1/2$ systematic code with $d_{min} = 9$.

(d) The best $R = 1/2$ nonsystematic code with $d_{free} = 9$.

REFERENCES

1. J. L. Massey, *Threshold Decoding*, MIT Press, Cambridge, Mass., 1963.
2. J. L. Massey and R. W. Liu, "Application of Lyapunov's Direct Method to the Error-Propagation Effect in Convolutional Codes," *IEEE Trans. Inf. Theory*, IT-10, pp. 248–250, July 1964.
3. J. P. Robinson, "Error Propagation and Definite Decoding of Convolutional Codes," *IEEE Trans. Inf. Theory*, IT-14, pp. 121–128, January 1968.
4. T. N. Morrissey, Jr., "A Unified Markovian Analysis of Decoders for Convolutional Codes," Technical Report No. EE-687, Department of Electrical Engineering, University of Notre Dame, Notre Dame, Ind., October 1968.
5. J. J. Bussgang, "Some Properties of Binary Convolutional Code Generators," *IEEE Trans. Inf. Theory*, IT-11, pp. 90–100, January 1965.
6. J. P. Robinson and A. J. Bernstein, "A Class of Binary Recurrent Codes with Limited Error Propagation," *IEEE Trans. Inf. Theory*, IT-13, pp. 106–113, January 1967.
7. J. L. Massey, "Uniform Codes," *IEEE Trans. Inf. Theory*, IT-12, pp. 132–134, April 1966.
8. D. D. Sullivan, "Error-Propagation Properties of Uniform Codes," *IEEE Trans. Inf. Theory*, IT-15, pp. 152–161, January 1969.
9. D. J. Costello, Jr., "Free Distance Bounds for Convolutional Codes," *IEEE Trans. Inf. Theory*, IT-20, pp. 356–365, May 1974.
10. L. D. Rudolph, "Generalized Threshold Decoding of Convolutional Codes," *IEEE Trans. Inf. Theory*, IT-16, pp. 739–745, November 1970.

14

Burst-Error-Correcting Convolutional Codes

Convolutional codes for correcting burst errors were first constructed by Hagelbarger [1]. More efficient codes of the same type were later constructed independently by Iwadare [2] and Massey [3]. These codes require shorter guard spaces than Hagelbarger's codes for the same burst-error-correcting capability. Convolutional codes for correcting phased-burst errors were first studied by Wyner and Ash [4]. Optimal codes of the same type were later discovered independently by Berlekamp [5] and Preparata [6]. Both the Iwadare–Massey codes and the Berlekamp–Preparata codes are presented in Section 14.2.

Interleaving can be used to convert convolutional codes for correcting random errors into burst-error-correcting codes. This is discussed in Section 14.3.

Three different approaches to correcting both burst and random errors with convolutional codes are presented in Section 14.4. The diffuse codes of Kohlenberg and Forney [7] and Massey [8] can be decoded with a simple majority-logic decoding rule. Gallager's [9] burst-finding scheme uses an adaptive decoder that separates bursts from random errors. Finally, Tong's [10] burst-trapping scheme applies the same idea to block codes, although the overall code remains convolutional because there is memory in the encoder.

We begin our discussion of burst-error-correcting convolutional codes in Section 14.1 by considering bounds on burst-error-correcting capability.

14.1 BOUNDS ON BURST-ERROR-CORRECTING CAPABILITY

Assume that $\mathbf{e} = (e_0, e_1, e_2, \ldots)$ represents the error sequence on a BSC.

Definition 14.1. A sequence of error bits $e_{l+1}, e_{l+2}, \ldots, e_{l+b}$ is called a *burst of length b* relative to a *guard space of length g* if:

1. $e_{l+1} = e_{l+b} = 1$.
2. The g bits preceding e_{l+1} and the g bits following e_{l+b} are all 0's.
3. The b bits from e_{l+1} through e_{l+b} contain no subsequence of g 0's.

Example 14.1

Consider the error sequence $\mathbf{e} = (\cdots\ 0\,0\,0\,0\,0\,0\,1\,0\,0\,1\,1\,1\,1\,0\,0\,0\,0\,1\,1\,1\,0\ 0\,1\,0\,0\,0\,1\,0\,0\,1\,1\,0\,1\,1\,0\,0\,0\,0\,0\,0\ \cdots)$. This sequence contains a burst of length $b = 28$ relative to a guard space of length $g = 6$. Alternatively, it contains two bursts, one of length 7 and the other of length 16, relative to a guard space of length 4, or three bursts, of lengths 7, 6, and 8, relative to a guard space of length 3. This example illustrates that the length of a burst is always determined relative to some guard space, and that the two cannot be specified independently.

Gallager [9] has shown that for any convolutional code of rate R that corrects all bursts of length b or less relative to a guard space of length g,

$$\frac{g}{b} \geq \frac{1+R}{1-R}.^{1} \tag{14.1}$$

The bound of (14.1) is known as the bound on complete burst-error correction. Massey [11] has also shown that if we allow a small fraction of the bursts of length b to be decoded incorrectly, the guard space requirements can be reduced significantly. In particular, for a convolutional code of rate R that corrects all but a fraction ϵ of bursts of length b or less relative to a guard space of length g,

$$\frac{g}{b} \geq \frac{R + [\log_2(1-\epsilon)]/b}{1-R} \approx \frac{R}{1-R} \tag{14.2}$$

for small ϵ. The bound of (14.2) is known as the bound on "almost all" burst-error correction.

Example 14.2

For $R = 1/2$, complete burst correction requires $g/b \geq 3$, whereas "almost all" burst correction only requires a g/b ratio of approximately 1, a difference of a factor of 3 in the necessary guard space.

14.2 BURST-ERROR-CORRECTING CONVOLUTIONAL CODES

In this section we discuss two classes of convolutional codes for correcting burst errors, the Berlekamp–Preparata codes and the Iwadare–Massey codes.

[1]Wyner and Ash [4] earlier obtained a special case of this bound for phased-burst-error correction.

Berlekamp–Preparata Codes

Consider designing an $(n, n-1, m)$ systematic convolutional code to correct bursts confined to a single block relative to a guard space of m error-free blocks (i.e., a burst can affect at most one block in a constraint length). Such a code would have a phased-burst-error-correcting capability of one block relative to a guard space of m blocks. To design such a code, we must assure that each correctable error sequence $[\mathbf{e}]_m = (\mathbf{e}_0, \mathbf{e}_1, \ldots, \mathbf{e}_m)$ results in a distinct syndrome $[\mathbf{s}]_m = (s_0, s_1, \ldots, s_m)$. This implies that each error sequence with $\mathbf{e}_0 \neq \mathbf{0}$ and $\mathbf{e}_l = \mathbf{0}, l = 1, 2, \ldots, m,$ must yield a distinct syndrome, and that each of these syndromes must be distinct from the syndrome caused by any error sequence with $\mathbf{e}_0 = \mathbf{0}$ and a single block $\mathbf{e}_l \neq \mathbf{0}$, $l = 1, 2, \ldots, m$. Under these conditions, the first error block $\mathbf{e}_0$ can be correctly decoded if the first constraint length contains at most one nonzero block, and assuming feedback decoding, each successive error block can be decoded in the same way.

An $(n, n-1, m)$ convolutional code is described by the set of generator polynomials $\mathbf{g}_1^{(n)}(D), \mathbf{g}_2^{(n)}(D), \ldots, \mathbf{g}_{n-1}^{(n)}(D)$. The parity-check matrix can be written as

$$[\mathbf{H}^T]_m = \begin{bmatrix} \mathbf{B}_0 \\ \mathbf{B}_1 \\ \cdot \\ \cdot \\ \cdot \\ \mathbf{B}_m \end{bmatrix}, \tag{14.3}$$

where

$$\mathbf{B}_0 = \begin{bmatrix} g_{1,0}^{(n)} & g_{1,1}^{(n)} & \cdots & g_{1,m}^{(n)} \\ \cdot & \cdot & & \cdot \\ \cdot & \cdot & & \cdot \\ \cdot & \cdot & & \cdot \\ g_{n-1,0}^{(n)} & g_{n-1,1}^{(n)} & \cdots & g_{n-1,m}^{(n)} \\ 1 & 0 & \cdots & 0 \end{bmatrix} \tag{14.4}$$

is an $n \times (m+1)$ matrix. For $0 < l \leq m$, $\mathbf{B}_l$ is obtained from $\mathbf{B}_{l-1}$ by shifting $\mathbf{B}_{l-1}$ one column to the right and deleting the last column. Mathematically, this can be expressed as

$$\mathbf{B}_l = \mathbf{B}_{l-1} \begin{bmatrix} 0 & 1 & 0 & \cdots & 0 \\ 0 & 0 & 1 & \cdots & 0 \\ \cdot & \cdot & \cdot & \cdot & \cdot \\ \cdot & \cdot & \cdot & \cdot & \cdot \\ \cdot & \cdot & \cdot & \cdot & \cdot \\ 0 & 0 & 0 & \cdots & 1 \\ 0 & 0 & 0 & \cdots & 0 \end{bmatrix} \triangleq \mathbf{B}_{l-1}\mathbf{T}, \tag{14.5}$$

where $\mathbf{T}$ is an $(m+1) \times (m+1)$ shifting matrix. Using this notation, the syndrome can be written as

$$\begin{aligned} [\mathbf{s}]_m = [\mathbf{e}]_m[\mathbf{H}^T]_m &= \mathbf{e}_0\mathbf{B}_0 + \mathbf{e}_1\mathbf{B}_1 + \mathbf{e}_2\mathbf{B}_2 + \cdots + \mathbf{e}_m\mathbf{B}_m \\ &= \mathbf{e}_0\mathbf{B}_0 + \mathbf{e}_1\mathbf{B}_0\mathbf{T} + \mathbf{e}_2\mathbf{B}_0\mathbf{T}^2 + \cdots + \mathbf{e}_m\mathbf{B}_0\mathbf{T}^m. \end{aligned} \tag{14.6}$$

From (14.1), for an optimum burst-correcting code, $g/b = (1 + R)/(1 - R)$. For the case above with $R = (n - 1)/n$ and $g = mn = mb$, this implies that

$$\frac{g}{b} = m = \frac{1 + [(n-1)/n]}{1 - [(n-1)/n]} = 2n - 1 \tag{14.7}$$

(i.e., $\mathbf{B}_0$ is an $n \times 2n$ matrix). We must now choose $\mathbf{B}_0$ such that the conditions for burst-error correction are satisfied.

If the first n columns of $\mathbf{B}_0$ are chosen to be the *skewed* $n \times n$ identity matrix

$$\begin{bmatrix} 0 & \cdots & 0 & 1 \\ 0 & \cdots & 1 & 0 \\ \cdot & & \cdot & \cdot \\ \cdot & \cdot & \cdot & \cdot \\ \cdot & & \cdot & \cdot \\ 1 & \cdots & 0 & 0 \end{bmatrix},$$

then (14.6) implies that each error sequence with $\mathbf{e}_0 \neq \mathbf{0}$ and $\mathbf{e}_l = \mathbf{0}$, $l = 1, 2, \ldots, m$, will yield a distinct syndrome. In this case, the estimate of $\mathbf{e}_0$ is obtained by simply reversing the first n bits in the $2n$-bit syndrome. In addition, for each $\mathbf{e}_0 \neq \mathbf{0}$, the condition

$$\mathbf{e}_0\mathbf{B}_0 \neq \mathbf{e}_l\mathbf{B}_0\mathbf{T}^l, \qquad l = 1, 2, \ldots, m, \tag{14.8}$$

must be satisfied for all $\mathbf{e}_l \neq \mathbf{0}$. This ensures that an error in some other block will not be confused for an error in block zero. The last n columns of $\mathbf{B}_0$ are chosen to be the $n \times n$ matrix

$$\begin{bmatrix} 0 & A & B & D & \cdots \\ 0 & 0 & C & E & \cdots \\ 0 & 0 & 0 & F & \cdots \\ \cdot & \cdot & \cdot & \cdot & \cdot \\ \cdot & \cdot & \cdot & & \cdot \\ \cdot & \cdot & \cdot & & \cdot \\ 0 & 0 & 0 & \cdots & 0 \end{bmatrix},$$

where $A, B, C, \ldots$ must be chosen so that (14.8) is satisfied.

First note that for any $\mathbf{e}_l \neq \mathbf{0}$ and $l \geq n$, the first n positions in the vector $\mathbf{e}_l\mathbf{B}_0\mathbf{T}^l$ must be all 0's, since $\mathbf{T}^l$ shifts $\mathbf{B}_0$ such that $\mathbf{B}_0\mathbf{T}^l$ has all 0's in its first l columns. On the other hand, for any $\mathbf{e}_0 \neq \mathbf{0}$, the vector $\mathbf{e}_0\mathbf{B}_0$ cannot have all 0's in its first n positions. Hence, condition (14.8) is automatically satisfied for $n \leq l \leq m = 2n - 1$, and we can replace (14.8) with the condition that for each $\mathbf{e}_0 \neq \mathbf{0}$,

$$\mathbf{e}_0\mathbf{B}_0 \neq \mathbf{e}_l\mathbf{B}_0\mathbf{T}^l, \qquad l = 1, 2, \ldots, n - 1, \tag{14.9}$$

must be satisfied for all $\mathbf{e}_l \neq \mathbf{0}$. Now note that:

1. The matrix $\mathbf{B}_0$ has rank n since it contains a skewed $n \times n$ identity matrix in its first n columns.
2. For $1 \leq l \leq n - 1$, the matrix $\mathbf{B}_0\mathbf{T}^l$ also has rank n since it also contains a skewed $n \times n$ identity matrix in some n columns.
3. Condition (14.9) is equivalent to requiring that the row spaces of $\mathbf{B}_0$ and $\mathbf{B}_0\mathbf{T}^l$ have only the vector $\mathbf{0}$ in common.

Hence, condition (14.9) can be replaced by the condition that

$$\text{rank}\begin{bmatrix}\mathbf{B}_0\mathbf{T}^l \\ \hdashline \mathbf{B}_0\end{bmatrix} = \text{rank}\,[\mathbf{B}_0\mathbf{T}^l] + \text{rank}\,[\mathbf{B}_0] = n + n = 2n. \tag{14.10}$$

Since $\begin{bmatrix}\mathbf{B}_0\mathbf{T}^l \\ \hdashline \mathbf{B}_0\end{bmatrix}$ is a $2n \times 2n$ matrix, condition (14.10) is equivalent to requiring that this matrix be nonsingular, that is, that

$$\det\begin{bmatrix}\mathbf{B}_0\mathbf{T}^l \\ \hdashline \mathbf{B}_0\end{bmatrix} \neq 0, \qquad l = 1, 2, \ldots, n-1. \tag{14.11}$$

We now show that $A, B, C, \ldots$ can always be chosen so that (14.11) is satisfied. First, for $n = 2$, (14.11) yields the condition

$$\det\begin{bmatrix} 0 & 0 & 1 & 0 \\ 0 & 1 & 0 & 0 \\ \hdashline 0 & 1 & 0 & A \\ 1 & 0 & 0 & 0 \end{bmatrix} = 1,$$

which is satisfied by choosing $A = 1$. For $n = 3$, (14.11) yields two conditions

$$\det\begin{bmatrix} 0 & 0 & 0 & 1 & 0 & 1 \\ 0 & 0 & 1 & 0 & 0 & 0 \\ 0 & 1 & 0 & 0 & 0 & 0 \\ \hdashline 0 & 0 & 1 & 0 & 1 & B \\ 0 & 1 & 0 & 0 & 0 & C \\ 1 & 0 & 0 & 0 & 0 & 0 \end{bmatrix} = 1, \qquad \det\begin{bmatrix} 0 & 0 & 0 & 0 & 1 & 0 \\ 0 & 0 & 0 & 1 & 0 & 0 \\ 0 & 0 & 1 & 0 & 0 & 0 \\ \hdashline 0 & 0 & 1 & 0 & 1 & B \\ 0 & 1 & 0 & 0 & 0 & C \\ 1 & 0 & 0 & 0 & 0 & 0 \end{bmatrix} = 1,$$

which are satisfied by choosing $B = 1$ and $C = 1$. It can be shown by induction that the missing elements of $\mathbf{B}_0$ can always be chosen so that (14.11) is satisfied (see Problem 14.1).

Example 14.3

For $n = 4$, the 4×8 matrix $\mathbf{B}_0$ is given by

$$\mathbf{B}_0 = \begin{bmatrix} 0 & 0 & 0 & 1 & 0 & 1 & 1 & 1 \\ 0 & 0 & 1 & 0 & 0 & 0 & 1 & 1 \\ 0 & 1 & 0 & 0 & 0 & 0 & 0 & 1 \\ 1 & 0 & 0 & 0 & 0 & 0 & 0 & 0 \end{bmatrix}.$$

This is a (4, 3, 7) systematic convolutional code with generator polynomials $\mathbf{g}_1^{(4)}(D) = D^3 + D^5 + D^6 + D^7$, $\mathbf{g}_2^{(4)}(D) = D^2 + D^6 + D^7$, and $\mathbf{g}_3^{(4)}(D) = D + D^7$, which is capable of correcting phased bursts of length 4 confined to a single block relative to a guard space of seven blocks (28 bits). Since $g/b = 28/4 = 7$ and $(1 + R)/(1 - R) = (7/4)/(1/4) = 7$, this code meets the Gallager bound of (14.1) and is optimal for phased-burst-error correction.

The construction discussed above, discovered independently by Berlekamp [5] and Preparata [6], always results in a code that meets the Gallager bound of (14.1). Hence, the *Berlekamp–Preparata codes* are optimum for phased-burst-error correction. This construction can also be extended to generate optimum phased-burst-error-correcting codes for $k < n - 1$. In addition, interleaving (see Section 14.3) can be used to convert any of these codes to phased-burst error-correcting codes which are capable of correcting bursts confined to λ blocks relative to a guard space of λm blocks, where λ is the degree of interleaving. Since the ratio of guard space to burst length remains the same, these interleaved codes are still optimum.

The Berlekamp–Preparata codes can be decoded using a general decoding technique for burst-correcting convolutional codes due to Massey [12]. We recall from (14.6) that the set of possible syndromes for a burst confined to block 0 is simply the row space of the $n \times 2n$ matrix $\mathbf{B}_0$. Hence, if $\mathbf{e}_0 \neq \mathbf{0}$ and $\mathbf{e}_l = \mathbf{0}$, $l = 1, 2, \ldots, m$, $[\mathbf{s}]_m$ is a code word in the $(2n, n)$ block code generated by $\mathbf{B}_0$. On the other hand, if $\mathbf{e}_0 = \mathbf{0}$ and a single block $\mathbf{e}_l \neq \mathbf{0}$ for some l, $1 \leq l \leq m$, condition (14.8) ensures that $[\mathbf{s}]_m$ is not a code word in the block code generated by $\mathbf{B}_0$. Therefore, $\mathbf{e}_0$ contains a correctable error pattern if and only if $[\mathbf{s}]_m$ is a code word in the block code generated by $\mathbf{B}_0$. This requires determining if $[\mathbf{s}]_m \mathbf{H}_0^T = \mathbf{0}$, where $\mathbf{H}_0$ is the $n \times 2n$ parity-check matrix corresponding to $\mathbf{B}_0$. Since

$$\mathbf{B}_0 = \left[\begin{array}{ccccc|cccccc} 0 & \cdots & 0 & 0 & 1 & 0 & A & B & D & \cdots & \\ 0 & \cdots & 0 & 1 & 0 & 0 & 0 & C & E & \cdots & \\ 0 & \cdots & 1 & 0 & 0 & 0 & 0 & 0 & F & \cdots & \\ \vdots & & & & \vdots & \vdots & & & & \ddots & \\ 1 & \cdots & 0 & 0 & 0 & 0 & 0 & 0 & \cdots & & 0 \end{array}\right], \tag{14.12}$$

the corresponding parity-check matrix is given by

$$\mathbf{H}_0 = \left[\begin{array}{ccccc|ccccc} 0 & \cdot & \cdots & & \cdot & 0 & \cdots & 0 & 0 & 1 \\ \vdots & & & & \vdots & 0 & \cdots & 0 & 1 & 0 \\ \vdots & & F & E & D & 0 & \cdots & 1 & 0 & 0 \\ 0 & \cdots & 0 & C & B & \vdots & & & & \vdots \\ 0 & \cdots & 0 & 0 & A & \vdots & & & & \vdots \\ 0 & \cdots & 0 & 0 & 0 & 1 & \cdots & 0 & 0 & 0 \end{array}\right]. \tag{14.13}$$

If $[\mathbf{s}]_m \mathbf{H}_0^T = \mathbf{0}$, the decoder must then find the correctable error pattern $\mathbf{e}_0$ which produces the syndrome $[\mathbf{s}]_m$. Since in this case $[\mathbf{s}]_m = \mathbf{e}_0 \mathbf{B}_0$, the estimate of $\mathbf{e}_0$ is obtained by simply reversing the first n bits in $[\mathbf{s}]_m$. For a feedback decoder, the syndrome must then be modified to remove the effect of $\mathbf{e}_0$. But for a correctable error pattern, $[\mathbf{s}]_m = \mathbf{e}_0 \mathbf{B}_0$ depends only on $\mathbf{e}_0$, and hence when the effect of $\mathbf{e}_0$ is removed, the syndrome will be reset to all zeros.

Example 14.4

For $n = 4$, the 4×8 parity-check matrix $\mathbf{H}_0$ is given by

$$\mathbf{H}_0 = \begin{bmatrix} 0 & 1 & 1 & 1 & 0 & 0 & 0 & 1 \\ 0 & 0 & 1 & 1 & 0 & 0 & 1 & 0 \\ 0 & 0 & 0 & 1 & 0 & 1 & 0 & 0 \\ 0 & 0 & 0 & 0 & 1 & 0 & 0 & 0 \end{bmatrix}.$$

A complete encoder/decoder block diagram for this code is shown in Figure 14.1. The

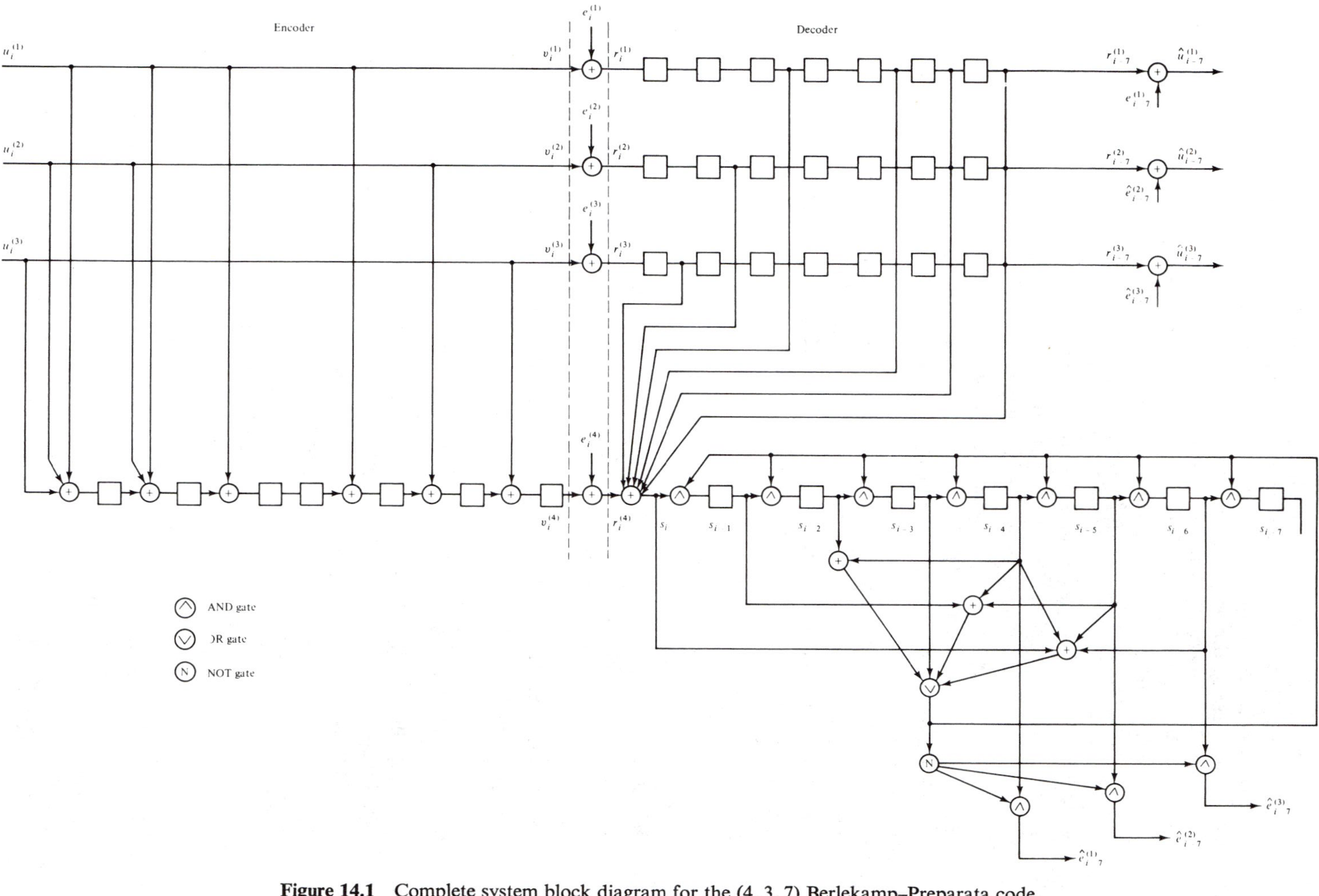

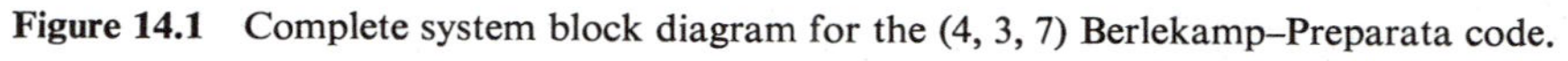
Figure 14.1 Complete system block diagram for the (4, 3, 7) Berlekamp–Preparata code.

encoder is implemented using the modified encoding circuit for systematic codes discussed in Example 10.13. The OR gate in the decoder has as inputs the four parity checks corresponding to the four rows of $\mathbf{H}_0$. The output of this OR gate is 0 if and only if the syndrome $[\mathbf{s}]_m$ corresponds to a correctable burst in $\mathbf{e}_0$. When this happens, the NOT gate activates the connections from s_3, s_2, and s_1 to provide the estimates $\hat{e}_0^{(1)}$, $\hat{e}_0^{(2)}$, and $\hat{e}_0^{(3)}$, respectively. In addition, the output of the OR gate is fed back to the AND gates between stages of the syndrome register. This resets the syndrome to all 0's after each burst is corrected.

Finally, we note that unlimited error propagation cannot occur in a decoder of the type described above. When the syndrome is reset, the only possible change is that some 1's are changed to 0's. Hence if the received sequence is error free for $2(m + 1)$ time units, the syndrome register must be cleared to all zeros. Since a guard space is m time units, this means that approximately two error-free guard spaces are needed to restore a decoder of this type to correct operation following a decoding error.

The phased-burst-error-correcting codes of Berlekamp and Preparata can also be used to correct bursts of length b relative to a guard space of length g, where b and g are not confined to an integral number of blocks. In general, though, the Gallager bound will not be met with equality in this case; that is, the codes will not achieve the optimum guard space-to-burst length ratio. However, for a code interleaved to degree λ, the shortest burst affecting $\lambda + 1$ blocks contains $(\lambda - 1)n + 2$ bits. Hence, the code will correct all bursts of length $b = (\lambda - 1)n + 1$ bits. Similarly, the longest guard space required to yield a guard space of λm blocks is $g = (\lambda m + 1)n - 1$ bits, and

$$\frac{g}{b} = \frac{(\lambda m + 1)n - 1}{(\lambda - 1)n + 1} = \frac{\lambda m + [(n - 1)/n]}{\lambda - [(n - 1)/n]} \approx m = \frac{1 + R}{1 + R} \tag{14.14}$$

when λ is large. Hence, the Berlekamp–Preparata codes are almost optimum for ordinary burst-error correction when the degree of interleaving λ is large.

Iwadare–Massey Codes

Another efficient class of convolutional codes for correcting burst errors of length b relative to a guard space of length g was discovered independently by Iwadare [2] and Massey [3]. These are called the *Iwadare–Massey codes.* For any n, a systematic $(n, n - 1, m)$ burst-error-correcting convolutional code can be constructed with the following parameters:

$$\begin{aligned} m &= (2n - 1)\lambda + 2n - 3 \\ b &= n\lambda \\ g &= n_A - 1 = n(m + 1) - 1, \end{aligned} \tag{14.15}$$

where λ is any positive integer. The $n - 1$ generator polynomials are given by

$$\mathbf{g}_{(i)}^{(n)}(D) = D^{a(i)} + D^{b(i)}, \qquad i = 1, 2, \ldots, n - 1, \tag{14.16}$$

where $a(i) \triangleq (\lambda + 1)(n - i) - 1$ and $b(i) \triangleq (\lambda + 1)(2n - 1) + i - 3$. The decoding of these codes can best be explained by an example.

Example 14.5

Consider the Iwadare–Massey code with $n = 3$ and $\lambda = 3$. In this case, $m = 18$, $b = 9$, and $g = 56$. The generator polynomials are

$$\mathbf{g}_1^{(3)}(D) = D^7 + D^{18} \qquad \text{and} \qquad \mathbf{g}_2^{(3)}(D) = D^3 + D^{15}.$$

The encoding circuit for this code is shown in Figure 14.2.

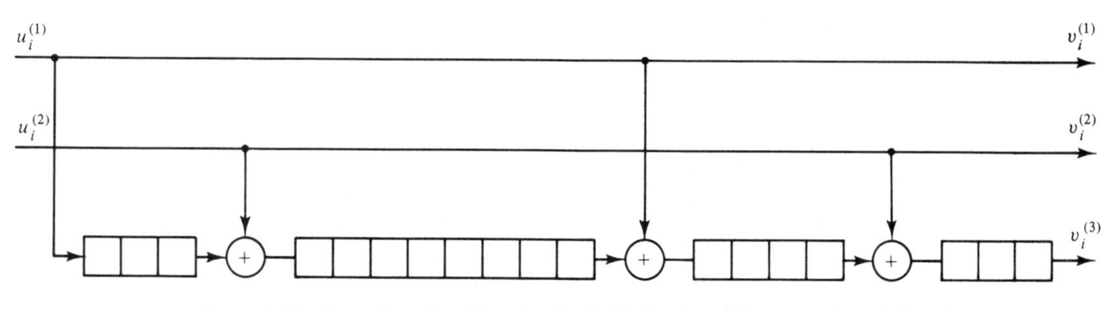

Figure 14.2 Encoding circuit for the (3, 2, 18) Iwadare–Massey code with $\lambda = 3$.

Assume that a burst error of length $b = 9$ begins with the first bit of block 0, that is,

$$\mathbf{e} = (e_0^{(1)} \ e_0^{(2)} \ e_0^{(3)} \ e_1^{(1)} \ e_1^{(2)} \ e_1^{(3)} \ e_2^{(1)} \ e_2^{(2)} \ e_2^{(3)} \ 0 \ 0 \ 0 \ \cdots).$$

Then from (13.24) the syndrome sequence $\mathbf{s}(D)$ is given by

$$\begin{aligned}\mathbf{s}(D) &= \mathbf{e}^{(1)}(D)\mathbf{g}_1^{(3)}(D) + \mathbf{e}^{(2)}(D)\mathbf{g}_2^{(3)}(D) + \mathbf{e}^{(3)}(D)\\ &= e_0^{(3)} + e_1^{(3)}D + e_2^{(3)}D^2 + e_0^{(2)}D^3 + e_1^{(2)}D^4 + e_2^{(2)}D^5\\ &\quad + e_0^{(1)}D^7 + e_1^{(1)}D^8 + e_2^{(1)}D^9 + e_0^{(2)}D^{15} + e_1^{(2)}D^{16} + e_2^{(2)}D^{17}\\ &\quad + e_0^{(1)}D^{18} + e_1^{(1)}D^{19} + e_2^{(1)}D^{20}\end{aligned}$$

or

$$\begin{aligned}\mathbf{s} = (&e_0^{(3)} \ e_1^{(3)} \ e_2^{(3)} \ e_0^{(2)} \ e_1^{(2)} \ e_2^{(2)} \ 0 \ e_0^{(1)} \ e_1^{(1)} \ e_2^{(1)} \ 0 \ 0 \ 0 \ 0\\ &0 \ e_0^{(2)} \ e_1^{(2)} \ e_2^{(2)} \ e_0^{(1)} \ e_1^{(1)} \ e_2^{(1)}).\end{aligned}$$

Examining the syndrome sequence above, we note that:

1. Each information error bit appears twice in $\mathbf{s}$.
2. $e_l^{(2)}$ appears in $\mathbf{s}$ before $e_l^{(1)}$, $l = 0, 1, 2$.
3. The number of positions between the two appearances of $e_l^{(j)}$ is $9 + j$, for $j = 1, 2$ and $l = 0, 1, 2$.

The integer $9 + j$ is called the *repeat distance* of $e_l^{(j)}$, and we see that the information error bits $e_l^{(1)}$ and $e_l^{(2)}$ have distinct repeat distances of 10 and 11, respectively. This fact will be useful in decoding.

If the burst starts with the second bit of block 0, then

$$\mathbf{e} = (0 \ e_0^{(2)} \ e_0^{(3)} \ e_1^{(1)} \ e_1^{(2)} \ e_1^{(3)} \ e_2^{(1)} \ e_2^{(2)} \ e_2^{(3)} \ e_3^{(1)} \ 0 \ 0 \ 0 \ \cdots)$$

and

$$\begin{aligned}\mathbf{s} = (&e_0^{(3)} \ e_1^{(3)} \ e_2^{(3)} \ e_0^{(2)} \ e_1^{(2)} \ e_2^{(2)} \ 0 \ 0 \ e_1^{(1)} \ e_2^{(1)} \ e_3^{(1)} \ 0 \ 0 \ 0 \ 0\\ &e_0^{(2)} \ e_1^{(2)} \ e_2^{(2)} \ 0 \ e_1^{(1)} \ e_2^{(1)} \ e_3^{(1)}).\end{aligned}$$

If the burst starts with the third bit of block 0,

$$\mathbf{e} = (0 \quad 0 \quad e_0^{(3)} \quad e_1^{(1)} \quad e_1^{(2)} \quad e_1^{(3)} \quad e_2^{(1)} \quad e_2^{(2)} \quad e_2^{(3)} \quad e_3^{(1)} \quad e_3^{(2)} \quad 0 \quad 0 \quad 0 \quad \cdots)$$

and

$$\mathbf{s} = (e_0^{(3)} \quad e_1^{(3)} \quad e_2^{(3)} \quad 0 \quad e_1^{(2)} \quad e_2^{(2)} \quad e_3^{(2)} \quad 0 \quad e_1^{(1)} \quad e_2^{(1)} \quad e_3^{(1)} \quad 0 \quad 0 \quad 0 \quad 0 \quad 0$$
$$e_1^{(2)} \quad e_2^{(2)} \quad e_3^{(2)} \quad e_1^{(1)} \quad e_2^{(1)} \quad e_3^{(1)}).$$

In each of these cases, the repeat distance for the information error bit $e_l^{(j)}$ is still $9 + j$.

The decoding circuit for this code is shown in Figure 14.3. It consists of an encoder for syndrome calculation and a correction circuit. The two inputs of AND gate A_2 are separated by 11 stages of the syndrome register [the repeat distance of $e_l^{(2)}$]. Two of the inputs of AND gate A_1 are separated by 10 stages of the syndrome register [the repeat distance of $e_l^{(1)}$]. The third input to A_1 ensures that both AND gates cannot have a 1 output at the same time. A careful examination of the syndromes above reveals that A_2 cannot be activated until error bit $e_0^{(2)}$ appears at both its inputs, and then its output will be the correct value of $e_0^{(2)}$. At the same time, the received information bit $r_0^{(2)}$ is at the output of the fifteenth stage of the buffer register that stores the received information sequence $\mathbf{r}^{(2)}$. Correction is then achieved by adding the output of A_2 to $r_0^{(2)}$. The output of A_2 is also fed back to reset the syndrome register. After one shift of the syndrome register, $e_1^{(2)}$ appears at both inputs of A_2. Hence, the output of A_2 will be the correct value of $e_1^{(2)}$, which is then used to correct the received information bit

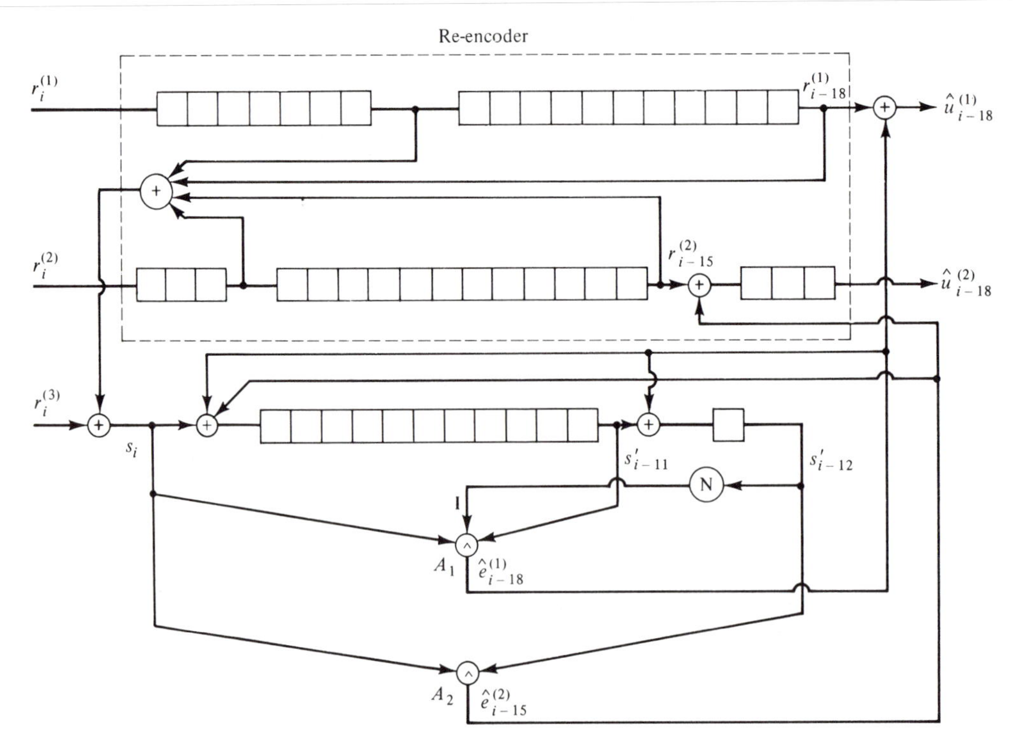

Figure 14.3 Decoding circuit for the (3, 2, 18) Iwadare–Massey code with $\lambda = 3$.

$r_1^{(2)}$. After the next shift of the syndrome register, the decoder estimates $e_2^{(2)}$ and corrects $r_2^{(2)}$ in exactly the same way.

Note that the output of the last stage of the syndrome register is inverted and fed into A_1 as input I. This prevents A_1 from making any erroneous estimates while $e_0^{(2)}$, $e_1^{(2)}$, and $e_2^{(2)}$ are being estimated by A_2. For example, when $e_0^{(2)}$ appears at both inputs of A_2, $e_0^{(2)}$ and $e_1^{(2)}$ appear as inputs of A_1. If input I is not provided, and if $e_0^{(2)} = e_1^{(2)} = 1$, the output of A_1 will be a 1, which would cause an erroneous correction at the output of the buffer register that stores the received information sequence $\mathbf{r}^{(1)}$.

After $e_2^{(2)}$ is corrected, the syndrome register is shifted once, and $e_0^{(1)}$ appears at two inputs of A_1. The last stage of the syndrome register contains a 0, and hence A_2 is prevented from making any erroneous estimates and input I of A_1 is a 1. Therefore, the output of A_1 is the correct value of $e_0^{(1)}$. At the same time, the received information bit $r_0^{(1)}$ is at the output of the last stage of the buffer register that stores the received information sequence $\mathbf{r}^{(1)}$. Correction is then achieved by adding the output of A_1 to $r_0^{(1)}$. The output of A_1 is also fed back to reset the syndrome register. Hence, after one shift, the last stage of the syndrome register again contains a 0, and the inputs of A_1 are 1, $e_1^{(1)}$, and $e_1^{(1)}$. Therefore, the output of A_1 is the correct value of $e_1^{(1)}$, which is then used to correct the received information bit $r_1^{(1)}$. After the next shift of the syndrome register, the decoder estimates $e_2^{(1)}$ and corrects $r_2^{(1)}$ in exactly the same way. Since a correctable error burst of length $b = 9$ must be followed by an error-free guard space of length $g = 56$, a careful examination of the error bursts and syndromes listed above reveals that the next burst cannot enter the decoder until after the decoder corrects $r_2^{(1)}$, at which time the syndrome register contains all zeros.

A general decoder for an Iwadare–Massey code is shown in Figure 14.4. Note that additional inverted syndrome outputs are fed into AND gates $A_1, A_2, \ldots, A_{n-2}$ to prevent any erroneous decoding estimates. It can be shown that unlimited error propagation cannot occur in a decoder of this type, and if a decoding error is followed by $n[m + (\lambda + 2)n - 1] - 1$ error-free bits, the syndrome will be cleared and error propagation terminated (see Problem 14.7).

For $k = n - 1$, the Gallager bound on complete burst-error correction is given by

$$\frac{g}{b} \geq \frac{1 + R}{1 - R} = \frac{1 + [(n-1)/n]}{1 - [(n-1)/n]} = 2n - 1. \tag{14.17}$$

For the Iwadare–Massey codes,

$$\frac{g}{b} = \frac{n(m+1) - 1}{n\lambda} = \frac{(2n-1)n\lambda + n(2n-2) - 1}{n\lambda}, \tag{14.18}$$

and we see that these codes require an excess guard space of $2n(n-1) - 1$ bits compared to an optimum code. Hence, they are very efficient for small values of n. For large n, however, the interleaved Berlekamp–Preparata codes require smaller guard spaces if the interleaving degree λ is large enough. But a comparison of decoder hardware shows that Iwadare–Massey codes are simpler to implement (see Problem 14.11). There is a second class of Iwadare–Massey codes, which requires a somewhat larger guard space than the class described above, but for large λ results in simpler encoding and decoding circuits (see Problem 14.6).

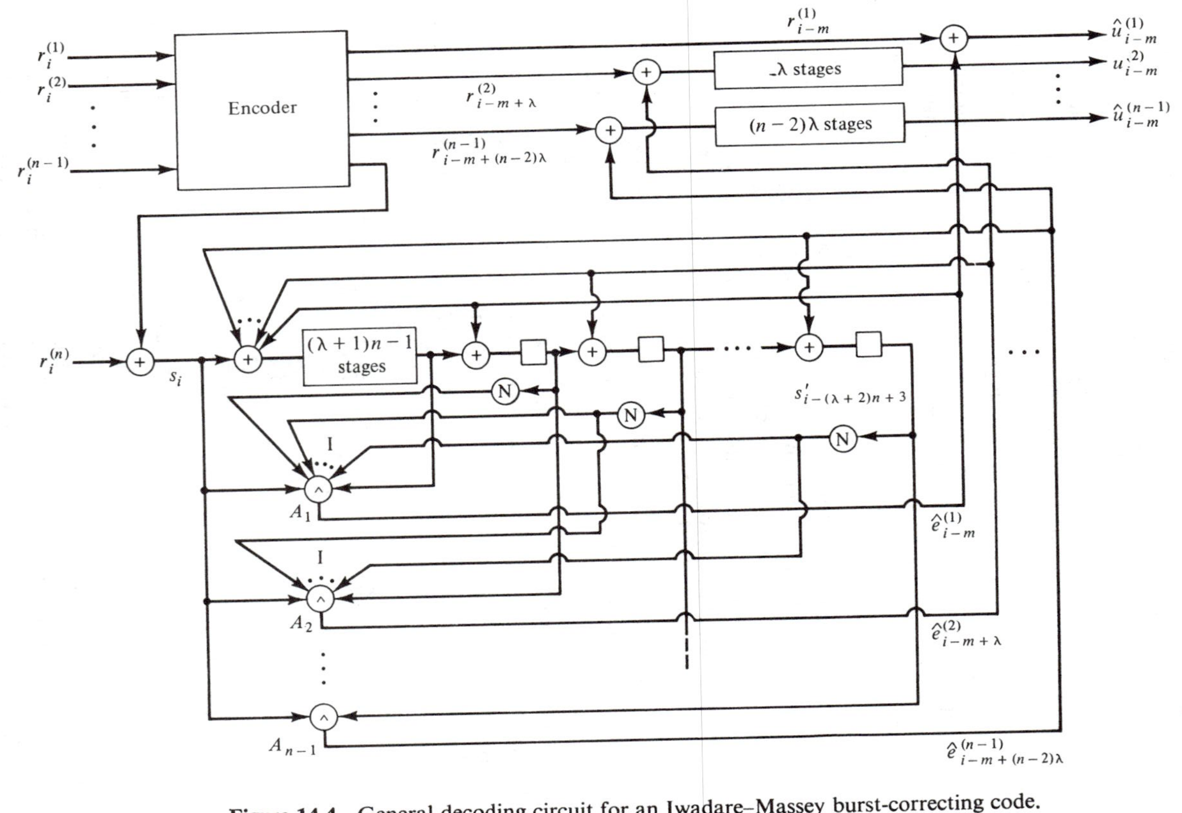

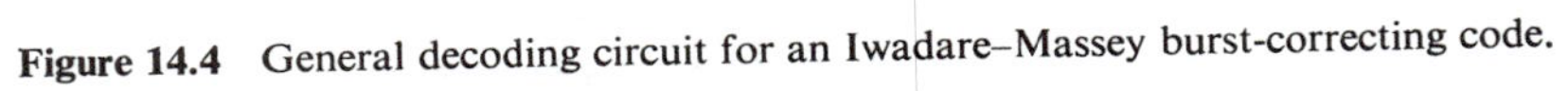

Figure 14.4 General decoding circuit for an Iwadare–Massey burst-correcting code.

14.3 INTERLEAVED CONVOLUTIONAL CODES

The technique of interleaving to obtain good long burst-error-correcting codes from good short burst-error-correcting codes, first discussed in Section 9.4 for block codes, can also be applied to convolutional codes. The idea of interleaving is simply to multiplex the outputs of λ separate encoders with constraint length n_A for transmission over the channel, where λ is the *interleaving degree*. The received bits are then demultiplexed and sent to λ separate decoders. A burst of length λ on the channel will then look like single errors to each of the separate decoders. Hence, if each decoder is capable of correcting single errors in a constraint length, then, with interleaving, all bursts of length λ or less relative to a guard space of length $(n_A - 1)\lambda$ will be corrected. Similarly, t bursts of length λ on the channel will look like weight-t error sequences to each of the separate decoders. Hence, if each decoder is capable of correcting t errors in a constraint length, then, with interleaving, all sequences of t or fewer bursts of length λ or less relative to a guard space of most $(n_A - t)\lambda$ will be corrected.[2] More generally, a burst of length $b'\lambda$ on the channel will look like bursts of length b' or less to each of the separate decoders. In this case if each decoder is capable of correcting bursts of length b' relative to a guard space g', the interleaved code will correct all bursts of length $b'\lambda$ or less relative to a guard space $g'\lambda$. In practice it is not necessary to use λ separate encoders and decoders, but only to implement one encoder and one decoder in such a way that their operation is equivalent to λ separate encoders and decoders.

An (n, k, m) convolutional coding system with interleaving degree λ is shown in Figure 14.5, where it is assumed that $\lambda - 1$ is a multiple of n. The interleaver is placed between the encoder and the multiplexer, and separates the n encoded bits in a block by $\lambda - 1$ intervening bits prior to transmission over the channel. In addition, the encoder is modified by replacing each delay unit with a string of λ delay units. This makes the encoder equivalent to λ separate encoders whose n-bit encoded blocks are formed in succession, and ensures that there will be $\lambda - 1$ intervening bits between the last bit in one block and the first bit in the next block corresponding to the same encoder. Hence, the encoder of Figure 14.5 achieves an interleaving degree of λ for the original convolutional code.

The interleaver of Figure 14.5 requires

$$\frac{\lambda - 1}{n} + 2\frac{\lambda - 1}{n} + \cdots + (n-1)\frac{\lambda - 1}{n} = \frac{(\lambda - 1)(n-1)}{2}$$

delay units. In addition, assuming a standard decoder containing k m-bit registers for reencoding and $(n - k)$ m-bit syndrome registers, the decoder in Figure 14.5 requires a total of λnm delay units. Hence, the total memory required in the interleaved decoder is given by $(\lambda - 1)(n - 1)/2 + \lambda nm$.

Example 14.6

Consider the (2, 1, 1) systematic convolutional code with $\mathbf{g}^{(2)}(D) = 1 + D$. This code can correct single errors in a constraint length of $n_A = n(m+1) = 4$ bits using the

[2]The actual guard space requirements between groups of t or fewer bursts depends on how the bursts are distributed. In essence, there can be no more than t bursts of length λ or less in any one constraint length of received bits (see Problem 14.8).

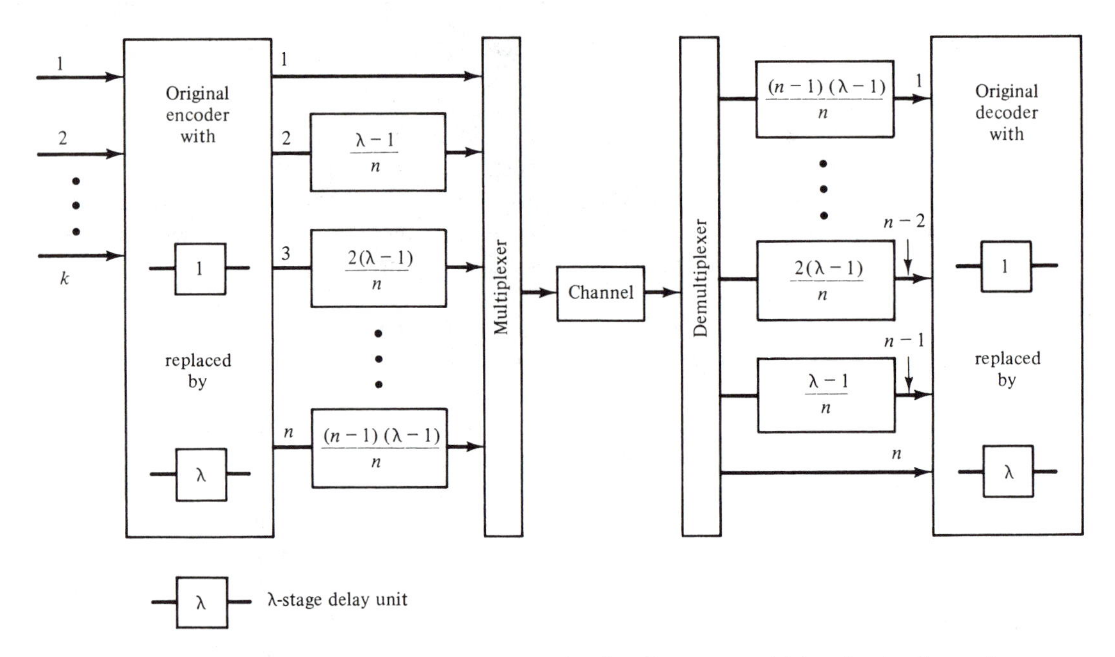

Figure 14.5 An (n, k, m) convolutional coding system with interleaving degree λ.

simple feedback decoding circuit shown in Figure 14.6(a). Hence when interleaved to degree λ, where $\lambda - 1$ must be a multiple of $n = 2$, this code will correct all bursts of length λ or less with a guard space of 3λ. Since $3\lambda/\lambda = 3$, the guard space-to-burst length ratio meets the Gallager bound for $R = 1/2$ codes, and this simple interleaved code is optimum for complete burst-error correction! The total memory required in the interleaved decoder is $(\lambda - 1)(n - 1)/2 + \lambda nm = (5\lambda - 1)/2$, which equals 12 for $\lambda = 5$. The complete interleaved convolutional coding system is shown in Figure 14.6(b) for $\lambda = 5$.

Example 14.7

Consider the (4, 3, 7) Berlekamp–Preparata code of Examples 14.3 and 14.4 and Figure 14.1, which is capable of correcting phased bursts of one block relative to a guard space of seven blocks. If this code is interleaved to degree λ, it can correct bursts confined to λ blocks relative to a guard space of 7λ blocks. Alternatively, it will correct all bursts of length $b = 4(\lambda - 1) + 1$ bits, relative to a guard space of $g = 4(7\lambda + 1) - 1$ bits, and for $\lambda = 5$, $g/b = 143/17 \approx 8.4$, which is about 20% above the Gallager bound of $g/b = 7$ for $R = 3/4$ codes.

14.4 BURST-AND-RANDOM-ERROR-CORRECTING CONVOLUTIONAL CODES

Several convolutional coding techniques are available for correcting errors on channels which are subject to a combination of random and burst errors. As noted in Section 14.3, interleaving a code with random-error-correcting capability t to degree λ results in a system that can correct groups of t or fewer bursts of length λ or less. This is called *multiple-burst-error correction,* and requires only that there be no more

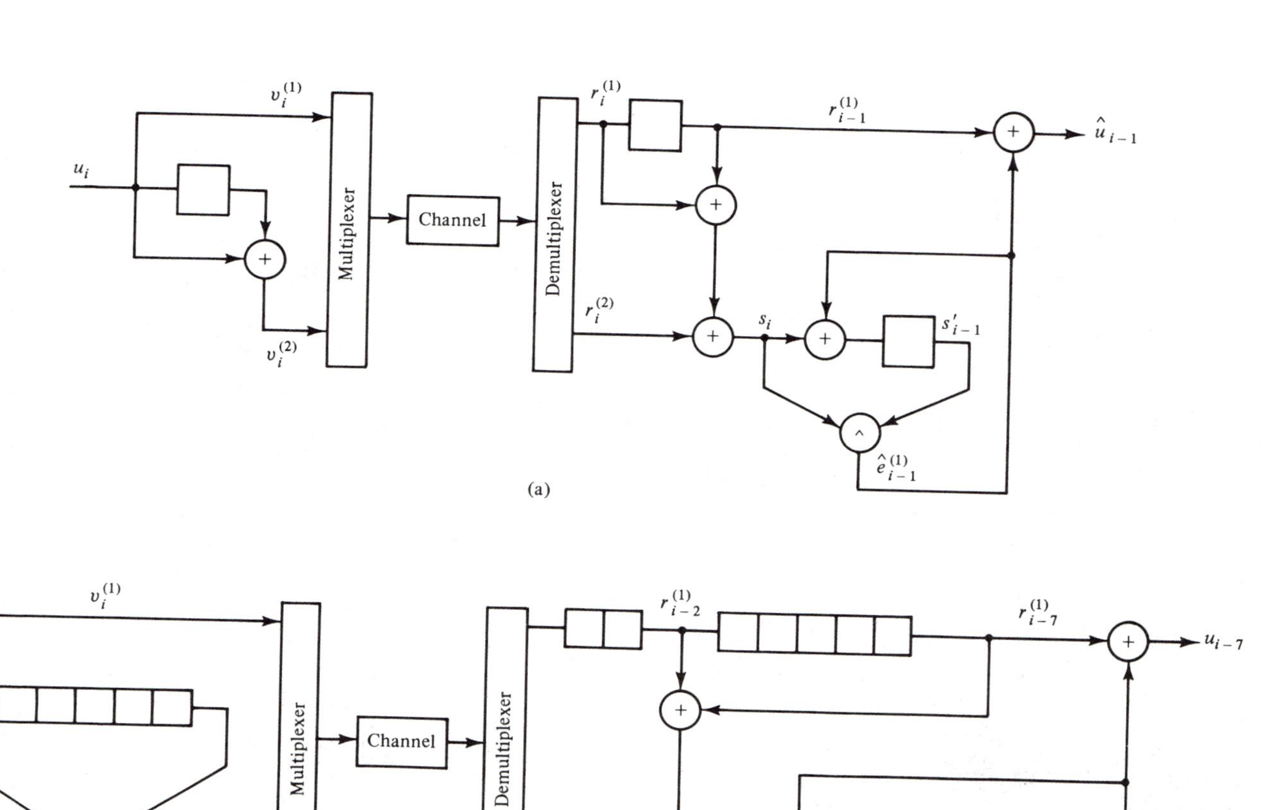

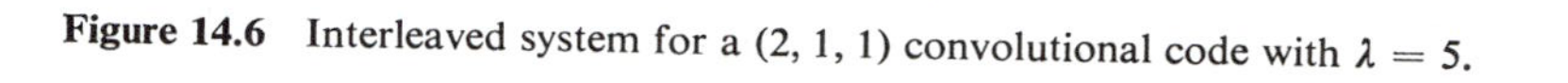
Figure 14.6 Interleaved system for a (2, 1, 1) convolutional code with $\lambda = 5$.

than t bursts of length λ or less in any λ constraint lengths of received bits. Since some of the bursts in a group may contain only scattered errors, this system in effect corrects a combination of burst and random errors.

Codes can also be constructed to correct a specific combination of burst and random errors. The *diffuse convolutional codes* of Forney and Kolenberg [7] and Massey [8], discussed in this section, are an example of this type of construction. Adaptive decoding algorithms can also be employed to determine which type of error pattern has been received and then switch to the appropriate correction circuit. Both Gallager's [9] *burst-finding* system and Tong's [10] *burst-trapping* system use an adaptive decoder, and are discussed in this section.

Diffuse Codes

Consider the (2, 1, m) systematic convolutional code with $m = 3\lambda + 1$ and $\mathbf{g}^{(2)}(D) = 1 + D^{\lambda} + D^{2\lambda} + D^{3\lambda+1}$, where λ is any positive integer greater than 1. The syndrome sequence is given by

$$\mathbf{s}(D) = \mathbf{e}^{(1)}(D)\mathbf{g}^{(2)}(D) + \mathbf{e}^{(2)}(D), \tag{14.19}$$

and four orthogonal check sums on $e_0^{(1)}$ can be formed as follows:

$$\begin{aligned} s_0 &= e_0^{(1)} & &+ e_0^{(2)} \\ s_\lambda &= e_0^{(1)} + e_\lambda^{(1)} & &+ e_\lambda^{(2)} \\ s_{2\lambda} + s_{3\lambda} &= e_0^{(1)} \quad + e_{3\lambda}^{(1)} & &+ e_{2\lambda}^{(2)} + e_{3\lambda}^{(2)} \\ s_{3\lambda+1} &= e_0^{(1)} + e_{\lambda+1}^{(1)} + e_{2\lambda+1}^{(1)} + e_{3\lambda+1}^{(1)} & &+ e_{3\lambda+1}^{(2)} \end{aligned} \tag{14.20}$$

Hence, if there are two or fewer errors among the 11 error bits checked in (14.20), they can be corrected by a majority-logic decoder.

Now suppose that a burst of length 2λ or less appears on the channel. If $e_0^{(1)} = 1$, the only other error bit in (14.20) that could have value 1 is $e_0^{(2)}$, since all the other error bits in (14.20) are at least 2λ positions away from $e_0^{(1)}$. On the other hand, if $e_0^{(1)} = 0$, a burst of length 2λ can affect at most two of the four check sums. In either case, the estimate $\hat{e}_0^{(1)}$ made by a majority logic decoder will be correct. Hence, the code corrects any $t_{\text{ML}} = 2$ or fewer random errors among the 11 error bits in (14.20) as well as bursts of length $b = 2\lambda$ or less with a guard space $g = 2(3\lambda + 1) = 6\lambda + 2$. With a feedback decoder, if all past decoding decisions have been correct, the same error-correcting capability applies to the decoding of all information error bits. A complete encoder/decoder block diagram for this code is shown in Figure 14.7. Note that for large λ,

$$\frac{g}{b} = \frac{6\lambda + 2}{2\lambda} \approx 3, \tag{14.21}$$

which is optimum for an $R = 1/2$ code according to the Gallager bound for complete burst-error correction.

The code of Figure 14.7 is an example of a diffuse convolutional code. A convolutional code is *λ-diffuse* and *t_{ML}-error-correcting* if $2t_{\text{ML}}$ orthogonal check sums can be formed on each block zero information error bit $e_0^{(i)}$ such that for each i, $1 \leq i \leq k$:

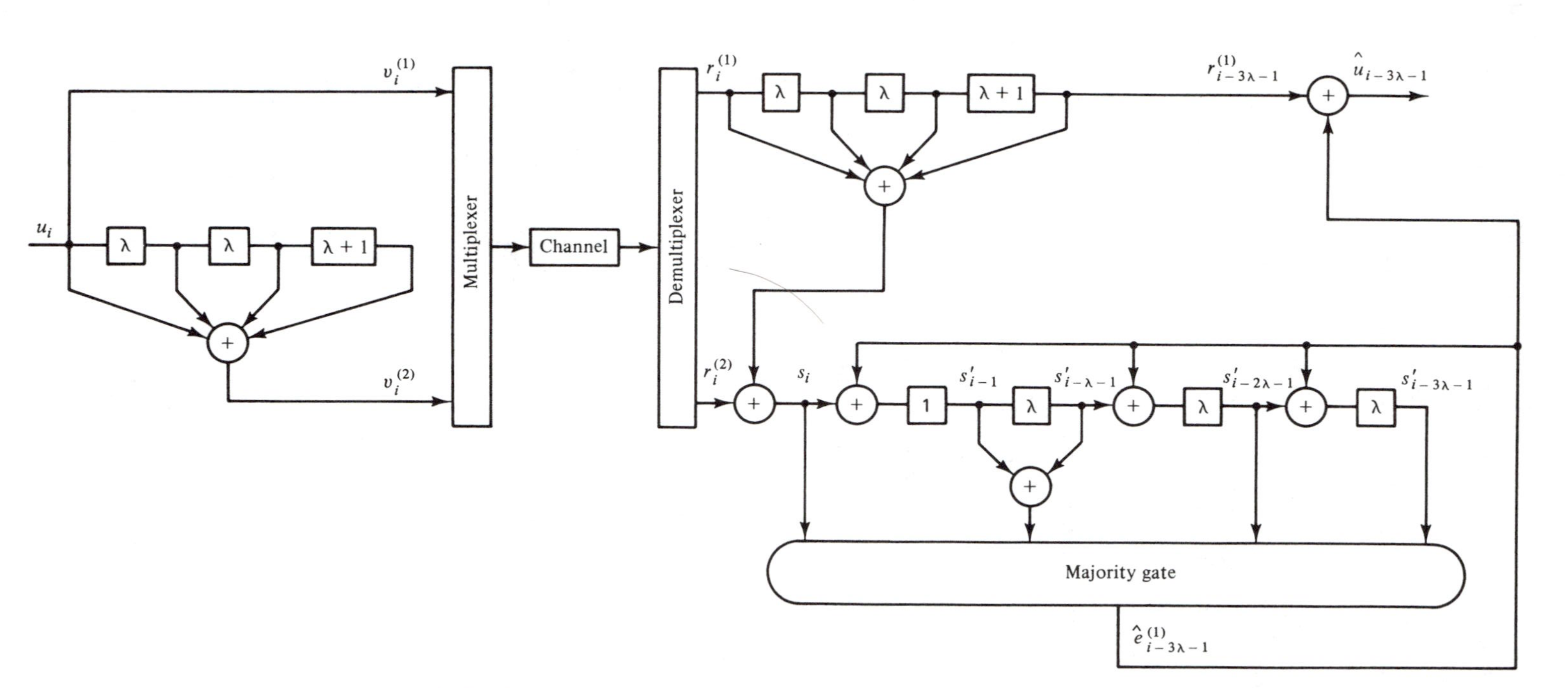

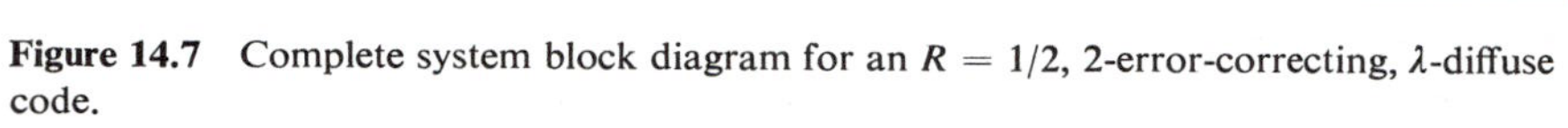

Figure 14.7 Complete system block diagram for an $R = 1/2$, 2-error-correcting, λ-diffuse code.

1. Error bits other than $e_0^{(i)}$ from a burst of length $n\lambda$ or less which starts in block 0 and includes $e_0^{(i)}$ are checked by no more than $t_{ML} - 1$ of the check sums orthogonal on $e_0^{(i)}$.
2. Error bits from a burst of length $n\lambda$ or less which starts anywhere after the ith position in block 0 are checked by no more than t_{ML} of the check sums orthogonal on $e_0^{(i)}$.

Hence, a majority-logic decoder will correctly estimate each information error bit $e_0^{(i)}$ when there are t_{ML} or fewer random errors in the $2t_{ML}$ orthogonal sums checking $e_0^{(i)}$, or when the first constraint length of received bits contains a burst of length $b = n\lambda$ or less with a guard space $g = n_A - t_{ML}$. With feedback decoding, the same error-correcting capability applies to all information error bits if all past decoding decisions have been correct. The error propagation properties of diffuse codes are examined in Problem 14.13.

For the special case when $t_{ML} = 1$ and $R = (n - 1)/n$, the Iwadare–Massey codes discussed in Section 14.2 form a class of λ-diffuse, single-error-correcting codes. For any t_{ML}, Ferguson [13] has constructed a class of λ-diffuse codes with $R = 1/2$ and $g/b \approx 4$, and Tong [14] has constructed a similar class of asymptotically optimum $R = 1/2$, λ-diffuse codes for which $g/b \longrightarrow 3$ as λ becomes large. Some of these diffuse codes are listed in Table 14.1. In addition, Tong [15] has constructed a

TABLE 14.1 $R = 1/2$, λ-DIFFUSE, t_{ML}-ERROR-CORRECTING ORTHOGONALIZABLE CODES

t_{ML}	m	λ_{min}*	$g_1^{(2)}$	Orthogonalization rules†
2	$3\lambda + 3$	2	$\{0, \lambda, 2\lambda + 2, 3\lambda + 3\}$	$(3\lambda + 3, 2\lambda + 1)$
3	$3\lambda + 12$	5	$\{0, 1, \lambda + 1, 2\lambda + 7, 3\lambda + 9, 3\lambda + 12\}$	$(3\lambda + 9, 2\lambda + 3)(3\lambda + 12, \lambda + 4)$
4	$3\lambda + 37$	17	$\{0, 2, 3, \lambda + 3, 2\lambda + 18, 2\lambda + 23, 3\lambda + 27, 3\lambda + 37\}$	$(2\lambda + 23, \lambda + 8)$ $(3\lambda + 27, 2\lambda + 12, 2\lambda + 7)$ $(3\lambda + 37, \lambda + 13)$
5	$3\lambda + 88$	44	$\{0, 3, 4, 5, \lambda + 5, 2\lambda + 40, 2\lambda + 54, 3\lambda + 60\ 3\lambda + 67, 3\lambda + 88\}$	$(5, 1)(2\lambda + 54, \lambda + 19)(2\lambda + 67, \lambda + 12)$ $(3\lambda + 60, 2\lambda + 11, 2\lambda + 25)$ $(3\lambda + 88, \lambda + 33, \lambda + 26)$
6	$3\lambda + 217$	120	$\{0, 2, 3, 7, 8, \lambda + 8, 2\lambda + 88, 2\lambda + 118, 2\lambda + 138, 3\lambda + 147, 3\lambda + 157, 3\lambda + 217\}$	$(8, 4, 5, 6)(2\lambda + 118, \lambda + 38)$ $(2\lambda + 138, \lambda + 58, \lambda + 28)$ $(3\lambda + 147, 2\lambda + 17, 2\lambda + 67, 2\lambda + 37)$ $(3\lambda + 157, \lambda + 18)$ $(3\lambda + 217, \lambda + 68, \lambda + 78)$
7	$3\lambda + 374$	223	$\{0, 6, 7, 9, 10, 11, \lambda + 11, 2\lambda + 141, 2\lambda + 154, 2\lambda + 245, 3\lambda + 257, 3\lambda + 296, 3\lambda + 322, 3\lambda + 374\}$	$(10, 3, 1)(11, 8, 4)(2\lambda + 154, \lambda + 24)$ $(2\lambda + 245, \lambda + 115, \lambda + 102)$ $(3\lambda + 257, 2\lambda + 23, 2\lambda + 127, \lambda + 114)$ $(3\lambda + 296, \lambda + 50)$ $(3\lambda + 322, \lambda + 76, \lambda + 37)$ $(3\lambda + 374, \lambda + 89, \lambda + 63)$

*The minimum value of λ for which these codes are λ-diffuse and t_{ML}-error-correcting.

†$(x, y, \ldots)$ indicates that the sum $s_x + s_y + \cdots$ forms an orthogonal check sum on $e_0^{(1)}$. Only those orthogonal equations that require a sum of syndrome bits are listed.

class of λ-diffuse, t_{ML}-error-correcting, self-orthogonal codes. These codes are easy to implement, have limited error propagation, and their g/b ratio, although much larger than the Gallager bound, is optimum within the class of self-orthogonal diffuse codes. A list of self-orthogonal diffuse codes is given in Table 14.2. Note that the g/b ratio for the diffuse codes in Table 14.1 is much less than for the corresponding $R = 1/2$ self-orthogonal diffuse codes in Table 14.2. However, the self-orthogonal diffuse codes are easier to implement and less sensitive to error propagation.

TABLE 14.2 λ-DIFFUSE, t_{ML}-ERROR-CORRECTING SELF-ORTHOGONAL CODES

(a) $R = 1/2$ codes

t_{ML}	m	λ_{min}*	$\mathbf{g}_1^{(2)}$
1	3λ	1	$\{0, \lambda, 3\lambda\}$
2	$4\lambda + 1$	2	$\{0, \lambda, 3\lambda, 4\lambda + 1\}$
3	$5\lambda + 4$	4	$\{0, 1, \lambda + 3, 3\lambda + 3, 4\lambda + 3, 5\lambda + 4\}$
4	$6\lambda + 10$	8	$\{0, 1, 3, \lambda + 7, 3\lambda + 7, 4\lambda + 7, 5\lambda + 8, 6\lambda + 10\}$
5	$7\lambda + 19$	13	$\{0, 1, 4, 6, \lambda + 12, 3\lambda + 12, 4\lambda + 12, 5\lambda + 13, 6\lambda + 15, 7\lambda + 19\}$

(b) $R = 2/3$ codes

t_{ML}	m	λ_{min}*	$\mathbf{g}_1^{(3)}$	$\mathbf{g}_2^{(3)}$
2	$8\lambda + 3$	3	$\{0, \lambda, 4\lambda, 8\lambda + 3\}$	$\{0, 2\lambda, 6\lambda + 1, 7\lambda + 2\}$
3	$10\lambda + 10$	7	$\{0, 1, \lambda + 2, 4\lambda + 4, 8\lambda + 5, 10\lambda + 10\}$	$\{0, 2, 2\lambda + 2, 6\lambda + 4, 7\lambda + 4, 9\lambda + 7\}$
4	$12\lambda + 26$	12	$\{0, 1, 4, \lambda + 7, 4\lambda + 11, 8\lambda + 12, 9\lambda + 13, 11\lambda + 24\}$	$\{0, 2, 7, 2\lambda + 7, 6\lambda + 11, 7\lambda + 11, 10\lambda + 20, 12\lambda + 26\}$

(c) $R = 3/4$ codes

t_{ML}	m	λ_{min}*	$\mathbf{g}_1^{(4)}$	$\mathbf{g}_2^{(4)}$	$\mathbf{g}_3^{(4)}$
2	$12\lambda + 5$	6	$\{0, \lambda, 6\lambda, 9\lambda + 2\}$	$\{0, 2\lambda, 11\lambda + 3, 12\lambda + 5\}$	$\{0, 4\lambda, 7\lambda, 8\lambda + 1\}$
3	$15\lambda + 12$	9	$\{0, 1, 2\lambda + 3, 11\lambda + 6, 12\lambda + 7, 15\lambda + 12\}$	$\{0, 2, 4\lambda + 3, 7\lambda + 3, 8\lambda + 5, 13\lambda + 7\}$	$\{0, 3, \lambda, 6\lambda, 9\lambda + 2, 14\lambda + 8\}$

(d) $R = 4/5$ codes

t_{ML}	m	λ_{min}*	$\mathbf{g}_1^{(5)}$	$\mathbf{g}_2^{(5)}$	$\mathbf{g}_3^{(5)}$	$\mathbf{g}_4^{(5)}$
2	$16\lambda + 8$	7	$\{0, \lambda, 10\lambda + 3, 12\lambda + 4\}$	$\{0, 2\lambda, 13\lambda + 5, 14\lambda + 7\}$	$\{0, 3\lambda, 7\lambda, 16\lambda + 8\}$	$\{0, 5\lambda, 8\lambda + 1, 9\lambda + 2\}$

*The minimum value of λ for which these codes are λ-diffuse and t_{ML}-error-correcting.

Example 14.8

Consider the (2, 1, 9) systematic code with $\mathbf{g}^{(2)}(D) = 1 + D^3 + D^7 + D^9$. The parity triangle for this code is given by

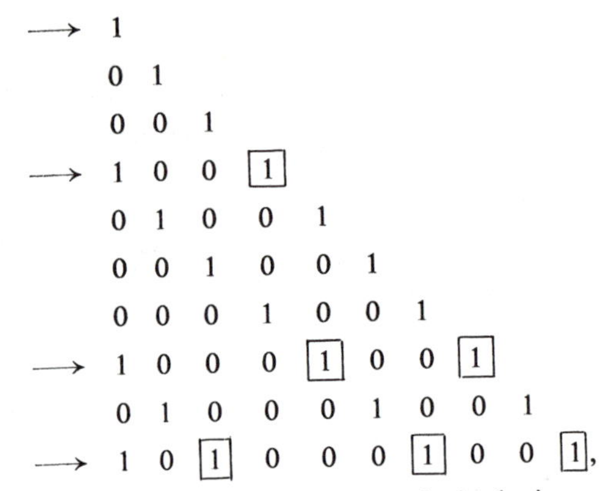

and we see that the code is self-orthogonal with majority-logic error-correcting capability $t_{ML} = 2$. The four syndrome bits that are orthogonal on $e_0^{(1)}$ are given by

$$\begin{aligned} s_0 &= e_0^{(1)} & & & & + e_0^{(2)} \\ s_3 &= e_0^{(1)} & + e_3^{(1)} & & & + e_3^{(2)} \\ s_7 &= e_0^{(1)} & + e_4^{(1)} & & + e_7^{(1)} & + e_7^{(2)} \\ s_9 &= e_0^{(1)} + e_2^{(1)} & & + e_6^{(1)} & & + e_9^{(1)} + e_9^{(2)}. \end{aligned}$$

The error bits other than $e_0^{(1)}$ which belong to a burst of length $b = 4$ or less, including $e_0^{(1)}$, can affect only syndrome bit s_0. In addition, error bits from a burst of length $b = 4$ or less which starts after $e_0^{(1)}$ can affect at most two of the syndrome bits orthogonal on $e_0^{(1)}$. Hence, this is a two-diffuse, two-error-correcting code which can correct any two or fewer random errors in a constraint length, or any burst of length $b = 4$ or less with a guard space $g = 18$.

The Burst-Finding System

Consider the $(2, 1, L + M + 5)$ systematic convolutional code with $\mathbf{g}^{(2)}(D) = 1 + D^3 + D^4 + D^5 + D^{L+M+5}$. In general, the Gallager burst-finding system corrects "almost all" bursts of length $b = 2(L - 5)$ or less with a guard space $g = 2(L + M + 5)$ as well as t'_{ML} or fewer random errors in a constraint length. In all cases $t'_{ML} < t_{ML}$, and in this example $t'_{ML} = 1$. Typically, L is on the order of 100's of bits, whereas M is on the order of 10's of bits. The encoder for the code above is shown in Figure 14.8.

The first five delay units in the encoder together with their associated connections form a set of $J = 4$ orthogonal check sums on each information error bit (see Example 13.2). By themselves, these orthogonal check sums could be used to correct $t_{ML} = 2$ or fewer random errors in a constraint length. The key to the Gallager burst-finding system is that only patterns of t'_{ML} or fewer random errors are corrected, where $t'_{ML} < t_{ML}$, and the additional error-correcting capability of the orthogonal check sums is used to detect bursts.

The decoding circuit for the code above is shown in Figure 14.9. To understand

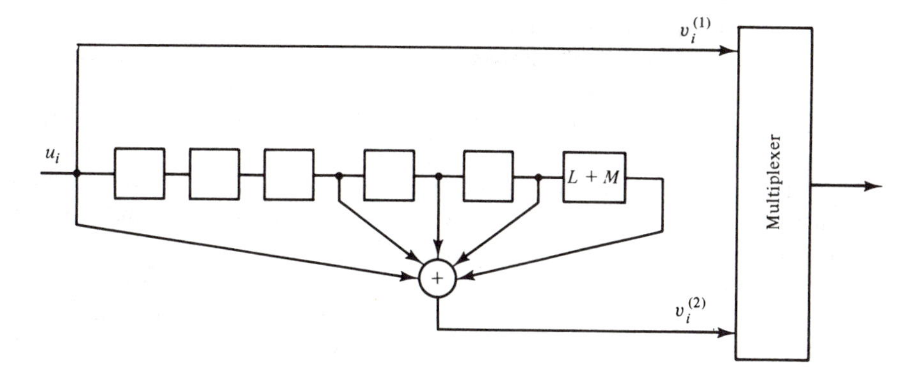

Figure 14.8 Encoding circuit for the $R = 1/2$ Gallager burst-finding code.

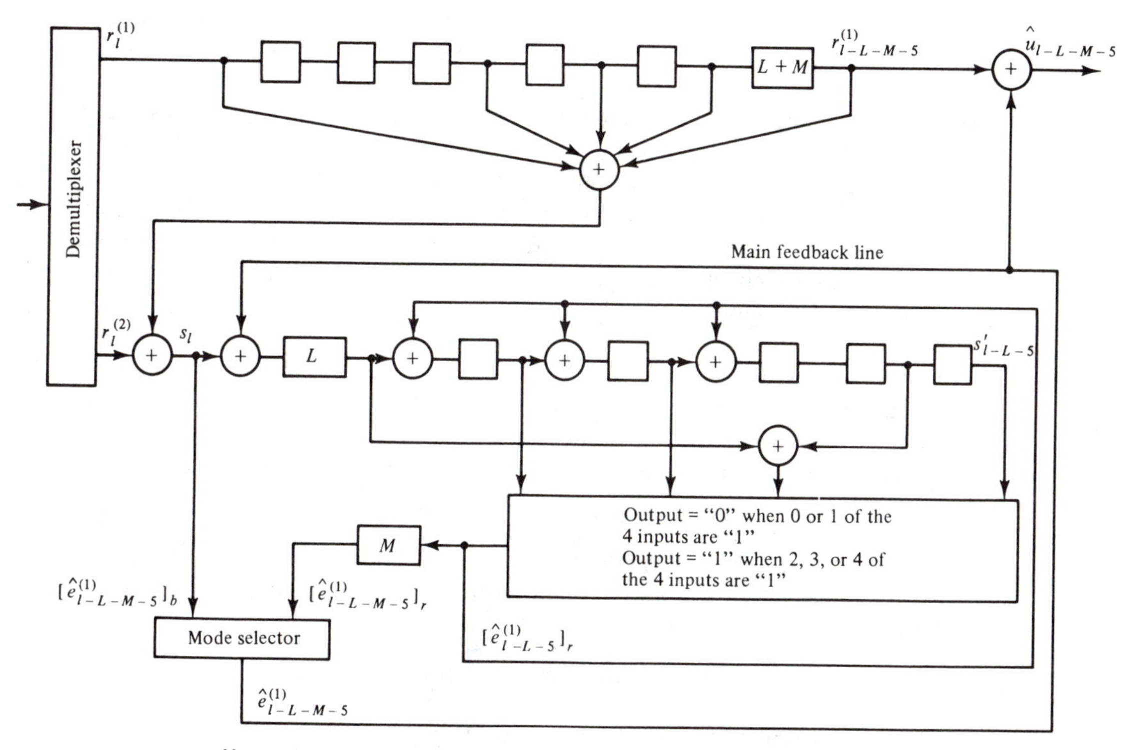

Notes: (1) Mode selector switches to "b-mode" when 2 or 3 of the 4 check sums have value "1"

(2) Mode selector switches to "r-mode" when M consecutive "r-mode" estimates have value "0"

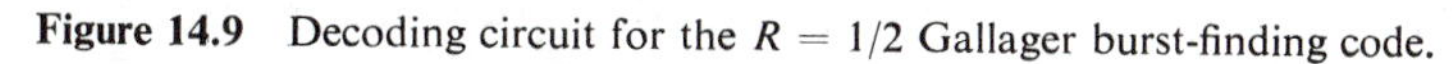

Figure 14.9 Decoding circuit for the $R = 1/2$ Gallager burst-finding code.

the functioning of this circuit, assume that the lth block has just been received, that all past decoding estimates have been correct, and that the decoder is in the "random mode" (or "r-mode") rather than the "burst mode" (or "b-mode"). In the r-mode, the input to the "mode selector" with subscript r is taken as the decoding estimate, whereas in the b-mode, the input with subscript b is taken as the decoding estimate.

Because all past estimates have been correct, the feedback of these estimates along the "main feedback line" removes the effect of $e_{l-L-M-5}^{(1)}$ from the syndrome, since at this time, $e_{l-L-M-5}^{(1)}$ is the error bit that is fed to the syndrome from the last stage of the encoder replica. Hence, the modified syndrome bits entering the syndrome register are precisely those that would enter if the last $M + L$ stages of the encoder replica and the associated connection were removed. This modified code is decoded just as if it were a random-error-correcting code, except that the syndrome bits are delayed by L time units before decoding begins.

Now assume that there are $t'_{\text{ML}} = 1$ or fewer errors in the $n_E = 11$ error bits affecting the $J = 4$ orthogonal check sums on $e_{l-L-5}^{(1)}$. Then if $e_{l-L-5}^{(1)} = 1$, all four check sums will equal 1, and if $e_{l-L-5}^{(1)} = 0$, at most one of the check sums will equal 1. Hence, the estimate $[\hat{e}_{l-L-5}^{(1)}]_r$, which is the decision accepted by the mode selector in the r-mode, will be correct, and the decoder will stay in the r-mode and decode correctly as long as there is at most one error in a constraint length. Note also that there is an additional M-time-unit delay before the r-mode decision is actually accepted by the mode selector.

Now suppose that at some time two or three of the four check sums have value 1. This will always occur when there are two or three errors in the $n_E = 11$ error bits affecting the check sums, and the r-mode estimate is incorrect. (Note that it is possible for two or three errors to cause zero, one, or all four of the check sums to equal 1, but in this case the estimate must be correct since two of the errors must have canceled their effect on the check sums.) When this occurs, the mode selector changes to the b-mode, and $[\hat{e}_{l-L-M-5}^{(1)}]_b$ is now chosen as the estimate. This causes the preceding M decisions of the r-mode decoder to be rejected, which ensures that when a burst occurs, the r-mode decoder will have detected the burst before any of its estimates of error bits in the burst are accepted by the mode selector. If the bits in a burst have probability 1/2 of being in error, then when the first bit in the burst reaches the end of the syndrome register, all $2^4 = 16$ possible outcomes for the four check sums are equally likely. Since $\binom{4}{2} + \binom{4}{3} = 10$ of these outcomes cause the mode selector to switch to the b-mode, the probability that the burst is not detected in time for the mode selector to switch to the b-mode before accepting an r-mode estimate for any error bit in the burst is

$$\Pr[\text{undetected burst}] \approx \left(\frac{6}{16}\right)^{M+1}. \tag{14.22}$$

Clearly, this probability is quite small for M greater than about 10.

The decoding estimate in the b-mode is given by

$$[\hat{e}_{l-L-M-5}^{(1)}]_b = s_l = e_{l-L-M-5}^{(1)} + e_{l-5}^{(1)} + e_{l-4}^{(1)} + e_{l-3}^{(1)} + e_l^{(1)} + e_l^{(2)}. \tag{14.23}$$

If $e_{l-L-M-5}^{(1)}$ is part of a burst, the other error bits in (14.23) must come from the guard space if the guard space has length $g = 2(L + M + 5)$ or more. Hence, the decoding estimate in the b-mode will be correct. Note that it is possible for a burst to cause the switch to the b-mode as soon as its first bit reaches the first input to a check sum in the syndrome register. This would result in up to $M + 5$ guard space error bits being estimated in the b-mode. For these estimates to be correct, the burst length should not exceed $L - 5$ blocks or $b = 2(L - 5)$ bits, since otherwise error bits from the burst would affect the decoding of these guard space error bits.

While the decoder is in the b-mode, the r-mode estimates continue to be monitored as an indicator of when the decoder should switch back to the r-mode. When M consecutive r-mode estimates have value 0, this is taken as evidence that the most recent $L + M + 5$ received blocks have been in a guard space, and the mode selector is returned to the r-mode. Since during a burst only $\binom{4}{0} + \binom{4}{1} = 5$ of the $2^4 = 16$ possible outcomes for the four check sums give a value of 0 for the r-mode estimate, the probability of a false return to the r-mode during a burst is

$$\Pr[\text{false return to } r\text{-mode}] \approx \left(\frac{5}{16}\right)^M. \tag{14.24}$$

Again, this probability is quite small for M greater than about 10.

The adaptive decoding scheme described above corrects "almost all" bursts of length $b = 2(L - 5)$ with a guard space $g = 2(L + M + 5)$, since the probabilities of an undetected burst and of a false return to the random mode, given by (14.22) and (14.24), respectively, can be made quite small by choosing M large enough. Since L can be chosen much greater than M, we see that

$$\frac{g}{b} = \frac{2(L + M + 5)}{2(L - 5)} \approx 1, \tag{14.25}$$

which meets the lower bound of (14.2) on "almost all" burst-error correction for $R = 1/2$ codes.

Gallager originally described his burst-finding system only for $R = (n - 1)/n$ codes, but Reddy [16] has since generalized this to any $R = k/n$. All of these schemes have the property that the lower bound of (14.2) on "almost all" burst-error correction is met with near equality. Note that (14.23) implies that when the decoder is in the b-mode, it is sensitive to errors in the guard space. Sullivan [17] has shown how the burst-finding system can be modified to provide some protection against errors in the guard space by lowering the code rate.

The Burst-Trapping System

Tong's burst-trapping system is similar to Gallager's burst-finding system, except that it is based on block codes rather than convolutional codes. The overall code remains convolutional, however, since there is memory in the encoder.

Consider an $(n = 3M, k = 2M)$ systematic block code with $R = 2/3$, error-correcting capability t, and generator matrix

$$\mathbf{G} = \left[\mathbf{I}_{2M} \;\vdots\; \begin{matrix}\mathbf{G}_1 \\ \hdashline \mathbf{G}_2\end{matrix}\right], \tag{14.26}$$

where $\mathbf{G}_1$ and $\mathbf{G}_2$ are $M \times M$ submatrices of $\mathbf{G}$. The code word $\mathbf{v} = [\mathbf{u}^{(1)}, \mathbf{u}^{(2)}]\mathbf{G} = [\mathbf{u}^{(1)}, \mathbf{u}^{(2)}, \mathbf{p}]$, where $\mathbf{u}^{(1)}$ and $\mathbf{u}^{(2)}$ each contain M information bits, and $\mathbf{p}$ is the M-bit parity vector given by

$$\mathbf{p} = \mathbf{u}^{(1)}\mathbf{G}_1 + \mathbf{u}^{(2)}\mathbf{G}_2. \tag{14.27}$$

The encoder for the Tong burst-trapping system based on this code is shown in Figure 14.10. Note that in addition to the operations required by the block encoder, memory has been added to the encoder for the burst-trapping system. This converts the block code into an $(n = 3M, k = 2M, m = 2L)$ convolutional code! The code is still systematic, and the encoding equations at time unit l are given by

$$\begin{aligned}\mathbf{v}_l^{(1)} &= \mathbf{u}_l^{(1)} \\ \mathbf{v}_l^{(2)} &= \mathbf{u}_l^{(2)} \\ \mathbf{v}_l^{(3)} &= \mathbf{u}_l^{(1)}\mathbf{G}_1 + \mathbf{u}_l^{(2)}\mathbf{G}_2 + \mathbf{u}_{l-L}^{(1)} + \mathbf{u}_{l-2L}^{(2)}.\end{aligned} \tag{14.28}$$

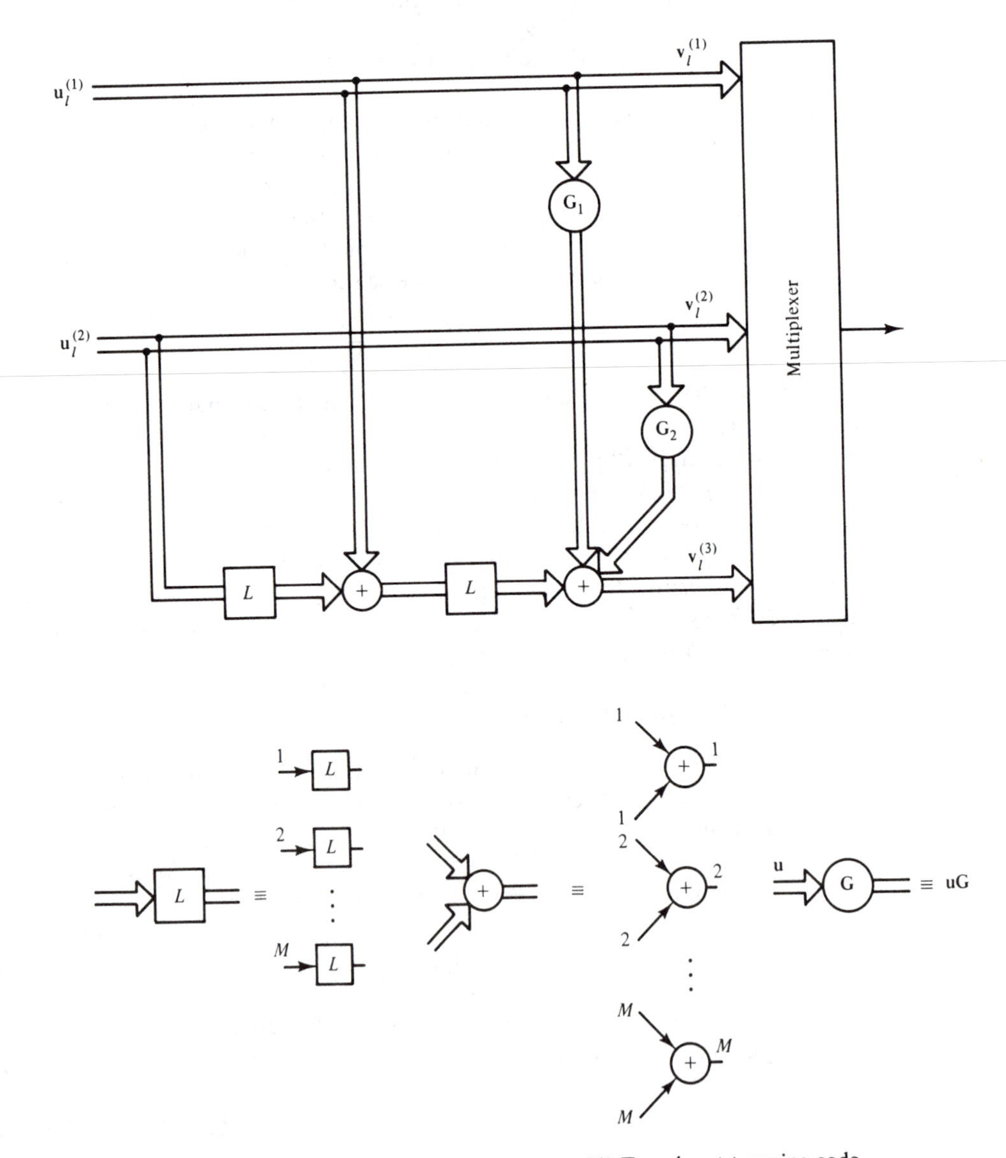

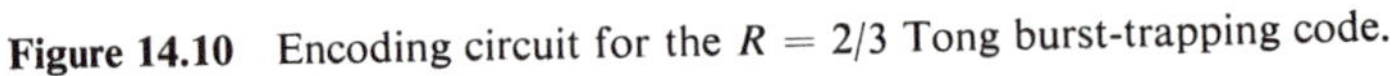

Figure 14.10 Encoding circuit for the $R = 2/3$ Tong burst-trapping code.

The decoder for the Tong burst-trapping system is shown in Figure 14.11. Assume that the lth block has just been received, that all past decoding decisions have been correct, and that the decoder is in the r-mode. The feedback of the decoding estimates $\hat{\mathbf{u}}_{l-L}^{(1)}$ and $\hat{\mathbf{u}}_{l-2L}^{(2)}$ to the parity input line of the block decoder removes the effect of past decisions from the encoding equations, so that the block decoder simply decodes the original block code. However, the block decoder is designed to correct only t' or fewer random errors, where $t' < t$, with the reserve error-correcting power of the code being used to detect patterns of $t' + 1$ or more errors. As long as the block decoder estimates t' or fewer errors in a block, it remains in the r-mode. Note

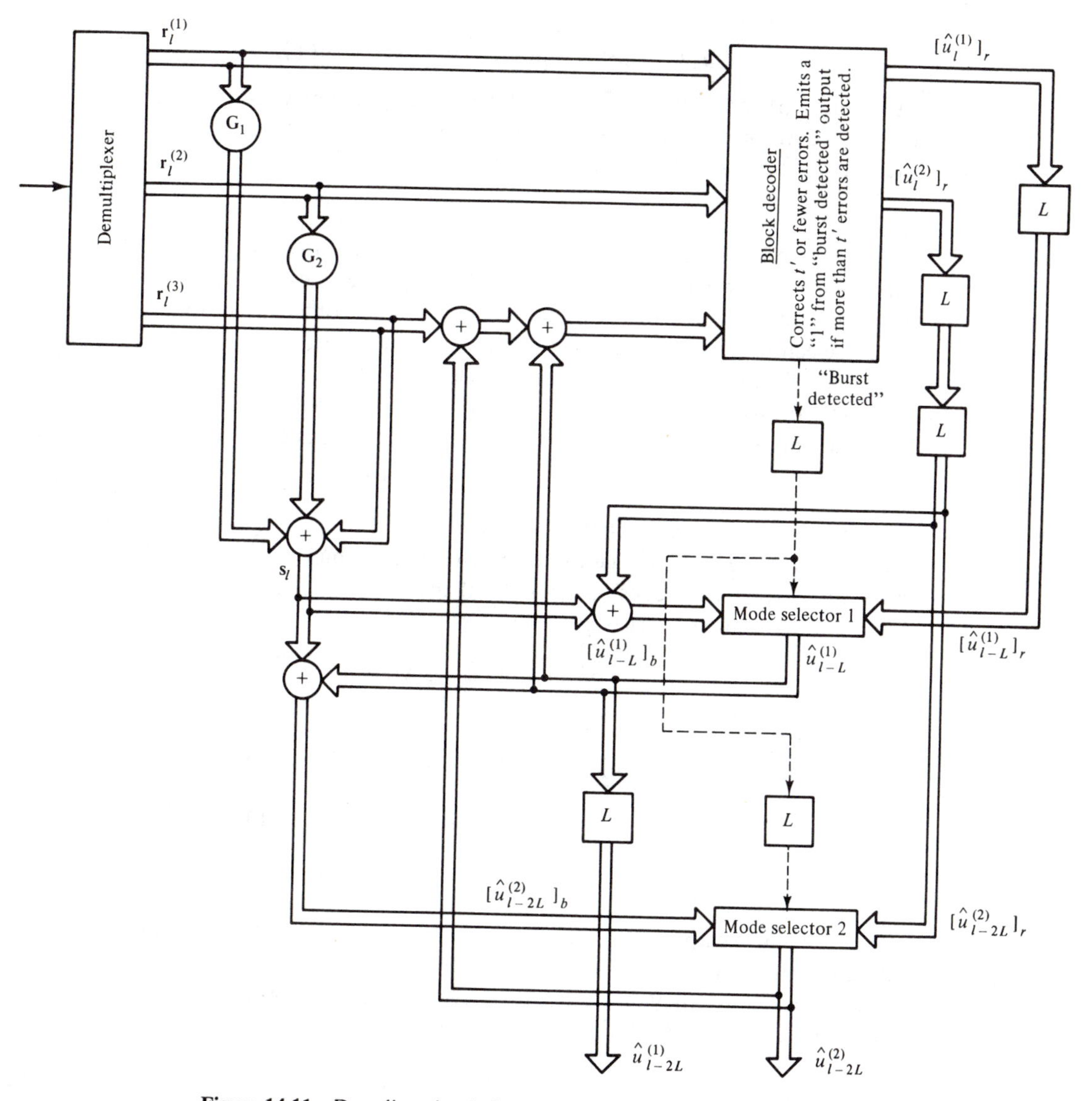

Figure 14.11 Decoding circuit for the $R = 2/3$ Tong burst-trapping code.

that, just as in the burst-finding system, the decoding estimates of the block decoder are delayed by $2L$ time units before they are accepted as final.

When the block decoder detects a pattern of $t' + 1$ or more errors, say at time unit $l - L$, it emits a single 1 from its "burst detected" output. L time units later, at time unit l, this 1 reaches the first mode selector in Figure 14.11 and causes it to switch to the b-mode for that block only. At this time, the output of the first mode selector is

$$\begin{aligned}\hat{\mathbf{u}}_{l-L}^{(1)} &= [\hat{\mathbf{u}}_{l-L}^{(1)}]_b = \mathbf{s}_l + [\hat{\mathbf{u}}_{l-2L}^{(2)}]_r \\ &= (\mathbf{u}_l^{(1)} + \mathbf{e}_l^{(1)})\mathbf{G}_1 + (\mathbf{u}_l^{(2)} + \mathbf{e}_l^{(2)})\mathbf{G}_2 + \mathbf{u}_l^{(1)}\mathbf{G}_1 + \mathbf{u}_l^{(2)}\mathbf{G}_2 \\ &\quad + \mathbf{u}_{l-L}^{(1)} + \mathbf{u}_{l-2L}^{(2)} + \mathbf{e}_l^{(3)} + [\hat{\mathbf{u}}_{l-2L}^{(2)}]_r \\ &= \mathbf{u}_{l-L}^{(1)} + \mathbf{u}_{l-2L}^{(2)} + [\hat{\mathbf{u}}_{l-2L}^{(2)}]_r + \mathbf{e}_l^{(1)}\mathbf{G}_1 + \mathbf{e}_l^{(2)}\mathbf{G}_2 + \mathbf{e}_l^{(3)}.\end{aligned} \tag{14.29}$$

From our earlier assumption that past decoding decisions were all correct, $[\hat{\mathbf{u}}_{l-2L}^{(2)}]_r = \mathbf{u}_{l-2L}^{(2)}$. Also assuming that time unit l comes from the error-free guard space following the burst at time unit $l - L$, it follows that

$$\hat{\mathbf{u}}_{l-L}^{(1)} = \mathbf{u}_{l-L}^{(1)}, \tag{14.30}$$

and the b-mode decoding estimate will be correct. A similar argument shows that at time unit $l + L$, when the "burst detected" output reaches the second mode selector and causes it to switch to the b-mode for that block only, the output of the second mode selector will be the correct estimate for $\mathbf{u}_{l-L}^{(2)}$, provided that time unit $l + L$ also comes from the error-free guard space.

The probability of failing to switch to the b-mode during a burst can be estimated as follows. Assume that an (n, k) block code is designed to correct t' or fewer errors. There are then a total of $N(t') \triangleq \binom{n}{0} + \binom{n}{1} + \cdots + \binom{n}{t'}$ correctable error patterns. Hence, for each of the 2^k code words, there are $N(t')$ received blocks that will be corrected and not cause a switch to the b-mode. Assuming that during a burst all 2^n received blocks are equally likely, the probability of failing to switch to the b-mode during a burst is given by

$$\Pr[\text{undetected burst}] \approx N(t')\frac{2^k}{2^n} = \frac{N(t')}{2^{n-k}}. \tag{14.31}$$

Example 14.9

Consider a (30, 20) shortened BCH code with $t' = 1$. Then $N(t') = 1 + 30 = 31$ and

$$\Pr[\text{undetected burst}] \approx \frac{31}{2^{10}} = 3.0 \times 10^{-2},$$

which implies about a 3% failure rate in detecting bursts.

The burst-trapping system described above corrects "almost all" bursts that can affect at most L consecutive received blocks. Hence, $b = (n - 1)L + 1$ bits. The guard space must include at least $2L$ consecutive error-free blocks following the burst. Hence, $g = 2nL + (n - 1)$. Therefore, for large L and n,

$$\frac{g}{b} = \frac{2nL + (n-1)}{(n-1)L + 1} \approx 2, \tag{14.32}$$

which meets the lower bound of (14.2) on "almost all" burst-error correction for $R = 2/3$ codes.

Although the discussion above concerned only $R = 2/3$ codes, the burst-trapping system can be easily generalized to any rate $R = \lambda M/(\lambda + 1)M$. These systems all meet the lower bound of (14.2) on "almost all" burst-error correction with near equality. Like the closely related burst-finding system, the burst-trapping system is sensitive to errors in the guard space when it is in the b-mode. Burton et al. [18] have shown, however, that the system can be modified to provide some protection against errors in the guard space by lowering the code rate.

PROBLEMS

14.1. Using mathematical induction, show that the unknown elements of the matrix $\mathbf{B}_0$ can always be chosen so that (14.11) is satisfied.

14.2. Show how to construct optimum phased-burst error correcting Berlekamp–Preparata codes with $k < n - 1$.

14.3. Consider the Berlekamp–Preparata code with $n = 3$.
- **(a)** Find m, b, and g for this code.
- **(b)** Find the $\mathbf{B}_0$ matrix.
- **(c)** Find the generator polynomials $\mathbf{g}_1^{(3)}(D)$ and $\mathbf{g}_2^{(3)}(D)$.
- **(d)** Find the $\mathbf{H}_0$ matrix.
- **(e)** Draw the complete encoder/decoder block diagram for this code.

14.4. Consider the Iwadare–Massey code with $n = 2$ and $\lambda = 4$.
- **(a)** Find m, b, and g for this code.
- **(b)** Find the generator polynomial $\mathbf{g}^{(2)}(D)$.
- **(c)** Find the repeat distance of the information error bit $e_l^{(1)}$.
- **(d)** Draw the complete encoder/decoder block diagram for this code.

14.5. A second class of Iwadare–Massey codes exists with the following parameters:

$$m = (2n - 1)\lambda + \frac{n^2 - n - 2}{2}$$

$$b = n\lambda$$

$$g = n(m + 1) - 1$$

The $n - 1$ generator polynomials are given by (14.16), where $a(i) \triangleq \frac{1}{2}(n - i)(4\lambda + n - i - 3) + n - 1$ and $b(i) \triangleq \frac{1}{2}(n - i)(4\lambda + n - i - 1) + n + \lambda - 2$. Consider the code with $n = 3$ and $\lambda = 3$.
- **(a)** Find m, b, and g for this code.
- **(b)** Find the generator polynomials $\mathbf{g}_1^{(3)}(D)$ and $\mathbf{g}_2^{(3)}(D)$.
- **(c)** Find the repeat distance of the information error bits $e_l^{(1)}$ and $e_l^{(2)}$.
- **(d)** Construct a decoding circuit for this code.

14.6. Construct a general decoding circuit for the class of Iwadare–Massey codes in Problem 14.5. For the two classes of Iwadare–Massey codes:
- **(a)** Compare the excess guard space required beyond the Gallager bound.
- **(b)** Compare the number of register stages required to implement a general decoder.

14.7. Show that, for the Iwadare–Massey code of Example 14.5, if $n[m + (\lambda + 2)n - 1] - 1 = 95$ consecutive error-free bits follow a decoding error, the syndrome will return to the all-zero state.

14.8. Consider the (2, 1, 5) double-error-correcting orthogonalizable code from Table 13.3 interleaved to degree $\lambda = 7$.

(a) Completely characterize the multiple-burst-correcting capability and the associated guard space requirements of this interleaved code.

(b) Find the maximum single-burst length that can be corrected, and the associated guard space.

(c) Find the guard space-to-burst length ratio for part (b).

(d) Find the total memory required in the interleaved decoder.

(e) Draw a block diagram of the complete interleaved system.

14.9. Consider the interleaved encoder shown in Figure 14.6(b). Assume that an information sequence $u_0, u_1, u_2, \ldots$ enters the encoder. Write down the string of encoded bits and verify that an interleaving degree of $\lambda = 5$ is achieved.

14.10. Consider the Berlekamp–Preparata code of Problem 14.3 interleaved to degree $\lambda = 7$.

(a) Find the g/b ratio, and compare with the Gallager bound.

(b) Draw a block diagram of the complete interleaved system.

14.11. Consider the $n = 3$ Berlekamp–Preparata code interleaved to degree $\lambda = 7$ and the $n = 3$ Iwadare–Massey code with $\lambda = 7$.

(a) Compare the g/b ratios of the two codes.

(b) Compare the number of register stages required to implement the decoder in both cases.

14.12. Consider the (2, 1, 9) systematic code with $\mathbf{g}^{(2)}(D) = 1 + D^2 + D^5 + D^9$.

(a) Is this code self-orthogonal? What is t_{ML} for this code?

(b) Is this a diffuse code? What is the burst-error-correcting capability b and its required guard space g?

(c) Draw a complete encoder/decoder block diagram for this code.

14.13. For the diffuse code of Figure 14.7, find the minimum number of error-free bits that must be received following a decoding error to guarantee that the syndrome returns to the all-zero state.

14.14. Consider using the (2, 1, 11) triple-error-correcting orthogonalizable code from Table 13.3 in the Gallager burst-finding system.

(a) Draw a block diagram of the encoder.

(b) Draw a block diagram of the decoder.

(c) With $t'_{ML} = 1$, choose M and L such that the probability of an undetected burst and of a false return to the r-mode are less than 10^{-2} and the g/b ratio is within 1% of the Gallager bound on "almost all" burst-error correction for $R = 1/2$ codes.

(d) Repeat part (c) for $t'_{ML} = 2$.

14.15. Consider the $R = 2/3$ burst-trapping code of Example 14.9.

(a) Choose L such that the g/b ratio is within 1% of the Gallager bound on "almost all" burst-error correction for $R = 2/3$ codes.

(b) Describe the generator matrix $\mathbf{G}$ of the (30, 20, $2L$) convolutional code.

REFERENCES

1. D. W. Hagelbarger, "Recurrent Codes: Easily Mechanized, Burst-Correcting, Binary Codes," *Bell Syst. Tech. J.*, 38, pp. 969–984, July 1959.
2. I. Iwadare, "On Type B1 Burst-Error-Correcting Convolutional Codes," *IEEE Trans. Inf. Theory*, IT-14, pp. 577–583, July 1968.

3. J. L. Massey, unpublished memorandum, July 1967.
4. A. D. Wyner and R. B. Ash, "Analysis of Recurrent Codes," *IEEE Trans. Inf. Theory*, IT-9, pp. 143–156, July 1963.
5. E. R. Berlekamp, "Note on Recurrent Codes," *IEEE Trans. Inf. Theory*, IT-10, pp. 257–258, July 1964.
6. F. P. Preparata, "Systematic Construction of Optimal Linear Recurrent Codes for Burst Error Correction," *Calcolo*, 2, pp. 1–7, 1964.
7. A. Kohlenberg and G. D. Forney, Jr., "Convolutional Coding for Channels with Memory," *IEEE Trans. Inf. Theory*, IT-14, pp. 618–626, September 1968.
8. J. L. Massey, "Advances in Threshold Decoding," in *Advances in Communication Systems*, Vol. 2, A. V. Balakrishnan, ed., Academic Press, New York, 1968.
9. R. G. Gallager, *Information Theory and Reliable Communication*, McGraw-Hill, New York, 1968.
10. S. Y. Tong, "Burst-Trapping Techniques for a Compound Channel," *IEEE Trans. Inf. Theory*, IT-15, pp. 710–715, November 1969.
11. J. L. Massey, "Coding for Everyday Channels," in *Coding Techniques for Digital Communication*, notes for a tutorial session, 1973 IEEE Int. Conf. Commun., Seattle, Wash., June 1973.
12. J. L. Massey, "Implementation of Burst-Correcting Convolutional Codes," *IEEE Trans. Inf. Theory*, IT-11, pp. 416–422, July 1965.
13. M. J. Ferguson, "Diffuse Threshold Decodable Rate 1/2 Convolutional Codes," *IEEE Trans. Inf. Theory*, IT-17, pp. 171–180, March 1971.
14. S. Y. Tong, "Construction of Asymptotically Optimal Rate 1/2 Diffuse Codes," unpublished manuscript, October 1970.
15. S. Y. Tong, "Systematic Construction of Self-Orthogonal Diffuse Codes," *IEEE Trans. Inf. Theory*, IT-16, pp. 594–604, September 1970.
16. S. M. Reddy, "Linear Convolutional Codes for Compound Channels," *Inf. Control*, 19, pp. 387–400, December 1971.
17. D. D. Sullivan, "A Generalization of Gallager's Adaptive Error Control Scheme," *IEEE Trans. Inf. Theory*, IT-17, pp. 727–735, November 1971.
18. H. O. Burton, D. D. Sullivan, and S. Y. Tong, "Generalized Burst-Trapping Codes," *IEEE Trans. Inf. Theory*, IT-17, pp. 736–742, November 1971.

15

Automatic-Repeat-Request Strategies

As we pointed out in Chapter 1, there are two categories of techniques for controlling transmission errors in data transmission systems: the forward-error control (FEC) scheme and the automatic-repeat-request (ARQ) scheme. In an FEC system, an error-correcting code is used. When the receiver detects the presence of errors in a received vector, it attempts to determine the error locations and then corrects the errors. If the exact locations of errors are determined, the received vector will be correctly decoded; if the receiver fails to determine the exact locations of errors, the received vector will be decoded incorrectly and erroneous data will be delivered to the user (or data sink). In an ARQ system, a code with good error-detecting capability is used. At the receiver, the syndrome of the received vector is computed. If the syndrome is zero, the received vector is assumed to be error-free and is accepted by the receiver. At the same time, the receiver notifies the transmitter, via a return channel, that the transmitted code vector has been successfully received. If the syndrome is not zero, errors are detected in the received vector. Then the transmitter is instructed, through the return channel, to retransmit the same code vector. Retransmission continues until the code vector is successfully received. With this system, erroneous data are delivered to the data sink only if the receiver fails to detect the presence of errors. Using a proper linear code, the probability of an undetected error can be made very small.

We have devoted many chapters in discussing various types of error-correcting codes and decoding methods for FEC. In this chapter we present various types of ARQ schemes and the combinations of ARQ and FEC, called *hybrid-ARQ schemes*. Major references on ARQ are listed at the end of this chapter.

15.1 BASIC ARQ SCHEMES

There are three basic types of ARQ schemes: the stop-and-wait ARQ, the go-back-N ARQ, and the selective-repeat ARQ. In a stop-and-wait ARQ data transmission system, the transmitter sends a code vector to the receiver and waits for an acknowledgment from the receiver as shown in Figure 15.1. A positive acknowledgment

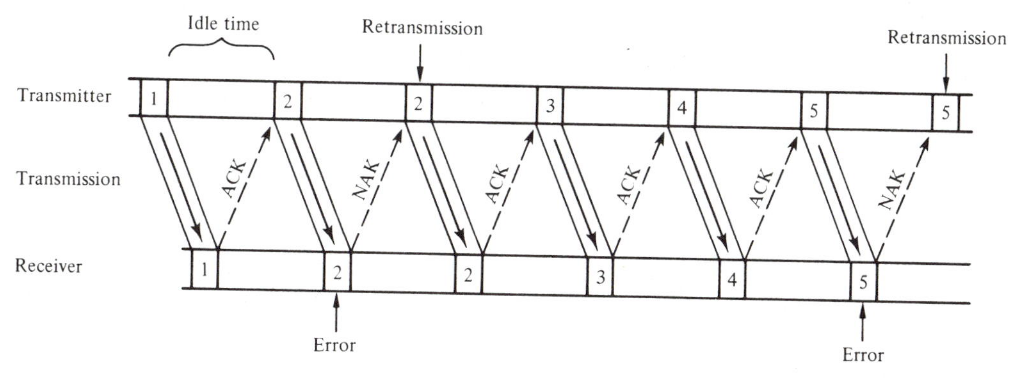

Figure 15.1 Stop-and-wait ARQ.

(ACK) from the receiver signals that the code vector has been successfully received (i.e., no errors being detected); and the transmitter sends the next code vector. A negative acknowledgment (NAK) from the receiver indicates that the received vector has been detected in error; the transmitter resends the code vector. Retransmissions continue until an ACK is received by the transmitter. The stop-and-wait ARQ scheme is simple and is used in many data communication systems; IBM's widely used Binary Synchronous Communication (BISYNC) protocol [1] is an example. However, the stop-and-wait ARQ scheme is inherently inefficient because of the idle time spent waiting for an acknowledgment for each transmitted code vector. Unless the code length n is extremely long, the fraction of idle time can be large. However, the use of a very long block length does not really provide a solution, since the probability that a block contains errors increases rapidly with the block length. Hence, using a long block length reduces the idle time but increases the frequency of retransmissions for each code vector. Moreover, a long block length may be impractical in many applications because of restrictions imposed by the data.

In a go-back-N ARQ system, code vectors are transmitted continuously. The transmitter does not wait for an acknowledgment after sending a code vector; as soon as it has completed sending one, it begins sending the next code vector as shown in Figure 15.2. The acknowledgment for a code vector arrives after a *round-trip delay*. The round-trip delay is defined as the time interval between the transmission of a code vector and the receipt of an acknowledgment for that code vector. During this interval, $N - 1$ other code vectors have also been transmitted. When a NAK is received, the transmitter backs up to the code vector that is negatively acknowledged and resends that code vector and $N - 1$ succeeding code vectors that were transmitted during the round-trip delay (i.e., the transmitter pulls back and resends N code

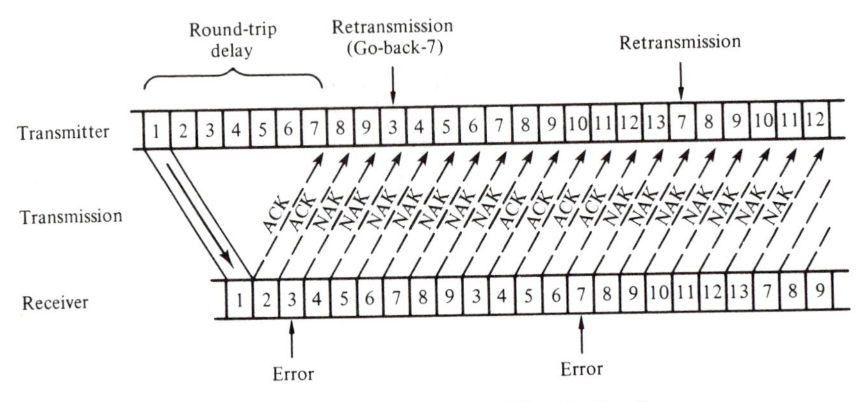

Figure 15.2 Go-back-N ARQ with $N = 7$.

vectors). Of course, buffer must be provided at the transmitter for these code vectors. At the receiver, the $N - 1$ received vectors following an erroneously received vector (a received vector detected in error) are discarded regardless of whether they are error-free or not. Therefore, the receiver needs to store only one received vector at a time. Because of the continuous transmission and retransmission of code vectors, the go-back-N ARQ scheme is more effective than the stop-and-wait ARQ; and it can be implemented at moderate cost. Communication protocols such as SDLC (Synchronous Data Link Control) [2] and ADCCP (Advanced Data Communications Control Procedure) [3] employ the go-back-N ARQ scheme. The go-back-N ARQ scheme becomes ineffective when the round-trip delay is large and data transmission rate is high. This inefficiency is caused by the retransmission of many error-free code vectors following a code vector detected in error. This inefficiency can be overcome by using the selective-repeat strategy.

In a selective-repeat ARQ system, code vectors are also transmitted continuously. However, the transmitter only resends (or repeats) those code vectors that are negatively acknowledged as shown in Figure 15.3. Since ordinarily code vectors must be delivered to the user in correct order, a buffer must be provided at the receiver to store the error-free received vectors following a received vector detected in error. When the first negatively acknowledged code vector is successfully received, the receiver than releases the error-free received vectors *in consecutive order* until the next erroneously received vector is encountered. Sufficient receiver buffer must be

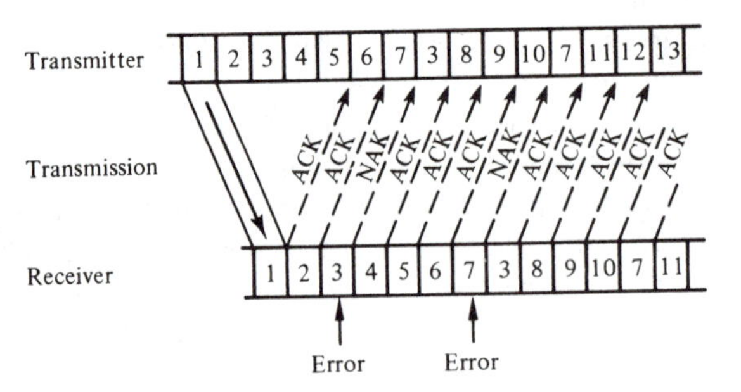

Figure 15.3 Selective-repeat ARQ.

provided; otherwise, buffer overflow may occur and data may be lost. The selective-repeat ARQ is the most efficient one among the three basic ARQ schemes; however, it is also the most complex one to implement.

In an ARQ system, the receiver commits a decoding error whenever it accepts a received vector with undetected errors. Such event is called an *error event*. Let $P(E)$ denote the probability of an error event. Clearly, for an ARQ system to be reliable, $P(E)$ should be very small. Therefore, the *reliability* of an ARQ system is measured by its error probability $P(E)$. Suppose that an (n, k) linear code C is used for error detection in an ARQ system. Let us define the following probabilities:

P_c = probability that a received vector contains no error;

P_d = probability that a received vector contains a detectable error pattern;

P_e = probability that a received vector contains an undetectable error pattern.

These probabilities add to 1 (i.e., $P_c + P_d + P_e = 1$). The probability P_c depends on the channel error statistics, the probabilities P_d and P_e depend on both the channel error statistics and the choice of the (n, k) error detecting code C. A received vector will be accepted by the receiver only if it either contains no error or contains an undetectable error pattern. Therefore, the probability $P(E)$ that the receiver commits an error is given by

$$P(E) = \frac{P_e}{P_c + P_e}. \tag{15.1}$$

The probability P_e can be made very small relative to P_c by choosing the code C properly (e.g., a long Hamming code or a long double-error correcting primitive BCH code). Consequently, the error probability $P(E)$ can be made very small. For a BSC with transition probability p, we have

$$P_c = (1 - p)^n. \tag{15.2}$$

In Section 3.6 we have shown that, for an average (n, k) linear code,

$$P_e \leq 2^{-(n-k)}[1 - (1 - p)^n]. \tag{15.3}$$

Combining (15.1), (15.2), and (15.3), we can obtain an upper bound on $P(E)$.

Example 15.1

Consider a BSC with transition probability $p = 10^{-3}$. Suppose that we use the double-error-correcting (1023, 993) BCH code for error detection in an ARQ system. Then

$$P_c = (1 - 10^{-3})^{1023} \approx 2^{-1.476},$$

and $P_e \leq 2^{-20}$ [double-error-correcting primitive BCH codes satisfy the bound $P_e \leq 2^{-(n-k)}$ for $p \leq 1/2$]. We see that $P_e \ll P_c$. Using (15.1), we find that

$$P(E) \leq 10^{-6}.$$

This illustrates that, using a relatively small number of parity-check digits, an ARQ system can achieve very high reliability.

Another measure of the performance of an ARQ system is its *throughput efficiency* (or simply throughput). The throughput is defined as the ratio of the average number of information digits successfully accepted by the receiver per unit of time to the total number of digits that could be transmitted per unit of time [4,5]. All three basic ARQ schemes achieve the same reliability; however, they have different throughput efficiencies. In the following we derive the throughput for each of the three basic ARQ schemes. For simplicity, we assume that the feedback channel is noiseless. The assumption may not be realistic, but the results give a good indication of the effectiveness of each ARQ scheme. For throughput analysis of various ARQ schemes with noisy feedback channel, the reader is referred to Reference 4.

We first derive the throughput of the selective-repeat ARQ scheme. We recall that, with this scheme, the transmitter sends code vectors to the receiver continuously and resends only those code vectors that are negatively acknowledged. The probability that a received vector will be accepted by the receiver is

$$P = P_c + P_e.$$

For the usual situation where $P_e \ll P_c$, then $P \approx P_c$. The probability that a code vector will be retransmitted is simply

$$P_d = 1 - P \approx 1 - P_c.$$

For a code vector to be successfully accepted by the receiver, the average number of retransmissions (including the original transmission) required is

$$\begin{aligned} T_{\text{SR}} &= 1 \cdot P + 2 \cdot P(1-P) + 3 \cdot P(1-P)^2 + \cdots + l \cdot P(1-P)^{l-1} + \cdots \\ &= \frac{1}{P}. \end{aligned}$$

Then the throughput of the selective-repeat ARQ is

$$\eta_{\text{SR}} = \frac{1}{T_{\text{SR}}}\left(\frac{k}{n}\right) = \left(\frac{k}{n}\right)P, \tag{15.4}$$

where k/n is the rate of the (n, k) code used in the system. We see that the throughput η_{SR} depends only on the channel error rate.

In a go-back-N ARQ system, when a code vector is negatively acknowledged, the transmitter resends that code vector and the $N - 1$ subsequent code vectors that were transmitted earlier. Therefore, for a code vector to be successfully accepted by the receiver, the average number of retransmissions (including the original transmission) required is

$$\begin{aligned} T_{\text{GBN}} &= 1 \cdot P + (N+1)P(1-P) + (2N+1)P(1-P)^2 + \cdots \\ &\quad + (lN+1)P(1-P)^l + \cdots \\ &= 1 + \frac{N(1-P)}{P}. \end{aligned}$$

Consequently, the throughput efficiency of a go-back-N ARQ system is

$$\eta_{\text{GBN}} = \frac{1}{T_{\text{GBN}}}\left(\frac{k}{n}\right) = \frac{P}{P + (1-P)N}\left(\frac{k}{n}\right). \tag{15.5}$$

We see that the throughput η_{GBN} depends on both channel block error rate $1 - P$ and the round-trip delay N. When the channel error rate is low, the effect of the round-trip delay, $(1 - P)N$, is insignificant and the throughput is high. However, the effect of $(1 - P)N$ becomes significant when the channel error rate is high, and the throughput will be sharply reduced. The go-back-N ARQ provides satisfactory throughput performance for systems where the round-trip delay is small and the data transmission rate is not too high. Otherwise, its throughput performance becomes inadequate.

In deriving the throughput of a stop-and-wait ARQ system, the idle time spent waiting for an acknowledgment for each transmitted code vector must be taken into consideration. Let D be the idle time from the end of transmission of one code vector to the beginning of transmission of the next. Let τ be the signaling rate of the transmitter in bits per second. In one round-trip delay time, the transmitter could transmit a total of $n + D \cdot \tau$ digits if it does not stay idle. During the interval from the beginning of transmission of one code vector to the receipt of a positive acknowledgment for that code vector, the average number of digits (including the idling effect) that the transmitter could have transmitted is

$$T_{\text{SW}} = (n + D\tau)P + 2(n + D\tau)P(1 - P) + 3(n + D\tau)P(1 - P)^2 + \cdots$$
$$= \frac{n + D\tau}{P}.$$

Therefore, the throughput efficiency of a stop-and-wait ARQ system is

$$\eta_{\text{SW}} = \frac{k}{T_{\text{SW}}} = \frac{P}{1 + D\tau/n}\left(\frac{k}{n}\right). \tag{15.6}$$

The factor $D\tau/n$ may be interpeted as the number of code vectors that could be transmitted during the idle time of the transmitter. We see that the throughput can never achieve the maximum value k/n even if the channel is noiseless ($P = 1$). For data transmission systems where the data transmission rate is low and the round-trip delay is short, $D\tau$ can be made relatively small compared to the code length n. In this case, the stop-and-wait ARQ provides satisfactory throughput performance. However, its throughput performance becomes unacceptable for systems where the data transmission rate is high and the round-trip delay is large, such as the satellite communication systems.

From the analysis of throughput performance just given, we see that the selective-repeat ARQ is the most efficient scheme, whereas the stop-and-wait ARQ scheme is the least efficient one. The throughput of the selective-repeat ARQ does not depend on the round-trip delay of the system; however, the throughputs of the other two ARQ schemes depend on the round-trip delay. In communication systems where the round-trip delay is large and data rate is high, the parameter N for the go-back-N ARQ and the parameter $D\tau/n$ for the stop-and-wait may become very large. In this case, the throughput for the go-back-N ARQ drops rapidly as the channel error rate increases, while the throughput of the stop-and-wait ARQ becomes unacceptable.

The high throughput of the selective-repeat ARQ is achieved at the expense of extensive buffering at the receiver and more complex logic at both transmitter and receiver. Theorically, infinite buffering is needed to achieve the efficiency $(k/n)P$. If finite buffer is used at the receiver, buffer overflow may occur which would reduce the throughput of the system. However, if sufficient buffer (say, a buffer that is capable of storing N code vectors) is used at the receiver, even with a reduction in throughput,

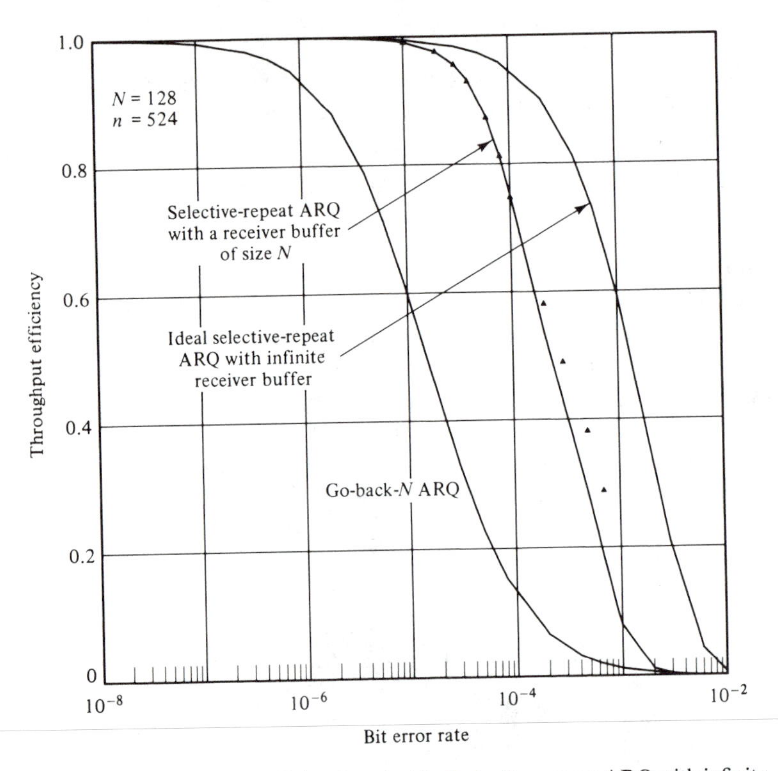

Figure 15.4 Throughput efficiencies: the ideal selective-repeat ARQ with infinite receiver buffer, the selective-repeat ARQ with a receiver buffer of size $N = 128$ (the solid triangles represent simulation results), and the go-back-N ARQ.

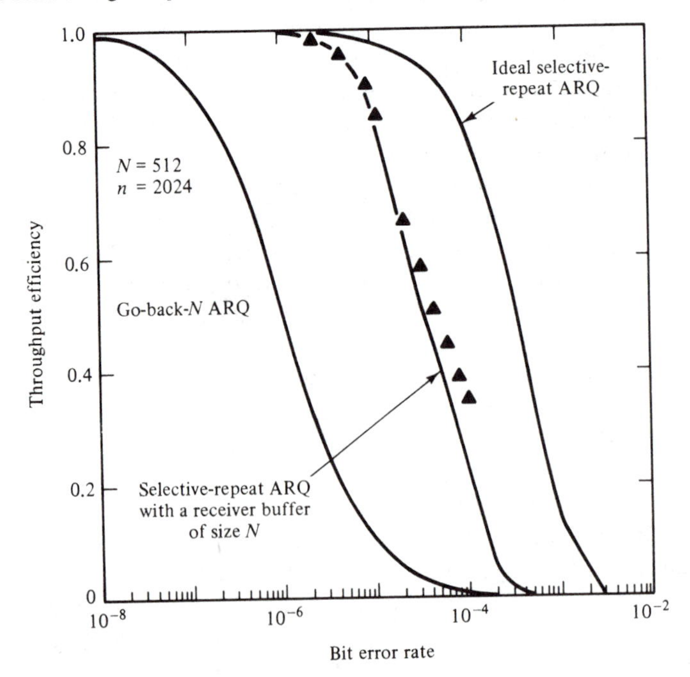

Figure 15.5 Throughput efficiencies: the ideal selective-repeat ARQ with infinite receiver buffer, the selective-repeat ARQ with a receiver buffer of size $N = 512$ (the solid triangles represent simulation results), and the go-back-N ARQ.

the selective-repeat ARQ still significantly outperforms the other two ARQ schemes in systems where data transmission rate is high and round-trip delay is large [6–11]. This is shown in the next section. Figures 15.4 to 15.6 show the throughput efficiencies of the selective-repeat ARQ and the go-back-N ARQ for various code lengths and round-trip delays. The channel is a BSC.

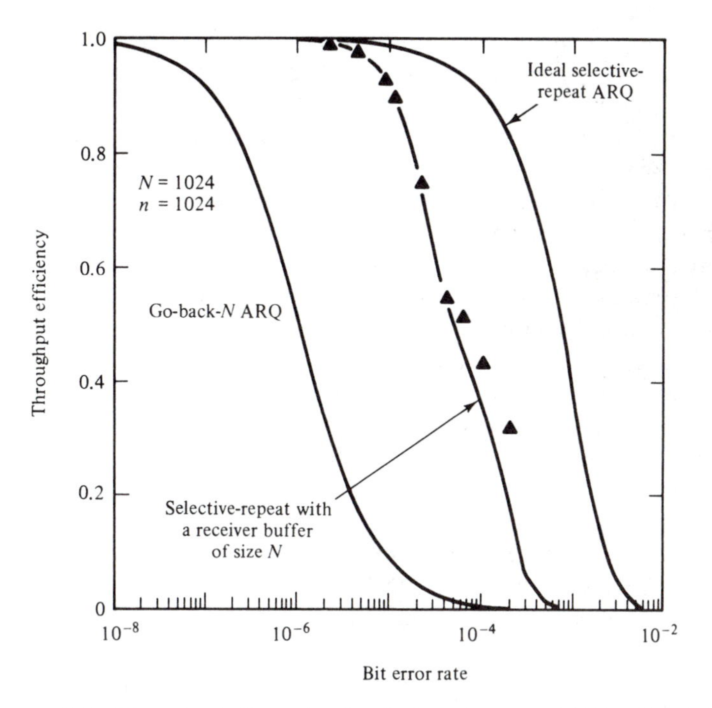

Figure 15.6 Throughput efficiencies: the ideal selective-repeat ARQ with infinite receiver buffer, the selective-repeat ARQ with a receiver buffer of size $N = 1024$ (the solid triangles represent simulation results), and the go-back-N ARQ.

Several variations of the go-back-N ARQ scheme have been proposed [12–15]. These variations improve the throughput efficiency; however, they are still less efficient than the selective-repeat ARQ scheme.

15.2 SELECTIVE-REPEAT ARQ SYSTEM WITH FINITE RECEIVER BUFFER

In this section we present a practical selective-repeat ARQ system in which the receiver employs a finite-size buffer [9,10]. Let N be the number of code vectors that can be transmitted during a round-trip delay period. For simplicity, we consider only the case for which the size of the receiver buffer is N (i.e., the receiver buffer is capable of storing N code vectors).

Each code vector to be transmitted has a sequence number. The range of the sequence number is set to $3N$ (i.e., code vectors are numbered from 1 to $3N$). These numbers will be used cyclically (i.e., a new code vector following a code vector with sequence number $3N$ is numbered with 1 again). For a receiver buffer of size N, if the range of the sequence numbers is $3N$, the receiver will be able to distinguish whether a received vector with sequence number q is a new vector or a vector that has been accepted and delivered [6]. This will be clear in what follows. Using the numbering system described above, $m = \lceil \log_2 3N \rceil$ bits must be appended to each code vector.

Suppose that an (n, k) linear code is used for error detection. When a vector is received, its syndrome is computed. If no errors are detected, the received vector is either delivered to the user or stored in the receiver buffer until it is ready to be delivered in the correct order. If the received vector is detected in error, it is discarded and a space in the receiver buffer is reserved for storing that vector at a later time.

When a code vector is ready for transmission, it is numbered and stored in the *input queue* of the transmitter. After its transmission, it is saved in the retransmission buffer of the transmitter until it is positively acknowledged. The acknowledgment will arrive after a round-trip delay. During this interval $N - 1$ other code vectors have been transmitted. A code vector is said to be a *time-out* vector if it has been transmitted (or retransmitted) for a round-trip delay time. When a code vector in the retransmission buffer becomes a time-out vector, it should be either positively acknowledged or negatively acknowledged. When an ACK for a code vector in the retransmission buffer is received, the code vector will be released from the retransmission buffer. However, when a NAK for a code vector in the retransmission buffer is received, a retransmission of that vector will be initiated. There is a third possibility that neither an ACK nor a NAK will be received when a code vector becomes a time-out vector (e.g., the acknowledgment is lost in the return channel). In this event, the transmitter will regard that the unacknowledged time-out vector was unsuccessfully transmitted or lost. Retransmission will be initiated for this vector.

Before we detail the transmission and retransmission process, we introduce another concept. Let q_0 be the sequence number of the earliest NAK'ed (negatively acknowledged) or unACK'ed (unacknowledged) code vector in the retransmission buffer. The *forward index* (FWX), denoted f_T, of a code vector with sequence number q in the retransmission buffer or input queue with respect to the earliest NAK'ed or unACK'ed code vector in the retransmission buffer is defined as the remainder resulting from dividing $q - q_0$ by $3N$. Mathematically, f_T is expressed as follows:

$$f_T \equiv q - q_0 \pmod{3N}. \tag{15.7}$$

Clearly, $0 \leq f_T < 3N$.

Transmission and Retransmission Procedure

When the transmitter is sending a code vector, it also computes the forward index f_T of the code vector in the retransmission buffer that is to become a time-out vector (we will refer to this vector as the *current time-out vector*). Then the transmitter decides whether the next code vector in the input queue is to be transmitted or a retransmission is to be initiated. The decision rules are:

1. If the current time-out vector is positively acknowledged and its forward index f_T is less than N (or there is no time-out vector in the retransmission

buffer), the first code vector in the input queue is to be transmitted. If the input queue is empty, the transmitter sits idle until a new code vector arrives or a retransmission is initiated.

If the current time-out vector, say $\mathbf{v}_j$, is either NAK'ed or unACK'ed and its forward index f_T is less than N, a retransmission for $\mathbf{v}_j$ is initiated. If the current NAK'ed (or unACK'ed) time-out vector is the earliest vector in the retransmission buffer that has not been positively acknowledged ($f_T = 0$), all the code vectors in the retransmission buffer with forward indices equal to or greater than N are moved back to the input queue for retransmission at a later time. These are the code vectors that have been transmitted; however, when they arrive at the receiver, the receiver buffer is full and has no space to store them (this event is referred to as *buffer overflow*). Therefore, these code vectors must be retransmitted (see Figure 15.7).

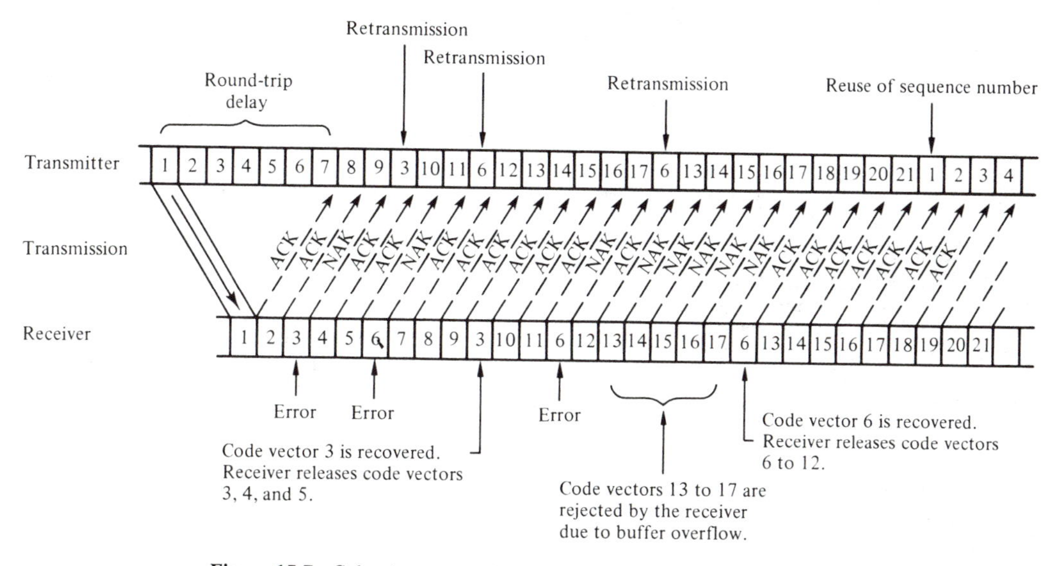

Figure 15.7 Selective-repeat ARQ with finite receiver buffer of size $N = 7$.

3. If the forward index f_T of the current time-out vector is equal to or greater than N, the first code vector in the input queue is the next to be transmitted (this vector may be a code vector that was moved back to the input queue from the retransmission buffer due to the receiver buffer overflow.)

The transmission and retransmission operations of the transmitter are illustrated in Figures 15.7 and 15.8.

Receiver's Operation and Error Recovery Procedure

Normally, the transmitter sends code vectors continuously to the receiver. The receiver checks the syndrome of each incoming received vector and sends an ACK to the transmitter for each successfully received vector. When the channel is quiet, the transmission proceeds smoothly, error-free code vectors are delivered to the user in

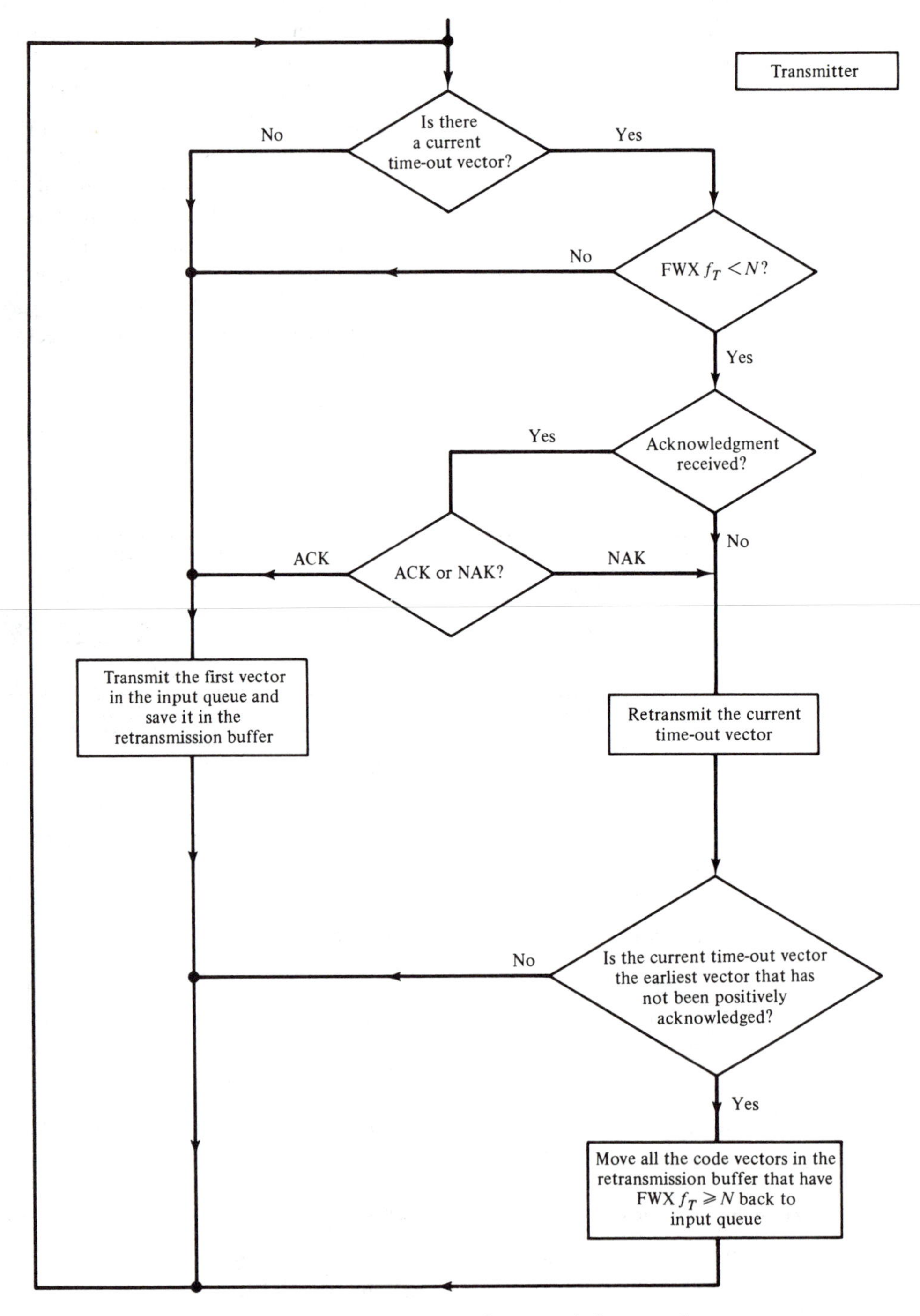

Figure 15.8 Transmission and retransmission procedure.

consecutive order and the receiver buffer is empty. The receiver is said to be in the *normal state* when the receiver buffer is empty and no space is reserved for any NAK'ed block.

If a received vector is detected in error or a received vector with an *out-of-order* sequence number is detected while the receiver is in the normal state, the receiver sends a NAK to the transmitter and enters the *blocked state*. In the blocked state, the receiver proceeds to check the syndromes of the incoming received vectors, it stores those vectors that have zero syndrome at the proper locations in the receiver buffer, and reserves proper locations for the vectors whose syndromes are not zero until the receiver buffer is full. No vectors are delivered to the user. When the retransmitted code vectors arrive after a round-trip delay, their syndromes are checked again. The vectors that are successfully received (zero syndrome) will be stored at the reserved locations in the receiver buffer. Once the earliest NAK'ed vector is successfully received, the receiver outputs that vector and all the subsequent consecutive zero-syndrome vectors (held in the receiver buffer) until the next NAK'ed vector is encountered (see Figure 15.7). If all the vectors in the receiver buffer are released and all the reserved locations are freed after the earliest NAK'ed vector has been successfully received, the receiver returns to the normal state.

If not all the vectors in the receiver buffer can be released to the user or if there are still reserved locations in the receiver buffer for the NAK'ed vectors, the receiver will continue its error recovery process until all the NAK'ed vectors are successfully received.

Next, we describe the detailed operation of the receiver in both the normal state and the blocked state.

Normal-State Operation

When a vector with sequence number q is received, the receiver checks its syndrome and computes its *forward index* (FWX), *denoted by* f_R, *with respect to the last accepted and delivered vector*. Let q_0 be the sequence number of the last accepted and delivered vector. The forward index f_R of the current received vector is defined as the remainder resulting from dividing $q - q_0$ by $3N$, that is,

$$f_R \equiv q - q_0 \pmod{3N}. \tag{15.8}$$

If no errors are detected in the current received vector and $f_R = 1$ ($f_R = 1$ indicates that the current received vector has the correct sequence number), the received vector is accepted and delivered to the user. The receiver proceeds to check the next incoming received vector.

However, if either the current received vector is detected in error or its forward index f_R is not equal to 1 but less than $N + 1$ ($1 < f_R \leq N$ indicates that the sequence number of the current received vector is out of order), the receiver enters the blocked state. There are three cases to be considered. In the first case for which the syndrome is not zero and $f_R = 1$, the receiver discards the erroneously received vector and reserves the first location of the receiver buffer for that vector. In the second case for which the syndrome is not zero and $f_R > 1$, $f_R - 1$ vectors between the last delivered vector and the current received vector are lost. The receiver rejects the current erroneously received vector and reserves the first f_R locations of the receiver buffer for the lost vectors and the rejected vector. In the third case for which the syndrome is zero but $f_R > 1$, $f_R - 1$ vectors between the last delivered vector and the current received vector are lost. The receiver then reserves the first $f_R - 1$ locations of the receiver buffer for the lost vectors and saves the current received vector at the f_Rth location of the receiver buffer.

If the forward index f_R of the current received vector is greater than N, it is regarded as a vector that was previously accepted and delivered, it is then ignored and an ACK is sent to the transmitter. The receiver stays in the normal state and proceeds to process the next incoming vector.

The flowchart shown in Figure 15.9 details the receiver operations in the normal state.

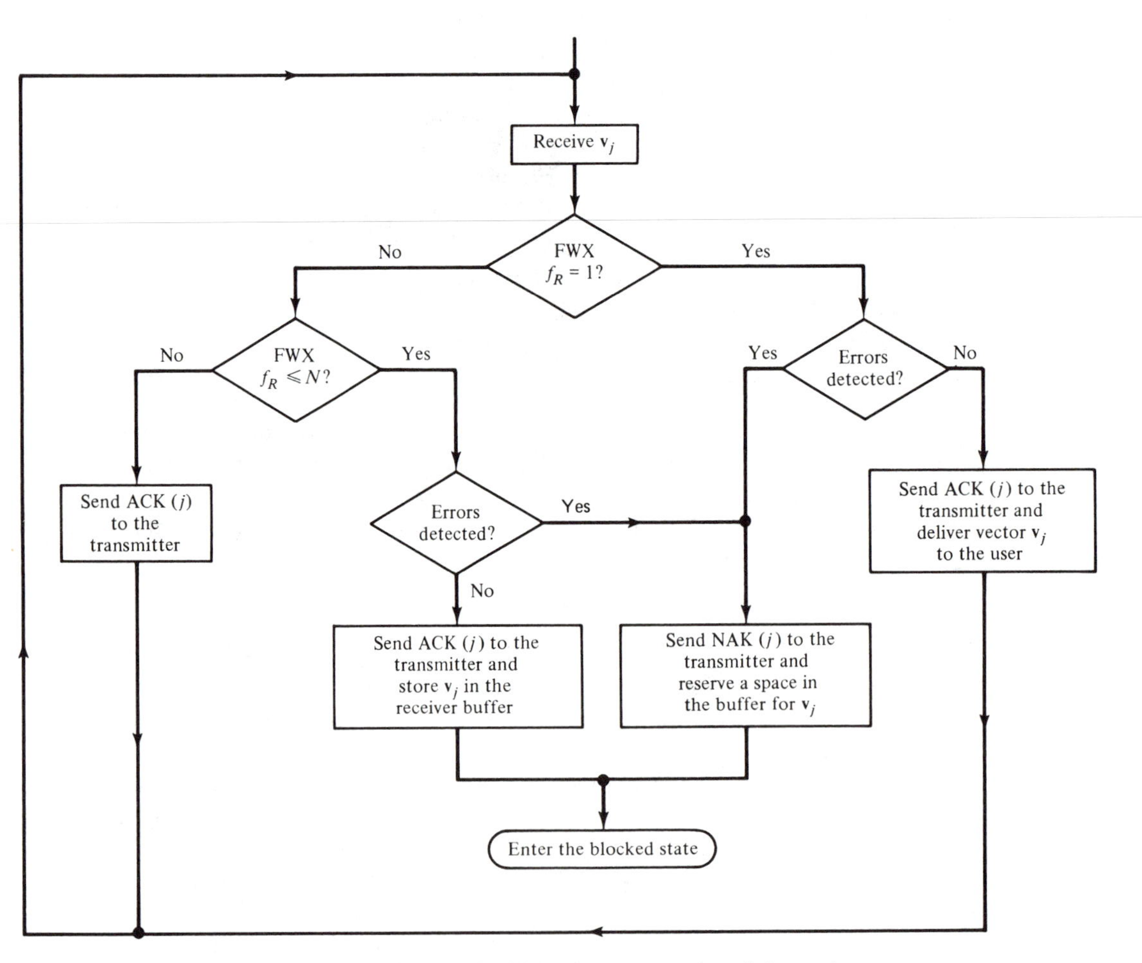

Figure 15.9 Normal-state operation of the receiver.

Blocked-State Operation

The receiver enters the blocked state when either a received vector is detected in error or a received vector with an out-of-order sequence number is detected. When the receiver enters the blocked state, each subsequent zero-syndrome received vector is held in the receiver buffer at a proper location and a space in the receiver buffer is reserved for each erroneously received or lost vector. When the buffer is full, all subsequent new received vectors will be rejected and buffer overflow occurs.

Let $\mathbf{v}_j$ be the earliest vector that has not been successively received. When a retransmitted copy of $\mathbf{v}_j$ is successfully received, the receiver releases $\mathbf{v}_j$ and the subsequent zero-syndrome vectors which are held in the receiver buffer (see Figure 15.7). Suppose that the receiver can release $L + 1$ consecutive vectors including $\mathbf{v}_j$ (i.e., $\mathbf{v}_j, \mathbf{v}_{j+1}, \ldots, \mathbf{v}_{j+L}$, where $0 \leq L < N$). Since the retransmission is selective and since $\mathbf{v}_{j+1}$ to $\mathbf{v}_{j+L}$ were successfully received before $\mathbf{v}_j$, the vectors following the retransmitted copy of $\mathbf{v}_j$ are new vectors $\mathbf{v}_{j+N}, \mathbf{v}_{j+N+1}, \ldots, \mathbf{v}_{j+N+L-1}$. When these L new vectors arrive, their syndromes are computed. The vectors that have zero syndrome are then temporarily stored in the receiver buffer until they are ready to be released, and proper locations in the receiver buffer are reserved for those vectors that have nonzero syndrome.

If the retransmitted copy of $\mathbf{v}_j$ is not successfully received, the zero-syndrome vectors held in the receiver buffer cannot be released to the user. Therefore, there are no places in the receiver buffer to store the new vectors, $\mathbf{v}_{j+N}, \mathbf{v}_{j+N+1}, \ldots$ (since the buffer size is N). In this event, buffer overflow occurs and the new vectors, $\mathbf{v}_{j+N}$, $\mathbf{v}_{j+N+1}, \ldots$, will be rejected no matter whether they are successfully received or not.

Let q_0 be the sequence number of the earliest vector that has not been successfully received. In the blocked state, when a vector with sequence number q is received, its forward index with respect to the earliest unsuccessfully received vector, denoted by l_f, is computed. This forward index l_f is defined as the remainder resulting from dividing $q - q_0$ by $3N$, that is,

$$l_f \equiv q - q_0 \pmod{3N}. \tag{15.9}$$

If the current received vector has zero syndrome and its forward index $l_f < N$, it is stored in the receiver buffer and an ACK is set to the transmitter. If the current received vector is detected in error and its forward index $l_f < N$, it is rejected and a space in the receiver buffer is reserved for this vector; also, a NAK is sent to the transmitter. If $l_f \geq N$, the receiver then computes the *backward index* (BWX) l_b of the received vector with respect to the last received vector that was stored in the receiver buffer or for which a space in the receiver buffer is reserved. This backward index l_b is defined as the remainder resulting from dividing $q_0 - q$ by $3N$, that is,

$$l_b \equiv q_0 - q \pmod{3N}, \tag{15.10}$$

where q_0 is the sequence number of the last received vector that was stored in the receiver buffer or for which a space in the receiver buffer is reserved and q is the sequence number of the current received vector. If $l_b < 2N$, the current received vector is regarded as a vector that was previously accepted and delivered. It is then ignored and an ACK is sent to the transmitter. If $l_b \geq 2N$, the current received vector is a new vector. However, since the receiver buffer is full, buffer overflow occurs. In this

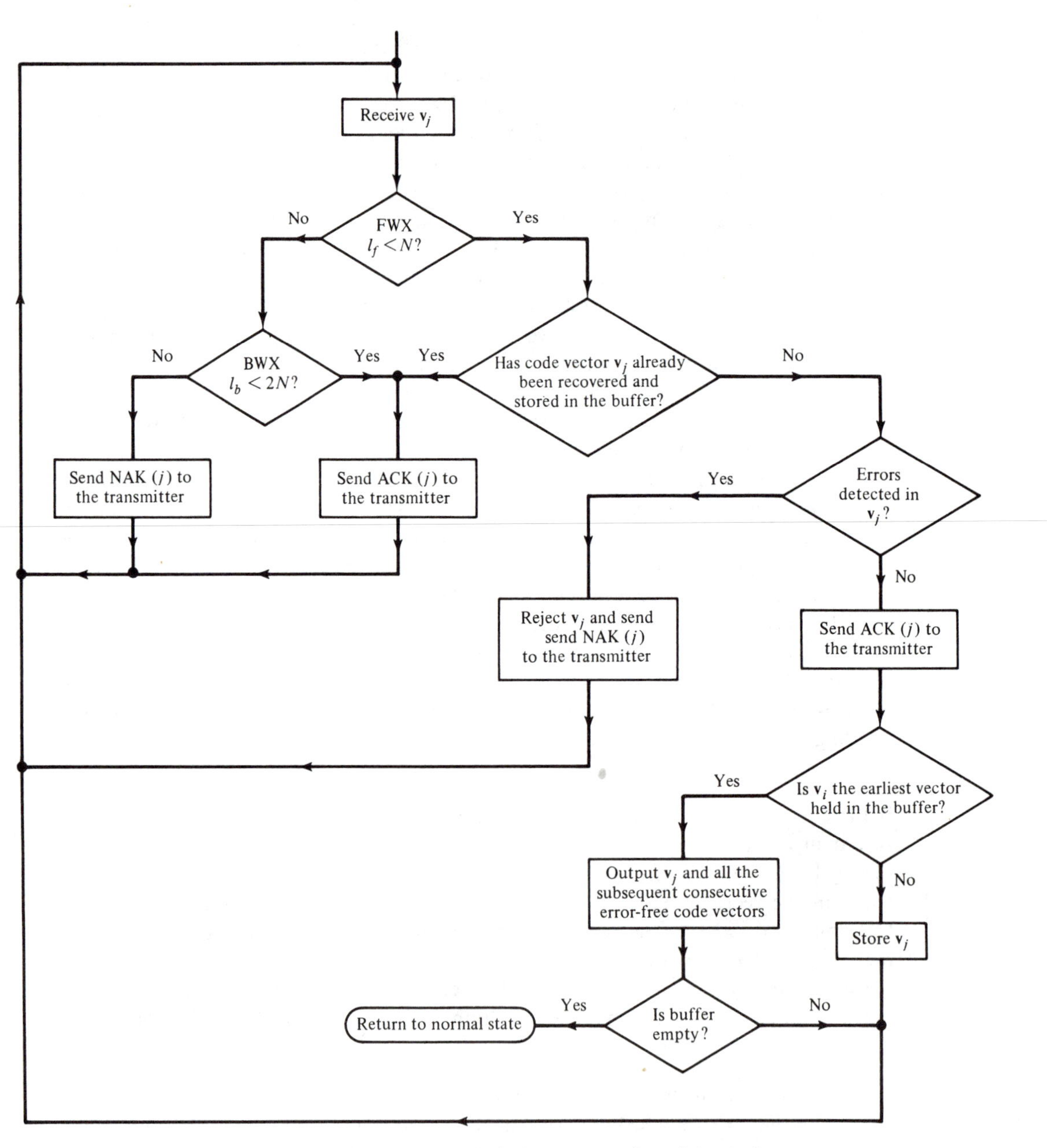

Figure 15.10 Blocked-state operation of the receiver.

event, the current received vector is rejected and a NAK is sent to the transmitter. The receiver's operation in the blocked state is detailed by the flowchart shown in Figure 15.10.

Throughput Efficiency

If we assume that the channel errors are randomly distributed (i.e., a BSC) and the feedback channel is noiseless, a lower bound on the throughput efficiency of the selective-repeat ARQ system above can be derived [10]. Suppose that a proper (n, k) code is used for error detection. Let p be the transition probability of the BSC. Then the probability P that a received vector will be accepted by the receiver is

$$P \approx P_c = (1 - p)^n.$$

Let

$$\phi_0 = 1 - (1 - P)^2, \qquad \phi_1 = 1 - (1 - P)^3, \qquad \phi_2 = 1 - (1 - P)^4.$$

These are the probabilities of success of the second, third, and fourth retransmissions of a code vector, respectively. Define

$$\lambda_0 = \frac{\phi_1}{1 - \phi_2}(1 - \phi_1\phi_2^{N-1}),$$

$$\lambda_1 = P^2\{\phi_0^{N-2} + (1 - P)\phi_1^{N-2} + (1 - P)^2\phi_2^{N-2}\},$$

$$\lambda_2 = 3 - P^2\phi_0^{N-2} - \phi_0^2\phi_1^{N-2} - \phi_1^2\phi_2^{N-2}.$$

Then the throughput efficiency of the selective ARQ with a receiver buffer of size N is lower bounded as follows:

$$\eta_{\text{SRF}} \geq \frac{\lambda_0}{\lambda_0 + \lambda_1 + \lambda_2 N}\left(\frac{k}{n}\right). \tag{15.11}$$

(Assuming that the number of bits used for sequence numbers is small compared to the code length n, its effect on the throughput is ignored.)

For various n and N, the lower bound on the throughput given by (15.11) is compared to the throughput of the ideal selective-repeat ARQ with infinite receiver buffer and the throughput of the go-back-N ARQ as shown in Figures 15.4 to 15.6. We see that the selective-repeat ARQ with a receiver buffer of size N is less efficient than the ideal selective-repeat ARQ because of the reduction in buffer size at the receiver. However, it significantly outperforms the go-back-N ARQ particularly for communication systems where the round-trip delay is large and data rate is high. If the round-trip delay is taken to be 700 ms (suitable for satellite channels), the data transmission rate corresponding to $n = 524$ and $N = 128$ (Figure 15.4) is 100 kbps, the data rate corresponding to $n = 2024$ and $N = 512$ (Figure 15.5) is 1.54 Mbps, and the data rate corresponding to $n = 1024$ and $N = 1024$ (Figure 15.6) is also 1.54 Mbps.

The selective-repeat ARQ described in this section can be extended for any receiver buffer of size greater than N, say $2N, 3N, \ldots$. Of course, the throughput efficiency increases as the buffer size increases. Figure 15.11 shows the throughput with buffer size equal to $2N$.

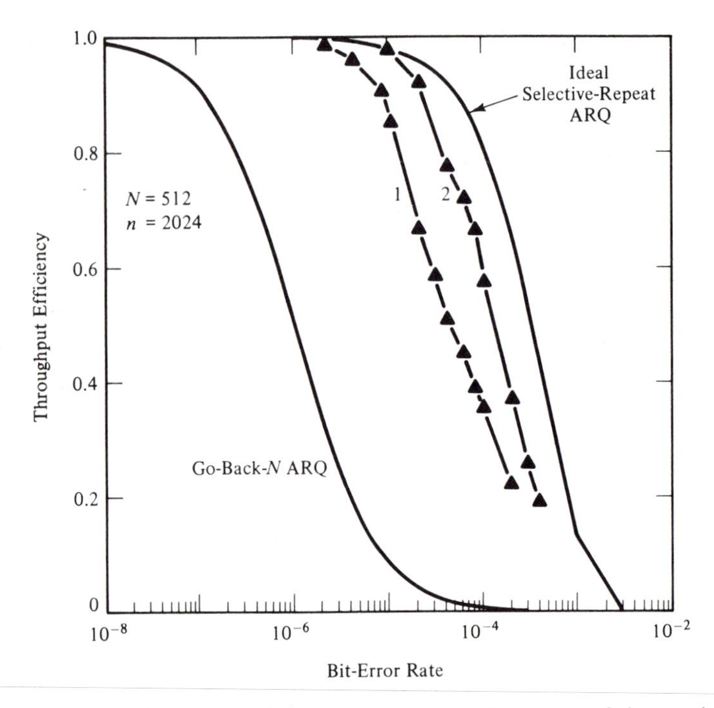

Figure 15.11 Throughout of the selective-repeat ARQ with a finite receiver buffer: (1) size N and (2) size $2N$ where $N = 512$.

Selective-repeat ARQ that operates with a finite receiver buffer and a finite range of sequence numbers was first studied by Metzner and Morgan [6, 16]. Other selective-repeat ARQ schemes employing finite receiver buffers can be found in References 7, 8, and 11.

15.3 ARQ SCHEMES WITH MIXED MODES OF RETRANSMISSION

It is possible to devise ARQ schemes with mixed modes of retransmission. One such scheme is the *selective-repeat plus go-back-N* (SR + GBN) ARQ [11]. When a code vector $\mathbf{v}$ in the retransmission buffer becomes the earliest negatively acknowledged vector, the transmitter resends $\mathbf{v}$ and other vectors in the selective-repeat (SR) mode (i.e., only $\mathbf{v}$ and other negatively acknowledged vectors are retransmitted). During the SR mode, the receiver stores those vectors that are successfully received. The transmitter remains in SR mode until either $\mathbf{v}$ is positively acknowledged or ν retransmissions of $\mathbf{v}$ have been made but the transmitter fails to receive an ACK for $\mathbf{v}$. In the first event, the transmitter proceeds to send new code vectors. When the second event occurs, the transmitter *switches* to the go-back-N (GBN) retransmission mode. The transmitter stops sending new code vectors, backs up to $\mathbf{v}$, and resends $\mathbf{v}$ and $N - 1$ succeeding vectors that were transmitted following the νth SR retransmission attempt for $\mathbf{v}$. The transmitter stays in the GBN mode until $\mathbf{v}$ is positively acknowledged. At the receiving end, when the νth retransmission of $\mathbf{v}$ is detected in error, all the subsequent $N - 1$ received vectors are discarded regardless of whether they are successfully received or not. The receiver continues this operation until $\mathbf{v}$ is successfully received. Then the receiver releases $\mathbf{v}$ and all the consecutive error-free vectors following $\mathbf{v}$ which were held in the receiver buffer. Using the SR + GBN ARQ, the receiver buffer must be able to store $\nu(N - 1) + 1$ vectors to prevent buffer overflow.

Figure 15.12 illustrates the transmission/retransmission procedure for SR + GBN with $\nu = 1$, namely one retransmission for a NAK'ed vector in SR mode and all subsequent retransmissions of the same vector being in GBN mode. Following the second consecutive NAK for a given vector (e.g., block 7 in Figure 15.12) the receiver simply discards all $N - 1$ subsequent received vectors until the successful reception of the erroneous vector. In the transmitter and the receiver, provision would be required for counters to keep check on the number of transmission/retransmission attempts made for each vector in order to determine whether an SR or GBN type of retransmission is required. If there is more than one vector stored in the transmitter retransmission buffer that have been transmitted twice unsuccessfully, they must be queued for successive independent go-back-N type retransmissions. The earliest double-NAK'ed vector, say $\mathbf{v}_j$, will be retransmitted in the GBN mode, that is, followed by $N - 1$ vectors which were transmitted previously after $\mathbf{v}_j$ and subsequently discarded by the receiver. This will be repeated until an ACK for $\mathbf{v}_j$ is received. Then the same procedure is repeated for any subsequent double-NAK'ed vectors in the retransmission buffer.

The throughput efficiency of the SR + GBN ARQ has been analyzed [11]. For $\nu = 1$, its throughput efficiency is

$$\eta_1 = \frac{P}{1 + (N - 1)(1 - P)^2}\left(\frac{k}{n}\right), \tag{15.12}$$

where P is the probability that a vector will be successfully received. For $\nu > 1$,

$$\eta_\nu = \frac{P}{1 + (N - 1)(1 - P)^{\nu+1}}\left(\frac{k}{n}\right). \tag{15.13}$$

For $\nu = 0$, the SR + GBN scheme becomes the conventional go-back-N ARQ and

$$\eta_0 = \frac{P}{P + (1 - P)N}\left(\frac{k}{n}\right).$$

Figure 15.13 demonstrates the throughput performance of the SR + GBN ARQ for $\nu = 1$ and 2. Throughput efficiencies of other ARQ schemes are included in the same figure for comparison. We see that SR + GBN ARQ significantly outperforms the conventional go-back-N ARQ in throughput for high bit error rate and large round-trip delay. We also see that, with the same receiver buffer size, the SR + GBN ARQ is inferior to the selective-repeat ARQ described in Section 15.2 for high channel error rate. However, SR + GBN ARQ requires simpler logic at the transmitter and receiver than the selective-repeat ARQ.

Other ARQ schemes with mixed modes of retransmission can be found in References 11 to 14. They are all simpler but less efficient than the selective-repeat ARQ with finite receiver buffer described in Section 15.2.

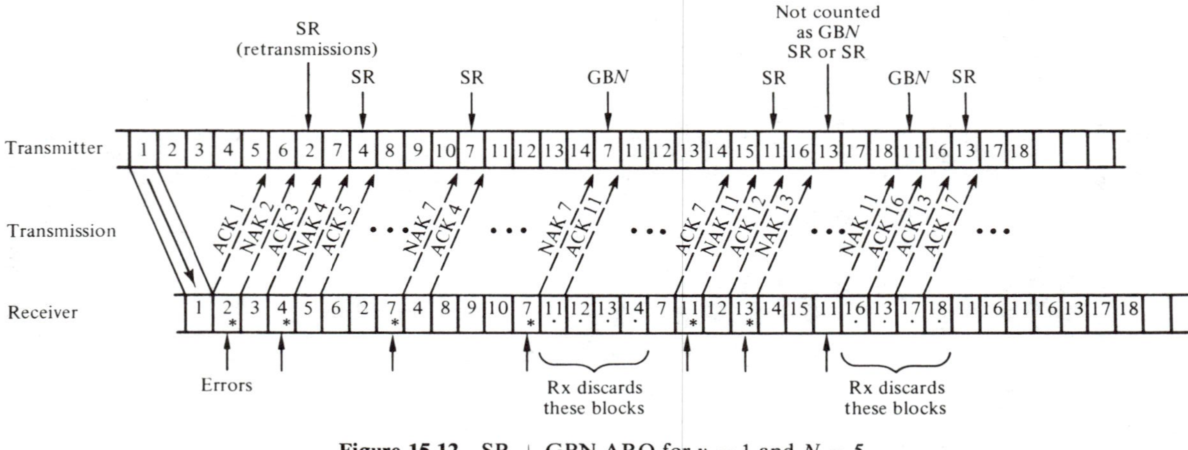

Figure 15.12 SR + GBN ARQ for $\nu = 1$ and $N = 5$.

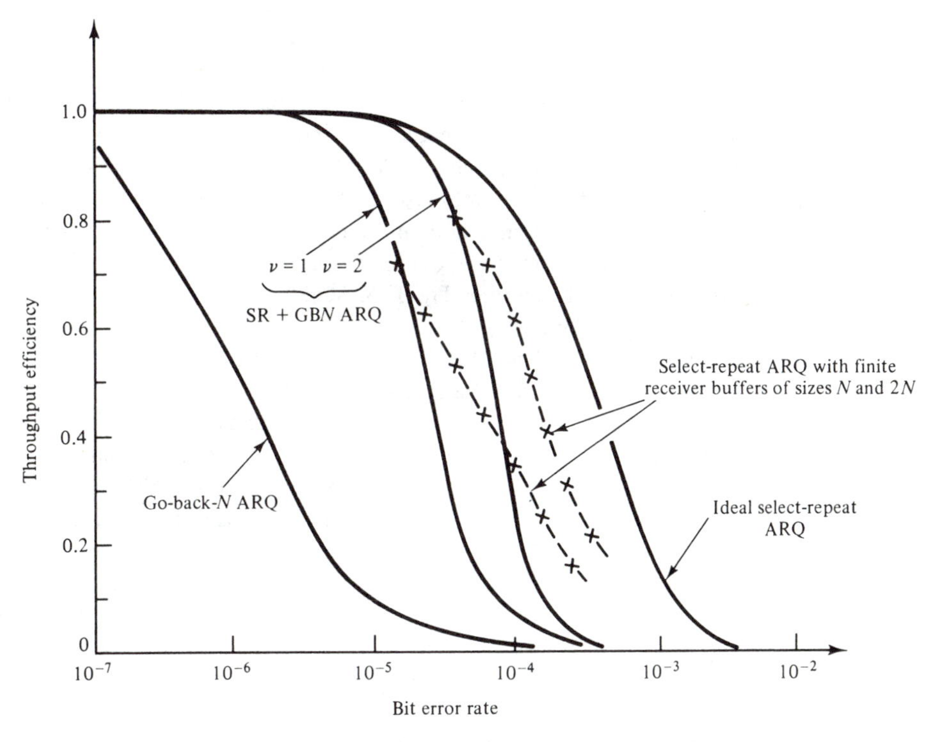

Figure 15.13 Throughput efficiencies of the SR + GBN with $\nu = 1$ and 2, $N = 512$ and $n = 2024$.

15.4 HYBRID ARQ SCHEMES

Comparing the two error control schemes, we see that ARQ is simple and provides high system reliability. However, ARQ systems have a severe drawback—their throughputs fall rapidly with increasing channel error rate. Systems using FEC maintain constant throughput (equal to the code rate $R = k/n$) regardless of the channel error rate. However, FEC systems have two drawbacks. First, when a received vector is detected in error, it must be decoded and the decoded message must be delivered to the user regardless of whether it is correct or incorrect. Since the probability of a decoding error is much greater than the probability of an undetected error, it is hard to achieve high system reliability with FEC. Second, to obtain high system reliability, a long powerful code must be used and a large collection of error patterns must be corrected. This makes decoding hard to implement and expensive. For these reasons, ARQ is often preferred over FEC for error control in data communication systems, such as packet-switching data networks and computer communication networks. However, in communication (or data storage) systems where return channels are not available or retransmission is not possible for some reason, FEC is the only choice.

The drawbacks in both ARQ and FEC could be overcome if two error control schemes are *properly* combined. Such a combination of the two basic error control schemes is referred to as a *hybrid* ARQ [5,17]. A hybrid ARQ system consists of an FEC subsystem contained in an ARQ system. The function of the FEC portion is to reduce the frequency of retransmission by correcting the error patterns that occur most frequently. This increases the system throughout. When a less frequent error pattern occurs and is detected, the receiver requests a retransmission rather than passing the unreliably decoded message to the user. This increases the system reliability. As a result, a proper combination of FEC and ARQ would provide higher reliability than an FEC system alone and a higher throughput than the system with ARQ only. Furthermore, since the decoder is designed to correct a small collection of error patterns, it can be simple. The FEC scheme can be incorporated with any of the three basic ARQ schemes.

A straightforward hybrid ARQ scheme is to use a code, say an (n, k) linear code, which is designed for simultaneous error correction and error detection. When a received vector is detected in error, the receiver first attempts to locate and correct the errors. If the number of errors (or the length of an error burst) is within the designed error-correcting capability of the code, the errors will be corrected and the decoded message will be passed to the user or saved in a buffer until it is ready to be delivered. If an uncorrectable error pattern is detected, the receiver rejects the received vector and requests a retransmission. When the retransmitted vector is received, the receiver again attempts to correct the errors (if any). If the decoding is not successful, the receiver again rejects the received vector and asks for another retransmission. This error-correction and retransmission process continues until the vector is successfully received or decoded. For example, one may use the (1023, 923) BCH code in a hybrid ARQ system. This code has a minimum distance 21; it can be used for correcting five or fewer errors and simultaneously detecting any combination of 15 or fewer errors (and many other error patterns). If an error pattern with five or fewer errors occurs, it will be detected and corrected. If an error pattern with more than five but less than 16 errors occurs, it will be detected. In this event, the receiver will request a retransmission of the erroneous vector.

The hybrid ARQ scheme described above is referred to as the *type I hybrid ARQ scheme*. Since a code used in this scheme must be able to correct a certain collection of error patterns and simultaneously detect other error patterns, more parity-check digits are needed. This increases the overhead for each transmission and retransmission. As a result, when the channel error rate is low, it has a lower throughput than its corresponding ARQ scheme. However, when the channel error rate increases, the throughput of the ARQ scheme drops rapidly and the hybrid-ARQ scheme provides higher throughput, as shown in Figure 15.14. The type I hybrid ARQ scheme is capable of maintaining significant high throughput over a wide range of channel error rates if the designed error-correcting capability of the code is sufficiently large.

Let P_t be the probability that a received vector will be decoded successfully. This probability depends on the designed error-correcting capability of the code being used. Replacing P by P_t in Equations (15.4), (15.5), (15.6), and (15.11), we obtain the throughput efficiencies for the type I hybrid selective-repeat ARQ with infinite receiver buffer, the type I hybrid go-back-N ARQ, the type I hybrid stop-

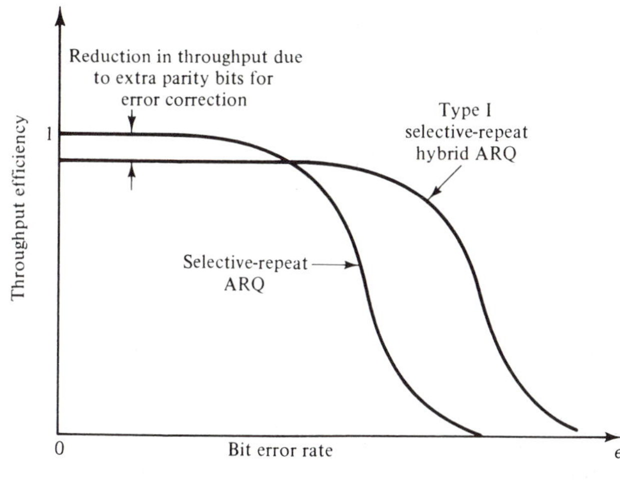

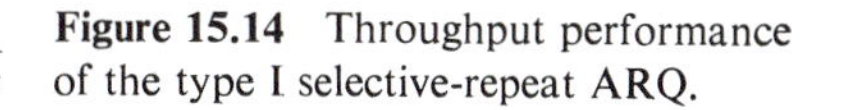
Figure 15.14 Throughput performance of the type I selective-repeat ARQ.

and-wait ARQ, and the type I hybrid selective-repeat ARQ with receiver buffer of size N, respectively.

The second type (or type II) of hybrid ARQ scheme is devised based on the concept that the parity-check digits for error correction are sent to the receiver only when they are needed [18,19]. Two linear codes are used in this type of scheme; one is a high-rate (n, k) code C_0 which is designed for error detection only, the other is an *invertible* half-rate $(2k, k)$ code C_1 which is designed for simultaneous error correction and error detection. A code is said to be invertible if, knowing only the parity-check digits of a code vector, the corresponding information digits can be uniquely determined by an *inversion process*. In the next section, a class of invertible half-rate linear codes is presented and we show that the inversion can be accomplished by a linear sequential circuit.

When a message $\mathbf{u}$ of k information digits is ready for transmission, it is encoded into a code vector $\mathbf{v} = (f(\mathbf{u}), \mathbf{u})$ of n digits based on the error-detecting code C_0, where $f(\mathbf{u})$ denotes the $n - k$ parity-check digits. The code vector $\mathbf{v} = (f(\mathbf{u}), \mathbf{u})$ is then transmitted. At the same time, the transmitter computes the k parity-check digits, denoted by $q(\mathbf{u})$, based on the message $\mathbf{u}$ and the half-rate invertible code C_1. Clearly, the $2k$-tuple $(q(\mathbf{u}), \mathbf{u})$ is a code vector in C_1. The k-bit parity block $q(\mathbf{u})$ is not transmitted but stored in the retransmission buffer of the transmitter for later use. Let $\tilde{\mathbf{v}} = (\tilde{f}(\mathbf{u}), \tilde{\mathbf{u}})$ denote the received vector corresponding to $\mathbf{v} = (f(\mathbf{u}), \mathbf{u})$. When $\tilde{\mathbf{v}}$ is received, the receiver computes the syndrome of $\tilde{\mathbf{v}}$ based on C_0. If the syndrome is zero, $\tilde{\mathbf{u}}$ is assumed to be error-free and will be accepted by the receiver. If the syndrome is not zero, errors are detected in $\tilde{\mathbf{v}}$. The erroneous message $\tilde{\mathbf{u}}$ is then *saved* in the receiver buffer and a NAK is sent to the transmitter. Upon receiving this NAK, the transmitter encodes the k-bit parity block $q(\mathbf{u})$ into a code vector $\mathbf{v}^* = (f[q(\mathbf{u})], q(\mathbf{u}))$ of n bits based on C_0, where $f[q(\mathbf{u})]$ denotes the $n - k$ parity-check digits for $q(\mathbf{u})$. This vector $\mathbf{v}^* = (f[q(\mathbf{u})], q(\mathbf{u}))$ is then transmitted (*here the retransmission is a parity vector*). Let $\tilde{\mathbf{v}}^* = (\tilde{f}[q(\mathbf{u})], \tilde{q}(\mathbf{u}))$ denote the received vector corresponding to $\mathbf{v}^* = (f[q(\mathbf{u})], q(\mathbf{u}))$. When $\tilde{\mathbf{v}}^*$ is received, the syndrome of $\tilde{\mathbf{v}}^*$ is computed based on C_0. If the syndrome is zero, $\tilde{q}(\mathbf{u})$ is assumed to be error-free and the message $\mathbf{u}$ is recovered from $\tilde{q}(\mathbf{u})$ by inversion. If the syndrome is not zero, $\tilde{q}(\mathbf{u})$ and the erroneous message $\tilde{\mathbf{u}}$ (stored in the receiver buffer) *together* are used for *error correction* based on the half-rate code C_1. If the errors in $(\tilde{q}(\mathbf{u}), \tilde{\mathbf{u}})$ form a correctable error pattern, they will be corrected. The decoded message $\mathbf{u}$ is then accepted by the receiver. If the errors in $(\tilde{q}(\mathbf{u}), \tilde{\mathbf{u}})$ form a detectable but not a correctable error pattern, $\tilde{\mathbf{u}}$ is discarded and the erroneous parity block $\tilde{q}(\mathbf{u})$ is stored in the receiver buffer; also, a NAK is sent to the transmitter.

Upon receiving the second NAK for the code vector $\mathbf{v} = (f(\mathbf{u}), \mathbf{u})$, the transmitter resends $\mathbf{v} = (f(\mathbf{u}), \mathbf{u})$. When $\tilde{\mathbf{v}} = (\tilde{f}(\mathbf{u}), \tilde{\mathbf{u}})$ is received, the syndrome of $\tilde{\mathbf{v}}$ is again computed based on C_0. If the syndrome is zero, $\tilde{\mathbf{u}}$ is assumed to be error-free and is accepted by the receiver, the erroneous parity block $\tilde{q}(\mathbf{u})$ is then discarded. If the syndrome is not zero, $\tilde{\mathbf{u}}$ and the erroneous parity block $\tilde{q}(\mathbf{u})$ (stored in the receiver buffer) together are used for error correction based on C_1. If the errors in $(\tilde{q}(\mathbf{u}), \tilde{\mathbf{u}})$ are corrected, the decoded message $\mathbf{u}$ is then accepted by the receiver and an ACK is sent to the transmitter. However, if the errors in $(\tilde{q}(\mathbf{u}), \tilde{\mathbf{u}})$ are detectable but not correctable, $\tilde{q}(\mathbf{u})$ is discarded, $\tilde{\mathbf{u}}$ is stored in the receiver buffer, and a NAK is sent to the transmitter. The next retransmission will be the parity vector $\mathbf{v}^* = (f[q(\mathbf{u})], q(\mathbf{u}))$. Therefore, the retransmissions are *alternating repetitions* of the parity code vector $\mathbf{v}^* = (f[q(\mathbf{u})], q(\mathbf{u}))$ and the information code vector $\mathbf{v} = (f(\mathbf{u}), \mathbf{u})$. The receiver stores the received message $\tilde{\mathbf{u}}$ and the received parity block $\tilde{q}(\mathbf{u})$ alternately. The retransmissions continue until $\mathbf{u}$ is finally recovered.

The most important feature of the type II hybrid ARQ is the *parity retransmission* for error correction based on a half-rate invertible code C_1. This parity-retransmission strategy can be incorporated with any of the three basic types of ARQ. It is particularly effective when it is used in conjunction with the selective-repeat ARQ. Because of the invertible property of C_1, the message $\mathbf{u}$ can be reconstructed uniquely from the parity block $q(\mathbf{u})$ by inversion. Hence, the parity block $q(\mathbf{u})$ contains the same amount of information as the message. As a result, the overhead per transmission or retransmission is simply the number of parity-check digits, $n - k$, needed for error detection based on the (n, k) code C_0, which is required by any ARQ scheme. Therefore, when the channel is quiet or the channel error rate is low, the type II hybrid ARQ has the same throughput efficiency as its corresponding ARQ scheme. When the channel error rate is high, the error correction provided by the half-rate code C_1 will maintain the throughput high. In Section 15.5 we present a type II hybrid ARQ that incorporates the parity-retransmission strategy with the selective-repeat ARQ with finite receiver buffer. We will show that, for a BSC, the throughput efficiency of the ideal selective-repeat ARQ with infinite receiver buffer can be achieved by a half-rate invertible code C_1 with a very small designed error-correcting capability, say $t = 3$ to 5. With a larger t, the type II hybrid selective-repeat ARQ with finite receiver buffer will be far superior to the ideal selective-repeat ARQ with infinite receiver buffer.

The decoding complexity for a type II hybrid ARQ is only slightly greater than that of a corresponding type I hybrid ARQ with the same designed error-correcting capability. The extra circuits needed for a type II hybrid ARQ scheme are an inversion circuit based on C_1, which is simply a linear sequential circuit and an error detection circuit based on C_0.

The disadvantage of the type I hybrid ARQ is that the overhead due to the extra parity-check digits for error correction must be included in each transmission or retransmission regardless of the channel error rate. When the channel is quiet, this represents a waste. However, the type II hybrid ARQ removes this disadvantage. It is an adaptive scheme. This scheme is particularly attractive for high-speed data communication systems where round-trip delay is large and error rate is nonstationary such as satellite communication systems.

Various hybrid ARQ schemes and their analysis can be found in References 17 to 29.

15.5 CLASS OF HALF-RATE INVERTIBLE CODES

In a type II hybrid ARQ system, C_1 is chosen as a half-rate invertible code. The invertible property facilitates the data recovery process. During a retransmission, if the parity block $q(\mathbf{u})$ is successfully received (no errors being detected), the message $\mathbf{u}$ can be reconstructed from $q(\mathbf{u})$ by a simple inversion process rather than by a more complicated decoding process. The inversion process also reduces the frequency of retransmission. For example, if the received message $\tilde{\mathbf{u}}$ contains more than t errors and the received parity block $\tilde{q}(\mathbf{u})$ is error-free, the decoding process based on $(\tilde{q}(\mathbf{u}), \tilde{\mathbf{u}})$ would not be able to recover the message $\mathbf{u}$. Hence, another retransmission would be required. However, taking the inverse of the error-free parity block $\tilde{q}(\mathbf{u})$, we will be able to recover $\mathbf{u}$ and thus avoid another retransmission.

In the following, a class of half-rate invertible block codes will be presented and we will show that inversion can be accomplished by a linear sequential circuit.

Let C be an (n, k) cyclic code with $n - k \leq k$. Let $\mathbf{g}(X)$ be the generator polynomial of C with the form

$$\mathbf{g}(X) = 1 + g_1X + g_2X^2 + \cdots + g_{n-k-1}X^{n-k-1} + X^{n-k}.$$

Let

$$\mathbf{v}(X) = v_0 + v_1X + v_2X^2 + \cdots + v_{n-1}X^{n-1}$$

be a code polynomial. In systematic form, the k leading high-order coefficients $v_{n-k}, v_{n-k+1}, \ldots, v_{n-1}$ are identical to k information digits, the $n - k$ low-order coefficients $v_0, v_1, \ldots, v_{n-k-1}$ are parity-check digits. Consider the set of those code vectors in C whose $2k - n$ leading high-order components $v_{2(n-k)}, v_{2(n-k)+1}, \ldots, v_{n-1}$ are zeros. There are 2^{n-k} such code vectors in C. If the $2k - n$ high-order zero components are removed from these code vectors, we obtain a set of 2^{n-k} vectors of length $2(n - k)$. These vectors form a half-rate $(2n - 2k, n - k)$ shortened cyclic code C_1 (see section 4.7). This shortened cyclic code has at least the same error-correcting capability as C. We have shown in Section 4.7 that the encoding and decoding of C_1 can be accomplished by the same circuits (or with a slight modification) as employed by C.

Next, we show that the shortened cyclic code C_1 has the invertible property. Let

$$\mathbf{u}(X) = u_0 + u_1X + \cdots + u_{n-k-1}X^{n-k-1}$$

be the message to be encoded. Dividing $X^{n-k}\mathbf{u}(X)$ by the generator polynomial $\mathbf{g}(X)$, we have

$$X^{n-k}\mathbf{u}(X) = \mathbf{a}(X)\mathbf{g}(X) + \mathbf{b}(X), \quad (15.14)$$

where $\mathbf{a}(X)$ and $\mathbf{b}(X)$ are the quotient and the remainder, respectively. The code vector for $\mathbf{u}(X)$ is then

$$\mathbf{w}(X) = \mathbf{b}(X) + X^{n-k}\mathbf{u}(X)$$

and $\mathbf{b}(X)$ is the parity-check portion. The following theorem proves the invertible property of C_1.

Theorem 15.1. No two code vectors in a half-rate shortened cyclic code C_1 have the same parity-check digits.

Proof. Let $\mathbf{u}_1(X)$ and $\mathbf{u}_2(X)$ be two distinct messages. Dividing $X^{n-k}\mathbf{u}_1(X)$ and $X^{n-k}\mathbf{u}_2(X)$ by the generator polynomial $\mathbf{g}(X)$, respectively, we obtain

$$X^{n-k}\mathbf{u}_1(X) = \mathbf{a}_1(X)\mathbf{g}(X) + \mathbf{b}_1(X), \quad (15.15)$$

$$X^{n-k}\mathbf{u}_2(X) = \mathbf{a}_2(X)\mathbf{g}(X) + \mathbf{b}_2(X). \quad (15.16)$$

Then the code vectors for $\mathbf{u}_1(X)$ and $\mathbf{u}_2(X)$ are

$$\mathbf{w}_1(X) = \mathbf{b}_1(X) + X^{n-k}\mathbf{u}_1(X),$$

$$\mathbf{w}_2(X) = \mathbf{b}_2(X) + X^{n-k}\mathbf{u}_2(X),$$

respectively. Suppose that

$$\mathbf{b}_1(X) = \mathbf{b}_2(X) = \mathbf{b}(X).$$

Adding (15.15) and (15.16), we obtain

$$[\mathbf{u}_1(X) + \mathbf{u}_2(X)]X^{n-k} = [\mathbf{a}_1(X) + \mathbf{a}_2(X)]\mathbf{g}(X).$$

Since $\mathbf{g}(X)$ and X^{n-k} are relatively prime, $\mathbf{u}_1(X) + \mathbf{u}_2(X)$ must be divisible by $\mathbf{g}(X)$. However, this is impossible since $\mathbf{u}_1(X) + \mathbf{u}_2(X) \neq 0$ and its degree is less than $n - k$ but the degree of $\mathbf{g}(X)$ is $n - k$. Therefore, $\mathbf{b}_1(X) \neq \mathbf{b}_2(X)$.

Q.E.D.

Since the remainder $\mathbf{b}(X)$ resulting from dividing $X^{n-k}\mathbf{u}(X)$ by $\mathbf{g}(X)$ is unique, Theorem 15.1 implies that there is one-to-one correspondence between a message $\mathbf{u}(X)$ and its parity check $\mathbf{b}(X)$. Therefore, knowing $\mathbf{b}(X)$, $\mathbf{u}(X)$ can be uniquely determined.

Next we want to show how to recover the message $\mathbf{u}(X)$ from the parity $\mathbf{b}(X)$. To see this, we multiply both sides of (15.14) by X^k:

$$X^n\mathbf{u}(X) = \mathbf{a}(X)\mathbf{g}(X)X^k + \mathbf{b}(X)X^k. \quad (15.17)$$

Rearranging (15.17), we obtain

$$\mathbf{u}(X)(X^n + 1) + \mathbf{u}(X) = \mathbf{a}(X)\mathbf{g}(X)X^k + \mathbf{b}(X)X^k. \quad (15.18)$$

Since $\mathbf{g}(X)$ is a factor of $X^n + 1$, (15.8) can be rewritten as

$$\mathbf{b}(X)X^k = [\mathbf{u}(X)\mathbf{h}(X) + \mathbf{a}(X)X^k]\mathbf{g}(X) + \mathbf{u}(X), \quad (15.19)$$

where $\mathbf{h}(X) = (X^n + 1)/\mathbf{g}(X)$. From (15.19) we see that the message $\mathbf{u}(X)$ is simply the remainder resulting from dividing $\mathbf{b}(X)X^k$ by the generator polynomial $\mathbf{g}(X)$. This can be achieved by using a division circuit with feedback connection based on $\mathbf{g}(X)$. The process of finding the message $\mathbf{u}(X)$ from its parity-check digits $\mathbf{b}(X)$ is called an inversion process.

A faster circuit for inverting $\mathbf{b}(X)$ can be implemented as follows. Dividing X^k by $\mathbf{g}(X)$, we have

$$X^k = \mathbf{c}(X)\mathbf{g}(X) + \boldsymbol{\rho}(X), \tag{15.20}$$

where the remainder

$$\boldsymbol{\rho}(X) = \rho_0 + \rho_1 X + \cdots + \rho_{n-k-2}X^{n-k-2} + \rho_{n-k-1}X^{n-k-1}. \tag{15.21}$$

Multiplying both sides of (15.20) by $\mathbf{b}(X)$ and using the equality of (15.19), we obtain

$$\mathbf{b}(X)\boldsymbol{\rho}(X) = [\mathbf{u}(X)\mathbf{h}(X) + \mathbf{a}(X)X^k + \mathbf{b}(X)\mathbf{c}(X)]\mathbf{g}(X) + \mathbf{u}(X). \tag{15.22}$$

The expression above suggests that the message $\mathbf{u}(X)$ can be obtained by multiplying the parity $\mathbf{b}(X)$ by $\boldsymbol{\rho}(X)$ and dividing the product $\mathbf{b}(X)\boldsymbol{\rho}(X)$ by $\mathbf{g}(X)$. This can be achieved with the circuit shown in Figure 15.15.

For example, consider the (1023, 523) BCH code. This code has minimum distance at least 111. We can use this code for simultaneous correction of all combinations of five or fewer errors and detection of all combinations of 105 or fewer errors. In this case, the probability of an undetected error will be extremely small. Shortening this code by 23-digits, we obtain a (1000, 500) invertible code.

15.6 TYPE II HYBRID SELECTIVE-REPEAT ARQ WITH FINITE RECEIVER BUFFER

The parity retransmission for error correction can be easily incorporated with the selective-repeat ARQ with finite receiver buffer described in Section 15.2. Again we consider only the case for which the size of the receiver buffer is N (the number of vectors transmitted in one round-trip delay time). The range of sequence numbers is still $3N$.

Let $\mathbf{u}$ be a k-bit message to be transmitted. There are three code vectors associated with this message:

1. The code vector $\mathbf{v} = (f(\mathbf{u}), \mathbf{u})$ based on an (n, k) error-detecting code C_0.
2. The code vector $\mathbf{w} = (q(\mathbf{u}), \mathbf{u})$ based on a half-rate invertible $(2k, k)$ code C_1.
3. The code vector $\mathbf{v}^* = (f[q(\mathbf{u})], q(\mathbf{u}))$ based on the k-bit parity block $q(\mathbf{u})$ and the code C_0.

For convenience, we call $\mathbf{v} = (f(\mathbf{u}), \mathbf{u})$ the *information code vector* of $\mathbf{u}$ and $\mathbf{v}^* = (f[q(\mathbf{u})], q(\mathbf{u}))$ the *parity code vector* of $\mathbf{u}$. We will use $\tilde{\mathbf{v}} = (\tilde{f}(\mathbf{u}), \tilde{\mathbf{u}})$ and $\tilde{\mathbf{v}}^* = (\tilde{f}[q(\mathbf{u})], \tilde{q}(\mathbf{u}))$ to denote the received vectors corresponding to $\mathbf{v}$ and $\mathbf{v}^*$, respectively. In transmission or retransmission, the information code vector $\mathbf{v}$ and the parity code vector $\mathbf{v}^*$ of the message $\mathbf{u}$ have the same sequence number.

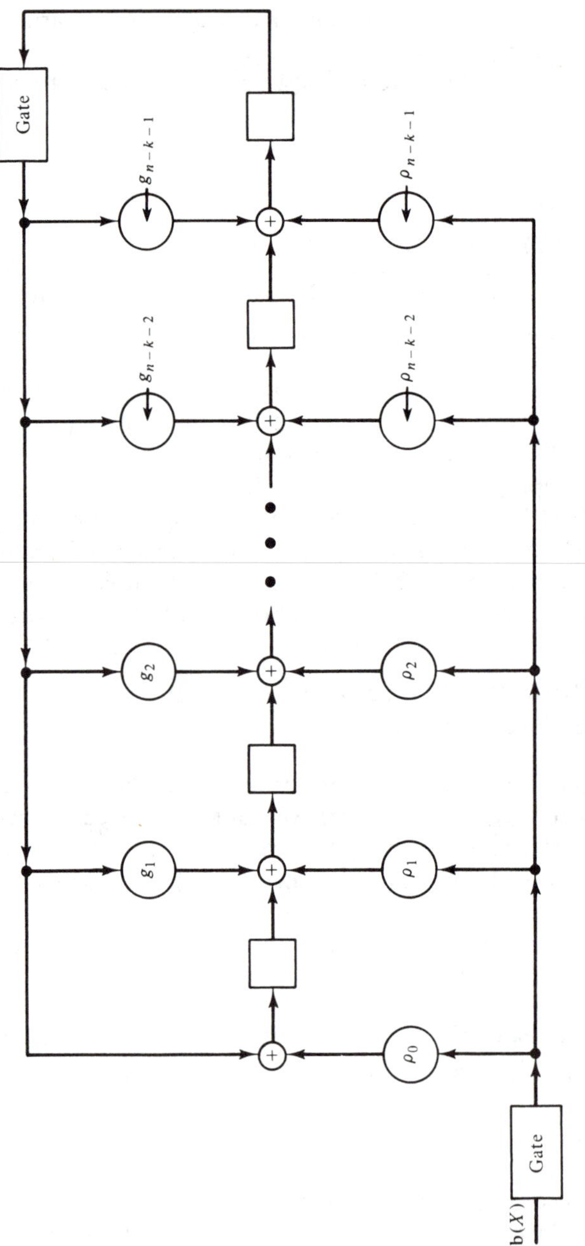

Figure 15.15 Inversion circuit.

When an information code vector $\mathbf{v}$ is ready for transmission, it is numbered and stored in the input queue of the transmitter. After its transmission, $\mathbf{v}$ and its corresponding parity code vector $\mathbf{v}^*$ are saved in the retransmission buffer until $\mathbf{v}$ is positively acknowledged. When an ACK is received after a round-trip delay, both $\mathbf{v}$ and $\mathbf{v}^*$ will be released. When a NAK (or no acknowlegdment) is received after a round-trip delay, $\mathbf{v}^*$ will be sent to the receiver for error correction (if necessary). After another round-trip delay, if an ACK is received, both $\mathbf{v}$ and $\mathbf{v}^*$ will be released; otherwise, $\mathbf{v}$ will be retransmitted. The transmitter resends $\mathbf{v}$ and $\mathbf{v}^*$ alternatively until $\mathbf{v}$ is positively acknowledged, as shown in Figure 15.16.

Transmission and Retransmission Procedure

When the transmitter is sending a code vector, information, or parity, it also computes the forward index f_T of the code vector in the retransmission buffer that is to become a time-out vector. Based on this forward index f_T, the transmitter decides whether the next information code vector in the input queue is to be transmitted or a retransmission is to be initiated. The decision rule is given as follows:

1. If the current time-out vector is positively acknowledged and $f_T < N$ (or if there is no current time-out vector), the first information code vector in the input queue is to be transmitted.
2. If the current time-out vector, say $\mathbf{v}_j$, is either negatively acknowledged or unacknowledged and $f_T < N$, a retransmission for $\mathbf{v}_j$ is initiated. The retransmission for $\mathbf{v}_j$ is the parity-code vector $\mathbf{v}_j^*$ if $\mathbf{v}_j$ was previously transmitted, and is a repetition of $\mathbf{v}_j$ if $\mathbf{v}_j^*$ was previously transmitted. If the current NAK'ed (or unACK'ed) time-out vector is the earliest vector in the retransmission buffer that has not been positively acknowledged ($f_T = 0$), all the information code vectors in the retransmission buffer with forward indices equal to or greater than N are moved back to the input queue for retransmission at a later time (these vectors are rejected by the receiver due to the receiver buffer overflow).
3. If $f_T \geq N$, the first information code vector in the input queue is the next to be transmitted.

The transmission and retransmission operations of the type II hybrid selective-repeat ARQ with receiver buffer of size N are detailed by the flowchart shown in Figure 15.17.

Receiver's Operation in the Normal State

In the normal state, the receiver receives information code vectors and the receiver buffer is empty. When an information code vector is received, the receiver computes its syndrome based on C_0 and its forward index f_R with respect to the last accepted and delivered information code vector. If the syndrome is zero and $f_R = 1$, the received information code vector is accepted and delivered to the user. If $f_R > N$, the received vector is regarded as an information code vector that was previously accepted and delivered; it is then ignored and an ACK is sent to the transmitter.

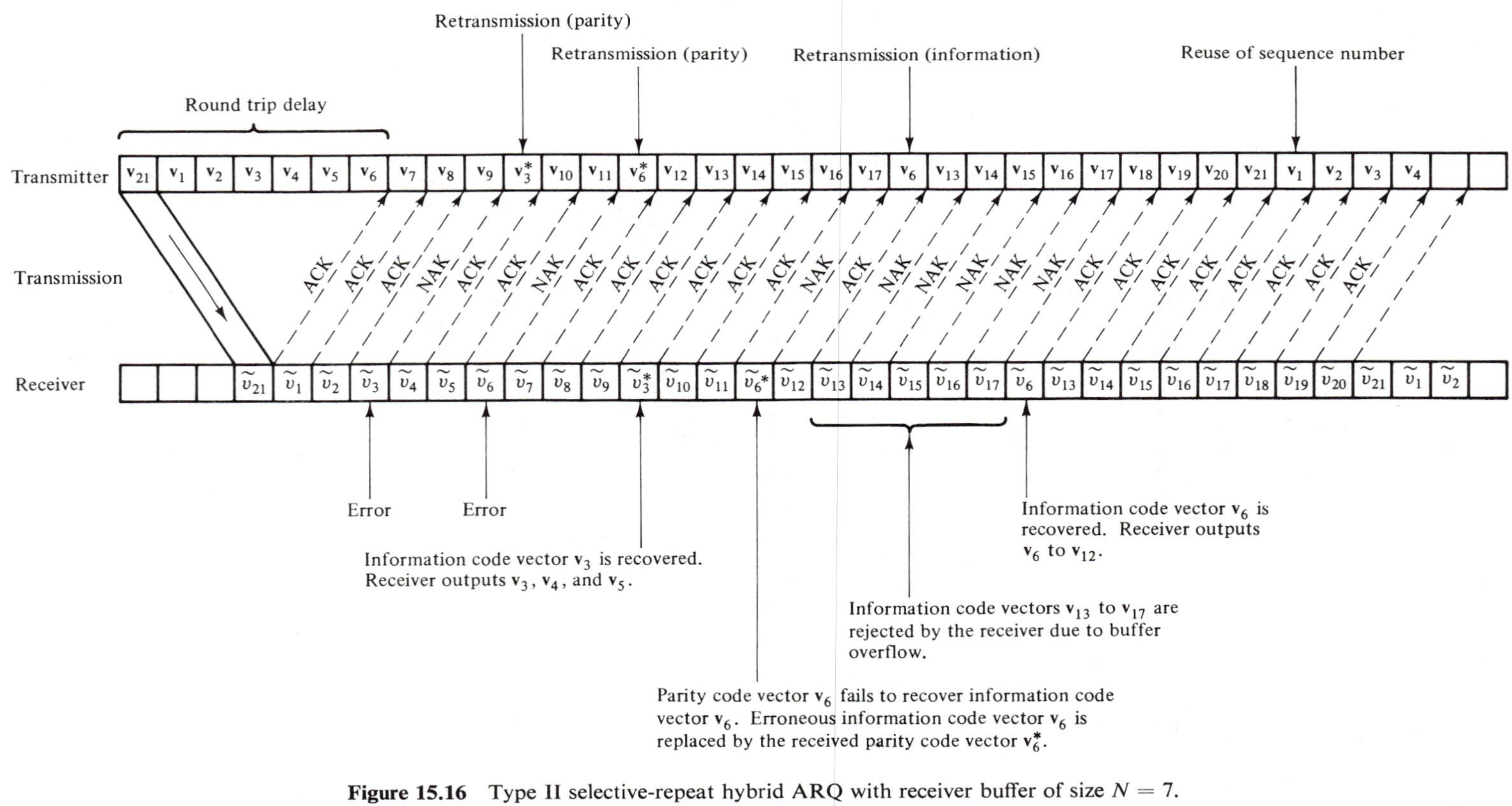

Figure 15.16 Type II selective-repeat hybrid ARQ with receiver buffer of size $N = 7$.

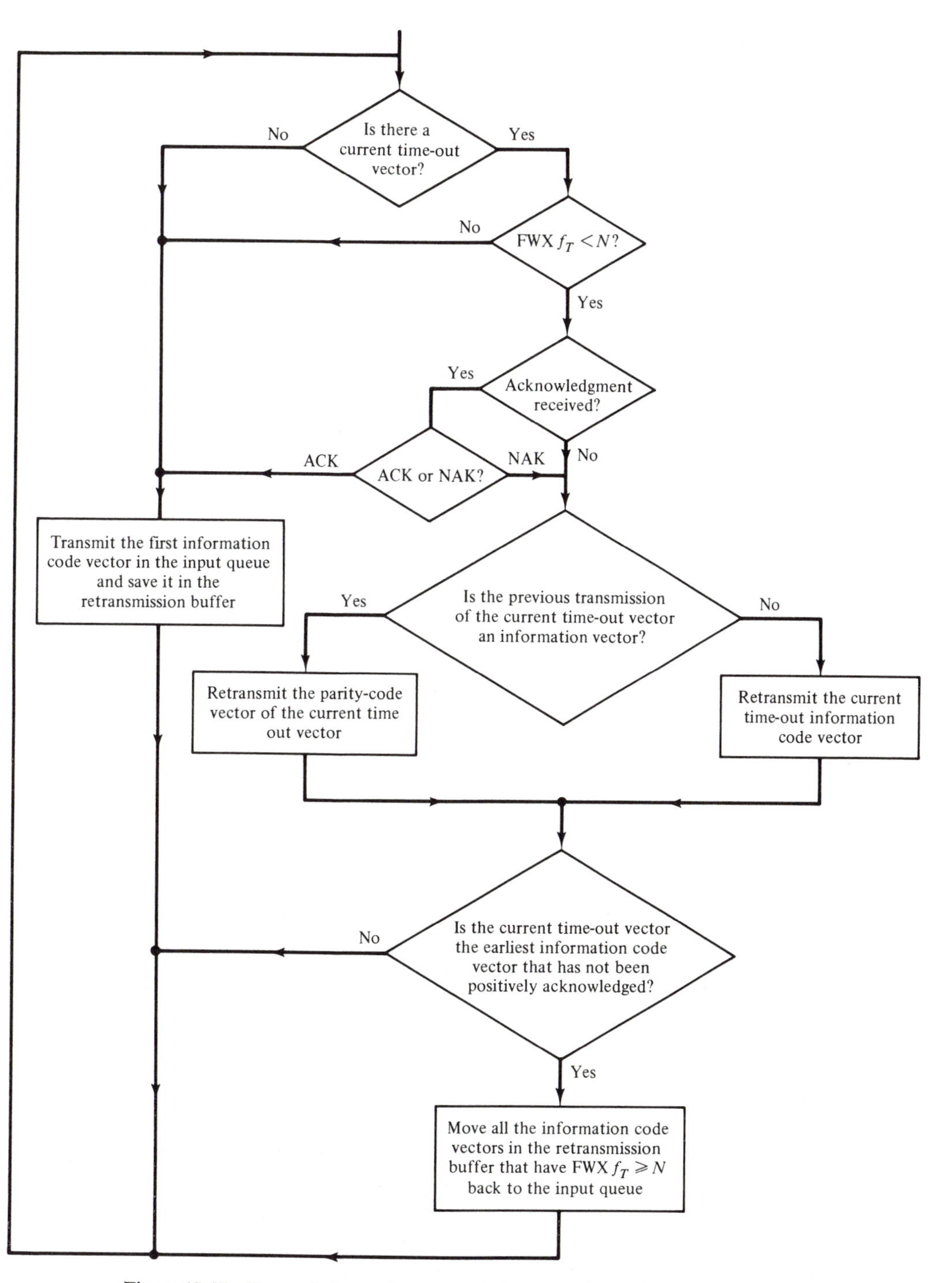

Figure 15.17 Transmission and retransmission procedure of the type II selective-repeat hybrid ARQ with receiver buffer of size N.

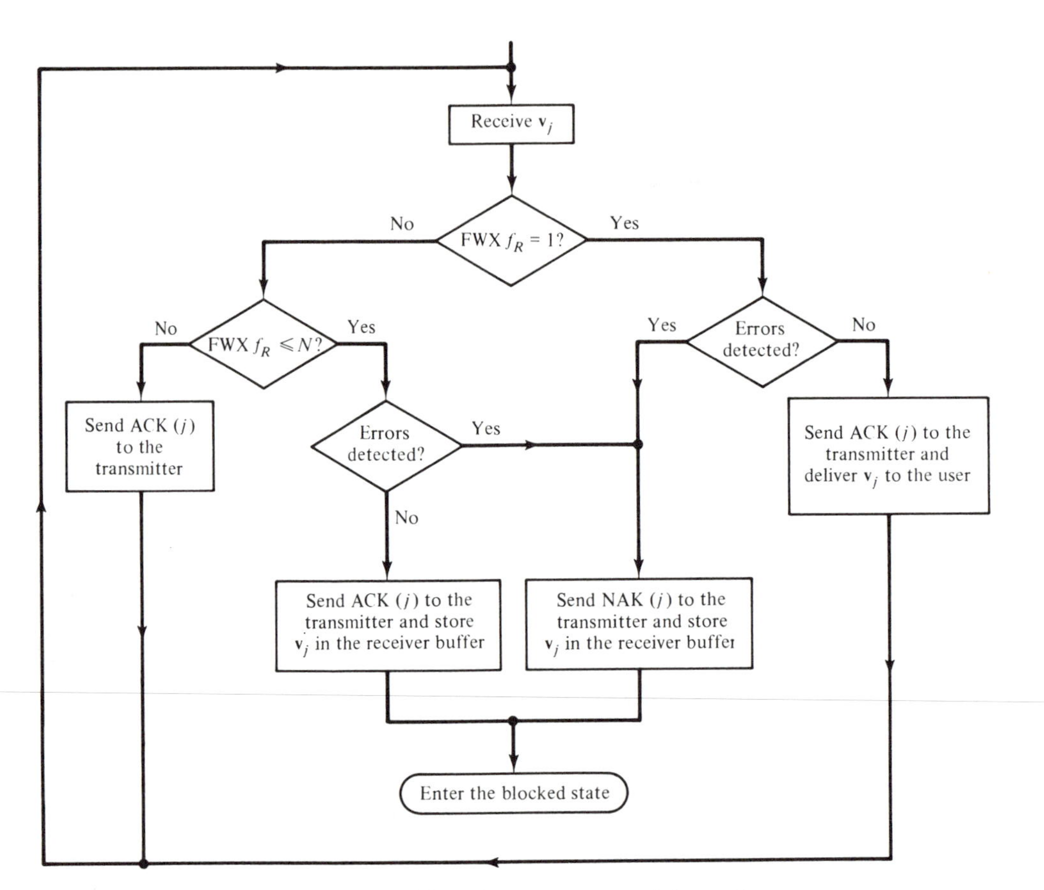

Figure 15.18 Normal-state operation of the receiver.

However, if either the received information code vector is detected in error or $1 < f_R \leq N$ (i.e., its sequence number is out of order), the receiver enters the blocked state. The receiver's operation in the normal state is detailed by the flowchart shown in Figure 15.18.

Receiver's Operation in the Blocked State

If the receiver enters the blocked state due to the detection of errors in a received information code vector $\tilde{\mathbf{v}} = (\tilde{f}(\mathbf{u}), \tilde{\mathbf{u}})$ with $f_R = 1$, the erroneous message $\tilde{\mathbf{u}}$ is then stored at the first location of the receiver buffer and a NAK is sent to the transmitter. The receiver proceeds to check the subsequent received information code vectors and stores them in the receiver buffer at the proper locations according to their sequence numbers until the buffer is full; also an appropriate acknowledgment is sent to the transmitter for each received vector. If the receiver enters the blocked state with $f_R > 1$, then $f_R - 1$ information code vectors between the last delivered information code vector and the current received information code vector are lost. The receiver then reserves the first $f_R - 1$ locations of the receiver buffer for the lost vectors and saves the current received vector (only the message part) at the (f_R)th location of the buffer. The subsequent received information code vectors are stored in the rest of the locations of the receiver buffer.

When the first retransmitted parity code vector, say $\tilde{\mathbf{v}}_j^* = (\tilde{f}[q(\mathbf{u}_j)], \tilde{q}(\mathbf{u}_j))$, is received, it is used to recover the earliest erroneously received message (or the earliest lost message) $\mathbf{u}_j$. If the syndrome of $\tilde{\mathbf{v}}_j^*$ is zero, $\mathbf{u}_j$ is recovered by taking the inversion of $\tilde{q}(\mathbf{u}_j)$. If the syndrome of $\tilde{\mathbf{v}}_j^*$ is not zero, $\tilde{q}(\mathbf{u}_j)$ and $\tilde{\mathbf{u}}_j$ (stored in the buffer) together are used for error correction based on the half-rate error-correcting code C_1. If the errors are correctable, $\mathbf{u}_j$ will be recovered. When $\mathbf{u}_j$ is recovered, the receiver releases $\mathbf{u}_j$ and the subsequent error-free (or zero-syndrome) messages in consecutive order, say $\mathbf{u}_j, \mathbf{u}_{j+1}, \ldots, \mathbf{u}_{j+L}$ with $0 \leq L < N$. Since the retransmission is selective and since $\mathbf{u}_{j+1}$ to $\mathbf{u}_{j+L}$ were successfully recovered, the vectors following $\tilde{\mathbf{v}}_j^*$ are new information vectors $\tilde{\mathbf{v}}_{j+N}, \tilde{\mathbf{v}}_{j+N+1}, \ldots, \tilde{\mathbf{v}}_{j+N+L-1}$. When these new information code vectors arrive, their syndromes are checked and are temporarily stored in the receiver buffer until they are ready to be released to the user or to be corrected.

If $\tilde{q}(\mathbf{u}_j)$ fails to recover $\mathbf{u}_j$, the receiver discards $\tilde{\mathbf{u}}_j$ and stores $\tilde{q}(\mathbf{u}_j)$ in the receiver buffer; also, a NAK is sent to the transmitter. Since error-free messages cannot be released, there are no places in the receiver buffer to store the new received information code vectors $\tilde{\mathbf{v}}_{j+N}$ to $\tilde{\mathbf{v}}_{j+N+L-1}$. In this event, buffer overflow occurs and these new information code vectors are rejected.

Upon receiving the second NAK for $\mathbf{v}_j = (f(\mathbf{u}_j), \mathbf{u}_j)$, the transmitter resends $\mathbf{v}_j$ to the receiver. Upon receiving $\tilde{\mathbf{v}}_j = (\tilde{f}(\mathbf{u}_j), \tilde{\mathbf{u}}_j)$, the receiver again attempts to recover $\mathbf{u}_j$. If the syndrome of $\tilde{\mathbf{v}}_j$ is zero, $\tilde{\mathbf{u}}_j$ is assumed to be error-free and will be released. If the syndrome of $\tilde{\mathbf{v}}_j$ is not zero, $\tilde{\mathbf{u}}_j$ and $\tilde{q}(\mathbf{u}_j)$ (stored in the receiver buffer) together are used for error correction based on C_1. If the receiver again fails to recover $\mathbf{u}_j$, then $\tilde{q}(\mathbf{u}_j)$ is discarded and $\tilde{\mathbf{u}}_j$ will be stored in the receiver buffer. Retransmissions for $\mathbf{v}_j$ (also for any other erroneously received information code vectors) continue until $\mathbf{u}_j$ is successfully recovered. The retransmissions are alternate repetitions of the parity vector $\mathbf{v}_j^*$ and the information vector $\mathbf{v}_j$, and the receiver stores $\tilde{\mathbf{u}}_j$ and $\tilde{q}(\mathbf{u}_j)$ alternately.

In the blocked state, when an information code vector is received, its forward index l_f is computed. If $l_f < N$, the vector is stored in the receiver buffer. If no errors are detected, an ACK is set to the transmitter; otherwise, a NAK is sent to the transmitter. If $l_f \geq N$, the receiver computes its backward index l_b. If $l_b < 2N$, the current received vector is regarded as an information code vector that was previously accepted and delivered. It is then ignored and an ACK is sent to the transmitter. If $l_b \geq 2N$, the received vector is a new information code vector; however, the buffer is full and buffer overflow occurs. In this event, the received vector will be rejected and a NAK is sent to the transmitter.

When a retransmission of a negatively acknowledged vector arrives, the receiver checks whether this received vector and its corresponding vector stored in the receiver buffer form *a message and parity pair*. If they do form such a pair, the receiver then attempts to recover the original message either by inversion or by decoding process. If they do not form a message and parity pair, the receiver checks whether the newly received vector and its corresponding vector in the buffer are both information vectors or are both parity vectors. If they are both information vectors, the receiver stores the one that is error-free; if both are erroneous, the receiver replaces the old

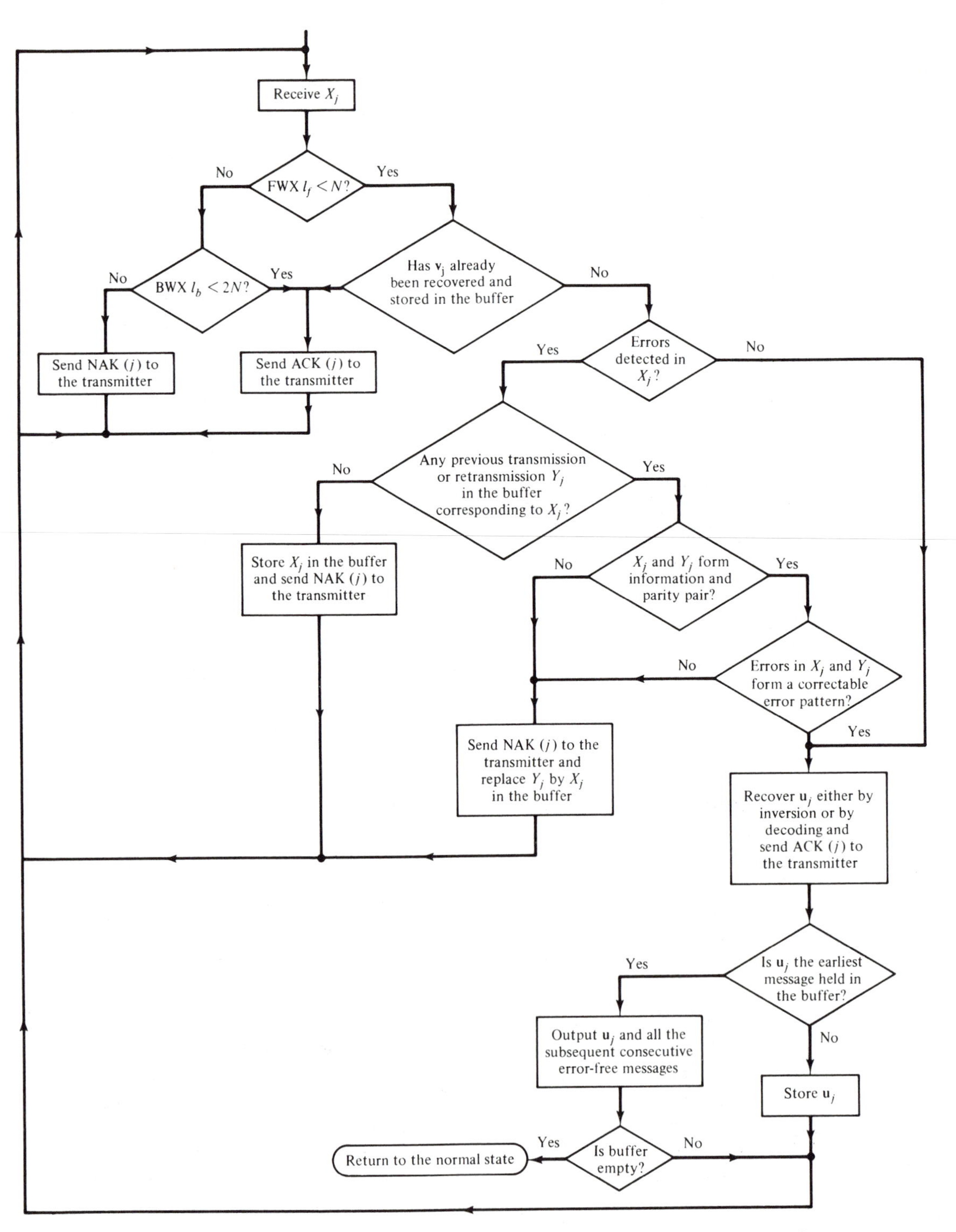

Figure 15.19 Blocked state of the receiver where X_j and Y_j represent either the information code vector $\mathbf{v}_j$ or the parity code vector $\mathbf{v}_j^*$.

one with the newly received one. If they are both parity vectors, the newly received parity vector (or its inversion) replaces the old one in the buffer. If the received vector corresponds to a lost vector, it will be stored at a reserved location in the receiver buffer.

The receiver's operation in the blocked state is detailed in the flowchart shown in Figure 15.19.

Throughput Efficiency

To analyze the throughput efficiency of the hybrid ARQ described above is very difficult. However, if we assume that the channel is a BSC and the return channel is noiseless, a lower bound on the throughput can be obtained. Again, let p be the channel transition probability and let P be the probability that a code vector, information or parity, will be received successfully (i.e., the syndrome based on C_0 is zero).

Suppose that C_1 is capable of correcting any combination of t or fewer errors and simultaneously detecting any combination d ($d > t$) or fewer errors. Let Q_1 be the conditional probability that a message $\mathbf{u}_j$ will be recovered from the first received parity block $\tilde{q}(\mathbf{u}_j)$ either by inversion or by decoding based on C_1, given that errors are detected in the received information code vector $\tilde{\mathbf{v}}_j = (\tilde{f}(\mathbf{u}_j), \tilde{\mathbf{u}}_j)$. Let Q_2 be the conditional probability that $\mathbf{u}_j$ will be successfully recovered from the second retransmission of $\mathbf{v}_j = (f(\mathbf{u}_j), \mathbf{u}_j)$, given that errors are detected in the first received parity code vector $\tilde{\mathbf{v}}_j^* = (\tilde{f}[q(\mathbf{u}_j)], \tilde{q}(\mathbf{u}_j))$ and the first received information code vector $\tilde{\mathbf{v}}_j = (\tilde{f}(\mathbf{u}_j), \tilde{\mathbf{u}}_j)$, and that the first received parity block $\tilde{q}(\mathbf{u}_j)$ fails to recover $\mathbf{u}_j$ but detects the presence of errors in $(\tilde{q}(\mathbf{u}_j), \tilde{\mathbf{u}}_j)$. Then it is possible to show that [29]

$$Q_1 = P + \frac{1}{1-P}\left\{\sum_{i=0}^{t}\binom{2k}{i}p^i(1-p)^{2k-i} + (1-p)^{2n} - 2(1-p)^n\left[(1-p)^k + \sum_{i=1}^{t}\binom{k}{i}p^i(1-p)^{k-i}\right]\right\}, \tag{15.23}$$

$$Q_2 \geq P + \frac{1}{(1-P)(1-Q_1)}\left\{\sum_{i=1}^{t-1}\Delta_i S_{t-i}(1 - \Delta_0 - S_{t-i})\right\} \tag{15.24}$$

where

$$\Delta_i = \binom{k}{i}p^i(1-p)^{k-i},$$

$$S_j = \sum_{l=1}^{j}\Delta_l.$$

Define

$$\theta_0 = 1 - (1-P)(1-Q_1),$$
$$\theta_1 = 1 - (1-P)(1-Q_1)(1-Q_2),$$
$$\theta_2 = Q_1 + (1-Q_1)Q_2.$$

Then the throughput of the type II hybrid selective-repeat ARQ with receiver buffer of size N is lower bounded as follows:

$$\eta_{\text{II}} \geq \frac{\delta_0}{\delta_0 + \delta_1 + \delta_2 N} \tag{15.25}$$

where

$$\delta_0 = \frac{\theta_2}{1 - \theta_1}(1 - \theta_2\theta_1^{N-1}),$$

$$\delta_1 = PQ_1\theta_0^{N-2} + P(1 - Q_1)Q_2\theta_1^{N-2},$$

$$\delta_2 = 2 - PQ_1\theta_0^{N-2} - \theta_0\theta_2\theta_1^{N-2}.$$

For various n, k, t, and N, the lower bound on throughput versus bit error rate above is plotted in Figures 15.20 to 15.22. We see that the throughput efficiency of the ideal selective-repeat ARQ with infinite receiver buffer can be achieved by the type II hybrid ARQ scheme described above with a *relatively small* designed error-correcting capability t ($t \leq 5$). For $t > 5$, throughput much higher than that of the ideal selective-repeat can be obtained.

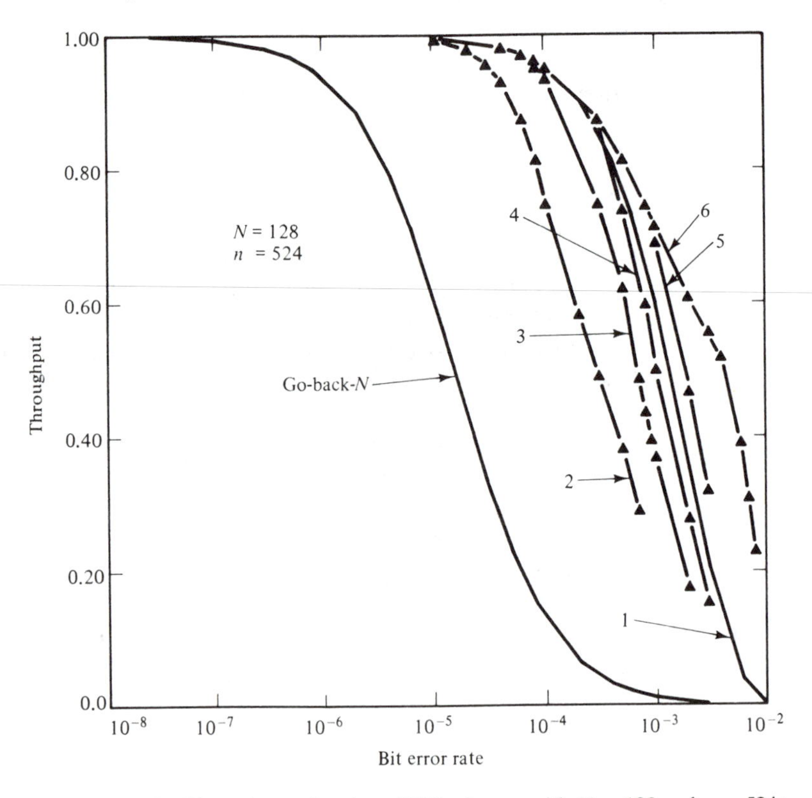

Figure 15.20 Throughput of various ARQ schemes with $N = 128$ and $n = 524$: (1) ideal selective-repeat with infinite receiver buffer; (2) and (3) selective-repeat with receiver buffer of size N and $2N$; (4), (5), and (6) type II selective-repeat hybrid ARQ with receiver buffer of size N and error-correction parameter $t = 3$, 5, and 10.

Reliability

If the error-detecting capability d of the half-rate code C_1 is sufficiently large, the type II hybrid ARQ scheme provides the same reliability as a pure ARQ scheme.

Let P_e be the probability that C_0 fails to detect the presence of errors. Let

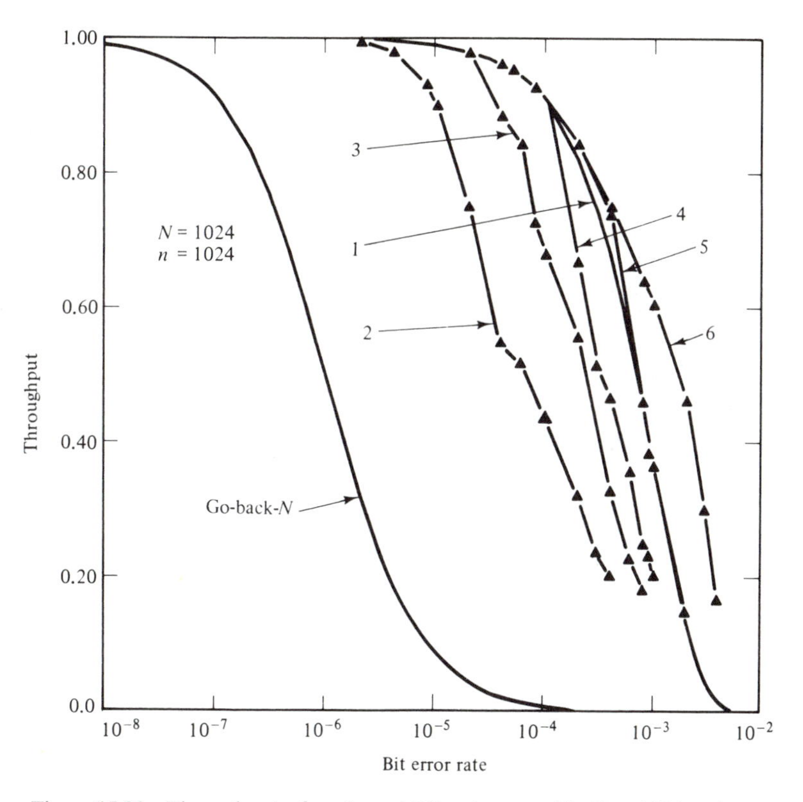

Figure 15.21 Throughput of various ARQ schemes with $N = 1024$ and $n = 1024$: (1) ideal selective-repeat with infinite receiver buffer: (2) and (3) selective-repeat with receiver buffer of size N and $2N$; (4), (5), and (6) type II selective-repeat hybrid ARQ with receiver buffer of size N and error correction parameter $t = 3$, 5, and 10.

$$\sigma = \sum_{i>d}^{2k} \binom{2k}{i} p^i(1-p)^{2k-i}, \tag{15.26}$$

which is the probability that the number of errors in $(\tilde{q}(\mathbf{u}), \tilde{\mathbf{u}})$ exceeds the designed error-detecting capability d of C_1. Then the probability that the receiver will commit a decoding error is bounded as follows [29]:

$$\frac{P_e}{P_e + P_c} \leq P(E) \leq \frac{P_e}{P_e + P_c} + \frac{\sigma}{P_e + P_c}, \tag{15.27}$$

where $P_c = (1 - p)^n$. If we choose d sufficiently large, we can make σ of the same order as P_e or even smaller than P_e. As a result, the probability of a decoding error, $P(E)$, is the same order as the error probability of a pure ARQ.

Example 15.2

Let C_0 be a properly chosen (524, 500) linear code. Let C_1 be the (1000, 500) shortened BCH code with $d_{\min} \geq 111$. Suppose that C_1 is used to correct $t = 10$ or fewer errors and simultaneously detect $d = 100$ or fewer errors. Let $p = 10^{-2}$. Then we find that

$$P_c = (1 - 10^{-2})^{524} = 5.16 \times 10^{-3},$$
$$\sigma \leq 1.8 \times 10^{-63}.$$

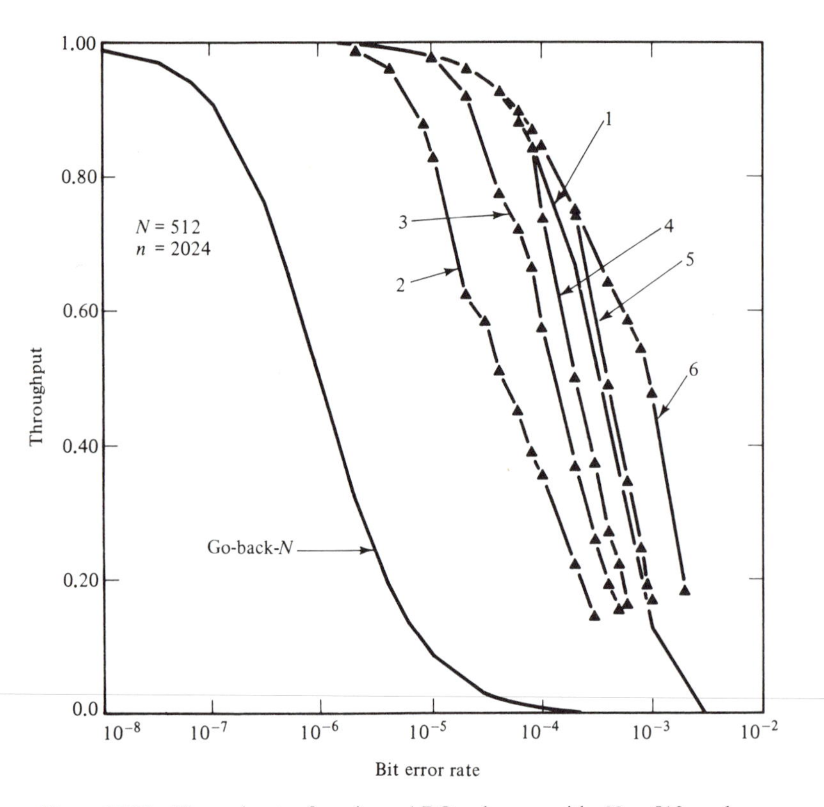

Figure 15.22 Throughput of various ARQ schemes with $N = 512$ and $n = 2024$: (1) ideal selective-repeat with infinite receiver buffer; (2) and (3) selective-repeat with receiver buffer of size N and $2N$; (4), (5), and (6) type II selective-repeat hybrid ARQ with receiver buffer of size N and error correction parameter $t = 3$, 5, and 10.

Since $n - k = 24$, it follows from (15.3) that

$$P_e \leq 6 \times 10^{-8},$$

We see that $\sigma \ll P_e$. Therefore, we have

$$P(E) \doteqdot \frac{P_e}{P_e + P_c} \leq 1.17 \times 10^{-5},$$

which is the same error probability for a pure ARQ scheme [see (15.1)].

So far we have considered only hybrid ARQ schemes that employ block codes for error correction. Hybrid ARQ schemes using convolutional codes for error correction can also be devised and are discussed in Section 17.5.

PROBLEMS

15.1. In (15.5) we see that the throughput of the go-back-N ARQ depends on the channel block error rate $P = 1 - (1 - p)^n$, where n is the code block length and p is the channel (BSC) transition probability. Let τ be the data rate in bits per second. Let T be round-trip delay time in seconds. Then $N = \tau \cdot T/n$. Suppose that p and k/n are fixed. Determine the block length n_0 that maximizes the throughput η_{GBN}. The block length n_0 is called the *optimal block length*. Optimal block lengths for the three basic ARQ schemes were investigated by Morris [30].

15.2. Consider a continuous ARQ scheme that operates as follows. When the transmitter receives a NAK for a particular vector **v** under the condition that the $N - 1$ vectors preceding **v** have been positively acknowledged, the transmitter stops sending new vectors and simply repeats the vector **v** continuously until an ACK for **v** is received. After receiving an ACK, the transmitter renews transmission of new vectors. At the receiver, when a received vector $\tilde{\mathbf{v}}$ is detected in error under the condition that all the vectors preceding $\tilde{\mathbf{v}}$ have been successfully received, the receiver rejects $\tilde{\mathbf{v}}$ and all the $N - 1$ subsequent received vectors until the first repetition of **v** arrives. Then the receiver checks the syndrome of $\tilde{\mathbf{v}}$ and the following repetitions of **v**. An ACK is sent to the transmitter as soon as one repetition of **v** has been successfully received.

(a) Derive the throughput of this scheme.

(b) Compare the throughput of this scheme and the throughput of the conventional go-back-N ARQ.

15.3. Suppose that we use the retransmission strategy described in Problem 15.2 but that a buffer of size N is provided at the receiver. When a received vector $\tilde{\mathbf{v}}$ is detected in error, the receiver stores the subsequent successfully received vectors. When a repetition of **v** is successfully received, the receiver will release $\tilde{\mathbf{v}}$ and the error-free vectors held in the receiver buffer in consecutive order until the next erroneous vector is encountered.

(a) Derive the throughput of this ARQ scheme.

(b) Compare its throughput with that of the conventional go-back-N ARQ.

15.4. We may shorten the (31, 16) BCH code to obtain a (30, 15) invertible code. Devise an inversion circuit for this code.

15.5. In a stop-and-wait ARQ system, suppose that the forward channel is a BSC with transition probability p_1 and the feedback channel is a BSC with transition probability p_2. Derive the throughput efficiency of this system.

15.6. Repeat Problem 15.5 for the go-back-N ARQ system.

15.7. Repeat Problem 15.5 for the ideal selective-repeat ARQ system.

REFERENCES

1. General Information: Binary Synchronous Communication, *IBM Publication GA27-3004*, IBM Corp., White Plains, N.Y., 1969.
2. J. R. Kersey, "Synchronous Data Link Control," *Data Commun.*, May/June 1974.
3. *Advanced Data Communication Control Procedures* (ADCCP), American National Standards Institute, Washington, D.C., 1977.
4. R. J. Benice and A. H. Frey, Jr., "An Analysis of Retransmission Systems," *IEEE Trans. Commun. Technol.*, COM-12, pp. 135–145, December 1964.
5. H. O. Burton and D. D. Sullivan, "Errors and Error Control," *Proc. IEEE*, 60(11), pp. 1293–1310, November 1972.
6. J. J. Metzner, "A Study of an Efficient Retransmission Strategy for Data Links," *NTC'77 Conf. Rec.*, pp. 3B:1-1 to 3B:1-5.

7. J. A. Lockitt, A. G. Gatfield, and R. R. Dobyns, "A Selective-Repeat ARQ System," Proc. Third Int. Conf. Digital Satellite Commun., Kyoto, Japan, November 1975.
8. M. C. Easton, "Efficient Transfer of Large Data Files over Satellite Links," Proc. 13th Hawaii Int. Conf. Syst. Sci., Honolulu, January 1979.
9. P. S. Yu and S. Lin, "An Efficient Selective-Repeat ARQ for Satellite Channels," *ICC'80 Conf. Rec.*, pp. 4.6.1–4.6.4, Seattle, Washington, June 1980.
10. P. S. Yu and S. Lin, "An Efficient Selective-Repeat ARQ Scheme for Satellite Channels and Its Throughput Analysis," *IEEE Trans. Commun.*, COM-29, pp. 353–363, March 1981.
11. M. J. Miller and S. Lin, "The Analysis of Some Selective-Repeat ARQ Schemes with Finite Receiver Buffer," *IEEE Trans. Commun.*, COM-29, pp. 1307–1315, September 1981.
12. A. R. K. Sastry, "Improving Automatic Repeat-Request (ARQ) Performance on Satellite Channels Under High Error Rate Conditions," *IEEE Trans. Commun.*, COM-23, pp. 436–439, April, 1975.
13. J. M. Morris, "On Another Go-Back-*N* ARQ Technique for High Error Rate Conditions," *IEEE Trans. Commun.*, COM-26, pp. 187–189, January 1978.
14. D. Towsley, "The Stutter Go-Back-*N* ARQ Protocol," *IEEE Trans. Commun.*, COM-27, pp. 869–875, June 1979.
15. S. Lin and P. S. Yu, "An Efficient Error Control Scheme for Satellite Communications," *IEEE Trans. Commun.*, COM-28, pp. 395–401, March 1980.
16. J. J. Metzner and K. C. Morgan, "Word Selection Procedures for Error-Free Communication Systems," New York University, New York, 2nd Sci. Rep., Contract AF19(628)-4321, June 1965.
17. K. Brayer, "Error Control Techniques Using Binary Symbol Burst Codes," *IEEE Trans. Commun.*, COM-16, pp. 199–214, April 1968.
18. D. M. Mandelbaum, "Adaptive-Feedback Coding Scheme Using Incremental Redundancy," *IEEE Trans. Inf. Theory*, IT-20(3), pp. 388–389, May 1974.
19. J. J. Metzner, "Improvements in Block Retransmission Schemes," *IEEE Trans. Commun.*, COM-27, pp. 524–532, February 1979.
20. E. Y. Rocher and R. L. Pickholtz, "An Analysis of the Effectiveness of Hybrid Transmission Schemes," *IBM J. Res. Dev.*, pp. 426–433, July 1970.
21. A. R. K. Sastry, "Performance of Hybrid Error Control Schemes on Satellite Channels," *IEEE Trans. Commun.*, COM-23, pp. 689–694, July 1975.
22. A. R. K. Sastry and L. N. Kanal, "Hybrid Error Control Using Retransmission and Generalized Burst-Trapping Codes," *IEEE Trans. Comun.*, COM-24, pp. 385–393, April 1976.
23. P. S. Sindhu, "Retransmission Error Control with Memory," *IEEE Trans. Commun*, COM-25, pp. 473–479, May 1977.
24. T. C. Ancheta, "Convolutional Parity Check Automatic Repeat Request," presented at the 1979 IEEE International Symposium on Information Theory, Grignano, Italy, June 25–29, 1979.
25. S. Lin and J. S. Ma, "A Hybrid ARQ System with Parity Retransmission for Error. Correction," *IBM Res. Rep.* 7478 (#32232), January 11, 1979.

26. S. Lin and P. S. Yu, "SPREC–An Effective Hybrid-ARQ Scheme," *IBM Res. Rep.* 7591 (#32852), April 4, 1979.
27. A. Drukarev and D. J. Costello, Jr., "ARQ Error Control Using Sequential Decoding," *ICC'80 Conf. Rec.*, pp. 4.7.1–4.7.5, Seattle, Wash., June 1980.
28. H. Yamamoto and K. Itoh, "Viterbi Decoding Algorithm for Convolutional Codes with Repeat Request," *IEEE Trans. on Inf. Theory*, IT-26, pp. 540–547, September 1980.
29. S. Lin and P. S. Yu, "A Hybrid ARQ Scheme with Parity Retransmission for Error Control of Satellite Channels," submitted to *IEEE Trans. Commun.*, 1981.
30. J. M. Morris, "Optimal Blocklengths for ARQ Error Control Schemes," *IEEE Trans. Commun.*, COM-27, pp. 488–493, February 1979.

16

Applications of Block Codes for Error Control in Data Storage Systems

In this chapter several applications of block coding for error control in data storage systems are discussed. In Section 16.1 a class of single-error-correcting and double-error-detecting (SEC-DED) codes is presented. This class of codes is widely used for error control in computer main/or control memories. In Sections 16.2 and 16.3, we describe two coding schemes that are used for error control in IBM magnetic tape storage systems. In Section 16.4 coding schemes used in various magnetic disk storage systems are discussed.

Block codes are also used for error detection/or error correction in other data storage systems (e.g., magnetic drums, photodigital storage systems, magnetic bubble memories etc.).

16.1 ERROR CONTROL FOR COMPUTER MAIN PROCESSOR AND CONTROL STORAGES

Error-correcting block codes are widely used to improve the system-level reliability of computer main storage or control storage. The IBM system 7030, built in 1961, was the first IBM computer system to use a single-error-correcting and double-error-detecting (SEC-DED) Hamming code for its core memory. Later, SEC-DED codes were used in IBM system 360 and computer systems of other computer makers for improving the reliability of core memories. However, core memories are very reliable especially since the technology has advanced to a mature state. In the 1970s, semiconductor memories were used to replace core memories. Semiconductor memories are faster than core memories in speed; however, they are less reliable than core memories due to their high density per chip and their exposure to radiation, which induces soft failures (errors). As a result, the use of error-correcting codes for improving semiconductor memory reliability becomes a standard design feature. The improvement in reliability is especially evident when the memory system is organized on 1-bit-per-card base (bit-oriented memory). With this organization, most error patterns (or multiple-bit failures caused by malfunction) on each card appear as if they were single errors. These errors can be effectively controlled by using SEC-DED codes. SEC-DED codes are used in many current IBM computer systems (e.g., System 370) as well as many computer systems of other computer makers in the United States and abroad.

Class of SEC-DED Codes

In the following we present a class of SEC-DED codes that are suitable and commonly used for improving computer memory reliability. This class of codes was discovered by Hsiao [1]. The most important feature of this class of codes is that it provides fast encoding and error detection in the decoding process, which are the most critical on-line processes in the memory operations.

It has been shown in Chapter 2 that an (n, k) linear code is uniquely specified by a parity-check matrix $\mathbf{H}$. This parity-check matrix $\mathbf{H}$ can be used to generate parity-check bits during encoding operation and to generate syndrome bits during decoding operation. The total number of 1's in each row of $\mathbf{H}$ relates to the number of logic levels necessary to generate the parity-check bit or the syndrome bit corresponding to that row. Let N_i be the number of 1's in the ith row of the $\mathbf{H}$ matrix. Let $L_i^{(p)}$ be the number of logic levels required to generate the ith parity-check bit with b-input modulo-2 adders (or b-input X-OR gate). Let $L_i^{(s)}$ be the number of logic levels required to generate the ith syndrome bit with b-input modulo-2 adders. Then we have

$$L_i^{(p)} = \lceil \log_b (N_i - 1) \rceil,$$
$$L_i^{(s)} = \lceil \log_b N_i \rceil,$$

where $\lceil X \rceil$ denotes the smallest integer greater than or equal to X. In practical applications, b is fixed for a given circuit family. Therefore, to minimize $L_i^{(p)}$ and $L_i^{(s)}$, minimum N_i is necessary. If all N_i for $i = 1, 2, \ldots, n - k$ are minimum and equal, we obtain the least propagation delay for generating parity-check bits or syndrome bits. Moreover, the code with minimum N_i requires less hardware for implementation. Less hardware implies lower cost and better reliability.

A class of SED-DED codes that has the features described above may be obtained from shortening Hamming codes. As we know, for every positive integer m there exists a Hamming code with the following parameters:

Code length:	$n = 2^m - 1$
Number of information symbols:	$k = 2^m - m - 1$
Number of parity-check bits:	$n - k = m$
Minimum distance:	$d_{\min} = 3.$

It is a single-error-correcting code. The parity-check matrix $\mathbf{H}$ consists of all the $2^m - 1$ nonzero m-tuples as columns. If we delete l columns from $\mathbf{H}$, we would obtain a new matrix $\mathbf{H}_0$ with $2^m - l - 1$ columns. This new parity-check matrix $\mathbf{H}_0$ will generate a linear code of length $2^m - l - 1$ and m parity-check bits. It is a $(2^m - l - 1, 2^m - l - m - 1)$ shortened Hamming code. Its minimum distance is at least 3. Now, we construct a class of shortened Hamming codes as follows. Delete columns from a Hamming code parity-check matrix $\mathbf{H}$ such that the new parity-check matrix $\mathbf{H}_0$ satisfies the following requirements:

1. Every column should have an odd number of 1's.
2. The total number of 1's in the $\mathbf{H}_0$ matrix should be a minimum.
3. The number of 1's in each row of $\mathbf{H}_0$ should be made equal, or as close as psssible, to the average number (i.e., the total number of 1's in $\mathbf{H}_0$ divided by the number of rows).

The first requirement guarantees the code generated by $\mathbf{H}_0$ has minimum distance at least 4. Therefore, it can be used for single-error-correction and double-error detection. The second and third requirements would yield minimum logic levels in forming parity or syndrome bits, and less hardware in implementation of the code. Hsiao [1] provided an algorithm to construct $\mathbf{H}_0$ and found some optimal SEC-DED codes. Several parity-check matrices for message (or data) lengths 16, 32, and 64 are given in Figure 16.1. The parameters of a list of Hsiao's codes are given in Table 16.1.

In computer applications, these codes are encoded and decoded in a parallel manner. In encoding, the message bits enter the encoding circuit in parallel and the parity-check bits are formed simultaneously. In decoding, the received bits enter the

$$\mathbf{H}_0 = \begin{bmatrix}
1&0&0&0&0&0&1&0&0&1&1&0&0&1&0&0&1&1&1&1&0&0\\
0&1&0&0&0&0&0&0&1&1&1&1&1&0&1&0&0&0&1&0&1&0\\
0&0&1&0&0&0&1&1&1&0&1&1&1&0&0&1&1&0&0&0&0&0\\
0&0&0&1&0&0&1&1&1&0&0&0&0&1&1&1&0&1&0&0&0&1\\
0&0&0&0&1&0&0&0&0&1&0&0&1&1&1&1&0&0&0&1&1&1\\
0&0&0&0&0&1&0&1&0&0&0&1&0&0&0&0&1&1&1&1&1&1
\end{bmatrix}$$

(a)

$$\mathbf{H}_0 = \begin{bmatrix}
1&0&0&0&0&0&0&1&0&0&0&1&0&1&0&1&0&0&0&0&0&1&0&0&0&0&0&1&1&1&1&0&0&0&1&1&0&1&1\\
0&1&0&0&0&0&0&0&0&0&1&0&0&0&0&0&0&0&1&1&1&1&1&0&1&1&1&0&0&0&1&0&1&1&0&0&0&0&1\\
0&0&1&0&0&0&0&0&0&0&1&0&1&1&0&1&1&1&1&0&0&0&0&1&0&0&1&0&0&1&0&1&0&1&0&0&1&1&0\\
0&0&0&1&0&0&0&1&1&1&1&1&1&1&1&0&0&0&0&0&0&0&1&1&0&1&0&0&1&0&0&0&1&0&0&0&1&0&0\\
0&0&0&0&1&0&0&0&1&1&0&1&1&0&0&1&1&1&1&1&1&1&1&0&0&0&0&1&0&0&0&0&0&0&0&1&0&0&0\\
0&0&0&0&0&1&0&0&0&1&0&0&0&0&1&0&0&1&0&0&1&0&0&1&1&1&1&1&1&1&1&1&0&0&1&0&0&0&0\\
0&0&0&0&0&0&1&1&1&0&0&0&0&0&1&0&1&0&0&1&0&0&0&0&0&1&0&0&0&0&0&0&1&1&1&1&1&1&1
\end{bmatrix}$$

(b)

Figure 16.1 (a) Parity-check matrix of an optimal (22, 16) SEC-DED code; (b) parity-check matrix of an optimal (39, 32) SEC-DED code; (c) parity-check matrix of an optimal (72, 64) SEC-DED code; (d) parity-check matrix of another optimal (72, 64) SEC-DED code.

$$
\mathbf{H}_0 = \left[\begin{array}{c|c|c|c|c|c|c|c|c}
10000000 & 11111111 & 00000111 & 00111000 & 11001000 & 00001000 & 00001001 & 10010010 & 01100100 \\
01000000 & 01100100 & 11111111 & 00000111 & 00111000 & 11001000 & 00001000 & 00001001 & 10010010 \\
00100000 & 10010010 & 01100100 & 11111111 & 00000111 & 00111000 & 11001000 & 00001000 & 00001001 \\
00010000 & 00001001 & 10010010 & 01100100 & 11111111 & 00000111 & 00111000 & 11001000 & 00001000 \\
00001000 & 00001000 & 00001001 & 10010010 & 01100100 & 11111111 & 00000111 & 00111000 & 11001000 \\
00000100 & 11001000 & 00001000 & 00001001 & 10010010 & 01100100 & 11111111 & 00000111 & 00111000 \\
00000010 & 00111000 & 11001000 & 00001000 & 00001001 & 10010010 & 01100100 & 11111111 & 00000111 \\
00000001 & 00000111 & 00111000 & 11001000 & 00001000 & 00001001 & 10010010 & 01100100 & 11111111
\end{array}\right]
$$

(c)

$$
\mathbf{H}_0 = \left[\begin{array}{c|c|c|c|c|c|c|c|c}
10000000 & 11111111 & 00001111 & 00001111 & 00001100 & 01101000 & 10001000 & 10001000 & 10000000 \\
01000000 & 11110000 & 11111111 & 00000000 & 11110011 & 01100100 & 01000100 & 01000100 & 01000000 \\
00100000 & 00110000 & 11110000 & 11111111 & 00001111 & 00000010 & 00100010 & 00100010 & 00100110 \\
00010000 & 11001111 & 00000000 & 11110000 & 11111111 & 00000001 & 00010001 & 00010001 & 00010110 \\
00001000 & 01101000 & 10001000 & 10001000 & 10000000 & 11111111 & 00001111 & 00000000 & 11110011 \\
00000100 & 01100100 & 01000100 & 01000100 & 01000000 & 11110000 & 11111111 & 00001111 & 00001100 \\
00000010 & 00000010 & 00100010 & 00100010 & 00100110 & 11001111 & 00000000 & 11111111 & 00001111 \\
00000001 & 00000001 & 00010001 & 00010001 & 00010110 & 00110000 & 11110000 & 11110000 & 11111111
\end{array}\right]
$$

(d)

Figure 16-1 *(cont.)*

TABLE 16.1 PARAMETERS OF A LIST OF HSIAO'S CODES

n	k	$n - k$	Total number of 1's in **H**	Average number of 1's per row
12	8	4	16	4
14	9	5	32	6.4
15	10	5	35	7
16	11	5	40	8
22	16	6	54	9
26	20	6	66	11
30	24	6	86	14.3
39	32	7	103	14.7
43	36	7	117	16.7
47	40	7	157	22.4
55	48	7	177	25.3
72	64	8	216	27
80	72	8	256	32
88	80	8	296	37
96	88	8	336	42
104	96	8	376	47
112	104	8	416	52
120	112	8	456	57
128	120	8	512	64
130	121	9	446	49.6
137	128	9	481	53.5

decoding circuit in parallel, the syndrome bits are formed simultaneously and the received bits are corrected in parallel. Single-error correction is accomplished by the table-lookup decoding as described in Example 3.9. Double-error detection is accomplished by examining the number of 1's in the syndrome vector $\mathbf{s}$. If the syndrome $\mathbf{s}$ contains an even number of 1's, then either a double-error pattern or a multiple-even-error pattern has occurred.

Consider the parity-check matrix of the (72, 64) SED-DED code given in Figure 16.1(c). Each row contains 27 ones. If 3-input X-OR gates are used to form syndrome bits, each syndrome bit is formed by a three-level X-OR tree with 13 gates. The eight X-OR trees for generating the 8 syndrome bits are identical. These provide uniform and minimum delay in the error-correction process.

Error Control for IBM System 370 Main Storage or Control Storage

Many of the IBM System 370 models (e.g., 145, 155, 165, 158, and 168) use a (72, 64) optimal SED-DED code for error control in main storage or control storage. Encoding and decoding are performed on a double-word (8-bytes) basis. Eight parity-check bits for error checking and correcting are used.

When a double-word as shown in Figure 16.2 is to be placed in processor storage by a program or in control storage during microprogram loading, the 8 parity-check bits, one from each byte, are removed (a single parity-check bit for each 8-bit data byte is used to check data validity for data not in the storage). Then the 64 data bits are encoded into a code word of 72 bits, as shown in Figure 16.3 and the resultant code word is stored in the memory. When a code word is fetched from the storage, the decoding circuit forms 8 syndrome bits. If the syndrome bits are all zero, the 64 data bits are assumed to be error-free. Then appropriate parity bit for each of the 8-bit byte is generated and the double-word is reformated to look as shown in Figure 16.2. If a single error is detected and located, it is then corrected. The corrected double-word is sent to the CPU but not back to storage. If a double-bit or multiple-

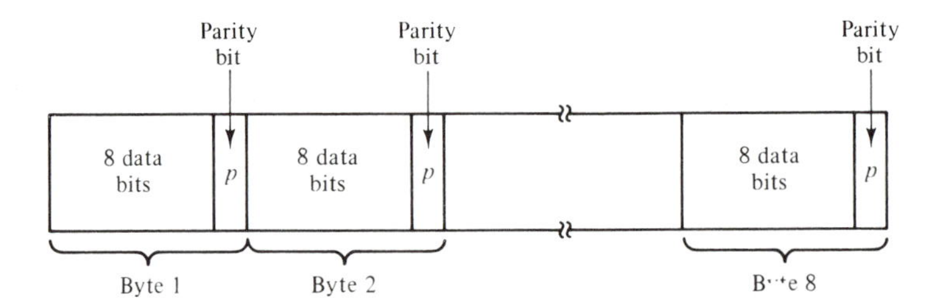

Figure 16.2 Data representation for double-word not in storage.

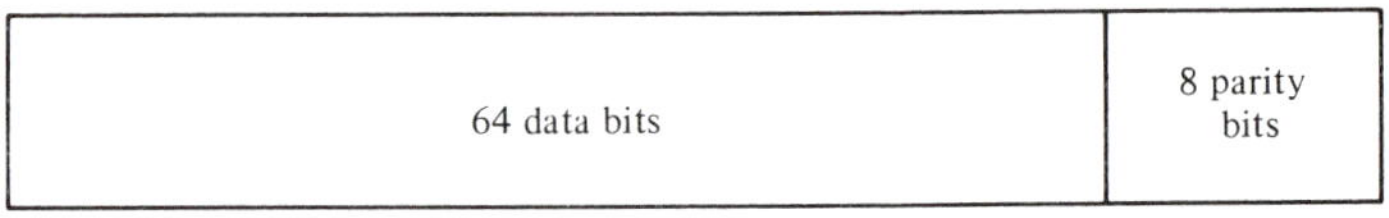

64 data bits	8 parity bits

Figure 16.3 Data representation for double-word in processor/or control storage.

bit error occurs during instruction execution, the instruction is retried, if possible. If the storage error is of an intermittent type and a single-bit error results after any one of these tries, the error is corrected as usual and processing continues.

Hsiao's codes are also implemented in IBM 3000 series and 4300 series for error control.

16.2 ERROR CONTROL FOR MAGNETIC TAPES

On the standard $\frac{1}{2}$-inch nine-track magnetic tapes, the most common errors are burst errors or erasures on tracks. These track errors or erasures are primarily caused by surface defects or variations in head-media separation in the presence of dust particles. These errors or erasures often affect as many as 100 bits at a time, depending on the density of recording. In this and next sections, we will describe two error-correcting schemes which are currently used in the IBM 3420 and 3850 tape systems for error control.

IBM 3420 Series Tape Units

First, we describe the error-correcting scheme used in the IBM 3420 series tape units with a recording density of 6250 bits per inch (bpi). The error-correcting scheme, devised by Patel and Hong [2], is capable of correcting any error pattern on a single track or any error patterns on two tracks provided that the erroneous tracks i and j are identified by some external pointers (the external pointers could be hardware indicators for inadequate amplitude or phase of the received signal corresponding to a particular track, or any other signals indicating errors in a track).

An IBM 3420 series tape unit writes characters in parallel across nine tracks on a $\frac{1}{2}$-inch tape as shown in Figure 16.4. Each character consists of 8 information bits and 1 overall parity-check bit. Let

$$\mathbf{B}_i = (b_{i0}, b_{i1}, b_{i2}, b_{i3}, b_{i4}, b_{i5}, b_{i6}, b_{i7})$$

Characters

Track no.												
0	b_{00}	b_{10}	b_{20}	b_{30}	b_{40}	b_{50}	b_{60}	b_{70}				
1	b_{01}	b_{11}	b_{21}	b_{31}	b_{41}	b_{51}	b_{61}	b_{71}				
2												
3												
4	$\mathbf{B}_0$	$\mathbf{B}_1$	$\mathbf{B}_2$	$\mathbf{B}_3$	$\mathbf{B}_4$	$\mathbf{B}_5$	$\mathbf{B}_6$	$\mathbf{B}_7$	$\mathbf{B}_0$	$\mathbf{B}_1$	$\mathbf{B}_3$	→ Tape motion
5												
6												
7	b_{07}	b_{17}	b_{27}	b_{37}	b_{47}	b_{57}	b_{67}	b_{77}				
Overall parity track 8	q_0	q_1	q_2	q_3	q_4	q_5	q_6	q_7				

|← Code array →|← Code array

Figure 16.4 Format of a code array on a IBM 3420 nine-track tape.

denote the 8 information bits of the ith character. For convenience, we call this 8-bit byte an *information byte*. The overall parity-check bit for $\mathbf{B}_i$, denoted q_i, is simply the modulo-2 sum of the bits in $\mathbf{B}_i$, that is,

$$q_i = \sum_{j=0}^{7} b_{ij}. \tag{16.1}$$

Thus, $\mathbf{B}_i$ and q_i together form a 9-bit character. The bit b_{ij} is recorded on the jth track with $0 \leq j \leq 7$ and the overall parity-check bit q_i is recorded on the eighth track.

Using Patel and Hong's coding scheme, a code word consists of eight characters (72 bits) arranged in a 9×8 rectangular array as shown in Figure 16.4, where $\mathbf{B}_7$, $\mathbf{B}_6, \ldots, \mathbf{B}_1$ are 8-bit information bytes and $\mathbf{B}_0 = (b_{00}, b_{01}, \ldots, b_{07})$ is a *parity-check byte*. The parity-check bits $b_{00}, b_{01}, \ldots, b_{07}$ are computed based on the 56 bits of the information bytes $\mathbf{B}_7, \mathbf{B}_6, \ldots, \mathbf{B}_1$ and the following irreducible polynomial over GF(2),

$$\mathbf{g}(X) = 1 + X^3 + X^4 + X^5 + X^8. \tag{16.2}$$

[Note that $\mathbf{g}(X)$ is self-reciprocal.] Let us represent the ith information byte in polynomial form as follows:

$$\mathbf{B}_i(X) = b_{i0} + b_{i1}X + b_{i2}X^2 + b_{i3}X^3 + b_{i4}X^4 + b_{i5}X^5 + b_{i6}X^6 + b_{i7}X^7.$$

Then the parity-check byte $\mathbf{B}_0(X)$ is simply the remainder resulting from dividing the polynomial

$$X\mathbf{B}_1(X) + X^2\mathbf{B}_2(X) + \cdots + X^7\mathbf{B}_7(X)$$

by the generator polynomial $\mathbf{g}(X)$. Therefore, $\mathbf{B}_0(X) + X\mathbf{B}_1(X) + \cdots + X^7\mathbf{B}_7(X)$ is divisible by $\mathbf{g}(X)$. After $\mathbf{B}_0(X)$ has been formed, an overall parity-check bit q_0 is appended to it to form a *parity-check character*. Clearly, the code is a (72, 56) linear code. The encoding can be carried out with a division circuit based on the generator polynomial $\mathbf{g}(X)$ as shown in Figure 16.5. The information bytes $\mathbf{B}_7, \mathbf{B}_6, \ldots, \mathbf{B}_1$ are successively shifted into the shift register with $\mathbf{B}_7$ first and $\mathbf{B}_1$ last (the 8 bits of each information byte enter the register in parallel). At the end of the first shift, the register contains the remainder resulting from dividing $X\mathbf{B}_7(X)$ by $\mathbf{g}(X)$. At the end of the second shift, the register contains the remainder resulting from dividing $X\mathbf{B}_6(X) + X^2\mathbf{B}_7(X)$ by $\mathbf{g}(X)$. At the end of the seventh shift, the register contains the parity-check byte $\mathbf{B}_0(X)$, which is then gated onto the tape (together with its overall parity-check digit q_0). The overall parity-check digits, $q_7, q_6, \ldots, q_0$ for $\mathbf{B}_7, \mathbf{B}_6, \ldots, \mathbf{B}_0$ are formed successively with an eight-input modulo-2 adder.

Before we can describe the decoding process of the code above, we need to develop some further properties of the code. Let $\mathbf{Z}_0, \mathbf{Z}_1, \ldots, \mathbf{Z}_7$ and $\mathbf{Q}$ denote the nine rows of a code array as shown in Figure 16.6. The row $\mathbf{Z}_j$ on the j-track with $0 \leq j \leq 7$ consists of the bits

$$\mathbf{Z}_j = (b_{0j}, b_{1j}, b_{2j}, b_{3j}, b_{4j}, b_{5j}, b_{6j}, b_{7j}). \tag{16.3}$$

The bottom row $\mathbf{Q}$ on the eight track consists of eight overall parity-check digits,

$$\mathbf{Q} = (q_0, q_1, q_2, q_3, q_4, q_5, q_6, q_7).$$

It follows from (16.1) and (16.3) that

$$\mathbf{Q} + \sum_{j=0}^{7} \mathbf{Z}_j = \mathbf{0}. \tag{16.4}$$

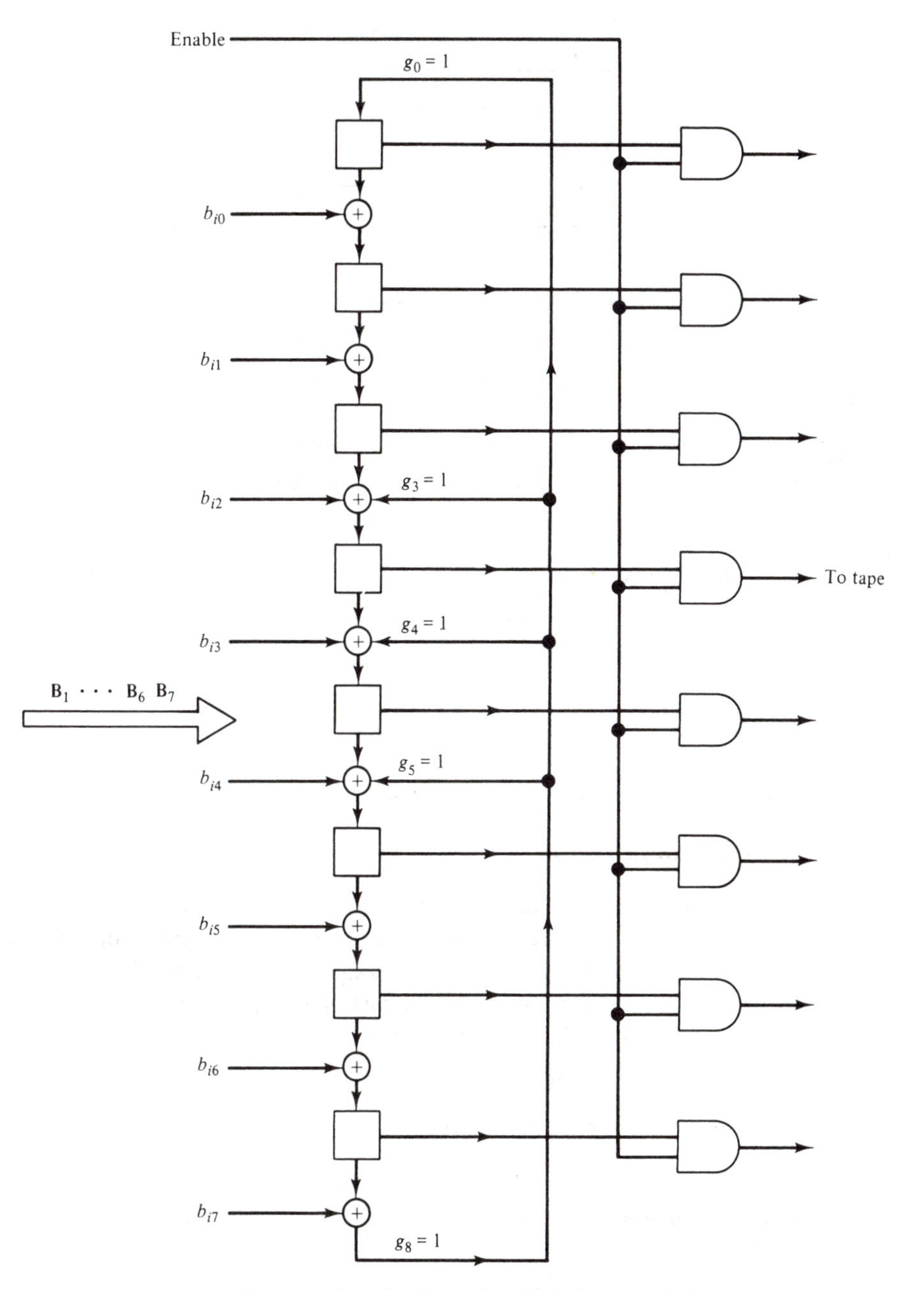

Figure 16.5 Encoding circuit for the IBM 3420 tape Unit.

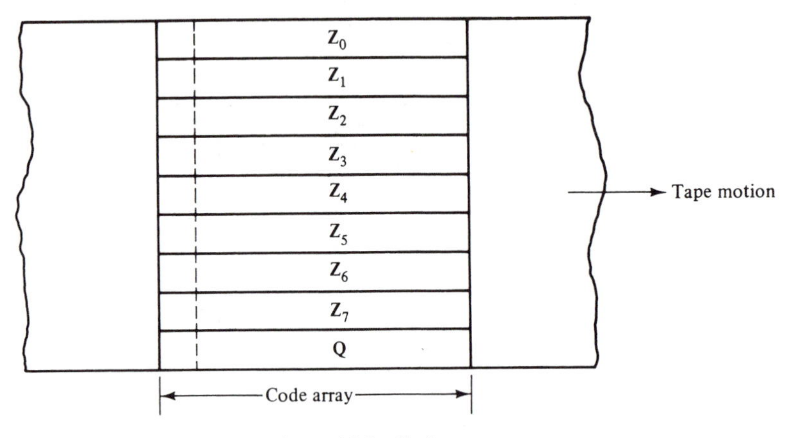

Figure 16.6 Code array.

Let us represent $\mathbf{Z}_j$ in polynomial form as follows:

$$\mathbf{Z}_j(X) = b_{0j} + b_{1j}X + b_{2j}X^2 + b_{3j}X^3 + b_{4j}X^4 + b_{5j}X^5 + b_{6j}X^6 + b_{7j}X^7. \tag{16.5}$$

Then, we can readily see that the following equality holds:

$$\sum_{j=0}^{7} X^j \mathbf{Z}_j(X) = \sum_{i=0}^{7} X^i \mathbf{B}_i(X). \tag{16.6}$$

Consequently,

$$\sum_{j=0}^{7} X^j \mathbf{Z}_j(X)$$

is divisible by the generator polynomial $\mathbf{g}(X)$.

Syndrome Computation

During the read operation, data are read from the tape character by character. Let $(\hat{B}_i, \hat{q}_i)$ be the ith character read from a code array. Let $\hat{Z}_j$ and $\hat{Q}$ denote the jth row and the overall parity-check row read from a code array, respectively. The syndrome of a code array read from the tape consists of two parts, $\mathbf{S}_1 = (\mathbf{S}_{10}, \mathbf{S}_{11}, \ldots, \mathbf{S}_{17})$ and $\mathbf{S}_2 = (\mathbf{S}_{20}, \mathbf{S}_{21}, \ldots, \mathbf{S}_{27})$, where

$$\mathbf{S}_1 = \hat{Q} + \hat{Z}_0 + \hat{Z}_1 + \cdots + \hat{Z}_7 \tag{16.7}$$

and $\mathbf{S}_2$ is the remainder resulting from dividing

$$\hat{B}_0(X) + X\hat{B}_1(X) + \cdots + X^7\hat{B}_7(X)$$

by the generator polynomial $\mathbf{g}(X)$. For convenience, we use the following notation:

$$\mathbf{S}_2 = \sum_{i=0}^{7} X^i \hat{B}_i(X) \text{ modulo } \mathbf{g}(X). \tag{16.8}$$

Since

$$\sum_{j=0}^{7} X^j \hat{Z}_j(X) = \sum_{i=0}^{7} X^i \hat{B}_i(X),$$

$\mathbf{S}_2$ is also the remainder resulting from dividing

$$\sum_{j=0}^{7} X^j \hat{Z}_j(X)$$

by $\mathbf{g}(X)$, that is,

$$\mathbf{S}_2 = \sum_{j=0}^{7} X^j \hat{Z}_j(X) \text{ modulo } \mathbf{g}(X). \tag{16.9}$$

The computation of $\mathbf{S}_1$ is accomplished by a nine-input modulo-2 adder iteratively as each character $(\hat{B}_i, \hat{q}_i)$ is read from the tape as shown in Figure 16.7. The second part of syndrome $\mathbf{S}_2$ can be computed by a division circuit based on $\mathbf{g}(X)$ as shown in Figure 16.8. The bytes $\hat{B}_7, \hat{B}_6, \ldots, \hat{B}_0$ are fed into a shift register as they are read off the tape. After $\hat{B}_0$ has been shifted in, the register contains the syndrome $\mathbf{S}_2$.

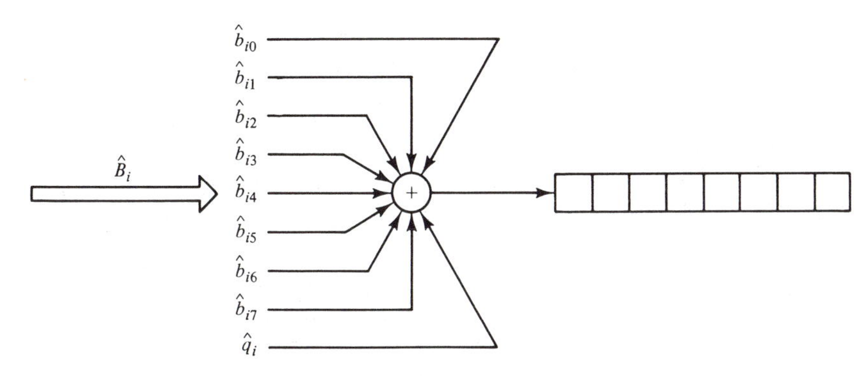

Figure 16.7 Circuit for computing $\mathbf{S}_1$ in the IBM 3420 Tape Unit.

It is possible to compute $\mathbf{S}_2$ by reading the bytes into a shift register in the order from $\hat{B}_0$ to $\hat{B}_7$ [i.e., $\hat{B}_0$ is read into the shift register first and $\hat{B}_7$ last (reverse order)]. This will result in a circuit that yields faster decoding operation, as we will see. First, we notice that the period of the generator polynomial $\mathbf{g}(X)$ is 17 [i.e., 17 is the smallest integer such that $\mathbf{g}(X)$ divides $X^{17} + 1$]. Using this fact, we can readily see that $\mathbf{S}_2(X)$ is also the remainder resulting from dividing

$$\sum_{i=0}^{7} X^{16(7-i)}[X^7 \hat{B}_i(X)]$$

by $\mathbf{g}(X)$, that is,

$$\mathbf{S}_2(X) = \sum_{i=0}^{7} X^{16(7-i)}[X^7 \hat{B}_i(X)] \text{ modulo } \mathbf{g}(X).$$

Let $\boldsymbol{\beta}_i(X) = \beta_{i0} + \beta_{i1}X + \cdots + \beta_{i7}X^7$ denote the remainder resulting from dividing $X^7\hat{B}_i(X)$ by $\mathbf{g}(X)$, that is,

$$\boldsymbol{\beta}_i(X) = X^7 \hat{B}_i(X) \text{ modulo } \mathbf{g}(X).$$

Then we can show that

$$\mathbf{S}_2(X) = \sum_{i=0}^{7} X^{16(7-i)} \boldsymbol{\beta}_i(X) \text{ modulo } \mathbf{g}(X) \tag{16.10}$$

[see Problem 16.2].

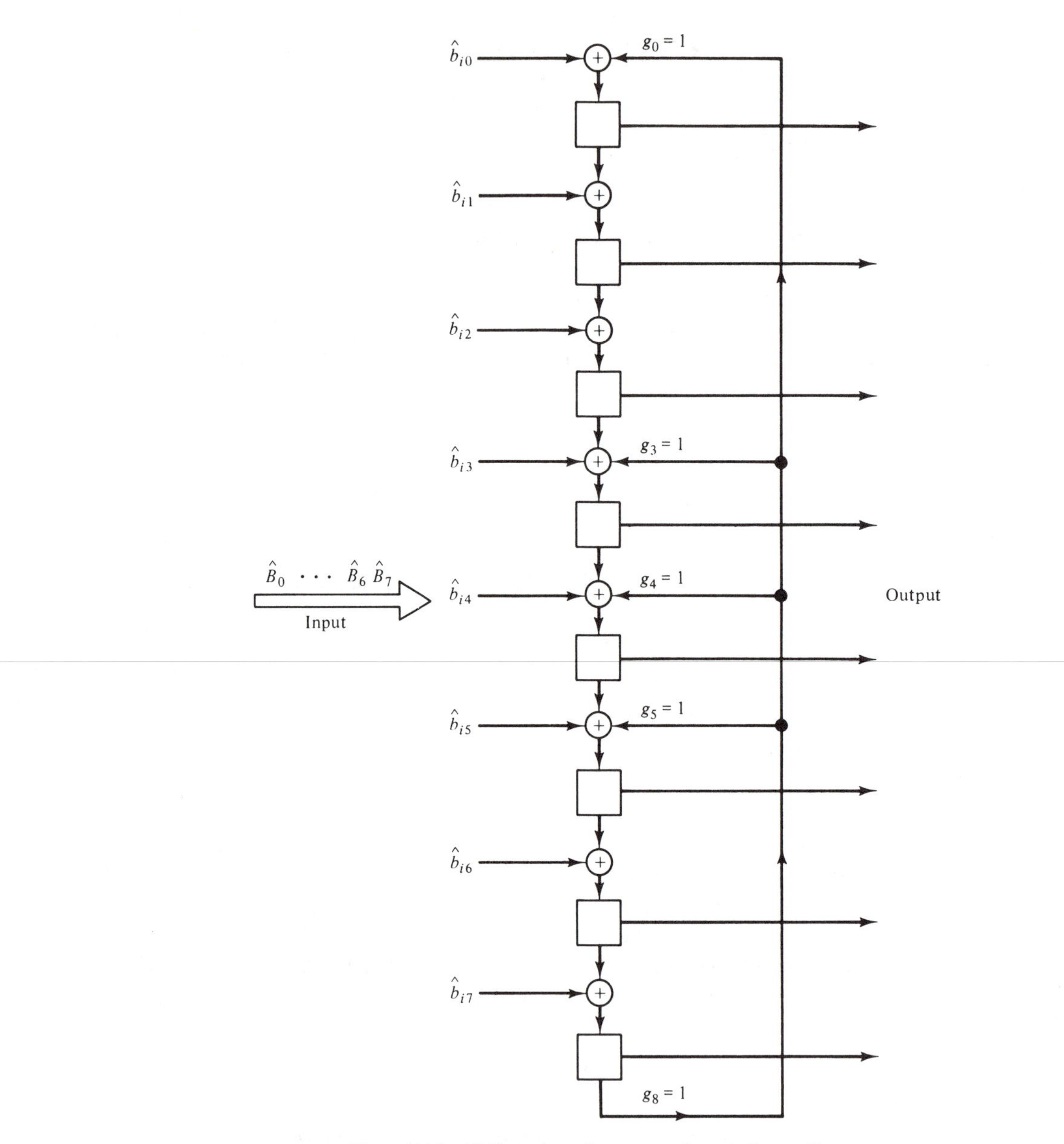

Figure 16.8 Shift register for computing syndrome $\mathbf{S}_2$.

To compute $\mathbf{S}_2(X)$, we first compute $\boldsymbol{\beta}_i(X)$ from $\hat{B}_i(X)$ for $i = 0, 1, \ldots, 8$. Then $\boldsymbol{\beta}_0(X), \boldsymbol{\beta}_1(X), \ldots, \boldsymbol{\beta}_7(X)$ are fed into a backward shift register based on $\mathbf{g}(X)$ as shown in Figure 16.9. Each shift of this backward shift register is equivalent to multiplying its contents by X^{16} and then dividing the product by $\mathbf{g}(X)$. Initially, the register contains zeros. At the end of the first shift, the register contains $\boldsymbol{\beta}_0(X)$. At the end of the second shift, the register contains

$$\boldsymbol{\rho}_2(X) = \boldsymbol{\beta}_1(X) + X^{16}\boldsymbol{\beta}_0(X) \text{ modulo } \mathbf{g}(X).$$

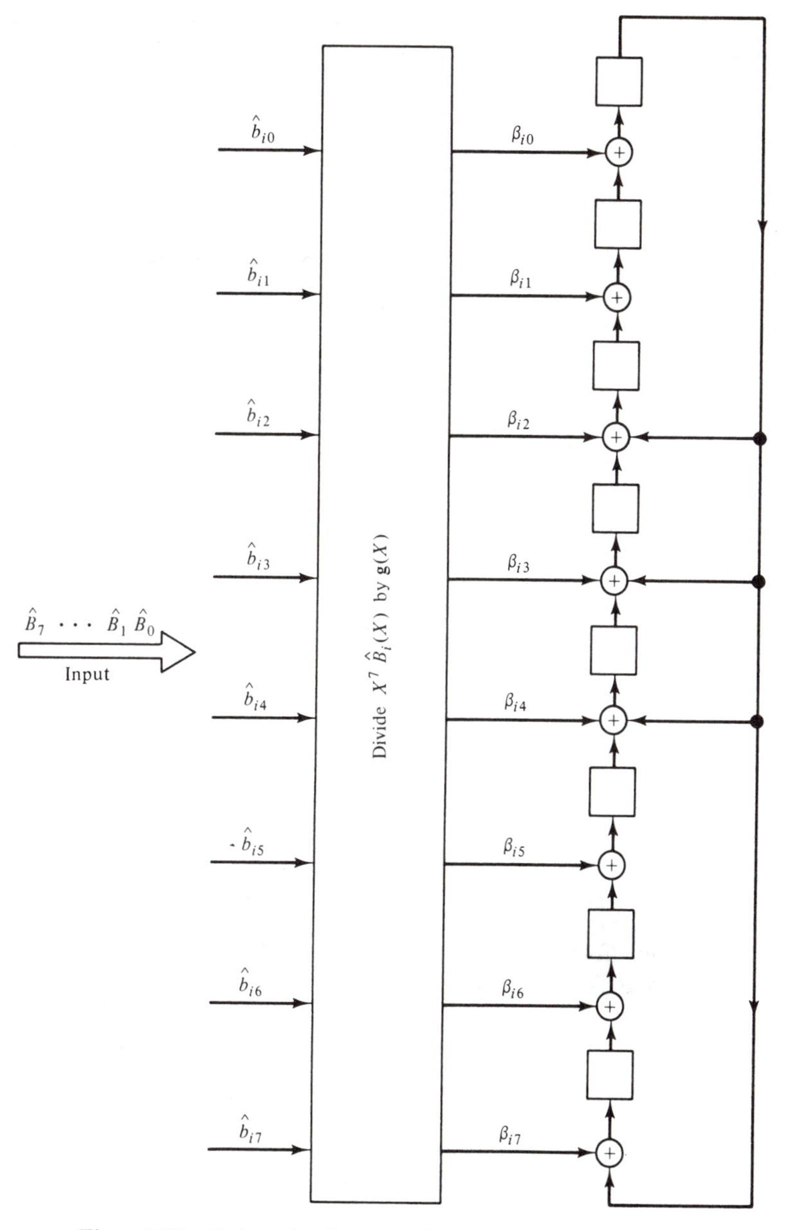

Figure 16.9 Backward shift register for computing syndrome $\mathbf{S}_2$.

At the end of the third shift, the register contains

$$\boldsymbol{\rho}_3(X) = \boldsymbol{\beta}_2(X) + X^{16}\boldsymbol{\beta}_1(X) + X^{2\times 16}\boldsymbol{\beta}_0(X) \text{ modulo } \mathbf{g}(X).$$

At the end of the eighth shift, the register will contain

$$\boldsymbol{\rho}_8(X) = \sum_{i=0}^{7} X^{16(7-i)}\boldsymbol{\beta}_i(X) \text{ modulo } \mathbf{g}(X),$$

which is the syndrome $\mathbf{S}_2(X)$.

Next, we show how to form $\boldsymbol{\beta}_i(X)$ from the byte $\hat{B}_i(X) = \hat{b}_{i0} + \hat{b}_{i1}X + \cdots + \hat{b}_{i7}X^7$. Dividing $X\hat{B}_i(X)$ by $\mathbf{g}(X)$, we obtain the following remainder:

$$\hat{b}_{i7} + \hat{b}_{i0}X + \hat{b}_{i1}X^2 + (\hat{b}_{i2} + \hat{b}_{i7})X^3 + (\hat{b}_{i3} + \hat{b}_{i7})X^4 + (\hat{b}_{i4} + \hat{b}_{i7})X^5 + \hat{b}_{i5}X^6 + \hat{b}_{i6}X^7.$$

We can readily check that the coefficients of the remainder above satisfy the following equality:

$$\begin{bmatrix} \hat{b}_{i7} \\ \hat{b}_{i0} \\ \hat{b}_{i1} \\ \hat{b}_{i2} + \hat{b}_{i7} \\ \hat{b}_{i3} + \hat{b}_{i7} \\ \hat{b}_{i4} + \hat{b}_{i7} \\ \hat{b}_{i5} \\ \hat{b}_{i6} \end{bmatrix} = \begin{bmatrix} 0 & 0 & 0 & 0 & 0 & 0 & 0 & 1 \\ 1 & 0 & 0 & 0 & 0 & 0 & 0 & 0 \\ 0 & 1 & 0 & 0 & 0 & 0 & 0 & 0 \\ 0 & 0 & 1 & 0 & 0 & 0 & 0 & 1 \\ 0 & 0 & 0 & 1 & 0 & 0 & 0 & 1 \\ 0 & 0 & 0 & 0 & 1 & 0 & 0 & 1 \\ 0 & 0 & 0 & 0 & 0 & 1 & 0 & 0 \\ 0 & 0 & 0 & 0 & 0 & 0 & 1 & 0 \end{bmatrix} \begin{bmatrix} \hat{b}_{i0} \\ \hat{b}_{i1} \\ \hat{b}_{i2} \\ \hat{b}_{i3} \\ \hat{b}_{i4} \\ \hat{b}_{i5} \\ \hat{b}_{i6} \\ \hat{b}_{i7} \end{bmatrix}.$$

Let

$$\mathbf{A} = \begin{bmatrix} 0 & 0 & 0 & 0 & 0 & 0 & 0 & 1 \\ 1 & 0 & 0 & 0 & 0 & 0 & 0 & 0 \\ 0 & 1 & 0 & 0 & 0 & 0 & 0 & 0 \\ 0 & 0 & 1 & 0 & 0 & 0 & 0 & 1 \\ 0 & 0 & 0 & 1 & 0 & 0 & 0 & 1 \\ 0 & 0 & 0 & 0 & 1 & 0 & 0 & 1 \\ 0 & 0 & 0 & 0 & 0 & 1 & 0 & 0 \\ 0 & 0 & 0 & 0 & 0 & 0 & 1 & 0 \end{bmatrix}. \tag{16.11}$$

With some effort, we can show that the coefficients of the remainder resulting from dividing $X^l\hat{B}_i(X)$ by $\mathbf{g}(X)$ with $0 \leq l \leq 7$ equal to the column entries of the column matrix

$$\mathbf{A}^l\hat{B}_i^T,$$

where $\mathbf{A}^l$ denotes the lth power of $\mathbf{A}$ and $\hat{B}_i^T$ denotes the transpose of $\hat{B}_i$ (see Problem 16.3). For $l = 7$, we find

$$\mathbf{A}^7 = \begin{bmatrix} 0 & 1 & 0 & 0 & 1 & 1 & 1 & 1 \\ 0 & 0 & 1 & 0 & 0 & 1 & 1 & 1 \\ 0 & 0 & 0 & 1 & 0 & 0 & 1 & 1 \\ 0 & 1 & 0 & 0 & 0 & 1 & 1 & 0 \\ 0 & 1 & 1 & 0 & 1 & 1 & 0 & 0 \\ 0 & 1 & 1 & 1 & 1 & 0 & 0 & 1 \\ 0 & 0 & 1 & 1 & 1 & 1 & 0 & 0 \\ 1 & 0 & 0 & 1 & 1 & 1 & 1 & 0 \end{bmatrix}.$$

Consequently, the coefficients of $\boldsymbol{\beta}_i(X) = \beta_{i0} + \beta_{i1}X + \cdots + \beta_{i7}X^7$ are equal to the column entries of the product matrix $\mathbf{A}^7\hat{B}_i^T$, that is,

$$\begin{bmatrix} \beta_{i0} \\ \beta_{i1} \\ \beta_{i2} \\ \beta_{i3} \\ \beta_{i4} \\ \beta_{i5} \\ \beta_{i6} \\ \beta_{i7} \end{bmatrix} = \begin{bmatrix} 0 & 1 & 0 & 0 & 1 & 1 & 1 & 1 \\ 0 & 0 & 1 & 0 & 0 & 1 & 1 & 1 \\ 0 & 0 & 0 & 1 & 0 & 0 & 1 & 1 \\ 0 & 1 & 0 & 0 & 0 & 1 & 1 & 0 \\ 0 & 1 & 1 & 0 & 1 & 1 & 0 & 0 \\ 0 & 1 & 1 & 1 & 1 & 0 & 0 & 1 \\ 0 & 0 & 1 & 1 & 1 & 1 & 0 & 0 \\ 1 & 0 & 0 & 1 & 1 & 1 & 1 & 0 \end{bmatrix} \begin{bmatrix} \hat{b}_{i0} \\ \hat{b}_{i1} \\ \hat{b}_{i2} \\ \hat{b}_{i3} \\ \hat{b}_{i4} \\ \hat{b}_{i5} \\ \hat{b}_{i6} \\ \hat{b}_{i7} \end{bmatrix}.$$

Multiplying out the product above, we obtain the following linear sums:

$$\begin{aligned} \beta_{i0} &= \hat{b}_{i1} + \hat{b}_{i4} + \hat{b}_{i5} + \hat{b}_{i6} + \hat{b}_{i7}, \\ \beta_{i1} &= \hat{b}_{i2} + \hat{b}_{i5} + \hat{b}_{i6} + \hat{b}_{i7}, \\ \beta_{i2} &= \hat{b}_{i3} + \hat{b}_{i6} + \hat{b}_{i7}, \\ \beta_{i3} &= \hat{b}_{i1} + \hat{b}_{i5} + \hat{b}_{i6}, \\ \beta_{i4} &= \hat{b}_{i1} + \hat{b}_{i2} + \hat{b}_{i4} + \hat{b}_{i5}, \\ \beta_{i5} &= \hat{b}_{i1} + \hat{b}_{i2} + \hat{b}_{i3} + \hat{b}_{i4} + \hat{b}_{i7}, \\ \beta_{i6} &= \hat{b}_{i2} + \hat{b}_{i3} + \hat{b}_{i4} + \hat{b}_{i5}, \\ \beta_{i7} &= \hat{b}_{i0} + \hat{b}_{i3} + \hat{b}_{i4} + \hat{b}_{i5} + \hat{b}_{i6}. \end{aligned} \tag{16.12}$$

Based on (16.12), $\mathbf{S}_2(X)$ can be computed by the circuit shown in Figure 16.10.

Decoding Operation

After $\mathbf{S}_1$ and $\mathbf{S}_2$ have been computed, the decoder proceeds to correct errors (if any). If $\mathbf{S}_1 = \mathbf{S}_2 = 0$, the code array read from the tape is assumed to be error-free. If $\mathbf{S}_1$ or $\mathbf{S}_2$ are not equal to zero, errors are detected. There are two modes of error recovery.

Mode I Correction of Single-Track Error

Suppose that the ith track is in error. Let $\mathbf{e}_i$ be the error vector. Then

$$\hat{Z}_i(X) = \mathbf{Z}_i(X) + \mathbf{e}_i(X)$$

and

$$\hat{Z}_j(X) = \mathbf{Z}_i(X) \qquad \text{for } j \neq i.$$

It follows from (16.7) and (16.9) that

$$\mathbf{S}_1(X) = \mathbf{e}_i(X) \tag{16.13}$$

and

$$\mathbf{S}_2(X) = \begin{cases} X^i\mathbf{e}_i(X) \text{ modulo } \mathbf{g}(X) & \text{if } 0 \leq i \leq 7 \\ 0 & \text{if } i = 8. \end{cases} \tag{16.14}$$

If $\mathbf{S}_1 \neq 0$ and $\mathbf{S}_2 = 0$, the errors are in the eighth track (the overall parity-check track). In this case, $\hat{B}_7, \hat{B}_6, \ldots, \hat{B}_1$ are error-free and are delivered to the user. If

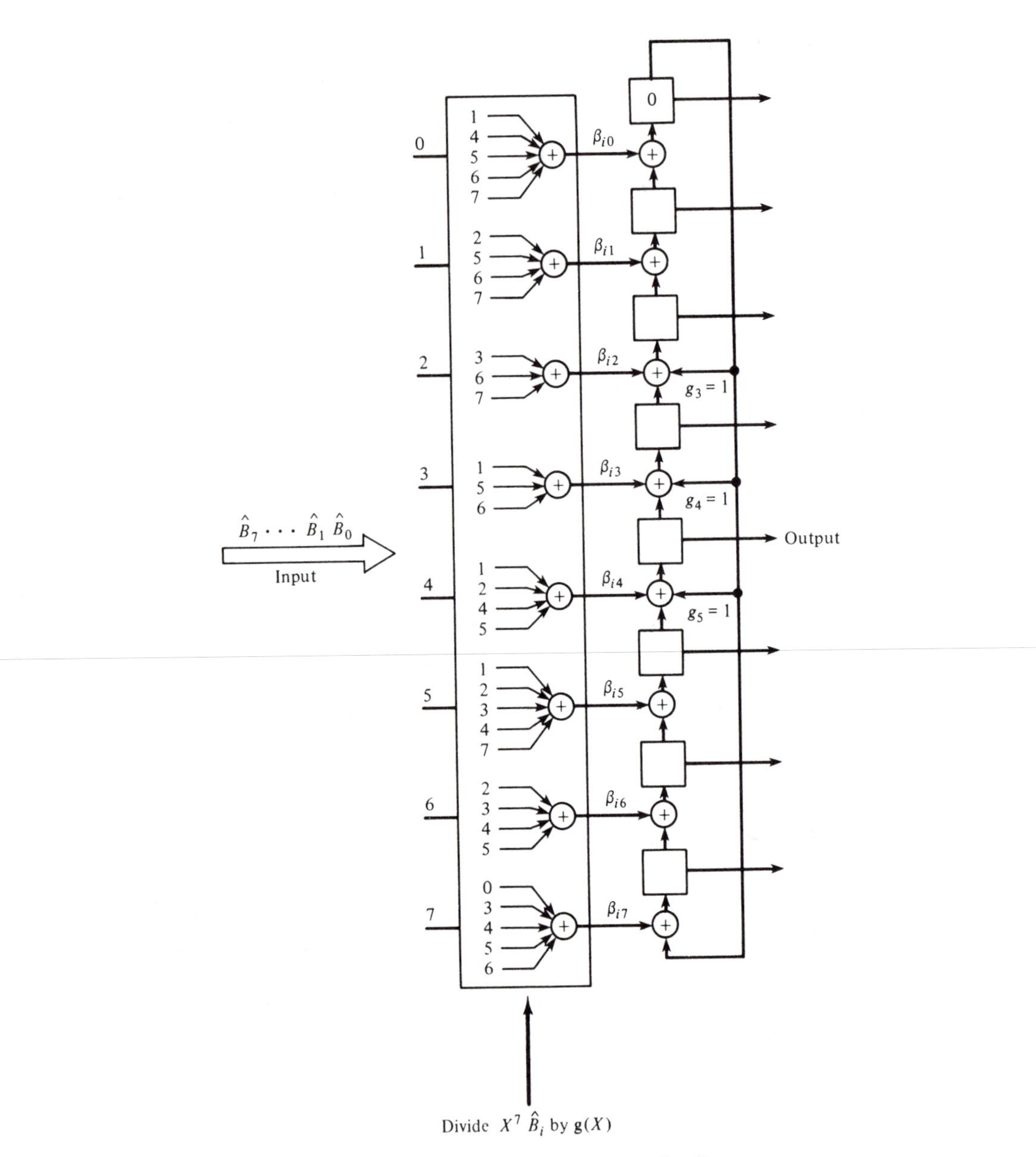

Figure 16.10 Circuit for computing $\mathbf{S}_2$.

$\mathbf{S}_1 \neq 0$ and $\mathbf{S}_2 \neq 0$, then $\mathbf{S}_1$ is identical to the error pattern $\mathbf{e}_i(X)$ and $\mathbf{S}_2$ tells which track contains the errors. If the circuit of Figure 16.8 is used to compute $\mathbf{S}_2(X)$, after $\mathbf{S}_2(X)$ has been formed, the shift register is then shifted without external inputs. After $17 - i$ shifts, the register should contain

$$\mathbf{e}_i(X) = X^{17-i}\,[X^i\mathbf{e}_i(X)] \text{ modulo } \mathbf{g}(X),$$

which matches $\mathbf{S}_1(X) = \mathbf{e}_i(X)$. When this event occurs, the location of the erroneous track is

$$17 - (17 - i) = i.$$

If $\mathbf{S}_2(X) \neq 0$ and the contents of the register do not match $\mathbf{S}_1$ after a maximum of 17 shifts, there are two or more tracks in error. In this event, uncorrectable errors are detected.

If the circuit of Figure 16.10 is used to compute $\mathbf{S}_2(X)$, a faster decoding operation can be obtained. After $\mathbf{S}_2(X)$ has been formed, the backward shift register is shifted without external inputs. After i shifts, the register should contain

$$\mathbf{e}_i(X) = X^{16i}[X^i \mathbf{e}_i(X)] \text{ modulo } \mathbf{g}(X),$$

which matches $\mathbf{S}_1(X) = \mathbf{e}_i(X)$. Then i is the location of the erroneous track. If $\mathbf{S}_2(X) \neq 0$ and if the contents of the register do not match $\mathbf{S}_1$ after a maximum of seven shifts, more than one track is in error. Therefore, the circuit of Figure 16.10 results in a faster decoding operation.

Mode II Correction of Double-Track Errors

Suppose that it is known (e.g., by a loss of signal in tape-reading head) that tracks i and j are in error (or are erased), where i and j are given by the track pointers. Assume that $i < j$. Let $\mathbf{e}_i(X)$ and $\mathbf{e}_j(X)$ denote the error patterns on track i and track j, respectively. Then it follows from (16.7) and (16.9) that

$$\mathbf{S}_1(X) = \mathbf{e}_i(X) + \mathbf{e}_j(X) \tag{16.15}$$

and

$$\mathbf{S}_2(X) = \begin{cases} X^i \mathbf{e}_i(X) + X^j \mathbf{e}_j(X) \text{ modulo } \mathbf{g}(X) & \text{if } j \leq 7 \\ X^i \mathbf{e}_i(X) \text{ modulo } \mathbf{g}(X) & \text{if } j = 8. \end{cases} \tag{16.16}$$

[If $\mathbf{e}_i(X) = 0$ or $\mathbf{e}_j(X) = 0$, we have the single-track error case.]

Suppose that the circuit shown in Figure 16.10 is used to compute $\mathbf{S}_2(X)$. After $\mathbf{S}_2(X)$ has been formed, the register is shifted i times without external inputs. Let $\mathbf{S}_2^{(i)}(X)$ denote the contents of the register. Then

$$\mathbf{S}_2^{(i)}(X) = \begin{cases} X^{16i}[X^i \mathbf{e}_i(X) + X^j \mathbf{e}_j(X)] \text{ modulo } \mathbf{g}(X) & \text{if } j \leq 7 \\ X^{17i} \mathbf{e}_i(X) \text{ modulo } \mathbf{g}(X) & \text{if } j = 8. \end{cases}$$

Since $X^{17i} + 1$ is divisible by $\mathbf{g}(X)$,

$$\mathbf{S}_2^{(i)}(X) = \begin{cases} \mathbf{e}_i(X) + X^{j-i} \mathbf{e}_j(X) \text{ modulo } \mathbf{g}(X) & \text{if } j \leq 7 \\ \mathbf{e}_i(X) & \text{if } j = 8. \end{cases} \tag{16.17}$$

Adding (16.15) and (16.17), we obtain

$$\mathbf{S}_2^{(i)}(X) + \mathbf{S}_1(X) = \begin{cases} (X^{j-1} + 1)\mathbf{e}_j(X) \text{ modulo } \mathbf{g}(X) & \text{if } j \leq 7 \\ \mathbf{e}_j(X) & \text{if } j = 8. \end{cases} \tag{16.18}$$

It can be shown that there is one and only one $\mathbf{e}_j(X)$ of degree 7 or less which satisfies the first relation of (16.17) (see Problem 16.4).

Since $j - i \leq 7$ and $\mathbf{g}(X)$ is an irreducible polynomial of degree 8, $X^{j-i} + 1$ and $\mathbf{g}(X)$ are relatively prime. Then there exist two polynomials, $\mathbf{c}_{j-i}(X)$ and $\mathbf{d}_{j-i}(X)$, over GF(2) such that

$$\mathbf{c}_{j-i}(X)(X^{j-i} + 1) + \mathbf{d}_{j-i}(X)\mathbf{g}(X) = 1. \tag{16.19}$$

Multiplying both sides of (16.19) by $\mathbf{S}_2^{(i)}(X) + \mathbf{S}_1(X)$ and rearranging it, we have

$$\begin{aligned}\mathbf{c}_{j-i}(X)[\mathbf{S}_2^{(i)}(X) + \mathbf{S}_1(X)](X^{j-i} + 1) = \mathbf{d}_{j-i}(X)[\mathbf{S}_2^{(i)}(X) + \mathbf{S}_1(X)]\mathbf{g}(X) \\ + [\mathbf{S}_2^{(i)}(X) + \mathbf{S}_1(X)].\end{aligned} \tag{16.20}$$

Therefore,

$$\mathbf{S}_2^{(i)}(X) + \mathbf{S}_1(X) = \mathbf{c}_{j-i}(X)[\mathbf{S}_2^{(i)}(X) + \mathbf{S}_1(X)](X^{j-i} + 1) \text{ modulo } \mathbf{g}(X). \tag{16.21}$$

Comparing (16.18) and (16.21), we obtain the following conclusions:

1. If the degree of $\mathbf{c}_{j-i}(X)[\mathbf{S}_2^{(i)}(X) + \mathbf{S}_1(X)]$ is less than 8,
$$\mathbf{e}_j(X) = \mathbf{c}_{j-i}(X)[\mathbf{S}_2^{(i)}(X) + \mathbf{S}_1(X)].$$
2. If the degree of $\mathbf{c}_{j-i}(X)[\mathbf{S}_2^{(i)}(X) + \mathbf{S}_1(X)]$ is 8 or greater, then
$$\mathbf{e}_j(X) = \mathbf{c}_{j-i}(X)[\mathbf{S}_2^{(i)}(X) + \mathbf{S}_1(X)] \text{ modulo } \mathbf{g}(X).$$

For different values of $j - i$, $\mathbf{c}_{j-i}(X)$'s are listed in Table 16.2.

TABLE 16.2

$j - i$	$\mathbf{c}_{j-i}(X)$
1	$X^3 + X^5 + X^6 + X^7$
2	$X^3 + X^4 + X^6$
3	$1 + X^2 + X^3 + X^7$
4	$X + X^5 + X^7$
5	$1 + X + X^4$
6	$X + X^2 + X^4 + X^5 + X^6$
7	$1 + X^2 + X^3 + X^4 + X^5$
8	1

Now we may summarize the error-correction process as follows. First, $\mathbf{S}_1(X)$ and $\mathbf{S}_2(X)$ are formed [the circuit of Figure 16.10 is used to form $\mathbf{S}_2(X)$]. Then shift $\mathbf{S}_2(X)$ i times to form $\mathbf{S}_2^{(i)}(X)$, and add $\mathbf{S}_2^{(i)}$ to $\mathbf{S}_1(X)$. If $j = 8$, then $\mathbf{e}_j(X) = \mathbf{S}_1(X) + \mathbf{S}_2^{(i)}(X)$. If $j \leq 7$, $\mathbf{S}_1(X) + \mathbf{S}_2^{(i)}(X)$ is multiplied by $\mathbf{c}_{j-i}(X)$. If the degree of $\mathbf{c}_{j-i}(X)[\mathbf{S}_1(X) + \mathbf{S}_2^{(i)}(X)]$ is less than 8,

$$\mathbf{e}_j(X) = \mathbf{c}_{j-i}(X)[\mathbf{S}_1(X) + \mathbf{S}_2^{(i)}(X)];$$

otherwise,

$$\mathbf{e}_j(X) = \mathbf{c}_{j-i}(X)[\mathbf{S}_1(X) + \mathbf{S}_2^{(i)}(X)] \text{ modulo } \mathbf{g}(X).$$

Once $\mathbf{e}_j(X)$ is found, then $\mathbf{e}_i(X) = \mathbf{S}_1(X) + \mathbf{e}_j(X)$. Finally, the errors are corrected by adding $\mathbf{e}_i(X)$ to track i and $\mathbf{e}_j(X)$ to track j.

A circuit for determining $\mathbf{e}_j(X)$ can be implemented by using the fact that the coefficients of the remainder resulting from dividing $[\mathbf{S}_1(X) + \mathbf{S}_2^{(i)}(X)]X^l$ by $\mathbf{g}(X)$

are equal to the column entries of the following product:

$$\mathbf{A}^l[\mathbf{S}_1 + \mathbf{S}_2^{(i)}]^T,$$

where $\mathbf{A}$ is given by (16.11) and $[\mathbf{S}_1 + \mathbf{S}_2^{(i)}]^T$ is the transpose of the vector $\mathbf{S}_1 + \mathbf{S}_2^{(i)}$. Let

$$\mathbf{c}_{j-i}(X) = \mathbf{c}_{j-i,0} + \mathbf{c}_{j-i,1}X + \cdots + c_{j-i,m}X^m.$$

Then the coefficients of the remainder resulting from dividing $\mathbf{c}_{j-i}(X)[\mathbf{S}_1(X) + \mathbf{S}_2^{(i)}(X)]$ by $\mathbf{g}(X)$ are equal to the column entries of the product

$$\mathbf{M}_{j-i}[\mathbf{S}_1 + \mathbf{S}_2^{(i)}]^T,$$

where

$$\mathbf{M}_{j-i} = c_{j-i,0}\mathbf{I} + c_{j-i,1}\mathbf{A} + \cdots + c_{j-i,m}\mathbf{A}^m \tag{16.22}$$

($\mathbf{I}$ is an 8×8 identity matrix). Table 16.3 gives the matrices $\mathbf{M}_{j-i}$ for all seven possible values of $j - i$. Consequently, we have

$$\mathbf{e}_j = \mathbf{M}_{j-i}[\mathbf{S}_1 + \mathbf{S}_2^{(i)}]^T, \tag{16.23}$$

where $\mathbf{M}_{j-i} = \mathbf{I}$ if $j - i = 8$. The right-hand side of (16.23) gives eight linear sums

TABLE 16.3 MODE II DECODING MATRICES

$$\mathbf{M}_1 = \begin{bmatrix} 0&1&1&1&1&1&1&1 \\ 0&0&1&1&1&1&1&1 \\ 0&0&0&1&1&1&1&1 \\ 1&1&1&1&0&0&0&0 \\ 0&0&0&0&0&1&1&1 \\ 1&1&1&1&1&1&0&0 \\ 1&1&1&1&1&1&1&0 \\ 1&1&1&1&1&1&1&1 \end{bmatrix} \quad \mathbf{M}_2 = \begin{bmatrix} 0&0&1&0&1&0&1&0 \\ 0&0&0&1&0&1&0&1 \\ 0&0&0&0&1&0&1&0 \\ 1&0&1&0&1&1&1&1 \\ 1&1&1&1&1&1&0&1 \\ 0&1&0&1&0&1&0&0 \\ 1&0&1&0&1&0&1&0 \\ 0&1&0&1&0&1&0&1 \end{bmatrix} \quad \mathbf{M}_3 = \begin{bmatrix} 1&1&0&0&1&0&0&1 \\ 0&1&1&0&0&1&0&0 \\ 1&0&1&1&0&0&1&0 \\ 1&0&0&1&0&0&0&0 \\ 0&0&0&0&0&0&0&1 \\ 0&1&0&0&1&0&0&1 \\ 0&0&1&0&0&1&0&0 \\ 1&0&0&1&0&0&1&0 \end{bmatrix}$$

$$\mathbf{M}_4 = \begin{bmatrix} 0&1&0&1&1&1&0&1 \\ 1&0&1&0&1&1&1&0 \\ 0&1&0&1&0&1&1&1 \\ 0&1&1&1&0&1&1&0 \\ 0&1&1&0&0&1&1&0 \\ 1&1&1&0&1&1&1&0 \\ 0&1&1&1&0&1&1&1 \\ 1&0&1&1&1&0&1&1 \end{bmatrix} \quad \mathbf{M}_5 = \begin{bmatrix} 1&0&0&0&1&0&0&0 \\ 1&1&0&0&0&1&0&0 \\ 0&1&1&0&0&0&1&0 \\ 0&0&1&1&1&0&0&1 \\ 1&0&0&1&0&1&0&0 \\ 0&1&0&0&0&0&1&0 \\ 0&0&1&0&0&0&0&1 \\ 0&0&0&1&0&0&0&0 \end{bmatrix} \quad \mathbf{M}_6 = \begin{bmatrix} 0&0&1&1&1&1&1&0 \\ 1&0&0&1&1&1&1&1 \\ 1&1&0&0&1&1&1&1 \\ 0&1&0&1&1&0&0&1 \\ 1&0&0&1&0&0&1&0 \\ 1&1&1&1&0&1&1&1 \\ 1&1&1&1&1&0&1&1 \\ 0&1&1&1&1&1&0&1 \end{bmatrix}$$

$$\mathbf{M}_7 = \begin{bmatrix} 1&0&0&1&1&1&0&0 \\ 0&1&0&0&1&1&1&0 \\ 1&0&1&0&0&1&1&1 \\ 1&1&0&0&1&1&1&1 \\ 1&1&1&1&1&0&1&1 \\ 1&1&1&0&0&0&0&1 \\ 0&1&1&1&0&0&0&0 \\ 0&0&1&1&1&0&0&0 \end{bmatrix} \quad \mathbf{M}_8 = \mathbf{I}$$

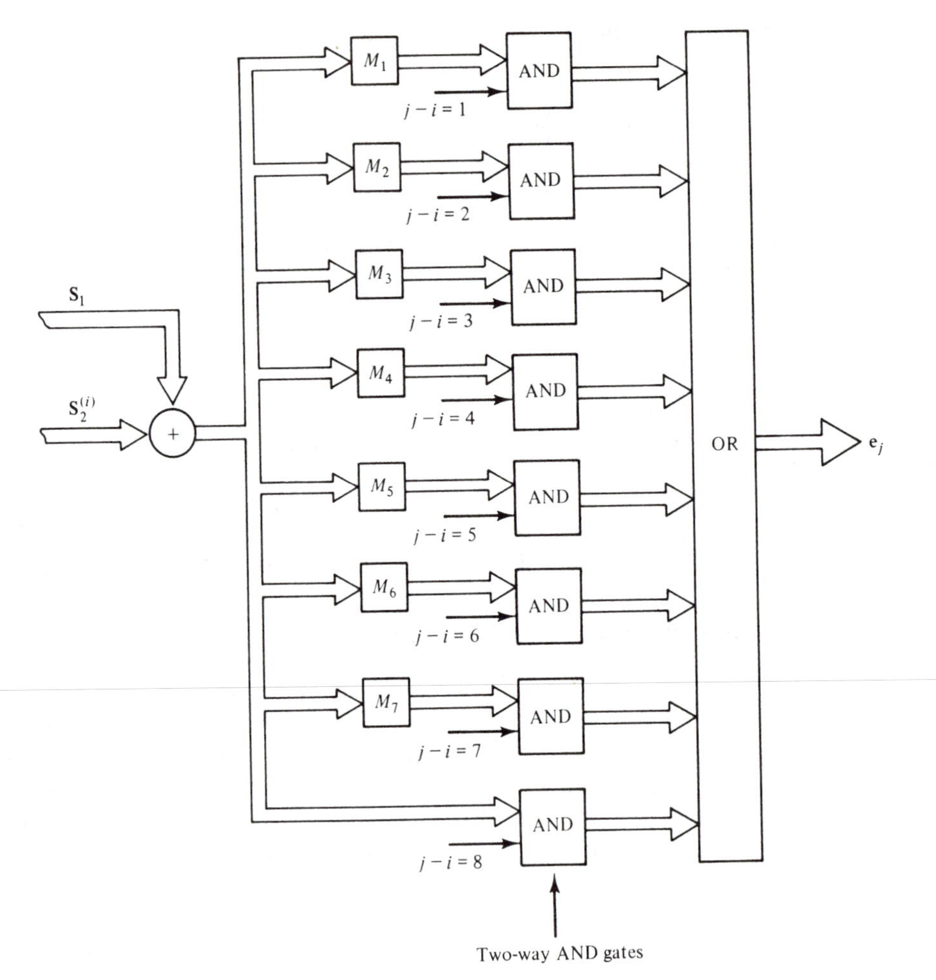

Figure 16.11 Double-erasure decoding circuit.

of the bits of $\mathbf{S}_1 + \mathbf{S}_2^{(i)}$, which can be implemented by eight modulo-2 adders. Figure 16.11 gives the block diagram of the circuit which computes $\mathbf{e}_i$ and $\mathbf{e}_j$ from $\mathbf{S}_1$ and $\mathbf{S}_2^{(i)}$.

Another interesting error control system for improving the reliability of tape storage was reported in Reference 3. This system is a software package and is implemented on the Univac 1108 computer.

16.3 ERROR CONTROL IN IBM 3850 MASS STORAGE SYSTEM

The IBM 3850 mass storage system (MSS) consists of an array of data cartridges, each cartridge is approximately 1.9 inches in diameter and 3.5 inches long and can contain up to 50.4 million 8-bit bytes of data [4]. Each cartridge holds a spool of magnetic tape of 2.7 inches wide and 64 feet long.

The 3850 MSS does not have the conventional parallel multitrack data format

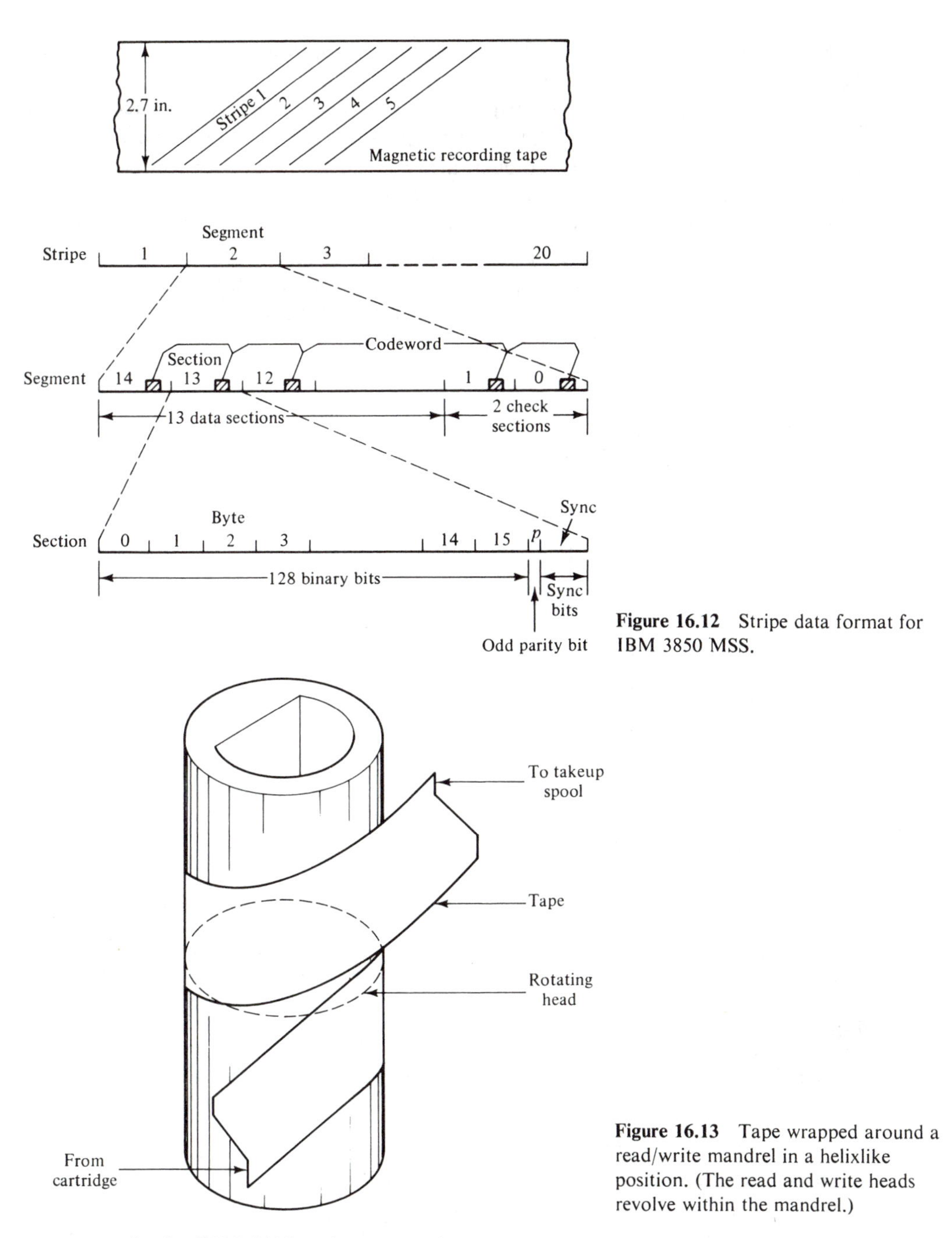

Figure 16.12 Stripe data format for IBM 3850 MSS.

Figure 16.13 Tape wrapped around a read/write mandrel in a helixlike position. (The read and write heads revolve within the mandrel.)

as in the IBM 3420 series tape units. Data are recorded in short slanted stripes across the tape as shown in Figure 16.12. The tape follows a helical path around a read/write mandrel and is stepped in position from one slanted stripe to the next over a circular slit as shown in Figure 16.13. The read and write heads revolve within the mandrel.

In this section we describe the error-correcting scheme used in the IBM 3850

MSS. The scheme was devised by Patel [5]. For other important features of the system, the reader is referred to References 4 and 5.

The data format of a 3850 stripe is illustrated in Figure 16.12. Each stripe is divided into 20 segments. The segment is a separate entity and can be decoded independently. Each segment is a rectangular array consisting of 13 data sections and two parity-check sections as shown in Figure 16.14. Each section consists of 16 8-bit bytes (a total of 128 bits). The 15 bytes in each column form a code word from a (15, 13) BCH code with symbols form the Galois field GF(2^8) [each 8-bit byte represents an element in GF(2^8)]. Therefore, each segment is a code array consisting of 16 interleaved code words. In the write operation, the 16 code words are first formed in a temporary storage and then recorded onto the tape section by section.

Codeword ↓ (column 2)

Section	0	1	2	3	4	--- k ---	14	15	
14	$B_{14\text{-}0}$	$B_{14\text{-}1}$	$B_{14\text{-}2}$	$B_{14\text{-}3}$	$B_{14\text{-}4}$	$B_{14\text{-}k}$	$B_{14\text{-}14}$	$B_{14\text{-}15}$	13 data sections
13	$B_{13\text{-}0}$	$B_{13\text{-}1}$	$B_{13\text{-}2}$	$B_{13\text{-}3}$	$B_{13\text{-}4}$	$B_{13\text{-}k}$	$B_{13\text{-}14}$	$B_{13\text{-}15}$	
12	$B_{12\text{-}0}$	$B_{12\text{-}1}$	$B_{12\text{-}2}$	$B_{12\text{-}3}$	$B_{12\text{-}4}$	$B_{12\text{-}k}$	$B_{12\text{-}14}$	$B_{12\text{-}15}$	
11	$B_{11\text{-}0}$	$B_{11\text{-}1}$	$B_{11\text{-}2}$	$B_{11\text{-}3}$	$B_{11\text{-}4}$	$B_{11\text{-}k}$	$B_{11\text{-}14}$	$B_{11\text{-}15}$	
⋮						⋮			
j	$B_{j\text{-}0}$	$B_{j\text{-}1}$	$B_{j\text{-}2}$	$B_{j\text{-}3}$	$B_{j\text{-}4}$	$B_{j\text{-}k}$	$B_{j\text{-}14}$	$B_{j\text{-}15}$	
⋮						⋮			
2	$B_{2\text{-}0}$	$B_{2\text{-}1}$	$B_{2\text{-}2}$	$B_{2\text{-}3}$	$B_{2\text{-}4}$	$B_{2\text{-}k}$	$B_{2\text{-}14}$	$B_{2\text{-}15}$	
1	$B_{1\text{-}0}$	$B_{1\text{-}1}$	$B_{1\text{-}2}$	$B_{1\text{-}3}$	$B_{1\text{-}4}$	$B_{1\text{-}k}$	$B_{1\text{-}14}$	$B_{1\text{-}15}$	2 parity-check sections
0	$B_{0\text{-}0}$	$B_{0\text{-}1}$	$B_{0\text{-}2}$	$B_{0\text{-}3}$	$B_{0\text{-}4}$	$B_{0\text{-}k}$	$B_{0\text{-}14}$	$B_{0\text{-}15}$	

Figure 16.14 A segment: 15 sections formed with 16 interleaved codewords. Sections are appended to each other to form a segment.

The Code

Consider the Galois field GF(2^8) which is constructed based on the primitive polynomial

$$\mathbf{p}(X) = 1 + X + X^3 + X^5 + X^8.$$

Let α be a primitive element in GF(2^8) such that

$$\mathbf{p}(\alpha) = 1 + \alpha + \alpha^3 + \alpha^5 + \alpha^8 = 0 \tag{16.24}$$

[i.e., $\mathbf{p}(X)$ is the minimal polynomial of α]. Then

$$\alpha^{2^8-1} = \alpha^{255} = 1.$$

Let

$$\gamma = \alpha^{68}.$$

Then the order of γ is 15 (i.e., $\gamma^{15} = 1$). We can readily check that

$$0, 1, \gamma, \gamma^2, \ldots, \gamma^{14}$$

form the field GF(2^4), a subfield of GF(2^8).

The code used in IBM 3850 MSS is the (15, 13) BCH code with symbols from GF(2^8) and is generated by

$$\begin{aligned}\mathbf{g}(X) &= (X + 1)(X + \gamma)\\ &= \gamma + (1 + \gamma)X + X^2.\end{aligned}$$

This code has minimum distance 3 and is capable of correcting any single-symbol error. Since each symbol in GF(2^8) represented by an 8-bit byte, the code is capable of correcting any 1-byte errors.

Let $\mathbf{B}_2, \mathbf{B}_3, \ldots, \mathbf{B}_{14}$ be the 13 data symbols to be encoded. Then the two parity symbols $\mathbf{B}_0$ and $\mathbf{B}_1$ are the coefficients of the remainder resulting from dividing the message polynomial

$$\mathbf{B}_2 X^2 + \mathbf{B}_3 X^3 + \cdots + \mathbf{B}_{14} X^{14}$$

by $\mathbf{g}(X)$. Thus, the code word is

$$\mathbf{v}(X) = \mathbf{B}_0 + \mathbf{B}_1 X + \mathbf{B}_2 X^2 + \cdots + \mathbf{B}_{14} X^{14},$$

which has 1 and γ as roots. Therefore,

$$\mathbf{v}(1) = \sum_{i=0}^{14} \mathbf{B}_i = 0, \tag{16.25}$$

$$\mathbf{v}(\gamma) = \sum_{i=0}^{14} \mathbf{B}_i \gamma^i = 0. \tag{16.26}$$

The encoding circuit for his code is shown in Figure 16.15. As soon as the 13 data symbols have been shifted into the register, the parity-check symbols $\mathbf{B}_0$ and $\mathbf{B}_1$ are in the low-order and high-order stages.

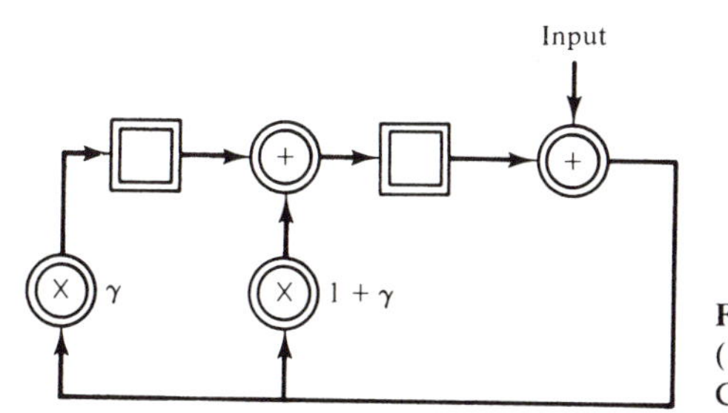

Figure 16.15 Encoding circuit for the (15, 13) BCH code with symbols from GF(2^8).

The multiplications of a field element β in GF(2^8) by γ and $1 + \gamma$ can be carried out in terms of modulo-2 operations. Any field element β in GF(2^8) can be expressed as the following sum:

$$\beta = b_0 + b_1\alpha + \cdots + b_7\alpha^7,$$

where $(b_0, b_1, \ldots, b_7)$ is the vector representation of β. Consider the multiplication of β by α,

$$\beta\alpha = b_0\alpha + b_1\alpha^2 + \cdots + b_7\alpha^8.$$

Using the identity of (16.24), we obtain the following:

$$\beta\alpha = b_7 + (b_0 + b_7)\alpha + b_1\alpha^2 + (b_2 + b_7)\alpha^3 + b_3\alpha^4 + (b_4 + b_7)\alpha^5 + b_5\alpha^6 + b_6\alpha^7.$$

Hence, the 8-bit byte representation of $\beta\alpha$ is

$$(b_7, b_0 + b_7, b_1, b_2 + b_7, b_3, b_4 + b_7, b_5, b_6).$$

We can readily see that it satisfies the following equalities:

$$\begin{bmatrix} b_7 \\ b_0 + b_7 \\ b_1 \\ b_2 + b_7 \\ b_3 \\ b_4 + b_7 \\ b_5 \\ b_6 \end{bmatrix} = \begin{bmatrix} 0 & 0 & 0 & 0 & 0 & 0 & 0 & 1 \\ 1 & 0 & 0 & 0 & 0 & 0 & 0 & 1 \\ 0 & 1 & 0 & 0 & 0 & 0 & 0 & 0 \\ 0 & 0 & 1 & 0 & 0 & 0 & 0 & 1 \\ 0 & 0 & 0 & 1 & 0 & 0 & 0 & 0 \\ 0 & 0 & 0 & 0 & 1 & 0 & 0 & 1 \\ 0 & 0 & 0 & 0 & 0 & 1 & 0 & 0 \\ 0 & 0 & 0 & 0 & 0 & 0 & 1 & 0 \end{bmatrix} \begin{bmatrix} b_0 \\ b_1 \\ b_2 \\ b_3 \\ b_4 \\ b_5 \\ b_6 \\ b_7 \end{bmatrix}.$$

Let

$$\mathbf{A} = \begin{bmatrix} 0 & 0 & 0 & 0 & 0 & 0 & 0 & 1 \\ 1 & 0 & 0 & 0 & 0 & 0 & 0 & 1 \\ 0 & 1 & 0 & 0 & 0 & 0 & 0 & 0 \\ 0 & 0 & 1 & 0 & 0 & 0 & 0 & 1 \\ 0 & 0 & 0 & 1 & 0 & 0 & 0 & 0 \\ 0 & 0 & 0 & 0 & 1 & 0 & 0 & 1 \\ 0 & 0 & 0 & 0 & 0 & 1 & 0 & 0 \\ 0 & 0 & 0 & 0 & 0 & 0 & 1 & 0 \end{bmatrix}. \tag{16.27}$$

Then the 8-bit byte representation of $\beta\alpha^{68} = \beta\gamma$, $(c_0, c_1, \ldots, c_7)$, satisfies the following equality:

$$\begin{bmatrix} c_0 \\ c_1 \\ c_2 \\ c_3 \\ c_4 \\ c_5 \\ c_6 \\ c_7 \end{bmatrix} = \mathbf{A}^{68} \begin{bmatrix} b_0 \\ b_1 \\ b_2 \\ b_3 \\ b_4 \\ b_5 \\ b_6 \\ b_7 \end{bmatrix}, \tag{16.28}$$

where

$$\mathbf{A}^{68} = \begin{bmatrix} 0 & 0 & 0 & 0 & 1 & 0 & 0 & 0 \\ 1 & 0 & 0 & 0 & 1 & 1 & 0 & 0 \\ 0 & 1 & 0 & 0 & 0 & 1 & 1 & 0 \\ 0 & 0 & 1 & 0 & 1 & 0 & 1 & 1 \\ 1 & 0 & 0 & 1 & 0 & 1 & 0 & 1 \\ 0 & 1 & 0 & 0 & 0 & 0 & 1 & 0 \\ 0 & 0 & 1 & 0 & 0 & 0 & 0 & 1 \\ 0 & 0 & 0 & 1 & 0 & 0 & 0 & 0 \end{bmatrix}. \tag{16.29}$$

From (16.28) and (16.29), we obtain the following modulo-2 sums:

$$c_0 = b_4,$$
$$c_1 = b_0 + b_4 + b_5,$$
$$c_2 = b_1 + b_5 + b_6,$$
$$c_3 = b_2 + b_4 + b_6 + b_7,$$
$$c_4 = b_0 + b_3 + b_5 + b_7,$$
$$c_5 = b_1 + b_6,$$
$$c_6 = b_2 + b_7,$$
$$c_7 = b_3.$$

Therefore, multiplication of β by $\gamma = \alpha^{68}$ can be accomplished by the circuit shown in Figure 16.16. Consequently, we obtain an encoding circuit for the (15, 13) BCH over GF(2^8) as shown in Figure 16.17.

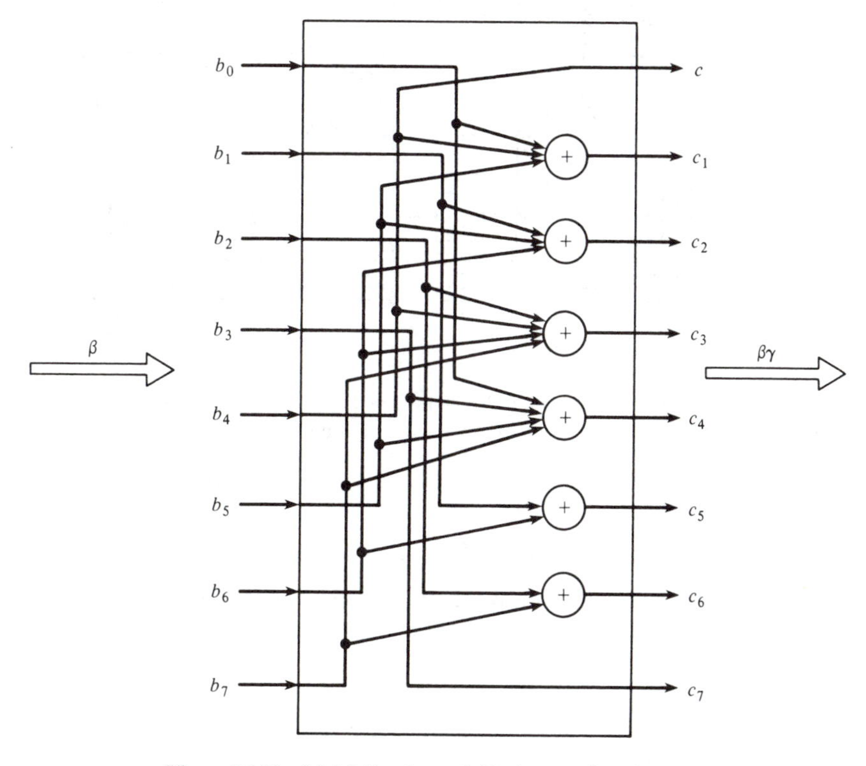

Figure 16.16 Multiplication a field element β in GF(2^8) by γ.

Syndrome Computation

When data are read from the tape, all 16 read words of a segment are stored in a temporary storage pending possible error correction. The decoding process is carried out by applying the decoding algorithm to each of the 16 read words independently. We will show that the code is capable of correcting any single-byte errors in unknown position and two-byte errors in indicated positions.

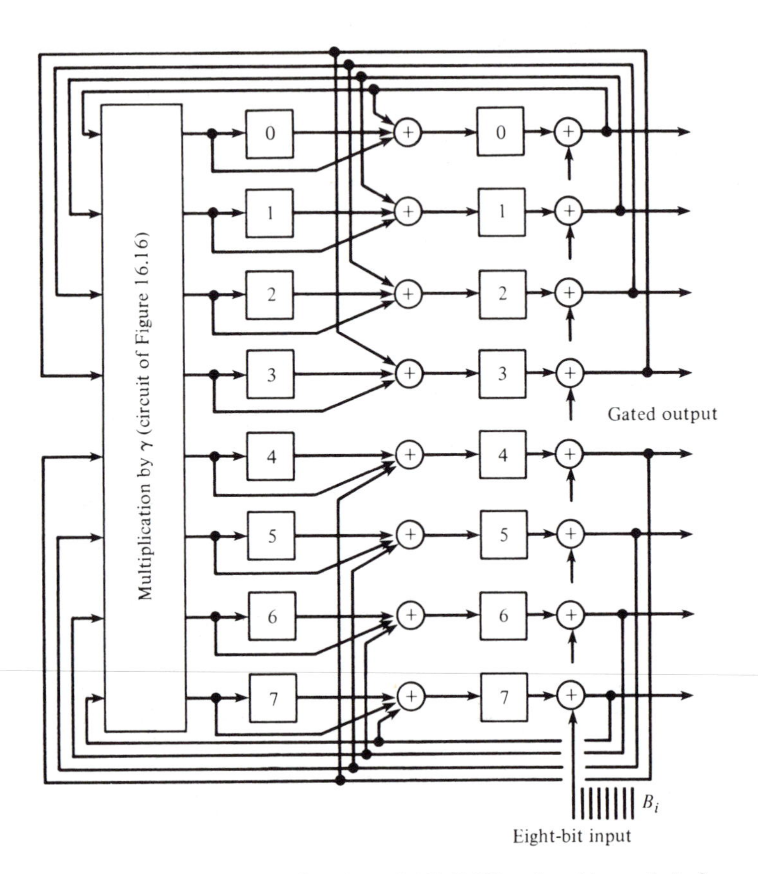

Figure 16.17 Encoding circuit for the (15, 13) BCH code with symbols from GF(2^8) (each symbol is represented by a 8-bit byte).

Let $\hat{B}_0, \hat{B}_1, \ldots, \hat{B}_{14}$ denote the 15 bytes of a read word. The syndrome of the read word

$$\mathbf{r}(X) = \hat{B}_0 + \hat{B}_1 X + \cdots + \hat{B}_{14} X^{14}$$

consists of the following components:

$$\mathbf{S}_1 = \mathbf{r}(1) = \sum_{i=0}^{14} \hat{B}_i, \tag{16.30}$$

$$\mathbf{S}_2 = \mathbf{r}(\gamma) = \sum_{i=0}^{14} \hat{B}_i \gamma^i. \tag{16.31}$$

These two components can be computed by the two shift registers shown in Figures 16.18 and 16.19, respectively.

Mode I: Correction of One-Byte Errors

Suppose that *i*th byte $\hat{B}_i$ is in error. Let $\mathbf{e}_i$ denote the error pattern Then

$$\begin{aligned} \hat{B}_i &= \mathbf{B}_i + \mathbf{e}_i, \\ \hat{B}_j &= \mathbf{B}_j \qquad \text{for } j \neq i. \end{aligned} \tag{16.32}$$

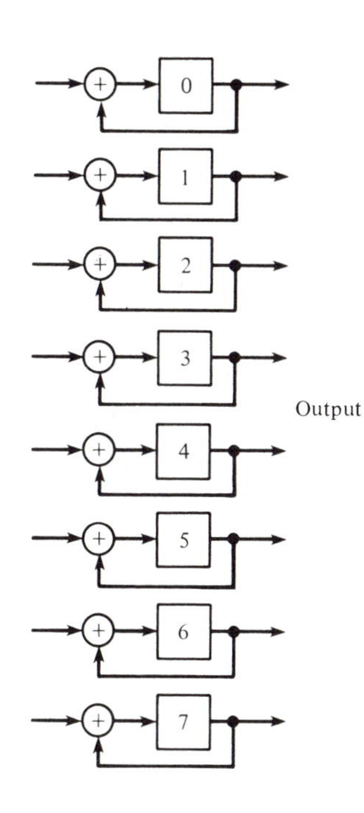

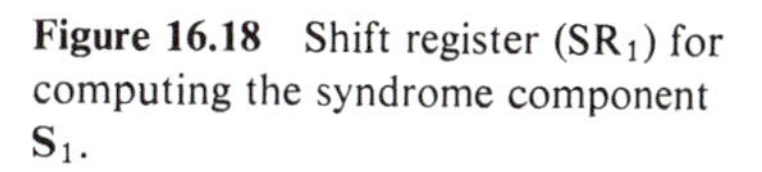

Figure 16.18 Shift register (SR_1) for computing the syndrome component $\mathbf{S}_1$.

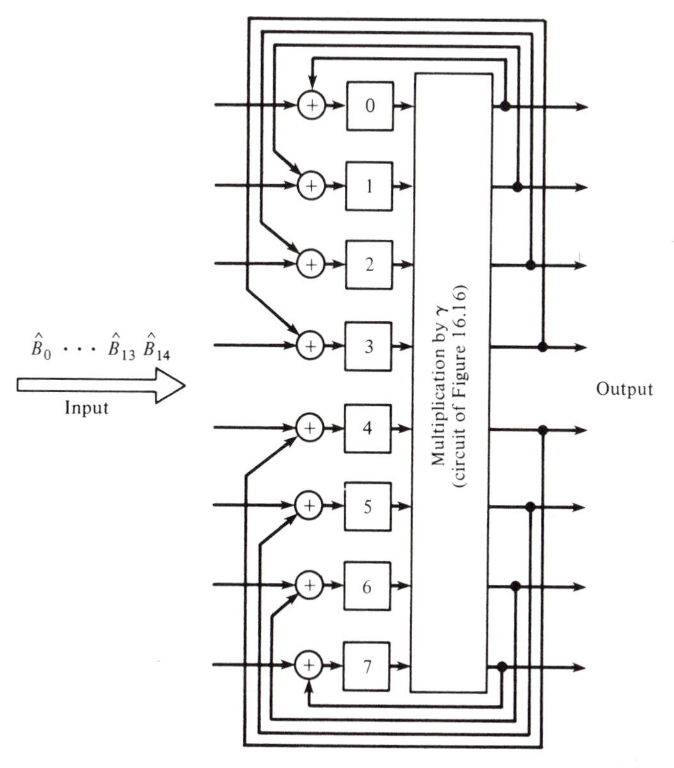

Figure 16.19 Shift register (SR_2) for computing the syndrome component $\mathbf{S}_2$.

It follows from (16.25), (16.26), (16.30), (16.31), and (16.32) that

$$\mathbf{S}_1 = \mathbf{e}_i,$$

$$\mathbf{S}_2 = \gamma^i \mathbf{e}_i.$$

Thus, the component $\mathbf{S}_1$ gives the error pattern $\mathbf{e}_i$ and $\gamma^i = \mathbf{S}_2/\mathbf{S}_1$ gives the index of the erroneous byte $\hat{B}_i$.

The decoding process is as follows. After $\mathbf{S}_2$ has been formed, the register of Figure 16.19 is shifted with no external inputs. A counter counts down from 15 with each shift. For each shift, the contents of the register are multiplied by $\gamma = \alpha^{68}$. After $15 - i$ shifts, the register contains

$$\gamma^{(15-i)}(\gamma^i \mathbf{e}_i) = \mathbf{e}_i$$

(since $\gamma^{15} = 1$), which matches $\mathbf{S}_1 = \mathbf{e}_i$. When this event occurs, the counter contains the index i of the erroneous byte $\hat{B}_i$. Error correction is then accomplished by adding $\mathbf{S}_1 = \mathbf{e}_i$ to $\hat{B}_i$. If the contents of the register of Figure 16.19 never match $\mathbf{S}_1$ after a maximal of 15 shifts, there are two or more bytes in error.

Mode II: Correction of Two-Byte Errors

Suppose it is known that byte i and byte j are in error (or being erased). Assume that $i < j$. Let $\mathbf{e}_i$ and $\mathbf{e}_j$ be the error patterns in $\hat{B}_i$ and $\hat{B}_j$, respectively. Then it follows from (16.30) and (16.31) that

$$\mathbf{S}_1 = \mathbf{e}_i + \mathbf{e}_j, \tag{16.33}$$

$$\mathbf{S}_2 = \gamma^i \mathbf{e}_i + \gamma^j \mathbf{e}_j. \tag{16.34}$$

Combining the two equalities above, we obtain

$$\mathbf{e}_j = (\mathbf{S}_1 + \mathbf{S}_2\gamma^{-i})(1 + \gamma^{j-i})^{-1}.$$

Since γ is a primitive element in the subfield GF(2^4), $\gamma^{-i} = \gamma^{(15-i)}$ and

$$(1 + \gamma^{j-i})^{-1} = \gamma^q$$

for some q. Consequently, we obtain the following:

$$\mathbf{e}_i = \mathbf{S}_1 + \mathbf{e}_j,$$

$$\mathbf{e}_j = \gamma^q[\mathbf{S}_1 + \gamma^{15-i}\mathbf{S}_2].$$

The values of q for all possible values of $j - i < 15$ are listed in Table 16.4.

TABLE 16.4 PARAMETER q AS A FUNCTION OF $j - i$

$j-i$	1	2	3	4	5	6	7	8	9	10	11	12	13	14
q	3	6	11	12	5	7	2	9	13	10	1	14	8	4

The decoding can be accomplished with the following steps:

1. After $\mathbf{S}_2$ has been formed, shift the register SR_2 of Figure 16.19 $15 - i$ times. This gives $\gamma^{15-i}\mathbf{S}_2$.
2. Add $\mathbf{S}_1$ to $\gamma^{15-i}\mathbf{S}_2$ and store $\mathbf{S}_1 + \gamma^{15-i}\mathbf{S}_2$ in register SR_2.

3. Shift SR_2 q times. This gives the error pattern $\mathbf{e}_j = \gamma^q[\mathbf{S}_1 + \gamma^{15-i}\mathbf{S}_2]$.
4. Add $\mathbf{e}_j$ to $\mathbf{S}_1$ in SR_1 of Figure 16.18. This results in the error pattern $\mathbf{e}_i$.
5. Add $\mathbf{e}_i$ to byte $\hat{B}_i$ and $\mathbf{e}_j$ to byte $\hat{B}_j$. This completes the error correction.

Using the coding scheme above, any error pattern that affects up to one full section of unknown position in a segment is correctable because such an error pattern affects at most one byte in each of the 16 code words in a segment. The coding scheme is also capable of correcting any error pattern which affects up to two full sections in a segment if the positions of the erroneous sections are known to the decoder. This type of error patterns affects at most two bytes of known positions in each of the 16 code words in a segment. Many other combinations of random and burst errors common to magnetic tapes are also correctable by the coding scheme above.

The IBM 3850 mass storage system has other error-checking features which are not included here. For a more complete description of the system, the reader is referred to References 4 and 5.

16.4 ERROR CONTROL FOR MAGNETIC DISKS

As with magnetic tape storage systems, burst errors predominate in magnetic disk storage systems. Again these errors are primarily caused by surface irregularities, such as defects or variations in head (read/write)-media separation in the presence of dust particles. A disk file consists of a number of tracks, but there is no coordination between the tracks and each track is accessed separately. Consequently, any coding scheme used for error control must serve a single track file containing long serial records. Moreover, fast decoding is a necessity for efficient operation.

Fire codes, variations of Fire codes, or Reed–Solomom (RS) codes are usually used for error control in magnetic disk storage systems. For example, a generalized Fire code is used in IBM 3330 disk systems; shortened Fire codes are used in IBM 3340 and 3350 disk systems; and a shortened RS code with symbols from GF(2^8) is used in IBM 3370 disk systems.

The generalized Fire code used in IBM 3330-compatible disk systems is generated by the following polynomial:

$$\mathbf{g}(X) = (X^{22} + 1)\mathbf{p}_1(X)\mathbf{p}_2(X)\mathbf{p}_3(X),$$

where

$$\mathbf{p}_1(X) = 1 + X + X^6 + X^7 + X^{11},$$

$$\mathbf{p}_2(X) = 1 + X + X^2 + X^3 + X^4 + X^5 + X^6 + X^7 + X^8 + X^9 + X^{10} + X^{11} + X^{12},$$

$$\mathbf{p}_3(X) = 1 + X + X^5 + X^6 + X^7 + X^9 + X^{11}$$

are irreducible polynomials. The periods of $\mathbf{p}_1(X)$, $\mathbf{p}_2(X)$, and $\mathbf{p}_3(X)$ are 89, 13, and 23, respectively. Therefore, the code length is

$$n = \text{LCM}(22, 89, 13, 23) = 585,422.$$

It has $22 + 11 + 12 + 11 = 56$ parity check digits and is capable of correcting any single burst of length 11 or less and detecting any single burst of length 22 or less.

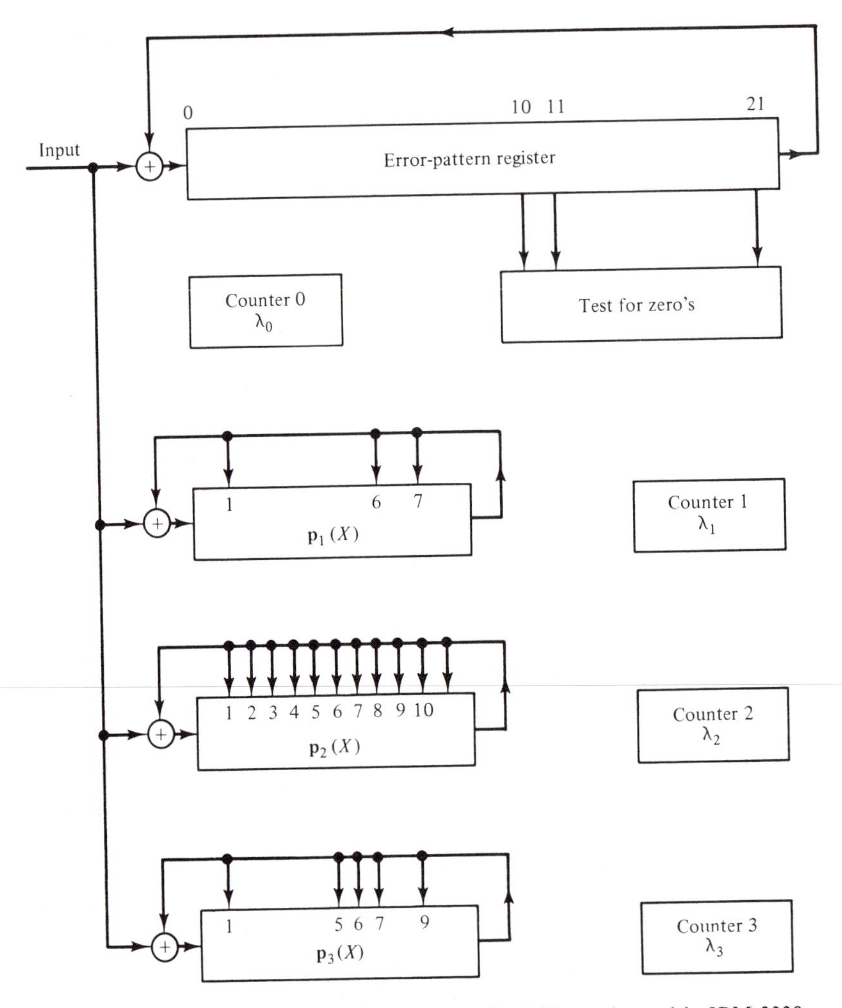

Figure 16.20 Decoding circuit for the generalized Fire code used in IBM 3330 disk systems.

This code is decoded by using Peterson–Chien fast error-trapping decoding algorithm as described in Section 9.3. The decoding circuit as shown in Figure 16.20 consists of four shift registers with feedback connections based on the four polynomials $X^{22} + 1$, $\mathbf{p}_1(X)$, $\mathbf{p}_2(X)$, and $\mathbf{p}_3(X)$, respectively. The register based on $X^{22} + 1$ is the error-pattern register. The other three registers are used to determine the error burst location. The decoding operation proceeds as follows. Entering the entire read word into four registers, we obtain four syndromes, $\mathbf{s}_0(X)$ in the error pattern register, $\mathbf{s}_1(X)$ in the $\mathbf{p}_1(X)$-register, $\mathbf{s}_2(X)$ in the $\mathbf{p}_2(X)$-register, and $\mathbf{s}_3(X)$ in the $\mathbf{p}_3(X)$-register. If all four syndromes are zero, the decoder assumes that the read word contains no errors. If some but not all four syndromes are zero, an uncorrectable error pattern is detected. If all four syndromes are nonzero, errors will be corrected provided that the error burst has length 11 or less. Shift the error-pattern register λ_0 time until the 11 high-order stages contain zeros. Then the burst is trapped in the 11 low-order stages of the error-pattern register. Shift the three error-location registers until their contents match the error pattern in the 11 low-order stages of the error-pattern register. Let λ_1, λ_2, and λ_3 be numbers of their shifts, respectively. Then we determine the positive integer q such that the remainders resulting from dividing q by 22, 89, 13, and 23 are λ_0, λ_1, λ_2, and λ_3, respectively. Then the error-location number is

$$n - q = 585442 - q.$$

Let $B(X)$ be the error pattern in the 11 low-order stages of the error-pattern register. Then the burst is $X^{n-q}B(X)$. The correction is done by adding $X^{n-q}B(X)$ to the read word. For decoding this code, no more than $22 + 89 = 111$ shifts are required.

Actually, q may be computed by using the Chinese remainder theorem as described in Section 9.3. Find four integers A_0, A_1, A_2 and A_3 such that the remainder resulting from dividing

$$22 \times A_0 + 89 \times A_1 + 13 \times A_2 + 23 \times A_3$$

by $n = 585{,}442$ is 1. The q is the remainder obtained from dividing

$$22 \times A_0 \times \lambda_0 + 89 \times A_1 \times \lambda_1 + 13 \times A_2 \times \lambda_2 + 23 \times A_3 \times \lambda_3$$

by n. For this code, we find that

$$\begin{aligned} 22 \times A_0 &= 452387, \\ 89 \times A_1 &= 72358, \\ 13 \times A_2 &= 315238, \\ 23 \times A_3 &= 330902. \end{aligned}$$

Shortened versions of a Fire code are used for error control in IBM 3340 and 3350 disk storage systems. The generator polynomial of this Fire code is

$$\begin{aligned} \mathbf{g}(X) &= (X^{13} + 1)(X^{35} + X^{23} + X^8 + X^2 + 1) \\ &= X^{48} + X^{36} + X^{35} + X^{23} + X^{21} + X^{15} + X^{13} \\ &\quad + X^8 + X^2 + 1. \end{aligned} \tag{16.35}$$

Since the period of $X^{35} + X^{23} + X^8 + X^2 + 1$ is $2^{35} - 1$, the natural length of the code is

$$\text{LCM}(13, 2^{35} - 1) \approx 4.47 \times 10^{11},$$

which far exceeds any practical record length on a disk file. Therefore, the code must be shortened for applications. The code is capable of correcting any error burst of length 7 or less; or it can be used for correcting any error burst or length b or less and simultaneously detecting any error burst of length $l \geq b$ with $b + l - 1 \leq 13$. In IBM 3340 disk systems, the code is used for correcting any error burst of length 3 or less and simultaneously detecting any error burst of length 11 or less. In IBM 3350 disk system, the code is used for correcting any error burst of length 4 or less and simultaneously detecting any error burst of length 10 or less.

The IBM 3340 and 3350 disk systems do not have a fixed-block data format. The length of a data record varies. The maximum record length in IBM 3340 disk systems is 70,320 bits long and the maximum record length in IBM 3350 disk systems is 152,552 bits long. As a result, the Fire code generated by (16.35) must be shortened to various lengths. The amount of shortening depends on the length of the data record to be encoded. This variable code length does not affect the encoding circuitry; however, it does affect the decoding process and circuitry.

In Section 4.7 we have shown that, when a cyclic code is shortened by l bits, it is possible to eliminate l preshifts of the syndrome register by modifying its feedback connections. The modification is based on the polynomial $\mathbf{\rho}(X)$, which is the remainder obtained from dividing X^l by the generator polynomial of the code (see Figure 4.15). However, if the amount of shortening varies from word to word (i.e., l is not fixed), this cannot be done. This difficulty can be overcome by processing the received word $\mathbf{r}(X) = \mathbf{r}_0 + \mathbf{r}_1 X + \cdots + \mathbf{r}_{n-1} X^{n-1}$ from the low-order end to the high-order end [i.e., to decode $\mathbf{r}(X)$ in *backward direction* from r_0 to r_{n-1}]. By doing this, the amount of shortening does not affect the decoding operation. In IBM 3340 and 3350 disk systems, conventional error-trapping decoding is used, but error correction is carried out in backward direction.

The IBM 3370 disk storage systems use a fixed-block data format. Data are processed in serial-by-byte form. Ech data block consists of 512 bytes (or 4096 bits). When a data block is recorded, 9 parity-check bytes are appended to it for error control during the read operation. The code used in the IBM 3370 disk systems is a shortened RS (Reed–Solomon) code with symbols from the field GF(2^8). The field GF(2^8) is generated by the primitive polynomial $\mathbf{p}(X) = 1 + X^2 + X^3 + X^4 + X^8$. The generator polynomial of the code is

$$\mathbf{g}(X) = (X + 1)(X + \alpha)(X + \alpha^{-1}), \tag{16.36}$$

where α is a primitive element of GF(2^8) and is a root of $\mathbf{p}(X) = 1 + X^2 + X^3 + X^4 + X^8$. The natural length of the code is 255 and its minimum distance is 4. Therefore, the code is capable of correcting any single-symbol (one-byte) error and simultaneously detecting any combination of double-symbol (two-byte) errors. In IBM 3370 disk systems, this code is shortened to 174 symbols long. Each symbol is represented by an 8-bit byte.

For the convenience of describing the coding scheme in IBM 3370 disk systems, we name the 512 bytes of a data block as in Figure 16.21, where a zero byte is added to the right end. For encoding, a data block is divided into three subblocks, $\mathbf{Z}_1$, $\mathbf{Z}_2$, and $\mathbf{Z}_3$, as shown in Figure 16.22; each consists of 171 bytes (the rightmost byte of $\mathbf{Z}_1$ is a zero byte to make it 171 bytes long). We represent these three subblocks in polynomial form as follows:

$$\mathbf{Z}_1(X) = B_{0,1} + B_{1,1}X + \cdots + B_{170,1}X^{170},$$
$$\mathbf{Z}_2(X) = B_{0,2} + B_{1,2}X + \cdots + B_{170,2}X^{170},$$
$$\mathbf{Z}_3(X) = B_{0,3} + B_{1,3}X + \cdots + B_{170,3}X^{170},$$

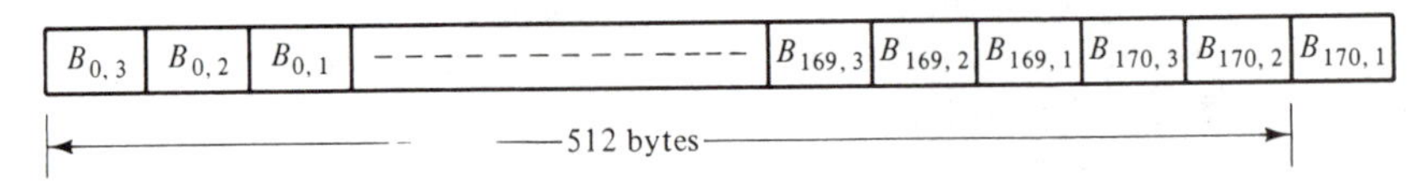

Figure 16.21 IBM 3370 data block format ($B_{170,1} = 0$).

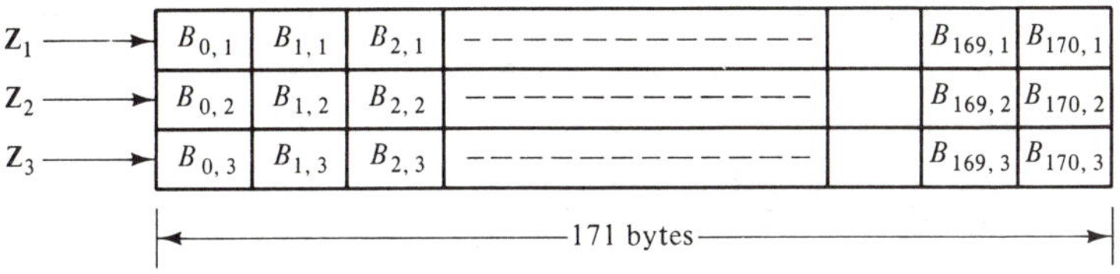

$Z_1 \rightarrow$	$B_{0,1}$	$B_{1,1}$	$B_{2,1}$	-------------		$B_{169,1}$	$B_{170,1}$
$Z_2 \rightarrow$	$B_{0,2}$	$B_{1,2}$	$B_{2,2}$	-------------		$B_{169,2}$	$B_{170,2}$
$Z_3 \rightarrow$	$B_{0,3}$	$B_{1,3}$	$B_{2,3}$	-------------		$B_{169,3}$	$B_{170,3}$

|←——— 171 bytes ———→|

Figure 16.22 Subblocks.

where each byte $B_{i,j}$ is regarded as a symbol in GF(2^8). Then each data subblock is encoded based on the shortened (174, 171) RS code generated by (16.36). The 3 parity-check bytes for $\mathbf{Z}_i$, denoted $C_{0,i}$, $C_{1,i}$, and $C_{2,i}$, are remainders resulting from dividing $X\mathbf{Z}_i(X)$ by $(X+1)$, $(X+\alpha)$, and $(X+\alpha^{-1})$, respectively. Note the difference between this encoding method and the method used for encoding cyclic (or shortened cyclic) codes, where all 3 parity-check bytes are computed by dividing $X^3\mathbf{Z}_i(X)$ by $\mathbf{g}(X) = (X+1)(X+\alpha)(X+\alpha^{-1})$. The circuits for computing the parity-check bytes $C_{0,i}$, $C_{1,i}$, and $C_{2,i}$ are shown in Figure 16.23. After the 3 parity-check bytes for $\mathbf{Z}_i$ have been formed, they are appended to $\mathbf{Z}_i$ to form a code word $\mathbf{W}_i$, as shown in Figure 16.24. Then the three code words $\mathbf{W}_1$, $\mathbf{W}_2$, and $\mathbf{W}_3$ are interleaved to form a coded block as shown in Figure 16.25. Therefore, the overall code is the (174, 171) RS code interleaved to degree 3. This interleaved RS code is capable of correcting any error burst confined to 3 consecutive bytes and detecting any error-burst confined to 6 consecutive bytes.

When a coded block is read from the disk, it is decomposed into three subwords, $\hat{W}_1$, $\hat{W}_2$, and $\hat{W}_3$, where

$$\hat{W}_i = (\hat{C}_{0,i}, \hat{C}_{1,i}, \hat{C}_{2,i}, \hat{Z}_i).$$

Any error burst confined to 3 bytes in the read block can affect at most one byte in each subword $\hat{W}_i$. Consequently, the error burst can be corrected by decoding each subword separately. Any error burst confined to 6 consecutive bytes in the read block can affect no more than two bytes in each subword $\hat{W}_i$. Thus, it can be detected by checking each subword separately.

For error correction, $\hat{C}_{0,i} + X\hat{Z}_i(X)$, $\hat{C}_{1,i} + X\hat{Z}_i(X)$, and $\hat{C}_{2,i} + X\hat{Z}_i(X)$ are divided by $X+1$, $X+\alpha$, and $X+\alpha^{-1}$, respectively. These divisions result in three syndromes, $S_{0,i}$, $S_{1,i}$, and $S_{2,i}$, which are symbols in GF(2^8). The circuits shown in Figure 16.23 can be used to compute these syndromes. If $S_{0,i} = S_{1,i} = S_{2,i} = 0$, the decoder assumes that the subword $\hat{W}_i$ contains no errors. If all three syndromes are nonzero, the decoder assumes that errors are confined to one data byte in $\hat{Z}_i$. In this case, $S_{0,i}$ gives the error pattern and $S_{1,i} = S_{0,i}\alpha^j$. Hence,

$$\frac{S_{1,i}}{S_{0,i}} = \alpha^j$$

gives the location of the erroneous data byte in $\hat{Z}_i$. More specifically, $\hat{B}_{j-1,i}$ is the erroneous byte. Other information regarding errors is given in Table 16.5. Further information regarding error control for IBM 3370 systems can be found in References 6 and 7.

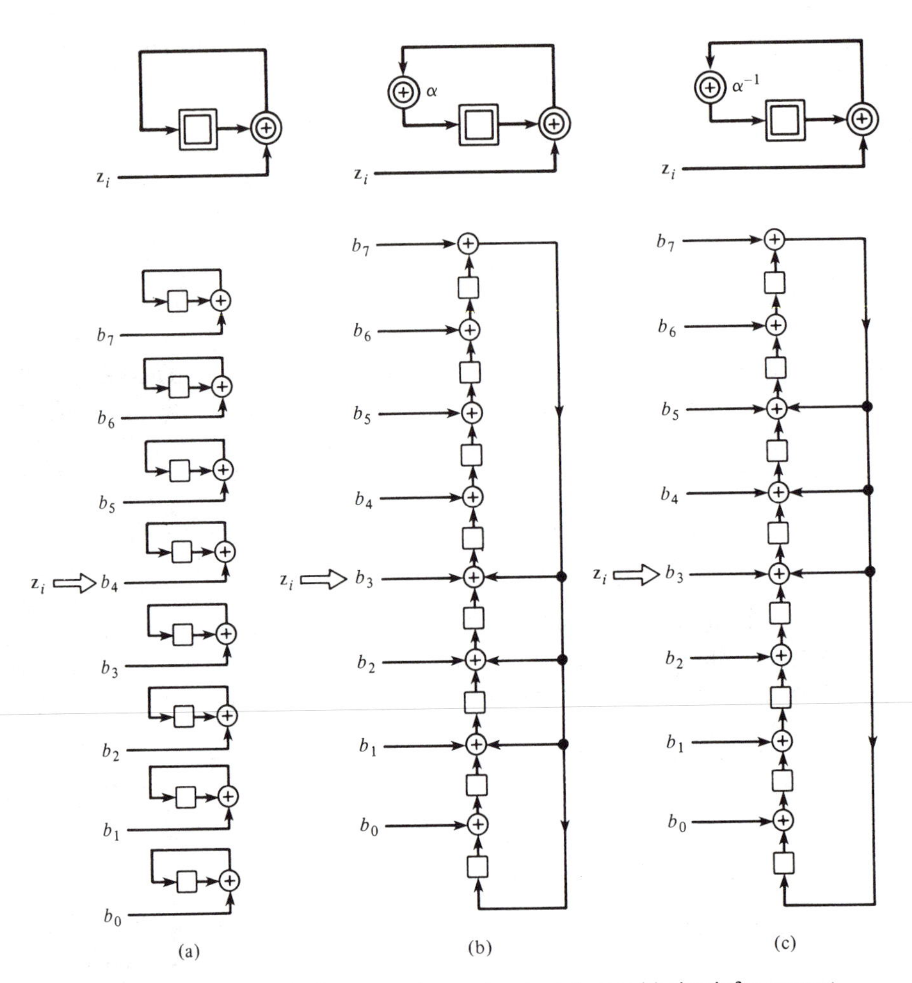

Figure 16.23 Circuits for computing parity-check bytes: (a) circuit for computing parity byte $C_{0,i}$; (b) circuit for computing parity byte $C_{1,i}$; (c) circuit for computing parity byte $C_{2,i}$.

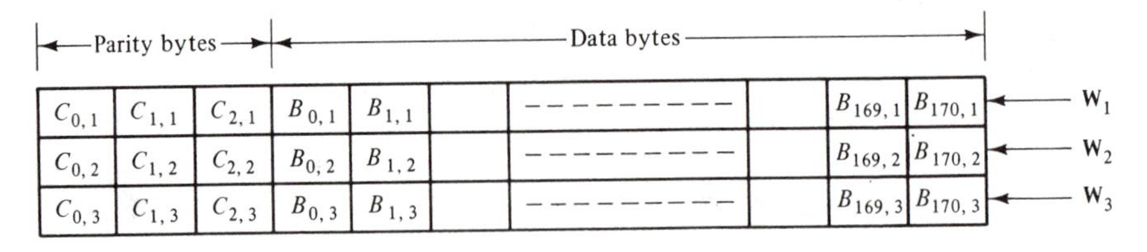

Figure 16.24 Subcodewords.

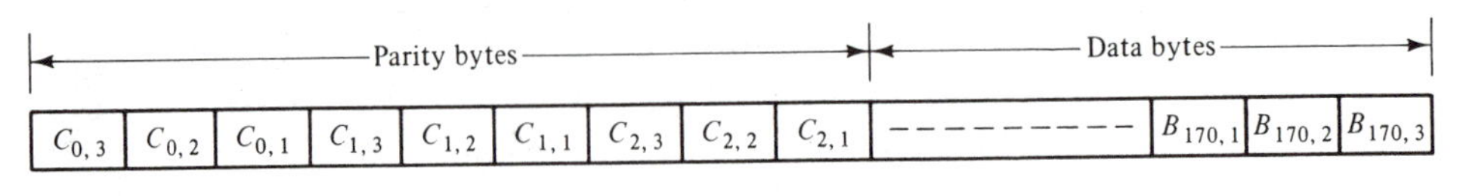

Figure 16.25 Overall codeword.

TABLE 16.5 ERROR INFORMATION BASED ON THE SYNDROMES

$S_{0,i}$	$S_{1,i}$	$S_{2,i}$	Error information in $\hat{W}_i$
0	0	0	Error-free
NZ	NZ	NZ	Single-byte errors
0	NZ	NZ	Double-byte errors
NZ	NZ	0	Errors in $\hat{C}_{0,i}$ and $\hat{C}_{1,i}$
NZ	0	NZ	Errors in $\hat{C}_{0,i}$ and $\hat{C}_{2,i}$
0	0	NZ	Errors in $\hat{C}_{2,i}$
0	NZ	0	Errors in $\hat{C}_{1,i}$
NZ	0	0	Errors in $\hat{C}_{0,i}$

A Fire code is also used for error control in DEC disks, RM and RP series. The generator polynomial of this code is

$$\mathbf{g}(X) = (X^{21} + 1)(X^{11} + X^2 + 1).$$

The natural length of the code is 42,987 bits, but it is shortened to $n = 4644$ bits. This code is capable of correcting any burst of length 11 or less.

16.5 ERROR CONTROL IN OTHER DATA STORAGE SYSTEMS

Block codes are also used for error control in other data storage systems, such as magnetic drums and photodigital storage systems. The IBM photodigital mass storage system known as Digital Cypress uses a (61, 50) shortened Reed–Solomon code with symbols from GF(2^6) for error control [8,9]. The field GF(2^6) is generated by using the primitive polynomial

$$\mathbf{p}(X) = X^6 + X + 1.$$

The generator polynomial of the code is

$$\begin{aligned}\mathbf{g}(X) &= \prod_{i=-5}^{5} (X - \alpha^i) \\ &= 1 + \alpha^{14}X + \alpha^{59}X^2 + \alpha^6X^3 + \alpha^{28}X^4 + \alpha^{54}X^5 \\ &\quad + \alpha^{54}X^6 + \alpha^{28}X^7 + \alpha^6X^8 + \alpha^{59}X^9 + \alpha^{14}X^{10} + X^{11},\end{aligned}$$

where α is a primitive element of GF(2^6) and is a root of $\mathbf{p}(X) = X^6 + X + 1$. The natural length of the code is 63 symbols, but it is shortened to 61 symbols in the IBM Digital Cypress. Each code symbol is represented by a 6-bit character. Therefore, in binary form, it is a (366, 300) code. The minimum distance of the code is 12; thus, it is capable of correcting any combination of independent and burst errors representable by five characters. A sixth character error, plus many other error patterns, can be detected.

Encoding and syndrome computations for this code are done by hardware with shift registers as discussed in Section 6.3. Error correction is implemented by a software program in a control processor which is time-shared with the control functions and other functions of the storage system. The error-correction procedure is designed to minimize the average decoding time. When a nonzero syndrome is detected, a retrial is called for first. If errors are still present, the program goes to a single-character partial-correction subroutine. If that procedure is unsuccessful in correcting the errors, a two-character partial-correction routine is called. The full-power correction routine is used only when both the single-character and double-character error-correction subroutines fail to correct the errors.

Error-correction and error-detection features are also to be included in the Burroughs' optical disk storage systems.

PROBLEMS

16.1. Find the parity-check matrix for the (12, 8) Hsiao code.

16.2. Prove the relation of (16.10).

16.3. Prove that the coefficients of the remainder resulting from dividing $X^l \hat{B}_i(X)$ by $\mathbf{g}(X) = 1 + X^3 + X^4 + X^5 + X^8$ with $0 \leq l \leq 7$ equal to the entries of the column matrix

$$\mathbf{A}^l \hat{B}_i^T,$$

where $\hat{B}_i(X) = \hat{b}_{i0} + \hat{b}_{i1}X + \hat{b}_{i2}X^2 + \hat{b}_{i3}X^3 + \hat{b}_{i4}X^4 + \hat{b}_{i5}X^5 + \hat{b}_{i6}X^6 + \hat{b}_{i7}X^7$ and $\mathbf{A}$ is the matrix given by (16.11).

16.4. Prove that there is one and only one $\mathbf{e}_j(X)$ of degree 7 or less which satisfies the first relation of (16.17).

REFERENCES

1. M. Y. Hsiao, "A Class of Optimal Minimum Odd-Weight-Column SEC–DED Codes," *IBM J. Res. Dev.*, 14, July 1970.
2. A. M. Patel and S. J. Hong, "Optimal Rectangular Code for High Density Magnetic Tapes," *IBM J. Res. Dev.*, 18, pp. 579–588, November 1974.
3. E. R. Berlekamp, "Algebraic Codes for Improving the Reliability of Tape Storage," *Proc. 1975 Nat. Comput. Conf.*
4. "Introduction to the IBM 3850 Storage System," IBM Systems GA 32-0028-1.
5. A. M. Patel, "Error Recovery Scheme for the IBM 3850 Mass Storage System," *IBM J. Res. Dev.*, 24(1), pp. 32–42, January 1980.
6. P. Hodges, W. J. Schaeuble, and P. L. Shaffer, "Error Correcting System for Serial by Byte Data," U.S. Patent No. 4,185,269, January 22, 1980.
7. "IBM 3370 Direct Access Storage Description," IBM Systems GA 26-1657-1, File No. 4300-07.
8. I. B. Oldham, R. T. Chien, and D. T. Tang, "Error Detection and Correction in a Photo-digital Storage System," *IBM J. Res. Dev.*, 12(6), pp. 422–430, November 1968.
9. R. T. Chien, "Memory Error Control: Beyond Parity," *IEEE Spectrum*, No. 7, pp. 18–23, July 1973.

17

Practical Applications of Convolutional Codes

Convolutional codes have found application in many diverse systems. In this chapter we review some of the most important of these applications. Naturally, because of the limitations of space, there are many applications of convolutional codes which we are not able to cover. The references listed at the end of the chapter provide numerous examples of the practical use of convolutional codes.

Section 17.1 deals with the applications of Viterbi decoding to space and satellite communication. The use of sequential decoding in similar applications is covered in Section 17.2. Applications of majority-logic decoding are discussed in Section 17.3. The application of some of the burst-error-correcting techniques from Chapter 14 to channels subject to both fading and interference is covered in Section 17.4. Finally, in Section 17.5, ARQ systems similar to those presented in Chapter 15, but employing convolutional codes instead of block codes, are discussed.

17.1 APPLICATIONS OF VITERBI DECODING

Convolutional coding with hard- and soft-decision Viterbi decoding has found application in many space and satellite communication systems. Several examples of such systems and the codes employed will now be briefly discussed.

In the 1977 launch of the Voyager space mission to Mars, Jupiter, and Saturn, NASA used a codec (encoder/decoder) designed by the Jet Propulsion Laboratory which can be operated with either of two codes: one a (2, 1, 6) code with

$$\begin{aligned} \mathbf{g}^{(1)}(D) &= 1 + D + D^3 + D^4 + D^6 \\ \mathbf{g}^{(2)}(D) &= 1 + D^3 + D^4 + D^5 + D^6 \end{aligned} \tag{17.1}$$

and $d_{\text{free}} = 10$, and the other a (3, 1, 6) code with

$$\begin{aligned} \mathbf{g}^{(1)}(D) &= 1 + D + D^3 + D^4 + D^6 \\ \mathbf{g}^{(2)}(D) &= 1 + D^3 + D^4 + D^5 + D^6 \\ \mathbf{g}^{(3)}(D) &= 1 + D^2 + D^4 + D^5 + D^6 \end{aligned} \tag{17.2}$$

and $d_{\text{free}} = 15$. These codes have come to be known as the NASA Planetary Standard Codes [1]. The hardware decoders employ 3-bit ($Q = 8$) soft quantization and operate at data rates up to 100 Kbps. The coding gains achievable by these codes for bit error probabilities of 10^{-5} are illustrated in Figure 17.1.

The Linkabit Corporation has designed and built convolutional encoder/Viterbi decoder codecs for a wide variety of applications. The (2, 1, 6) code described above with a $Q = 8$ soft-decision hardware decoder has been deployed throughout the Defense Satellite Communication System (DSCS) Network. These decoders are capable of operation at 10 Mbps data rates [2]. An interesting variation on this basic coding system allows the user to select either the (2, 1, 6) code or a higher rate (4, 3, 2) code with $K = 6$, which is derived from the basic code by deleting two bits from every three branches of the code trellis (see Problem 17.2). Both codes are

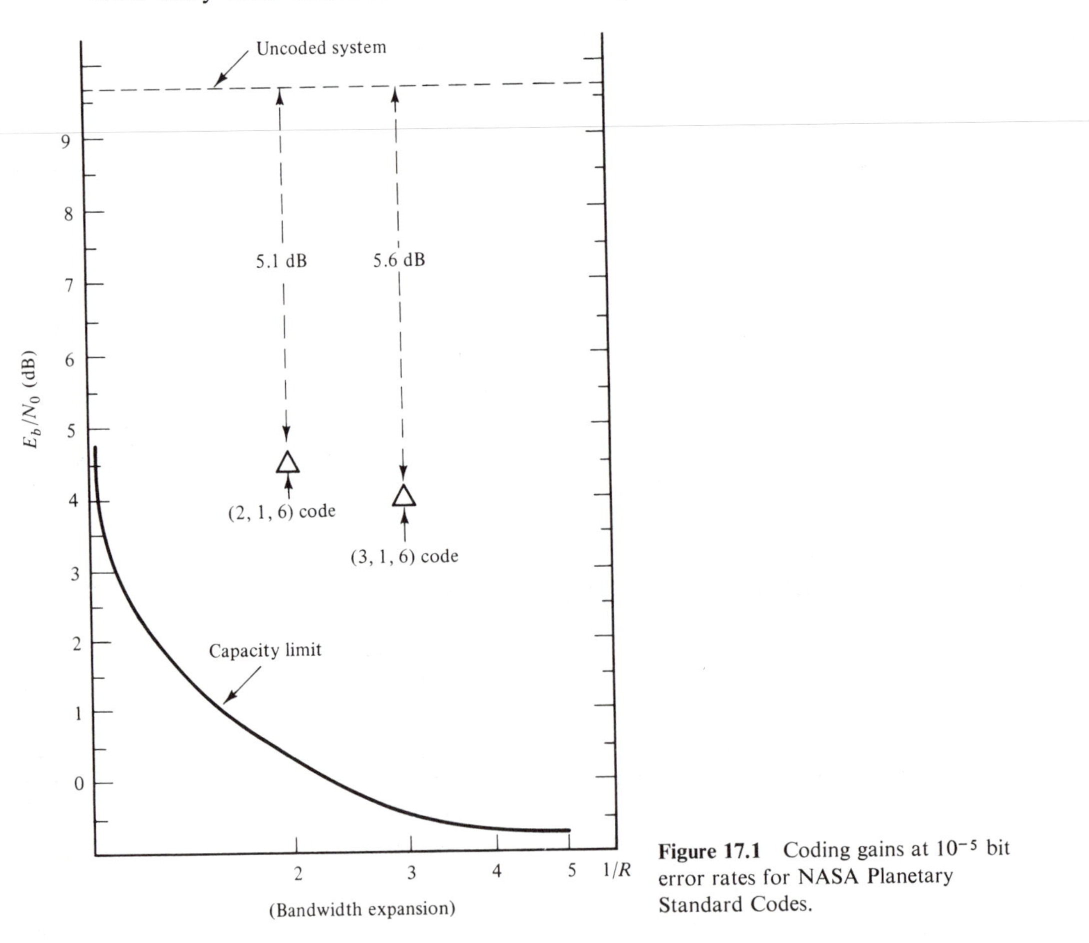

Figure 17.1 Coding gains at 10^{-5} bit error rates for NASA Planetary Standard Codes.

decoded using the same decoder, which is capable of operating at 8 Mbps data rates [2]. This multiple-rate coding system is presently being used on NASA's CTS satellite system. The coding gain of the (4, 3, 2) code for a bit error rate of 10^{-5} is also illustrated in Figure 17.1.

The Harris Corporation has developed a high-speed hardware decoder employing 3-bit ($Q = 8$) soft quantization for the (2, 1, 6) code described above. This system operates up to 16 10 Mbps decoders in parallel, thereby achieving a maximum decoding speed of 160 Mbps. This decoder has been developed for the Tracking and Data Relay Satellite System (TDRSS) program [3] commissioned by NASA.

Another approach to high-speed Viterbi decoding has recently been suggested by Acampora and Gilmore [4]. They have reported decoding speeds of 50 Mbps with an experimental hardware decoder using analog components to process the unquantized demodulator outputs. Decoding speeds of up to 200 Mbps may be possible using this approach.

Viterbi decoding of convolutional codes can also be used in a concatenated coding scheme with a Reed–Solomon block code [5,6]. In one such system, an $(n_1, 1, m)$ convolutional code with Viterbi decoding is used as the inner code along with an (n_2, k_2) Reed–Solomon outer code with symbols in GF(2^J), where $n_2 = 2^J - 1$. A block diagram of the system is shown in Figure 17.2. At the encoder, k_2 bytes of J information bits each are encoded by the outer code into an n_2-byte codeword. λ consecutive codewords are then stored by row in a $\lambda \times (2^J - 1)$-byte interleaving buffer, and are read out by column in bytes. These interleaved bytes are then converted to J-bit vectors for encoding by the inner code. The received sequence is hard-quantized and decoded by a 2^m-state truncated Viterbi decoder. The decoded information bits are then grouped into J-bit bytes, stored by column in a $\lambda \times (2^J - 1)$-byte de-interleaving buffer, and read out by row in bytes. These de-interleaved bytes are then decoded by the Reed–Solomon decoder. The byte size J is chosen to correspond to the length of a typical error event at the output of the Viterbi decoder. This typical error event, if properly aligned, would then cause one symbol error in

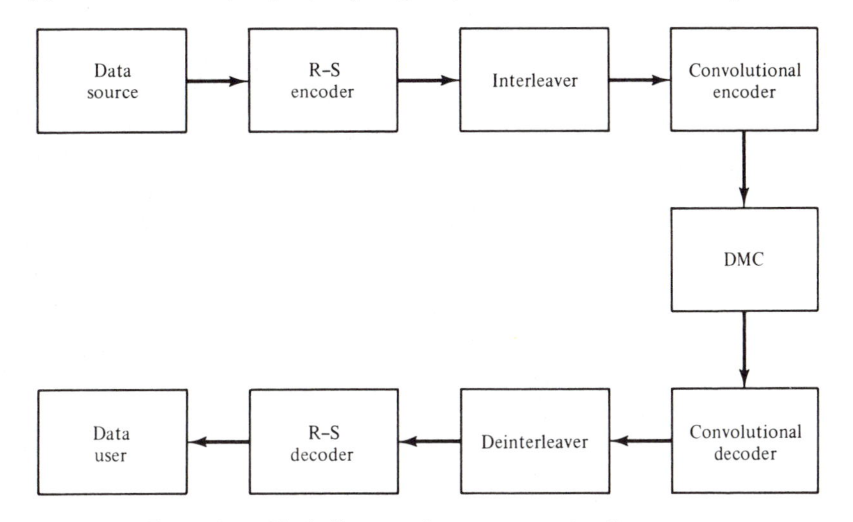

Figure 17.2 Block diagram of a concatenated coding system.

the Reed–Solomon code. Since some error events may extend beyond J bits and affect several consecutive bytes, interleaving is required to distribute these erroneous bytes among different Reed–Solomon codewords. It is customary to choose $J \approx m + 1$, since error events must be at least $m + 1$ bits long and much larger byte sizes would make the Reed–Solomon code too long.

Odenwalder [5] has investigated the performance of such a system. An example of his results for the (2, 1, 6) convolutional inner code with $d_{\text{free}} = 10$ described above and an $n_2 = 2^J - 1$, $k_2 = n_2 - 2t = 2^J - 2t - 1$ Reed–Solomon outer code with error-correcting capability t is shown in Figure 17.3 for various values of J and t. The overall code rate in this case is $R = k_2/2n_2 = (2^J - 2t - 1)/(2^{J+1} - 2)$. The performance is improved by increasing t up to a point, but further increases in t cause the performance to degrade, as illustrated in Figure 17.4 for the case $J = 7$.

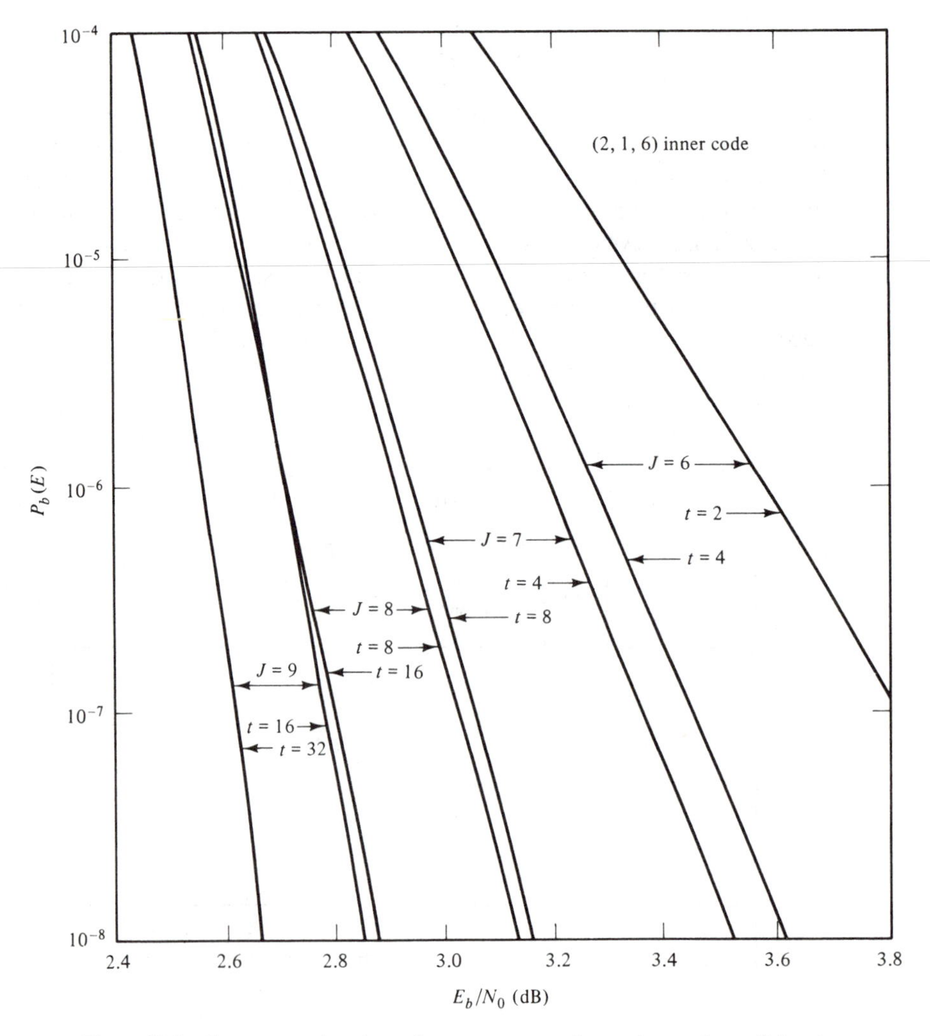

Figure 17.3 Concatenated code performance curves for various values of J and t.

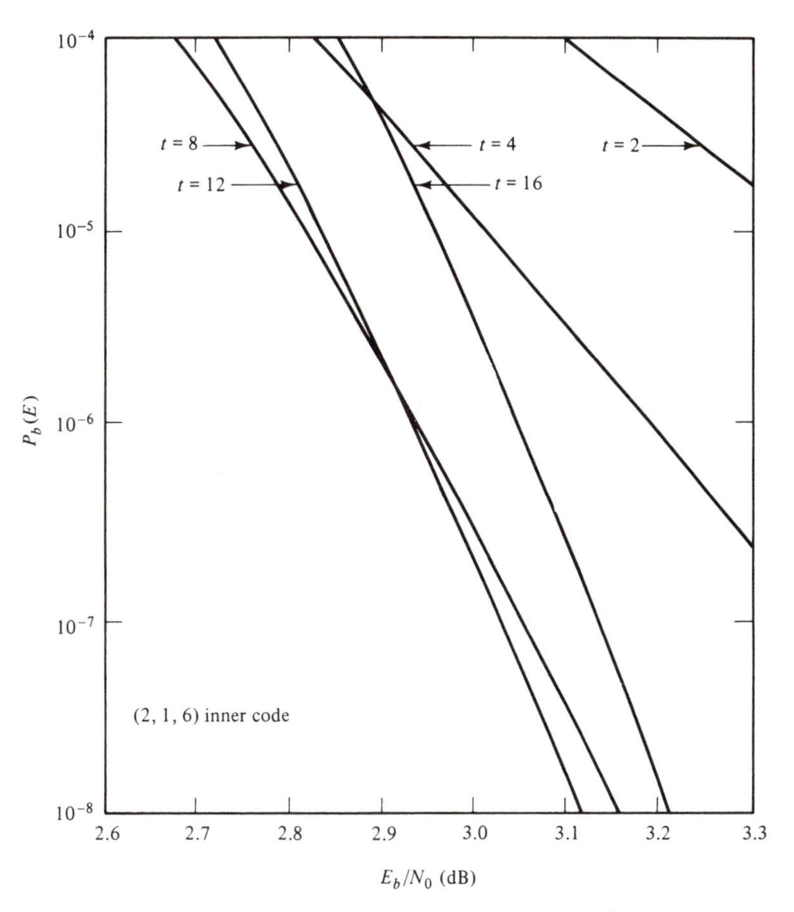

Figure 17.4 Concatenated code performance curves for $J = 7$.

This is due to the fact that the reduced rate is not compensated for by the increased error-correcting capability of the Reed–Solomon code.

A similar system has been proposed by Lee [6]. In this system an $(n_1, k_1, 1)$ convolutional inner code is used along with an (n_2, k_2) Reed–Solomon outer code with symbols in GF(2^{k_1}); that is, the byte size of the outer code is equal to the number of information bits encoded per unit time by the unit-memory inner code. This provides a natural interface between the inner and outer portions of the encoder and decoder.

The performance of an (18, 6, 1) inner code with $d_{\text{free}} = 16$ along with a t-error-correcting length 63 outer code with symbols in GF(2^6) is shown in Figure 17.5. For comparison, the performance of the same outer code with a (3, 1, 6) inner code with $d_{\text{free}} = 15$ and a (3, 1, 7) inner code with $d_{\text{free}} = 16$ is also shown. The overall code rate in each case is $R = k_2/3n_2 = (63 - 2t)/189$.[1] It is seen that the unit-memory

[1] In Figure 17.5, only one data point was taken for each value of t. This corresponded to an E/N_0 ratio of -3.5 dB. Each different value of t gave a different overall code rate, resulting in distinct values of E_b/N_0.

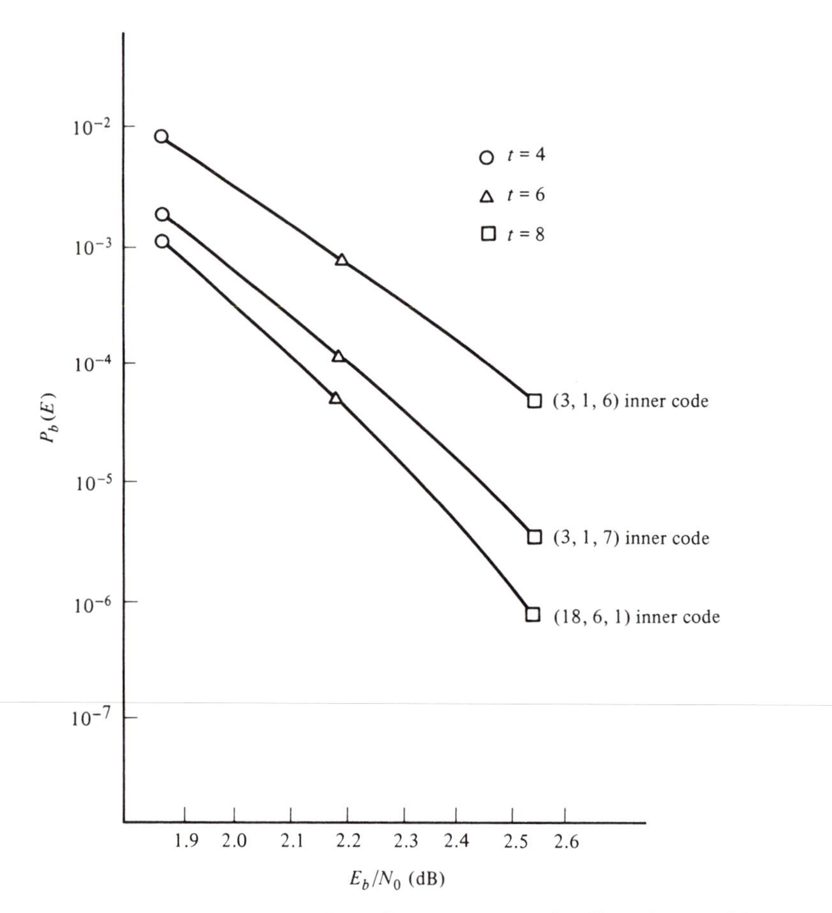

Figure 17.5 Performance comparison of two concatenated coding schemes with $J = 6$.

code outperforms both of the other codes. Its larger free distance accounts for the improvement over the (3, 1, 6) code. The improvement over the (3, 1, 7) code, which has the same free distance, is attributed to the byte-oriented structure of the unit-memory code.

Many variations are possible on the basic concatenation schemes covered above. Lee [6] has described several of these, including using a modified decoder for the unit-memory inner code which minimizes the byte error probability and provides the outer decoder with reliability information on each decoded byte, and using feedback from the output of the outer decoder to aid the inner decoder in making decisions. Zeoli [7] has also described a modification of Odenwalder's basic concatenation scheme in which reliability information is passed from the inner decoder to the outer decoder, and feedback from the outer decoder is used to aid the inner decoder. The high reliability and modest complexity of concatenated coding systems make them prime candidates for many of the space and satellite applications of the 1980s.

17.2 APPLICATIONS OF SEQUENTIAL DECODING

Convolutional coding with hard- and soft-decision sequential decoding has been applied in many space communication systems. Several examples of such systems and the codes employed will now be briefly discussed.

NASA's Pioneer 9 solar orbit space mission, launched in 1968, used a modification of the (2, 1, 20) systematic code constructed by Lin and Lyne [8]. The generator sequences of this code are given in octal notation by

$$\begin{aligned} \mathbf{g}^{(1)} &= 4000000 \\ \mathbf{g}^{(2)} &= 7154737. \end{aligned} \tag{17.3}$$

At the time this code was proposed, the minimum distance $d_{\min}$ was still considered the best measure of code performance. The code above has $d_{\min} = 10$. The actual code used on Pioneer 9 was a (2, 1, 24) code with four zeros appended to each of the generator sequences above. This code has $d_{\min} = 11$. The software decoder used the Fano algorithm with 3-bit ($Q = 8$) soft quantization and operated at data rates up to 512 bps. Erased frames were recorded and later decoded off-line.

Pioneer 9 was the first deep-space mission to make use of error-correcting codes. The coding system yielded a coding gain of about 3 dB compared to the bit error rate of an uncoded system.

Several later deep-space missions used a (2, 1, 31) nonsystematic, quick-look-in (QLI) code constructed by Massey and Costello [9]. The generator sequences of this code in octal notation are given by

$$\begin{aligned} \mathbf{g}^{(1)} &= 73353367672 \\ \mathbf{g}^{(2)} &= 53353367672. \end{aligned} \tag{17.4}$$

This code has $d_{\min} = 11$ and $d_{\text{free}} = 23$. Software decoding using the Fano algorithm with 3-bit soft quantization was again selected. The Pioneer 10 Jupiter fly-by mission, launched in 1972, and the Pioneer 11 Saturn fly-by mission, launched in 1973, both operated at data rates up to 2 Kbps. A high-speed (1 Mbps) hardware decoder was also developed by the Jet Propulsion Laboratory for this project [10]. The Helios A and Helios B West German solar orbiter missions, as well as the Pioneer 12 Venus orbiter, also used this same coding system operating at data rates up to 4 Kbps.

Because of its many applications, the code of (17.4) is probably the most widely used code in deep-space missions. However, more recent work by Massey [11] and Johannesson [12] has suggested a better (2, 1, 31) nonsystematic QLI code for use in future space missions. This code has generator sequences

$$\begin{aligned} \mathbf{g}^{(1)} &= 74042402072 \\ \mathbf{g}^{(2)} &= 54042402072, \end{aligned} \tag{17.5}$$

$d_{\min} = 13$, $d_{\text{free}} = 21$, and is an optimum distance profile (ODP) code. A comparison of the column distance function (CDF) d_i for $i \leq 71$ and the computational distribution of the codes of (17.4) and (17.5) is shown in Figure 17.6. This clearly illustrates that the ODP code of (17.5) is superior from a computational point of view, and hence is

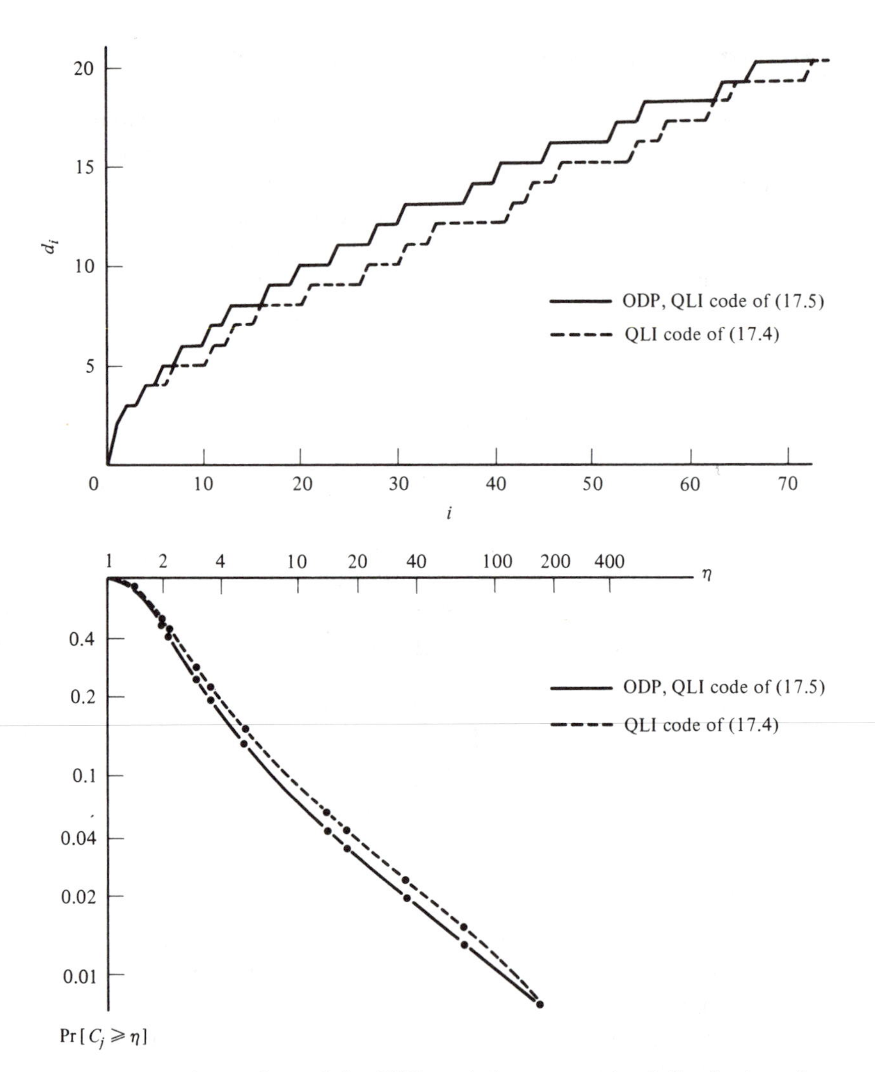

Figure 17.6 Comparison of the CDFs and the computational distribution of two (2, 1, 31) QLI codes.

capable of achieving higher decoding speeds. Although the free distance of the OPD code of (17.5) is smaller, simulations appear to show that it has a better error performance than the code of (17.4) [11].

The Linkabit Corporation designed and built a number of hardware sequential decoders for NASA applications. One of these, a 50 Mbps hard-decision Fano decoder built in 1970 remains the fastest sequential decoder ever constructed, and is used on a NASA space station-to-ground telemetry link. Another more flexible Fano decoder is capable of handling memory orders from 7 to 47, systematic or nonsystematic codes, variable frame lengths, and hard or soft decisions, and operates at a speed of 3 million computations per second. This unit, built in 1975, has been used in a variety of NASA applications, including the TELOPS program, the International Ultraviolet Explorer (IUE) telemetry system, and the International Sun–Earth Explorer (ISEE) program.

Massey [13] has conducted a thorough study of (2, 1, m) codes with an *equivalent memory order*[2] $m_E = 23$ for these and future applications in NASA's deep-space network. Since in a systematic (2, 1, m) code, the tail bits corresponding to the systematic generator must be all 0's, it is not necessary to actually transmit these bits. Hence, a (2, 1, m) systematic code transmits the same number of tail bits as a (2, 1, $m/2$) nonsystematic code, and can be considered to have an equivalent memory order $m_E = m/2$. The complete list of codes studied by Massey is listed below.

Code 1. A (2, 1, 23) QLI code constructed by Massey and Costello [9].

Code 2. A (2, 1, 23) ODP-QLI code constructed by Johannesson [14].

Code 3. A second (2, 1, 23) ODP-QLI code constructed by Johannesson [14].

Code 4. A (2, 1, 23) complementary code constructed by Bahl and Jelinek [15].

Code 5. A (2, 1, 23) code constructed by Massey et al. [16] from an (89, 44) quadratic residue cyclic code. [This code actually has $m = 22$, since $\mathbf{g}_{23}^{(1)} = \mathbf{g}_{23}^{(2)} = 0$. See Table 11.4A.]

Code 6. A (2, 1, 23) code constructed by Massey [13].

Code 7. A (2, 1, 47) QLI code with $m_E = 23$ constructed by Massey and Costello [9].

Code 8. A (2, 1, 47) ODP-QLI code with $m_E = 23$ constructed by Johannesson [12].

Code 9. A (2, 1, 46) extension by Forney [17] of Lin and Lyne's [8] (2, 1, 20) systematic code. Since this code is systematic, $m_E = m/2 = 23$. [This code actually has $m = 44$ since $\mathbf{g}_{45}^{(2)} = \mathbf{g}_{46}^{(2)} = 0$.]

Code 10. A (2, 1, 46) ODP systematic code with $m_E = m/2 = 23$ constructed by Johannesson [12].

Code 11. A (2, 1, 23) ODP code constructed by Johannesson and Paaske [18].

A summary of the distance properties of these 11 codes is given in Table 17.1. The two $m = 47$ nonsystematic codes (codes 7 and 8) are expected to have the best free distances, although d_{free} is not known exactly for either of these codes. The nonsystematic ODP-QLI code with $m = 47$ (code 8) and the systematic ODP code with $m = 46$ (code 10) have the best distance profile [13].

Massey [13] performed extensive computer simulations on these 11 codes. A summary of his recommendations regarding code selection for future deep-space missions now follows. The best code would be the (2, 1, 47) ODP-QLI code (code 8) encoded with an undersized tail of only $m_E = 23$ branches. This code had the best simulated computational performance and was among the best in decoding error probability. In addition, it has the "quick-look" property, and can be adapted for

[2]A code with memory order m is said to have an equivalent memory order $m_E \leq m$ if each data frame is terminated with a tail of m_E rather than m branches.

TABLE 17.1 DISTANCE PROPERTIES OF EQUIVALENT MEMORY ORDER 23 CODES

Code number	$g^{(1)}$	$g^{(2)}$	d_{23}	d_{free}
1	73353367	53353367	9	17
2	74042417	54042417	11	18
3	74041567	54041567	11	19
4	51202215	66575563	10	24
5	77441232	54502376	10	20
6	75105323	55105323	10	19
7	73353367	53353367	9	≥ 32
	67373553	67373553		
8	74042402	54042402	11	≥ 33
	07121635	07121635		
9	40000000	71547370	9	19
	00000000	13174650		
10	40000000	67114545	11	17
	00000000	75564666		
11	55346125	75744143	11	25

use with any tail length up to $m_E = 47$. If the undersized tail option is not acceptable,[3] the next best code would be the (2, 1, 23) ODP code (code 11). This code has the best d_{free} ($= 25$), and hence the lowest decoding error probability, of any of the $m = 23$ codes, and its computational performance was nearly as good as code 8. However, it is not a QLI code. The best QLI code with $m = 23$ was the second (2, 1, 23) ODP-QLI code (code 3).

In 1969, the Codex Corporation developed a hardware hard-decision Fano decoder for the U.S. Army Satellite Communication Agency's TACSAT channel [19]. The code employed was the (2, 1, 46) systematic code constructed by Forney [17] listed as code 9 in Table 17.1. The decoder used backsearch limiting to avoid erasures, and was able to achieve 5-dB coding gains at data rates of 5 Mbps. A complete discussion of the performance of this system was included in Section 12.3.

In 1971, the Harris Corporation designed and built a flexible hardware hard-decision Fano decoder for NASA's digital television test set [20]. This decoder could operate at data rates up to 9 Mbps and could be used with any $(2, 1, m)$ systematic code with $m \leq 39$.

As a final example of the application of sequential decoding, we discuss Lincoln Laboratory's software implementation of a soft-decision Fano decoder for the U.S. Navy's Project Sanguine [21]. This system, field tested in 1972, was designed to operate in the extremely low frequency (ELF) band in order to facilitate communication with submarines around the world. The limited power that can be generated at ELF frequencies of around 76 Hz required very low data rates (≈ 0.03 bps) and

[3] One disadvantage of encoding an undersized tail ($m_E < m$) is that the contents of the encoder register must be reset to zero after each frame is encoded.

a highly sophisticated modulation/coding technique to meet the performance specifications at the receiver.

The code selected was an $R = 1/6$ code interleaved so that all the bits in $\mathbf{v}^{(1)}$ are sent first, then all the bits in $\mathbf{v}^{(2)}$, and so on. This allows the sequential decoder to immediately attempt decoding the first two received sequences ($\mathbf{r}^{(1)}$ and $\mathbf{r}^{(2)}$) as an $R = 1/2$ code. If decoding is unsuccessful (an erasure), the decoder waits for the next received sequence ($\mathbf{r}^{(3)}$) and then attempts decoding as an $R = 1/3$ code, and so on. The probability of successful decoding increases with each decoding attempt. Different receivers, with different received signal-to-noise ratios (SNRs), will then take different amounts of time to decode a message. In other words, the system adapts so that receivers with large SNRs decode much faster than receivers with low SNRs.

A conservative strategy is employed by the decoder to ensure an extremely low undetected error probability. This strategy combines the use of a long-constraint-length code with three rules for declaring an erasure. An erasure is declared if (1) the computational limit is exceeded, (2) the final metric falls below some threshold, or (3) at any time in the decoding process, the metrics of all paths examined fall below some reject level. The parameters involved in this strategy allow considerable flexibility in designing the system for any given undetected error probability and erasure probability specifications.

17.3 APPLICATIONS OF MAJORITY-LOGIC DECODING

Most of the early applications of convolutional codes in the 1960s involved some form of majority-logic decoding due to the relatively low cost of decoder implementation. In this section we discuss several more recent applications of majority-logic decoding.

The Communications Satellite Corporation has used convolutional codes with majority-logic decoding in several different applications [22]. In 1969, a majority-logic decoder for an (8, 7, 146) self-orthogonal code was designed and built for a digital television system (DITEC) which transmits digitized color TV signals at a data rate of 33.6 Mbps over the INTELSAT IV satellite system. The generator polynomials of the code are given by

$$\begin{aligned}
\mathbf{g}_1^{(8)}(D) &= 1 + D^2 + D^8 + D^{32} + D^{88} + D^{142} \\
\mathbf{g}_2^{(8)}(D) &= 1 + D^3 + D^{19} + D^{52} + D^{78} + D^{146} \\
\mathbf{g}_3^{(8)}(D) &= 1 + D^{11} + D^{12} + D^{62} + D^{85} + D^{131} \\
\mathbf{g}_4^{(8)}(D) &= 1 + D^{21} + D^{25} + D^{39} + D^{82} + D^{126} \\
\mathbf{g}_5^{(8)}(D) &= 1 + D^5 + D^{20} + D^{47} + D^{84} + D^{144} \\
\mathbf{g}_6^{(8)}(D) &= 1 + D^{58} + D^{96} + D^{106} + D^{113} + D^{141} \\
\mathbf{g}_7^{(8)}(D) &= 1 + D^{41} + D^{77} + D^{108} + D^{117} + D^{130}
\end{aligned} \tag{17.6}$$

This code forms $J = 6$ orthogonal parity checks on each information error bit, has $d_{\min} = 7$, and is triple-error correcting. The measured performance of this code,

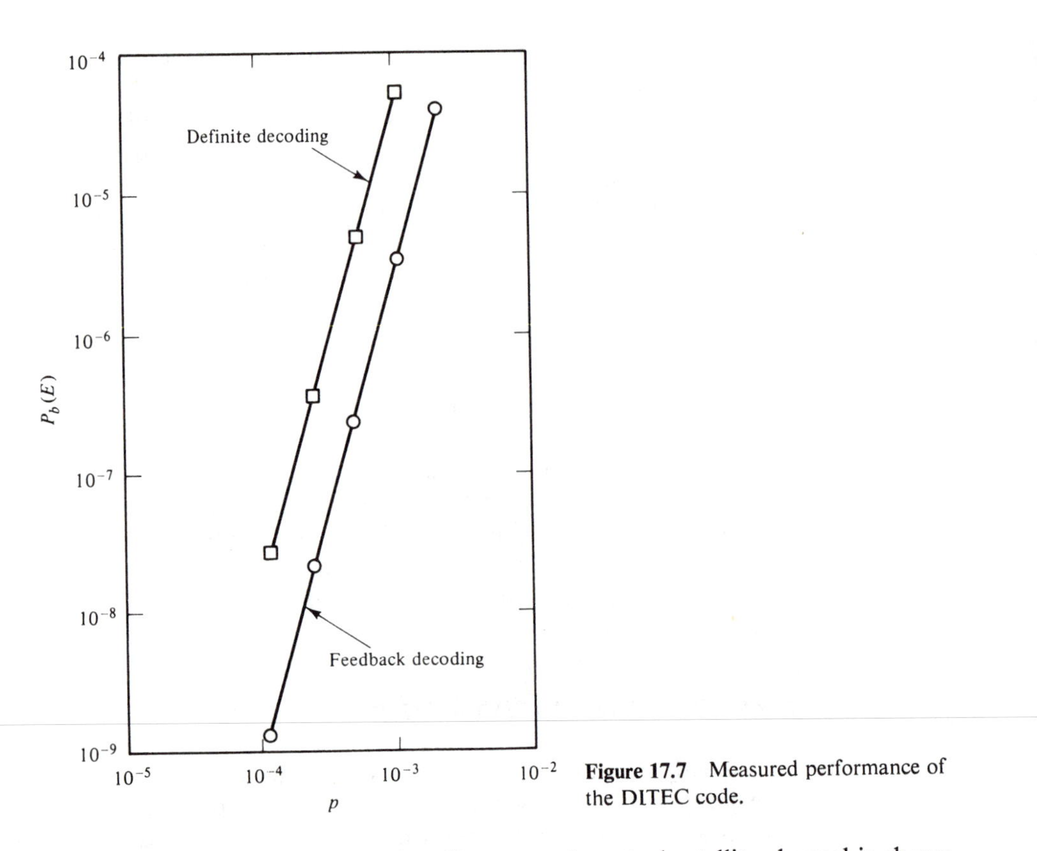

Figure 17.7 Measured performance of the DITEC code.

with both definite and feedback decoding, over the actual satellite channel is shown in Figure 17.7. Note that channel error rates of 10^{-4} are reduced to bit error rates of 10^{-9} with feedback decoding, but only to 5×10^{-8} with definite decoding.

In 1970, a majority-logic decoder for a (4, 3, 19) self-orthogonal code was designed and built for a demand-assignment PCM/FDMA system (SPADE) which transmits data at 40.8 Kbps over regular voice channels of the INTELSAT system. The generator polynomials of the code are given by

$$\begin{aligned} \mathbf{g}_1^{(4)}(D) &= 1 + D^3 + D^{15} + D^{19} \\ \mathbf{g}_2^{(4)}(D) &= 1 + D^8 + D^{17} + D^{18} \\ \mathbf{g}_3^{(4)}(D) &= 1 + D^6 + D^{11} + D^{13}. \end{aligned} \tag{17.7}$$

This code forms $J = 4$ orthogonal parity checks on each information error bit, has $d_{\min} = 5$, and is double-error-correcting. The measured performance of this code with feedback decoding over the actual satellite channel is shown in Figure 17.8. Note that channel error rates of 10^{-4} are reduced to bit error rates of about 5×10^{-9} after decoding.

In 1975, a majority-logic decoder for an (8, 7, 47) self-orthogonal code was designed and built for the 64.0 Kbps INTELSAT single-channel-per-carrier (SCPC) system [23]. The generator polynomials of the code are given by

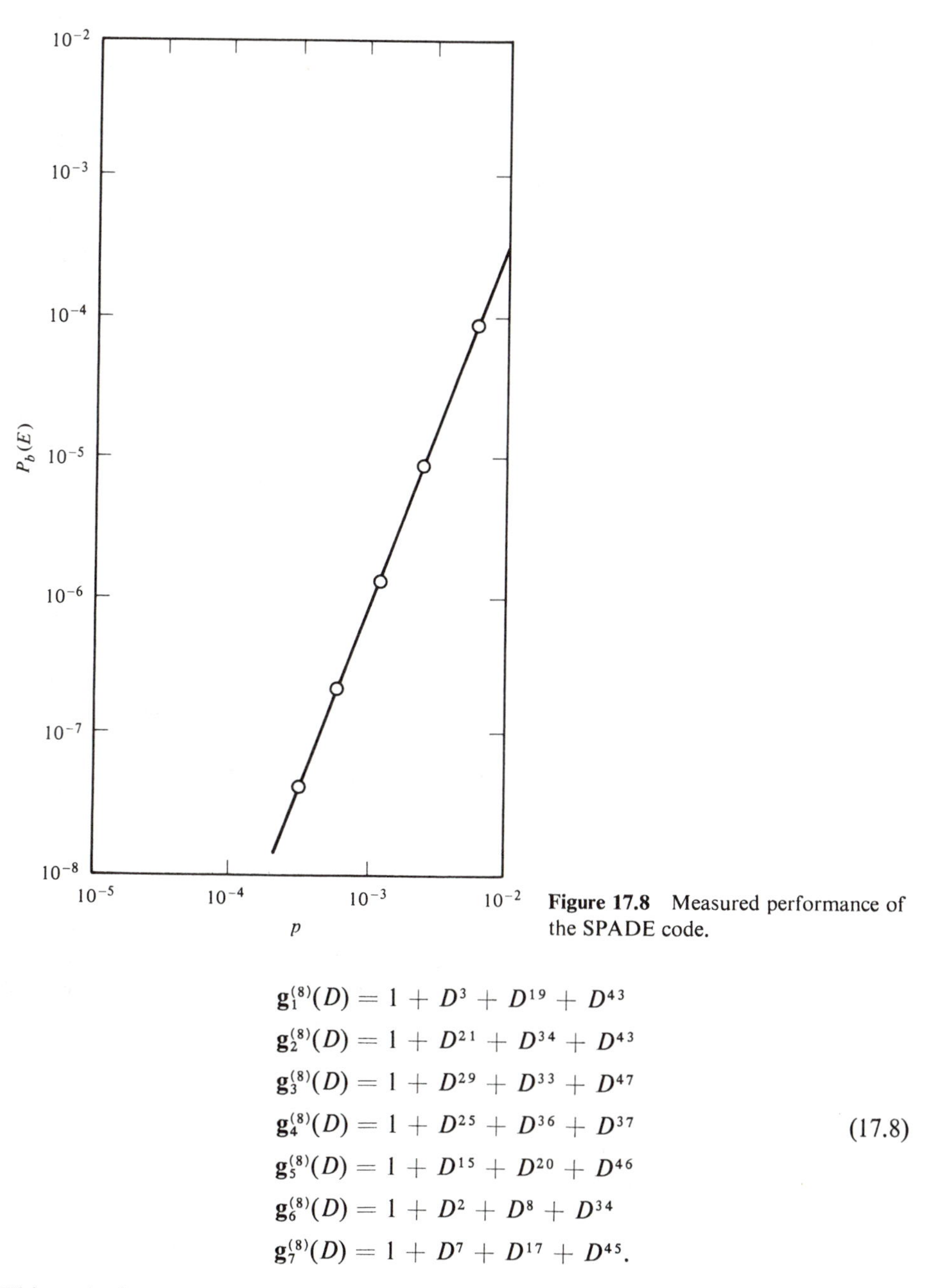

Figure 17.8 Measured performance of the SPADE code.

$$
\begin{aligned}
\mathbf{g}_1^{(8)}(D) &= 1 + D^3 + D^{19} + D^{43} \\
\mathbf{g}_2^{(8)}(D) &= 1 + D^{21} + D^{34} + D^{43} \\
\mathbf{g}_3^{(8)}(D) &= 1 + D^{29} + D^{33} + D^{47} \\
\mathbf{g}_4^{(8)}(D) &= 1 + D^{25} + D^{36} + D^{37} \\
\mathbf{g}_5^{(8)}(D) &= 1 + D^{15} + D^{20} + D^{46} \\
\mathbf{g}_6^{(8)}(D) &= 1 + D^{2} + D^{8} + D^{34} \\
\mathbf{g}_7^{(8)}(D) &= 1 + D^{7} + D^{17} + D^{45}.
\end{aligned}
\tag{17.8}
$$

This code forms $J = 4$ orthogonal parity checks on each information error bit, has $d_{\min} = 5$, and is double-error-correcting. The measured performance of this code with feedback decoding over the actual satellite channel is shown in Figure 17.9. Note that channel error rates of 10^{-4} are reduced to bit error rates of about 5×10^{-8} after decoding.

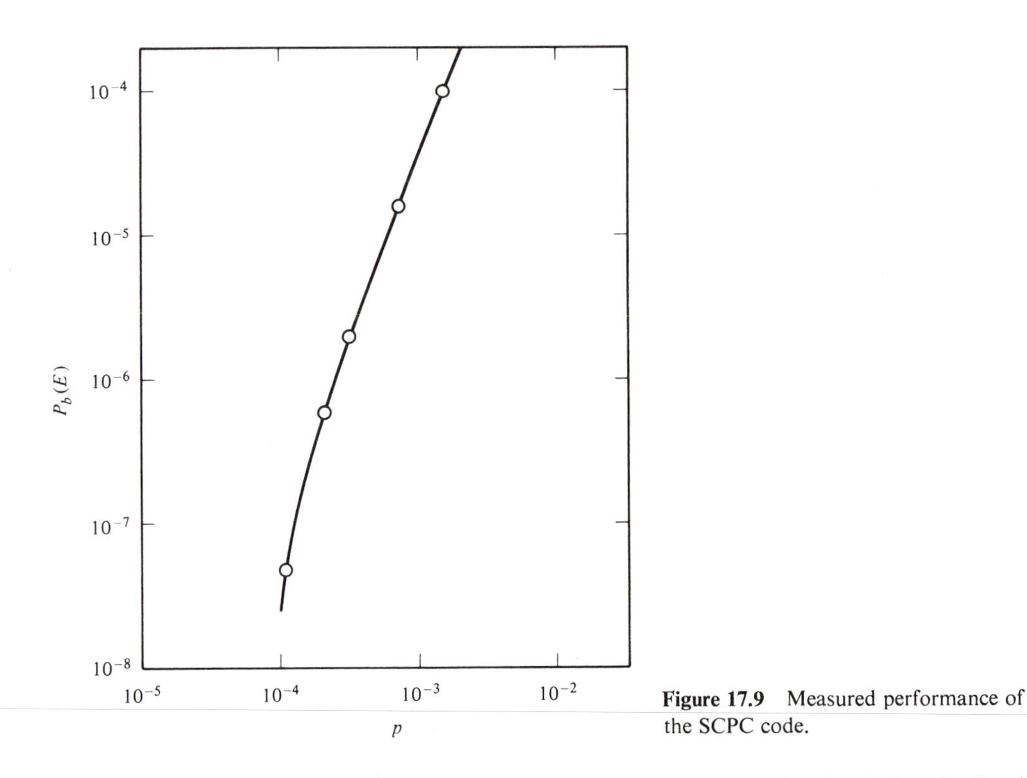

Figure 17.9 Measured performance of the SCPC code.

In 1971, the Linkabit Corporation designed and built a hard-decision feedback majority-logic decoder for a (2, 1, 9) self-orthogonal diffuse code with generator polynomial

$$\mathbf{g}^{(2)}(D) = 1 + D^3 + D^7 + D^9. \tag{17.9}$$

This code forms $J = 4$ orthogonal parity checks on each information error bit, has $d_{\text{min}} = 5$, is double-error-correcting, and corrects many error patterns of higher weight. It is combined with 256 bits of interleaving for protection against burst errors, and operates at data rates up to 1 Mbps. Similar feedback decoder systems with rates $R = 2/3$ and $R = 3/4$ have also been built by Linkabit. These systems have been used in applications ranging from airborne satellite communication to terrestial telephone lines [2].

In the early 1970s, the Harris Corporation designed and built a soft-decision majority-logic decoder for a (2, 1, 12) orthogonalizable diffuse code with generator polynomial

$$\mathbf{g}^{(2)}(D) = 1 + D^3 + D^8 + D^{12}. \tag{17.10}$$

This code forms $J = 4$ orthogonal parity checks on each information error bit, has $d_{\text{min}} = 5$, and is decoded using an approximation to Massey's [24] APP decoding. The decoder was designed for use as an integral part of an Army PSK modem, operates at a 5 Mbps data rate, and provides a 2.6 dB coding gain at a bit error rate of 10^{-5}.

17.4 APPLICATIONS TO BURST-ERROR CORRECTION

The first practical applications of convolutional codes were for burst-error correction in the mid-1960s. Some examples of these early applications will now be discussed, together with some more recent applications of convolutional codes for protection against radio-frequency interference (RFI).

The Codex Corporation pioneered the use of convolutional codes for error correction in the mid-1960s [25]. Among the early systems designed and built by Codex are the following:

1. An $R = 1/2$ Massey–Kohlenberg diffuse convolutional coding system with data rates up to 1.2 Kbps. In the presence of random errors of probability p, the probability of incorrect decoding is $102p^3$. The burst-correcting capability varies from 12 bits with a 44-bit guard space to 2400 bits with a 7208-bit guard space, depending on the application. Applications included troposcatter radio, HF radio, telegraph terminals, telephone line terminals, and the Air Force's Ballistic Missile Early Warning System (BMEWS).
2. An $R = 2/3$ Massey–Kohlenberg diffuse convolutional coding system with a 1.6 Kbps data rate for telephone line applications. For random errors of probability p, the probability of incorrect decoding is $11p^2$. The burst-correcting capability is 270 bits with a guard space of 1623 bits.
3. An $R = 3/4$ Massey–Kohlenberg diffuse convolutional coding system with data rates up to 3.6 Kbps for telephone line and submarine cable terminals. For random errors of probability p, the probability of incorrect decoding is $19p^2$. The burst-correcting capability is 240 bits with a 2404-bit guard space or 384 bits with a 3844-bit guard space.
4. An $R = 1/2$ Gallager burst-finding system with a 75 bps data rate for telegraph and military applications. For random errors of probability p, the probability of incorrect decoding is $102p^3$. The burst-correcting capability is 733 bits with a guard space of only about 20 bits more than the burst length.

These systems were all developed by 1965 and are the first known applications of convolutional codes for error correction.

In 1973, the Linkabit Corporation designed and built an interleaved convolutional encoder/Viterbi decoder for the U.S. Army Satellite Communications Agency [26]. This system uses the (2, 1, 6) code of (17.1) with either hard or 3-bit ($Q = 8$) soft-decision decoding. It was designed to correct burst errors on a troposcatter channel. The interleaver is capable of dispersing bursts up to 1024 bits in length such that no two bits in a burst will appear at the decoder closer than 64 bits apart. With hard-decision decoding and an effective channel error rate (after interleaving) of 10^{-2}, this system can achieve a bit error rate of 5×10^{-7}.

In the early 1970s, the Harris Corporation developed a 2 Mbps coding system for the Naval Electronics Systems Command for protection of a shipboard satellite terminal against RFI bursts caused by radar [27]. This system used the (2, 1, 5) code with

$$\mathbf{g}^{(1)}(D) = 1 + D^2 + D^4 + D^5$$
$$\mathbf{g}^{(2)}(D) = 1 + D + D^5 \tag{17.11}$$

and $d_{\text{free}} = 7$. The decoder used the Viterbi algorithm with 3-bit ($Q = 8$) soft quantization. The system provided a 4.7 dB coding gain at a bit error rate of 10^{-5} in a Gaussian noise environment.

Protection against RFI bursts was provided by an interleaver which dispersed 200 bit bursts 20 bits apart over a 4000-bit block. Assuming RFI which affects 100 data bits out of every 2000 (a 5% duty cycle) in addition to Gaussian background noise, the performance of this system is illustrated in Figure 17.10. Note that the system retains about a 1.4 dB coding gain at a 10^{-5} error rate against a combination of noise and interference.

A more sophisticated coding system for a similar application has recently been implemented by Harris. This system also employs convolutional codes with either hard- or soft-decision Viterbi decoding. A variable code rate of 1/2, 2/3, or 3/4 is achieved by deleting bits from an $R = 1/2$ code. The basic code is a (2, 1, 8) code with

$$\mathbf{g}^{(1)}(D) = 1 + D^2 + D^3 + D^4 + D^8$$
$$\mathbf{g}^{(2)}(D) = 1 + D + D^2 + D^3 + D^5 + D^7 + D^8 \tag{17.12}$$

and $d_{\text{free}} = 12$. A (3, 2, 4) code with $K = 8$ is derived from the basic code by deleting one bit from every two branches of the code trellis, and a (4, 3, 3) code with $K = 8$ is

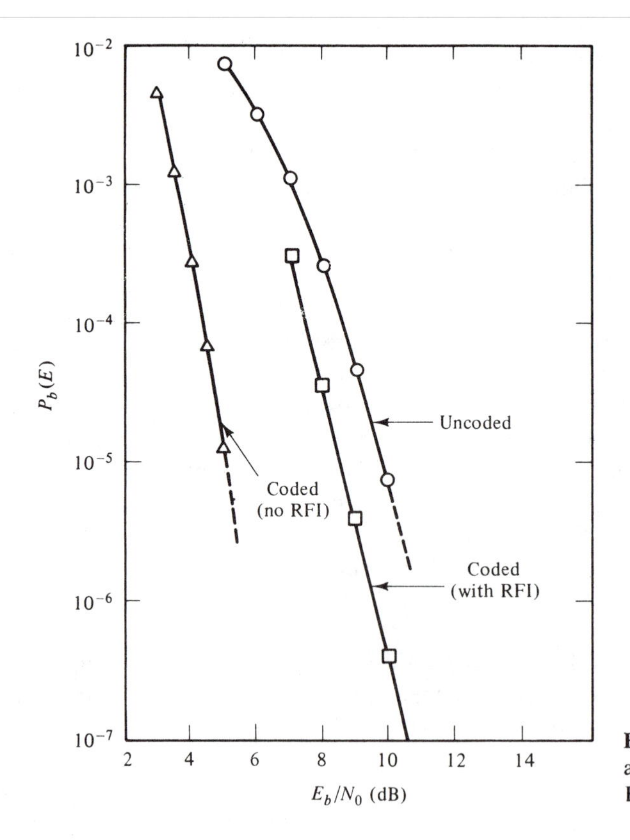

Figure 17.10 Performance curves for a (2, 1, 5) code in Gaussian noise and RFI.

derived by deleting two bits from every three branches of the trellis. Since the same basic trellis describes all three codes, they can all be decoded by the same decoder. The relatively low data rate of 32 Kbps makes the use of these longer-constraint-length ($K = 8$) codes feasible with Viterbi decoding.

The interleaver used in this system disperses bits over a 1000-bit block and orders them according to a pseudorandom pattern. This pseudorandom interleaving provides robust protection against a variety of RFI duty cycles. Predicted coding gains of this system with soft-decision decoding at a bit error rate of 10^{-5} are shown in Table 17.2 [3].

TABLE 17.2 PREDICTED CODING GAINS (IN dB) OF HARRIS RFI SYSTEM WITH 3-BIT ($Q = 8$) SOFT-DECISION DECODING AT A BIT ERROR RATE OF 10^{-5}

Code	No RFI	5% RFI	10% RFI
(2, 1, 8)	5.6	5.1	4.5
(3, 2, 4)	4.9	4.3	3.2
(4, 3, 3)	4.3	3.5	—

A similar system was developed in the mid-1970s by Lincoln Laboratory for military satellite communication systems affected by radar-induced RFI [28]. A block diagram of the system is shown in Figure 17.11. It is assumed that the system has the capability to detect RFI pulses. The demodulated channel symbols received during an RFI pulse are replaced by erasures prior to de-interleaving and decoding.

Four different codes with rates 1/2, 2/3, 3/4, and 4/5 were selected. The codes used are listed in Table 17.3 together with their free distances. Each code has $K = 6$, and hence requires a 64-state Viterbi decoder. The decoder used 3-bit ($Q = 8$) soft quan-

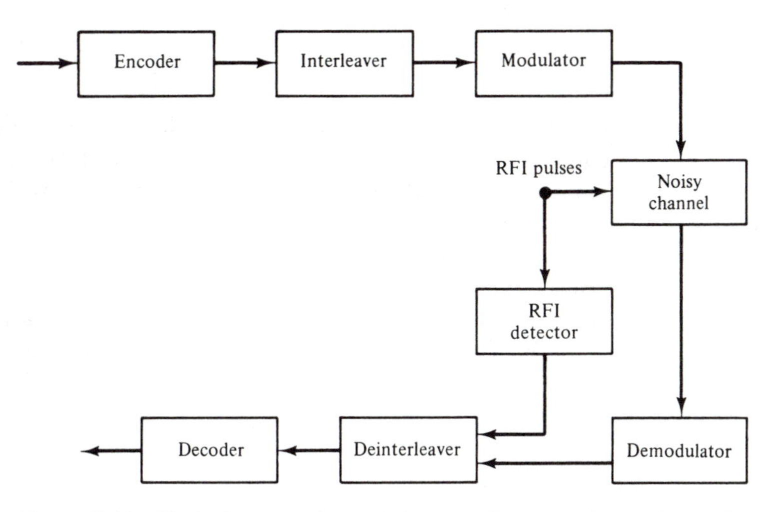

Figure 17.11 Block diagram of a coded system for protection against radar-induced RFI.

TABLE 17.3 CODES USED IN LINCOLN LABORATORY RFI SYSTEM

Code	Generator sequences				d_{free}
(2, 1, 6)	$\mathbf{g}^{(1)} = 1\ 1\ 1\ 1\ 0\ 0\ 1$				10
	$\mathbf{g}^{(2)} = 1\ 0\ 1\ 1\ 0\ 1\ 1$				
(3, 2, 3)	$\mathbf{g}_1^{(1)} = 1\ 1\ 0\ 1$	$\mathbf{g}_2^{(1)} = 0\ 1\ 1\ 0$			7
	$\mathbf{g}_1^{(2)} = 0\ 1\ 1\ 0$	$\mathbf{g}_2^{(2)} = 1\ 1\ 0\ 1$			
	$\mathbf{g}_1^{(3)} = 1\ 1\ 0\ 1$	$\mathbf{g}_2^{(3)} = 1\ 1\ 1\ 1$			
(4, 3, 2)	$\mathbf{g}_1^{(1)} = 1\ 1\ 0$	$\mathbf{g}_2^{(1)} = 0\ 1\ 1$	$\mathbf{g}_3^{(1)} = 0\ 1\ 0$		6
	$\mathbf{g}_1^{(2)} = 0\ 0\ 1$	$\mathbf{g}_2^{(2)} = 1\ 0\ 0$	$\mathbf{g}_3^{(2)} = 0\ 1\ 1$		
	$\mathbf{g}_1^{(3)} = 0\ 0\ 0$	$\mathbf{g}_2^{(3)} = 0\ 0\ 1$	$\mathbf{g}_3^{(3)} = 1\ 1\ 1$		
	$\mathbf{g}_1^{(4)} = 1\ 1\ 1$	$\mathbf{g}_2^{(4)} = 1\ 1\ 0$	$\mathbf{g}_3^{(4)} = 1\ 0\ 0$		
(5, 4, 2)	$\mathbf{g}_1^{(1)} = 1\ 1\ 0$	$\mathbf{g}_2^{(1)} = 1\ 0\ 1$	$\mathbf{g}_3^{(1)} = 1\ 1\ 0$	$\mathbf{g}_4^{(1)} = 1\ 1\ 0$	5
	$\mathbf{g}_1^{(2)} = 1\ 0\ 0$	$\mathbf{g}_2^{(2)} = 0\ 0\ 0$	$\mathbf{g}_3^{(2)} = 0\ 1\ 0$	$\mathbf{g}_4^{(2)} = 1\ 0\ 0$	
	$\mathbf{g}_1^{(3)} = 0\ 1\ 1$	$\mathbf{g}_2^{(3)} = 1\ 0\ 1$	$\mathbf{g}_3^{(3)} = 0\ 0\ 0$	$\mathbf{g}_4^{(3)} = 1\ 0\ 0$	
	$\mathbf{g}_1^{(4)} = 0\ 1\ 0$	$\mathbf{g}_2^{(4)} = 0\ 0\ 1$	$\mathbf{g}_3^{(4)} = 1\ 1\ 0$	$\mathbf{g}_4^{(4)} = 0\ 0\ 0$	
	$\mathbf{g}_1^{(5)} = 0\ 0\ 1$	$\mathbf{g}_2^{(5)} = 0\ 0\ 0$	$\mathbf{g}_3^{(5)} = 0\ 0\ 0$	$\mathbf{g}_4^{(5)} = 1\ 0\ 0$	

TABLE 17.4 LINCOLN LABORATORY RFI SYSTEM PERFORMANCE

			Degradation (dB)		
Code	RFI	E_b/N_0	Predicted	Simulated	Measured
(2, 1, 6)	0%	2.7 dB	—	—	—
	10%		0.8	0.9	0.9
	20%		1.8	1.8	2.0
(4, 3, 2)	0%	3.6 dB	—	—	—
	5%		0.7	0.8	0.9
	10%		1.6	2.0	2.1

tization, had a path memory of 64 data bits, and was designed to decode four different code rates. A summary of system performance is presented in Table 17.4. This table shows the degradation in decibels compared to code performance with no RFI at a bit error rate of 10^{-3}. The E_b/N_0 values given in the table are those required to achieve a 10^{-3} error rate for the two code rates examined. Two RFI erasure rates are considered for each code. The erasures are assumed to be randomly distributed due to ideal interleaving. Three degradation estimates are given for each case, one a prediction based on analysis, one from computer simulations, and one from actual measurements taken on a hardware decoder. Results tended to support the following conclusion. At a given $P_b(E)$ and E_b/N_0, each lower code rate could tolerate an additional 5% of RFI. For example, the $R = 4/5$ code achieves $P_b(E) = 10^{-3}$ at $E_b/N_0 \approx 6$ dB with 5% RFI, and the rate $R = 3/4$ code achieves $P_b(E) = 10^{-3}$ at $E_b/N_0 \approx 6$ dB with 10% RFI.

17.5 APPLICATIONS OF CONVOLUTIONAL CODES IN ARQ SYSTEMS

In Chapter 15 we discussed the application of block codes to ARQ systems. Block codes are a natural choice for many ARQ systems which require high-rate codes used for error detection only. However, in a hybrid-ARQ system which combines error correction with error detection, thereby requiring lower code rates, convolutional codes can also be used. In this section we present several examples of the application of convolutional codes to hybrid-ARQ systems.

In the late 1970s, the Linkabit Corporation developed a hybrid-ARQ system for a packet radio application of the Defense Advanced Research Projects Agency (DARPA) [29]. The system uses convolutional codes with a Fano sequential decoder, and is capable of operation at data rates up to 300 Kbps. Flexibility is achieved with multiple code rates (1/4, 1/2, 3/4, 7/8) and the ability to handle hard-quantized, hard-quantized with erasures, or 2-bit ($Q = 4$) soft-quantized demodulator outputs.

The retransmission strategy employed by this system is quite simple. If the decoding of a packet of data exceeds some prespecified time limit, an erasure is declared and a retransmission of the packet is requested. This strategy is referred to as the *time-out* (TO) *algorithm*. Some advantages of the TO algorithm are as follows:

1. The undetected error probability can be made very low with almost no increase in decoding time by choosing a long-constraint-length code. The basic system is designed to work with any (2, 1, 23) code, such as those discussed in Section 17.2. For packet lengths on the order of 1000 data bits, the fractional rate loss incurred by adding a 23-bit tail to each data packet is negligible.
2. High throughput can be achieved because most packets are decoded quickly.
3. Design flexibility can be obtained by varying the decoder time limit, the code rate, and the demodulator quantization. A larger time limit will result in a longer average decoding time per received packet, but fewer retransmissions. A smaller time limit, on the other hand, will result in a shorter average decoding time per received packet, but more retransmissions. The code rate options make the system adaptable to a variety of channel conditions. The demodulator is designed to operate in two basic modes: hard quantization or 2-bit ($Q = 2$) soft quantization. In the hard-decision mode, a retransmitted packet presents the decoder with two separate versions of the received sequence. Those positions that do not agree are then treated as erasures, and the decoder switches to a hard-decision with erasures mode, thereby reducing the probability of another decoding failure.

This hybrid-ARQ system can be used with any of the standard ARQ protocols: stop-and-wait (SW), go-back-N (GBN), or selective repeat (SR). Depending on the protocol employed and the round-trip delay, the TO algorithm described above has an optimum *time limit* T to achieve the highest *throughput efficiency* η for a given choice of code and demodulator quantization. Drukarev and Costello [30] have performed a random coding analysis of the TO algorithm and determined the optimum

T as a function of packet length, decoder speed (average number of computations per second), E_b/N_0, and code rate. They also performed computer simulations and determined the optimum T experimentally for code rates of 1/2, hard demodulator decisions (a BSC), and an SR protocol (no delays). The results of these simulations for three different packet lengths are shown in Figure 17.12. The code used was an ODP (2, 1, 8) code with optimum free distance (OFD). The BSC transition probability was $p = 0.045$. The time limit T was expressed in terms of computations per bit, which is equivalent to an absolute time limit for a given decoder speed and packet length.

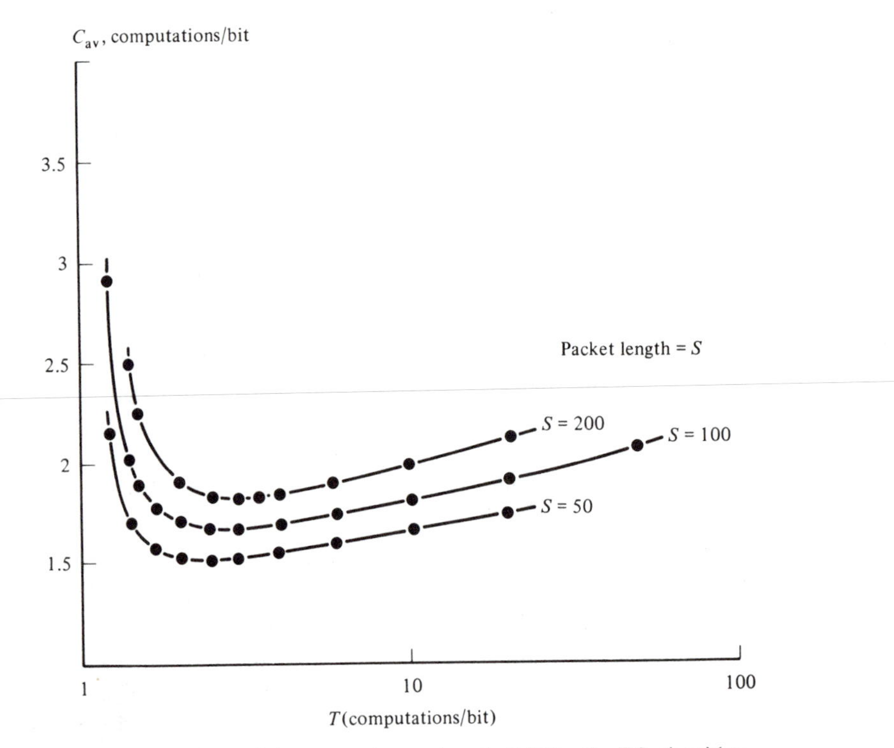

Figure 17.12 Determining an optimum time limit T for the TO algorithm.

The throughput efficiency is then given by

$$\eta = \frac{1}{C_{av}}, \tag{17.13}$$

where C_{av} is the average number of computations per decoded information bit. Note that in each case a definite minimum for C_{av} is achieved, since if T is too low, too many retransmissions are requested, and if T is too high, the average time to decode a packet is too large. In Figure 17.13, η is plotted as a function of undetected error probability $P_b(E)$ with T as a parameter for several different code constraint lengths. By connecting the points of maximum throughput on each curve, we see that undetected error probability can be reduced sharply with almost no decrease in throughput by using longer code lengths! This property makes sequential decoding of convo-

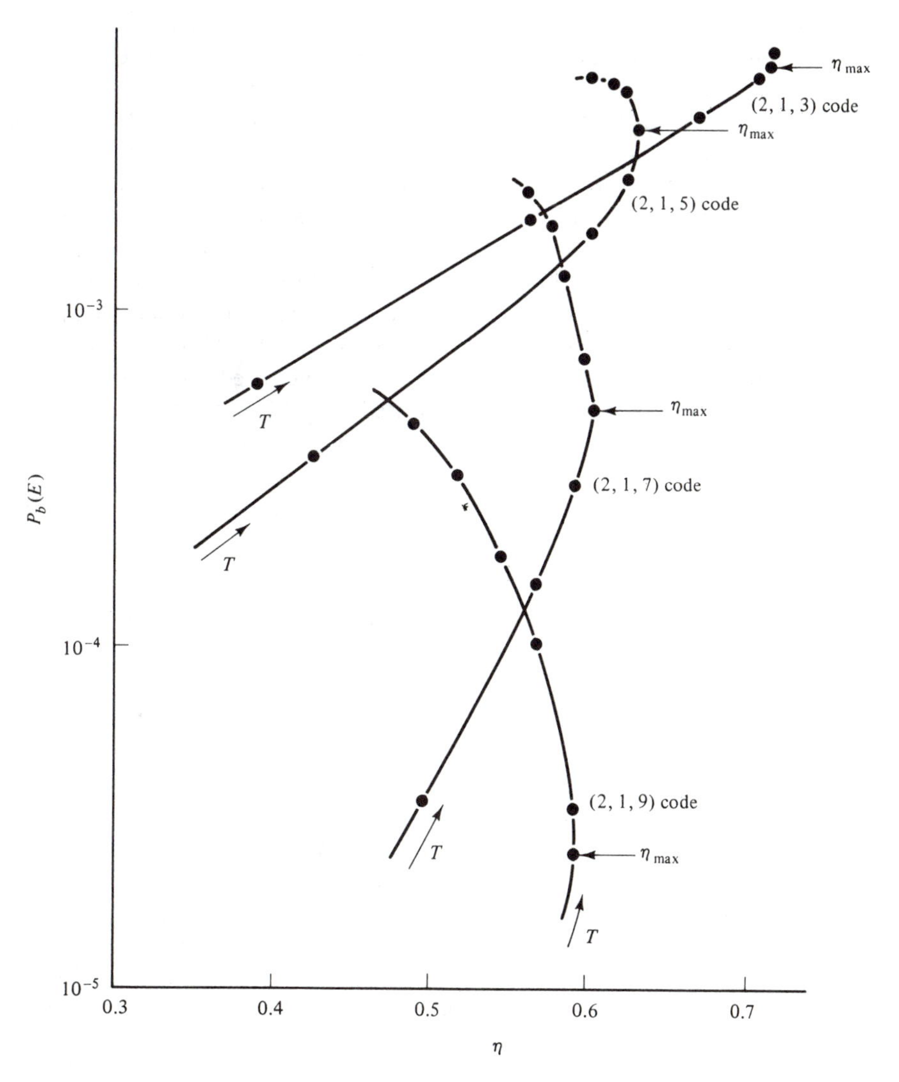

Figure 17.13 η versus $P_b(E)$ for the TO algorithm.

lutional codes a very attractive option in hybrid-ARQ systems requiring very low undetected error probabilities.

Drukarev and Costello [30] have proposed a more efficient retransmission strategy for the hybrid-ARQ system described above. Their strategy is based on the stack algorithm, and is designed to predict the occurrence of a time-out in advance by monitoring the metric of the path on top of the stack. In particular, if the metric of the best path has fallen by an amount more than some (negative) threshold Γ over the last W branches (i.e., if its slope over the last W branches is less than Γ/W), this is taken as an indication that the packet is noisy, and a retransmission is requested immediately. This results in less wasted time trying to decode noisy packets before the time limit is reached, and hence yields improved throughput, particularly on noisy channels. This strategy is called the *slope control algorithm* (SCA).

A random coding analysis was also performed for the SCA to obtain optimum values of Γ and W [30]. It was found that an optimum value for Γ exists, but that η is relatively insensitive to W. In other words, it is the slope Γ/W which determines the efficiency of the algorithm. The results of computer simulations for an ODP, OFD (2, 1, 10) code on a BSC with $p = 0.045$ and an SR protocol are shown in Figure 17.14. Note that, as in the TO algorithm, a definite minimum in C_{av}, and hence a maximum η, is achieved for a certain value of Γ. The throughput advantage of the SCA compared to the TO algorithm is illustrated in Figure 17.15 for a (2, 1, 6) code, where the TO curve has T as a parameter and the SCA curve has Γ as a parameter. The throughput advantage of the SCA becomes more pronounced on noisier channels.

Viterbi decoding can also be used in a hybrid-ARQ mode. Yamamoto and

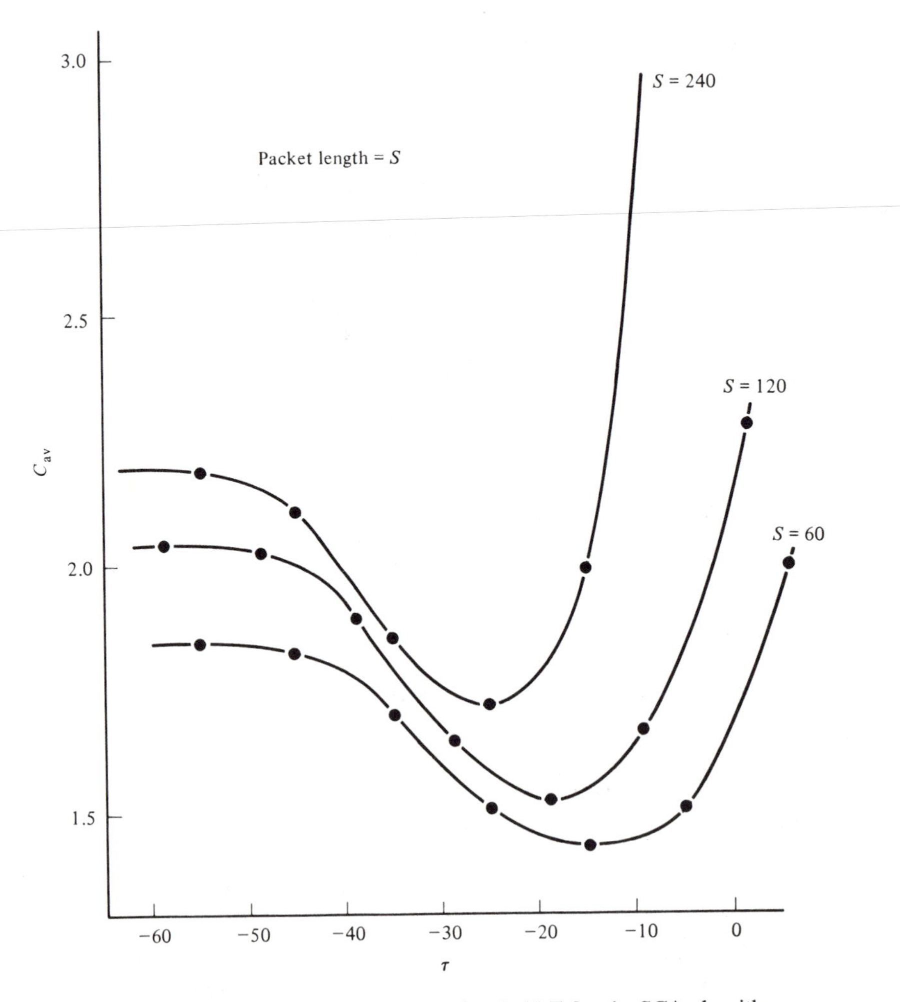

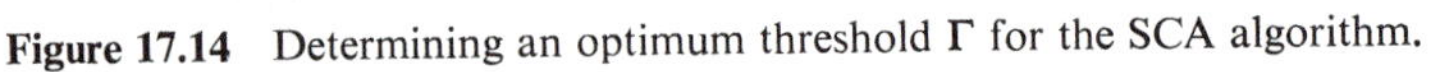

Figure 17.14 Determining an optimum threshold Γ for the SCA algorithm.

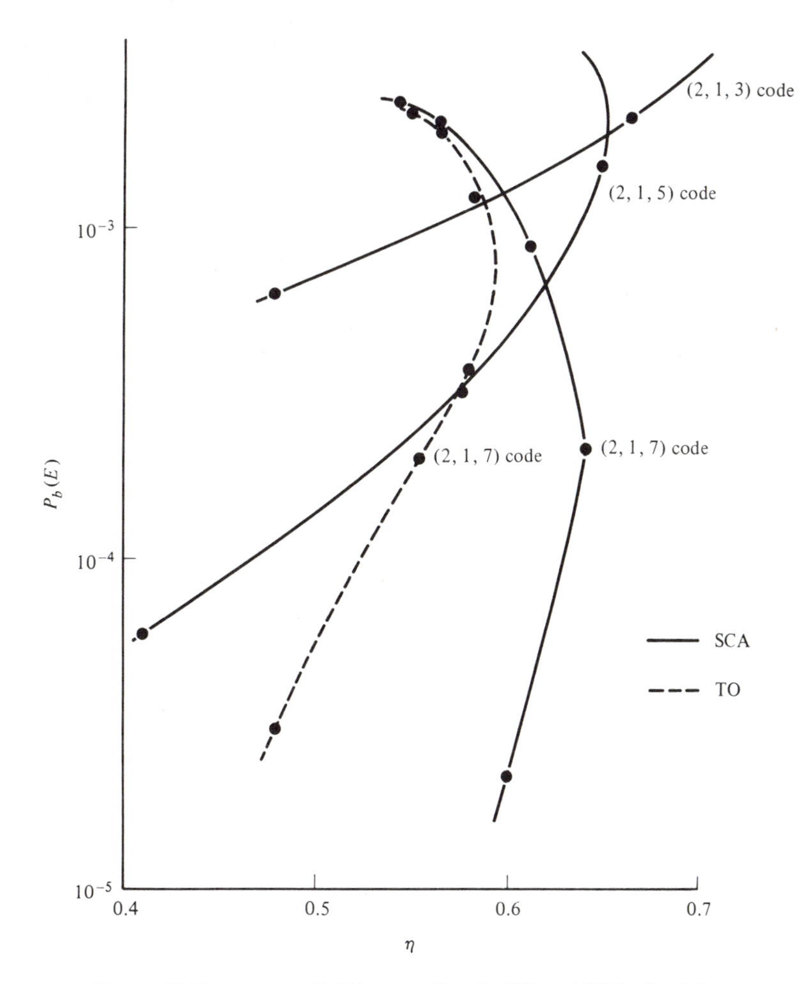

Figure 17.15 η versus $P_b(E)$ curves for the TO and SCA algorithms.

Itoh [31] have described a system, similar in concept to the one described above for sequential decoding, which uses the Viterbi algorithm. A retransmission request is generated according to the following strategy. If the difference in the metrics of the survivor and the next best path is less than some prespecified value A, the survivor is labeled with an X. The X label remains on that path as long as it continues to be a survivor. If, at any level of the trellis, all survivors are labeled with an X, a retransmission is requested. In other words, a survivor is considered "unreliable" if, in any one of its comparisons, some other path had a metric close to it. When all survivors are considered "unreliable," a retransmission request is generated.

The performance of this system was analyzed using both random coding arguments and the code-generating-function approach, and computer simulations were also performed [31]. At a bit error rate of 10^{-5}, the hybrid-ARQ system was able to achieve a savings of about 1.0 to 1.5 dB in E_b/N_0 compared to ordinary Viterbi decoding, with only a minor loss in throughput. Compared to the sequential decoding system described above, the Viterbi ARQ system achieves a higher throughput for moderate error rates (short constraint lengths), but the sequential decoding system achieves a higher throughput for low error rates (long constraint lengths). This is because in order to achieve a low error rate, the Viterbi ARQ system must use a large K, and therefore requires a long time to decode each packet, resulting in a low throughput. The decoding time of a sequential decoder is relatively insensitive to K, however, and hence most packets are decoded quickly even for large K, and a high throughput is maintained. A qualitative comparison of the undetected error probability versus throughput characteristics of both ARQ systems is shown in Figure 17.16. Hence for high-performance hybrid-ARQ systems using convolutional codes, sequential decoding appears to be the best choice.

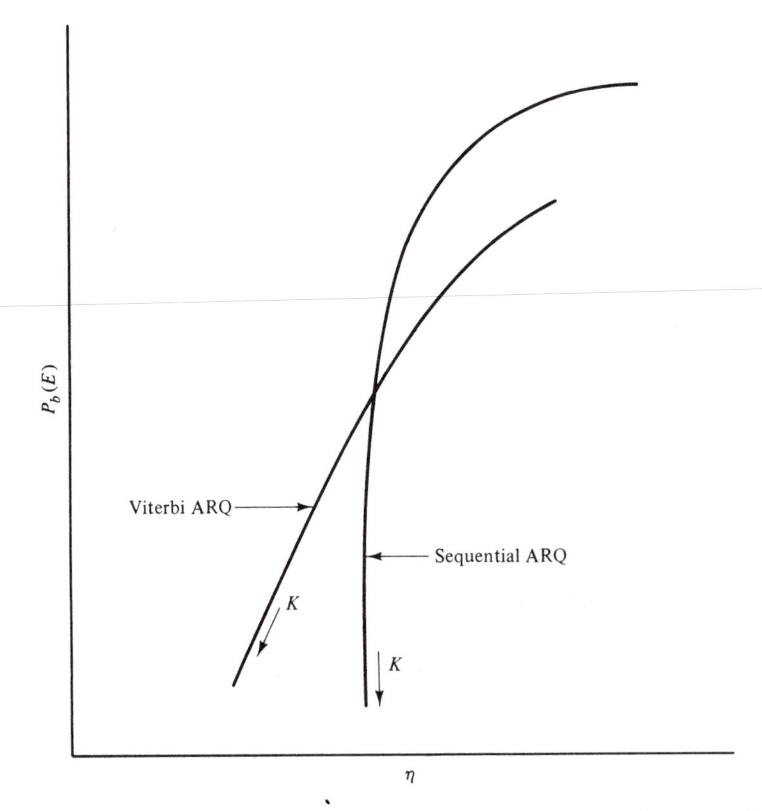

Figure 17.16 Comparison of hybrid-ARQ systems using sequential and Viterbi decoding.

PROBLEMS

17.1. Consider the two NASA Planetary Standard Codes described in Section 17.1.

(a) Draw a block diagram of the encoding circuits.

(b) Find the asymptotic coding gains for a hard-quantized and an unquantized demodulator output.

(c) Compare with the coding gains shown in Figure 17.1.

17.2. Show how bits can be deleted from the code words of the (2, 1, 6) code of (17.1) to convert it to a (4, 3, 2) code. Find d_{free} for this new code. Describe how both codes can be decoded using the same decoding trellis.

17.3. Consider the (3, 1, 6) code, the (3, 1, 7) code, and the (18, 6, 1) code described with reference to Figure 17.5. Compare the complexity of a Viterbi decoder for these three codes and discuss their relative performance.

17.4. The (2, 1, 31) code with generator sequences

$$\mathbf{g}^{(1)} = 42545013236$$

$$\mathbf{g}^{(2)} = 70436206116$$

is known to have $d_{free} = 28$. Calculate its CDF d_i for $i \leq 7$ and compare with the CDFs for the two (2, 1, 31) codes shown in Figure 17.6. Order the three codes from the point of view of computational performance with sequential decoding. What can you say about error performance?

17.5. Consider the seven (2, 1, 23) codes listed in Table 17.1. Develop a computer program, based on the flowchart of Figure 12.14 or one of your own design, to compute the CDF d_i of the codes. Using this program, find d_{free} for each of these codes, and verify the entries in Table 17.1.

17.6. Consider the (4, 3, 19) code used with the SPADE system described in Section 17.3.

(a) Verify that the positive difference sets of its generator polynomials are full and disjoint.

(b) Find the constraint length n_A and the effective constraint length n_E.

(c) Draw a block diagram of the decoder.

17.7. Consider the (2, 1, 12) code of (17.10). Find a set of $J = 4$ orthogonal parity checks on $e_0^{(1)}$ for this code, and show that the code is a three-diffuse code.

17.8. Determine the generator polynomials of the three diffuse codes discussed in Section 17.4.

17.9. Consider the (2, 1, 5) code of (17.11). Using the approximate expressions for bit error probability $P_b(E)$ developed in Chapter 11, find the coding gain when $P_b(E) = 10^{-5}$ for both a hard-quantized and an unquantized demodulator output. Compare your results with the 4.7 dB coding gain claimed for 3-bit ($Q = 8$) soft quantization.

17.10. Design an interleaver to space the bits in a 200-bit burst 20 bits apart over a 4000-bit block for the (2, 1, 5) code of (17.11).

17.11. For $W = 3$ and $\Gamma = -9$, apply the SCA to Examples 12.4 and 12.5. In each case, determine if a retransmission will be requested. (Apply the slope test only to paths that extend at least $W = 3$ branches into the tree.)

17.12. For $A = 5$, apply the Viterbi ARQ algorithm described in Section 17.5 to Example 11.1. Determine if a retransmission will be requested.

REFERENCES

1. M. K. Simon and J. G. Smith, "Alternate Symbol Inversion for Improved Symbol Synchronization in Convolutionally Coded Systems," *IEEE Trans. Commun.*, COM-28, pp. 228–237, February 1980.
2. J. P. Odenwalder and A. J. Viterbi, "Overview of Existing and Projected Uses of Coding

in Military Satellite Communications," *NTC Conf. Rec.*, pp. 36:4.1–36:4.2, Los Angeles, Calif., December 1977.

3. J. M. Geist, private communication, 1978.
4. A. S. Acampora and R. P. Gilmore, "Analog Viterbi Decoding for High Speed Digital Satellite Channels," *IEEE Trans. Commun.*, COM-26, pp. 1463–1470, October 1978.
5. J. P. Odenwalder, "Optimal Decoding of Convolutional Codes," Ph.D. thesis, University of California, Los Angeles, 1970.
6. L. N. Lee, "Concatenated Coding Systems Employing a Unit-Memory Convolutional Code and a Byte-Oriented Decoding Algorithm," *IEEE Trans. Commun.*, COM-25, pp. 1064–1074, October 1977.
7. G. W. Zeoli, "Coupled Decoding of Block-Convolutional Concatenated Codes," *IEEE Trans. Commun.*, COM-21, pp. 219–226, March 1973.
8. S. Lin and H. Lyne, "Some Results on Binary Convolutional Code Generators," *IEEE Trans. Inf. Theory*, IT-13, pp. 134–139, January 1967.
9. J. L. Massey and D. J. Costello, Jr., "Nonsystematic Convolutional Codes for Sequential Decoding in Space Applications," *IEEE Trans. Commun. Technol.*, COM-19, pp. 806–813, October 1971.
10. J. W. Layland and W. A. Lushbaugh, "A Flexible High-Speed Sequential Decoder for Deep Space Channels," *IEEE Trans. Commun. Technol.*, COM-19, pp. 813–820, October 1971.
11. J. L. Massey, "A Recommended $R = 1/2$, $K = 32$, Quick-Look-In Convolutional Code for NASA Use," Technical Report No. EE-751, University of Notre Dame, Notre Dame, Ind., April 1975.
12. R. Johannesson, "Some Long Rate One-Half Binary Convolutional Codes with an Optimum Distance Profile," *IEEE Trans. Inf. Theory*, IT-22, pp. 629–631, September 1976.
13. J. L. Massey, "Comparison of Rate One-Half, Equivalent Constraint Length 24, Binary Convolutional Codes for Use with Sequential Decoding on the Deep-Space Channel," Technical Report No. EE-762, University of Notre Dame, Notre Dame, Ind., April 1976.
14. R. Johannesson, "Robustly Optimal Rate One-Half Binary Convolutional Codes," *IEEE Trans. Inf. Theory*, IT-21, pp. 464–468, July 1975.
15. L. R. Bahl, and F. Jelinek, "Rate 1/2 Convolutional Codes with Complementary Generators," *IEEE Trans. Inf. Theory*, IT-17, pp. 718–727, November 1971.
16. J. L. Massey, D. J. Costello, Jr., and J. Justesen, "Polynomial Weights and Code Constructions," *IEEE Trans. Inf. Theory*, IT-19, pp. 101–110, January 1973.
17. G. D. Forney, Jr., "Use of a Sequential Decoder to Analyze Convolutional Code Structure," *IEEE Trans. Inf. Theory*, IT-16, pp. 793–795, November 1970.
18. R. Johannesson and E. Paaske, "Further Results on Binary Convolutional Codes with an Optimum Distance Profile," *IEEE Trans. Inf. Theory*, IT-24, pp. 264–268, March 1978.
19. G. D. Forney, Jr., and E. K. Bower, "A High-Speed Sequential Decoder: Prototype Design and Test," *IEEE Trans. Commun. Technol.*, COM-19, pp. 821–835, October 1971.
20. G. C. Clark, Jr., "Error Correction Coding at Harris ESD," Harris Corp. Internal Report, 1974.
21. S. L. Bernstein, D. A. McNeill, and I. Richer, "A Signaling Scheme and Experimental Receiver for Extremely Low Frequency (ELF) Communication," *IEEE Trans. Commun.*, COM-22, pp. 508–528, April 1974.

22. W. W. Wu, "New Convolutional Codes—Part I," *IEEE Trans. Commun.*, COM-23, pp. 942–956, September 1975.
23. W. W. Wu, "New Convolutional Codes—Part II," *IEEE Trans. Commun.*, COM-24, pp. 19–33, January 1976.
24. J. L. Massey, *Threshold Decoding*, MIT Press, Cambridge, Mass., 1963.
25. A. Kohlenberg, and G. D. Forney, Jr., "Convolutional Coding for Channels with Memory," *IEEE Trans. Inf. Theory*, IT-14, pp. 618–626, September 1968.
26. J. E. Quigley, private communication, 1978.
27. G. C. Clark, Jr. and R. C. Davis, "Two Recent Applications of Error-Correction Coding to Communications Systems Design," *IEEE Trans. Commun. Technol.*, COM-19, pp. 856–863, October 1971.
28. S. L. Bernstein, H. M. Heggestad, S. Y. Mui, and I. Richer, "Variable-Rate Viterbi Decoding in the Presence of RFI," *NTC Conf. Rec.*, pp. 36.6.1–36.6.5, Los Angeles, Calif., December 1977.
29. R. E. Kahn, S. A. Gronemeyer, J. Burchfiel, and R. E. Kunzelman, "Advances in Packet Radio Technology," *Proc. IEEE,* 66, pp. 1468–1497, November 1978.
30. A. Drukarev, and D. J. Costello, Jr., "ARQ Error Control Using Sequential Decoding," *ICC Conf. Rec.*, pp. 4.7.1–4.7.5, Seattle, Wash., June 1980.
31. H. Yamamoto and K. Itoh, "Viterbi Decoding Algorithm for Convolutional Codes with Repeat Request," *IEEE Trans. Inf. Theory*, IT-26, pp. 540–547, September 1980.

APPENDIX A

Tables of Galois Fields

Galois fields GF(2^m) for $3 \le m \le 10$ are given. Each element in GF(2^m) is expressed as a power of some primitive element α and as a linear sum of $\alpha^0, \alpha^1, \ldots, \alpha^{m-1}$. The minimal polynomial of α is used to generate the field. For every m, the generating polynomial of the field is given. Each element is represented by the following notation:

$$i \qquad (a_0, a_1, \ldots, a_{m-1}),$$

where the integer i represents α^i and the binary m-tuple $(a_0, a_1, \ldots, a_{m-1})$ represents

$$\alpha^i = a_0 + a_1\alpha + \cdots + a_{m-1}\alpha^{m-1}.$$

Galois Fields of Order 2^m

1. GF(2^3) generated by $\mathbf{p}(X) = 1 + X + X^3$

-	000
0	100
1	010
2	001
3	110
4	011
5	111
6	101

2. $GF(2^4)$ generated by $\mathbf{p}(X) = 1 + X + X^4$

-	0000
0	1000
1	0100
2	0010
3	0001
4	1100
5	0110
6	0011
7	1101
8	1010
9	0101
10	1110
11	0111
12	1111
13	1011
14	1001

3. $GF(2^5)$ generated by $\mathbf{p}(X) = 1 + X^2 + X^5$

-	00000	15	11111
0	10000	16	11011
1	01000	17	11001
2	00100	18	11000
3	00010	19	01100
4	00001	20	00110
5	10100	21	00011
6	01010	22	10101
7	00101	23	11110
8	10110	24	01111
9	01011	25	10011
10	10001	26	11101
11	11100	27	11010
12	01110	28	01101
13	00111	29	10010
14	10111	30	01001

4. $GF(2^6)$ generated by $\mathbf{p}(X) = 1 + X + X^6$

-	000000	31	101001
0	100000	32	100100
1	010000	33	010010
2	001000	34	001001
3	000100	35	110100
4	000010	36	011010
5	000001	37	001101
6	110000	38	110110
7	011000	39	011011
8	001100	40	111101
9	000110	41	101110
10	000011	42	010111
11	110001	43	111011
12	101000	44	101101
13	010100	45	100110
14	001010	46	010011
15	000101	47	111001
16	110010	48	101100
17	011001	49	010110
18	111100	50	001011
19	011110	51	110101
20	001111	52	101010
21	110111	53	010101
22	101011	54	111010
23	100101	55	011101
24	100010	56	111110
25	010001	57	011111
26	111000	58	111111
27	011100	59	101111
28	001110	60	100111
29	000111	61	100011
30	110011	62	100001

5. GF(2^7) generated by $\mathbf{p}(X) = 1 + X^3 + X^7$

–	0000000
0	1000000
1	0100000
2	0010000
3	0001000
4	0000100
5	0000010
6	0000001
7	1001000
8	0100100
9	0010010
10	0001001
11	1001100
12	0100110
13	0010011
14	1000001
15	1101000
16	0110100
17	0011010
18	0001101
19	1001110
20	0100111
21	1011011
22	1100101
23	1111010
24	0111101
25	1010110
26	0101011
27	1011101
28	1100110
29	0110011
30	1010001
31	1100000
32	0110000
33	0011000
34	0001100
35	0000110
36	0000011
37	1001001
38	1101100
39	0110110
40	0011011
41	1000101
42	1101010
43	0110101
44	1010010
45	0101001
46	1011100
47	0101110
48	0010111
49	1000011
50	1101001
51	1111100
52	0111110
53	0011111
54	1000111
55	1101011
56	1111101
57	1110110
58	0111011
59	1010101
60	1100010
61	0110001
62	1010000
63	0101000
64	0010100
65	0001010
66	0000101
67	1001010
68	0100101
69	1011010
70	0101101
71	1011110
72	0101111
73	1011111
74	1100111
75	1111011
76	1110101
77	1110010
78	0111001
79	1010100
80	0101010
81	0010101
82	1000010
83	0100001
84	1011000
85	0101100
86	0010110
87	0001011
88	1001101
89	1101110
90	0110111
91	1010011
92	1100001
93	1111000
94	0111100
95	0011110
96	0001111
97	1001111
98	1101111
99	1111111
100	1110111
101	1110011
102	1110001
103	1110000
104	0111000
105	0011100
106	0001110
107	0000111
108	1001011
109	1101101
110	1111110
111	0111111
112	1010111
113	1100011
114	1111001
115	1110100
116	0111010
117	0011101
118	1000110
119	0100011
120	1011001
121	1100100
122	0110010
123	0011001
124	1000100
125	0100010
126	0010001

6. GF(2^8) generated by $\mathbf{p}(X) = 1 + X^2 + X^3 + X^4 + X^8$

–	00000000	63	10000101
0	10000000	64	11111010
1	01000000	65	01111101
2	00100000	66	10000110
3	00010000	67	01000011
4	00001000	68	10011001
5	00000100	69	11110100
6	00000010	70	01111010
7	00000001	71	00111101
8	10111000	72	10100110
9	01011100	73	01010011
10	00101110	74	10010001
11	00010111	75	11110000
12	10110011	76	01111000
13	11100001	77	00111100
14	11001000	78	00011110
15	01100100	79	00001111
16	00110010	80	10111111
17	00011001	81	11100111
18	10110100	82	11001011
19	01011010	83	11011101
20	00101101	84	11010110
21	10101110	85	01101011
22	01010111	86	10001101
23	10010011	87	11111110
24	11110001	88	01111111
25	11000000	89	10000111
26	01100000	90	11111011
27	00110000	91	11000101
28	00011000	92	11011010
29	00001100	93	01101101
30	00000110	94	10001110
31	00000011	95	01000111
32	10111001	96	10011011
33	11100100	97	11110101
34	01110010	98	11000010
35	00111001	99	01100001
36	10100100	100	10001000
37	01010010	101	01000100
38	00101001	102	00100010
39	10101100	103	00010001
40	01010110	104	10110000
41	00101011	105	01011000
42	10101101	106	00101100
43	11101110	107	00010110
44	01110111	108	00001011
45	10000011	109	10111101
46	11111001	110	11100110
47	11000100	111	01110011
48	01100010	112	10000001
49	00110001	113	11111000
50	10100000	114	01111100
51	01010000	115	00111110
52	00101000	116	00011111
53	00010100	117	10110111
54	00001010	118	11100011
55	00000101	119	11001001
56	10111010	120	11011100
57	01011101	121	01101110
58	10010110	122	00110111
59	01001011	123	10100011
60	10011101	124	11101001
61	11110110	125	11001100
62	01111011	126	01100110

GF (2^8) Continued

127	00110011
128	10100001
129	11101000
130	01110100
131	00111010
132	00011101
133	10110110
134	01011011
135	10010101
136	11110010
137	01111001
138	10000100
139	01000010
140	00100001
141	10101000
142	01010100
143	00101010
144	00010101
145	10110010
146	01011001
147	10010100
148	01001010
149	00100101
150	10101010
151	01010101
152	10010010
153	01001001
154	10011100
155	01001110
156	00100111
157	10101011
158	11101101
159	11001110
160	01100111
161	10001011
162	11111101
163	11000110
164	01100011
165	10001001
166	11111100
167	01111110
168	00111111
169	10100111
170	11101011
171	11001101
172	11011110
173	01101111
174	10001111
175	11111111
176	11000111
177	11011011
178	11010101
179	11010010
180	01101001
181	10001100
182	01000110
183	00100011
184	10101001
185	11101100
186	01110110
187	00111011
188	10100101
189	11101010
190	01110101
191	10000010
192	01000001
193	10011000
194	01001100
195	00100110
196	00010011
197	10110001
198	11100000
199	01110000
200	00111000
201	00011100
202	00001110
203	00000111
204	10111011
205	11100101
206	11001010
207	01100101
208	10001010
209	01000101
210	10011010
211	01001101
212	10011110
213	01001111
214	10011111
215	11110111
216	11000011
217	11011001
218	11010100
219	01101010
220	00110101
221	10100010
222	01010001
223	10010000
224	01001000
225	00100100
226	00010010
227	00001001
228	10111100
229	01011110
230	00101111
231	10101111
232	11101111
233	11001111
234	11011111
235	11010111
236	11010011
237	11010001
238	11010000
239	01101000
240	00110100
241	00011010
242	00001101
243	10111110
244	01011111
245	10010111
246	11110011
247	11000001
248	11011000
249	01101100
250	00110110
251	00011011
252	10110101
253	11100010
254	01110001

7. $GF(2^9)$ generated by $\mathbf{p}(X) = 1 + X^4 + X^9$

-	000000000
0	100000000
1	010000000
2	001000000
3	000100000
4	000010000
5	000001000
6	000000100
7	000000010
8	000000001
9	100010000
10	010001000
11	001000100
12	000100010
13	000010001
14	100011000
15	010001100
16	001000110
17	000100011
18	100000001
19	110010000
20	011001000
21	001100100
22	000110010
23	000011001
24	100011100
25	010001110
26	001000111
27	100110011
28	110001001
29	111010100
30	011101010
31	001110101
32	100101010
33	010010101
34	101011010
35	010101101
36	101000110
37	010100011
38	101000001
39	110110000
40	011011000
41	001101100
42	000110110
43	000011011
44	100011101
45	110011110
46	011001111
47	101110111
48	110101011
49	111000101
50	111110010
51	011111001
52	101101100
53	010110110
54	001011011
55	100111101
56	110001110
57	011000111
58	101110011
59	110101001
60	111000100
61	011100010
62	001110001
63	100101000
64	010010100
65	001001010
66	000100101
67	100000010
68	010000001
69	101010000
70	010101000
71	001010100
72	000101010
73	000010101
74	100011010
75	010001101
76	101010110
77	010101011
78	101000101
79	110110010
80	011011001
81	101111100
82	010111110
83	001011111
84	100111111
85	110001111
86	111010111
87	111111011
88	111101101
89	111100110
90	011110011
91	101101001
92	110100100
93	011010010
94	001101001
95	100100100
96	010010010
97	001001001
98	100110100
99	010011010
100	001001101
101	100110110
102	010011011
103	101011101
104	110111110
105	011011111
106	101111111
107	110101111
108	111000111
109	111110011
110	111101001
111	111100100
112	011110010
113	001111001
114	100101100
115	010010110
116	001001011
117	100110101
118	110001010
119	011000101
120	101110010
121	010111001
122	101001100
123	010100110
124	001010011
125	100111001
126	110001100

GF (2^9) Continued

127	011000110
128	001100011
129	100100001
130	110000000
131	011000000
132	001100000
133	000110000
134	000011000
135	000001100
136	000000110
137	000000011
138	100010001
139	110011000
140	011001100
141	001100110
142	000110011
143	100001001
144	110010100
145	011001010
146	001100101
147	100100010
148	010010001
149	101011000
150	010101100
151	001010110
152	000101011
153	100000101
154	110010010
155	011001001
156	101110100
157	010111010
158	001011101
159	100111110
160	010011111
161	101011111
162	110111111
163	111001111
164	111110111
165	111101011
166	111100101
167	111100010
168	011110001
169	101101000
170	010110100
171	001011010
172	000101101
173	100000110
174	010000011
175	101010001
176	110111000
177	011011100
178	001101110
179	000110111
180	100001011
181	110010101
182	111011010
183	011101101
184	101100110
185	010110011
186	101001001
187	110110100
188	011011010
189	001101101
190	100100110
191	010010011
192	101011001
193	110111100
194	011011110
195	001101111
196	100100111
197	110000011
198	111010001
199	111111000
200	011111100
201	001111110
202	000111111
203	100001111
204	110010111
205	111011011
206	111111101
207	111101110
208	011110111
209	101101011
210	110100101
211	111000010
212	011100001
213	101100000
214	010110000
215	001011000
216	000101100
217	000010110
218	000001011
219	100010101
220	110011010
221	011001101
222	101110110
223	010111011
224	101001101
225	110110110
226	011011011
227	101111101
228	110101110
229	011010111
230	101111011
231	110101101
232	111000110
233	011100011
234	101100001
235	110100000
236	011010000
237	001101000
238	000110100
239	000011010
240	000001101
241	100010110
242	010001011
243	101010101
244	110111010
245	011011101
246	101111110
247	010111111
248	101001111
249	110110111
250	111001011
251	111110101
252	111101010
253	011110101
254	101101010

GF (2^9) Continued

255	010110101
256	101001010
257	010100101
258	101000010
259	010100001
260	101000000
261	010100000
262	001010000
263	000101000
264	000010100
265	000001010
266	000000101
267	100010010
268	010001001
269	101010100
270	010101010
271	001010101
272	100111010
273	010011101
274	101011110
275	010101111
276	101000111
277	110110011
278	111001001
279	111110100
280	011111010
281	001111101
282	100101110
283	010010111
284	101011011
285	110111101
286	111001110
287	011100111
288	101100011
289	110100001
290	111000000
291	011100000
292	001110000
293	000111000
294	000011100
295	000001110
296	000000111
297	100010011
298	110011001
299	111011100
300	011101110
301	001110111
302	100101011
303	110000101
304	111010010
305	011101001
306	101100100
307	010110010
308	001011001
309	100111100
310	010011110
311	001001111
312	100110111
313	110001011
314	111010101
315	111111010
316	011111101
317	101101110
318	010110111
319	101001011
320	110110101
321	111001010
322	011100101
323	101100010
324	010110001
325	101001000
326	010100100
327	001010010
328	000101001
329	100000100
330	010000010
331	001000001
332	100110000
333	010011000
334	001001100
335	000100110
336	000010011
337	100011001
338	110011100
339	011001110
340	001100111
341	100100011
342	110000001
343	111010000
344	011101000
345	001110100
346	000111010
347	000011101
348	100011110
349	010001111
350	101010111
351	110111011
352	111001101
353	111110110
354	011111011
355	101101101
356	110100110
357	011010011
358	101111001
359	110101100
360	011010110
361	001101011
362	100100101
363	110000010
364	011000001
365	101110000
366	010111000
367	001011100
368	000101110
369	000010111
370	100011011
371	110011101
372	111011110
373	011101111
374	101100111
375	110100011
376	111000001
377	111110000
378	011111000
379	001111100
380	000111110
381	000011111
382	100011111

GF (2^9) Continued

383	110011111
384	111011111
385	111111111
386	111101111
387	111100111
388	111100011
389	111100001
390	111100000
391	011110000
392	001111000
393	000111100
394	000011110
395	000001111
396	100010111
397	110011011
398	111011101
399	111111110
400	011111111
401	101101111
402	110100111
403	111000011
404	111110001
405	111101000
406	011110100
407	001111010
408	000111101
409	100001110
410	010000111
411	101010011
412	110111001
413	111001100
414	011100110
415	001110011
416	100101001
417	110000100
418	011000010
419	001100001
420	100100000
421	010010000
422	001001000
423	000100100
424	000010010
425	000001001
426	100010100
427	010001010
428	001000101
429	100110010
430	010011001
431	101011100
432	010101110
433	001010111
434	100111011
435	110001101
436	111010110
437	011101011
438	101100101
439	110100010
440	011010001
441	101111000
442	010111100
443	001011110
444	000101111
445	100000111
446	110010011
447	111011001
448	111111100
449	011111110
450	001111111
451	100101111
452	110000111
453	111010011
454	111111001
455	111101100
456	011110110
457	001111011
458	100101101
459	110000110
460	011000011
461	101110001
462	110101000
463	011010100
464	001101010
465	000110101
466	100001010
467	010000101
468	101010010
469	010101001
470	101000100
471	010100010
472	001010001
473	100111000
474	010011100
475	001001110
476	000100111
477	100000011
478	110010001
479	111011000
480	011101100
481	001110110
482	000111011
483	100001101
484	110010110
485	011001011
486	101110101
487	110101010
488	011010101
489	101111010
490	010111101
491	101001110
492	010100111
493	101000011
494	110110001
495	111001000
496	011100100
497	001110010
498	000111001
499	100001100
500	010000110
501	001000011
502	100110001
503	110001000
504	011000100
505	001100010
506	000110001
507	100001000
508	010000100
509	001000010
510	000100001

8. $GF(2^{10})$ generated by $\mathbf{p}(X) = 1 + X^3 + X^{10}$

-	0000000000
0	1000000000
1	0100000000
2	0010000000
3	0001000000
4	0000100000
5	0000010000
6	0000001000
7	0000000100
8	0000000010
9	0000000001
10	1001000000
11	0100100000
12	0010010000
13	0001001000
14	0000100100
15	0000010010
16	0000001001
17	1001000100
18	0100100010
19	0010010001
20	1000001000
21	0100000100
22	0010000010
23	0001000001
24	1001100000
25	0100110000
26	0010011000
27	0001001100
28	0000100110
29	0000010011
30	1001001001
31	1101100100
32	0110110010
33	0011011001
34	1000101100
35	0100010110
36	0010001011
37	1000000101
38	1101000010
39	0110100001
40	1010010000
41	0101001000
42	0010100100
43	0001010010
44	0000101001
45	1001010100
46	0100101010
47	0010010101
48	1000001010
49	0100000101
50	1011000010
51	0101100001
52	1011110000
53	0101111000
54	0010111100
55	0001011110
56	0000101111
57	1001010111
58	1101101011
59	1111110101
60	1110111010
61	0111011101
62	1010101110
63	0101010111
64	1011101011
65	1100110101
66	1111011010
67	0111101101
68	1010110110
69	0101011011
70	1011101101
71	1100110110
72	0110011011
73	1010001101
74	1100000110
75	0110000011
76	1010000001
77	1100000000
78	0110000000
79	0011000000
80	0001100000
81	0000110000
82	0000011000
83	0000001100
84	0000000110
85	0000000011
86	1001000001
87	1101100000
88	0110110000
89	0011011000
90	0001101100
91	0000110110
92	0000011011
93	1001001101
94	1101100110
95	0110110011
96	1010011001
97	1100001100
98	0110000110
99	0011000011
100	1000100001
101	1101010000
102	0110101000
103	0011010100
104	0001101010
105	0000110101
106	1001011010
107	0100101101
108	1011010110
109	0101101011
110	1011110101
111	1100111010
112	0110011101
113	1010001110
114	0101000111
115	1011100011
116	1100110001
117	1111011000
118	0111101100
119	0011110110
120	0001111011
121	1001111101
122	1101111110
123	0110111111
124	1010011111
125	1100001111
126	1111000111

GF (2^{10}) Continued

127	1110100011
128	1110010001
129	1110001000
130	0111000100
131	0011100010
132	0001110001
133	1001111000
134	0100111100
135	0010011110
136	0001001111
137	1001100111
138	1101110011
139	1111111001
140	1110111100
141	0111011110
142	0011101111
143	1000110111
144	1101011011
145	1111101101
146	1110110110
147	0111011011
148	1010101101
149	1100010110
150	0110001011
151	1010000101
152	1100000010
153	0110000001
154	1010000000
155	0101000000
156	0010100000
157	0001010000
158	0000101000
159	0000010100
160	0000001010
161	0000000101
162	1001000010
163	0100100001
164	1011010000
165	0101101000
166	0010110100
167	0001011010
168	0000101101
169	1001010110
170	0100101011
171	1011010101
172	1100101010
173	0110010101
174	1010001010
175	0101000101
176	1011100010
177	0101110001
178	1011111000
179	0101111100
180	0010111110
181	0001011111
182	1001101111
183	1101110111
184	1111111011
185	1110111101
186	1110011110
187	0111001111
188	1010100111
189	1100010011
190	1111001001
191	1110100100
192	0111010010
193	0011101001
194	1000110100
195	0100011010
196	0010001101
197	1000000110
198	0100000011
199	1011000001
200	1100100000
201	0110010000
202	0011001000
203	0001100100
204	0000110010
205	0000011001
206	1001001100
207	0100100110
208	0010010011
209	1000001001
210	1101000100
211	0110100010
212	0011010001
213	1000101000
214	0100010100
215	0010001010
216	0001000101
217	1001100010
218	0100110001
219	1011011000
220	0101101100
221	0010110110
222	0001011011
223	1001101101
224	1101110110
225	0110111011
226	1010011101
227	1100001110
228	0110000111
229	1010000011
230	1100000001
231	1111000000
232	0111100000
233	0011110000
234	0001111000
235	0000111100
236	0000011110
237	0000001111
238	1001000111
239	1101100011
240	1111110001
241	1110111000
242	0111011100
243	0011101110
244	0001110111
245	1001111011
246	1101111101
247	1111111110
248	0111111111
249	1010111111
250	1100011111
251	1111001111
252	1110100111
253	1110010011
254	1110001001

GF (2^{10}) Continued

255	1110000100
256	0111000010
257	0011100001
258	1000110000
259	0100011000
260	0010001100
261	0001000110
262	0000100011
263	1001010001
264	1101101000
265	0110110100
266	0011011010
267	0001101101
268	1001110110
269	0100111011
270	1011011101
271	1100101110
272	0110010111
273	1010001011
274	1100000101
275	1111000010
276	0111100001
277	1010110000
278	0101011000
279	0010101100
280	0001010110
281	0000101011
282	1001010101
283	1101101010
284	0110110101
285	1010011010
286	0101001101
287	1011100110
288	0101110011
289	1011111001
290	1100111100
291	0110011110
292	0011001111
293	1000100111
294	1101010011
295	1111101001
296	1110110100
297	0111011010
298	0011101101
299	1000110110
300	0100011011
301	1011001101
302	1100100110
303	0110010011
304	1010001001
305	1100000100
306	0110000010
307	0011000001
308	1000100000
309	0100010000
310	0010001000
311	0001000100
312	0000100010
313	0000010001
314	1001001000
315	0100100100
316	0010010010
317	0001001001
318	1001100100
319	0100110010
320	0010011001
321	1000001100
322	0100000110
323	0010000011
324	1000000001
325	1101000000
326	0110100000
327	0011010000
328	0001101000
329	0000110100
330	0000011010
331	0000001101
332	1001000110
333	0100100011
334	1011010001
335	1100101000
336	0110010100
337	0011001010
338	0001100101
339	1001110010
340	0100111001
341	1011011100
342	0101101110
343	0010110111
344	1000011011
345	1101001101
346	1111100110
347	0111110011
348	1010111001
349	1100011100
350	0110001110
351	0011000111
352	1000100011
353	1101010001
354	1111101000
355	0111110100
356	0011111010
357	0001111101
358	1001111110
359	0100111111
360	1011011111
361	1100101111
362	1111010111
363	1110101011
364	1110010101
365	1110001010
366	0111000101
367	1010100010
368	0101010001
369	1011101000
370	0101110100
371	0010111010
372	0001011101
373	1001101110
374	0100110111
375	1011011011
376	1100101101
377	1111010110
378	0111101011
379	1010110101
380	1100011010
381	0110001101
382	1010000110

GF (2^{10}) Continued

383	0101000011
384	1011100001
385	1100110000
386	0110011000
387	0011001100
388	0001100110
389	0000110011
390	1001011001
391	1101101100
392	0110110110
393	0011011011
394	1000101101
395	1101010110
396	0110101011
397	1010010101
398	1100001010
399	0110000101
400	1010000010
401	0101000001
402	1011100000
403	0101110000
404	0010111000
405	0001011100
406	0000101110
407	0000010111
408	1001001011
409	1101100101
410	1111110010
411	0111111001
412	1010111100
413	0101011110
414	0010101111
415	1000010111
416	1101001011
417	1111100101
418	1110110010
419	0111011001
420	1010101100
421	0101010110
422	0010101011
423	1000010101
424	1101001010
425	0110100101
426	1010010010
427	0101001001
428	1011100100
429	0101110010
430	0010111001
431	1000011100
432	0100001110
433	0010000111
434	1000000011
435	1101000001
436	1111100000
437	0111110000
438	0011111000
439	0001111100
440	0000111110
441	0000011111
442	1001001111
443	1101100111
444	1111110011
445	1110111001
446	1110011100
447	0111001110
448	0011100111
449	1000110011
450	1101011001
451	1111101100
452	0111110110
453	0011111011
454	1000111101
455	1101011110
456	0110101111
457	1010010111
458	1100001011
459	1111000101
460	1110100010
461	0111010001
462	1010101000
463	0101010100
464	0010101010
465	0001010101
466	1001101010
467	0100110101
468	1011011010
469	0101101101
470	1011110110
471	0101111011
472	1011111101
473	1100111110
474	0110011111
475	1010001111
476	1100000111
477	1111000011
478	1110100001
479	1110010000
480	0111001000
481	0011100100
482	0001110010
483	0000111001
484	1001011100
485	0100101110
486	0010010111
487	1000001011
488	1101000101
489	1111100010
490	0111110001
491	1010111000
492	0101011100
493	0010101110
494	0001010111
495	1001101011
496	1101110101
497	1111111010
498	0111111101
499	1010111110
500	0101011111
501	1011101111
502	1100110111
503	1111011011
504	1110101101
505	1110010110
506	0111001011
507	1010100101
508	1100010010
509	0110001001
510	1010000100

GF (2^{10}) Continued

511	0101000010
512	0010100001
513	1000010000
514	0100001000
515	0010000100
516	0001000010
517	0000100001
518	1001010000
519	0100101000
520	0010010100
521	0001001010
522	0000100101
523	1001010010
524	0100101001
525	1011010100
526	0101101010
527	0010110101
528	1000011010
529	0100001101
530	1011000110
531	0101100011
532	1011110001
533	1100111000
534	0110011100
535	0011001110
536	0001100111
537	1001110011
538	1101111001
539	1111111100
540	0111111110
541	0011111111
542	1000111111
543	1101011111
544	1111101111
545	1110110111
546	1110011011
547	1110001101
548	1110000110
549	0111000011
550	1010100001
551	1100010000
552	0110001000
553	0011000100
554	0001100010
555	0000110001
556	1001011000
557	0100101100
558	0010010110
559	0001001011
560	1001100101
561	1101110010
562	0110111001
563	1010011100
564	0101001110
565	0010100111
566	1000010011
567	1101001001
568	1111100100
569	0111110010
570	0011111001
571	1000111100
572	0100011110
573	0010001111
574	1000000111
575	1101000011
576	1111100001
577	1110110000
578	0111011000
579	0011101100
580	0001110110
581	0000111011
582	1001011101
583	1101101110
584	0110110111
585	1010011011
586	1100001101
587	1111000110
588	0111100011
589	1010110001
590	1100011000
591	0110001100
592	0011000110
593	0001100011
594	1001110001
595	1101111000
596	0110111100
597	0011011110
598	0001101111
599	1001110111
600	1101111011
601	1111111101
602	1110111110
603	0111011111
604	1010101111
605	1100010111
606	1111001011
607	1110100101
608	1110010010
609	0111001001
610	1010100100
611	0101010010
612	0010101001
613	1000010100
614	0100001010
615	0010000101
616	1000000010
617	0100000001
618	1011000000
619	0101100000
620	0010110000
621	0001011000
622	0000101100
623	0000010110
624	0000001011
625	1001000101
626	1101100010
627	0110110001
628	1010011000
629	0101001100
630	0010100110
631	0001010011
632	1001101001
633	1101110100
634	0110111010
635	0011011101
636	1000101110
637	0100010111
638	1011001011

GF (2^{10}) Continued

639	1100100101
640	1111010010
641	0111101001
642	1010110100
643	0101011010
644	0010101101
645	1000010110
646	0100001011
647	1011000101
648	1100100010
649	0110010001
650	1010001000
651	0101000100
652	0010100010
653	0001010001
654	1001101000
655	0100110100
656	0010011010
657	0001001101
658	1001100110
659	0100110011
660	1011011001
661	1100101100
662	0110010110
663	0011001011
664	1000100101
665	1101010010
666	0110101001
667	1010010100
668	0101001010
669	0010100101
670	1000010010
671	0100001001
672	1011000100
673	0101100010
674	0010110001
675	1000011000
676	0100001100
677	0010000110
678	0001000011
679	1001100001
680	1101110000
681	0110111000
682	0011011100
683	0001101110
684	0000110111
685	1001011011
686	1101101101
687	1111110110
688	0111111011
689	1010111101
690	1100011110
691	0110001111
692	1010000111
693	1100000011
694	1111000001
695	1110100000
696	0111010000
697	0011101000
698	0001110100
699	0000111010
700	0000011101
701	1001001110
702	0100100111
703	1011010011
704	1100101001
705	1111010100
706	0111101010
707	0011110101
708	1000111010
709	0100011101
710	1011001110
711	0101100111
712	1011110011
713	1100111001
714	1111011100
715	0111101110
716	0011110111
717	1000111011
718	1101011101
719	1111101110
720	0111110111
721	1010111011
722	1100011101
723	1111001110
724	0111100111
725	1010110011
726	1100011001
727	1111001100
728	0111100110
729	0011110011
730	1000111001
731	1101011100
732	0110101110
733	0011010111
734	1000101011
735	1101010101
736	1111101010
737	0111110101
738	1010111010
739	0101011101
740	1011101110
741	0101110111
742	1011111011
743	1100111101
744	1111011110
745	0111101111
746	1010110111
747	1100011011
748	1111001101
749	1110100110
750	0111010011
751	1010101001
752	1100010100
753	0110001010
754	0011000101
755	1000100010
756	0100010001
757	1011001000
758	0101100100
759	0010110010
760	0001011001
761	1001101100
762	0100110110
763	0010011011
764	1000001101
765	1101000110
766	0110100011

GF (2^{10}) Continued

767	1010010001
768	1100001000
769	0110000100
770	0011000010
771	0001100001
772	1001110000
773	0100111000
774	0010011100
775	0001001110
776	0000100111
777	1001010011
778	1101101001
779	1111110100
780	0111111010
781	0011111101
782	1000111110
783	0100011111
784	1011001111
785	1100100111
786	1111010011
787	1110101001
788	1110010100
789	0111001010
790	0011100101
791	1000110010
792	0100011001
793	1011001100
794	0101100110
795	0010110011
796	1000011001
797	1101001100
798	0110100110
799	0011010011
800	1000101001
801	1101010100
802	0110101010
803	0011010101
804	1000101010
805	0100010101
806	1011001010
807	0101100101
808	1011110010
809	0101111001
810	1011111100
811	0101111110
812	0010111111
813	1000011111
814	1101001111
815	1111100111
816	1110110011
817	1110011001
818	1110001100
819	0111000110
820	0011100011
821	1000110001
822	1101011000
823	0110101100
824	0011010110
825	0001101011
826	1001110101
827	1101111010
828	0110111101
829	1010011110
830	0101001111
831	1011100111
832	1100110011
833	1111011001
834	1110101100
835	0111010110
836	0011101011
837	1000110101
838	1101011010
839	0110101101
840	1010010110
841	0101001011
842	1011100101
843	1100110010
844	0110011001
845	1010001100
846	0101000110
847	0010100011
848	1000010001
849	1101001000
850	0110100100
851	0011010010
852	0001101001
853	1001110100
854	0100111010
855	0010011101
856	1000001110
857	0100000111
858	1011000011
859	1100100001
860	1111010000
861	0111101000
862	0011110100
863	0001111010
864	0000111101
865	1001011110
866	0100101111
867	1011010111
868	1100101011
869	1111010101
870	1110101010
871	0111010101
872	1010101010
873	0101010101
874	1011101010
875	0101110101
876	1011111010
877	0101111101
878	1011111110
879	0101111111
880	1011111111
881	1100111111
882	1111011111
883	1110101111
884	1110010111
885	1110001011
886	1110000101
887	1110000010
888	0111000001
889	1010100000
890	0101010000
891	0010101000
892	0001010100
893	0000101010
894	0000010101

GF (2^{10}) Continued

895	1001001010
896	0100100101
897	1011010010
898	0101101001
899	1011110100
900	0101111010
901	0010111101
902	1000011110
903	0100001111
904	1011000111
905	1100100011
906	1111010001
907	1110101000
908	0111010100
909	0011101010
910	0001110101
911	1001111010
912	0100111101
913	1011011110
914	0101101111
915	1011110111
916	1100111011
917	1111011101
918	1110101110
919	0111010111
920	1010101011
921	1100010101
922	1111001010
923	0111100101
924	1010110010
925	0101011001
926	1011101100
927	0101110110
928	0010111011
929	1000011101
930	1101001110
931	0110100111
932	1010010011
933	1100001001
934	1111000100
935	0111100010
936	0011110001
937	1000111000
938	0100011100
939	0010001110
940	0001000111
941	1001100011
942	1101110001
943	1111111000
944	0111111100
945	0011111110
946	0001111111
947	1001111111
948	1101111111
949	1111111111
950	1110111111
951	1110011111
952	1110001111
953	1110000111
954	1110000011
955	1110000001
956	1110000000
957	0111000000
958	0011100000
959	0001110000
960	0000111000
961	0000011100
962	0000001110
963	0000000111
964	1001000011
965	1101100001
966	1111110000
967	0111111000
968	0011111100
969	0001111110
970	0000111111
971	1001011111
972	1101101111
973	1111110111
974	1110111011
975	1110011101
976	1110001110
977	0111000111
978	1010100011
979	1100010001
980	1111001000
981	0111100100
982	0011110010
983	0001111001
984	1001111100
985	0100111110
986	0010011111
987	1000001111
988	1101000111
989	1111100011
990	1110110001
991	1110011000
992	0111001100
993	0011100110
994	0001110011
995	1001111001
996	1101111100
997	0110111110
998	0011011111
999	1000101111
1000	1101010111
1001	1111101011
1002	1110110101
1003	1110011010
1004	0111001101
1005	1010100110
1006	0101010011
1007	1011101001
1008	1100110100
1009	0110011010
1010	0011001101
1011	1000100110
1012	0100010011
1013	1011001001
1014	1100100100
1015	0110010010
1016	0011001001
1017	1000100100
1018	0100010010
1019	0010001001
1020	1000000100
1021	0100000010
1022	0010000001

APPENDIX B

Minimal Polynomials of Elements in GF(2^m)

A list of all minimal polynomials of elements in GF(2^m) with $m \leq 10$ is given. Each element in GF(2^m) is expressed as a power of a primitive element α in GF(2^m). The minimal polynomial $\phi_i(X)$ of α^i is given by the powers of its nonzero terms. For example, let $m = 6$. The notation

5 (0, 1, 2, 5, 6)

represents the field element α^5 in GF(2^6) whose minimal polynomial is

$$\phi_5(X) = 1 + X + X^2 + X^5 + X^6.$$

For each m, the field GF(2^m) is generated by the first minimal polynomial [i.e., $\phi_1(X)$]. For example, GF(2^6) is generated by

$$\phi_1(X) = 1 + X + X^6.$$

Also, α^i and all its conjugates have the same minimal polynomial. Therefore, for $m = 6$,

3 (0, 1, 2, 4, 6)

represents that

$$\phi_3(X) = 1 + X + X^2 + X^4 + X^6$$

is the minimal polynomial of

$$\alpha^3,\quad (\alpha^3)^2 = \alpha^6,\quad (\alpha^3)^{2^2} = \alpha^{12},\quad (\alpha^3)^{2^3} = \alpha^{24},\quad (\alpha^3)^{2^4} = \alpha^{48},\quad (\alpha^3)^{2^5} = \alpha^{33}.$$

Minimal Polynomials of Elements in GF(2^m) *with* $1 < m \leq 10$

1. $m = 2$

   1 (0, 1, 2)

2. $m = 3$

   1 (0, 1, 3) 3 (0, 2, 3)

3. $m = 4$

1	(0, 1, 4)	3	(0, 1, 2, 3, 4)
5	(0, 1, 2)	7	(0, 3, 4)

4. $m = 5$

1	(0, 2, 5)	3	(0, 2, 3, 4, 5)
5	(0, 1, 2, 4, 5)	7	(0, [illegible] 2, 3, 5)
11	(0, 1, 3, 4, 5)	15	(0, 3, 5)

5. $m = 6$

1	(0, 1, 6)	3	(0, 1, 2, 4, 6)
5	(0, 1, 2, 5, 6)	7	(0, 3, 6)
9	(0, 2, 3)	11	(0, 2, 3, 5, 6)
13	(0, 1, 3, 4, 6)	15	(0, 2, 4, 5, 6)
21	(0, 1, 2)	23	(0, 1, 4, 5, 6)
27	(0, 1, 3)	31	(0, 5, 6)

6. $m = 7$

1	(0, 3, 7)	3	(0, 1, 2, 3, 7)
5	(0, 2, 3, 4, 7)	7	(0, 1, 2, 4, 5, 6, 7)
9	(0, 1, 2, 3, 4, 5, 7)	11	(0, 2, 4, 6, 7)
13	(0, 1, 7)	15	(0, 1, 2, 3, 5, 6, 7)
19	(0, 1, 2, 6, 7)	21	(0, 2, 5, 6, 7)
23	(0, 6, 7)	27	(0, 1, 4, 6, 7)
29	(0, 1, 3, 5, 7)	31	(0, 4, 5, 6, 7)
43	(0, 1, 2, 5, 7)	47	(0, 3, 4, 5, 7)
55	(0, 2, 3, 4, 5, 6, 7)	63	(0, 4, 7)

7. $m = 8$

1	(0, 2, 3, 4, 8)	3	(0, 1, 2, 4, 5, 6, 8)
5	(0, 1, 4, 5, 6, 7, 8)	7	(0, 3, 5, 6, 8)
9	(0, 2, 3, 4, 5, 7, 8)	11	(0, 1, 2, 5, 6, 7, 8)
13	(0, 1, 3, 5, 8)	15	(0, 1, 2, 4, 6, 7, 8)
17	(0, 1, 4)	19	(0, 2, 5, 6, 8)
21	(0, 1, 3, 7, 8)	23	(0, 1, 5, 6, 8)
25	(0, 1, 3, 4, 8)	27	(0, 1, 2, 3, 4, 5, 8)
29	(0, 2, 3, 7, 8)	31	(0, 2, 3, 5, 8)
37	(0, 1, 2, 3, 4, 6, 8)	39	(0, 3, 4, 5, 6, 7, 8)
43	(0, 1, 6, 7, 8)	45	(0, 3, 4, 5, 8)
47	(0, 3, 5, 7, 8)	51	(0, 1, 2, 3, 4)
53	(0, 1, 2, 7, 8)	55	(0, 4, 5, 7, 8)
59	(0, 2, 3, 6, 8)	61	(0, 1, 2, 3, 6, 7, 8)
63	(0, 2, 3, 4, 6, 7, 8)	85	(0, 1, 2)
87	(0, 1, 5, 7, 8)	91	(0, 2, 4, 5, 6, 7, 8)
95	(0, 1, 2, 3, 4, 7, 8)	111	(0, 1, 3, 4, 5, 6, 8)
119	(0, 3, 4)	127	(0, 4, 5, 6, 8)

8. $m = 9$

1	(0, 4, 9)	3	(0, 3, 4, 6, 9)
5	(0, 4, 5, 8, 9)	7	(0, 3, 4, 7, 9)
9	(0, 1, 4, 8, 9)	11	(0, 2, 3, 5, 9)
13	(0, 1, 2, 4, 5, 6, 9)	15	(0, 5, 6, 8, 9)
17	(0, 1, 3, 4, 6, 7, 9)	19	(0, 2, 7, 8, 9)
21	(0, 1, 2, 4, 9)	23	(0, 3, 5, 6, 7, 8, 9)

25	(0, 1, 5, 6, 7, 8, 9)	27	(0, 1, 2, 3, 7, 8, 9)
29	(0, 1, 3, 5, 6, 8, 9)	31	(0, 1, 3, 4, 9)
35	(0, 8, 9)	37	(0, 1, 2, 3, 5, 6, 9)
39	(0, 2, 3, 6, 7, 8, 9)	41	(0, 1, 4, 5, 6, 8, 9)
43	(0, 1, 3, 6, 7, 8, 9)	45	(0, 2, 3, 4, 5, 6, 9)
47	(0, 1, 3, 4, 6, 8, 9)	51	(0, 2, 4, 6, 7, 8, 9)
53	(0, 2, 4, 7, 9)	55	(0, 2, 3, 4, 5, 7, 9)
57	(0, 2, 4, 5, 6, 7, 9)	59	(0, 1, 2, 3, 6, 7, 9)
61	(0, 1, 2, 3, 4, 6, 9)	63	(0, 2, 5, 6, 9)
73	(0, 1, 3)	75	(0, 1, 3, 4, 5, 6, 7, 8, 9)
77	(0, 3, 6, 8, 9)	79	(0, 1, 2, 6, 7, 8, 9)
83	(0, 2, 4, 8, 9)	85	(0, 1, 2, 4, 6, 7, 9)
87	(0, 2, 5, 7, 9)	91	(0, 1, 3, 6, 9)
93	(0, 3, 4, 5, 6, 7, 9)	95	(0, 3, 4, 5, 7, 8, 9)
103	(0, 1, 2, 3, 5, 7, 9)	107	(0, 1, 5, 7, 9)
109	(0, 1, 2, 3, 4, 5, 6, 8, 9)	111	(0, 1, 2, 3, 4, 8, 9)
117	(0, 1, 2, 3, 6, 8, 9)	119	(0, 1, 9)
123	(0, 1, 2, 7, 9)	125	(0, 4, 6, 7, 9)
127	(0, 3, 5, 6, 9)	171	(0, 2, 4, 5, 7, 8, 9)
175	(0, 5, 7, 8, 9)	183	(0, 1, 3, 5, 8, 9)
187	(0, 3, 4, 6, 7, 8, 9)	191	(0, 1, 4, 5, 9)
219	(0, 2, 3)	223	(0, 1, 5, 8, 9)
239	(0, 2, 3, 5, 6, 8, 9)	255	(0, 5, 9)

9. $m = 10$

1	(0, 3, 10)	3	(0, 1, 2, 3, 10)
5	(0, 2, 3, 8, 10)	7	(0, 3, 4, 5, 6, 7, 8, 9, 10)
9	(0, 1, 2, 3, 5, 7, 10)	11	(0, 2, 4, 5, 10)
13	(0, 1, 2, 3, 5, 6, 10)	15	(0, 1, 3, 5, 7, 8, 10)
17	(0, 2, 3, 6, 8, 9, 10)	19	(0, 1, 3, 4, 5, 6, 7, 8, 10)
21	(0, 1, 3, 5, 6, 7, 8, 9, 10)	23	(0, 1, 3, 4, 10)
25	(0, 1, 5, 8, 10)	27	(0, 1, 3, 4, 5, 6, 8, 9, 10)
29	(0, 4, 5, 8, 10)	31	(0, 1, 5, 9, 10)
33	(0, 2, 3, 4, 5)	35	(0, 1, 4, 9, 10)
37	(0, 1, 5, 6, 8, 9, 10)	39	(0, 1, 2, 6, 10)
41	(0, 2, 5, 6, 7, 8, 10)	43	(0, 3, 4, 8, 10)
45	(0, 4, 5, 9, 10)	47	(0, 1, 2, 3, 4, 5, 6, 9, 10)
49	(0, 2, 4, 6, 8, 9, 10)	51	(0, 1, 2, 5, 6, 8, 10)
53	(0, 1, 2, 3, 7, 8, 10)	55	(0, 1, 3, 5, 8, 9, 10)
57	(0, 4, 6, 9, 10)	59	(0, 3, 4, 5, 8, 9, 10)
61	(0, 1, 4, 5, 6, 7, 8, 9, 10)	63	(0, 2, 3, 5, 7, 9, 10)
69	(0, 6, 7, 8, 10)	71	(0, 1, 4, 6, 7, 9, 10)
73	(0, 1, 2, 6, 8, 9, 10)	75	(0, 1, 2, 3, 4, 8, 10)
77	(0, 1, 3, 8, 10)	79	(0, 1, 2, 5, 6, 7, 10)
83	(0, 1, 4, 7, 8, 9, 10)	85	(0, 1, 2, 6, 7, 8, 10)
87	(0, 3, 6, 7, 10)	89	(0, 1, 2, 4, 6, 7, 10)
91	(0, 2, 4, 5, 7, 9, 10)	93	(0, 1, 2, 3, 4, 5, 6, 7, 8, 9, 10)
95	(0, 2, 5, 6, 10)	99	(0, 1, 2, 4, 5)
101	(0, 2, 3, 5, 10)	103	(0, 2, 3, 4, 5, 6, 8, 9, 10)
105	(0, 1, 2, 7, 8, 9, 10)	107	(0, 3, 4, 5, 6, 9, 10)
109	(0, 1, 2, 5, 10)	111	(0, 1, 4, 6, 10)

115	(0, 1, 2, 4, 5, 6, 7, 8, 10)
119	(0, 1, 3, 4, 6, 9, 10)
123	(0, 4, 8, 9, 10)
127	(0, 1, 2, 3, 4, 5, 6, 7, 10)
149	(0, 2, 4, 9, 10)
155	(0, 3, 5, 7, 10)
159	(0, 1, 2, 4, 5, 6, 7, 9, 10)
167	(0, 1, 4, 5, 6, 7, 10)
173	(0, 1, 2, 3, 4, 6, 7, 9, 10)
179	(0, 3, 7, 9, 10)
183	(0, 1, 2, 3, 8, 9, 10)
189	(0, 1, 5, 6, 10)
205	(0, 1, 3, 7, 10)
213	(0, 1, 3, 4, 7, 8, 10)
219	(0, 3, 4, 5, 7, 8, 10)
223	(0, 2, 5, 9, 10)
235	(0, 1, 2, 3, 6, 9, 10)
239	(0, 1, 2, 4, 6, 8, 10)
247	(0, 1, 6, 9, 10)
253	(0, 5, 6, 8, 10)
341	(0, 1, 2)
347	(0, 1, 6, 8, 10)
363	(0, 2, 5)
375	(0, 2, 3, 4, 10)
383	(0, 2, 7, 8, 10)
447	(0, 3, 5, 7, 8, 9, 10)
495	(0, 1, 2, 3, 5)
117	(0, 3, 4, 7, 10)
121	(0, 1, 2, 5, 7, 9, 10)
125	(0, 6, 7, 9, 10)
147	(0, 2, 3, 5, 6, 7, 10)
151	(0, 5, 8, 9, 10)
157	(0, 1, 3, 5, 6, 8, 10)
165	(0, 3, 5)
171	(0, 2, 3, 6, 7, 9, 10)
175	(0, 2, 3, 7, 8, 10)
181	(0, 1, 3, 4, 6, 7, 8, 9, 10)
187	(0, 2, 7, 9, 10)
191	(0, 4, 5, 7, 8, 9, 10)
207	(0, 2, 4, 5, 8, 9, 10)
215	(0, 5, 7, 8, 10)
221	(0, 3, 4, 6, 8, 9, 10)
231	(0, 1, 3, 4, 5)
237	(0, 2, 6, 7, 8, 9, 10)
245	(0, 2, 6, 7, 10)
251	(0, 2, 3, 4, 5, 6, 7, 9, 10)
255	(0, 7, 8, 9, 10)
343	(0, 2, 3, 4, 8, 9, 10)
351	(0, 1, 2, 3, 4, 5, 7, 9, 10)
367	(0, 2, 3, 4, 5, 8, 10)
379	(0, 1, 2, 4, 5, 9, 10)
439	(0, 1, 2, 4, 8, 9, 10)
479	(0, 1, 2, 4, 7, 8, 10)
511	(0, 7, 10)

APPENDIX C

Generator Polynomials of Binary Primitive BCH Codes of Length up to $2^{10}-1$

The generator polynomials of all the binary primitive BCH codes of length up to $2^{10} - 1$ are given. They are given in an octal representation. Each digit in the representation is coded as follows:

$$0 \longleftrightarrow 000 \qquad 2 \longleftrightarrow 010 \qquad 4 \longleftrightarrow 100 \qquad 6 \longleftrightarrow 110$$

$$1 \longleftrightarrow 001 \qquad 3 \longleftrightarrow 011 \qquad 5 \longleftrightarrow 101 \qquad 7 \longleftrightarrow 111.$$

When the octal representation of a generator polynomial is expanded in binary, the binary digits are the coefficients of the polynomial, with the high-order coefficients at the left. For example, consider the (63, 45) BCH code in the table. Its generator polynomial in octal form is

$$1701317.$$

Expanding this in binary using the code above, we obtain

$$001 \quad 111 \quad 000 \quad 001 \quad 011 \quad 001 \quad 111$$

The generation polynomial is then

$$\mathbf{g}(X) = X^{18} + X^{17} + X^{16} + X^{15} + X^{9} + X^{7} + X^{6} + X^{3} + X^{2} + X + 1.$$

Primitive Binary BCH Codes of Length up to $2^{10} - 1$

n	k	t	Generator polynomial
7	4	1	13
15	11	1	23
15	7	2	721
15	5	3	2467
31	26	1	45
31	21	2	3551
31	16	3	107657
31	11	5	5423325
31	6	7	313365047
63	57	1	103
63	51	2	12471
63	45	3	1701317
63	39	4	166623567
63	36	5	1033500423
63	30	6	157464165547
63	24	7	17323260404441
63	18	10	1363026512351725
63	16	11	6331141367235453
63	10	13	472622305527250155
63	7	15	5231045543503271737
127	120	1	211
127	113	2	41567
127	106	3	11554743
127	99	4	3447023271
127	92	5	624730022327
127	85	6	130704476322273
127	78	7	26230002166130115
127	71	9	6255010713253127753
127	64	10	1206534025570773100045
127	57	11	335265252505705053517721
127	50	13	54446512523314012421501421
127	43	14	17721772213651227521220574343
127	36	15	3146074666522075044764574721735
127	29	21	403114461367670603667530141176155

n	k	t	Generator polynomial
127	22	23	123376070404722522435445626637647043
127	15	27	22057042445604554770523013762217604353
127	8	31	7047264052751030651476224271567733130217
255	247	1	435
255	239	2	267543
255	231	3	156720665
255	223	4	75626641375
255	215	5	23157564726421
255	207	6	16176560567636227
255	199	7	7633031270420722341
255	191	8	2663470176115333714567
255	187	9	52755313540001322236351
255	179	10	22624710717340432416300455
255	171	11	15416214212342356077061630637
255	163	12	7500415510075602551574724514601
255	155	13	3757513005407665015722506464677633
255	147	14	1642130173537165525304165305441011711
255	139	15	461401732060175561570722730247453567445
255	131	18	215713331471510151261250277442142024165471
255	123	19	12061405224206600371721032651614122627250626 7
255	115	21	60526665572100247263636404600276352556313472737
255	107	22	22205772322066256312417300235347420176574750154441
255	99	23	1065666725347317422274141620157433225241107643230343 1
255	91	25	675026503032744417272363172473251107555076272072434456 1
255	87	26	110136763414743236435231634307172046206722545273311721317
255	79	27	66700035637657500020270344207366174621015326711766541342355
255	71	29	2402471052064432151555417211233116320544425036255764322170603 5
255	63	30	107544750551635443253152173577070036661117264552676136567025433 01
255	55	31	731542520350110013301527530603205432541 4

n	k	t	Generator polynomial
			326755010557044426035473617
255	47	42	2533542017062646563033041377406233175123 334145446045005066024552543173
255	45	43	1520205605523416113110134637642370156367 0024470762373033202157025051541
255	37	45	5136330255067007414177447245437530420735 706174323432347644354737403044003
255	29	47	3025715536673071465527064012361377115342 242324201174114060254657410403565037
255	21	55	1256215257060332656001773153607612103227 341405653074542521153121614466513473725
255	13	59	4641732005052564544426573714250066004330 6774454765614031746772135702613446050054 7
255	9	63	1572602521747246320103104325535513461416 2367212044074545112766115547705561677516 057
511	502	1	1021
511	493	2	1112711
511	484	3	1530225571
511	475	4	1630256304641
511	466	5	1112724662161763
511	457	6	1142677410335765707
511	448	7	1034122337164372224005
511	439	8	1561350064670543777423345
511	430	9	1727400306127620173461431627
511	421	10	1317711625267264610360644707513
511	412	11	1337530164410305712316173767147101
511	403	12	1573436303657311762726657724651203651
511	394	13	1102510344130333354270407474305341234033
511	385	14	1775546025777712372455452107300530331444 031
511	376	15	1116744706521725532227162607146216210106 733203
511	367	17	1126657202505666323017001652245562614435 511600655
511	358	18	1574456154545450414733416176535156070037 760411373255
511	349	19	1455012234675753242074501555657377006165 521557376050313
511	340	20	1036075427062474623220662047122611677363 511364110105517777

n	*k*	*t*	Generator polynomial
511	331	21	1540644721106050570342772405117177453215 13666367746164145732 1
511	322	22	1153614675506111211374366675624670755236 52364506267706177073507 3
511	313	23	1037043005346411102774516447047070735602 32722463742153673625151743 7
511	304	25	1510736212272133353502352536320703410147 2252730643370771600352540473 51
511	295	26	1112630477530033170044524747727767532752 0466126030774720522476717444670 35
511	286	27	1317463403265645042064775326044775737416 7140717560167145236500227345054014 71
511	277	28	1111170752122547100341422773660302256230 3177512454137173036077374264015263260 45
511	268	29	1241160247151367165615372023170221264427 2264376531630435034363106314253017352056 01
511	259	30	1121314111162101532370722243711014463333 4772560250516566143547137606623504332146 46117
511	250	31	1007276607431444342624513277572752237465 2773105153763257577607352353353361571636 37347245
511	241	36	1142630652622415305026460545443230406701 2432776321666056566417103173036006026112 51467341721
511	238	37	1266115765104357454721732420217755547411 1377076177455240605755033452532207221321 266342055373
511	229	38	1441474125377630723466544074104363527262 5110602507742556745735375653737245201524 721767740003105
511	220	39	1330350650041246137404371541263763570720 1431575333227603262602752030077406436030 760454644342665055
511	211	41	1537262671145756636471271211466023057434 5664477051253665031617124007140032144713 400007755712030376003
511	202	42	1355074061631664750650654005233224126373 5501617417747326444652077100470767607227 335404174253126577344477
511	193	43	1151502313317462115717012400332054150716 2357615552060114767351204573257260045473 425675673673505075470425055
511	184	45	1320551776766374340772233460740727550552 2041306675275253552627101444370345136765 164253644172265100216633505671
511	175	46	1202260564674312433743724246412313100673 3703766567307723446714611616650062423137

n	k	t	Generator polynomial
			2142007622705777576135715734456503
511	166	47	1105764456740764631742144442303723134662 4415551173770646017645044577670522052736 107022447017356700151227221443336113
511	157	51	1710171171400604264434731140711521370753 6440012114236665475350701102415515624310 174161400004517373603747715514145565503
511	148	53	1414131546120371255541603030602601124642 7727350174267055345645550670707547107070 5355774504056620750463641124465774123024 61
511	139	54	1775336143053070047771371572323076147655 0747705134625235605330765134143000233634 1267612204553536475125547007426342706405 51763
511	130	55	1126210420631670733262507445577025266400 0744725674510475154245554654156641313571 4050203537624232123030520531734566124772 13206321
511	121	58	15412713576557725255604106044116552170135 4434750270764625323125324327677742325121 2771735535703734317405374311756764357501 74233577257
511	112	59	1264236440177126545461720120275756614321 2522620050735005241425745450416057231261 2131152145444635107565754242126117753372 24522460653365
511	103	61	1267502203437327636237202040115050677105 7315147230644142256367333406167475350432 5307201333612360340675162226650265473307 65357307372010437
511	94	62	1034624554522076136543376246276501412751 3662235012007006252201651305170664740707 5134565753153573011560335354474211375326 02636772420754543335
511	85	63	1311423305612705436042312400305355360613 5536677204454060166477652423477032137140 3157621106144232601520207000161502003022 44744512244665460344315
511	76	85	1266305771250774445612456156722760607324 4460173371073371716611544036044377532354 6736213462571106045203515637171742621252 52705066142632374105237105
511	67	87	15721237117653466537773251761251635002412 6410254565007065546756015043572302057573 40105470100445777303456773650454016441676 52645603000655246474332003041
511	58	91	10575217600004167050454306543214444576237 1110323251120106217732446025515645663300 5122223411465454201703211427513246445624 30356677543412533230060230330601
511	49	93	14003731252226245616434770021451434774715 0430652515175422234333765344626250514664 2652723355123643633615716343345655723373

n	k	t	Generator polynomial
			56773023742177503076702013524172673
511	40	95	1065643260601413014144161647626313166664 6125563104544564702042267572604002367722 2071247202762236702540237554475163650230 54261717244400242403714177455740276623
511	31	109	1004032013504641301465031022773733377231 2030141207564005615007277456377116231511 6355123767005745123620162301722500173755 1261347775206565523421255764676722014574 5
511	28	111	1506424217016105771311126531671315537006 62227135624440407472104067534437240506204 0406060362307553060252105071101510111346 0677710774225141620173656560126201213355 31
511	19	119	1323204574570661164212403733264114023404 6117057401637365534154143137744473020704 4764741324776540766245335511016032261140 7052420104522110303361617354750363145550 46713
511	10	127	1702172311027431733521206643077042605327 7252024574422247642034145450713500264726 3477463246601133310064536566115023625306 7473677673467052234355271140607221533731 42701037
1023	1013	1	2011
1023	1003	2	4014167
1023	993	3	12052210423
1023	983	4	30135372217233
1023	973	5	67441634100257771
1023	963	6	155441273452021342255
1023	953	7	321370747475547513070313
1023	943	8	760744225715270200004506345
1023	933	9	1323526661245521113217162255031
1023	923	10	2023237633202230444160563331425623
1023	913	11	77555357222250615754561135410204703015
1023	903	12	1762347256322045077757166370232714216725 7
1023	893	13	3040745710762554654627001731571615206417 2421
1023	883	14	4773627447536521222606523723122700031460 4273713
1023	873	15	1373673270027255300005045713741010077651 22140012433
1023	863	16	344027215636044423517306353401033533501 14020277143115

n	k	t	Generator polynomial
1023	858	17	1322555466146551135464235057214741470461 7135735657240141
1023	848	18	3542752065222573717143677007420570562622 0205536512744500263
1023	838	19	5210617676532176677755466752642074636403 1144034344342361121165
1023	828	20	1213222404544636466003710275014422027057 75313007201145723077423013
1023	818	21	2214325742216310053650417376464144271657 51274313300617216324271557407
1023	808	22	5552750420617757113515375401021454054142 43020552307423532731034607546117
1023	798	23	1652724250022427704512246362640211603371 646303263710572663702515300240421537
1023	788	24	2246066255654012757523064141151344242436 4324674165464024461711130753671556534665
1023	778	25	7403630677220307364435256070064531622417 2156747445532621743456011004302553363211 61
1023	768	26	1420057410715701076671164654273645645222 5510712657147620420136563512515532547606 604067
1023	758	27	3544663745163634242516250415043615446630 1242677700521404233564674545716503565046 015304675
1023	748	28	5160035350014031172374207074650122660340 6725265542413700106041273652611227613754 613777751617
1023	738	29	1761052465717574451310664156407472743161 2025501044536741540070065741445436720677 4215713707625277
1023	728	30	2705040500317205454745436524651543310405 0017465005222512321066476651072114503137 3602071775011137127
1023	718	31	6421001547326360050675717264533767251455 4755764143772111200712325604323701627400 3637473623253276573051
1023	708	34	1221657751474373730117767342335416045271 4315315367757451645726466677117331525450 50454154042571133265734145
1023	698	35	2234207617247627652572131446233344733763 6432733672331007455154713117134375547477 57210324464635077421557061245
1023	688	36	6062331567761473445053573264026122102305 3201440006656062450307555544233405300271 40456114002742376254456531124177
1023	678	37	1173340210112354267372717132007746772117 2430242075060060272435267356231765714270 070343261342147115315606445423035675

n	k	t	Generator polynomial
1023	668	38	2726674363712255027500744177372151741454 5645433073762151476755130543142214506356 405532675477541702642035025443744706413
1023	658	39	4526553271372400673644374714763070706341 1031171237616540024630303133511353737123 3160732343127163117047027565041001443541 05
1023	648	41	1053012656042512406741405367113520216337 3335151361445051301722215370653734053400 0442526605347576637101417411054700433515 223273
1023	638	42	3760125774555145576316166607000752402515 4405641372077333177067640612307077706604 5666510673072260546441216265142152656464 533676375
1023	628	43	6421516056355057002552057134724545361312 2357027016417012220102012742464575330112 6724317410001044430017517040452131500121 350470437063
1023	618	44	14367672275404102542246475266313531611600 5027177636020644031621570413456443331322 7260670220051221512570532647243752525477 5754664246713353
1023	608	45	3205632662501303326372612360531506221660 1642671653635152412144657627370612401323 5612466273254637300536451672472606153046 2161570470103566541
1023	598	46	5175076274225613532026031146621612144714 3656373757601671144563673511345562033377 2363115242327614411151057011653316234767 2420714406612507752125
1023	588	47	1425717410513115113640172157603722701046 3011170047424744303123121316566534441253 5602227567425313454763311477014361615330 04527036262433614645152063
1023	578	49	3177430361201520403652734206337256777357 1514665321603455147706422374540553332214 5746177477620442053330430305772517145376 36407425537275744133016141137
1023	573	50	1355677017526536535146475314700031707456 3026607614213012666657500272520767133533 7304265646727506154132227032714377707023 2341637207202520573521417420615
1023	563	51	2772361504424321676653141026604362002113 2553454106501612101536430475161545700052 0475027073570114161614606247142254455044 6255557776515151167236264235327161
1023	553	52	6042334623077055525431324656261417515014 2274125133435506177502357332260547747255 3162040767437444775076764454714640345671 0027561254427647334263053360460303115
1023	543	53	1055532301605012610131175414035550021456 2763276437435575056231126247124212102427 6704204717425615057004044214740647640450 0651662113626166471166577340552720505314

n	k		Generator polynomial
			3
1023	533	54	30112522710240036012563556144101106312 41 4476043760457564721111127466077665562665 4217667430146671642613617516247055413742 6636741420630626224221121140152405654631 6753
1023	523	55	6173100556046353561760536040447255721247 4310077565636320561601165471476057132554 2322473307203230122443324706157714547610 2204305664132377402106560521763116372354 2140361
1023	513	57	1442056607323310761242261470600262276017 5101364127621557726777406532331057077546 1527721765575312220547404020757305657546 5457253141504105143444175702256556213675 47711533103
1023	503	58	3526111340256374262000600572256610352600 0577617220337007505520131512037473367771 3400541207320422505347425362014026650010 0004257043231653414626355376624351306710 02311565250731
1023	493	59	7535405755220775623403553444573506243560 7256600605234131432506166510202242405073 3356624720337475566117114661531204767172 3111426350630167625357243023251630010175 4522605016124 5001
1023	483	60	1010227406455200076236416372024031175567 5534053006400654636246221506521030256417 4635764603621630460402150663270555030262 2471552611215057442745721707226536376112 32507162644766772 4133
1023	473	61	3275255734777512716766130324236300561602 3266541073200606432722454332243070233312 1056616662644203305305353270024721414270 6252777265242372623345602661643601551362 36011471202207156352 2541
1023	463	62	4257710770354177520700120310175131741452 2646316040516304114052431351313706134426 2044226555241044230276332312417700133134 5011373217527767452370777645631335625621 27540107152554561702 7552161
1023	453	63	1531237560135257466053045644313751237706 0514513604361025341247434430553661067700 1052325072756200027154257505607050746074 2347574503151051424660227041423302376426 02570366025705161467606361 53261
1023	443	73	3051462775160646577510534745450470242247 4353060560266203436041336276663363011462 7247152315630145650707511617214040605320 4277716062400570267044607301501125151777 33745355177447267466240511176145 57
1023	433	74	6551077013564477321042777730023704151601 7521127762157242770232322533724016627637 1251547755210341250226270224512023346531 5173725652335621405465526246704034012300 70732711270754577541552732457757450 13

n	k	t	Generator polynomial
1023	423	75	1371403545740325650734320662641445605572 0647077714165030001606304253676672714102 5356347426042441521146714214012236710237 4164612444130330470102450320103726066270 4604725046607414253305030001515377207422 7
1023	413	77	3130514551013371502077510305732557321401 1760033161356760432560144747314167124256 5376173247642602317357010625427257776456 77342665446545120037332774475061557524550 1423614017265745456752500432104234734310 7567
1023	403	78	6530503067625562124141564675424734756257 5145101333262333026444047152033731026412 7462440055716211016277744037230663055316 4172765324526335350412724676631061133755 7637416161530226355450072447426761051420 0676657
1023	393	79	1653506701401422711244424506213255341202 7657471014506751046044362763202645334655 6625534505264745765143400727723477524642 3717565120014071373273475432524117405663 5606472676113552676751152426134176727371 57011323251
1023	383	82	2101556350316274636721774002521346374755 0416046710713235213260000173163311010751 1524536177002664305516503446336030163303 2543716222246044772213606523671240711261 7657615110133267371365102061275103775727 57105036743057
1023	378	83	1271727232475673032645416341027644322054 4742747152233707376724600063056046415216 0615441501471172331343574553307577366042 6310001651125246345406555466665316732504 0134511004147527430357700216203600436576 3562314467231257
1023	368	85	2667110244306717554116670535772352255702 0775135077721751514126771355024621053205 1314022567463732172361241171201412026160 3653333750471661055152633432534522430740 0302522242423267144130672542411721301423 3216317412603200011
1023	358	86	7435121372564560416164423405512531131273 4223333111503244154771550003234404757442 6171255246153400574361641566444466317375 4160134125055131104230756443674102100730 5155725322757546453441553326405634665130 7314476261042544230245
1023	348	87	1152150570436510265766676136141303474274 4526012535312616667574523032536722700050 0625207035201332737114107013243424325560 6164445352775311630144736763442471124475 7750150011616262637717177075102213217170 43446340305445612706534703
1023	338	89	3435404744265677602320351555134172304633 0127646153113200255460143043276470736246 6125401015341073237073746771541102307733 4025666153620557042677053703237045526065 4431637224726547347532253255127643611341

n	k	t	Generator polynomial
			66201344577712444057365034127
1023	328	90	4376075566465464447725523310167055725051 3364752460175061537135014226525551535275 6 1554472635775452747645066645517076050542 7007214641140377770232661301737600047265 0644621575075623315742537655651745704130 72547436036455673176766776021557
1023	318	91	1706570374315130744566215341745357673544 6506517074057307606757741716571770230670 4666124321175656663111343514112511526132 5767234612127537336563140520412006236673 0652025365545623610266465373576416357076 757337145232060114113270376256424571
1023	308	93	2642120320047717622231140633167326340740 4654622717462634441140442435052126716431 0571545025406673462177740705177336340321 1544225553064525153700121577654557715476 2747463245571152471573560120403030036251 611371571363536743575475627255612375247
1023	298	94	7607316125761433642526304450363545626472 5611714372152073305634315715022554055231 6076751112611277717145131135317527677046 6732556644427514047164046110636365565566 4550340671546270704215331125127241135267 1670301124644115710644070316271466331106 73
1023	288	95	1701604066633416774726444725261650551067 2124410746002141416267465201122102455167 6755376270304423013326176347761574061225 2161051122456127543465052666631743005425 2266655317553460754531732325667620546756 4036613660503377260207743030117536367621 203555
1023	278	102	2561112413204004672512333427015121540424 3237123071101755037216065252775302223547 4065127152454745331164320065161070571534 4371247324607717027041446261403024476234 2751113123413544231643217420335257714150 0236421250632332705126052420172332450755 310424635
1023	268	103	5654035737756644460334414133030111146310 0146152621621754746144205012055330527077 5303715417371641000727253627610744441051 5703446316265332602512753027230261252257 0230566667675120633312503010076171344377 7330061102755376503137425117533736240637 177561306317
1023	258	106	1472716560654553504476114427330200150115 1173472111661146472710536457264074616730 6212707640513421202420267426242134553637 0427517704053463140321453024446377474011 2076150643076734714642657240376712557330 2203556104577153413672763550007757247113 0110646647103743
1023	248	107	3461477710376543241414375231636516574724 5065316230137456120461323574541525726041 2556573562616526111127656576320525763510 5644024312145652627132756063506201454560 0503657215441172100402476306531430623526

n	k	t	Generator polynomial
			5426422633275763410271325014165632012203 2414745504624453615
1023	238	109	6107055505572007262714470615665443601727 4773321535432655566365546074301200532116 2555655214623411066175714545116221410540 1214402533154265163574524537271755667277 4742133671045121217224375055532423613465 2547057726572740725403741750507323711445 52267222531336165554655
1023	228	110	1645505272602350646313312172433135453243 4317660426124673504670142363510702632530 5465440350706502341177157432562253614237 0371356404544253565207222015007240167503 4540743040342334543505710424074123577104 0704601774071102570072722602051723521347 64771653006051570762272065
1023	218	111	2532727377067146456122304005405563345245 7027263257630231533014541245300437451413 4422124564661526271012077252745371352727 5724274535530277606021744662377454417475 5417200575157107713300760220650710444522 5676334024215714747554147237530037305312 55325367612674645741101010215
1023	208	115	7662425544271246134245023532655061777064 3113564744501532047502013441735363751643 4611742014356270417547440327560724415144 7613713075166362462555105072064606436604 7372533427761377771313714611060116225324 7273756021533523511473763263374430301047 41312067517077073657202575266431
1023	203	117	2600721761002761741110660757163553757530 6525157537267407357336666131163647123236 6275645264765562251403631501114054010175 7224017277322017716776500255740570067411 7120457710072403775723015475413276153570 1153221436207760470741233214262705415704 6022354153273653525227756023726523
1023	193	118	7053315727030267005076537662446626601361 2061777171150123703517602323634330163122 0440504733611235242564514652123365275225 5244134100153077772243144440563401766701 5257142421560325755421566155040177772450 5676452071661107471162522137406175403173 3166505125521364032352227514145713341
1023	183	119	1347473474775270044720557004055543253641 5414261074530641366160044747074270440647 6351725463337066164237425741310003010151 3453374232473772046203334555337543714510 7054072166471561270450770527154501716302 2332610013551710764724143151161146625775 6310407013234622100223740720176570172124 5
1023	173	122	2361302433102534366303174067425000312611 5407714446113241270367016250545512127107 4146077421244535275152145570435377754120 5555475532041352234100662235537577223261 4457553535540403471102147116764306321760 6032315053403217330370624061030766417314 5401136667040305764271011336376561455315 7553

n	k	t	Generator polynomial
1023	163	123	4113171033252014342001147750344201677441 7325213737663530655624555754314734523735 5425411305340760252512234263304567336743 6776416515774627717565175764750357722307 2217561502535205126601021474136200301335 5457103141643542557151144254001725320301 7436544220547403671156214574372425103751 6460607
1023	153	125	1473037520540655305664534611401067646510 6566466240750272645247101532717001327355 1255250401446241777221035154710077244137 3023726106475461742104112674543042527464 7451775407521677507671036742125536233317 6644217425306265727330713311536600712505 2432044341706551447006426647662422226300 32444750511
1023	143	126	2705672534137533656134605524750654047625 4275423257512563102361416067617730642771 1460013664347574610040617626371543266057 3554365713174265772110167734047217705111 3733747446147746503062302477575037740315 1165171334262046730256770526522266602575 2346240531211431754131515171115505251076 23337477461525
1023	133	127	4617241122502533137567475531722511566715 7425701644602354172410473477746724542114 5544443464632270001337721543511324751670 2616154503645720574203625447407506343636 6627145456220533275233402326143257055011 6720123731574052034715703301044271032476 7607251200167144164775070153251671274256 43765453321156665
1023	123	170	1606122475314452440364114065506377666425 5343610702760377310024522705343600423320 2424333105146144027327126143431675314002 3036402653162631026523102264715527745315 0374162211376360615537273444653255610566 1614203104347443540533723112046663543521 1253420231107716037536204725255752061446 310672727204716406065
1023	121	171	5222675662147725741314764210323377001541 2253272514721374170157274532253203570061 5555001733462474144045642452515062144017 1131414121534317141271717416540247672143 1364537676373322240234442776120501273102 5245611735266752042303471766360032052267 6522561717727552103333636452501626224763 172044444636551422213
1023	111	173	1522057216652345524151261131756011401100 2417457353411556263253430644235445523722 3141414362643623756035136777711661732620 2500037150531207635215557034127213515066 4410720607604354463431406515431477631373 2040303636642717670427347615762362355036 7600521462074250472554257242343544575670 3446634044261764644465517
1023	101	175	3545472772647733506027327007274644770353 7076474016112034433402242207212433412503 4212535626065370142605376134717357361465 6417762011214767217653655554653572111747

n	k	t	Generator polynomial
			4551152615662076355010304237412255115177 6272636505203446412116131004765425047506 2051240151667124466614627030314263307135 5576456772101677262472416421
1023	91	181	4055177762504121412577610205372003777645 6673574005261545600620674271417162613012 1506256105242557423640764336401204101741 7126672067376365721700727456241223231302 2756306103623641555001215465766126430341 7306703214450577041500501573212167321536 3434543304701503707164511142547260441612 1735616377403703647762331023117
1023	86	183	2264761771543214471675702410753621637726 1160424142731011126327103065573247413075 4432257227721764404464742501644540205701 3267260210053106372523361731350152677206 0076334324656464215710544105353577351675 3364012464542355412231261607767455076427 5143702025133630600744625136760211406353 453422170361560266523166234073223
1023	76	187	5504413226010767557023243061737064215533 4421615324723723124722403164751617036710 2332107021677452321433245440364276246075 0627166653450647114065506704152554300353 0021311306574650326702510673440306147402 0127524575753074316742263020463722341655 5043324233205722150256144415372750304032 050275143511316413151005306045676027
1023	66	189	1335505434441741406157525754573170243027 3424173625471123363551344271423420240635 4431211746027242072506552240651527643442 5225735404126141635066623161414334637002 3253213327457376454172114672756045050327 1457414742355472655525107236667037135445 6044605352522350777052251520016127100460 0523064264130454576437037741704275570603
1023	56	191	3533660424156746206407772444526075003401 3327044430111755764054213671261403637721 7365150066452035503620722031617013704506 1451154164441245647372600477550413342313 4332421423334727666016525177347306160053 4367723121360245640211624352467665254225 6204245534234357413520302403642764362211 3551667237042145430541554174713125325173 331
1023	46	219	6301541625665623020521356602744545517117 5526531114735670265141755744713461033137 0637137663547260635501352643104443676250 6773370422474424106014660455576033053653 6303137074076236722167541537206044032770 3604223201460001112124251137371265705531 3174645762574576401050005015160506604517 2730226274076512210521522004216362565324 376075
1023	36	223	1152711551526005136004274504217435002475 0437221435664073005171237220151140064707 5441034055747331016655123301566414331316 3711704233561771130764134237244667223150 7613166546645026325127605716064161523643 7407144501652522375660250700303477777350 5457120771071520002217212553614046136364

n	k	t	Generator polynomial
			3211657443425105061615376162061404232341 2027502143
1023	26	239	3476655146625052723141431762032154723323 2124522103643311017642323320202416451725 3130645425737600246350051466006303723670 4734745250307710106433015376231566674173 2473210524736630110637274120466644500552 3530736347265641711514677043326744476035 7766775101312115727255145732774036010471 3655061034476144410065213573347702152554 6455751231633
1023	16	247	6036262767731160101141105063701021363565 3762274052211432564657262162121174115766 0750767504220703662612124577754243717505 4060630501002657505104156311506734744442 5411743547305156755340240153001037104363 0513040170535057375012745423660541522550 5633323652623572617766360711322000255022 0251544067462673641704456634322534227212 4736702723114761
1023	11	255	3435423242053413257500125205705563224605 1630065263153013075623353102124314210612 6137022171665514567226426351765550103471 0474154323571003071265736644050721570112 5641143017433246415076511474510565005614 4327137466737311303124750456366004341761 1165663371330203524572775110303560116122 5714476377111212075267515476733757766662 624063406704036177

Index

Abramson, N.M., 257
Acampora, A.S., 535
ACK (*see* Acknowledgement)
Acknowledgement, 13
  negative, 13
  positive, 13
ADCCP (advanced data communications control procedure), 460
Adder, modulo-2, 56
Addition, 16
  modulo-2, 16
  modulo-m, 17
  vector, 41
Additive white Gaussian noise, 6, 327-29
Advanced data communications control procedure (ADCCP), 460
Affine permutation, 194
Analog-to-digital (A/D) conversion, 2
Arithmetic:
  binary field, 24-29
  Galois field, 39-40
ARQ, 13, 359, 458-94, 551-56
  continuous, 13
  go-back-N, 13, 459-60, 462-63, 551
  hybrid, 13, 477-81, 551-56
  mixed mode, 474-77
  selective-repeat, 13, 460-61, 462, 551-52, 554
  selective-repeat with finite receiver buffer, 465-74
  selective-repeat plus go-back-N, 474-77
  stop-and-wait, 13, 459, 463, 551
  type-I hybrid, 478
  type-II hybrid, 479
  type-II hybrid selective-repeat, 483-94
Ash, R.B., 429
Asymptotic coding gain, 328-29
Automatic repeat request (*see* ARQ)
AWGN (*see* Additive white Gaussian noise)

Backward index, 471
Bahl, L.R., 309, 332, 541
Bandwidth, 8, 343
Basis, 44
BCH bound, 151
BCH codes, 40, 141-80
  decoding, 151-60
  description, 142-51
  designed distance, 150
  error correction, 159-60
  error detection, 177-80
  generator polynomial, 142, 150, 151
  narrow-sense, 142
  nonbinary, 170-77
  nonprimitive, 150
  parity-check matrix, 148
  primitive, 142
  syndrome, 152
  syndrome computation, 167-69
  weight distribution, 177-80
BCH decoder, implementation, 167-70
Berlekamp, E.R.R., 141, 155, 429, 433
Bernstein, A.J., 414, 418
Binary operation, 15
Binary-phase-shift-keyed modulation, 6, 327
Binary symmetric channel (BSC), 7, 320, 389
Binary synchronous communication (BISYNC) protocol, 459
Birkhoff G., 15
Bit error probability, 326-29, 339-46, 371-73, 383, 413-14
Block codes, 3, 4, 52
  BCH (*see* BCH codes)
  binary, 51
  burst-error-correcting, 12, 257-74
  burst-and-random-error-correcting, 12, 274-82
  Burton, 273-74
  concatenated, 278
  cyclic, 85-124
  difference-set, 203-9
  direct product, 274 (*see also* Product codes)
  DTI, 194-201
  dual, 55
  EG, 228-40
  Euclidean geometry (*see* EG)
  finite geometry, 223-53
  Fire, 261-69
  Golay, 80, 134-38
  Goppa, 176-77
  Hamming, 79-82, 111-16
  Hsiao, 500, 501
  inner, 279
  interleaved, 271-72
  invariant, 194
  invertible, 479, 481-83
  linear, 51-84
  linear systematic, 54
  majority-logic decodable, 184-253
  maximum-length, 201-3
  modified Fire, 280-82
  outer, 279
  perfect, 80 (*see also* Golay and Hamming codes)
  PG, 242-45
  phased-burst-error-correcting, 272-74
  polynomial, 116 (*see also* Shortened cyclic codes)
  product, 274-78
  projective geometry (*see* PG)
  q-ary, 170
  quasi-cyclic, 221
  random-error-correcting, 11
  Reed-Muller, 233
  Reed-Solomon, 171-76
  RM (*see* Reed-Muller codes)
  RS (*see* Reed-Solomon codes)
  SEC-DED, 498, 499-502
  self-dual, 83
  (72, 64) SEC-DED, 502
  shortened cyclic, 116-21
  single-error-correcting and double-error-detecting (*see* SEC-DED codes)
  systematic, 51
  twofold EG, 235-39
Blocked state, 469
Block length, 4, 52
Bose, R.C., 141
Bounds:
  BCH, 151
  Hamming, 83
  Plotkin, 83
  Reiger, 258
  Varsharmov-Gilbert, 84
Bower, E.K., 370
BPSK modulation (*see* Binary-phase-shift-keyed modulation)
BSC (*see* Binary symmetric channel)
Buffer overflow, 463, 467
Burst, 12, 102, 257
  end-around, 102
  phased, 272
Burst-correcting efficiency, 259
Burst error (*see* Burst)
Burst error correction, 429-55, 547-50
  almost all, 430
  complete, 430
  multiple-,443-44
  phased-, 433
Burst length, 430
Burton, H.O., 141, 273, 455
Bussgang, J.J., 412, 414
BWX (*see* Backward Index)
Byte, 278
  information, 504
  parity, 504

Cain, J.B., 346
Catastrophic error propagation, 308
Channel, 2
  additive white gaussian noise, 327-29
  binary symmetric, 7, 320, 389
  burst-error, 12
  coding, 3
  compound, 12
  discrete memoryless, 7, 316-17, 353
  feedback, 462
  full-duplex, 13
  half-duplex, 13
  with memory, 8, 11
  random-error, 11
Channel capacity, 10, 366
Channel decoder, 2
Channel encoder, 2
Character, 503
  parity-check, 504
Characteristic of field, 22
Chaudhuri, D.K. Ray-, 141
Check sum (*see* Parity-check, sum)
Chen, C.L., 158, 159, 233, 240
Chevillat, P.R., 378, 381
Chien, R.T., 141, 159, 269, 526
Chien search, 159-60, 169-70
Cocke, J., 380
Code length, 4
Code polynomial, 86
Code rate, 4, 292
Code tree, 351-352
  incorrect subset, 364-66
Code vector, 52 (*see also* Code word)
Code word, 2, 3, 52, 289-95
Codex Corporation, 542, 547
Coding gain, 328, 347
Column distance function (CDF), 309-11, 378-79
Combinational logic circuit, 74
Communications Satellite Corporation, 543

digital television system (DITEC), 543-44
single-channel-per-carrier (SCPC) system, 544-46
SPAPE system, 544-45
Complementary error function, 7
Complete orthogonalization, 186, 210
in L steps, 210
in one step, 186
Conan, J., 346
Concatenated codes, 278, 535-38
Conjugates, 34
Constraint length, 11
Convolutional codes, 3-5
Berlekamp-Preparata, 431-36
burst-error-correcting, 429-55
burst-finding, 448-51
burst-and-random-error-correcting, 442-55
burst-trapping, 451-55
catastrophic, 308, 311-12, 334, 342, 378
complementary, 332
completely orthogonalizable, 410, 420-23
constraint length, 291
construction of, 329-37, 374-79, 414-24, 446-47
diffuse, 444-48
distance profile, 374-78
distance properties, 308-12
encoded (output) sequence, 288-95
equivalent memory order, 541
generating function, 297-303, 339-41
generator matrix, 289-92
generator polynomial, 293-95
generator sequence, 288-90
information (input) sequence, 288-95
Iwadare-Massey, 436-40
memory order, 291
NASA Planetary Standard, 533-34
noncatastrophic, 308, 311-12, 332-36, 378
nonsystematic, 304, 333-34
optimum distance profile, 374-78
optimum minimum distance, 414-16
parity-check matrix, 390
parity sequence, 304-5
parity transfer function matrix, 398
performance, 343-45
performance bounds, 322-29
quick-look-in, 333-34
self-orthogonal, 393, 414-20
structural properties, 295-308
systematic, 303-8, 333-34
transfer function matrix, 294-95
uniform, 423-24
unit memory, 537-38
Coset, 69
Coset leader, 69
weight distribution of, 71
Costello, D.J., Jr., 333, 381, 539, 541, 551, 553
Covering polynomials, 133
Cramer's rule, 25, 40
Current time-out vector, 466
Cyclic codes, 85-124, 335-37
decoding, 103-10
description, 85-91
encoding, 95-98
error detection, 102-3
generator matrix, 92-94
Meggit decoder, 106
parity-check matrix, 93-94
parity-polynomial, 93
shortened, 116-21
syndrome, 98-102
systematic, 90-91
Cyclic product codes, 276-77
Cyclic shifts, 85

Data rate, 370-73
Data storage system, 2, 498
Data transmission system, 2
DEC disks, 531
RM series, 531
RP series, 531
Decibel loss, 329
Decoder buffer, 359-60, 370-73
Decoder memory, 337
Decoder speed, 345, 359-60, 373, 384, 425-426
Decoder storage, 360-73
Decoder synchronization, 343, 370-71, 407, 420, 424
Decoders, 2
adaptive, 448-55
analog, 6, 329, 535
channel, 2, 3
digital, 6, 329
hybrid, 380
for linear codes, 74-75
maximum likelihood, 10
Meggit, 106
minimum distance, 10
parallel, 345
source, 3
syndrome, 390
truncated, 338-42
Decoding, 2
BCH codes, 151-60
bit-by-bit, 346
burst-error-correcting codes, 259-61
cyclic codes, 103-10
cyclic Hamming codes, 112-15
definite, 406-8
error-trapping, 125-38
feedback, 389-405, 410
Fire codes, 263-69
Golay code, 135-38
Hamming codes, 80
hard-decision, 7
IBM 3850 mass storage system, 521-25
IBM 3420 Series tape units, 511-16
linear block codes, 72-75
majority-logic, 184-219, 388-426
maximum likelihood, 8-11, 317
minimum distance, 10, 70
sequential , 350-384
sequential-code-reduction, 248
shortened cyclic codes, 116-21
soft-decision, 8
syndrome, 51, 68-75
table-lookup, 72
threshold (*see* Majority-logic)
Viterbi, 315-46
Decoding delay, 426
Decoding error, 2, 9, 59
Decoding rule, 9
Defense Advanced Research Projects Agency (DARPA), 551
Defense Satellite Communication System (DSCS), 534
Delsarte, P., 245
Demodulation, 5-8
Demodulator, 2
Descendant, 195
Difference equation, 97
Difference set:
perfect, 203
perfect simple, 205
positive, 414-17
Difference-set-codes, 203-9
Digital Cypress, 531
Digital-to-analog (D/A) conversion, 3
Digital sink, 3
Dimension, 44
Discrepancy, 155
Discrete memoryless channel (DMC), 316-17
symmetric, 353

Distance, 10
column, 309, 342, 378-79
designed, 150
free, 309-12, 376-78
Hamming, 63, 320, 322
minimum, 51, 63-65, 310, 408-14
truncation, 342
Distance bounds (*see* Bounds)
DMC (discrete memoryless channel), 8
Double-track error, 513
Doubly transitive invariant (DTI) property, 195
Drukarev, A., 551-553
DTI codes,
type 0, 197-99
type 1, 200-201
Dual code, 55
Dual space, 45

Effective constraint length, 393
EG codes, 228-34
twofold, 235-39
Elementary symmetric functions, 154
Elias, P., 287
Encoder inverse, 306-8, 333-34
Encoder memory, 296, 306
Encoders, 2
channel, 2
for convolutional codes, 4, 288-95, 305-6
for cyclic codes, 95-98
for linear block codes, 56-58
source, 2
Encoder state diagram, 295-303, 339-41
Encoding, 2
concatenated codes, 279
convolutional codes, 4, 288-95
cyclic codes, 95-98
IBM 3850 mass storage system 518-19
IBM3420 series tape units, 504, 505
IBM 3370 disk storage systems, 528-29
interleaved codes, 272
linear block codes, 56-58
product codes, 276
Reed-Solomon codes, 171-73
shortened cyclic codes, 116
Encoding equations:
for convolutional codes, 288-95
Energy per information bit, 328-29
Energy per transmitted symbol, 327-28
Erasure probability, 359, 368
Erasures, 359, 381-82
Error control, 1, 12-14
automatic-repeat-request (ARQ), 13-14
for data storage systems, 498-532
forward error correction (FEC), 12-13
Error-correcting capability, 65
burst, 258, 430
majority-logic, 186, 391
maximum, 409
phased-burst, 431
random, 67
Error-detecting capability, 65
burst, 102-103
random, 65-66
Error-detection, 13
with BCH codes, 177-80
with cyclic codes, 98-103
with Hamming codes, 81-82, 116
with linear block codes, 58-63, 65-66
Error location numbers, 153, 159-60, 169-70
Error location polynomial, 153-59, 169
Error patterns, 58
correctable, 70
detectable, 65
most probable, 62
undetectable, 59, 65

Error probability amplification factor, 333-34
Error propagation, 406-8, 436
Error sequence, 389
Errors, 11
bit, 326-28
burst, 12
event, 322-28
post-decoding, 406
random, 11
transmission, 58
truncation, 339-42
undetected, 359
Error vectors, 58 (*see* Error patterns)
Euclidean geometry, 223, 227
Euclidean geometry codes, 228-40 (*see* EG codes)
Euclid's division algorithm, 27
Euclid's theorem, 18
Event error probability, 322-28, 370
Exclusive-OR (X-OR) gate, 5
Expansion of an integer, 195, 228
radix-2, 195
radix-2^s, 228
Extended Reed-Solomon codes, 176
Extension of a code, 82, 194
Extremely low frequency (ELF), 542

Falconer, D.D., 380
Fano algorithm, 360-64, 370, 378
metric threshold, 361-63
number of computations, 362-64
threshold increment, 362-64
Fano, R. M., 350, 353
FEC (forward error correction), 12-13 (*see* Error control)
Ferguson, M. J., 380, 446
Field, 19-40
binary, 21
characteristic of, 22
definition, 19
extension, 22
finite, 20
Galois, 20, 29-40
ground, 32
prime, 21
subfield, 23
Field element, 19
order of, 23
primitive, 24
Finite field (*see* Galois field)
Finite geometries, 223-27, 240-42
Euclidean, 223-27
projective, 240-42
Fire, P., 257, 261
Flat, of an Euclidean geometry, 225-26
of a projective geometry, 242
Flip-flop, 56
Forney, G. D., 141, 278, 306, 315, 341, 343, 370, 378, 444, 541, 542
Forward error correction (FEC), 12, (*see* Error control)
Forward index, 466, 469, 471
Fractional rate loss, 293
Fraleigh, J. B., 15
Frame, 235
Free distance, 309-12, 376-78
bounds, 312, 335-37
FWX (*see* Forward index)

Gallager bound, 433
Gallager function, 366
Gallager, R.G., 261, 429, 430, 444
Galois field, 22
arithmetic, 24-29, 39-40
basic properties of, 22-24, 34-39
construction of, 21-22, 29-33
representations of, 32-33
subfield, 23, 32, 226
tables, 33, 145-46, 250, 560-79
Galois field arithmetic, 24-29, 39-40
implementation, 161-67
GCD (greatest common divisor), 206, 306
GF (*see* Galios field)
GF(2), 21 (*see also* Binary field)
GF(2^m), 29 (*see also* Galois field)
GF(p), 21
GF(P^m), 22
GF(q), 22
Generalized stack algorithm, 380
Generating orthogonal polynomial, 249
Generator matrix, 51
block codes, 53-54
convolutional codes, 289-92
cyclic codes, 92-94
Generator polynomial, 89
BCH codes, 142, 147, 150, 151, 584
Burton codes, 273
convolutional codes, 293-95, 335-36
cyclic codes, 87-90
cyclic product codes, 276
difference-set codes, 206-9
DTI codes, 197, 200
EG codes, 229
Fire codes, 261
Golay code, 135
Hamming codes, 111
maximum length codes, 202
PG codes, 243
Reed-Muller codes, 233
Reed-Solomon codes, 171
twofold EG codes, 235
Gilbert, E.N., 84
Gilmore, R.P., 535
Goethals, J.M., 245
Golay codes, 134-38
Golay, M.J.E., 80, 134, 135
Goppa, V.D., 176
Gorenstein, D., 141
Graham, R.L., 206
Ground field, 32
Group, 15-19
additive, 18
commutative, 16
cyclic, 24
finite, 16
multiplicative, 19
order of, 16
Guard space, 430

Haccoun, D., 346, 380
Hagelbarger, D. W., 429
Hamming, R. W., 79, 111
Hamming bound, 83
Hamming codes, 79-82, 111-16
weight distribution, 81, 115
Hamming distance, 63, 320-22
Hamming weight, 63
Hard decisions, 7, 322, 328-29, 331-32, 343-45, 370-73
Harris Corporation, 535, 542, 546, 547-49
Hartmann, C. R. P., 246,346
Helios space missions, 539
Heller, J. A., 343
Hirasawa, S., 280
Hocquenghem, A., 141
Hong, S. J., 503
Hsiao, M. Y., 499
Hybrid ARQ schemes, 13, 477-81, 483-94, 551-56
Hyperplane, 225 (*see also* Flat)

IBM (International Business Machines), 498
Digital Cypress, 531-32
system 7030, 498
system 370, 499, 502
system 360, 498
3850 mass storage system, 516-25
3420 series tape units, 503-16
3350 disk system, 527-28
3340 disk system, 527-28
3370 disk system, 528-31
3330 disk system, 525-27
Identity element, 16
Identity matrix, 48
Implementation:
of BCH decoders, 167-70
of cyclic code decoders, 104-6
of Fano algorithm, 364
of majority-logic decoders, 426
of multiple stack algorithm, 384
of stack algoithm, 359-60
of Viterbi decoders, 337-45
Incidence vector, 227
Information code vector, 483
Information error sequence, 389
Information sequence, 2
Information source, 1
Inner code, 279
Inner product, 45
Input queue, 466
INTELSAT satellite system, 543, 544
Interleaved codes, 271-72
Interleaving, 271, 434, 441-42, 549
Interleaving degree, 271, 441
Itoh, K., 555
Invariant codes, 194
Inverse, 16
additive, 20
multiplicative, 20
Inversion process, 479
Invertible codes, 479, 481-83
Irreducible polynotial, 28
Iterative algorithm, 155-59
Iwadare, Y., 429, 436

Jacobs, I.M., 343
Jelinek, F., 332, 350, 360, 380, 541
Jet Propulsion Laboratory, 533, 539
Johannesson, R., 376, 539, 541
Justesen, J., 280, 336

Kasami, T., xv, 125, 131, 133, 135, 155, 176, 177, 233
Kasami decoder, 135, 138
Kasahara, M., (*see* Hirasawa, S.)
Kohlenberg, A., 429, 444
Kolesnik, V. D., 233

Larsen, K. J., 309
LCM (least common multiple), 142
Lee, L.N., 537, 538
Lin, S., 176, 539, 541
Lincoln Laboratory, 542, 549-50
Line, 224
Linear block codes, 51-84
combination, 43
feed forward shift register, 288, 306-8
sequential circuit, 287, 306
Linearly dependent vectors, 44
Linearly independent vectors, 44
Linkabit Corporation, 534, 540, 546, 547, 551
Location numbers, 194
Logic-AND, 75
Logic-Complement, 75
Log-likelihood function, 10, 317
L-step majority-logic decoders, 216-19
L-step majority-logic decoding, 210
L-step orthogonalizable code, 210
Lyne, H., 539, 541

Maclane, S., 15
MacWilliams, F. J., 77, 135, 155, 177, 206, 233, 245
MacWilliams identity, 77

Majority-logic decodable codes:
one-step, 194-209
L-step, 227-39, 242-45
Majority-logic decoding, 184-221, 388-426, 543-46
multiple-step, 209-19
one-step, 184-94
Majority-logic decoding rule, 391
error probability, 408-14
Majority-logic gate, 188, 398
Mann, H. B., 233, 245
Masons' gain formula, 297-99
Massey, J. L., xv, 141, 184, 287, 308, 333, 335, 388, 400, 414, 421, 426, 429, 430, 434, 436, 444, 539, 541, 546
Matrix, 46
generator, 51, 53
identity, 46
parity-check, 51, 53
transpose, 45
Maximum length codes, 201-3
Maximum likelihood decoding, 8-11, 317
Maximum likelihood path, 317-18
Meggit, J. E., 106
Meggit decoder, 106, 397
Memory, 498
main, 498
semiconductor, 498, 499
Memory order, 4
Message, 3, 51
Metric:
bit, 317-20, 353-54, 370-73
branch, 317-20
Fano, 353-54, 368-69
integer, 318-19, 354-55, 368-69
partial path, 317, 353
path, 317
Metric table, 319, 354-55, 368-69
Metzner, J. J., 474
Minimal polynomial, 35-38
tables of, 38, 146, 580-83
Minimum distance, 63
of BCH codes, 142
of concatenated codes, 279
of convolutional codes, 310, 408-14
of difference-set codes, 206
of Hamming codes, 79, 80, 113
of linear block codes, 63-65
of maximum length codes, 201
of product codes, 275
of Reed-Muller codes, 233
of Reed-Solomon codes, 237
of twofold EG codes, 237
Mironchikov, E. T., 233
Mitchell, M. E., 125
MLD (maximum likelihood decoder), 8-11
Modified syndrome, 394
Modulation, 5-8
binary-phase-shift keyed, 6, 327
Modulator, 2
Modulo-m addition, 17
Modulo-p multiplication, 18
Modulo-2 addition, 16
Modulo-2 multiplication, 21
Morgan, K. C., 474
Morries, J. M., 495
Morrissey, T. N., 408
Muller, D. E., 184, 233
Multiple stack algorithm, 381-84
computational distribution, 382-83
error probability, 383-83
storage requirements, 383-84
Multiplication, 18
field, 19, 29
modulo-p, 18
scalar, 41

NAK (negative acknowledgement), 13
Namekawa, T. (*see* Hirasawa, S)
NASA, 533-35, 539-42
CTS satellite system, 535
digital television test set, 542
International Sun-Earth Explorer (ISEE), 541
International Ultraviolet Explorer (IUE), 541
Pioneer space missions, 539
TELOPS program, 541
Tracking and Data Relay Satellite System (TDRSS), 535
Voyager space missions, 533
Negative acknowledgement (*see* NAK)
Newton's identities, 154
(n, k) block code, 4, 52
(n, k, m) convolutional code, 4
Noise, 2
Noisy channel coding theorem, 10
Nonbinary BCH codes, 170-77
Nonorthogonal parity-check sums, 253
Nonprimitive BCH codes, 150
Normal state, 469
n-tuple, 3, 41
Null space, 45

Odenwalder, J. P., 536
Omura, J. K., 315, 328
One-step majority-logic decodable codes, 194-209
One-step majority-logic decoders, 188, 191-92
One-step majority-logic decoding, 184-94
Optimal block length, 495
Optimal burst-error-correcting codes, 258
Order:
of a field, 19
of a field element, 23
of a group, 16
Ordered differences, 203
Orthogonal check sum, 391
Orthogonalizable codes, 186, 210
completely one-step, 186
completely L-step, 210
Orthogonalization, 210
Orthogonal vectors, 185
Outer code, 279

Paske, E., 376, 541
Pareto distribution, 366
Pareto exponent, 366-68
Parity-check digit, 54
overall, 82, 194
Parity-check equation, 154, 184
Parity-check matrix, 51
BCH codes, 148
block codes, 55
cyclic codes, 92-94
SEC-DED codes, 499-501
Parity-check sums, 185, 391
orthogonal, 186, 189
Parity code vector, 483
Parity error sequence, 384
Parity parallelogram, 423
Parity polynomials:
cyclic code, 93
cyclic product code, 277
difference-set code, 206
Parity retransmission, 480
Parity triangle, 392
Patel, A. M., 503, 518
Path memory, 337-42
Perfect codes, 80, 134
Perfect simple difference set, 205
Period of an irreducible polynomial, 261
Peterson, W. W., xv, 141, 155, 159, 176, 269, 526

PG codes, 242-45
Phased burst, 272
Phased-burst-error-correcting codes, 272-74
Plotkin, M., 83
Plotkin bound, 83
Polynomials, 26-29
irreducible, 28
minimal, 35-38, 146
primitive, 28, 29
Positive acknowledgement (ACK), 13
Power spectral density, 6, 327-29
Power-sum symmetric functions, 153
Prange, E., 85
Preparata, F. P., 429, 433
Prime numbers, 18
Primitive BCH codes, 142 (*see also* BCH codes)
Primitive element, 24
Primitive polynomials, 28, 29
Probability, transition, 7
Probability of decoding error, 3, 9, 10, 67, 71, 362
Probability of undetected burst, for convolutional codes, 450, 454
Probability of undetected error, 65
for BCH codes, 178-80
for convolutional codes, 322-29, 368
for Hamming codes, 81-82, 116
for hybrid-ARQ systems, 552-53
for linear block codes, 65-66, 76-79
Product codes, 274-77
cyclic, 276-77
Projective Geometry, 240-42
codes, 242-45

q-ary codes, 170
Quasi-cyclic codes, 221
Quotient, 27

Radio frequency interference (RFI), 547-50
Random coding, 11, 312, 329, 341, 343, 350, 364, 368-69, 425, 551, 554-55
Random-error channels, 11
Random-error-correcting codes, 11
Random errors, 11
Rate, 4
baud, 8
code, 4, 292
data, 8, 370-73
information transmission rate (data), 8
symbol transmission (baud), 8
Reading unit, 2
Received information sequence, 389
n-tuple, 9
parity sequence, 389
sequence, 2, 316-17, 320
vector, 58
Reciprocal polynomial, 49
Reddy, S. M., 451
Redundant bits, 4
Reed, I. S., 141, 171, 184, 233
Reed-Muller codes, 233
Reed-Solomon codes, 171-76, 535-37
codes derived from, 278
decoding, 172-76
encoding, 171-72
Registers, 56
buffer, 104
error-location, 263
error-pattern, 263
message, 56
shift, 56, 95
syndrome, 104, 395
Reiger, S. H., 258
Reiger bound, 258
Reliability, 461
of hybrid ARQ's, 478

Reliability (*cont.*)
of pure ARQ's, 461
of a type-II hybrid ARQ, 492-94
Remainder, 17, 27
Representations of Galois fields, 32-33
Retransmission, 13
buffer, 466
parity, 480
RM codes (*see* Reed-Muller codes)
Robinson, J. P., 407, 414, 418
Roots of polynomial, 27, 34-36
Rosenberg, W. J., 308
Round-trip delay, 13, 459
Row operations, elementary, 46
Row space, 46
Rudolph, L. D., 125, 206, 223, 233, 245, 246, 346, 426

Sain, M. K., 308
Satellite channels, 473
Scalar multiplication, 41
Scalars, 41
Schalkwijk, J.P.M., 346
SCR (sequential-code-reduction) decoding, 248
SDLC (synchronous data link control), 460
SEC-DED (single-error-correcting and double-error-detecting) codes, 498, 499-502
Segment, 518
Selective-repeat ARQ, 13, 460
with finite receiver buffer, 465-77
hybrid, 483-94
ideal, 464, 465
with infinite receiver buffer, 462, 463-64
reliability of, 461
throughput efficiency of, 462, 464, 465, 473-74, 475, 477
Selective-repeat ARQ with finite receiver buffer, 465-74
buffer overflow, 467
throughput efficiency, 473-74
transmission and retransmission procedure, 466-67
Selective-repeat plus go-back-N ARQ, 474-77
throughput efficiency, 475, 477
Sequence, 2
encoded, 2, 288-95
estimated, 2
information, 2, 288-95
received, 2, 316-17, 320
Sequence numbers, 456-66
Sequential-code-reduction decoding (*see* SCR decoding)
Sequential decoding, 350-84, 539-43
backsearch limit, 370-73
computational cutoff rate, 367-68, 380
computational distribution, 364-71
error probability, 368-74
performance characteristics, 364-73
Sequential-logic circuit, 4
Set, 15
Shannon, C. E., 1, 10
Shift, cyclic, 85
Shortened cyclic codes, 116-21
decoding, 116-21
encoding, 116
invertible, 479, 481-83
Shortened Hamming codes, 80
Signal flow graph, 297-303
Singer, J., 205
Single-error-correcting and double-error-detecting codes (*see* SEC-DED codes)
Single-track error, 511
Sloane, N.J.A., 135, 155
Slope control algorithm, 554-56
Smith, K.J.C., 233, 245
Soft decision, 8, 322, 329, 332, 337, 343-45, 373, 383, 426
Solomon, G., 141, 171 (*see also* Reed-Solomon codes)
Source, 1
decoder, 3
digital, 3
encoder, 2
information, 1
Space, 40
dimension of, 44
dual, 45
null, 45
row, 46
vector, 40-46
Span, 44
SR + GBN ARQ (*see* Selective-repeat plus go-back-N)
Stack algorithm, 350-60, 378
number of computations, 356-60
Stack-bucket algorithm, 360-64
Standard array, 51, 68-75
Storage media, 2
core memories, 2
disk files, 2
drums, 2
magnetic bubble memories, 498
magnetic tapes, 2
optical memory units, 2
semiconductor memories, 2
Stripe, 517
Subcode, 113
Subfield, 23
Subgroup, 19
Submatrix, 48
Subspace, 42
Suiyama, Y. (*see* Hirasawa, S.)
Sullivan, D. D., 424, 451
Switching functions, 74
Synchronous data link control (*see* SDLC)
Syndrome, 51
BCH codes, 152-55
block codes, 58-63
computation for BCH codes, 167-69
convolutional codes, 346
cyclic codes, 98-102
decoding, 72
Syndrome forming circuit, 390
Syndrome register, 395
Syndrome sequence, 389
Syndrome vector, 398
Systematic codes, 54
Systematic search decoder for Golay code, 138
Systems, 1
ARQ, 13, 459-465
data storage, 1-3, 498-532
data transmission, 1-3
FEC, 12-13
one-way, 12
two-way, 13

Threshold decoding (*see* Majority-logic decoding)
Throughput efficiency, 461, 551
go-back-N ARQ, 462-63
selective-repeat plus go-back-N ARQ, 475, 477
selective-repeat ARQ with finite receiver buffer, 473-74
selective-repeat ARQ with infinite receiver buffer, 462
slope control algorithm, 554-55
stop-and-wait ARQ, 463
time-out algorithm, 551-53
type-I hybrid ARQs, 478-79
type-II hybrid ARQ, 480, 491-92, 493, 494
Time-out, 466
Time-out (TO) algorithm, 551-55
Time-out vector, 466
Tong, S. Y., 429, 444, 446
Transition probabilities, 7, 317, 320, 353
Transmission errors, 58
Transmission rates:
information, 8
symbol, 8
Transpose of a matrix, 48
Trellis diagram, 315-16
Truncated sequence, 309
Truncation distance, 342
Truncation length, 340-42

Undetectable error patterns, 65
Union bound, 324
Unit element of a field, 19
U.S. Air Force Ballistic Missile Early Warning System (BMEWS), 547
U.S. Army Satellite Communication Agency, 542, 547
TACSAT channel, 542
U.S. Navy, 542, 547
Electronic Systems Command, 547
Project Sanguine, 542

Vandermonde determinant, 150
Varsharmov, R. R., 84
Varsharmov-Gilbert bound, 84
Vectors, 41
incidence, 227
linearly dependent, 44
linearly independent, 44
orthogonal, 185
time-out, 466
Vector addition, 41
Vector space, 40-46
Vinck, A. J., 384
Viterbi, A. J., 287, 315, 328
Viterbi decoding, 315-46, 373, 376-78, 382-84, 533-38
error probability, 322-29, 339-46, 373, 382-83
hybrid-ARQ, 554-56
number of computations, 345, 356-58, 370, 373, 383
storage requirements, 337, 373, 383

Weight, Hamming, 63
minimum, 63
2^s -, 228
Weight distributions, 65, 76
of BCH codes, 117-80
of convolutional codes, 296-303
of Golay code, 138
of Hamming codes, 81, 115-16
of Reed-Solomon codes, 180
Weight enumerators, 77 (*see also* Weight distributions)
Weldon, E. J., Jr., 155, 206, 233
Wolf, J. K., 176
Wozencraft, J. M., 287, 350
Writing unit, 2
Wyner, A. D., 429

Yamamoto, H., 554

ZJ algorithm (*see* Stack algorithm)
Zeoli, G. W., 538
Zero element of a field, 19
Zierler, N. , 141
Zigangirov, K., 350